The Limnology, Climatology
and Paleoclimatology
of the East African Lakes

The Limnology, Climatology and Paleoclimatology of the East African Lakes

Edited by

Thomas C. Johnson

Large Lakes Observatory
University of Minnesota, Duluth

and

Eric O. Odada

Department of Geology
University of Nairobi, Kenya

with the assistance of Katherine T. Whittaker

CRC Press
Taylor & Francis Group
Boca Raton London New York

CRC Press is an imprint of the
Taylor & Francis Group, an **informa** business

First published 1996 by Gordon and Breach Publishers

Published 2019 by CRC Press
Taylor & Francis Group
6000 Broken Sound Parkway NW, Suite 300
Boca Raton, FL 33487-2742

© 1996 by Taylor & Francis Group, LLC
CRC Press is an imprint of Taylor & Francis Group, an Informa business

First issued in papeback 2019

No claim to original U.S. Government works

ISBN 13: 978-0-367-45590-3 (pbk)
ISBN 13: 978-2-88449-234-8 (hbk)

**Visit the Taylor & Francis Web site at
http://www.taylorandfrancis.com**

**and the CRC Press Web site at
http://www.crcpress.com**

British Library Cataloguing in Publication Data

The limnology, climatology and paleoclimatology of the East
    African Lakes
    1.Lakes – Africa, East   2.Limnology – Africa, East
    3.Freshwater biology – Africa, East   4.Paleoclimatology –
    Africa, East
    I.Johnson, Thomas C.   II.Odada, Eric O.
    551.4'82'09676

*This volume is dedicated to our friend and colleague
the late Dr. William F. Kudhongania,
former director of the Uganda Freshwater
Fisheries Research Organization in Jinja*

CONTENTS

Aquatic Chemistry

Food Webs and Fisheries

Sedimentary Processes and Deciphering the Past in the Large Lakes

INTRODUCTION

The large lakes of the East African Rift Valley are among the oldest on Earth, and are vital resources for the people of their basins. They are unique among the large lakes of the world in terms of their sensitivity to climatic change, rich and diverse populations of endemic species, circulation dynamics and water-column chemistry, and long, continuous records of past climatic change.

More than four kilometers of sediment underlie Lakes Tanganyika and Malawi, and their ages are estimated to be on the order of 10–15 million years based on models of sedimentation and compaction in rift basins. Some of the lakes of the Rift Valley are closed basins (i.e., without outlets) and their surface levels fluctuate dramatically both seasonally and interannually in response to rainfall variability. Even the open-basin lakes such as Tanganyika, Victoria, and Malawi lose 80–90% of their water by evaporation. These lakes have fluctuated between closed- and open-basin status frequently in response to varying rainfall and evaporation, causing their levels and water chemistry and biota to shift significantly with climatic change.

The largest lakes contain hundreds of endemic species of fish and invertebrates, providing ample opportunity for the study of evolutionary biology within confined systems. The African Great Lakes have unique physical qualities that affect water circulation. Coriolis Force is weaker in equatorial lakes than in lakes at higher latitudes. The Rift Valley channels the trade winds so effectively that wind forcing of the lake circulation is remarkably unidirectional. Temperature profiles are nearly isothermal, yet contain fascinating but subtle structure that greatly impacts vertical circulation. Water chemistry reflects a variety of biological, physical, and chemical processes. The deep lakes are anoxic in their hypolimnions; nutrient cycling is complex; and inorganic reverse weathering reactions may occur in early diagenesis of lake floor sediments that have major impact on the major ion composition of the overlying water. Changes in sediment composition in response to changing climate and lake level are easily discerned on a time scale resolvable to decades, if not individual years, because of rapid sedimentation rates and a lack of bioturbation in the deep basins.

A comprehensive study of the large African lakes is long overdue. The scientific justification for such an effort is noted in the previous paragraph and is illustrated in great detail in this volume. Societal need for the sustainable utilization of these lakes offers an even more compelling reason for examination of biological food webs, water quality, and past climate variability in East Africa. The lakes provide the most important source of protein for the people of the African Rift Valley, and fish populations are shifting dramatically in response to fishing pressure, introduction of exotic species, land use impact on water quality, and perhaps climatic change. Current estimates of primary productivity, the underpinning of the food resource, are extremely crude and based on only a few spot measurements.

Although evidence for shifting water quality is very localized in Lakes Tanganyika and Malawi, Lake Victoria has undergone tremendous change in the past three decades. Dominance in primary production has shifted from diatoms to blue-green algae. Secchi depth readings have shifted from an average of 12 to 2 meters. Bottom water anoxia has shifted from seasonal to permanent. At present nobody has a good explanation for these recent changes. Can the lake return to its earlier, more desirable state? Are the other large African lakes facing similar changes in water quality?

The African lakes are impacted significantly by climate change, as are agriculture and other aspects of the East African national economies. The country of Malawi, for example, derives over 80% of its electric power from a dam on the Shire River which drains Lake Malawi. The river was dry for the first three decades of this century because lake level had dropped below the outlet depth as a result of slightly more arid conditions at that time. It is likely that such arid events recur relatively frequently. An improved concept of past natural climate variability in East Africa will favorably impact the development of new strategies for the management of fisheries, agriculture, and other natural resources.

The purpose of this volume is to present results of scientific investigations on the East African lakes that were presented at the IDEAL (International Decade for the East African Lakes) Symposium on the Limnology, Climatology and Paleoclimatology of the East African Lakes, at Jinja, Uganda, in February 1993. More than 100 aquatic scientists from North America, Europe, Africa, Asia, and New Zealand attended the symposium. Issues were limited to five topical areas: climatology, physical limnology, geochemistry, biological sciences, and paleoclimatology. While every effort was made to provide a balance of the various disciplines presented, some were covered more extensively than others.

IDEAL was first conceived at a workshop supported by the United States National Science Foundation and the Swiss National Climate Program in Bern, Switzerland, in March 1990. Although the original scientific focus of IDEAL was on the paleoclimatic record archived in the bottom sediments of the Rift Valley lakes, it has evolved into an investigation of biogeochemical processes in the large lakes of East Africa as well. IDEAL also carries a major commitment to training in the aquatic sciences for African scientists, students, and technicians. The intention is to combine the talents and expertise of a multi-disciplinary group of limnologists and oceanographers to undertake a sustained and comprehensive study of the large African lakes over the coming decade. We hope that the papers which follow will excite curiosity and interest among our colleagues to apply their talents to some of the most intriguing lakes on Earth.

ACKNOWLEDGMENTS

We thank many people and organizations for completion of this volume. The authors of the articles are to be commended for their dedication to the collection, analysis, and dissemination of data on the East African lakes, given the logistical challenges that accompany such efforts. We express our sincere gratitude to the government officials of the East African countries bordering the lakes, who were instrumental in granting research permits, easing the passage of personnel and equipment across borders, and providing logistical support that has made research on the lakes of East Africa possible. The IDEAL Symposium held in Jinja, Uganda, was managed locally by the late Dr. William F. Kudhongania, former director of the Fisheries Research Institute in Jinja. He and his staff were instrumental in the symposium's success.

George Kitaka of the International Hydrology Program of UNESCO, based in Nairobi, donated much of his time to obtaining financial and logistical support for African participants to the IDEAL Symposium. Andy Cohen, Kerry Kelts, John Lehman, Sharon Nicholson, Bob Spigel, Mike Talbot, and Ray Weiss provided valuable suggestions for the IDEAL Science and Implementation Plan that resulted from the Jinja meeting. Much of the introduction to this volume is derived from that report.

We thank Katherine T. Whittaker, Doug Ricketts, Keith Sturgeon, and Mary Plante of the Large Lakes Observatory at the University of Minnesota for their assistance in the editing of this volume. Thomas C. Johnson expresses sincere thanks to the Fulbright Commission and to the CNRS Laboratoire de Geologie du Quaternaire at Luminy, France, for partial support while much of the editing of this volume took place at Luminy in 1993–1994. Financial support for the symposium was provided by the United States National Science Foundation, the International Hydrology Program of UNESCO, the American Society for Limnology and Oceanography, and the Societas Internationalis Limnologiae. We greatly appreciate the efforts of all the following reviewers, who substantially improved the quality of the papers published in this book.

Paul A. Baker, Charles E. Barton, David S. Brown, Louis P. Brzuzy, Eddie C. Carmack, Thure E. Cerling, Mark Chandler, Robert B. Cook, George Coulter, Jack Dymond, Steven J. Eisenreich, Daniel Engstrom, Everett J. Fee, Bruce P. Finney, Joel R. Gat, A.T. Grove, John D. Halfman, Robert E. Hecky, Sylvie Joussaume, George Kling, John Largier, Guy Lister, and Dan A. Livingstone.

W. Barry Lyons, Sally MacIntyre, John M. Melack, Philip A. Meyers, Sharon P. Nicholson, R.B. Owen, David K. Rea, Robin Renaut, Christopher A. Scholz, Frederick H. M. Semazzi, Joseph Shapiro, Val H. Smith, J. Curt Stager, Robert W. Sterner, Michael R. Talbot, Denis Tweddle, Hassan Virji, Ray F. Weiss, Jacques White, Joan D. Willey, Martin Williams, Alfred Wüest, and Richard Yuretich.

Tectonic Setting of the East African Lakes

Tectonic Controls on the Development of Rift-Basin Lakes and Their Sedimentary Character: Examples from the East African Rift System

W.A. WESCOTT   *Amoco Production Company, Houston, Texas, United States*

C.K. MORLEY   *Department of Petroleum Geoscience, University of Brunei, Darussalam, Brunei*

F.M. KARANJA   *Shell Exploration and Production Kenya, Nairobi, Kenya*

Abstract — Tectonic style and history exerted a strong influence on the development of East African Rift lakes. Fundamental rift-basin geometry is controlled, in large part, by extension rates and the angles of bounding faults. Where the ratio of vertical to horizontal displacement is large and climatic conditions result in fluvial discharge into the lake greater than evaporation, conditions are favorable for the development of deep, long-lived lacustrine systems. Where this ratio is lower, climatic conditions may only periodically support the formation of relatively shallow lakes. The timing and magnitude of rift margin uplift also exert an influence upon drainage patterns and therefore affect the discharge of water and sediment into the basin. Rift lakes with high-angle, high-relief margins tend to be filled by relatively fine-grained sediments with coarse clastics confined to rift margins and in the rift axis only at rift segment terminations or as basin plain turbidites. Basins with low-angle, low-relief boundary faults are commonly dominated by coarse clastics that may be deposited throughout the rift lake. Lakes in relatively short rift segments with well developed axial drainage systems flowing into them tend to be shallower and coarser-grained than those without them.

Lake Tanganyika is a long, deep, relatively fine-grained lake, in a humid environment, characterized by steep, high-angle boundary faults with high vertical-to-horizontal displacement ratios and limited axial input. On the other hand, the southwestern Turkana basins (Upper Oligocene–Middle Miocene) are examples of half-grabens characterized by low-angle boundary faults and relatively low vertical-to-horizontal displacement ratios. They are dominated by coarse-grained fluvial deposits with inter-bedded shales which were deposited during episodic, shallow lacustrine conditions. The present Lake Turkana developed during two separate rifting episodes and has a complex history. Lake Rukwa appears to be a hybrid of the end members and serves as an example of how rift-lake systems can evolve with time. As is typical of continental rifts, its early history (Neogene) was dominated by coarse-clastic deposition. During the Plio–Pleistocene, lacustrine conditions, characterized by the deposition of muddy sediments, prevailed. The modern lake, which is in an arid setting, is dominated by well developed axial drainage systems which maintained sedimentation rates at levels greater than or equal to tectonic subsidence rates and is consequently shallow and sand-dominated.

INTRODUCTION

The integration of structural and sedimentological data is necessary to understand how basins form and fill. The comparison of theoretical and predictive methods with field observations from modern basins and their ancient counterparts allows us to develop a model for tectonic controls on the sedimentary character of rift basins. For example, Morley (1989) suggested that changes in fault geometry with time can influence sedimentation style in rift basins. From the sedimentology side, unique information about provenance, tectonics, and sediment dispersal patterns may be preserved in the detritus filling a basin (Dickinson et al., 1986), even where the present geological setting bears little or no resemblance to the setting in which the sediments were deposited. In this study we discuss how spatial and temporal variations of structures at different scales may influence sedimentation in rifts, particularly with respect to the East African rift system.

TECTONIC CONTROLS ON DRAINAGE AND SEDIMENTATION PATTERNS

The structural and tectonic effects of rifting on depositional patterns in rifts occur at both temporal and spatial levels. These can be further divided according to the scale of the effect. The largest spatial control is the size and length of individual rift segments (which range from tens to hundreds of kilometers long) and the area of the drainage basins that feed the segments. The arrangement of rift segments affects the location, pattern, and flow direction of major drainage systems. Next in scale are the locations of transfer zones, which may actually link rift systems, or link half-grabens within rift segments. Depending upon the topographic relief generated at transfer zones, they may either form barriers to drainage systems into the rift or provide points of entry into the rift (e.g., Ebinger et al., 1984; Rosendahl et al., 1986; Rosendahl, 1987, Crossley, 1984; Frostick and Reid, 1987; Leeder and Gawthorpe, 1987; Cohen, 1990; Lambiase, 1990, Morley et al., 1990).

Commonly preexisting structural fabric may strongly influence the location of transfer zones and localise coarse-clastic input into rifts. Once an entry point for significant quantities of coarse sediment has been established it is further dispersed across the rift under the influence of gravity (commonly by fluvial systems or as turbidity currents) in tectonically created mazes of topographic highs and lows associated with en-echelon arrays of titled fault blocks. The following sections look at the effects on sedimentation of different sizes of structural elements, beginning with the largest.

Rift Segments

Rift segments are composed of one or more half-graben or full-graben elements. The main linking geometries of rift segments involve either line segments (e.g., gaps and splays) or en-echelon rift segments (e.g., jumps, offsets and passes) (Nelson et al., 1992; Fig. 1). These different rift segment interactions give rise to different large-scale variations in drainage patterns. Where rift segments are continuously linked by offsets there may be little opportunity for axial input of large quantities of sediment except at the extreme terminations of the rift. Major drainage systems have to enter from the rift margins through areas of relatively low relief. Such areas may be due to preexisting structural trends, regions where fluvial erosion has kept pace with uplift, and at transfer zones (e.g., Cohen 1990, see next section). Rift gaps and jumps provide the opportunity for axial sediment input into discontinuous rift segments (Fig. 2). Streams flowing down the rift axis generally drain larger areas and have lower gradients than marginal systems, especially short-headed systems along fault margins. Their deltas are constrained by rift structure and they contribute large amounts of

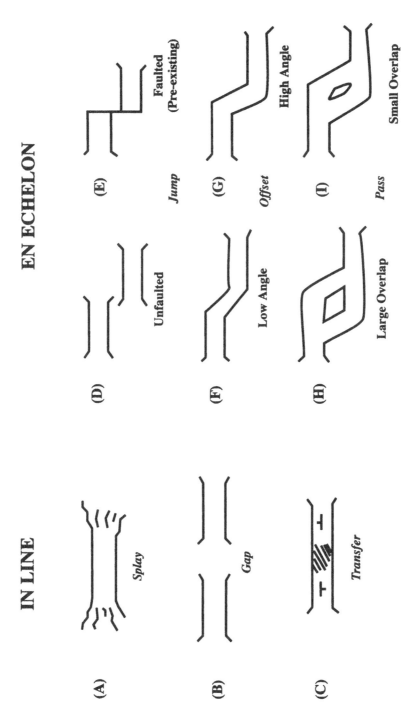

Figure 1. Large-scale geometries of rift segments (from Nelson et al., 1992).

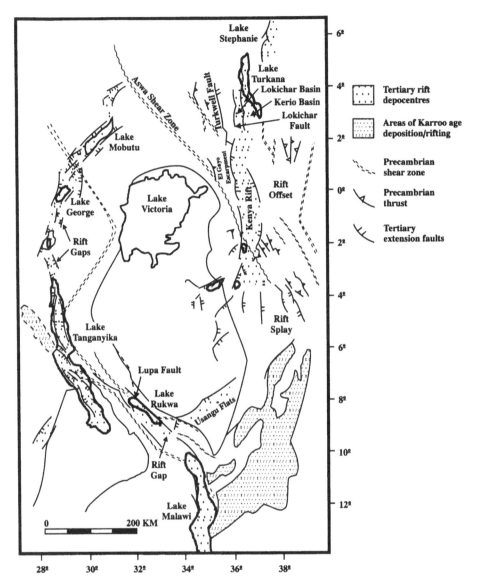

Figure 2. Generalized map of the East African Rift System showing the locations of rift segment interactions.

sand-sized sediment into the basin (Frostick and Reid, 1986; Cohen, 1989; Morley et al., 1992a).

In the Western Branch of the East African Rift System the rift is immature, and the various rift segments have not propagated together to form a continuous rift. Consequently this rift branch contains many examples of rift gaps and jumps (Nelson et al., 1992), where the main lakes lie in individual rift segments and the intervening basement highs or volcanic regions form the gaps or jumps.

The length of the rift segment has a significant impact on how effective the axial drainage is in affecting sediment distribution and sedimentation rates within a segment, as can be seen by examining Lake Tanganyika (700 km long) and Lake Rukwa (250 km long, present day). Axial systems in these rifts are very sand rich (e.g., Rusizi plain, Lake Tanganyika). However, the great length of the Tanganyika Rift means that the effect of axial drainage is limited to the northernmost part of the rift segment. The shorter Lake Rukwa Rift segment in contrast is greatly affected by axial drainage both from the NW and SE. Well information (Wescott et al., 1991) suggests that deep lacustrine conditions were rarely, if ever, attained in the Upper Miocene–Recent history of the Rukwa Rift despite large (up to a 7.5 km throw) displacement on the boundary fault. This suggests that, unlike Lake Tanganyika, sedimentation kept pace with subsidence, which in turn can be attributed to the large axial input into a relatively short rift segment.

Transfer Zones

Isostatic footwall uplifts occur in response to displacement on major faults (Buck, 1988; Kusznir and Egan, 1989). Unlike thermal uplifts, this mechanical uplift is permanent (until it is eroded). Footwall uplifts are greater where fault displacement is greatest and decrease as fault displacement decreases. The uplifts are generated during rifting and consequently can significantly affect synrift sedimentation patterns. Footwall uplifts create barriers that inhibit large river systems from entering the rift (e.g., Crossley, 1984; Ebinger et al., 1987; Cohen, 1989). As faults lose displacement (i.e., at transfer zones) the footwall uplift topography also lessens; consequently many transfer zones provide entry points for drainage systems into the rift (Morley et al., 1991).

Three main types of transfer zones associated with boundary faults can be considered: faults that dip toward each other (convergent), faults that dip away from each other (divergent), and faults that dip in the same direction (synthetic). Transfer zones are the same as the features referred to as accommodation zones by Rosendahl et al. (1986) and Rosendahl (1987). However, these previous classification schemes were not scale-independent as they focused on the rift unit-size structures associated with alternating half-grabens. The classification used here (after Morley et al., 1990) is scale-independent. In addition to including the scoop-shaped half-graben interactions described by Rosendahl et al. (1986), this broader classification includes geometries involving linear and zig-zag fault traces. The classification scheme of Morley et al. (1990) also allows for the description of transfer zones as they evolve through time.

The interplay between transfer zone geometry and sediment input is summarized in Fig. 3. The convergent type causes a switch in polarity of the basin along the rift axis. Such transfer zones may commonly mark the switch of major drainage from one side of the rift to the other. The overlapping diverging transfer zones tend to be characterised by a horst zone that separates basins and acts as a barrier to sedimentation (e.g., Lambiase, 1990). Synthetic transfer zones may provide a break in the topographic barrier that is commonly present along boundary fault margins. They may allow larger fluvial systems to enter the rift from the boundary fault margin (Crossley, 1984).

All the different types of transfer zone are present in the East African Rift System (e.g., Morley et al., 1990). To illustrate the role played by transfer zones on sedimentation patterns, the drainage systems surrounding Lake Tanganyika are shown in Fig. 4A. In Fig. 4B the individual drainage basins have been mapped. From these figures it is apparent that the drainage areas along the lake side of the major boundary faults are relatively small because the drainage divide is close to the lake. Streams on the more gently dipping backside of the boundary fault blocks develop larger drainage basins. Lower order streams in these systems initially flow away from the lake and trunk streams may flow parallel to the lake in valleys established between tilted fault blocks (Cohen 1990). These rivers frequently reenter the rift through topographic lows created at transfer zones. In general, the different mechanical uplift configurations caused by boundary fault interplay at transfer zones can either cause drainage barriers (e.g., conjugate, divergent, overlapping) or lows through which rivers may pass (e.g., conjugate, collinear) (Fig. 3).

Individual Fault Geometries

1. Fault block geometries

The large-scale features discussed above are also repeated at smaller scales within rifts. The well-exposed Kenya Rift provides an excellent example of the mazes of drainage that are generated in association with swarms of minor tilted fault blocks (e.g., Crossley, 1984). The uplifted footwall block can be a relative high on the rift floor onlapped by sediments. Under lacustrine conditions this may commonly lead to deposition of muds on the highs, with coarser material onlapping the highs. However it is possible for coarse material to be deposited on highs, and this may be a result of lake floor or sea floor currents being particularly active on the highs leading to winnowing of fine-grained sediments, unconformities, and channeling. Such submarine processes may be mistaken as evidence for subareal erosion and large-scale changes in lake level. Subareal erosion would be of a more regional and continuous nature than erosion by currents.

2. Boundary fault dip

Changes in major fault dip amount with time or spatially may exert an important control on sedimentation style in rift basins. It is clear from seismic reflection data in the East African Rift that boundary faults control basin geometries and act as large growth faults. Since for a fixed amount of extension steeply dipping faults have a greater throw than more gently dipping faults, fault dip should be expected to have some impact of subsidence rates. Basins with steep boundary faults and relatively high subsidence rates, such as Lake Tanganyika (Fig. 5), are prone to form deep lakes predominantly filled with fine-grained sediments (Livingstone, 1965; Degens et al., 1971; LeFournier et al., 1985). Boundary faults with lower dips and/or shallow depths to detachment appear to be less likely to form basins which undergo rapid synrift subsidence, unless strain rates are high. High strain rates may increase the potential for large block rotations and synsedimentary erosion. These rifts are more likely to be dominated by coarse-clastic deposits, with more episodic development of lacustrine conditions (Chapman et al., 1978; Morley, 1989).

Rifts that evolve under constant strain rates and experience shallowing of the detachment with time and/or a trend towards lower-angle boundary faults with time will concomitantly have a decrease in the amount of tectonic subsidence with time. Under these conditions an evolving continental rift basin might pass from deep lacustrine to shallow lacustrine to fluvial environments through time. This can occur simply as a result of changing detachment levels and not necessarily because of declining strain rates or increasing rates of sedimentation (Morley, 1989).

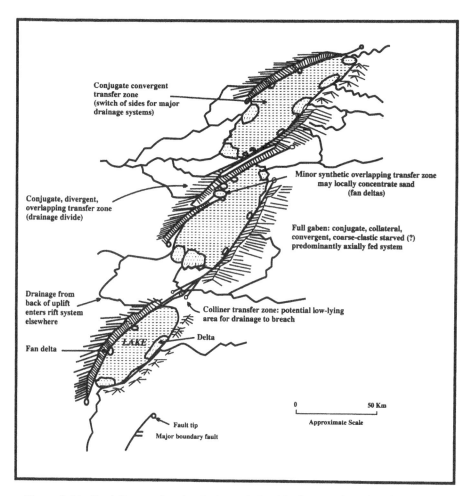

Figure 3. Idealized diagram showing the interrelationships between boundary fault geometries, transfer zones, drainage, and sand distribution in rift systems. Major boundary faults are mechanically uplifted and relatively small drainage basins develop on the footwall side. These basins deliver immature coarse-grained sediments into the lake via alluvial fans and fan deltas. Areas of the rift not bounded by major faults are relatively low and provide a sites for rivers with larger drainages to enter the rift and deposit more texturally mature sediments (from Morley et al., 1990).

The extensional histories of three basins in the Eastern and Western arms of the East African Rift System are summarized in Table 1. In attempting to quantify extension rates a large potential inaccuracy arises from estimates of the duration of rifting. Field work and drilling in the Lake Rukwa Rift has established a Late Miocene–Recent age for the synrift section (Ebinger et al., 1989; Wescott et al., 1991; Morley et al., 1992b). Ages from 9 to 25 Ma have been calculated for the Lake Tanganyika Rift (Rosendahl and Livingstone, 1983; Morley, 1988; Ebinger, 1989; Tiercelin and Mondeguer, 1991; Cohen et al., 1993). In the northern Kenya Rift sedimentary sections have been dated using vertebrate fossils, palynology, and radiometric dates from igneous rocks (Dixey, 1945; Hooijer, 1966, 1973; Maglio,

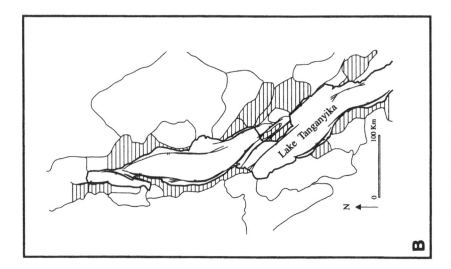

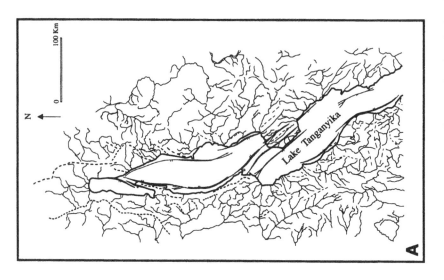

Figure 4. A) Map of the Lake Tanganyika area showing the locations of the fluvial systems that drain into the lake. B) Size and locations of individual drainage basins that feed Lake Tanganyika. Patterned areas represent basins that drain the footwall of major boundary faults; they are relatively small compared to the basins (unpatterned areas) developed by streams draining the more gently dipping backsides of the fault blocks.

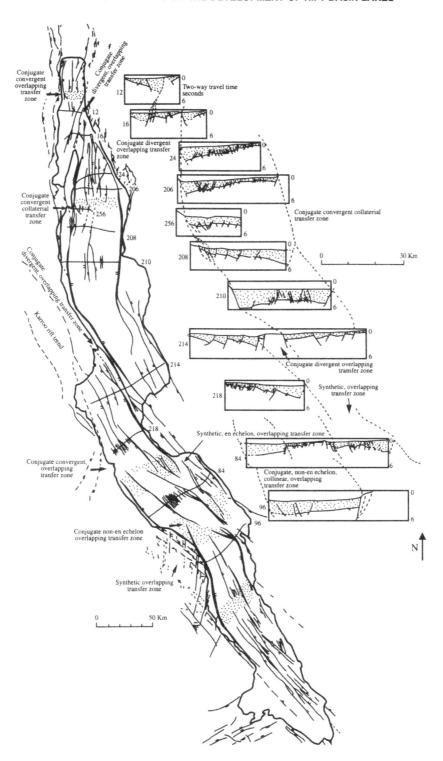

Figure 5. Changes in rift geometry and transfer zones in the Tanganyika Rift. The cross-sections are based upon seismic reflection profiles plotted in two-way travel time in seconds (from Morley, 1988).

Table 1. Calculation of Rates of Motion on Boundary Faults in the East African Rift

	Maximum synrift basin fill thickness, km	Maximum depth to Precambrian basement, km	Duration of rifting, Ma	Extension, km	Vertical throw on boundary fault (during Tertiary, km	Time-averaged extension rate, mm/yr	Time-averaged throw rate, mm/yr
Lake Tanganyika	7	7	9–25	4.5	7	0.5–0.18	0.78–0.28
Lake Rukwa	7.5	10.5	7	6	7.5	0.85	1.07
Northern Kenya Rift (Lokichar)	7	7	13	12	7	0.92	0.54

1969; Walsh and Dodson, 1969; Zanettin et al., 1983; Williamson and Savage, 1986; Boschetto, 1988; Feibel, 1988; Morley et al., 1992b). Thicknesses, extension, and throw rates in Table 1 were calculated from seismic lines. In most of the East African rift basins the synrift fill thickness corresponds to the depth to Precambrian basement and may include both sedimentary and volcanic (if present) rocks. In Lake Rukwa this value was calculated to the top of the Karroo Super Group (Fig. 6).

For the East African Rift the greatest time-averaged extension rates occur in the northern Kenya Rift and the lowest in Lake Tanganyika (Fig. 7). However, the throw rate in Lake Tanganyika is greater than that of the northern Kenya Rift. This supports the concept that, although some rifts have higher amounts of extension than others, they have not necessarily undergone tectonic subsidence either faster or by greater amounts than those rifts with lower amounts of extension.

The Western Arm is characterized by large, deep lakes that commonly occupy the entire width of the rift. The Eastern Arm contains smaller lakes that do not fill the entire width of the rift. This is, in part, because of climatic conditions. The Eastern Arm is much more arid than the Western rift. However, even under the same climatic conditions as Lake Tanganyika, Lake Turkana, the largest lake in the Eastern Arm, would not be deep and narrow, but rather become very wide and relatively shallow. This difference between the two basins is probably, at least partly, related to the different styles of the boundary faults, which in turn reflect the different thermal histories of the two branches.

3. Evolution of rifts with time

Rifts dominated by large half-grabens may evolve in several ways. The activity of boundary faults may change so that a full-graben geometry is actually a composite of half-grabens of switching polarity (e.g., Rosendahl et al., 1986). This results in switching of drainage patterns and sedimentary facies patterns as the rift margins alternate between boundary-fault margins and flexural margins. Another change that is commonly found in rifts is that the minor faults in a half-graben commonly deactivate prior to cessation of activity on the major boundary fault (e.g., line 24 Lake Tanganyika, Morley, 1988). Consequently towards the end of a rift cycle the rift floor may become smoother and less of a maze of minor tilted fault blocks; hence sediment dispersal patterns may become simpler.

The linkage of major faults may also change the half-graben architecture. There seems to be some indication that faults of a certain polarity may become dominant over distances of hundreds of kilometres. In the northern Kenyan Rift for example the original half-graben pattern in the early Tertiary may have been a series of half-grabens that alternated in polarity; by the middle Miocene, however, the half-grabens were dominated by easterly dipping boundary faults (Morley et al., 1992b).

An example of the complex structural evolution is provided by the northern Kenya Rift. During the latest Oligocene to early Miocene several west-thickening half-grabens formed in the area of the present-day Lokichar and Kerio basins. Although extensive areas of volcanic rocks were present to the north-northwest of these grabens, they were initially a depositional site for basement derived arkosic grits. A Precambrian basement terrain located to the south and drained by northeasterly flowing rivers was probably the source for this early basin-fill. Morley et al. (1992b) have suggested that this depositional system may have extended as far south as the Tugen Hills (Kamego Formation).

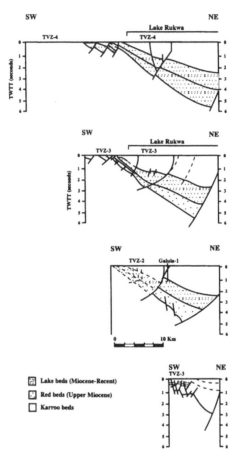

Figure 6. Cross-sections through the northwestern portion of the Rukwa Rift based upon seismic reflection profiles. Line locations are shown in Fig. 7 (from Morley et al., 1992a).

Volcanism shifted to the southeast during the Middle Miocene and formed an elongate north–south band of volcanic centers which extended from the Tugen Hills into the Napedet Hills and terminated in the Lothidok area. During this time volcaniclastic fluvial–lacustrine sediments were deposited in the basins and were interbedded with the volcanics (Boschetto, 1988).

At the end of the Miocene and on into the early Pliocene, the volcanic trend reoriented along a NNE–SSW line. Rifting trended N–S to NNE–SSW and extended from the northern half of the present-day Lake Turkana onshore into the Kerio Basin (Morley et al., 1992b). Data from seismic profiles, as well as outcrops on islands in Lake Turkana, indicate that this area experienced some volcanic activity during this time, in addition to being the primary depo-center for clastic sediments (Dunkleman, 1986; Dunkleman et al., 1988, 1989). Paleocurrent analyses from Lothagam Hill indicate that, during the Upper Miocene, fluvial sediment transport was initially from south to north, then reversed and flowed north to south

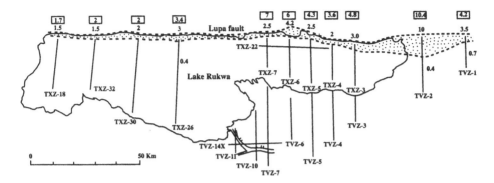

Figure 6, cont'd

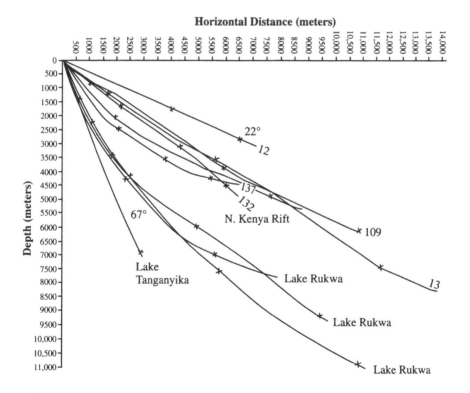

Figure 7. Comparison of fault profiles from seismic lines from some East African rift basins (from Morley, 1989). Points where the depth to seismic reflectors were calculated are indicated by X's. The variations between stacking velocities used in the calculations and actual rock velocities may result in errors of up to 10% in estimated depths. The Lake Tanganyika profile crosses the western boundary fault (line 210; Rosendahl et al., 1986). All three Lake Rukwa profiles cross the Lupa fault (see Fig. 9). Northern Kenya Rift profiles: lines 13 and 109 cross the late-Oligocene–early to mid-Miocene Lokichar Basin western boundary fault; lines 132, 137, and 12 cross boundary faults of the middle Miocene to Pliocene Kerio and Napedet Hills basins.

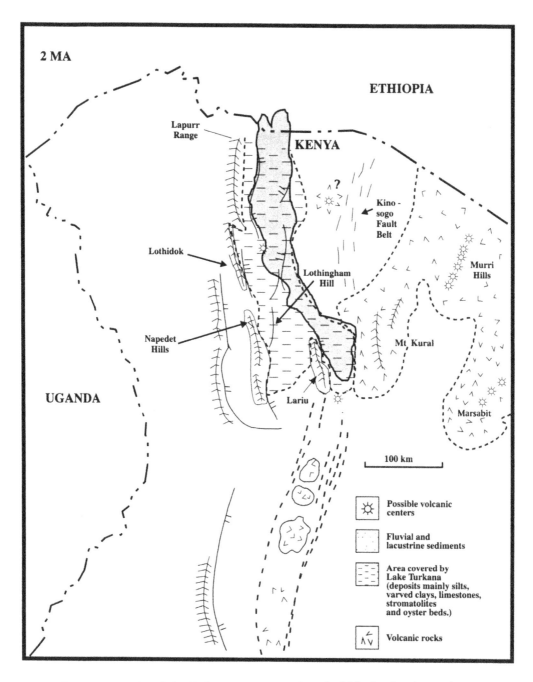

Figure 8. Map of the Lake Turkana area at approximately 2 Ma showing the maximum extent of lacustrine conditions based upon distribution of Plio-Pleistocene lake beds (after Feibel, 1988; Morley et al., 1992).

(Morley et al., 1992b). This indicates that a long-lived lacustrine environment had not been established in Lake Turkana at that time.

During the last 3 Ma, conditions in present-day Lake Turkana have alternated several times between lacustrine and fluvial environments (Feibel, 1988). The widest extent of lacustrine conditions as mapped from outcrops of lacustrine sediments is depicted in Fig. 8. The southern Lake Turkana Rift was probably also established at this time. The southern portion of the lake coincides with a zone of late minor faults. The minor fault swarms are the characteristic late deformation of the rift and extend along the entire length of the rift. Towards the north they lie to the east of the earlier half-grabens. This structural style marks a major change in the rift deformation, with abandonment of the large half-graben structural style. The change in style may be due to the intrusion of large volumes of magma into the crust and the rising of the geothermal gradient. The change in structural style marked the end of the potential for creating large (half-graben) rift lakes in the Kenya Rift. The numerous minor fault blocks have set up smaller lacustrine systems and created their own drainage systems. The Kenya Rift displays various stages of transition between severe erosion of the old rift topography, where remnants of the old half-graben drainage systems are partially preserved (southwestern Turkana), to the initial stages of modification (e.g., Elgayo Escarpment-Tugen Hills area), where much of the synrift topography is still preserved. This old topography lies to the east of the narrow trough created by the latest extension.

CONCLUSIONS

In this brief review we have tried to illustrate how structural styles at different scales can influence sedimentation patterns in rifts, and how some of those structural styles may change with time. Whilst regional doming may occur due to thermal effects or magmatic underplating, important uplifts may also be due to the isostatic responses due to faulting. The effects of faulting are more widespread in rifts and lead to certain predictable patterns of basin geometry. At the largest scale rift segment geometry affects drainage patterns over hundreds of kilometres, and represents the rift basin geometry at the largest scale. Within rift segments the geometry of transfer zones associated with boundary faults may affect one or several large drainage systems and may provide crucial entry points for large volumes of clastic material into the rift segment. At the smallest scale considered here individual fault blocks will affect the thicknesses and the detailed distribution of fine and coarse-clastic material within a basin.

Tectonism has been a major control on the development of lakes in the East African Rift System. The timing and style of tectonic movements has influenced drainage pattern development, hydrological budget, and the amount and type of sediment supplied to the rifts. Boundary fault geometries are the major control on the thicknesses of syntectonic basin-fill (Buck, 1988; Morley, 1989). Variations in extension (horizontal motions) and maximum synrift basin fill (vertical motion) between rifts can be characterized by values between high syntectonic sediment thickness to extension ratios that are associated with steep faults and deep depths to detachment (greater than 15 km), and those with low syntectonic sediment thickness to extension ratios that are characteristic of relatively low-angle faults and shallow depths to detachment (less than 15 km). The differences between these rift end members are summarized in Fig. 9.

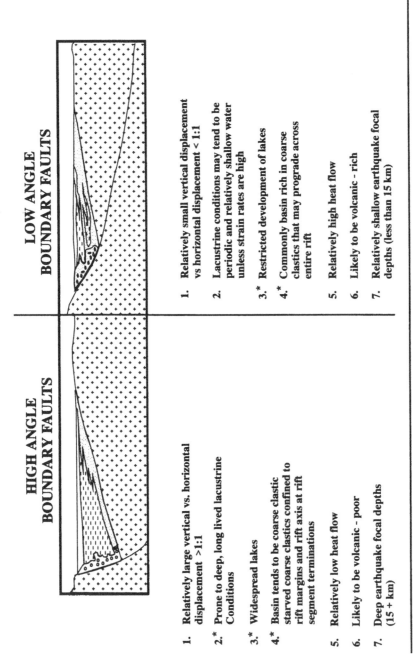

LOW ANGLE BOUNDARY FAULTS

1. Relatively small vertical displacement vs horizontal displacement < 1:1

2.* Lacustrine conditions may tend to be periodic and relatively shallow water unless strain rates are high

3.* Restricted development of lakes

4.* Commonly basin rich in coarse clastics that may prograde across entire rift

5. Relatively high heat flow

6. Likely to be volcanic – rich

7. Relatively shallow earthquake focal depths (less than 15 km)

HIGH ANGLE BOUNDARY FAULTS

1. Relatively large vertical vs. horizontal displacement >1:1

2.* Prone to deep, long lived lacustrine Conditions

3.* Widespread lakes

4.* Basin tends to be coarse clastic starved coarse clastics confined to rift margins and rift axis at rift segment terminations

5. Relatively low heat flow

6. Likely to be volcanic – poor

7. Deep earthquake focal depths (15 + km)

* Other factors, particularly climate may considerably modify these conditions arising from tectonic influence

Figure 9. Characteristics of rift basins bounded by steeply dipping, deep detachment faults versus rifts bounded by low-angle, shallow detachments (after Morley, 1989). Characteristics are generalized and other factors may also influence rift geometry and sedimentation. Also, rifts may evolve from one style to the other during their history and result in composite forms.

The evolution of faulting in rifts can follow numerous different paths. Some of the more important evolutionary patterns are as follows: 1) during the initial stages of rifting, lateral propagation and linkage of initially separate faults and basins, 2) deactivation of large faults and replacement of narrow deep half-grabens with broader, more symmetric thermal sag basins (slower subsidence rates), 3) a change from a half-graben style to minor fault swarms (Kenya Rift), and 4) changes from higher- to lower-angle boundary faults with time or vice versa.

REFERENCES

Boschetto, H. B., 1988, Geology of the Lothidok Range, northern Kenya, Unpublished MS Thesis, University of Utah, Salt Lake City, Utah.

Buck, R. W., 1988, Flexural rotation of normal faults: Tectonics, v. 7, pp. 959–973.

Chapman, G. R., Lippard, S. J., and Martyn, J. E., 1978, The stratigraphy and structure of the Kamasia Range, Kenya Rift Valley: Journal of the Geological Society, London, v. 135, pp. 265–281.

Cohen, A. S., 1989, Facies relationships and sedimentation in large rift lakes and implications for hydrocarbon exploration — examples from lakes Turkana and Tanganyika: Palaeogeography, Palaeoclimatology, Palaeoecology, v. 70, pp. 65–80.

Cohen, A. S., 1990, Tectono-stratigraphic model for sedimentation in Lake Tanganyika, Africa, *in* Katz, B. (ed.), Lacustrine Basin Exploration: case studies and modern analogues: American Association of Petroleum Geologists Memoir 50, pp. 137–150.

Cohen, A. S., Soreghan, M. J., and Scholz, C. A., 1993, Estimating the age of formation of lakes: An example from Lake Tanganyika, East African Rift system: Geology, v. 21, pp. 511–514.

Crossley, R., 1984, Controls of sedimentation in the Malawi rift valley, central Africa: Sedimentary Geology, v. 40, pp. 33–50.

Degens, E. T., von Herzen, R. P., and Wong, H. K., 1971, Lake Tanganyika: Water chemistry, sediments, geological structure: Naturwissenschaften, v. 58, pp. 229–241.

Dickinson, W. R., Lawton, T. F., and Inman, K. F., 1986, Sandstone detrital modes, central Utah foreland region: stratigraphic record of Cretaceous–Paleogene tectonic evolution: Journal of Sedimentary Petrology, v. 56, pp. 276–293.

Dixey, F., 1945, Miocene sediments in south Turkana: Journal of the East African Natural History Society, v. 18, pp. 13–14.

Dunkelman, T. J., 1986, The structural and stratigraphic evolution of Lake Turkana, Kenya, as deduced from a multichannel seismic survey: Unpublished MS Thesis, Duke University, Durham, North Carolina.

Dunkelman, T. J., Karson, J. A., and Rosendahl, B. R., 1988, Structural style of the Turkana Rift, Kenya: Geology, v. 16, pp. 258–261.

Dunkelman, T. J., Rosendahl, B. R., and Karson, J. A., 1989, Structure and stratigraphy of the Turkana Rift from seismic reflection data: Journal of African Earth Sciences (and the Middle East), v. 8, pp. 489–510.

Ebinger, C. J., 1989, Tectonic development of the western branch of the East African rift system: Geological Society of America Bulletin, v. 101, pp. 885–903.

Ebinger, C. J., Crow, M. J., Rosendahl, B. R., Livingstone, D. A., and LeFournier, J., 1984, Structural evolution of Lake Malawi, Africa: Nature, v. 308, pp. 627–629.

Ebinger, C. J., Rosendahl, B. R., and Reynolds, D. J., 1987, Tectonic model of the Malawi rift Africa: Tectonophysics, v. 141, pp. 215–235.

Ebinger, C. J., Deino, A., Drake, B., and Tesha, A. L., 1989, Chronology of volcanism and rift basin propagation: Rungwe volcanic province, East Africa: Journal of Geophysical Research, v. 94, pp. 85–803.

Feibel, C. S., 1988, Paleoenvironments of the Koobi Fora Formation, Turkana Basin, northern Kenya: Unpublished PhD Dissertation, University of Utah, Salt Lake City, Utah.

Frostick, L. E., and Reid, I., 1986, Evolution and sedimentary character of lake deltas fed by ephemeral rivers in the Turkana basin, northern Kenya, in Frostick, L. E., Renault, R. W., Reid, I., and Tiercelin, J. J. (eds.), Sedimentation in the African Rifts: Geological Society Special Publication No. 25, pp. 113–126.

Frostick, L. E., and Reid, I., 1987, Tectonic controls of desert sediments in rift basins, ancient and modern, in Frostick, L. E., and Reid, I. (eds.), Desert Sediments: Ancient and Modern: Geological Society Special Publication No. 35, pp. 53–68.

Hooijer, D. A., 1966, Miocene rhinoceroses of East Africa, in Fossil Mammals of Africa: British Museum, London, v. 21, pp. 117–190.

Hooijer, D. A., 1973, Additional Miocene to Pleistocene rhinoceroses of Africa: Zoology Meded., v. 46, pp. 149–177.

Kusznir, N. J., and Egan, S. S., 1990, Simple-shear and pure-shear models of extensional sedimentary basin formation: application to the Jeanne d'Arc Basin, Grand Banks of Newfoundland: American Association of Petroleum Geologists Bulletin, Memoir 46, pp. 305–322.

Lambiase, J. J., 1990, A model for tectonic control of lacustrine stratigraphic sequences in continental rift basins, in Katz, B. (ed.), Lacustrine Basin Exploration: case studies and modern analogues: American Association of Petroleum Geologists Memoir 50, pp. 265–276.

Leeder, M. R., and Gawthorpe, R. L., 1987, Sedimentary models for extensional tilt-block (half-graben) basins, in Coward, M. P., Dewey, J. F., and Hancock, P. L. (eds.), Continental Extensional Tectonics: Geological Society Special Publication No. 28, p. 139–152.

LeFournier, J., Chorowicz, J., Thouin, C., Balzer, F., Chenet, P. V., Henriet, J. P., and Masson, D., 1985, The Lake Tanganyika basin: tectonic and sedimentary evolution: Comptes Rendus de l'Acadamie des Sciences. v. 301, pp. 1053–1057.

Livingstone, D. A., 1965, Sedimentation and the history of water level change in Lake Tanganyika: Limnology and Oceanography, v. 10, pp. 607–611.

Maglio, V. J., 1969, A shovel tusked gomphothere from the Miocene of Kenya: Museum of Comparative Zoology Breviora, v. 310, pp. 1–10.

Morley, C. K., 1988, Variable extension in Lake Tanganyika: Tectonics, v. 7, pp. 785–801.

Morley, C. K., 1989, Extension, detachments, and sedimentation in continental rifts (with particular reference to East Africa): Tectonics, v. 8, pp. 1175–1192.

Morley, C. K., Nelson, R. A., Patton, T. L., and Munn, S. G., 1990, Transfer zones in the East Africa Rift System and their relevance to hydrocarbon exploration in rifts: American Association of Petroleum Geologists Bulletin, v. 74, pp. 1234–1253.

Morley, C. K., Cunningham, S. M., Harper, R. M., and Wescott, W. A., 1992a, Geology and geophysics of the Rukwa Rift, East Africa: Tectonics, v. 11, pp. 69–81.

Morley, C. K., Wescott, W. A., Stone, D. M., Harper, R. M., Wigger, S. T., and Karanja, F. M., 1992b, Tectonic evolution of the northern Kenyan Rift: Journal of the Geological Society, London, v. 149, pp. 333–348.

Nelson, R. A., Patton, T. L., and Morley, C. K., 1992, Rift–segment interaction and its relation to hydrocarbon exploration in continental rift systems: American Association of Petroleum Geologists, v. 76, pp. 1153–1169.

Rosendahl, B. R., 1987, Architecture of continental rifts with special reference to East Africa: Annual Review of Earth and Planetary Science Letters, v. 15, pp. 445–503.

Rosendahl, B. R., and Livingstone, D. A., 1983, Rift lakes of East Africa: New seismic data and implications for future research: Episodes, v. 83, pp. 14–19.

Rosendahl, B. R., Reynolds, D. J., Lorber, P. M., Burgess, D. F., McGill, J., Scott, D. L., Lambiase, J. J., and Derksen, S. J., 1986, Structural expressions of rifting: lessons from Lake Tanganyika, Africa, *in* Frostick, L. E. Renault, R. W., Reid, I., and Tiercelin, J. J. (eds.), Sedimentation in the African Rifts: Geological Society Special Publication No. 25, pp. 113–125.

Tiercelin, J. J., and Mondeguer, A., 1991, The geology of the Tanganyika Trough, *in* Coulter, G. (ed.), Lake Tanganyika and Its Life: New York, Natural History Museum, pp. 7–48.

Walsh, J., and Dodson, R. G., 1969, Geology of northern Kenya: Report of the Geological Survey of Kenya No. 82.

Wescott, W. A., Krebs, W. N., Englehardt, D. W., and Cunningham, S. M., 1991, New biostratigraphic age dates from the Lake Rukwa rift basin in western Tanzania: American Association of Petroleum Geologists Bulletin, v. 75, pp. 1255–1263.

Williamson, P. G., and Savage, R. J. G., 1986, Early rift sedimentation in the Turkana Basin, northern Kenya, *in* Frostick, L. E., Renault, R. W., Reid, I., and Tiercelin, J. J. (eds.), Sedimentation in the African Rifts: Geological Society Special Publication No. 25, pp. 267–283.

Zanettin, B., Justin-Visentin, E., Bellieni, G., Piccirillo, E. M., and Rita, F., 1983, Le volcanisme du basin du nord-Turkana (Kenya): age, succession et evolution structurale: BCREDP, No. 7, pp. 249–255.

East African Climate

A Review of Climate Dynamics and Climate Variability in Eastern Africa

S.E. NICHOLSON   *Department of Meteorology, Florida State University, Tallahassee, Florida, United States*

Abstract — This paper presents an overview of regional climate and spatial and temporal aspects of climate variability in eastern Africa. Causal mechanisms of climate variability are also considered. In general, widely diverse climates, ranging from desert to forest, exist over a relatively small area. Rainfall seasonality is quite complex, changing within tens of kilometers. Despite this diversity, interannual fluctuations of rainfall are markedly uniform over the region and appear to be governed by the same factors. In much of the region the annual cycle of rainfall is bimodal, with a main rainy season in March-to-May and "short rains" of October–November. The main rains are less variable, so that interannual variability is related primarily to fluctuations in the short rains. These are also linked more closely to large-scale, as opposed to local, atmospheric and oceanic factors. Rainfall fluctuations show strong links to the El Niño–Southern Oscillation (ENSO) phenomenon, with rainfall tending to be above average during ENSO years. The relationship appears to be indirect, however. The main causal mechanisms appear to be fluctuations in sea-surface temperatures in the tropical Atlantic and Indian Oceans, which in turn are loosely coupled to ENSO. For the region as a whole, the Atlantic seems to affect rainfall more than the Indian Ocean.

1. INTRODUCTION

Meteorologically, equatorial eastern Africa is one of the most complex sectors of the African continent. The large-scale tropical controls, which include several major convergence zones, are superimposed upon regional factors associated with lakes, topography and the maritime influence. As a result, the climatic patterns are markedly complex and change rapidly over short distances.

This is dramatically illustrated by the patterns of rainfall which prevail in the region (Fig. 1). The seasonality changes significantly within distances on the order of tens of kilometers; within eastern Africa are regions with one, two and three rainfall maxima in the seasonal cycle. The transition from desert, with rainfall less than 200 mm per year, to forests with annual rainfall exceeding 2000 mm, occurs over relatively small distances and altitudes.

On the surface, this complexity implies that the climate is to a large extent regionally controlled and that few generalities can be made concerning the region as a whole or its link to global climate. Fortunately, the complexity of the region's climate is underlain by rather simple and spatially coherent patterns of temporal variability. Likewise, while mean patterns are closely linked to regional factors, rainfall variability is clearly dominated by changes on the large-scale with clear links to the global tropics. These characteristics, which are demonstrated in this chapter, facilitate the study of the climate change in the region and its interpretation in terms of global climate.

For the IDEAL project, the main concern is water balance; therefore this chapter will be limited to rainfall variability, the prime determinant of the region's water balance. Emphasis will be on the region generally referred to as East Africa (i.e., the countries of Uganda,

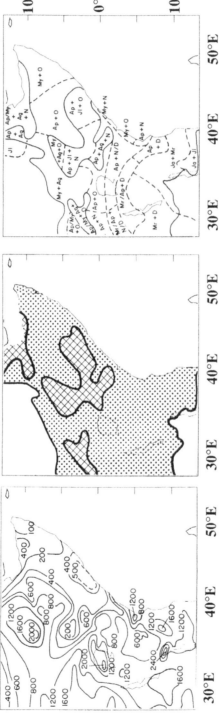

Figure 1. Map of rainfall characteristics over East Africa (from Nicholson et al., 1988a). Left: mean annual rainfall; Center: distribution of unimodal (no-shading), bimodal (dots), and trimodal (hatching) rainy seasons. Right: months of peak rainfall.

Tanzania and Kenya). In some analyses, a larger region is considered which includes Rwanda, Burundi, and small sectors of eastern Zaire, Ethiopia, Somalia and the Sudan. The chapter will start out with a general description of the large-scale climatic controls and the precipitation climatology. Then interannual variability of rainfall and its causes are considered.

2. CLIMATIC CONTROLS IN THE GENERAL ATMOSPHERIC CIRCULATION

The wind and pressure patterns governing the region's climate include three major air streams and three convergence zones (Fig. 2). The air streams are Congo air with westerly and southwesterly flow, the northeast monsoon and the southeast monsoon. Both monsoons, unlike the SW monsoon of Asia, are thermally stable and associated with subsiding air; they are therefore relatively dry. In contrast, the flow from the Congo is humid, convergent and thermally unstable, and therefore associated with rainfall. These air streams are separated by two surface convergence zones, the ITCZ and the Congo Air Boundary; the former separates the two monsoons, the latter, the easterlies and westerlies. A third convergence zone aloft separates the dry, stable, northery flow of Saharan origin and the moister southerly flow.

At low levels, the southeast and northeast monsoons prevail during the high and low sun seasons, respectively. The NE monsoon is made up of primarily of a dry air stream which has traversed the eastern Sahara, but during the NE monsoon season relatively humid currents from the Atlantic Ocean occasionally penetrate the region. The SE monsoon often splits into two streams when it encounters the coast, one continuing westward and the other turning abruptly northward to parallel the Somali coast. One reason for the relative dryness of both monsoons is that they flow more or less parallel to the shore. In such a situation the frictional contrast between the shore and the water induces subsidence of the air (Bryson and Kuhn, 1961). Another factor in their dryness is that, for complex reasons related to Coriolis force, easterly flow which moves equatorwards also tends to sink. Therefore, in most of eastern Africa, rainfall is highest during the transition seasons, when the strongly onshore flow is forced by topography and coastal friction to ascend.

A number of aspects of the climate of eastern Africa are poorly understood. One is the cause of the relatively arid conditions in much of the region. No primary cause can be identified; rather, many factors probably play a role (Trewartha, 1961; Anyamba 1984a). These would include the thermal and dynamic stability of the two monsoons, as described above. Another factor is that the moist air streams in the region are relatively shallow; there is dry, stable easterly air aloft, which originates from anticyclones over the Sahara and Arabia and also from the Mascarene High. Also, there is anticyclonic flow aloft because the fringes of high pressure cells over the Sahara and Saudi Arabia reach relatively low latitudes during much of the year.

Several other factors probably play a localized role. One is the highlands. Those of the Rift Valley block the moist, unstable westerly flow of Congo air from reaching coastal areas, especially in Ethiopia and Somalia. Also, leeward rainshadows produce subsiding, dry air. This leads to an extremely complex pattern of rainfall and aridity over the Ethiopian highlands, for example, with pockets of humid climates alternating with arid ones within a few tens of kilometers. An additional factor, only recently discovered, is the low-level Turkana jet (Kinuthia, 1992, Fig. 3). Such jet streams can produce patterns of descending air which enhance aridity. The cold upwelled water along the Somali coast likewise tends to

January Circulation

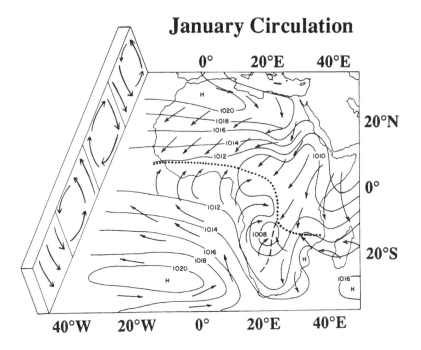

July/August Circulation

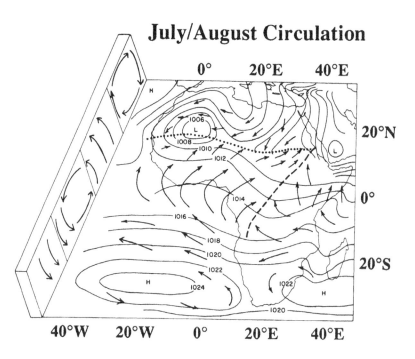

Figure 2. Schematic of the general patterns of winds, pressure and convergence over Africa (adapted from Nicholson et al., 1988a). Dotted lines indicate the Intertropical Convergence Zone, dashed lines, other convergence zones.

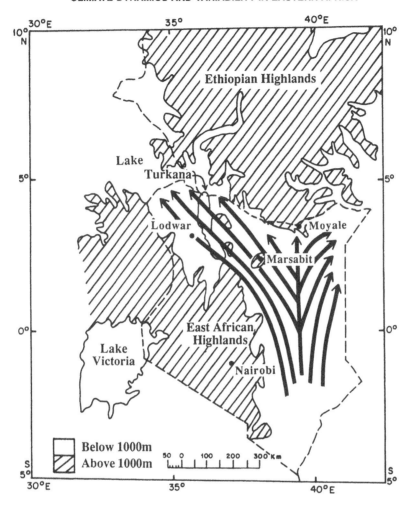

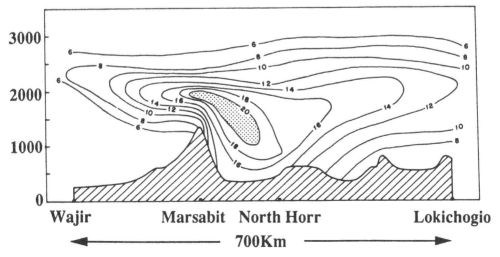

Figure 3. a) Small-scale topography over eastern Africa and the location of the Turkana jet (modified from Kinuthia, 1992). b) Typical easterly wind speeds of the Turkana jet as a function of altitude (in meters msl) (shading, speeds exceeding 20 m s^{-1}; modified from Kinuthia 1992).

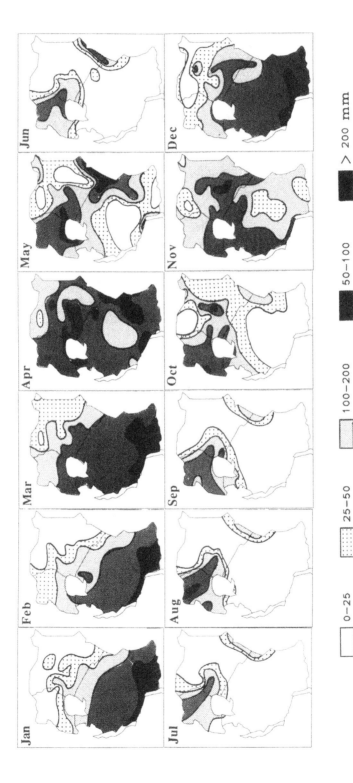

Figure 4. Mean monthly rainfall in mm during the period 1931–85.

suppress rainfall. Another factor is the low-level Somali jet just offshore, which, as part of the SE monsoon, parallels the coast and enhances the frictionally induced subsidence (Flohn, 1965; Lettau, 1978). Divergent flow resulting from regional pressure patterns and the extreme heating of the Ethiopian highlands enhance the summer aridity.

What brings rainfall (i.e., the general synoptic situations governing day to day weather) is unfortunately just as obscure as the causes of aridity. In general, incursions of the humid, unstable westerly Congo air stream bring high rainfall (Thompson, 1957; Griffiths, 1959; Trewartha, 1961). However, many of the phenomena associated with rainfall occurrences elsewhere in the tropics appear to be lacking here. For example, although rainfall in the region occurs as part of organized patterns of convection, clear linkages to either large-scale disturbances or to the convergence zones are apparent only in the far north.

Part of the reason for the limited knowledge of this region is that until recently it was believed most rainfall in the tropics originated as random thermal convection, i.e., isolated thunderstorms not associated with any large-scale disturbances. This myth has been dispelled for most of the tropics but for East Africa the picture has been slow to change. There is evidence of organized weather systems in the region (Forsdyke, 1949; Oliver, 1956; Thompson, 1957; Trewartha, 1961; Bhalotra, 1973), but these have never been analyzed and described, partly because the effects of the highlands mask the systems and make analysis difficult. The type of organized system so common elsewhere in tropical Africa, the wave perturbations such as the Easterly Waves of West Africa, have not been unambiguously identified in the region or linked to rainfall. However, a few studies have demonstrated their existence over the western Indian Ocean and near the East African coast (Freming, 1970; Gichuiya, 1970; Zangvil, 1975; Njau, 1982; Lyons, 1991).

3. RAINFALL CLIMATOLOGY

Mean annual rainfall for a large sector of eastern Africa is shown in Fig. 1. The areas with the most rainfall are those with the longest seasons, e.g., the northwest, the areas of highest relief surrounding Mt. Kenya and Mt. Kilimanjaro, and the rugged terrain northeast of Lake Malawi, northwest of Lake Victoria and in Ethiopia. Driest regions are in the northeast, where mean annual rainfall is generally less than 400 mm. In the desert core it is less than 200 mm. It exceeds 800 mm in only small portions of Kenya, but exceeds 1200 mm in nearly all of Uganda. Over most of Tanzania, mean annual rainfall is on the order of 800 to 1200 mm. In general, the highest rainfall is in the mountainous regions.

Much of the region experiences a bimodal seasonal distribution of rainfall, with maxima occurring in the two transition seasons. In large areas of Kenya and a few other regions, there is a third maximum, usually occurring in July or August. In the northern and southern extremes of the region, the seasonality is unimodal, with the maximum occurring during the high-sun season of the respective hemisphere.

Mean monthly rainfall for East Africa is depicted in Fig. 4. It is readily seen that the dominant pattern is a seasonal north–south movement of the main rainbelt, a consequence of the dominant influence of the Intertropical Convergence Zone (ITCZ). Heavy rainfall occurs in southern-hemisphere sectors from December to April, while intense rains occur from March to May north of the equator and again in October to November. From June to September the region is nearly rainless except for an area north and west of Lake Victoria and a small coastal strip from c. 1 to 7°S. These two areas receive rainfall nearly year-round.

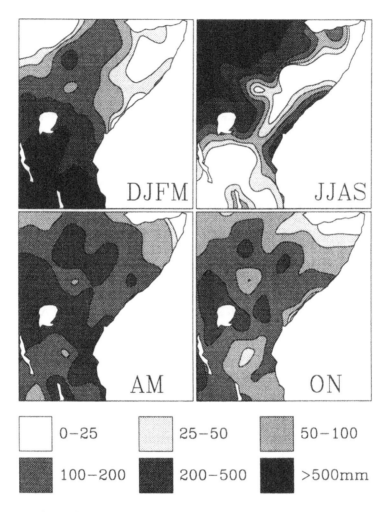

Figure 5. Mean seasonal rainfall in mm during the period 1931–85.

The only months in which significant rainfall occurs throughout the analysis sector are March and April and, to a lesser extent, November and December.

Figure 4 suggests that logical climatological seasons might be December-to-March, April–May, June-to-September and October–November. The means for these periods are shown in Fig. 5. The four months December-to-March contribute over 50% of the annual rainfall in the south (Fig. 6a). In equatorial regions where the seasonal distribution is strongly bi-modal, April–May (or sometimes March-to-May) is traditionally considered to be the "long-rains" while the October–November period is called in "short-rains" and is considered to be of secondary importance. The short rains make a large contribution to annual rainfall only in eastern equatorial latitudes.

It is interesting to note, however, that the October–November season contributes disproportionately to the inter-annual variability of rainfall. Correlation between annual totals and the October–November total exceeds 0.5 over nearly half of the analysis region and exceeds

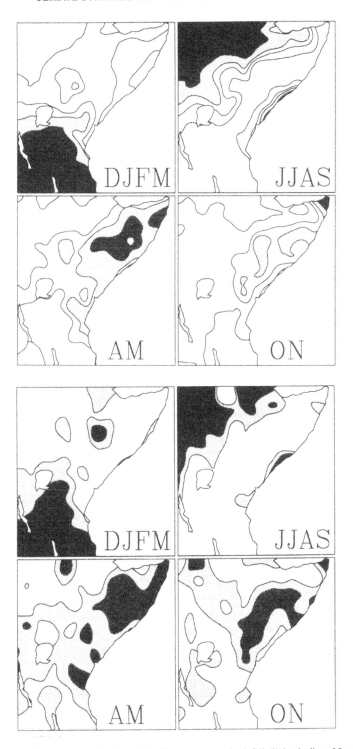

Figure 6. a) (top) Seasonal contribution (%) to the mean annual rainfall (light shading, 35 to 50%; dark shading, >50%). b) (bottom) Coefficient of correlation between the seasonal and annual rainfall departure series (approximately equal to the percent variance explained by each season; light shading, 50 to 70%; dark shading, >70%).

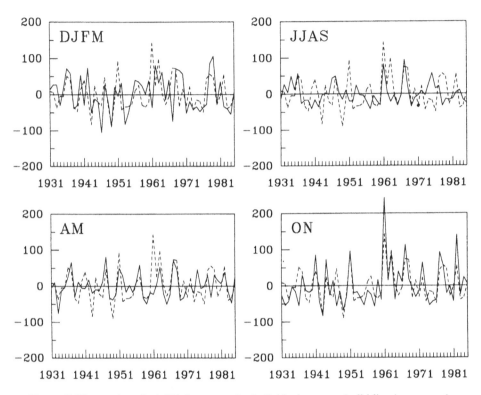

Figure 7. Time series of rainfall departures for individual seasons (solid lines) compared with the annual rainfall departure series. Data are for the time series R_T, representing eastern Africa as a whole, and are expressed as a percent standard departure.

0.7 over much of it (Fig. 6b). The correlations are somewhat higher during April–May, and the area of high correlation more extensive, but a much higher proportion of the rain occurs in these sectors during these two months. The importance of the "short rains" for interannual variability is underscored by Fig. 7, which compares annual time series of rainfall for the region as a whole with the corresponding time series for the four seasons defined above. A visual comparison shows that the similarity is clearly strongest with October–November rainfall. This is confirmed by linear correlation coefficients: the correlation between October–November departures and annual departures is 0.71, compared to 0.53 between April–May and annual rainfall departures.

4. SPATIAL PATTERNS OF RAINFALL VARIABILITY

In a previous study (Nyenzi, 1988), principal component analysis was used to identify dominant spatial patterns of rainfall variability in the region. The analysis was carried out for the period 1931 to 1982 using in one case three month averages and in a second, annual values. In the current chapter, the first eigenvector is of greatest relevance. This eigenvector is in-phase throughout the analysis sector in both cases (Fig. 8). The loadings are negative and fairly uniform throughout, although somewhat higher in the north than in the south.

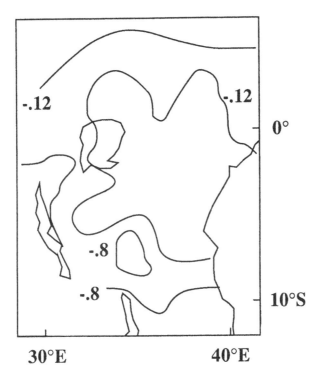

Figure 8. Eigenvector of the first principal component of rainfall variability (based on Nyenzi, 1988).

It is interesting to note the high percent variance explained by this eigenvector: 30% of the variance of seasonal data and 36% of the variance of annual data. For the "short rains" of October–November, it explains over 50% of the variance. This suggests that, despite the inhomogeneity of mean rainfall conditions over the region, interannual variability of rainfall is remarkably coherent within the region. This in turn implies that the primary factors governing interannual variability are large-scale and affect almost all of the region in the same way. These facts can greatly simplify studies of rainfall variability in East Africa and facilitate their interpretation.

The large-scale coherence implied by the uniformity of loadings of the first eigenvector and the high percent variance it explains is further substantiated by comparing rainfall time series for various subregions with that for the eastern African region as a whole. This is done in Fig. 9, using a regionalization given in Nicholson et al. (1988a) and consisting of the seven areas shown in Fig. 10. The time series for each of the seven regions is remarkably similar to the time series averaged for the all seven, despite the diverse climatologies of the smaller regions (see also Nicholson and Nyenzi, 1990).

The correlation coefficients between the pairs of rainfall time series in Fig. 9 indicate that the time series R_T captures for each region at least 50% of the variance on annual time scales and in four regions more than 80%. This demonstrates a high degree of coherence throughout the domain, consistent with the strong dominance of the first eigenvector (Fig. 8), which depicts a similar coherence.

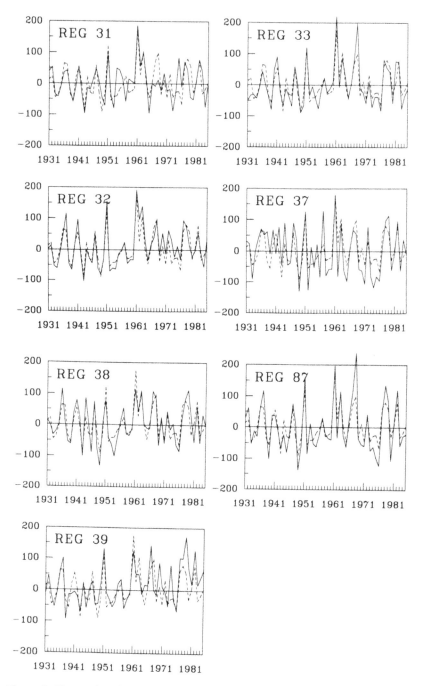

Figure 9. Time series of rainfall departures for the seven regions shown in Fig. 10 (solid lines) compared to the series R_T for eastern Africa as a whole (dashed lines). Values are expressed as a percent standard departure from the long-term mean.

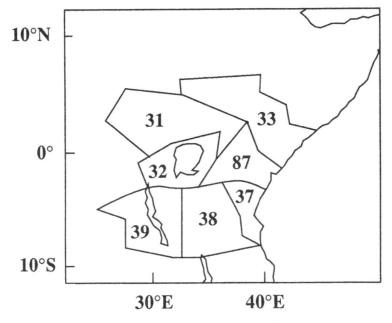

Figure 10. Seven rainfall regions of eastern Africa (numbers correspond to those in Nicholson et al., 1988a; rough seasonality of rainfall is November to April for regions 32, 38, 39 and 87, March to November for region 31, April/May and October/November for region 33 and March to May plus November/December for region 37).

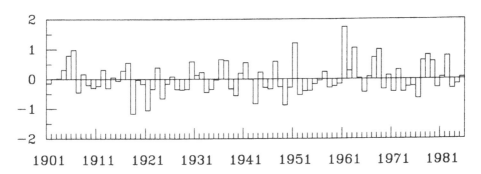

Figure 11. Time series of rainfall departures for eastern Africa as a whole for the period 1901 to 1985. Values are expressed as a percent standard departure from the long-term mean.

5. INTERANNUAL VARIABILITY

Interannual variability is described using the spatially averaged series R_T for the entire domain (Fig. 11). The series has been extended back to 1901 and forward to 1985, using the available station records. It should be noted that most of the peaks in rainfall correspond to Pacific ENSO years, e.g., 1941, 1951, 1957, 1963, 1968, 1972, 1978, and 1982. This linkage is further described in §6. Notable fluctuations are the extremely dry years within the period

Table 1. November Rainfall (mm) at Select Stations in Kenya in 1961 Compared to Mean Rainfall for the Month

	Mean	1961
Lodwar	16	190
Lamu	35	212
Lokitaung	39	302
Eldama	48	402
Makindu	48	478
Mandera	52	193
Malindi	54	238
Kitale	57	365
Wajir	58	612
Nakuru	60	280
Rumuruti	79	350
Garissa	81	412
Moyale	86	362
Mombasa	99	217
Marsabit	143	612
Murango	187	610
Embu	189	605
Machakos	189	600
Kitui	307	682
Meru	317	875

1918 to 1923, the relatively dry 1950s, the abnormally wet conditions in the 1960s and late 1970s, and the dry period c. 1968 to 1974.

A second characteristic of the interannual variability is its extreme magnitude in individual years. This is illustrated by the conditions of 1961, a year in which Lake Victoria rose several meters and reached levels unattained since the nineteenth century. The heaviest rainfall occurred in October and November (Flohn, 1987; Nicholson, 1989) (see Table 1). In northern Kenya, where fluctuations of the Turkana jet appear to intensify rainfall variability, the anomaly was most pronounced. The stations of Wajir, Eldama and Lokitaung received 612, 402 and 302 mm, respectively, in November, compared with monthly means of 58, 48 and 39 mm. November rainfall approached or exceeded annual means in the more arid sectors. Similar, but not quite so extreme, conditions occurred again in 1963.

A number of studies have utilized spectral analysis to demonstrate that these fluctuations are not completely random, but rather occur on preferred time scales of three to five years (Rodhe and Virji, 1976; Ogallo, 1979; Nicholson and Entekhabi, 1986). The spectrum for rainfall for the region as a whole is shown in Fig. 12. It is dominated by a strong peak at 5

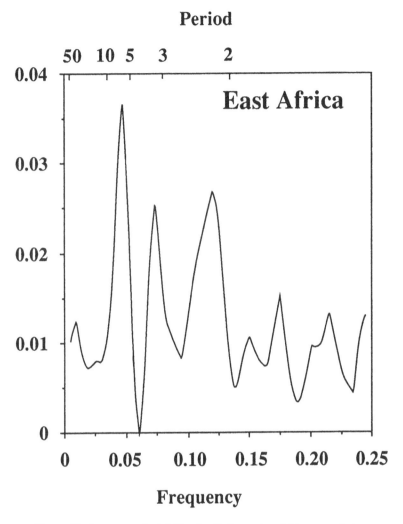

Figure 12. Spectrum of rainfall series R_T for eastern Africa as a whole.

to 6 years, but significant peaks at about 3.5 and 2.3 years are also evident. It is evident in spectra of both the long and short rains.

The 5–6 year peak is also the dominant one in the spectrum of the first principal component. This suggests that some forcing mechanism acting quasi-periodically with a time scale of 5–6 years is responsible for most of the interannual variability of rainfall in East Africa and for the relative coherence throughout the domain.

6. TELECONNECTIONS TO RAINFALL IN EASTERN AFRICA

Conditions of anomalous rainfall in eastern Africa are generally associated with anomalous conditions on a continental scale. Annual rainfall anomalies are generally well described by one of four preferred patterns. These include two patterns with anomalies of the same sign

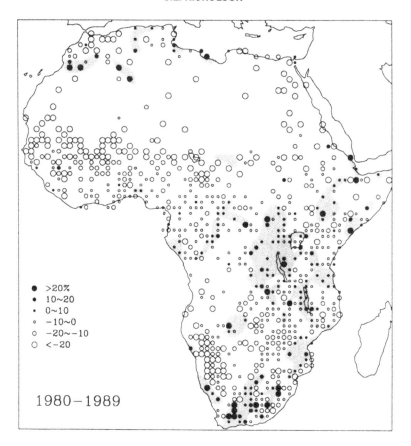

Figure 13. Mean rainfall for the decade 1980–89 (expressed as a percent departure from the long-term mean, with station data averaged over 1° squares).

over most of Africa, i.e., wetter or drier conditions occurring roughly synchronous over the continent. The latter case is illustrated by the mean anomaly map for the 1980s (Fig. 13). The remaining two patterns show strong opposition between the equatorial and subtropical latitudes. This is illustrated by the mean decadal anomalies for the 1950s (Fig. 14).

These configurations seem to be inherent features of rainfall variability in the region. They are apparent during recent historical periods (Fig. 15). They are also apparent in reconstructions of paleoclimates since the late Pleistocene (Fig. 16). These long-term fluctuations tend to occur during times of globally anomalous climate. Any factors explaining the interannual variability of rainfall in the region must be able to account for the existence of such patterns and the global teleconnections.

7. CAUSES OF INTERANNUAL VARIABILITY

7.1. General Aspects

The lack of a complete meteorological understanding of eastern Africa, as discussed in §2, makes it difficult to evaluate the causes of the interannual variability of rainfall in eastern

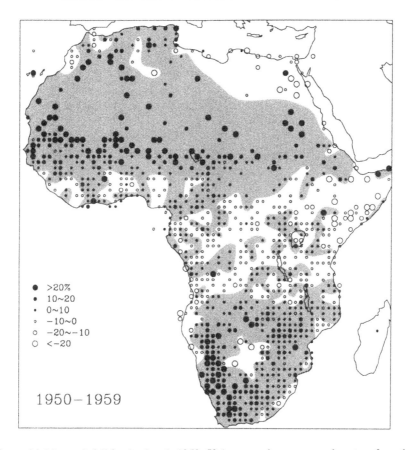

Figure 14. Mean rainfall for the decade 1950–59 (expressed as a percent departure from the long-term mean, with station data averaged over 1° squares).

Africa. A number of factors have been examined in various studies. Some, such as the position of convergence zones, are relatively local; others, such as sea-surface temperature variability, are large-scale and closely linked to the atmospheric general circulation. In most cases, the regional and large-scale factors are to some extent interrelated.

Numerous studies have focused on the importance of westerly flow in the region (Johnson and Mörth, 1960; Nakamura, 1969; Davies et al., 1985). Enhanced westerly flow with incursion of humid Congo air played a role in the 1961/62 floods (Anyamba, 1984b) and in the wet conditions of 1977/78 (Minja, 1985). However, there are examples of such westerlies being associated with anomalously dry conditions, especially over eastern Kenya (Nakamura, 1969; Kiangi and Temu, 1984).

Other studies have focused on the influence of the convergence zones which affect the region, namely the Inter-tropical Convergence Zone and the Zaire Air Boundary. These may be intensified in response to changes in the subtropical highs, the westerly flow, or other dynamic factors (Anyamba and Ogallo, 1985).

A number of recent studies have analyzed the interannual variability of rainfall in the region and have demonstrated some interesting associations with atmospheric and oceanic

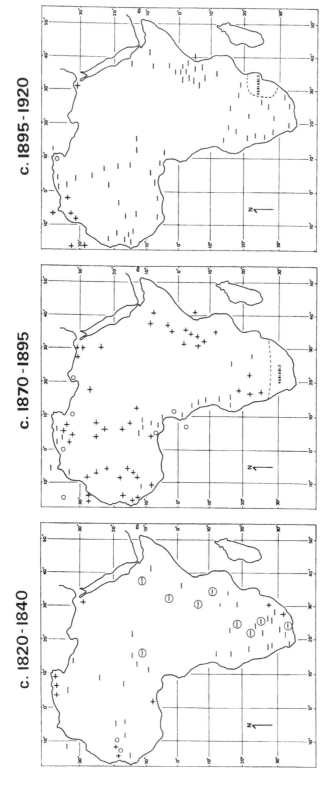

Figure 15. African rainfall anomalies for three historical periods (minus signs denote evidence of drier conditions; plus signs denote evidence of above-average rainfall; zeroes, near normal conditions; circled symbols denote regional integrators, such as lakes or rivers) (from Nicholson, 1989).

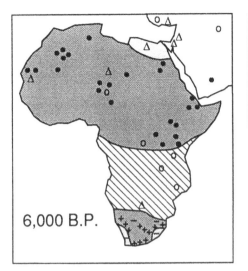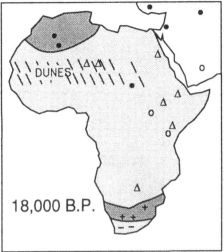

Figure 16. African paleoclimates c. 18,000 B.P. and 6,000 B.P. (rainfall indicators: shading or plus signs represnt generally greater rainfall than at present, minus signs indicate drier than at present, hatching represents decreasing rainfall; lake indicators: • = high stands, o = intermediate stands, Δ = low stands; from Nicholson, 1989; based on Street and Grove, 1979; Nicholson and Flohn, 1980; Street-Perrott and Roberts, 1983; Cockcroft et al., 1987; and others).

phenomena. Nicholson and Entekhabi (1986), Ropelewski and Halpert (1987), Ogallo (1987) and Farmer (1988) have all demonstrated a relationship between East African rainfall and the El Niño/Southern Oscillation (ENSO) phenomenon described in §7.3. Nicholson and Entekhabi (1987) have shown an even stronger association with sea-surface temperatures (SSTs) in the upwelling sector along the Benguela coast. Abnormally high SSTs in both the Atlantic and Indian Oceans (Fig. 17) were clearly linked to the tremendous floods in October and November, 1961 (Flohn, 1987; Nicholson, 1989). SSTs remained high throughout the very wet period from 1961 to 1963. Teleconnections to Pacific and global sea-surface temperatures are also apparent (Ogallo et al., 1988). It is interesting to note that the spectra (Fig. 18) of SSTs throughout the tropical Atlantic and Indian Oceans (Nicholson and Nyenzi, 1990) and that of the Southern Oscillation are all dominated by the 5 to 6 year spectral peak clearly evident in the rainfall spectrum for the region (Fig. 12). This hints at interrelationships in all three phenomena.

In sections 7.2 and 7.3, some of the evidence for the link to both sea-surface temperatures and the Pacific ENSO phenomenon is described. The analyses presented show that interannual variability in eastern Africa is clearly linked to the fluctuations in the global tropics. This conclusion is reinforced by the paleo- and historical fluctuations described in §6. Thus, knowledge of paleoclimate in eastern Africa will make a large contribution to the understanding of climate history of the global tropics, in particular, but also the earth as a whole. The discerned links will also help to draw conclusions concerning tropical response to future global climate change.

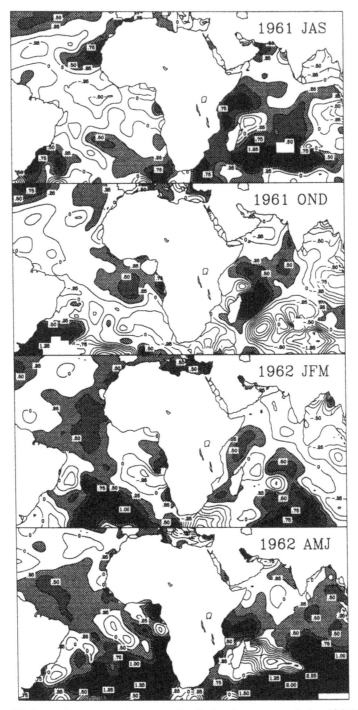

Figure 17. Sea-surface temperature anomalies during four seasons during 1961/62 (units are standard departures from mean; positive anomalies shaded; contours for each 0.25 units).

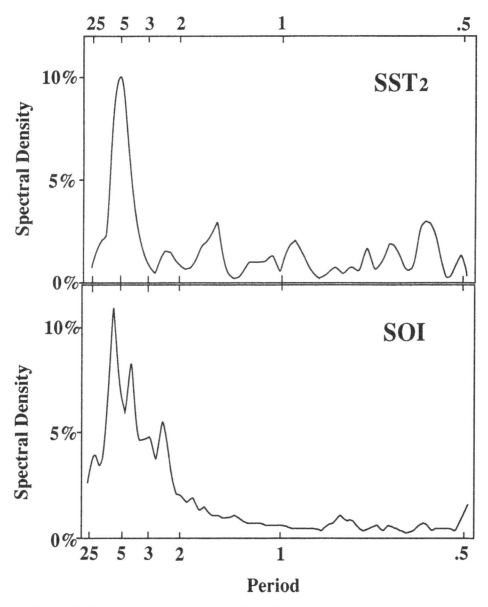

Figure 18. Power spectra of the Southern Oscillation (normalized sea-surface pressure difference between Tahiti and Darwin, Australia) and sea-surface temperatures along the Benguela coast of the southeastern Atlantic (from Nicholson, 1989). In lower diagram, dashed line indicates 95% significance level.

7.2. The Relationship Between Rainfall and Sea-Surface Temperatures

Figures 19 and 20 show the patterns of sea-surface temperature anomalies associated with conditions of anomalous rainfall. In each case, the temperature anomaly is the difference

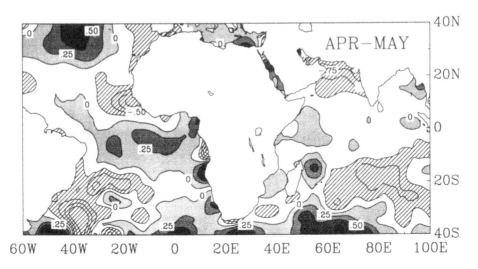

Figure 19. Seasonal SST composite differences between five wettest and five driest seasons for the long rains (April–May) (from Nicholson and Zheng, 1996; positive implies higher temperatures during wet years). SST anomalies are normalized by the standard deviation, with one standard deviation roughly equalling 1°C; contours are at intervals of 0.25 standard deviations, with positive departures shaded. Hatching indicates negative anomalies exceeding 0.25 standard deviations.

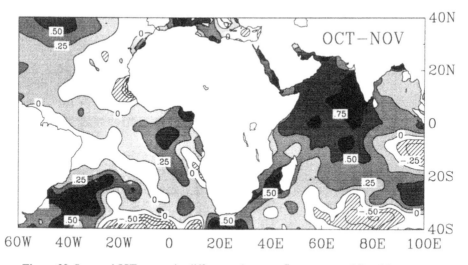

Figure 20. Seasonal SST composite differences between five wettest and five driest seasons for the short rains (October-to-November) (from Nicholson and Zheng, 1996; as in Fig. 19).

between SSTs in the five wettest and five driest years between 1946 and 1979, with the long and short rains being analyzed separately.

For the "long rain" season, the wet-minus-dry SST composite for April–May indicates negative SST anomalies in most of the Indian Ocean down to c. 35°S, positive anomalies further south, and a banded structure in most of the Atlantic, with positive anomalies in the

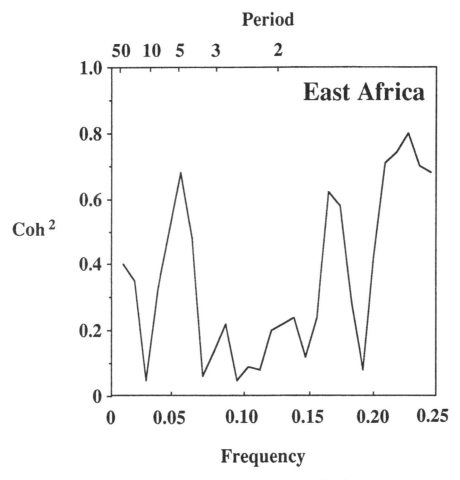

Figure 21. Coherence-squared between October-to-December rainfall and concurrent SSTs in Atlantic and Indian Ocean sectors (99% confidence limit = 0.72; 95% confidence limit = 0.61; from Nicholson and Zheng, 1996).

equatorial region and mid-latitudes, but negative anomalies in the subtropics. Consistent SST patterns appear to develop well in advance of the rainy season.

For the short rains, the SSTs of the October–November wet-minus-dry composite are nearly the opposite (Fig. 20). SST differences are predominantly positive in most of the Indian Ocean, especially in the west. In the Atlantic, SST differences are similar to those of the long rains, but with additional areas of positive anomalies. These patterns indicate that high rainfall in the short rainy season is associated with enhanced SST gradients in the Indian Ocean and, to a lesser extent, in the Atlantic. As with the long rains, anomalous SSTs are evident several seasons in advance (Nicholson and Zheng, 1996).

The general patterns apparent in Figs. 19 and 20 are confirmed using spectral analysis. An example, in Fig. 21, shows the coherence-squared of SSTs in several sectors of the Atlantic and Indian Oceans and rainfall in the October-to-November season. The coherence squared is particularly strong on time scales of about 4 to 7 years. Similar coherence is

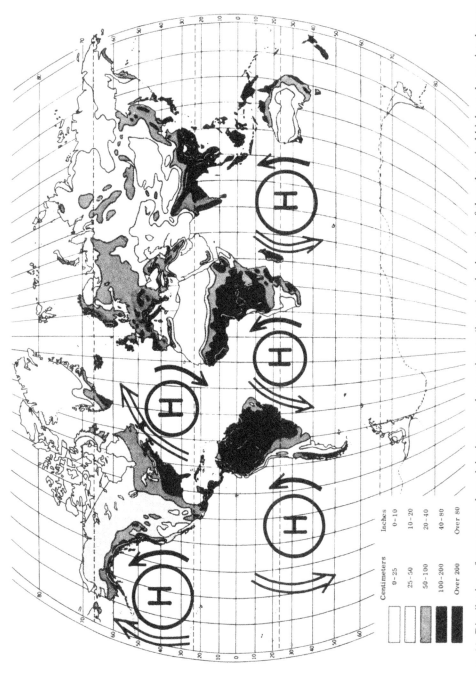

Figure 22. Distribution of mean annual precipitation over the continents (in cm) and schematic of the subtropical highs and associated ocean currents (blank arrows, warm currents; solid arrows, cold currents; adapted from Critchfield, 1983).

evident for numerous sectors of the tropical Atlantic and western Indian Oceans (Nicholson and Zheng, 1996).

Overall, the links to SSTs are stronger for the short rains than for the long rains and, in some cases, stronger for the Atlantic than for the Indian Oceans (Nicholson and Entekhabi, 1987). Likewise, they are strongest and most extensive on time scales of about 4 to 7 years, as illustrated with the example in Fig. 21. This is the same time scale as the ENSO phenomenon described in the next section. At least part of the rainfall–SST and rainfall–ENSO relationships described in the next section is due to the influence of ENSO on SSTs in these two ocean regions, but the ENSO influence on rainfall in East Africa may be secondary to the more local SST influence.

7.3. The Relationship Between Rainfall in Eastern Africa and ENSO

The El Niño–Southern Oscillation (ENSO) phenomenon is a quasi-global climatic fluctuation related to an asymmetric east–west tropical circulation linked to the subtropical highs. On the eastern portion of these highs is generally cold, upwelled water and strong subsidence, two characteristics which strongly promote aridity (Fig. 22). Hence, major coastal deserts are situated in these regions, along the western shores of low latitude continents. The air on the western portion of the subtropical highs is considerably more unstable and forced to ascend, partly for dynamic reasons and, over some continents, partly for orographic reasons. Here, along eastern margins of continents, the climate is relatively humid. The result is a series of vertical circulation cells, producing in the low latitudes a patchwork of arid and humid climates. These east–west oriented cells, termed the Walker circulation (Fig. 23), prescribe, in general, rising motion and relatively humid conditions on the eastern sides of continents and sinking motion and aridity on the west coasts.

Many years ago it was noted that along the desert west coast of South America, the cold water occasionally disappeared, replaced by warm water and heavy rainfall. The phenomenon was referred to as the "El Niño," or Christ Child, as it often occurred around Christmas. "El Niño" is now understood to be a global climatic phenomenon, marked by a general weakening of the global Walker circulation (the so-called Southern Oscillation phenomenon). Although most intense in the tropical Pacific, there are teleconnections to mid-latitude climates and evidence of quasi-synchronous warmings in the Atlantic and Indian Oceans as well (Fig. 24). ENSO is monitored using a parameter called the Southern Oscillation Index (Fig. 18), based on the pressure difference between Tahiti and Darwin (Australia), two stations on opposite sides of the Pacific Walker cell.

ENSO is linked to rainfall fluctuations in many areas of Africa, but these may be more directly a response to SST fluctuations in the Atlantic and Indian Oceans which occur in the context of ENSO. Fig. 25 shows annual rainfall departure series for several sections of eastern and Southern Africa; in Fig. 25a, ENSO years are shaded, but in Fig. 25b, the year following ENSO is shaded. These figures show a tendency for rainfall to be above average in most of these regions during ENSO years and for drought to occur during the following year. There is also a seasonal preference for the ENSO-related rainfall anomalies, as shown in Fig. 26. In eastern Africa, the positive anomaly tends to be during the short rains of the ENSO year, the drought during the long rains of the following year and short rains of the year prior to ENSO. In areas of southern Africa, rainfall is anomalous during these same seasons, which are not the core of the rainy season in southern Africa.

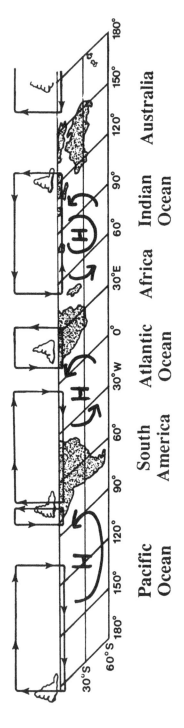

Figure 23. Schematic of the Walker circulation (modified from Newell, 1979)

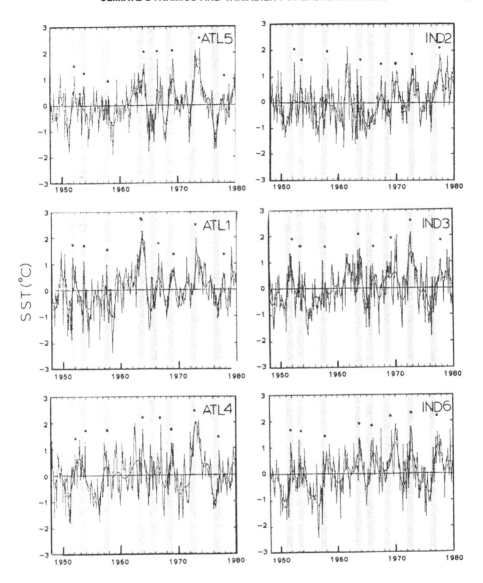

Figure 24. Time series of sea-surface temperatures (units: standard departures from the long-term mean) for select sectors of the tropical Atlantic and Indian Oceans (from Nicholson et al., 1988b). A1, A4, and A5, respectively, are Atlantic sectors along the Guinea Coast, east of northeast Brazil and along the Benguela coast of southern Africa. I2, I3 and I6 are Indian Ocean sectors just east of Madagascar, around the Somali current and in the Arabian Sea. Shading indicates ENSO years.

In eastern Africa, the strongest effect of ENSO appears to be during the short rains. Earlier it was noted that the short rains also have the strongest association with SSTs in the Atlantic and Indian Ocean. Figure 27 demonstrates that the ENSO-related SST anomalies of October-to-December of the ENSO year show essentially the same pattern as those during the wet-minus-dry composite of Fig. 20. Clearly the SST and ENSO influences in the region are

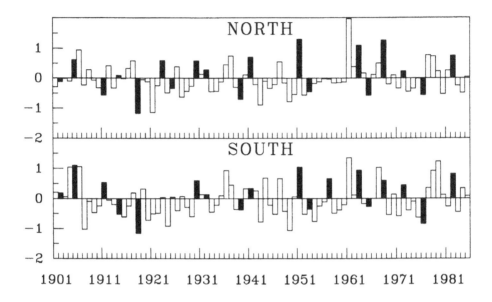

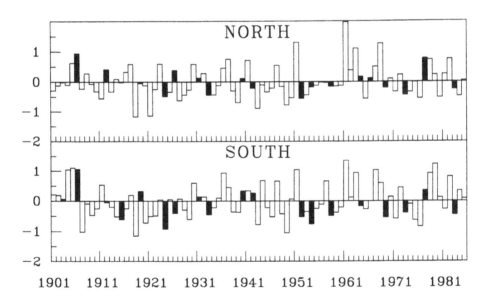

Figure 25. a) (top) Annual rainfall over eastern Africa 1901–85, with ENSO years shaded (rainfall expressed in units of standard departures from the long-term mean, for northern (regions 37, 38, 39) and southern sectors (regions 31, 32, 33, 87) of the eastern Africa (see Fig. 10). b) (bottom) Annual rainfall over eastern Africa 1901–85, with ENSO + 1 years shaded (rainfall expressed in units of standard departures from the long-term mean, for northern (regions 37, 38, 39) and southern sectors (regions 31, 32, 33, 87) of the eastern Africa (see Fig. 10).

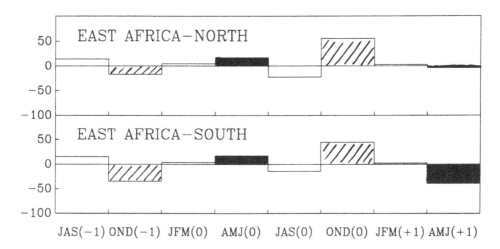

JAS(−1) OND(−1) JFM(0) AMJ(0) JAS(0) OND(0) JFM(+1) AMJ(+1)

Figure 26. Time evolution of seasonal rainfall anomalies during the two-year ENSO cycle (composite of eight episodes: 1951, 1953, 1957, 1963, 1965, 1968, 1972, 1976; units: percent standard departure; main rains shaded, short rains hatched). Seasons designated as −1 for year prior to ENSO, 0 for the ENSO year, and +1 for the year after ENSO.

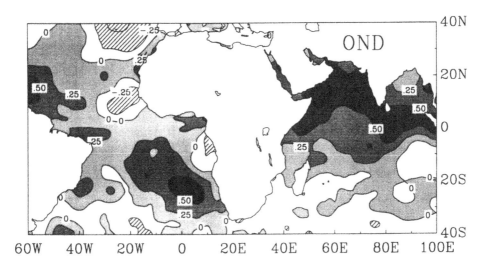

Figure 27. SST during the short rains (October-to-December season) for a composite of eight ENSO episodes (1951, 1953, 1957, 1963, 1965, 1968, 1972, 1976). SST anomalies are normalized by the standard deviation, with one standard deviation roughly equalling 1°C; contours are at intervals of 0.25 standard deviations, with positive departures shaded.

interrelated. The important point, however, is that the eastern Africa rainfall fluctuations are clearly linked to large-scale features of the general atmospheric circulation and marine sectors.

8. CONCLUSIONS

The interannual variability of rainfall is remarkably coherent throughout most of eastern Africa, despite quite diverse climatic mean conditions. The largest portion of this variability is accounted for by the "short rains" of the October-to-December season. Rainfall variability in the region shows strong teleconnections to the rest of Africa and to the global tropics.

Rainfall in eastern Africa is strongly quasi-periodic, with a dominant time scale of variability if 5 to 6 years. This is also the dominant time scale for the El Niño–Southern Oscillation phenomenon and for SST fluctuations in the equatorial Indian and Atlantic Oceans. Rainfall variability is closely linked to both ENSO and SSTs, and it tends to be enhanced in eastern Africa during ENSO years. These relationships, and the teleconnections discussed in §6, indicate that long-term rainfall fluctuations in eastern Africa are linked to quasi-global climate fluctuations. Thus, an understanding of paleoclimates in the region can contribute much to our understanding of global climate variability.

ACKNOWLEDGMENTS

This work was carried out with the support of NOAA Grant NO. NA89AA-D-AC203 and NSF Grant ATM-9024340.

REFERENCES

Anyamba, E.K., 1984a, Some aspects of the origin of rainfall deficiency in East Africa: Proceedings of the WMO Regional Scientific Conference on GATE, WAMEX and Tropical Meteorology, Dakar, Senegal, pp. 110–112.

Anyamba, E.K., 1984b, On the monthly mean lower tropospheric circulation and the anomalous circulation during the 1961/62 floods in East Africa: MSc Thesis, Department of Meteorology, University of Nairobi, Kenya.

Anyamba, E.K., and Ogallo, L.J., 1985, Anomalies in the windfield over East Africa during the East African rainy season of 1983/84: Proceedings of the first WMO workshop on diagnosis and prediction of monthly and seasonal atmospheric variations over the globe, Long-range forecasting research report series no. 6, v. 1, WMO/TD. No. 87, pp. 128–133.

Bhalotra, Y.P.R., 1973, Disturbances of the Summer Season Affecting Zambia: Zambian Meteorological Department, Technical Memorandum, no.3, 7 pp.

Bryson, R.A., and Kuhn, P.M., 1961, Stress-differential induced-divergence with application to littoral precipitation: Erdkunde, v. 15, pp. 287–294.

Cockcroft, M.J., Wilkinson, M.J., and Tyson, P.D., 1987, The application of a present-day climatic model to the Late Quaternary in Southern Africa: Climatic Change, v. 10, pp. 161–181.

Critchfield, H.J., 1983, General Climatology: Prentice Hall, Englewood Cliffs, N.J., 453 pp.

Davies, T.D., Vincent, C.E., and Beresford, A.K.C., 1985, July–August rainfall in West-Central Kenya: Journal of Climatology, v. 5, pp. 17–33.

Farmer, G., 1988, Seasonal forecasting of the Kenya coast short rains, 1901–84: Journal of Climatolology, v. 8, pp. 489–497.

Flohn, H., 1965, Studies on the Meteorology of Tropical Africa: Bonner Meteorologische Abhandlungen, v. 5, 57 pp.

Flohn, H., 1987, East African rains of 1961/62 and the abrupt change of the White Nile discharge: Paleoecology of Africa and the Surrounding Islands, v. 18, pp. 3–18.

Fordsdyke, A.G., 1949, Weather forecasting in tropical regions: Great Britain Meteorolological Office, Geophysical Memorandum, v. 10, no. 82, 32 pp.

Fremming, D., 1970, Notes on an Easterly Disturbances Affecting East Africa 5–7 Sept. 1967: East African Meteorological Department, Memoires, v. 13, 13 pp.

Gichuiya, S.N., 1970, Easterly disturbances in the Southeast monsoon: Proceedings of the Symposium on Tropical Meteorology, Honolulu, Hawaii.

Griffiths, J.F., 1959, The variability of annual rainfall in East Africa: Bulletin of the American Meteorological Society, v. 40, pp. 361–362.

Johnson, D.H., and Mörth, H.T., 1960, Forecasting research in East Africa, in Bargman, D.J., ed., Tropical Meteorology in Africa: Munitalp, pp. 56–137.

Kiangi, P.M.R., and Temu, J.J., 1984, Equatorial westerlies in Kenya. Are they always rain-laden?: Proceedings of the WMO Regional Scientific Conference on GATE, WAMEX and Tropical Meteorology, Dakar, Senegal, pp. 144–146.

Kinuthia, J.K., 1992, Horizontal and vertical structure of the Lake Turkana jet: Journal of Applied Meteorology, v. 31, pp. 1248–1274.

Lettau, H.H., 1978, Explaining the world's driest climate, in Lettau, H.H., and Lettau, K., eds., Exploring the World's Driest Climate, University of Wisconsin, Madison, pp. 182–248.

Lyons, S.W., 1991, Origins of convective variability over equatorial Southern Africa during austral summer: Journal of Climate, v. 4, pp. 23–39.

Minja, W.E.S., 1985, A comparative investigation of weather anomalies over East Africa during the 1972 drought and 1977–78 wet periods: MSc Thesis, Department of Meteorology, University of Nairobi.

Nakamura, K., 1969, Equatorial westerlies over East Africa and their climatological significance: Japanese Progress in Climatology, pp. 9–27.

Newell, R.E., 1979, Climate and the ocean: American Scientist, vol. 67, pp. 405–416.

Nicholson, S.E., 1989, African drought: Characteristics, casual theories and global teleconnections, in Berger, A., Dickinson, R.E., and Kidson, J.W., eds., Understanding Climate Change: American Geophysical Union, Washington, DC, pp. 79–100.

Nicholson, S.E., and Flohn, H., 1980, African environmental and climatic changes and the general atmospheric circulation in late Pleistocene and Holocene: Climatic Change, v. 2, pp. 313–348.

Nicholson, S.E., and Entekhabi, D., 1986, The quasi-periodic behavior of rainfall variability in Africa and its relationship to the Southern Oscillation: Journal of Climate and Applied Meteorology, p. 34, pp. 311–348.

Nicholson, S.E., and Entekhabi, D., 1987, Rainfall variability in equatorial and southern Africa: relationships with sea-surface temperatures along the southwestern coast of Africa: Journal of Climate and Applied Meteorology, v. 26, pp. 561–578.

Nicholson, S.E., Kim, J., and Hoopingarner, J., 1988, Atlas of African Rainfall and Its Interannual Variability: Florida State University, 252 pp

Nicholson, S.E., Nyenzi, B.S., and Cooper, H., 1988: Interannual variability in the Atlantic and Indian Oceans and its relationship to the Southern Oscillation. Twelfth Climate Diagnostics Workshop, NOAA, pp. 114–125.

Nicholson, S.E., and Nyenzi, B.S., 1990, Temporal and spatial variability of SSTs in the tropical Atlantic and Indian Oceans: Archives for Meteorology, Geophysics, and Bioclimatology, Ser. A, pp. 138–146.

Nicholson, S.E., and Zheng, H., 1996, The relationship between sea-surface temperatures and rainfall variability in eastern Africa: In preparation.

Njau, L.N., 1982, Tropospheric wave disturbances in East Africa: MSc Thesis, Department of Meteorology, University of Nairobi.

Nyenzi, B.S., 1988, Mechanisms of East African rainfall variability: PhD Thesis, Department of Meteorology, Florida State University.

Ogallo, L., 1979, Rainfall variability in Africa: Monthly Weather Review, v. 107, pp. 1133–1139.

Ogallo, L., 1987, Relationships between seasonal rainfall in East Africa and the Southern Oscillation: Journal of Climatolology, v. 7, pp. 1–13.

Ogallo, L.J., Janowiak, J.E., and Halpert, M.S., 1988, Teleconnection between seasonal rainfall over East Africa and global sea surface temperature anomalies: Journal of the Meteorological Society of Japan, v. 66, pp. 807–821.

Oliver, Mildred B., 1956, Some aspects of the rainfall in the Kenyan highlands: Geografiska Annaler, v. 38, pp. 102–111.

Rodhe, H., and Virji, H., 1976, Trends and periodicities in East Africa rainfall data: Monthly Weather Review, v. 104, pp. 307–315.

Ropelewski, C.F., and Halpert, M.S., 1987, Global and regional scale precipitation and temperature patterns associated with El Niño/Southern Oscillation: Monthly Weather Review, v. 115, pp. 1606–1626.

Street, F.A., and Grove, A.T., 1979, Global maps of lake-level fluctuations since 30,000 BP: Quaternary Research, v. 12, pp. 83–118.

Street-Perrott, F.A., and Roberts, N., 1983, Fluctuations in closed-basin lakes as an indicator of past atmospheric circulation patterns, in Street-Perrott, A., Beran, M., and Ratcliffe, R., eds., Variations in the Global Water Budget: Reidel, Dordrecht, pp. 331–345.

Thompson, B.W., 1957, Some reflections on equatorial and tropical forecasting: East African Meteorolological Department, Technical Memorandum, no. 7, 14 pp.

Trewartha, C., 1961, The Earth's Problem Climate: Methuen and Co., Ltd., 334 pp.

Zangvil, A., 1975, Temporal and spatial behaviour of large-scale disturbances in tropical cloudiness deduced from satellite brightness data: Monthly Weather Review, v. 103, pp. 904–920.

Sensitivity of Subtropical African and Asian Climate to Prescribed Boundary Condition Changes: Model Implications for the Plio-Pleistocene Evolution of Low-Latitude Climate

P.B. deMENOCAL   *Lamont-Doherty Earth Observatory, Palisades, New York, United States*

D. RIND   *NASA Goddard Space Flight Center, Institute for Space Studies, New York, United States*

Abstract — Climate model sensitivity experiments are used to assess the sensitivity of African and Asian climate to changes in solar insolation, high-latitude glacial ice cover, North Atlantic sea surface temperatures, and South Asian orography boundary conditions. Model climate responses to these boundary condition changes are discussed in terms of their response magnitude, regional extent, and response season signatures. By region, West African climate was most sensitive to cool North Atlantic SSTs which promoted annually cooler and drier surface conditions and stronger trade winds. East African/Arabian climate experienced annually cooler and drier conditions due to downstream effects of high-latitude ice sheets; the region was very sensitive to South Asian orographic changes as well. South Asian climate was very sensitive to variations in monsoon intensity resulting from orbital insolation variations and changes in Tibetan/Himalayan orography. These model results stress the importance of high-latitude boundary condition changes in affecting low-latitude African and Asian climate. Plio-Pleistocene African paleoclimate can be understood in terms of the relative influences of high- and low-latitude forcing factors.

INTRODUCTION

General circulation model (GCM) sensitivity tests are used to explore patterns of low-latitude climate variability attributable to changes in solar insolation, high-latitude ice sheet size, North Atlantic sea surface temperatures, and Asian orography. The primary objective of this study is to use climate model experiments to understand how seasonal components of Asian and African climate respond to specific boundary condition changes. This chapter is derived from results presented in deMenocal and Rind [1993] and the reader is directed to that reference for a complete discussion of the model experiments and results. The climatic signatures of each boundary condition change are evaluated by differencing two simulations which are configured identically except for the prescribed boundary condition change. The results are used to understand the dynamics which control the response of a given region to a given boundary condition change. The model results are then compared to available paleoclimate data to determine whether the sense of change indicated by the model results is consistent with the paleoclimate record.

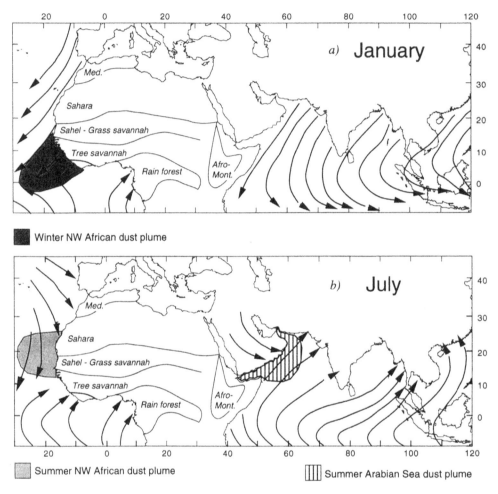

Figure 1. Modern surface wind fields for the eastern equatorial Atlantic and equatorial Indian Oceans for January (a) and July (b) [after Hastenrath, 1985]. Winter and summer dust plume trajectories from African and Asian dust source areas are shown [after Pye, 1987].

Modern Asian and African Climatology

Monsoonal circulation results from the differing heat capacities of land and water: Sensible heating warms land surfaces much more rapidly than the ocean mixed layer. During winter months, the South Asian landmass cools relative to the Indian Ocean and a broad high pressure cell develops over the Tibetan Plateau. General anticyclonic circulation over the Tibetan Plateau interacts with the regional Hadley circulation to produce dry and variable NE trade wind flow over the South Asian region from October to April (Fig. 1a). Sensible heating during Northern Hemisphere summer develops a strong low pressure cell over the Tibetan Plateau and regional cyclonic circulation prevails over South Asia from May to September [Fig. 1b; Hastenrath and Lamb, 1979]. The West African monsoon (Fig. 1a,b) develops in a similar fashion although it is less energetic. The strength of the summer Asian monsoon has been tied to the orographic effects of the Himalayas which focus the convergence of moist convection and latent heat at mid-tropospheric levels [Ramage, 1965; Hahn and Manabe, 1975].

East African climate is considerably more complex owing to its great topographic variability and resulting microclimates (see S. Nicholson, this volume). It has neither a purely monsoonal nor Mediterreanan-style climate; the main rains occur semi-annually and are associated with the overhead passage of the equatorial rain belt. March–May is the main rainy season ("long rains"), whereas the more variable "short rains" occur in October–November. Parts of the Ethiopian and Kenyan highlands receive copious rainfall throughout much of the year; mountain runoff feeds the numerous large lakes situated within the East African rift valley. Despite these rainy seasons much of low-lying NE Africa (below 1000 m) remains semiarid. Dry conditions are maintained by the combined effects of (1) high East African topography which blocks eastward penetration of Atlantic moisture, (2) the presence of cool, upwelled water near the Somali coast, and (3) frictionally induced subsidence of the Somali jet which parallels the East African coast (Flohn, 1965).

Asian-African Precipitation Regimes

The seasonality and magnitude of precipitation can be used to define climatological zones. Hsu and Wallace [1976] compiled precipitation data from 700 climatological stations distributed worldwide. Rainfall time series from each station were averaged into monthly means and then subjected to harmonic analysis which defined the phase and amplitude of the annual precipitation cycle. The phase and amplitude of the second harmonic was also calculated where a semiannual cycle was present.

Results for the African-Asian sector (Fig. 2a,b) demonstrate that African and Asian precipitation patterns fall into one of four broad categories: Equatorial, tropical, monsoonal, and Mediterranean [see Hsu and Wallace, 1976; Kendrew, 1961]. The annual cycle results are shown in Fig. 2a, the semiannual harmonics are shown in Fig. 2b. The equatorial pattern occurs where abundant rain occurs throughout the year as two distinct rainfall maxima when the sun is overhead. The tropical regime is defined as modest rains which occur during the hottest months and there is a pronounced dry winter season. The monsoon regime is defined as regions with very heavy summer rainfall and a pronounced winter dry season. The Mediterranean regime is broadly defined as wet winters and dry summers. The dominant semiannual precipitation pattern of the "equatorial" and "tropical" regimes occurs mainly over equatorial and East Africa and parts of Indonesia. The "monsoon regime" is clearly delineated for subtropical Africa, India, Indonesia, and northern Australia. The "Mediterranean regime" characterizes the Mediterranean periphery, but also extends over broad areas of northernmost Africa, the Middle East, and South Africa. As demonstrated in the following climate model experiments these climatic zones shift dramatically in response to changed boundary conditions.

THE MODEL

The GCM used in these experiments is the Goddard Institute for Space Studies (GISS) model II with $8° \times 10°$ medium grid resolution which produces a generally good simulation of the annual cycle of global climate [see Hansen et al., 1983]. The model solves the equations for conservation of mass, energy, momentum, and moisture for nine atmospheric layers. It calculates cloud cover, snow cover, soil moisture, and full radiative processes with

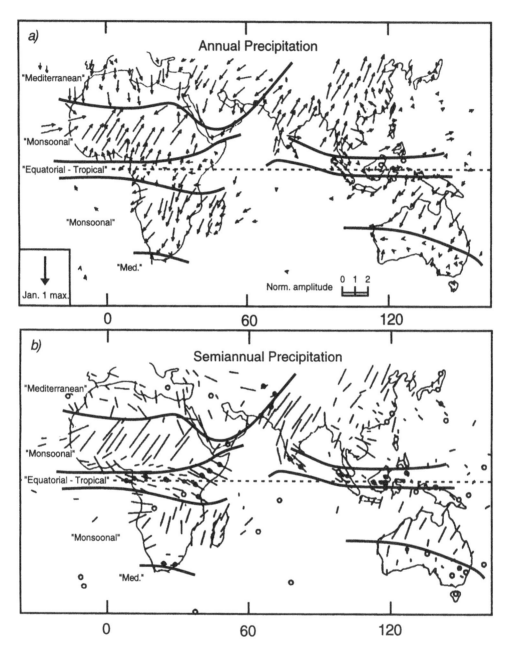

Figure 2. Annual (a) and semiannual (b) harmonics of precipitation for Africa, Asia, and Australia [from Hsu and Wallace, 1976]. Vector length is shown as normalized amplitude, where the maximum rainfall is divided by the monthly average rainfall (see text). Vector direction indicates month of maximum precipitation. Southward-pointing vectors indicate January 1 rainfall maxima, westwars vectors indicate April 1 maxima. Filled symbols in the semiannual plot indicate stations where the semiannual amplitude is greater than the annual. Open symbols indicate normalized amplitudes less than 0.075. "Equatorial-tropical," "Monsoonal," and "Mediterranean" rainfall regimes are shown based on classifications originally defined by Kendrew [1961] (see text).

Table 1. GCM Configuration Summary

GISS#	Title	Comments
822	CONT	Medium grid control run, modern SST, insolation and ice extent. 5-year average.
945	CONT.SST	Medium grid control run with cool 18 ka B.P. CLIMAP SSTs north of 25°N. 3-year average.
969	11K.FULL	Medium grid, modern SST, 11 ka B.P. insolation and ice extent. 3-year average.
969B	11K.NOICE	Medium grid, modern SST, 11 ka B.P. insolation, modern ice extent. 3-year average.
903	18K.FULL	Medium grid, 18 ka B.P. CLIMAP SST, insolation, and ice extent. 3-year average.
903E	18K.NOICE	Medium grid, 18 ka B.P. CLIMAP SST, and insolation, modern ice extent. 4-year average.
949	NOMTNS	Medium grid control with Tibetan Plateau elevation reduced to 500 msl. 2-year average.

a diurnal and seasonal cycle. The control (CONT) run has realistic topography and produces generally realistic temperature and precipitation fields when modern sea-surface temperatures (SSTs) are prescribed [Hansen et al., 1983]. Precipitation is calculated for both subgrid-scale convection, and also large-scale supersaturation which occurs when relative humidity exceeds 100%; the moisture condensed in both cases is allowed to reevaporate in unsaturated atmospheric layers below. SST fields are prescribed and non-interactive in all runs.

Comparisons between the model and observed climates are discussed by Druyan and Rind [1988]. The model also produces too much winter precipitation in South Asia (up to 4 mm/day) and there is excess winter precipitation in the Bay of Bengal (~12°N). The summer monsoon rains do not penetrate sufficiently northward into southeast Asia [Hansen et al., 1983]. The standard deviations of Northern Hemisphere January and July precipitation for the 5-year control run were 0.11 and 0.05 mm/day, respectively. The control run standard deviations of Northern Hemisphere January and July surface air temperatures were 0.14 and 0.06°C, respectively. The run diagnostics are based on multi-year averages of equilibrium runs (Table 1).

Experiment Configurations

Four experiments were selected to examine the low-latitude seasonal climate response to major changes in latest Cenozoic climatic boundary conditions (ca. last 5 Ma). This chapter is a summary of the model results presented in deMenocal and Rind [1993] and the reader is directed to that reference for a more thorough discussion of this work. The model experiments specifically examine the climatic effects of changes in the seasonal distribution of solar insolation, increased glacial ice cover, cold North Atlantic sea surface temperatures, and reduced Asian orography (Table 1).

To test the effects of orbital seasonal insolation changes on the monsoonal circulation, a model run with 11 ka B.P. insolation [Berger, 1978] and modern SST and ice distribution (11K.NOICE) is compared to the model control (CONT) run. The only difference between these runs is the seasonal distribution of insolation. The Earth's orbital configuration at 11 ka B.P. was such that perihelion occurred in Northern Hemisphere summer and the obliquity component (tilt) was higher than today, allowing ~8% more insolation during Northern Hemisphere summer and ~8% less during winter [Berger, 1978]. All model run configurations are summarized in Table 1.

Significant expansion of high-latitude ice sheets in Antarctica, North America, northern Europe occurred near 2.8 Ma, followed by a second period of enhanced glacial amplitude near 1 Ma [e.g., Shackleton et al., 1984; Ruddiman et al., 1989a]. A model configured with full Last Glacial Maximum (LGM; 18 ka B.P. radiocarbon years) boundary conditions (18K.FULL: LGM ice extent [Denton and Hughes, 1981], CLIMAP SSTs [CLIMAP project members, 1981], 120 m lower sea-level, and 18 ka B.P. orbital configurations [Berger, 1978]) with an identically configured run except that the ice sheets were restored to their modern extent (18K.NOICE). The 18 ka B.P. runs do not include the reduced glacial atmospheric CO_2 concentrations.

The effects of cooler glacial North Atlantic SSTs on low-latitude climate were investigated by comparing the control (CONT) run against an identically configured run with cooler 18 ka B.P. SSTs (CONT.SST) prescribed in the North Atlantic sector above 25°N. Sea ice was set at its modern latitudinal limit.

Ruddiman et al [1989b] summarized evidence regarding development of the Tibetan-Himalayan complex over the last 40 Ma, with particularly rapid uplift during the last 5 Ma. To evaluate orographic effects on the monsoon climate, the control run (CONT) is compared to an identically configured run except that Himalayan and Tibetan Plateau elevation is reduced to 500 m (NM).

RESULTS

1. Orbital Variations in Seasonal Insolation

Winter Climate. African and South Asian climate were relatively unaffected by the ~8% reduction in Northern Hemisphere winter insolation (Fig. 3a). The differenced surface wind vectors do not indicate any consistent change in direction or intensity of the trade wind circulation, nor are there significant changes in precipitation, except that South Africa became drier. Interestingly, central Europe and Asia became markedly warmer. An increased temperature gradient between North America and the adjacent Atlantic intensified the cyclonic Icelandic low (Fig. 3a) which, in turn, advected oceanic heat (and moisture) over northern Europe and Asia. A reverse analogy of this effect would be the dramatic Younger Dryas cooling of Europe and Asia resulting from cold North Atlantic SSTs [Rind et al., 1986].

December–February is the warm, wet monsoon season in South Africa (see Fig. 2), and the lower precipitation there most probably reflects a reduction in the summer South African monsoon intensity. Insolation variations due to orbital precession are out of phase between hemispheres for any given season.

Summer Climate. The summer SW Asian monsoon was greatly enhanced due to the 8% increase in summer insolation in the 11K.NOICE run; the West African monsoon was only

Model climate anomalies due to
10 kyr BP orbital insolation change

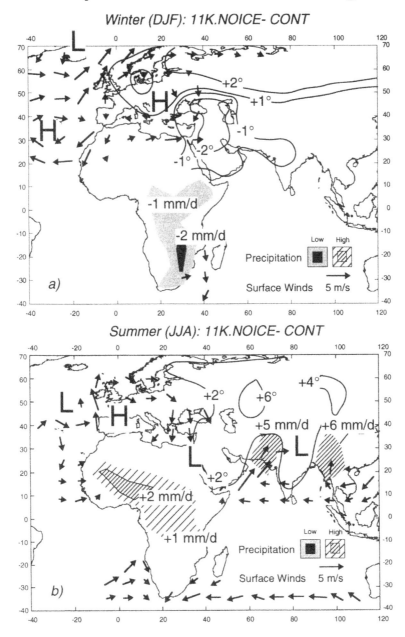

Figure 3. Effects of 11 ka B.P. solar insolation on the winter (DJF; a) and summer (JJA, b) model climate (11K.NOICE minus CONT). Climate patterns shown reflect anomalies attributable to the prescribed boundary condition change (see deMenocal and Rind [1993] for details).

moderately strengthened. Surface wind field differences (11K.NOICE-CONT) indicate that the summer SW monsoon circulation was enhanced by ~20% over the Arabian Sea and eastern Asia (2 m/s; Fig. 3b). The wind speed increases reflect an increase in the surface pressure gradient between South Asia and the Indian Ocean. Precipitation over South Asia increased by 2–6 mm/day (Fig. 3b). The northward shift of the pressure, precipitation, and temperature anomalies demonstrate that the stronger summer monsoon circulation penetrated more deeply into the Asian interior.

The African monsoon was less responsive to insolation changes than the Asian monsoon in this model (Fig. 3b). There was no characteristic deepening of the heat low over the Sahara and the wind field data do not indicate a net increase in the monsoonal inflow into the NW African interior. Increased precipitation (1–2 mm/day; Fig. 3b) in several West African gridboxes may indicate enhanced moisture convergence and moist convection. The weaker West African response to orbital insolation forcing may be due to the same reason that the actual African monsoon is less intense than the Asian monsoon: the high elevation and broad expanse of the Tibetan Plateau and the Himalayas promotes highly efficient focusing of sensible and latent heat which drives the monsoonal circulation [Ramage, 1965; Hahn and Manabe, 1975; Chen et al., 1985; Prell and Kutzbach, 1987].

Comparison with Other Model Results. A number of previous studies have demonstrated that the summer monsoon circulation is highly responsive to orbital insolation changes [Kutzbach, 1981; Kutzbach and Otto-Bliesner, 1982; Rind et al., 1986; Prell and Kutzbach, 1987; Kutzbach and Street-Perrott, 1985; Kutzbach and Guetter, 1984]. Prell and Kutzbach [1987] considered a series of GCM experiments in which summer insolation was varied from –6% to +12% of its modern value using the National Center for Atmospheric Research-Community Climate Model (NCAR-CCM). They noted that the intensity of the monsoon winds and precipitation was approximately linearly related to summer insolation intensity. The summer monsoon is so responsive because it is effectively a heat engine which responds directly (sensible heating) and indirectly (latent heating) to increased summer insolation. Their model suggests that an 8% increase in summer insolation produces a ~2 m/s (ca. 20%) increase in SW winds over the Arabian Sea (see Fig. 6d in Prell and Kutzbach, 1987), which is in good agreement with our results. However, contrary to the present findings, they suggest that the African monsoon is more responsive to orbital insolation variations than the Asian monsoon.

2. High-Latitude Glacial Ice Cover

Winter Climate. Increased glacial ice cover over North America, northern Europe, and Antarctica enhanced winter trade wind circulation over South Asia and reduced precipitation over Arabia, NE Africa, and South Asia. In contrast, West Africa was relatively unaffected. Although the main dynamic responses are centered at high latitudes, subtropical climate was sensitive to these changes.

The 18K results for DJF indicate much stronger trade wind circulation and broad temperature and precipitation decreases over South Asia, the Mediterranean, and East Africa (Fig. 4a). Surface wind anomalies in the 18K.FULL-18K.NOICE demonstrate that the anticyclonic circulation over South Asia is strengthened by 3–6 m/s (Fig. 4a). The zonal (16°N) mean u-wind velocities increased by +2.3 standard deviations over control values. The stronger winds are attributable to an intensification of the winter high pressure cell over South Asia. The eastern Mediterranean, South Asian, and East African regions are drier by

Model climate anomalies due to high-latitude glacial ice cover

Winter (DJF): 18KFULL-18KNOICE

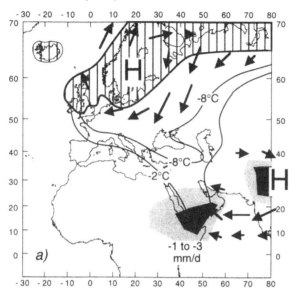

Summer (JJA): 18KFULL-18KNOICE

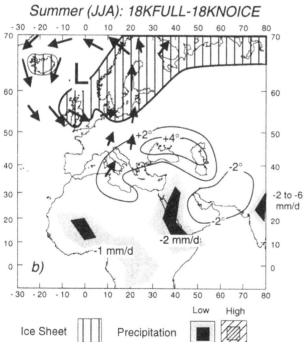

Figure 4. Effects of 18 ka B.P. glacial ice cover on the winter (DJF; a) and summer (JJA; b) model climate (18K.FULL minus 18K.NOICE). Climate patterns shown reflect anomalies attributable to the prescribed boundary condition change (see deMenocal and Rind [1993] for details).

up to 4 mm/day which, in some gridboxes, is equivalent to the entire annual precipitation. Surface temperature differences indicate that the largest cooling occurs over the same regions experiencing reduced precipitation: southern Asia and the Arabian Peninsula (Fig. 3a). A well-defined tongue of increased precipitation over the northern Arabian Sea/Persian Gulf is associated with enhanced ocean moisture advection due to the stronger easterly trade circulation.

The indicated precipitation decreases are important because winter is the precipitation season in these regions. Winter season cooling and drying in these regions is associated with the downstream advection of cooler and drier air from the high-latitude ice sheets. The high-latitude cooling is ultimately related to the radiative and sensible heat losses associated with the increased albedo and elevation of the ice sheets. Aside from the increased reflectivity of the ice, the elevation of the 18K.FULL Fennoscandian ice sheet (2.5 km) reduces the optical thickness of the overlying atmosphere which significantly diminishes the atmospheric retention of outgoing longwave radiation (greenhouse capacity) [Rind, 1987]. There is also a small ice sheet (<500 m thickness) over the Himalayan/Tibetan region in the 18K.FULL experiment [Denton and Hughes, 1981; Rind, 1986], and although it is neither high nor broad, it clearly plays a role in affecting the circulation and climate of the South Asian region.

Summer Climate. High-latitude glacial ice cover weakened the summer Asian monsoon and reduced southwest Asian and subtropical African precipitation (up to 6 mm/day; Fig. 4b). West African monsoonal climate was less sensitive to this boundary condition change than either East Africa/Arabia or South Asia. Surface temperature anomalies show that the ice sheet causes significant summer season warming (~10°C) over central Europe and the eastern Mediterranean (Fig. 4b). As noted by Rind [1987], summer warming in this region is due to downstream subsidence and consequent (dry) adiabatic warming of air flowing down off the Fennoscandian ice sheet (Fig. 4b).

The inclusion of full 18 ka B.P. ice cover and topography reduces the summer SW Asian monsoon winds, and there are new, "real" decreases in precipitation over a broad swath including West and East Africa, Arabia, and southwest Asia (Fig. 4b). The differenced 18K.FULL minus 18K.NOICE surface wind field indicates reduced cyclonic monsoonal circulation over South Asia (lower by 1–5 m/s; Fig. 4b). Comparison of the surface pressure anomalies for this experiment with the normal position of the summer heat low in the control run indicates that the ice cover reduces the development of the summer heat low over South Asia. This is the reason for the greatly reduced precipitation over East Africa and South Asia (–2 to –6 mm/day), and increased rainfall over Indonesia (Fig. 4b). The precipitation reductions over India are roughly 20–30% of the summer control summer precipitation over this region; reductions over the Arabian Peninsula approach ~50% of the control summer rainfall (see Fig. 4b).

Reduction of the monsoon circulation and rainfall is attributable to the direct cooling effects of high-albedo ice cover and the downstream dynamical effects of the high-elevation Fennoscandian ice sheet. Increased snow/ice cover over South Asian topography inhibits sensible and latent heating which drive the monsoon. This was studied explicitly by Rind [1987; see his Fig. 11] who studied the albedo and elevation effects of glacial ice cover using the GISS GCM. Emplacement of low-elevation ice (10m elevation only, but in full glacial locations) produced significant precipitation reductions over India and Arabia (2 mm/day). These reductions must be attributed to albedo and associated cooling effects only. Emplacement of the full elevation glacial ice cover (2–3 km) superimposed further precipitation reductions over the already drier regions of West India and Arabia [Rind, 1987].

Comparison with Other Model Results. Weakening of the summer Asian monsoon due to full glacial boundary conditions has been noted in many GCM studies [Gates, 1976; Manabe and Hahn, 1977; Rind and Peteet, 1985; Rind, 1987; Prell and Kutzbach, 1987]. When full glacial boundary conditions are considered, most of these models show reduced Asian precipitation. Few of these studies have isolated the effects of increased ice cover alone on the climate of the summer or winter monsoon, so it is difficult to determine a common response to glacial ice cover. Full glacial ice extent and elevation were responsible for South Asian precipitation decreases in the GISS model [Rind, 1987]. Prell and Kutzbach [1987] observed that the inclusion of glacial ice to their NCAR-CCM simulation of 18 ka B.P. climate caused a 20% decrease in the South Asian precipitation as well as a 2 m/s reduction in Arabian Sea winds.

Many of these regions have Mediterranean-type (winter-wet, summer-dry) rainfall patterns (see Fig. 2a), so the decreases in winter season rainfall due to expanded ice cover have important implications regarding glacial climate sensitivity of this low-latitude region. Specifically, the model results suggest that expanded high-latitude ice cover leads to enhanced aridification of the Asian monsoon dust source areas in Arabia and East Africa.

3. North Atlantic Sea-Surface Temperature

Winter Climate. Prescribing cooler (CLIMAP) 18 ka B.P. North Atlantic SSTs north of 25°N caused dramatic cooling and increased surface wind speeds over West Africa and the eastern equatorial Atlantic (Fig. 5a). Arabian and South Asian climate were relatively unaffected. Increased anticyclonic circulation over the cold North Atlantic strengthens the NE trade winds over West Africa (Fig. 5a). The anticyclonic circulation also advects cold central European air equatorward, cooling northwestern Africa by –2 to –4°C. These West African climate anomalies can be related directly to the dynamical effects associated with the anticyclonic circulation developed over the cold North Atlantic SSTs.

Summer Climate. Cooler North Atlantic SSTs reduced monsoonal surface winds into North Africa (1–2 m/s) and surface air temperatures decreased markedly (–2 to –6°C; Fig. 5b) and precipitation (up to 1 mm/day; Fig. 5b). The persistent anticyclonic circulation over the North Atlantic counteracted and reduced cyclonic monsoonal inflow into West Africa, reducing regional precipitation. No corresponding changes in Asian monsoonal climate were observed (Fig. 5b), with the exception of a limited area of reduced precipitation over East African and Arabia (1–2 mm/day). Summer precipitation decreases over Africa and Arabia can be related to enhanced surface air outflow from the respective continents due to the thermal effects of the cooler North Atlantic SSTs.

Comparison with Other Model Results. Rind et al. [1986] considered the effects of cooler North Atlantic SSTs on European and Asian climate and they concluded, as we do, that the largest amplitude changes are found at higher latitudes, whereas the low-latitudes are only moderately affected. Closer inspection of these model runs suggest that subtropical West African climate is responsive to cooler North Atlantic SSTs.

GCM experiments have been used to understand the origin of sub-Saharan drought. The surface climate of West Africa is very sensitive to Atlantic SST anomalies. Using a GCM with an interactive ocean, Druyan [1987] compared the SST distributions associated with anomalously wet and dry conditions in the sub-Sahara and Sahel regions of NW Africa. His

Model climate anomalies due to cool North Atlantic SSTs

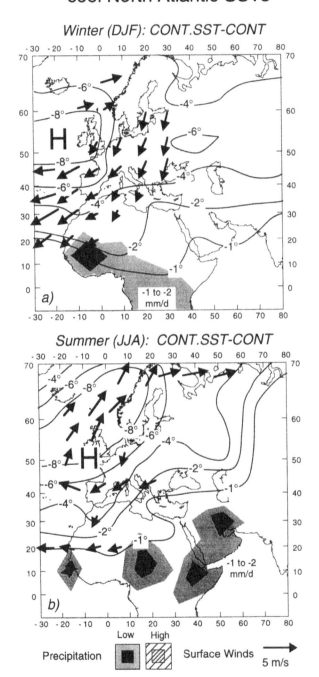

Figure 5. Effects of cold North Atlantic SSTs on the winter (DJF, a) and summer (JJA; b) model climate. CLIMAP SSTs imposed above 25°N in North Atlantic only (CONT.SST minus CONT). Climate patterns shown reflect anomalies attributable to the prescribed boundary condition change (see deMenocal and Rind [1993] for details).

results demonstrate that arid conditions prevail when North Atlantic SSTs (30–45°N) are relatively cool and South Atlantic SSTs (10–30°S) are relatively warm. The drier conditions were attributed to a weakening of the land-sea pressure gradient from the eastern South Atlantic and a strengthening of the high pressure cell over the North Atlantic, both of which act to decrease the advection of moisture from the adjacent Atlantic to the NW African interior. Similar correlations between historical recurrences of Sahelian aridity (drought) and Atlantic SST anomalies have been described using meteorological data [Lamb, 1978; Lough, 1986; see deMenocal et al., 1993].

4. South Asian Orography

Winter Climate. Reduced Himalayan-Tibetan orography enhanced winter anticyclonic circulation over southern Asia by up to +10 m/s (Fig. 6a). This regional increase in circulation increases the zonal average mean winds at 16°N by 0.74 m/s (+3 standard deviations). Warming (up to 12°C) and high SLP anomalies over South Asia can be attributed to increased atmospheric mass and thermal opacity (greenhouse capacity) of the atmosphere over the reduced elevation area (Fig. 6a), as well as adiabatic warming due to subsidence. Precipitation reductions (–2 to –4 mm/day) over South Asia coincide with outward surface air flow from the continental interior (Fig. 6a). The winter climate response to changed orography is limited to the reduced-topography region.

Increased Hadley circulation is indicated by the vertically integrated stream function diagnostic at 16°N, which increases by –6 × 10^9 kg/s (1 standard deviation) relative to the control value. The orographic influence of the Tibetan Plateau imparts regional scale eddy effects (lee cyclogenesis) upon the mean Hadley circulation [Murakami, 1981; Manabe and Terpstra, 1974]. Lower Tibetan Plateau elevation reduces the transient eddy interference with the mean Hadley circulation, enabling a stronger mean NE trade circulation to persist.

Summer Climate. Reducing the elevation of the Tibetan Plateau weakens, but does not completely eliminate Asian monsoonal cyclonic flow. The strong anticylonic pattern of the NOMTNS-CONT differenced surface wind plot illustrates the marked reduction in the normally cyclonic flow of the summer Asian monsoon (–4 m/s; Fig. 6b). The weakened circulation results from a weaker sea-land pressure gradient which dramatically changes the precipitation pattern as well (Fig. 6b): Broad areas to the east and southeast of the Himalayan-Tibetan complex have sharply reduced precipitation (–4 to –8 mm/day), whereas broad areas to the west and northwest of the complex have increased precipitation (up to 4 mm/day).

The model results indicate that uplift makes East Africa and Arabia drier and makes SE Asia wetter. As discussed by Ruddiman and Kutzbach [1989], high topography acts to draw moist maritime precipitation on its southeastern flanks and advect cool, dry continental air from the north over its western flank during the summer seasons (in the northern hemisphere). Broad, high-elevation regions enhance monsoonal, cyclonic circulation in the summer and enhance anticyclonic circulation in the winter.

Comparison with Other Model Results. Murakami et al. [1970], Godbole [1973] and Hahn and Manabe [1975] have investigated the role of the Tibetan Plateau in the South Asian summer monsoon circulation. Their results are in qualitative agreement with our findings, namely that the high topography is critically important to the development of an intense heat–low pressure cell through efficient focusing of moist convection and latent heat release. Strong convection over the plateau is driven initially by sensible heating but is maintained and strengthened by the release of latent heat at mid-tropospheric levels due to the forced

Model climate anomalies due to
reduced South Asian orography

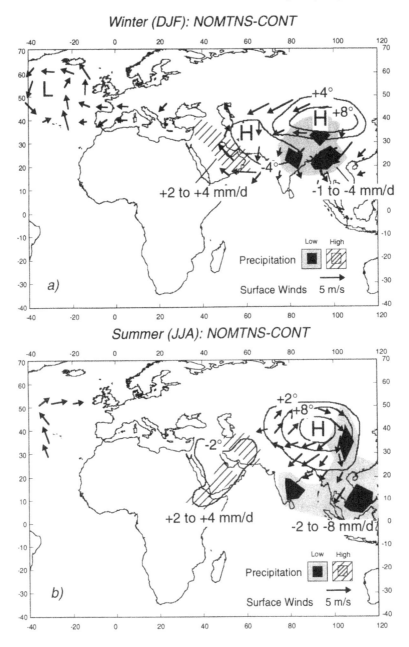

Figure 6. Effects of reduced South Asian orography (Tibetan-Himalayan elevations set at 500 msl) on the winter (DJF; a) and summer (JJA; b) model climate. Climate patterns shown reflect anomalies attributable to the prescribed boundary condition change (see deMenocal and Rind [1993] for details).

Table 2. Relative Sensitivities of Asian and African Climate to Imposed Boundary Condition Changes

Boundary condition	South Asia		West Africa		East Africa/Arabia	
	Winter	Summer	Winter	Summer	Winter	Summer
Insolation	−	+++	−	+	−	+
Ice sheets	++	++	−	+	++	+++
N. Atlantic SST	−	−	+++	++	−	+
Asian orography	+++	++	−	−	−	++

Least sensitive to most sensitive: −, + , ++ , +++.

ascent of moisture-laden air [Hahn and Manabe, 1975]. In the absence of such high topography this upward convection is replaced by downward subsidence of dry air over the Tibetan Plateau. The summer monsoon all but disappears in their no-mountain model. The summer monsoon is considerably weakened but still intact in the no-mountain GISS GCM.

Ruddiman and Kutzbach [1989] summarized model climate anomalies due to changes in Tibetan Plateau and American southwest orography. Concerning the tropics and subtropics outside of South Asia, increased orography causes: (1) reduced summer precipitation over the Mediterranean and East Africa, (2) increased summer precipitation over the sub-Sahara, and (3) annual cooling over North Africa. Our results indicate that while East Africa and Arabia are somewhat drier during summer due to uplift, the climate of the Mediterranean region is relatively unaffected (Fig. 6b). We see slightly *reduced* summer precipitation and perhaps *warmer* surface air temperatures over the sub-Sahara (item 2) (Figs. 6b).

DISCUSSION

A qualitative ranking of the winter and summer model climate of South Asia, West Africa, and East Africa to the various boundary condition changes is summarized in Table 2. We assess the relative sensitivities by considering the region in question and its sensitivity to a particular boundary condition.

South Asian summer and winter climate were most sensitive to changes in Himalayan-Tibetan topography; the changed orography experiment emphasized the critical importance of high topography in generating the seasonal circulation and precipitation characteristics of the Asian monsoon. Also influencing South Asian climate were variations in solar insolation and ice sheet size. Variations in seasonal insolation were important in modulating the intensity of the Asian summer monsoon, whereas high-latitude ice sheets tended to both reduce the intensity of the summer monsoon circulation, increase winter trade wind circulation, and promote subtropical drying during both seasons. South Asian climate was relatively insensitive to changes in North Atlantic SSTs.

West African climate was most sensitive to variations in North Atlantic SSTs. Cool SSTs acted to reduce the continental penetration of the summer African monsoon rains and increase winter trade wind circulation. Both of these responses tended to reduce the annual supply of precipitation. West African climate was only moderately sensitive to changes in ice sheet size or solar insolation; West Africa was effectively insensitive to changes in South Asian topography alone.

East African and Arabian climate were most sensitive to changes in ice sheet size. A regional swath encompassing East Africa, eastern Mediterranean, Arabia, and southwest Asia exhibited winter season cooling and drying; summers were drier as well. The precipitation decreases over the eastern Mediterranean and Arabia are important climatologically because winter is the rainy season (Fig. 2a), so a reduction in winter rainfall may indicate aridification. East African climate was also sensitive to changes in Asian orography: the high Asian topography acts to maintain aridity over the entire East African-southwest Asian region. African climate was least sensitive to changes in solar insolation and North Atlantic SSTs.

Comparison with Paleoclimate Data. Asian and African Monsoon Intensity. The bulk of available monsoon paleointensity indices suggest that both African and Asian monsoonal climate has followed seasonal insolation variability related to orbital precession variations. As summarized by Prell and Kutzbach [1987], several indicators of African and Asian summer monsoon paleointensity vary at 23–19 kyr periodicities corresponding to orbital precession. Over South Asia, indices of Arabian Sea upwelling [Prell, 1984; Prell and van Campo, 1986; Clemens et al., 1991], SW monsoon wind speed [Clemens and Prell, 1990; Clemens et al., 1991], and Arabian and West Indian lake levels [Singh et al., 1972; McClure, 1976; Bryson and Swain, 1981; Swain et al., 1983] attain their maximum values during maximum precessional summer insolation forcing. Anderson and Prell [1993] have suggested that monsoon intensity is perhaps forced by some component of high-latitude climate. Records of subtropical West African lake levels [Street and Grove, 1979], Niger River outflow to the Atlantic [Pastouret et al., 1978], Nile River outflow to the Mediterranean [Rossignol-Strick, 1983] indicate maximum monsoon precipitation coincident with precessional forcing. Similarly, precessional control of African monsoon intensity is suggested by marine records of freshwater diatoms from desiccated lake beds [Pokras and Mix, 1985; deMenocal et al., 1993], and monsoon-related upwelling in the eastern equatorial Atlantic [McIntyre et al., 1989; Molfino and McIntyre, 1991].

Terrestrial West African Paleoclimate. Many paleoclimate records from subtropical West and East Africa and Arabia indicate significantly more arid conditions and perhaps stronger trade wind intensities during glacial maxima. Increased concentrations of eolian dust are found in glacial-age sediments of the adjacent eastern equatorial Atlantic [Hays and Perruzza, 1972; Curry and Lohmann, 1990, 1990; Bloemendal and deMenocal, 1989; François et al., 1990; deMenocal et al., 1993; Kolla et al., 1979; Sarnthein et al., 1981; Tiedemann et al., 1989, 1994]. Marine pollen records off NW Africa suggest that African vegetation zones migrated southward during glacial maxima over the last 730 ka [Dupont and Hoogheimstra, 1989; Dupont et al., 1989; Lezine, 1991; Leroy and Dupont, 1994]; these results may also indicate increased trade wind vigor. Concentrations of wind-borne savannah grass cuticles (phytoliths) in eastern equatorial Atlantic sediments increased severalfold during the last glacial maximum [Pokras and Mix, 1985; deMenocal et al., 1993]. Several studies have also proposed stronger boreal winter NE trade wind velocities during glacial maxima based on increased eolian grain sizes [Parkin and Shackleton, 1973] and higher concentrations of pollen borne by the winter NE trade winds [Hoogheimstra et al., 1987], and increased trade-wind-driven upwelling [Müller and Suess, 1979] .

Climatological studies have tied interannual-to-decadal occurrences of Sahelian drought to cold SST anomalies in the North Atlantic and relatively warm SST anomalies in the South and Equatorial Atlantic [Lamb, 1976; Folland et al., 1986]. deMenocal et al. [1993] compared the long-term coevolution of African aridity and North Atlantic SST records and demonstrated that variations in eolian dust and grass phytolith supply are both coherent and in-phase with a record of North Atlantic (41°N) SST variability. Similar connections between African paleoclimate and Atlantic SSTs have been described by Street-Perrott and Perrott [1990] and Gasse et al. [1989].

Arabian and NE African Paleoclimate. Numerous studies have demonstrated that East African and Arabian climate was drier and perhaps cooler during glacial maxima [Kolla and Biscaye, 1977; van Campo et al., 1982; Bloemendal and deMenocal, 1989; Clemens and Prell, 1990; Sirocko and Sarnthein, 1989; deMenocal et al. 1991]. These studies cite greatly increased accumulations of eolian dust in Arabian Sea sediment cores which covary with high-latitude ice volume [Clemens and Prell, 1990]. Palynological evidence indicates glacial expansion of saline littoral and steppe environments in Arabian and East African dust source areas [van Campo et al., 1982].

Based on long and continuous dust records from West and East Africa/Arabia deMenocal et al. [1993, in press] proposed that African climate was fundamentally altered following the expansion of high-latitude ice sheets after 2.8 Ma. African climate responded to direct precessional insolation forcing of monsoonal climate prior to 2.8 Ma when high-latitude ice sheets were small and relatively invariant, but after 2.8 Ma when ice sheets grew such that large glacial cycles were sustained African climate became partially dependent upon remote high-latitude forcing. Specifically, the data and model results suggest that African climate became periodically cooler and drier after 2.8 Ma due to dynamical effects related to the development of cold glacial North Atlantic SSTs; this effect was apparently amplified after 1 Ma following the increase in duration and magnitude of high-latitude glacial cycles. Onset of glacial arid conditions in East Africa near 2.5–3.0 Ma is suggested by some palynological and macrofaunal data [Bonnefille, 1983; Cerling, 1992; Vrba et al., 1989].

South Asian Orography. Ruddiman et al. [1989b] have presented a thorough review of paleoclimate evidence for accelerated Tibetan Plateau (and American West) uplift rates during the latest Neogene. Tibet apparently attained half or more of its present elevation in the last 10 Ma. Citing a variety of paleoclimate and geological evidence from Eurasia, Ruddiman and Kutzbach [1989] indicate that late Cenozoic uplift was responsible for: 1) intensification of the Asian monsoon and increased South Asian precipitation [see also Prell and Kutzbach, 1992], 2) cooler annual temperatures over northern Europe and Asia, and 3) development of dry summer conditions over the Mediterranean and parts of the Eurasian interior.

SUMMARY

These results indicate that Asian and African model climate exhibit unique climatic signatures associated with prescribed changes in orbital insolation, glacial ice volume, North Atlantic SSTs, and South Asian orography. As summarized in Table 2, these boundary condition changes include:

1. Asian summer monsoonal circulation and precipitation were greatly increased by the ~8% orbital increase in summer insolation. The African monsoon system was less responsive to insolation changes in this model.

2. Expanded high-latitude glacial ice cover produced generally cooler and drier conditions over East Africa, Arabia, and southwest Asia, particularly during winter months. Increased glacial ice cover enhanced winter trade wind speeds and reduced the Asian summer monsoon intensity. African climate was relatively unresponsive.

3. Cool North Atlantic sea-surface temperatures caused significant cooling and drying and increased winter trade wind speeds over West Africa; summer African monsoon intensity was moderately reduced. South Asian and East African climates were largely unaffected.

4. Reduced Asian orography produced the most dramatic climate responses of all model runs. Winter trade wind circulation over Asia was enhanced in the reduced-elevation experiment, and the reduced orography also limited the interior penetration of the summer monsoon. West African climate was relatively unaffected in this experiment, however East Africa and Arabia exhibited significantly higher precipitation.

5. Many of the modeled climate anomaly patterns are consistent the paleoclimate record.

ACKNOWLEDGMENTS

This research was supported by the National Science Foundation (OCE-9203936). This is Lamont-Doherty Earth Observatory publication number 5460.

REFERENCES

Anderson, D. M., and Prell, W. L., A 300 kyr record of upwelling off Oman during the late Quaternary: Evidence of the Asian southwest monsoon, Paleoceanography, 8/2: 193–208, 1993.

Berger, A. L., Long-term variations of caloric solar insolation resulting from earth's orbital variations. Quat. Res., 12: 139–167, 1978.

Bonnefille, R., Evidence for a cooler and drier climate in the Ethiopian Uplands towards 2.5 Myr ago, Nature, 303: 487–491, 1983.

Bloemendal, J., and deMenocal, P. B., Evidence for a change in the periodicity of tropical climate cycles at 2.4 Myr from whole-core magnetic susceptibility measurements. Nature, 342: 897–900, 1989.

Bryson, R. A., and Swain, A. M., Holocene variations in monsoon rainfall in Rajasthan. Quat. Res., 16: 135–145, 1981.

Cerling, T. E., Development of grasslands and savannas in East Africa during the Neogene, Paleogeog., Paleoclimatol., Paleoecol., 97: 241–247, 1992.

Chen, L. R., Reiter, E., and Feng, Z., The atmospheric heat source over the Tibetan Plateau: May-August 1979. Mon. Weath. Rev., 113: 1771–1790, 1985.

Clemens, S., Prell, W., Murray, D., Shimmield, G., and Weedon, G., Forcing mechanisms of the Indian Ocean monsoon, Nature, 353: 720–725, 1992.

Clemens, S., and Prell, W. L., Late Pleistocene variability of Arabian Sea summer monsoon winds and continental aridity: Eolian records from the lithogenic component of deep-sea sediments. Paleoceanography, 5, 109–146, 1990.

Clemens, S., and Prell., W., Late Quaternary forcing of Indian Ocean summer-monsoon winds: A comparison of fourier model and general circulation model results., J. Geophys. Res., 96/D12: 22683–22700, 1991.

CLIMAP project members, Seasonal Reconstruction of the Earths Surface at the Last Glacial Maximum, Geol. Soc. Amer., Map Chart Ser., MC-36, Boulder, CO, 1981.

Curry, W. B., and Lohmann, G. P., Reconstructing past particle fluxes in the tropical Atlantic Ocean, Paleoceanography, 5: 487–505, 1990.

deMenocal, P. B., Bloemendal, J., and King, J. W., A rock-magnetic record of monsoonal dust deposition to the Arabian Sea: Evidence for a shift in the mode of deposition at 2.4 Ma. In Proc. of Ocean Drill. Prog., Scientific Results (Prell, W. L., and Niitsuma, N., et al., eds.), Ocean Drilling Program, College Station, TX, pp. 389–407, 1991.

deMenocal, P. B., Ruddiman, W. F., and Pokras, E. M., Influences of high- and low-latitude processes on African climate: Pleistocene eolian records from equatorial Atlantic Ocean Drilling Program Site 663, Paleoceanography, 8/2: 209–242, 1993.

deMenocal, P.B., and Rind, D., Sensitivity of Asian and African climate to variations in seasonal insolation, glacial ice cover, sea surface temperature, and Asian orography. J. Geophys. Res., 98/D4: 7265–7287, 1993.

deMenocal, P. B., and Bloemendal, J., Plio-Pleistocene subtropical African climate variability and the paleoenvironment of hominid evolution: A combined data-model approach. *In* Paleoclimate and Evolution with Emphasis on Human Origins, (Vrba, E., Denton, G., Burckle, L., and Partridge, T., eds.), Yale University Press, New Haven, in press.

Denton, G., and Hughes, T., The Last Great Ice Sheets. Wiley, New York, NY, 1981.

Druyan, L. M., GCM studies of African summer monsoon. Climate Dynamics, 2: 117–126, 1987.

Druyan, L. M., and Rind, D., Verification of Regional Climates of GISS GCM-Part I: Winter. NASA Technical Memorandum 100695, 1988.

Dupont, L. M., Beug, H.-J., Stalling, H., and Tiedemann, R., First palynological results from Site 658 at 21°N off northwest Africa: Pollen as climate indicators. *In* Proceed. of the Ocean Drilling Program, Scientific Results (Ruddiman, W. F. Sarnthein, M., et al., eds.), Ocean Drilling Program, College Station, TX, pp. 93–111, 1989.

Dupont, L. M., and Hoogheimstra, H., The Saharan-Sahelian boundary during the Brunhes chron, Acta Bot. Neerl., 38: 405–415, 1989.

Flohn, H., Studies on the Meteorology of Tropical Africa, Bonner Meteorologische Abhandlungen, vol. 5, 1965, 57 pp.

Folland, C., Palmer, T., and Parker, D., Sahel rainfall and worldwide sea temperatures, Nature, 320: 602–607, 1986.

François, R., Bacon, M. P., and Suman, D. O., Thorium 230 profiling in deep-sea sediments: High-resolution records of flux and dissolution of carbonate in the equatorial Atlantic during the last 24,000 years, Paleoceanogr., 5: 761–787, 1990.

Gasse, F., Lédée, V., Massualt, M., and Fontes, J.-C., Water-level fluctuations of Lake Tanganyika in phase with oceanic changes during the last glaciation and deglaciation. Nature, 342: 57–59, 1989.

Gates, W. L., Modelling the Ice Age climate. Science, 191: 1138–1144, 1976.

Godbole, R. V., Numerical simulation of the Indian summer monsoon. Indian J. Meteor. Geophys., 24: 1–13, 1973.

Hahn, D. G., and Manabe, S., The role of mountains in the South Asian monsoon circulation. J. Atmos. Sci., 32: 1515–1541, 1975.

Hansen, J., Russell, G., Rind, D., Stone, P., Lacis, A., Lebedeff, S., Ruedy, R., and Travis, L., Efficient three-dimensional global models for climate studies: Models I and II. Mon. Weath. Rev., 111: 609–662, 1983.

Hastenrath, S., Climate and the Circulation of the Tropics. D. Reidel Publishing Co., Boston, MA, 1985.

Hastenrath, S., and Lamb, P. J., Climate Atlas of the Indian Ocean. "Surface climate and atmospheric circulation." University of Wisconsin Press, Madison, Wisconsin, 1979.

Hoogheimstra, H., Bechler, A., and Beug, H.-A., Isopollen maps for 18,000 years B.P. of the Atlantic offshore of Northwest Africa: Evidence for paleowind circulation, Paleoceanogr., 2: 561–582, 1987.

Hsü, C.-P.F., and Wallace, J. M., The global distribution of the annual and semiannual cycles in precipitation. Month. Weath. Rev., 104/9: 1093–1101.

Kendrew, W. G., The Climates of the Continents, 5th Edition, Clarendon Press, 1961.

Kolla, V., and Biscaye, P., Distribution and origin of quartz in the sediments of the Indian Ocean. J. Sediment. Petrol., 47: 642–649, 1977.

Kolla, V., Kostecki, J. A., Robinson, F., Biscaye, P. E., and Ray, P. K., Distribution and origins of clay minerals and quartz in surface sediments of the Arabian Sea. J. Sed. Petrol., 51: 563–569, 1981.

Kutzbach, J. E., Monsoon climate of the early Holocene: Climate experiment with the Earth's orbital parameters for 9,000 years ago. Science, 214: 59–61, 1981.

Kutzbach, J. E., and Street-Perrott, F. A., Milankovitch forcing of fluctuations in the level of tropical lakes from 18 to 0 kyr BP. Nature, 317: 130–134, 1985.

Kutzbach, J. E., and Guetter, P. J., The sensitivity of monsoon climates to orbital parameter changes for 9000 yr B.P: Experiments with the NCAR GCM. In Milankovitch and Climate (Berger, A., and Imbrie, J., eds.), D. Reidel, Dordrecht, 1984.

Kutzbach, J. E., and Otto-Bliesner, B. L., The sensitivity of the African-Asian monsoonal climate to orbital parameter changes for 9000 years B.P. in a low-resolution general circulation model. J. Atmos. Sci., 39: 1177–1188, 1982.

Lamb, P. J., Case studies of tropical Atlantic surface circulation patterns during recent sub-Saharan weather anomalies: 1967 and 1968. Mon. Weather Rev., 106: 482, 1978.

Leroy, S., and Dupont, L., Development of vegetation and continental aridity in northwestern Africa during the late Pliocene: the pollen record of ODP Site 658, Paleogeo., Paleoclim., Paleoecol., 109: 295–316, 1994.

Lezine, A., West African paleoclimates during the last climatic cycle inferred from an Atlantic deep-seas pollen record, Quat. Res., 35: 456–463, 1991.

Lough, J., Tropical Atlantic sea surface temperatures and rainfall variations in sub-Saharan Africa. Mon. Weather Rev., 114: 561–570, 1986.

Manabe, S., and Hahn, D. G., Simulation of the tropical climate of an Ice Age. J. Geophys. Res., 82: 3889–3911, 1977.

Manabe, S., and Terpstra, T., The effects of mountains on the general circulation of the atmosphere as identified by numerical experiments. J. Atmos. Sci., 31: 3–42, 1974.

McClure, H. A., Radiocarbon chronology of the Quaternary lakes in the Arabian Desert. Nature, 263: 755, 1976.

McIntyre, A., Ruddiman, W. F., Karlin, K., and Mix, A. C., Surface water response of the equatorial Atlantic Ocean to orbital forcing. Paleoceanography, 4: 19–55, 1989.

Molfino, B., and McIntyre, A., Nutricline variation in the equatorial Atlantic coincident with the Younger Dryas. Paleoceanography, 5: 997–1008, 1990.

Müller, P. J., and Suess, E., Productivity, sedimentation rate, and sedimentary organic matter in the oceans. Part I, Organic carbon preservation. Deep-Sea Res., 26: 1347–1362, 1979.

Murakami, T., Orographic influence of the Tibetan Plateau on the Asiatic winter monsoon circulation, Part I: Large-scale aspects. J. Meteor. Soc. of Japan, 59: 40–65, 1981.

Murakami, T., Godbole, R. V., and Kelkar, R. R., Numerical simulation of the monsoon along 80°E. In Proceedings Conference Summer Monsoon of Southeast Asia. Navy Weather Research Facility, Norfolk, Virginia, 1970.

Nicholson, S. E., A review of climate dynamics and climate variability in eastern africa, In The Limnology, Climatology and Paleoclimatology of the East African Lakes (Johnson, T. C., and Odada, E. O., eds.), Gordon and Breach, Toronto, pp. 23–54, 1996.

Parkin, D. W., and Shackleton, N. J., Trade wind and temperature correlations down a deep-sea core off the Saharan coast. Nature, 245: 455–457, 1973.

Pastouret, L., Chamley, H., Delibrias, G., Duplessy, J.-C., and Theide, J., Late Quaternary climatic changes in western tropical Africa deduced from deep-sea sedimentation off the Niger delta. Oceanol. Acta, 1: 217–232, 1978.

Pokras, E. M., and Mix, A. C., Eolian evidence for spacial variability of late Quaternary climates in tropical Africa. Quat. Res., 24: 137–149, 1985.

Prell, W. L., and Kutzbach, J. E., Sensitivity of the Indian monsoon to forcing parameters and implications for its evolution, Nature, 360: 647–652, 1992.

Prell, W. L., Monsoonal climate of the Arabian Sea during the late Quaternary; a response to changing solar radiation. *In* Milankovitch and Climate (Berger, A., Imbrie, J., Hays, J., Kukla, G., and Satzman, B., eds.), D. Reidel, Hingham, Mass, 1984.

Prell, W. L., and Kutzbach, J. E., Monsoon variability over the past 150, 000 years. J. Geophys. Res. 92, 1987.

Prell, W. L., and van Campo, E., Coherent response of Arabian Sea upwelling and pollen transport to late Quaternary monsoonal winds. Nature, 323: 526–528, 1986.

Pye, K., Aeolian Dust and Dust Deposits., Academic Press, London, 1987.

Ramage, C. S., The summer atmospheric circulation over the Arabian Sea. J. Atmos. Sci., 23, 144, 1965

Rind, D., Components of the Ice Age circulation. J. Geophys., Res., 92: 4241–4281, 1987.

Rind, D., The dynamics of warm and cold climates. J. Atmos., Sci., 43: 3–24, 1986.

Rind, D., Peteet, D., Broecker, W., McIntyre, A., and Ruddiman, W., The impact of cold North Atlantic sea surface temperatures on climate: Implications for the Younger Dryas cooling (11–10 kyr). Clim. Dyn., 1: 3–34, 1986.

Rossignol-Strick, M., African monsoons, an immediate climatic response to orbital insolation. Nature, 303: 46–49, 1983.

Ruddiman, W. F., and Kutzbach, J. E., Forcing of late Cenozoic Northern Hemisphere climate by plateau uplift in southern Asia and the American West. J. Geophys. Res., D1594: 18409–18427, 1989.

Ruddiman, W. F., Raymo, M. E., Martinson, D. G., Clement, B. M., and Backman, J., Pleistocene evolution of northern hemisphere climate. Paleoceanography, 4: 353–412, 1989a.

Ruddiman, W. F., Prell, W. L., and Ratmo, M. E., Late Cenozoic uplift of southern Asia and the American West: Rationale for general circulation modeling experiments. J. Geophys. Res., 94: 18379–18391, 1989b.

Shackleton, N. J., et al., Oxygen isotope calibration of the onset of ice-rafting and history of glaciation in the North Atlantic region, Nature, 307: 620–623, 1984.

Singh, G., Joshi, R. D., and Singh, A. B., Stratigraphic and radiocarbon evidence for the age and development of three salt lake deposits in Rajasthan, India. Quat. Res., 2: 496–505, 1972.

Sirocko, F., and Sarnthein, M., Wind-borne deposits in the northwestern Indian Ocean: Record of Holocene sediments versus modern satellite data. *In* Paleoclimatology and Paleometeorology: Modern and Past Patterns of Global Atmospheric Transport (Leinen, M., and Sarnthein, M., eds.), NATO ASI Ser., 1989.

Street-Perrott, F. A., and Perrott, R. A., Abrupt climate fluctuations in the tropics: the influence of Atlantic Ocean circulation, Nature, 343: 607–612, 1990.

Street, A. E., and Grove, A. T., Global maps of lake-level fluctuations since 30,000 years BP. Quat. Res., 12: 83–118, 1979.

Swain, A. M., Kutzbach, J. E., and Hastenrath, S., Estimates of Holocene precipitation for Rajasthan, India based on pollen and lake-level data. Quat. Res. 19: 1–17, 1983.

Tiedemann, R., Sarnthein, M., and Shackleton, N. J., Astronomic timescale for the Pliocene Atlantic $\delta^{18}O$ and dust flux records of ODP Site 659, Paleoceanography, 9/4: 619–638, 1994.

Tiedemann, R. Sarnthein, M., and Stein, R., Climatic changes in the western Sahara: Aeolo-marine sediment record of the last 8 million years (Sites 657–661), In Proceedings of the Ocean Drilling Program, Scientific Results, Ruddiman, W. F. Sarnthein, M., et al., eds., Ocean Drilling Program, College Station, TX, 1989.

van Campo, E., Duplessy, J. C., and Rossignol-Strick, M., Climatic conditions deduced from a 150-kyr oxygen isotope-pollen record from the Arabian Sea. Nature, 296: 56–59, 1982.

Vrba, E. S., Denton, G. H., and Prentice, M. L., Climatic influences on early Hominid behavior, Ossa, 14: 127–156, 1989.

Isotope Patterns of Precipitation in the East African Region

K. ROZANSKI, L. ARAGUAS-ARAGUAS and
R. GONFIANTINI   *International Atomic
Energy Agency, A-1400 Vienna, Austria*

Abstract — The International Atomic Energy Agency (IAEA), in cooperation with the World Meteorological Organization (WMO), has been conducting since the early sixties a worldwide survey of hydrogen and oxygen isotopic composition of precipitation. The East African region is represented in the IAEA database by six stations: Addis Ababa (Ethiopia), Kericho (Kenya), Entebbe (Uganda), Dar Es Salaam (Tanzania), Ndola (Zambia) and Harare (Zimbabwe). The Indian Ocean, which is the major source of precipitation for this part of the African continent, is represented by two stations: Antananarivo (Madagascar) and Diego Garcia Island (USA). Four out of eight stations listed above are still actively collecting monthly precipitation for isotope analyses.

The isotopic composition of precipitation in the East African region reflects the regional circulation patterns. Seasonal fluctuations of deuterium and ^{18}O content of precipitation are well pronounced at almost all stations of the region and coincide with the seasonal displacement of the intertropical convergence zone (ITCZ). Isotopically depleted rains are in general associated with the rainy period(s). The amplitude of seasonal variations is relatively high in the northern and southern parts of the region (5 to 7‰ for δ^{18}O at Antananarivo, Harare, Ndola and Addis Ababa). The annual weighted mean values of δ^{18}O lie between –6.5‰ for the stations located in the south, and –1.3‰ for Addis Ababa. The average deuterium excess values vary between 12.5 and 15.5‰ for the inland stations, whereas the maritime stations (Diego Garcia and Dar Es Salaam) reveal deuterium excess values close to 10. Relatively large year-to-year variability of the isotopic composition of precipitation is observed. The region experiences recurrent events of extremely low deuterium and ^{18}O isotope composition of precipitation, often persisting for more than one month, which most probably is associated with unusually strong convective activity within the air column, associated with passage of the ITCZ.

INTRODUCTION

The International Atomic Energy Agency (IAEA), in cooperation with the World Meteorological Organization (WMO), initiated in 1961 a worldwide survey of the isotope composition of monthly precipitation. The programme was launched with the primary objective of collecting systematic data on the isotopic composition of precipitation on a global scale (deuterium, oxygen-18 and tritium), characterizing their spatial and temporal variability and, consequently, providing basic isotope data for the use of environmental isotopes in hydrogeological investigations. The collected data are also very useful in other water-related ·fields such as oceanography, hydrometeorology and climatology. About 220,000 isotope and meteorological data representing more than 420 locations all over the world have been accumulated to date in the IAEA database.

The isotope and meteorological data gathered by the network are published regularly by the IAEA in the form of data books (IAEA, 1969, 1970, 1971, 1973, 1975, 1979, 1983, 1986, 1990) and are also available on floppy disks. Basic statistical treatment of the data accumulated up to 1987 is available as a separate volume (IAEA, 1992). Temporal and spatial distribution patterns of deuterium and ^{18}O in modern global precipitation were recently discussed by Rozanski et al. (1993), based mainly on the network data.

The paper reviews available meteorological and isotope data collected by six stations of the IAEA/WMO network located in the East African region: Addis Ababa (Ethiopia), Kericho (Kenya), Entebbe (Uganda), Dar es Salaam (Tanzania), Ndola (Zambia) and Harare (Zimbabwe). The Indian Ocean, which is the major potential source of precipitation for this part of the African continent, is represented by two stations: Antananarivo (Madagascar) and Diego Garcia Island (USA). Spatial and temporal variations of stable isotope composition of precipitation in the region are discussed in the context of atmospheric circulation over East Africa.

METHODS

The isotope data stored in the IAEA database consist of tritium, deuterium and ^{18}O isotope concentration in total rainfall accumulated during any given month. The meteorological data comprise the amount of monthly precipitation, mean monthly surface air temperature and water vapour pressure. It should be kept in mind that, whereas the reported meteorological parameters represent average weather conditions during the given month, the isotope data are inherently related only to rainy periods.

The deuterium and ^{18}O content of precipitation is reported in conventional delta notation (δ), as parts per thousand deviation of the isotope ratios $^2H/^1H$ or $^{18}O/^{16}O$ relative to the standard V-SMOW (Vienna Standard Mean Ocean Water). The deuterium excess, defined as $d = \delta^2H - 8 \cdot \delta^{18}O$, describes the location of individual data points on the $\delta^2H - \delta^{18}O$ plot with respect to the so-called Global Meteoric Water Line (GMWL) defined by the equation: $\delta^2H = 8 \cdot \delta^{18}O + 10$. Seasonal and long-term means of δ^2H and $\delta^{18}O$ were calculated by arithmetic averaging of individual monthly data or by weighing by amount of monthly precipitation (weighted means).

The long-term seasonal variability of the isotopic composition of precipitation recorded in the stations of the region is presented in the form of box-and-whisker plots. In this presentation, the central box covers the middle 50% of the isotope data available for the given month, between the lower and upper quartiles. The "whiskers" extend out to the extremes (minimum and maximum values), while the central line is at the median. The outliers occurring far away from the bulk of data are plotted as separate points. The whiskers extend only to those points that are within 1.5 times the interquartile range.

RESULTS AND DISCUSSION

Table 1 summarizes long-term mean values of isotope and meteorological data calculated for eight stations of the region. Geographical location, elevation above sea level and the period of operation are also listed for each station. Four out of eight stations are still actively collecting monthly precipitation for isotope analyses. Unfortunately, due to numerous gaps and missing data, the available isotope and meteorological records are not always satisfactory.

The local meteoric water lines derived from monthly $\delta^{18}O$ and δ^2H data available for the analysed stations of the region are summarized in Table 2. Least square linear regression analysis was used to derive the parameters listed in Table 2.

Table 1. Long-Term Mean Values of Isotope (δ^2H, δ^{18}O, deuterium excess) and Meteorological Data (amount of precipitation, surface air temperature, vapour pressure) for Eight Stations of the IAEA/WMO Network, Located in the East African Region

Station name	Oxygen-18 (‰ vs. VSMOW)			Deuterium (‰ vs. VSMOW)			Deuterium excess $d = \delta^2H - 8\delta^{18}O$			Precip. (mm)	Temp. (°C)	Vapour press. (mb)
	Mean	W. mean	n	Mean	W. mean	n	Mean	W. mean	n			
Addis Ababa (1961–87)* 9.00 N, 38.73 E, 2360 m	−0.36	−1.31	102	8.5	1.8	95	11.6	13.2	93	1181	16.3	11.0
Entebbe (1960–76) 0.05 N, 32.45 E, 1155 m	−2.26	−2.91	120	−5.8	−11.2	115	12.6	12.4	105	1613	21.6	20.5
Kericho (1967–70) 0.37 S, 35.35 E, 2130 m	−2.90	−3.83	24	−10.0	−16.3	19	11.5	13.4	19	1037	–	–
Dar es Salaam (1961–73) 6.88 S, 39.20 E, 55 m	−2.03	−2,83	126	−7.1	−13.3	119	8.5	9.3	119	1139	25.7	25.6
Ndola (1968–85)* 13.00 S, 28.65 E, 1331 m	−6.03	−6.59	79	−35.1	−39.5	69	12.5	13.3	68	1172	20.4	14.8
Harare (1961–83)* 17.83 S, 31.02 E, 1471 m	−3.88	−6.14	175	−16.1	−32.4	110	12.1	13.4	106	809	18.0	12.9
Antananarivo (1961–75)* 18.90 S, 47.53 E, 1300 m	−4.95	−6.41	71	−26.4	−36.0	67	13.9	15.4	66	1390	17.8	15.8
Diego Garcia (1962–87)* 7.32 S, 72.40 E, 1 m	−3.46	−4.03	182	−20.0	−23.7	170	9.1	9.6	159	2195	26.9	29.0

Both arithmetic and weighted means (weighting by amount of precipitation) are reported for the isotope data. Stations marked by asterisks continue to collect isotope and meteorological data.

January   July

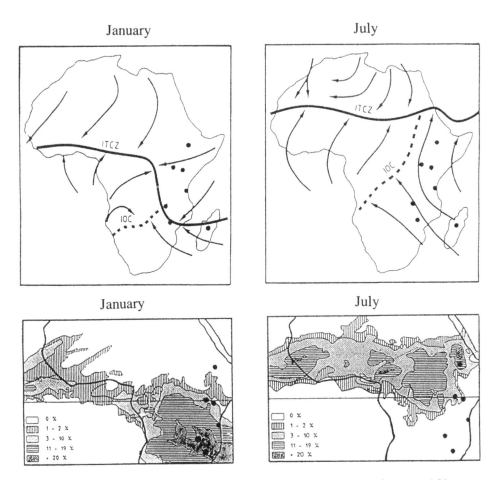

January   July

Figure 1. Upper graphs: General circulation patterns in the lower troposphere over Africa, after Lacaux et al. (1992). The seasonal displacement of the intertropical convergence zone (ITCZ) and the interoceanic confluence (IOC) determines the pluviometric regime of eastern Africa. Heavy dots indicate the location of the stations belonging to the IAEA/WMO network "Isotopes in Precipitation," which are discussed in the text. Lower graphs: Occurrence of cloud tops (in percent) with a temperature of less than –40°C, based on IR satellite images (Lahuec and Carn, 1988). This parameter can be used as a measure of the convective activity within the air column and the associated rainfall.

The seasonal variability of the isotopic composition of precipitation in the region and the selected climatic parameters (temperature, amount of precipitation) is illustrated in Figs. 2 and 3. Figure 4 shows individual $\delta^2H - \delta^{18}O$ plots for each of the analysed stations.

Figure 5 illustrates the long-term changes of the isotopic composition and the amount of rainfall during the major rainy period, recorded at selected locations in the region. A more detailed insight into the variability of the isotope signal in precipitation on a monthly basis is given by Fig. 6 which presents monthly $\delta^{18}O$ data and the amount of rainfall for selected periods when abnormally low ^{18}O content of precipitation was recorded.

Regional Circulation Patterns

The pluviometric regime of the East African region is controlled on an annual basis primarily by the presence of an intertropical convergence zone (ITCZ). Passage of ITCZ is associated with intensified convective activity within the air column, which usually leads to abundant rainfall (cf. Fig. 1). Thus, stations located in the southern part of the East African region (Harare, Ndola, Antananarivo), which is also the region of maximum southward displacement of ITCZ, reveal only one rainy period lasting from November to March (Fig. 2). On the other hand, equatorial stations of the region (Dar es Salaam, Kericho, Entebbe) experience two rainy periods (March–May and October–December) associated with northward and southward passage of the ITCZ, respectively. Finally, the Addis Ababa station, located in the region of maximum northward displacement of the ITCZ, has one rainy period lasting from June to September. The entire region is under the prevailing influence of easterly circulation bringing moisture from the southwestern Indian Ocean during the boreal summer monsoon period (June–September) and from the equatorial Indian Ocean and the Arabian Sea during the boreal winter months.

Although the presence of the ITCZ seems to provide an important control of the seasonal variability of rainfall, the causes of the interannual variability, and consequently of climate variability in East Africa are far more complex and remain a subject of scientific debate (e.g., Flohn, 1964; Nicholson, 1986; Chen and van Loon, 1987; Ropelewski and Halpert, 1987). Recently, it has been suggested that fluctuations in sea-surface temperature in the tropical Atlantic and Indian Oceans, which, in turn are coupled with the El Niño–Southern Oscillation (ENSO) phenomenon, impose a strong control on the interannual variability of rainfall in the region (S.E. Nicholson, this volume).

Seasonality

The seasonal changes of oxygen isotope composition of precipitation, recorded at the analysed stations of the region, are shown in Fig. 3, in the form of box-and-whisker plots. The patterns of $\delta^{18}O$ changes vary considerably from place to place, generally reflecting seasonal distribution of rainfall.

The most regular behaviour of $\delta^{18}O$ with the season is observed in the southern part of the region (Harare, Ndola). Here, rainfall is strongly depleted in ^{18}O during the rainy period; values between –6 and –7‰ are usually recorded between December and March. The austral winter values, represented in fact by a very small amount of rainfall, are relatively enriched in ^{18}O (values close to 0‰). Seasonal changes of $\delta^{18}O$ observed at Harare and Ndola esentially mirror the variations observed at Antananarivo station, located in the vicinity of the major sources of moisture (western Indian Ocean). The amplitude of seasonal changes of $\delta^{18}O$ in precipitation reaches in the southern part of the region about 6‰, with a substantially higher apparent variability of isotopic composition of rainfall during the rainy period. An analogous effect, most probably associated with the passage of the ITCZ over the given area, was also reported for the equatorial region of the South American continent (Matsui et al., 1983).

The stations located in the equatorial part of East Africa (Dar es Salaam, Kericho, Entebbe) reveal a more complex seasonal distribution of $\delta^{18}O$ of precipitation, with two small minima corresponding to two rainy periods in the region (March–May, October–December). At Kericho, the minimum in $\delta^{18}O$ seems to be more pronounced during "short

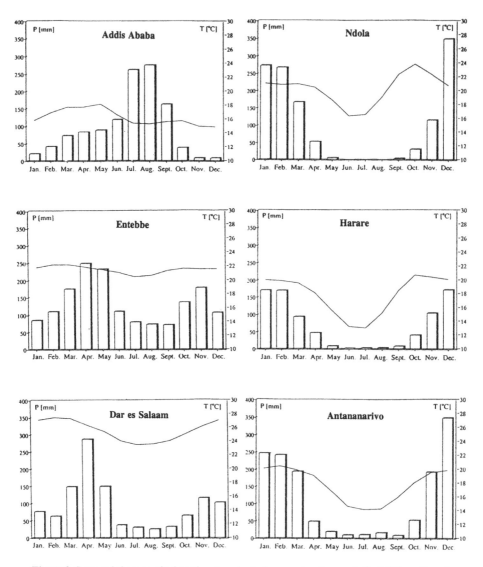

Figure 2. Seasonal changes of selected meteorological parameters (amount of rainfall, surface air temperature) for seven stations of the IAEA/WMO network, located in the East African region.

rains" (November–December). The significance of this effect cannot, however, be assessed due to the relatively short record of data available for this station (three years). In general, the isotopic composition of precipitation during the months of abundant rainfall is substantially more variable that during the dry season, the effect, being apparent also in the southern part of the region.

The northern part of the East African region is represented by only one station, Addis Ababa. It experiences one rainy period, lasting from June to September. The $\delta^{18}O$ of precipitation during months of abundant rainfall is in general more depleted than during the rest of the year. Interestingly, low $\delta^{18}O$ values persisting also during October and November

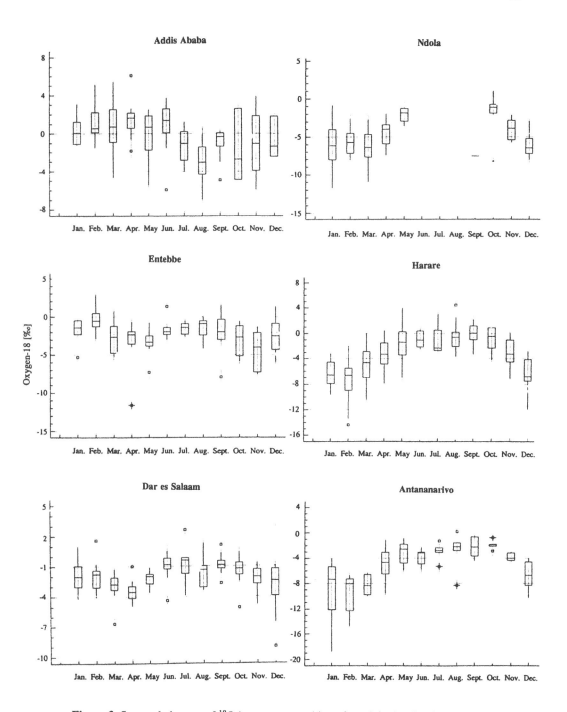

Figure 3. Seasonal changes of [18]O isotope composition of precipitation for six stations of the IAEA/WMO network, located in the East African region. Seasonal changes of $\delta^{18}O$ are shown in the form of box-and-whisker plots (see text).

which are characterized by a relatively low rainfall. It should be noted that Addis Ababa station, although situated at the highest elevation among the analysed stations (2360 m a.s.l.), reveals on the average the highest $\delta^{18}O$ and δ^2H values in precipitation. The data points cluster along the Global Meteoric Water Line (Fig. 4) suggesting a lack of substantial evaporative enrichment of rain collected at this station. Dansgaard in his pioneering work (Dansgaard, 1964) suggested that light rains at Addis Ababa undergo partial evaporation and isotope exchange with near-ground re-evaporated moisture characterized by a high value of the d-excess (indicative of fast evaporation of surface water to a relatively dry atmosphere), in order to explain the slope of the $\delta D - \delta^{18}O$ relationship for Addis Ababa and Entebbe, which was higher than 8. However, his analysis was based on a very limited set of data (only five months of observation for Addis Ababa). The presently available set of data (Table 2, Fig. 4) does not confirm this hypothesis.

It has been suggested (Sonntag et al., 1979) that the rain forest of the Congo Basin may represent an important source of moisture for the regions situated north and northeast of the basin, thus leading to relatively high $\delta^{18}O$ and δ^2H values of Addis Ababa precipitation. Because the transpiration process proceeds without isotope differentiation (Zimmermann et al., 1967), the moisture released in this process will be isotopically much heavier than the atmospheric water vapor of maritime origin. Consequently, rain produced with a substantial contribution of such recycled moisture can easily reach positive $\delta^{18}O$ and δ^2H values, which often happens at Addis Ababa between December and May. On the other hand, during this period the ITCZ is located south of Addis Ababa and, consequently, this station may receive a substantial portion of moisture from the equatorial Indian Ocean and the Arabian Sea, characterized by relatively high sea surface temperatures. Therefore, positive $\delta^{18}O$ values at Addis Ababa can be also explained assuming that the rain collected between December and May represents in fact the first condensation stage of the maritime moisture brought from the Indian Ocean north of the equator (Joseph et al., 1992).

$\delta^2H - \delta^{18}O$ Relationship

The $\delta^2H - \delta^{18}O$ plots of the monthly isotope data for the analysed stations are shown in Fig. 4. The data points were split into rainy season (full triangles) and dry season (open squares). In general, the data points cluster along the Global Meteoric Water Line, with the spread varying from station to station, being relatively high for Dar es Salaam and Harare. Least squares linear regression analysis of the monthly data yields the equations listed in Table 2. There are only minor differences among local meteoric water lines defined by these equations. Regional studies of isotopic composition of rainfall in the Kenya Rift Valley, based on isotope analyses of individual rain events, resulted in slightly different $\delta^2H - \delta^{18}O$ relationships than those reported in Table 2 for Entebbe and Kericho (Allen et al., 1989; Darling et al., 1990).

The deuterium excess derived from the long-term weighted mean $\delta^{18}O$ and δ^2H values vary between 9.3‰ (Dar es Salaam) and 15.4‰ (Antananarivo). In general, the average deuterium excess is higher than 10, with the exception of Dar es Salaam and Diego Garcia Island. The Antananarivo data reveal an apparent seasonal dependence on the d value, with generally lower deuterium excess during the rainy period (IAEA, 1992). Such a seasonal fluctuation of deuterium excess was observed also for the Reunion Island, about 1000 km east of Antananarivo, where a difference up to 7‰ between austral summer and winter rains

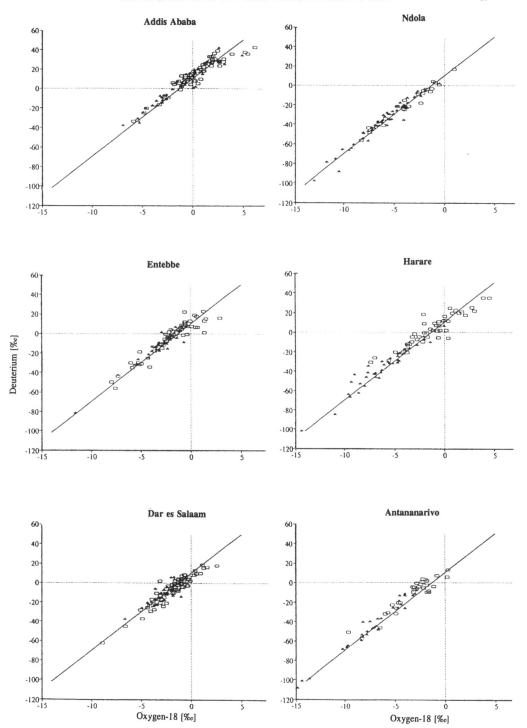

Figure 4. The δ^2H–δ^{18}O diagrams based on monthly isotope data available for six stations of the IAEA/WMO network, located in the East African region. Full triangles represent major rainy period at each station: June to September for Addis Ababa, March to May for Entebbe and Dar es Salaam, and November to March for Ndola, Harare and Antananarivo. The lines indicate the position of the Global Meteoric Water Line. δ^2H $= 8 \cdot \delta^{18}$O $+ 10$. δ^2H and δ^{18}O are expressed in per mil vs. V-SMOW standard.

Table 2. Local Meteoric Water Lines Derived from Availbale Monthly δ^2H and δ^{18}O Data for Eight Stations of the IAEA/WMO Network, Located in the East African Region

Station	δ^2H = A·δ^{18}O + B		
	A	B	r^2
Addis Ababa	6.95 ± 0.22	11.51 ± 0.58	0.918
Entebbe	7.38 ± 0.24	10.78 ± 0.78	0.902
Kericho	7.96 ± 0.35	11.35 ± 0.96	0.912
Dar es Salaam	7.01 ± 0.28	6.83 ± 0.72	0.845
Ndola	7.66 ± 0.16	9.29 ± 1.09	0.971
Harare	7.03 ± 0.80	9.24 ± 0.93	0.933
Antananarivo	8.06 ± 0.18	14.00 ± 1.17	0.969
Diego Garcia	6.93 ± 0.20	4.66 ± 0.81	0.880

was found (Grunberger, 1989). This difference may result from specific conditions over the western Indian Ocean during the monsoon period: strong winds and relatively low relative humidity will likely lead to enhanced values of the deuterium excess in the evaporated vapor and, consequently, in the austral winter rains. This seasonality disappears, however, over the continent (station Harare), suggesting that the original d-excess signature of the maritime moisture is no longer preserved in precipitation. This lack of preservation of the d-excess may be due to several reasons: (i) partial evaporation of light rains in the unsaturated atmosphere below the cloud base; (ii) mixing with air masses of different origin and/or rainout history; (iii) substantial admixture of re-evaporated moisture of continental origin (surface water bodies and/or soil water). The latter effect should be noticeable downwind of the large lakes in the region.

Long-Term Trends

The interannual variability of the isotope composition of rainfall in the region is presented in Fig. 5. The figure shows mean weighted δ^{18}O and cumulative amount of precipitation during the major rainy period for four selected stations (Addis Ababa, Entebbe, Dar es Salaam, Harare). Unfortunately, the available record is relatively short and incomplete. The correlation of δ^{18}O with the cumulative rainfall is, in general, very poor or nonexistent (correlation coefficients generally below 0.5). This suggests that the control of the long-term fluctuations of ^{18}O content in precipitation, via local rainout mechanism (amount effect), is less important and that the regional processes seem to play a dominant role. They may include changes in the source region(s) of the vapor, varying contribution of different sources of moisture and/or modified circulation patterns in the region.

The data presented in Fig. 5 reveal substantial differences in the long-term behaviour of the isotope signal in precipitation among different parts of the region. For instance, the Addis

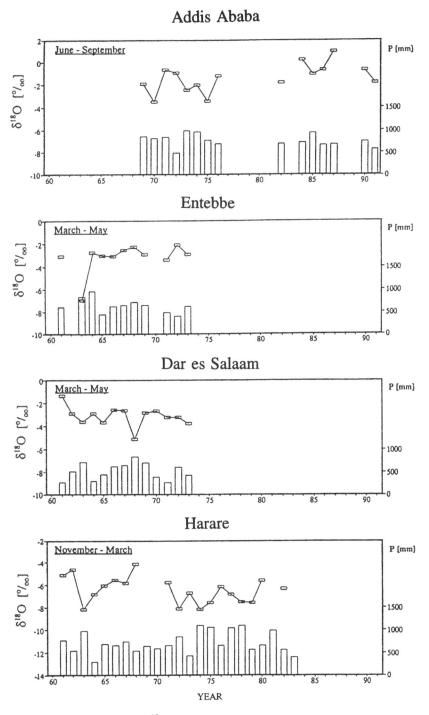

Figure 5. Interannual changes of ^{18}O isotope composition of precipitation for selected stations of the East African region. Weighted mean δ^{18}O values are shown (weighing by amount of precipitation) together with a cumulative amount of rainfall for the major rainy period at the given location.

Ababa data reveal an apparent shift of approximately 1.5‰ towards more positive $\delta^{18}O$ values between the seventies and the eighties. Substantial gaps in the latter part of the record do not allow any definite conclusions with regard to the significance of this shift. At Entebbe and Dar es Salaam the year-to-year changes of $\delta^{18}O$ are relatively small except for anomalously depleted $\delta^{18}O$ values recorded in April 1963 at Entebbe and in December 1969 at Dar es Salaam (see discussion below). The $\delta^{18}O$ record for Harare reveals the highest interannual variability among the stations analysed, with the amplitude of changes up to 4‰. No clear long-term trend can be seen in this case.

Extreme Events

An interesting feature of the long-term isotopic variability of rainfall in the region is the appearance of recurrent events characterized by abnormally low ^{18}O and deuterium content in precipitation. Two such events are presented in more detail in Fig. 6. The upper diagram of this figure shows monthly means of $\delta^{18}O$ and the amount of rainfall recorded at Entebbe between October 1962 and June 1964. A very low $\delta^{18}O$ value of −11.6‰ was recorded in April 1963 at this station, followed by still lower than normal value in May 1963 (−7.5‰). The amount of rainfall during these two months was higher than average, with more than 20 and 18 rainy days in April and May, respectively. Although during April and May 1964 the rainfall was also higher than normal, its isotopic composition was at that time almost 10‰ heavier. Abnormally low $\delta^{18}O$ values of −11.0 and −13.6‰ were also recorded two months earlier (February 1963) at Harare and Antananarivo, respectively. This suggests unusually strong convective activity within the air column during the spring months of 1963, possibly associated with disturbances during the passage of the ITCZ over the area. Interestingly, the deuterium content in non-exchangable hydrogen in a tree rings of Juniperus procera from Kenya (Krishnamurthy and Epstein, 1985) also reveal exceptionally low δD values during this period, pointing to a regional character of this phenomenon.

The second extreme event is apparent from the record of monthly $\delta^{18}O$ data for Ndola station, collected between November 1968 and April 1970 (lower graph in Fig. 6). In December 1968, exceptionally low $\delta^{18}O$ values were recorded at Ndola and Harare (−13.1 and −10.2‰, respectively), followed by slightly higher values in January 1969 (−8.1 and −10.0‰, respectively). In February 1969, the rainfall at Ndola was extremely depleted in ^{18}O ($\delta^{18}O$ = −17.5‰) and about 40% higher than the long-term average for this month. On the other hand, less than 50% of the average amount of rainfall was recorded at Harare, with enriched ^{18}O concentration ($\delta^{18}O$ = −4.2‰).

Anomalously low $\delta^{18}O$ values in precipitation, associated with passage of the ITCZ have also been recorded in other parts of the tropics (Salati et al., 1979; Matsui et al., 1983). For instance, 815 mm of rain with $\delta^{18}O$ value of −15.5‰ was recorded in April 1984 at the coastal station Belém, Brazil, which has to be compared with the long-term monthly means of 378 mm and −4.8‰, respectively. This exceptional disturbance within the ITCZ was propagated westward and reached Manaus, located about 1500 km inland, during the following month. The $\delta^{18}O$ value recorded in May 1984 at Manaus was close to −12.7‰, when compared with the long-term mean value of −7.2‰.

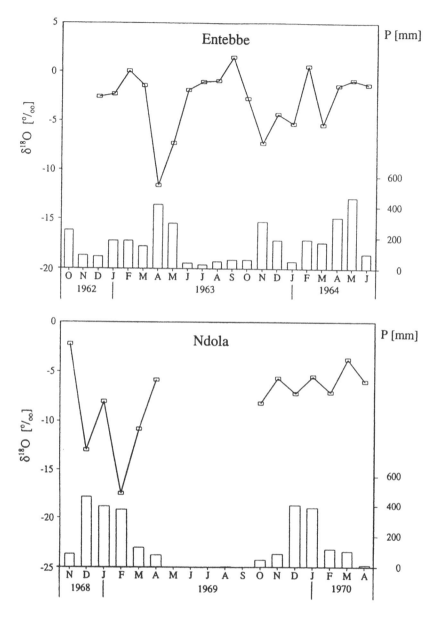

Figure 6. Two examples of events of abnormally low deuterium and ^{18}O contents in precipitation, recorded by the East African stations of the IAEA/WMO network.

CONCLUSIONS

The isotopic composition of precipitation in the East African region reflects the regional circulation patterns. Seasonal fluctuations of deuterium and ^{18}O content in monthly precipitation are well pronounced throughout the entire region. In general, isotopically depleted rains are associated with the rainy period(s). The annual weighted mean $\delta^{18}O$ values in precipitation increase northward, from about $-6.5‰$ in Zimbabwe and Zambia to about $-1.3‰$ in Ethiopia. The amplitude of seasonal changes of $\delta^{18}O$ also varies considerably, being relatively high in the south (approximately $6‰$) and dumped to about $3‰$ in the equatorial zone.

Whereas the seasonal changes of deuterium and ^{18}O isotope composition of precipitation in the region are in general well correlated with the amount of rainfall, for the interannual changes this correlation is very poor or nonexistent. This suggests that the control of long-term changes of the heavy isotope content of precipitation, via local rainout mechanism, is of minor importance, whereas the regional processes seem to play a dominant role. They may include changes in the source regions of the vapour, varying contribution of different sources of moisture and/or modified circulation patterns in the region.

An interesting feature of temporal variability of the isotopic composition of precipitation in the region is the occurrence of recurrent events of extremely low $\delta^{18}O$ and δ^2H values, often persisting for more than a month. Presumably, these events are induced by unusually strong convective activity within the intertropical convergence zone. These events may often be strong enough to leave their characteristic isotope signature in the freshly formed lake sediments or in the cellulose of the growing trees in the region (Krishnamurthy and Epstein, 1985). Thus, certain caution is required when past climatic and environmental conditions are reconstructed from isotope records preserved in such archives.

REFERENCES

Allen, D.J., Darling, W.G., and Burgess, W.G., 1989, Geothermics and hydrogeology of the southern part of the Kenya Rift Valley with emphasis on the Magadi-Nakuru area: British Geological Survey Research Report SD/89/1, pp. 30–33.

Chen, T.C., And Van Loon, H., 1987, Interannual variation of the Tropical Easterly Jet: Monthly Weather Review, v. 115, pp. 1739–1759.

Dansgaard, W., 1964, Stable isotopes in precipitation: Tellus, v. 16, pp. 436–468.

Darling, W.G., Allen, D.J., and Armannsson, H., 1990, Indirect detection of subsurface outflow from a Rift Valley lake: Journal of Hydrology, v. 113, pp. 297–305.

Flohn, H., 1964, Investigation on the Tropical Easterly Jet: Bonner Meteorologische Abhandlungen, v. 26, 83 pp.

Grunberger, O., 1989, Etude géochimique et isotopique de l'infiltration sous climat tropical contraste-massif du Piton des Neiges, Ile de la Réunion, Thése, Université de Paris-Sud.

IAEA, 1969, 1970, 1971, 1973, 1975, 1979, 1983, 1986, 1990, World Survey of Isotope Concentrations in Precipitation: Technical Report Series Nos. 69, 117, 129, 147, 165, 192, 226, 264, 311, International Atomic Energy Agency, Vienna.

IAEA, 1992, Statistical Treatment of Data on Environmental Isotopes in Precipitation: Technical Report Series No. 331, International Atomic Energy Agency, Vienna, 784 pp.

Joseph, A., Frangi, J.P., and Aranyossy, J.F., 1992, Isotope characteristics of meteoric water and groundwater in the Sahelo-Sudanese zone: Journal of Geophysical Research, v. 97, pp. 7543–7551.

Krishnamurthy, R.V., and Epstein, S., 1985, Tree ring D/H ratio from Kenya, East Africa and its palaeoclimatic significance: Nature, v. 317, pp. 160–162.

Lacaux, J.P., Delmas, R., Kouadio, G., Cros, B., and Andreae, M.O., 1992, Precipitation chemistry in the Mayombé forest of equatorial Africa: Journal of Geophysical Research, v. 97, pp. 6195–6206.

Lahuec, J.P., and Carn, M., 1988, Convergence intertropicale: L'intensité de la convection de juillet á septembre 1988: Veille Clim. Satell., v. 24, pp. 11–20.

Matsui, E., Salati, E., Ribeiro, M.N.G., Reis, C.M., Tancredi, A.C., and Gat, J.R., 1983, Precipitation in the Central Amazon Basin: The isotopic composition of rain and atmospheric moisture at Belém and Manaus: Acta Amazonica, v. 13, pp. 307–369.

Nicholson, S.E., 1986, The spatial coherence of African rainfall anomalies: interhemispheric teleconnections: Journal of Climate and Applied Meteorology, v. 25, pp. 1365–1381.

Nicholson, S.E., 1996, A review of climate dynamics and climate variability in eastern africa, in: The Limnology, Climatology and Paleoclimatology of the East African Lakes Johnson, T.C., and Odada, E.O. (Eds.), Gordon and Breach, Toronto, pp. 23–54.

Ropelewski, C.F., and Halpert, M.S., 1987, Global and regional scale precipitation and temperature patterns associated with El Niño/Southern Oscillation: Monthly Weather Review, v. 115, pp. 1606–1626.

Rozanski, K., Araguás-Araguás, L., and Gonfiantini, R., 1993, Isotope patterns in modern global precipitation, in: Climate Change in Continental Isotopic Records, P.K., Swart, K.C., Lohmann, J., McKenzie, and S., Savin (Eds.), Geophysical Monograph 78, American Geophysical Union, pp. 1–37.

Salati, E., Dall'Olio, A., Matsui, E., and Gat, J.R., 1979, Recycling of water in the Amazon Basin: An isotopic study: Water Resour. Res., v. 15, pp. 1250–1258.

Sonntag, Ch., Klitzsch, E., Löhnert, E.P., El-Shazly, E.M., Münnich, K.O., Junghans, Ch., Thorweihe, U., Weistroffer, K., and Swailem, F.M., 1979, Palaeoclimatic information from deuterium and oxygen-18 in carbon-14 dated north Saharian groundwaters, in: Isotope Hydrology 1978, Vol. II, International Atomic Energy Agency, Vienna, pp. 569–581.

Zimmermann, U., Ehhalt, D.H., and Münnich, K.O., 1967, Soil water movement and evapotranspiration: Changes in the isotopic composition of water, in: Isotopes in Hydrology, International Atomic Energy Agency, Vienna, pp. 567–585.

African River Discharges and Lake Levels in the Twentieth Century

A.T. GROVE *Department of Geography, Downing Place, Cambridge, England*

Abstract — Instrumental records show that African lake levels and river discharges have varied considerably from one decade to another in this century. The sequences of events in Equatorial, South Tropical and North Tropical Africa have differed. Most striking was the increase in lake levels and river discharges in both the Equatorial and North Tropical regions in 1961–62, and the decrease in all three regions since 1980. A question arises as to whether there have been times in earlier centuries when Lake Victoria ceased to overflow?

When African lake levels and river discharges are plotted over time, as in Fig. 1, two main features appear. Most striking is the rise in lake levels and river flows that took place in the early 1960s. Almost as impressive is the decline in river discharges and lake levels that has taken place since 1980. Evaporation and potential transpiration losses from water and vegetated surfaces are everywhere high in Africa, evaporation from open water being of the order of 1500–2500 mm a year. Records show little sign of a warming trend though, from the retreat of the Lewis Glacier in the course of this century, Kruss (1983) has deduced a temperature rise of about 0.3°C, which might be taken to be sufficient to give an increase in open water evaporation of about 30 mm a year (Dunne and Leopold 1978, p. 119). Except where there has been drastic interference in the form of dam construction and extraction for irrigation, as on the lower Nile, variations from year to year in values of river discharges and lake levels are primarily indicators of variations in precipitation over their catchments.

The rise in levels of the East African Great Lakes took place quite suddenly, late in 1961. Until that time, fluctuations in level this century had taken place on a modest scale with a pseudo-periodicity of about 11 years evident in the Lake Victoria record. Heavy rains towards the end of 1961 had a dramatic effect, the levels of Lake Victoria and Lake Tanganyika both rising by a metre at the end of1961 and by another metre the following year. The much smaller Lake Naivasha rose by about two metres (Vincent et al. 1979). All three lakes eventually peaked in 1964. By this time, Lake Victoria stood less than half a metre below the maximum it had reached in the 1870s, a level that may not have been exceeded since 3,720 ± 120 years BP (Y-688). Sediments which had accumulated on the floor of Hippo Cave by that time would not have survived the action of waves in a lake reaching up to a higher level (Bishop 1969, p. 107).

The greatest departures from average rainfall in 1961 were experienced in the semi-desert areas of northern Kenya which received in November more than six times their mean rainfall. The catchments of the Tana and Athi rivers received four times their mean monthly values in September, October and November 1961. The discharge of the Tana at Garissa in November averaged 1862 m^3 sec^{-1}, 5 to 10 times the usual values for the month (UNESCO 1971).

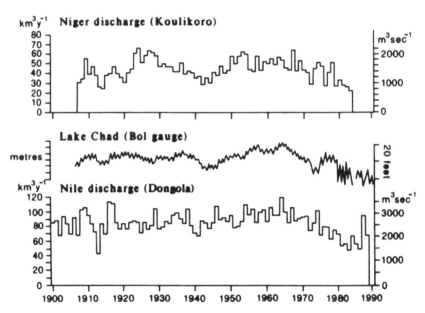

Figure 1. Fluctuations in African lake levels and river discharges in the twentieth century.

As Flohn (1987) has shown, these heavy 1961 rains were part of a great perturbation that extended from East Africa far out over the Indian Ocean, SE Tanzania receiving five times its normal rainfall in July, and the Seychelles more than three times their average rainfall in August and September 1961. Unusually high sea surface temperatures extended along the

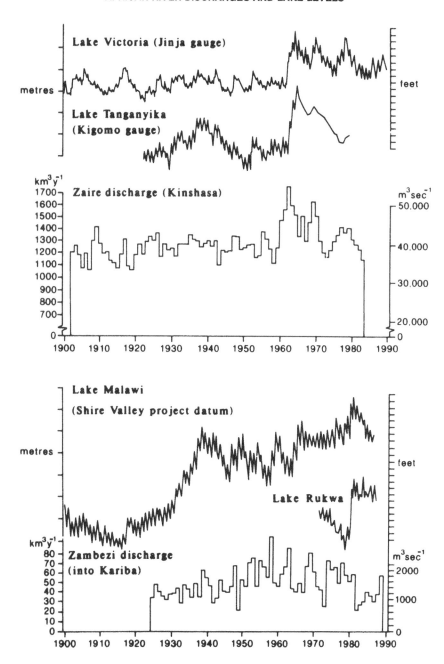

equator from the Somali Coast to the Bay of Bengal. Further east in contrast, approaching Sumatra about 90°E, skies were unusually cloudfree and sea surface temperatures unusually low.

Flohn described the excessive rainfall anomaly centred in October–December 1961 as extending from 30°E, about the longitude of the Western Rift, to 70°E, somewhat to the west of peninsular India. The longitudinal extent of the high rainfall anomaly for 1961 seems to have been even greater. The precipitation over the west, north–west and centre of India,

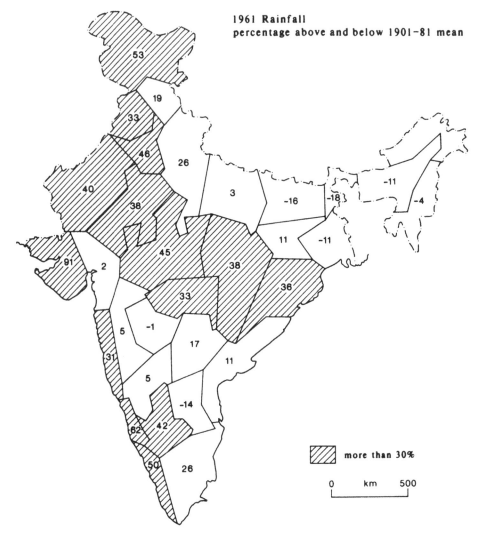

Figure 2. Rainfall over the Indian subcontinent in 1961 expressed as percentages above and below the mean (derived from Shukla 1987).

between 75 and 80°E, was more than 130 per cent of the mean (Fig. 2, based on Shukla 1987). In no other year this century has such a large positive anomaly been so widespread over that part of the subcontinent. In Africa, river discharge records (UNESCO 1969, 1971, 1974, 1979) show that high rainfall totals in 1961 extended far to the west of the Western Rift. The mean discharge of the Chari into Lake Chad was the highest November value for the period 1951–64 at least, being 1912 m³ sec⁻¹ compared with a mean of 887 m³ sec⁻¹. In Gabon, the mean discharge of the Ogoowi in November 1961 was 11,000 m³ sec⁻¹ as compared with a mean of 7458 m³ sec⁻¹; the highest recorded in any month over the period 1930–64. The Ubangui's mean discharge of 12,200 m³ sec⁻¹ in November 1961 compares with a 1911–64 November mean of 8500 m³ sec⁻¹; its discharge values for September,

October, November and December 1961, and January and February 1962 were the highest recorded since 1916 for each of those months. The 1961 discharges of both the Ogoowi and Ubangui were their highest annual values on record. The discharge of the Zaire at Kinshasa in November 1961 was the highest monthly value recorded between 1910 and 1983; furthermore, the great river's mean discharge in 1962, 54,000 m^3 sec^{-1}, gave the highest annual total ever recorded; 1609 km^3 as compared with a mean of about 1262 km^3 (Bultot and Dupriez 1987). It would therefore seem that the 1961 event extended at least as far west as longitude 10°E, well into that part of the continent normally under the influence of Atlantic weather systems.

After the heavy rains of November 1961 there was above average precipitation in 1962, 1963 and 1964 in the Lake Victoria region. The Equatorial Lakes, Edward and Albert as well as Victoria, continued to rise in the early 1960s. Over a period of twenty years, from 1961 until 1981, the White Nile delivered on average twice as much water to the Sudd as it had done in the first half of the century. The flow of the lower Nile in northern Sudan Republic and Egypt was higher in the early 1960s than it had been in the first half of the century, but this was not because of additional water from the Great Lakes; most of that was evaporated from the Sudd where rainfall was somewhat below the mean. The discharge of the lower river, shown on the diagram at Dongola, was maintained by rainfall totals being higher than usual over the Sobat and Blue Nile catchments throughout the early 1960s (Evans 1990, Conway and Hulme 1993).

The second feature which appears on these diagrams is the decline in river discharges and lake levels since about 1965. This did not affect Africa in the southern hemisphere. Lakes Malawi and Rukwa reached their highest levels around 1980 and then declined (UNDP 1986). The discharge of the Zambezi at the point where it enters Kariba was high from 1973 until 1980, about 60 km^3 annually; its mean annual discharge since 1949 had been some 40% higher than in the 25 years preceding 1949 (personal communication from Simon Bailey, Sir Alexander Gibb and Partners 1992). In 1981/2, its flow was much reduced again; it recovered over the next decade, and then came the great southern African drought of 1991/2 when it shrank more than had ever been recorded.

It has long been recognized that the smoothed values of the annual discharges of the great rivers of the western Sudan/Sahel have all varied in phase, in this century at least, with a periodicity of about three decades (UNESCO 1969, 1971, 1974, 1979, O.R.S.T.O.M. 1986). Gac and Faure (1981, 1986), when they plotted the seven-year running means of the Senegal, Niger and Chari, were so impressed by the regularity of the oscillations that they were tempted to predict on this basis that the Sahelian drought might come to an end in 1985. But it continued.

In spite of the early 1960s having been relatively wet, the precipitation in the Sahel/Sudan over the 30 year period 1961–90 was about 25% lower than for the period 1931–60 (Hulme 1992). One of the most spectacular results of the increased aridity was the shrinkage of Lake Chad from its highest level and greatest extent this century, about 25,000 km^2 in the 1960s, to about one tenth that area in the mid 1980s. Its decline and fall, as spectacular as that of the Sea of Aral, has not been the result of excessive abstraction of water for irrigation, but of diminished rainfall north of the Zaire watershed.

One might wonder about the chances of the East African lakes continuing to fall. What can be learned from the past? Tanganyika's discharge to the Zaire/Congo by the Lukuga channel has been interrupted from time to time over the last century. It is well-known that Lake Malawi had no overflow to the Shire and Zambezi between 1915 and 1935. Owen et al. (1990, p. 541) find evidence in cores and in historical and archaeological data that Lake Malawi fell "by at least 121 m for part of the time between 1390 and 1860." What of Lake Victoria? A reduction in the rainfall over its basin of 20 per cent would probably reduce the flow of the White Nile below Jinja to a trickle within a decade. Perhaps that has happened in recent times, for example during the Nyarabunga drought which is believed to have preceded a great famine about 1616–23, ""the worst drought ever to afflict the interlacustrine region within the recall of oral tradition" (Webster 1979, p. 152). At any rate it might be worth looking carefully in cores for evidence of such events within the last few centuries; they might be easier to identify than those rare occasions when the lake was as high as it was in the early 1960s.

REFERENCES

Bishop, W.W., 1969, Pleistocene Stratigraphy in Uganda. Geological Survey Uganda, Memoir 10. Entebbe.

Bultot, F., and Dupriez, G.L., 1987, Niveaux et débits du fleuve Zaire à Kinshasa. Mémoires de l'Académie royale des sciences d'Outre-Mer, Bruxelles, 4: 1–49.

Conway, D., and Hulme, M., 1993, Recent fluctuations in precipitation and runoff over the Nile sub-basins and their impact on main Nile discharge. Climatic Change 25: 127–151.

Dunne, T., and Leopold, L., 1978, Water in Environmental Planning. Freeman, San Francisco.

Evans, T.E., 1990, History of Nile Flows. In P.P. Howell and J.A. Allan (eds.), The Nile: resource evaluation, resource management, hydropolitics and legal issues. Centre of Near Eastern and Middle Eastern Studies, S.O.A.S., London University.

Faure, H., and Gac, J.Y., 1981, Will the Sahelian drought end in 1985? Nature 291: 475–478.

Flohn, H., 1987, East African rains of 1961/2 and the abrupt change of the White Nile discharge. Palaeoecology of Africa 18: 3–18.

Gac, J.Y., and Faure, H., 1986, Le "vrai" retour à l'humide au Sahel — est-il pour demain? C.R. Acad. Sci. (Paris), Ser. 2, 305: 777–781.

Hulme, M., 1992, Rainfall changes in Africa: 1931–60 to 1961–90. International Journal of Climatology 12: 685–699.

Itandala, Buluda, 1979, Ilembo, Nkanda and the girls: establishing a chronology of the Babinza. In J.B. Webster (ed.), 1979 Chronology, Migration and Drought in Interlacustrine Africa. Longman and Dalhousie University Press. Pp. 147–173.

Kruss, P.D., 1983, Climate change in East Africa: a numerical simulation from the 100 years of terminus record at Lewis Glacier, Mount Kenya. Zeitschrift für Gletscherkund und Glazialgeologie 19: 43–60.

Owen, R.B., Crossley, R., Johnson, T.C., Tweddle, D., Kornfield, R., Davison, S., Eccles, D.H., and Engstrom, D.E., 1990, Major low levels of Lake Malawi and their implications for speciation rates in cichlid fishes. Proceedings of the Royal Society London B240: 519–553.

Shukla, J., 1987, Interannual variability of monsoons. In J.S. Fein and P.L. Stephens (eds.), Monsoons. Wiley: New York, Chichester, Brisbane, Toronto, Singapore. Pp. 399–464.

U.N.D.P., 1986, Republic of Malawi, National Water Resources Master Plan.

U.N.E.S.C.O., 1969, 1971, 1974, 1979, Discharge of selected rivers of the World, Vols I, II, III, Paris.

Vincent, C.E., Davies, T.D., and Beresford, A.K.C., 1979, Recent changes in the level of Lake Naivasha, Kenya, as an indicator of equatorial westerlies over East Africa. Climatic Change 2: 175–189.

Physical Limnology

Comparison of Hydrology and Physical Limnology of the East African Great Lakes: Tanganyika, Malawi, Victoria, Kivu and Turkana (with reference to some North American Great Lakes)

R.H. SPIGEL   *Department of Civil Engineering, University of Canterbury, Christchurch, New Zealand*

G.W. COULTER   *Taupo, New Zealand*

INTRODUCTION

Extending from the southern tip of Lake Malawi to the north shore of Lake Turkana, the East African Great Lakes (Fig. 1) span a range of latitude from 14°30'S to 4°35'N and a north–south distance of over 2100 km, almost as great as that between Ottawa, Ontario, and Fort Lauderdale, Florida. The East African lakes are all found at medium to high altitudes (from 427 m asl for Lake Turkana to 1460 m asl for Lake Kivu) and are subject to a wide range of climatic conditions. The lake basins exhibit a variety of types, including the very deep, elongated and steep-sided troughs of Lakes Tanganyika and Malawi, and the more nearly circular, relatively shallow basin on Lake Victoria, with its highly complex, indented shoreline.

Hydrologic and hydrodynamic data from these lakes are scarce, fragmentary and often not easily accessible. We acknowledge the work of early investigators for much of the information presented here. Work published in the 1960s and 1970s by (among others) Talling on Victoria, Coulter on Tanganyika, Eccles on Malawi, Hopson on Turkana and Degens on Kivu forms the basis for much of what we know about these lakes. A resurgence of work is now underway over a wide range of disciplines; at the same time the lakes are coming under severe ecological pressure. It was therefore felt to be an opportune time to review what is known about the physical limnology of these lakes, and to bring into focus the factors that distinguish these remarkable lakes from other large lakes, particularly those in temperate zones.

While much of this paper is a review, we will have occasion to refer to some theoretical models for circulation and mixing in lakes. While the applicability of these models is by no means certain, they do provide both a framework for existing observations and working hypotheses to help organise future investigations. In keeping with the comparative nature of the review, reference will also be made to the extensive work that has been published on the physical limnology of the North American Great Lakes, especially that carried out on Lake Ontario in 1972–73 in connection with the International Field Year on the Great Lakes (IFYGL). It will be seen, however, that significant differences in climate, latitude and basin shape make the direct application of some of the IFYGL results to the East African lakes questionable.

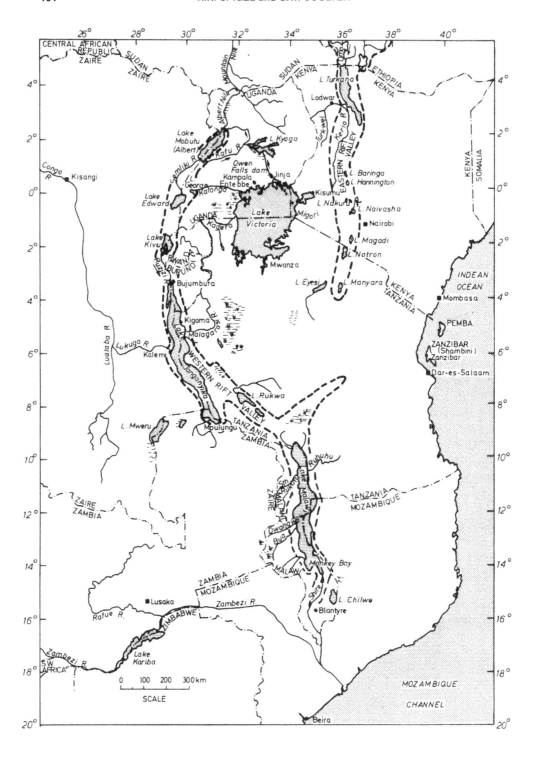

Figure 1. East Africa and the major lakes.

BASIN SHAPES: FEATURES OF SIGNIFICANCE FOR WATER MOVEMENTS

Table 1 summarises some basin and catchment parameters for the East African lakes and two of the North American lakes; bathymetric features are shown in Figs. 2 and 3. The surface area of Lake Victoria is greater than that of all the North American lakes except Lake Superior. Victoria is, however, relatively shallow compared with both the North American lakes and the other East African lakes. Lakes Tanganyika, Malawi and Turkana occupy elongated, narrow basins, those of Tanganyika and Malawi being of exceptional depth with extremely steep sides. Very detailed bathymetry has been published for Lake Tanganyika by Tiercelin (1991) and for Lake Malawi by Johnson and Ng'ang'a (1990). Lake Tanganyika is separated into two main basins by the Kalemi shoal (mean water depth over the sill of 500 m, maximum water depth over the sill between 500 and 550 m); the main basins on either side of this sill have relatively flat bottoms with depths greater than 1000 m and steep ends and sides, the bed slope of the southern end being somewhat more gradual. In contrast Lake Malawi consists of a single major basin with a more gradual slope from the southern end to a maximum depth greater than 700 m. The sides of Lake Turkana are not as steep as those of Malawi or Tanganyika, and its two basins are of unequal size and depth, the southernmost being relatively small and deep. Kivu occupies a deep and steep-sided basin, divided in two by the large Idjwi Island. Aside from its very shallow and broad aspect ratios, Lake Victoria is distinguished from the other four lakes by its intricate and highly indented shoreline, with the presence of numerous islands, shallow bays and connecting channels, and extensive areas of wetlands. These features are significant for exchange of water between the littoral and pelagic regions of the lake.

The North American Great Lakes, although elongated, have larger width to length aspect ratios than the deep, narrow African rift lakes, and more gently sloping sides. These features influence the seasonal pattern of stratification and the coastally trapped waves and currents observed in the North American lakes. Otherwise the difference in catchment and basin parameters among the African lakes is greater than between African and North American lakes. For example, the ratio of lake area to total catchment area varies from 0.067 to 0.26 for the African lakes (Table 1), the relatively large tributary area of Lake Turkana being an indication of the aridity of the climate and importance of the River Omo inflow for that lake. The higher ratios of lake/catchment areas are more typical of large lakes.

WATER BALANCE AND CLIMATE

A lake's water balance equates rate of change in lake volume with inflows minus outflows:

$$\Delta V/\Delta t = (P - E)A_s + \Sigma Q_{in} + Q_{sro} - Q_{out} + Q_{gw}, \qquad (1)$$

where P is rainfall on the lake surface, E is evaporation from the lake surface, A_s is lake surface area, ΣQ_{in} is the sum of all river inflows, Q_{sro} is storm runoff (other than that included in ΣQ_{in}, i.e., flows from non-point sources) from the surrounding catchment, Q_{out} is river outflow (if any), Q_{gw} is net groundwater inflow to the lake, ΔV is the change in volume of water stored in the lake, and Δt is the time interval over which the various inputs and outputs are averaged. For relatively small changes in storage ΔV is related to changes in lake surface elevation ΔZ_s by

$$\Delta V = A_s \, \Delta Z_s. \qquad (2)$$

Table 1. Catchment and Basin Parameters

	East Africa					North America	
	Victoria	Tanganyika	Malawi	Turkana	Kivu	Ontario	Superior
Elevation, m	1,134	774	475	427	1,460	74.6	183
A_s, lake surface area, km^2	68,800	32,600	22,490	8,860	2,370	19,520	82,100
A_t, tributary area, km^2	195,000	198,400	75,300	123,300		70,400	128,000
$A_s/(A_t + A_s)$	0.26	0.14	0.23	0.067		0.22	0.39
V, volume, km^3	2,760	18,900	6,140	251	569	1,640	12,200
$\bar{H}$, mean depth V/A_s, m	40	580	273	28.3	240	84	149
H_{max}, maximum depth, m	92	1,471	706	73 (106)[b]	480	244	406
L, transect[a] length, km	309	659	569	256	84	297	697
B, transect[a] width, km	212	58	55	29	28	67	204
$\bar{B}$, mean width A/L, km	223	49	40	35	28	66	118
B_{max}, maximum width, km	240	76	75	52	48	92	253

[a]Transect lengths and widths are those for the transects shown on the maps of Figure 2. [b]Maximum depths for central basin and southern basin (in brackets) for Lake Turkana.
Data from: Johnson et al. 1990, Johnson 1980, Johnson and Ng'ang'a 1990, Degens et al. 1973, Hecky and Bugenyi 1992, Hecky and Kling 1987, Richards and Aubert 1981, Dobson 1984, Talling 1966, Tiercelin 1991, Eccles 1974, Eccles 1984, Ferguson and Harbott 1982, Källqvist et al. 1988, Coulter and Spigel 1991, Quinn 1992.

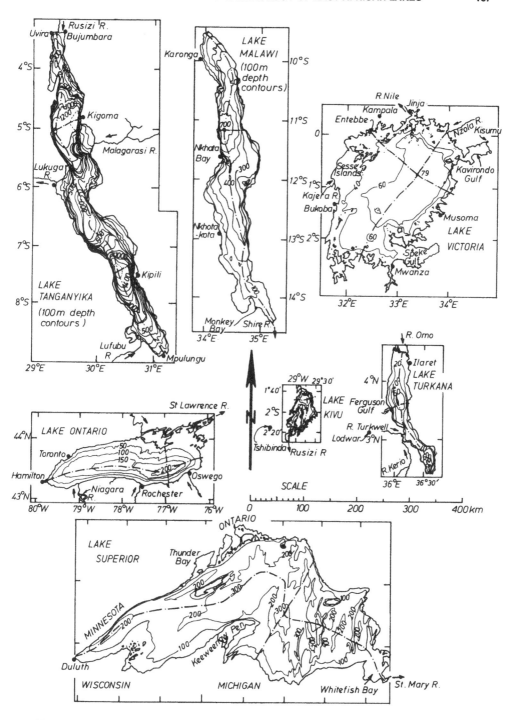

Figure 2. Bathymetry of the five East African lakes included in this study. North American Lakes Ontario and Superior also included for comparison. All lakes are drawn to the same horizontal scale but depth contour intervals differ. The depth contour visible in the central basin of Lake Kivu is 400 m. Map for Lake Tanganyika based on Tiercelin and Mondeguer (1991); Malawi, Johnson and Ng'ang'a (1990); Victoria, Talling (1966); Turkana, Ferguson and Harbott (1982); Kivu, Degens et al. (1973); Ontario, Dobson (1984); Superior, Johnson (1980) and Johnson (pers. comm.).

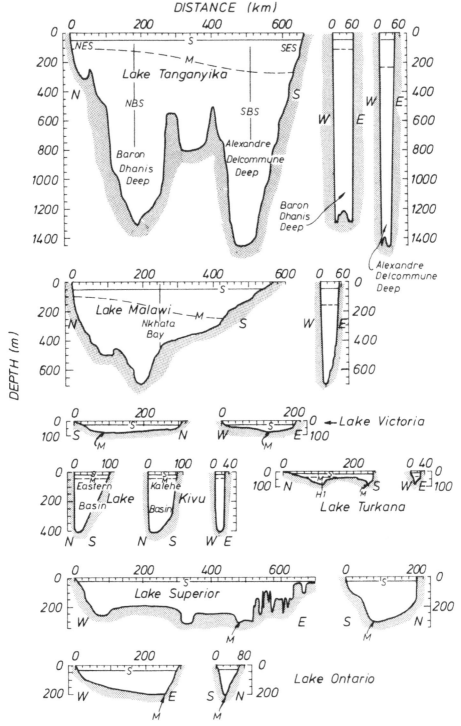

Figure 3. Bottom profiles through the lakes along the transects shown in Figure 2. All profiles are drawn to the same horizontal and vertical scales. S = approximate depth of summer stratification; M = maximum depth of mixing.

If the averaging time Δt is such that lake levels are the same at the beginning and end of the time interval, then $\Delta V/\Delta t = 0$ and all the inflows and outflows must balance.

Estimated annual average components for water balances of the African lakes (except Kivu, for which we could not find sufficient data) are listed in Table 2, as well as the exceptionally well documented 1972–73 figures for Lake Ontario from IFYGL and mean annual data for Lake Superior. The figures for the African lakes are at best approximations, and the balances have all been forced to close by assuming $\Delta V/\Delta t = 0$, thereby allowing some of the figures to be determined by subtraction. $\Delta V/\Delta t = 0$ implies no net long-term change in water level, which is not the case over tens or hundreds of years. However, on an annual time scale the average annual values of lake-level change are generally much smaller than the other water balance components. Groundwater flows have been neglected in the balances for all African lakes, and have been incorporated into the inflow term for Lake Ontario. As far as we know all of the precipitation figures for the African lakes are from land-based stations. In contrast, rainfall measurements for Lake Ontario were made at eighteen meteorological buoys spread evenly over the lake as well as at several land-based stations. Evaporation estimates for Lake Ontario were made by a number of methods, while those for the African lakes are based mainly on evaporation pan data, again from a few land-based stations. Rainfall and evaporation data for Lake Superior, while not as thoroughly documented as those for Lake Ontario, are based on data from many stations measured over extended periods of time (Bennett 1978). For purposes of comparison all terms have been expressed both as fluxes per unit lake surface area (mm/yr) and as rates of change in volume (km^3/yr).

The water balances of Lakes Tanganyika, Malawi and Victoria are dominated by rainfall on the lakes and evaporation, with river inflow and outflow making minor contributions. The balance of Lake Turkana is primarily between evaporation and inflow from the River Omo, which according to Ferguson and Harbott (1982) supplies 90% of all water inputs. Only Lakes Malawi and Victoria have significant river outflows, Malawi to the Shire River and Victoria to the Nile. Even these outflows are small compared to the St Lawrence River outflow from lake Ontario. The relatively minor contribution of river flows to the water balances of most of the East African Great Lakes, when coupled with their large volumes, leads to very long residence or flushing times for these lakes, exceptionally so for Lake Tanganyika. Of all the North American Great Lakes only Lake Superior has a flushing time in excess of 100 years (Quinn 1992). The longer its flushing time, the more vulnerable a lake is to damage from the effects of human activities and development in its catchment. Moreover, the longer the residence time, the less likely it is that damage can be reversed once it occurs. The very large volumes of Lakes Tanganyika and Malawi may provide temporary buffers against deterioration of water quality in these lakes. In Lake Victoria, however, with its much smaller volume, investigators have recently documented increased eutrophication compared with conditions of the 1960s (Hecky 1993).

The large proportion of rainfall and evaporation in the balances of the African lakes makes their levels particularly sensitive to climate change. Relatively small changes in rainfall and evaporation may lead to shifting between closed-and open-basin status as has happened in historic times for Lake Tanganyika. Closed-basin lake levels tend to fluctuate widely in response to changes in rainfall and evaporation. The level of lake Turkana, at present the only closed-basin lake among those considered here, declined by approximately 5 m from 1972 to 1988 (Källqvist et al. 1988).

Table 2. Water Balance[a] (all components mm/yr and km³/yr (in brackets))

	East Africa (mean annual)				North America	
	Victoria	Tanganyika	Malawi	Turkana	Ontario (1972–73)	Superior (mean annual)
P, precipitation on lake	1450 (100)	1050 (35)	1350 (30)	360 (3.2)	1038 (20.3)	696 (57.1)
E, evaporation from lake	1370 (94.6)	1530 (50)	1610 (36)	2340 (21)	744 (14.5)	470 (38.6)
$\Sigma Q_{in} + Q_{sro}$, river inflows + storm runoff	260 (18)	560 (18)	650 (15)	1980 (17.5)	13,718 (267.8)	657 (53.9)
Q_{out}, river outflow	340 (23.4)	83 (2.7)	390 (8.8)	0 (0)	13,332 (260.2)	883 (72.5)
$\Delta V/\Delta t$, change in storage[a]	0	0	0	0	+680 (13.3)	0
Flushing time, V/Q_{out}, years	120	7,000	700	∞	6.3	168
Refill time, $V/(\Sigma Q_{in} + Q_{sro} + PA_s)$, years	24	360	140	14	6.1	110

[a]$\Delta V/\Delta t = (P - E) A_s + \Sigma Q_{in} + Q_{sro} - Q_{out}$. For East African lakes, mean annual water balances have been forced to balance with $\Delta V/\Delta t = 0$. Groundwater flow has been neglected for all East African lakes, but has been incorporated in the tributary inflow term for Lake Ontario.

Data from: Rzoska 1976, Newell 1960, Coulter and Spigel 1991, Griffiths 1972a,b, Bultot and Griffiths 1972, Bultot 1972, Torrance 1972, Ferguson and Harbott 1982, Källqvist et al. 1988, Hecky and Bugenyi 1992, Eccles 1984, Eccles 1974, DeCooke and Witherspoon 1981, Bennett 1978, Phillips 1974.

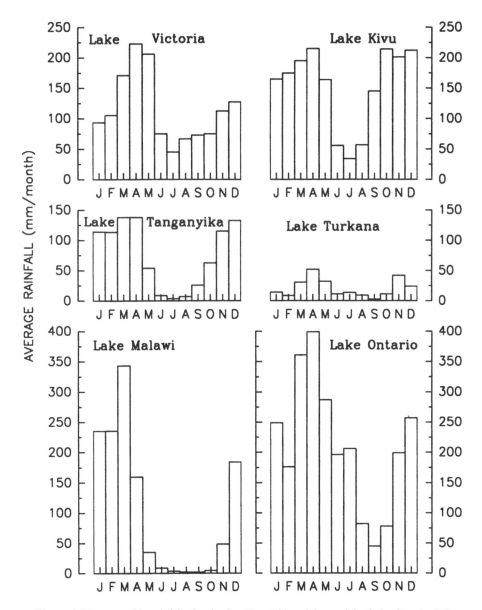

Figure 4. Mean monthly rainfalls for the five East African lakes and for Lake Ontario. Lake Victoria values are averages for Entebbe, Kisumu, Mwanza and Bukoba (Griffiths 1972b, Rzoska and Wickens 1976); Lake Kivu, Tshibinda (Bultot and Griffiths 1972); Lake Tanganyika, averages for Bujumbura and Kigoma (Griffiths 1972b); Lake Turkana, averages for Lodwar and Ilaret (Ferguson and Harbott 1982); Lake Malawi, averages for Karonga and Nkhotakota (Torrance 1972); Lake Ontario from IFYGL (DeCooke and Witherspoon 1981).

Seasonal influences on the East African lakes are dominated by the annual cycle of monsoon winds and the accompanying changes in air temperature and humidity. Rainfall (Fig. 4), humidity and evaporation data (Fig. 5) indicate that a dry season extends from May or June through August, September or October, with wetter conditions persisting for the

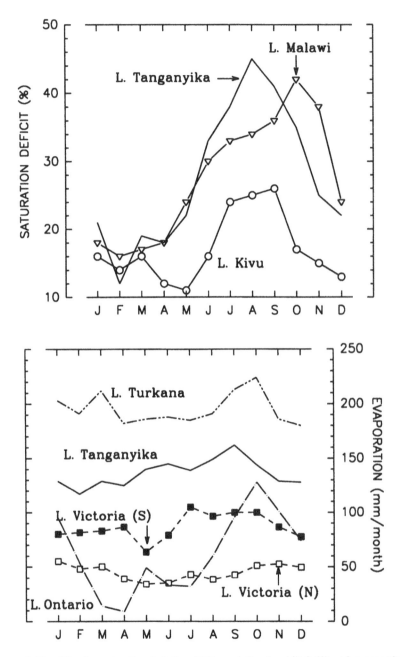

Figure 5. Humidity (as saturation deficit = [100 – relative humidity] %) and evaporation; illustrating generally drier conditions during the period May–September. Decreasing evaporation from south to north on Lake Victoria probably results from the gain in humidity in the atmospheric boundary layer over the lake in predominantly southerly winds. Data for Lake Tanganyika are from Bujumbura (Griffiths 1972a); Lake Malawi, average for Karonga and Nkhotakota (Torrance 1972); Lake Kivu, Tshibinda (Bultot and Griffiths 1972); Lake Turkana, Lodwar (Källqvist et al. 1988); Lake Victoria: northern, average Jinja and Entebbe; southern, average Musoma and Mwanza (Newell 1960); Lake Ontario, IFYGL (Quinn and den Hartog 1981).

remainder of the year. A double-peaked wet season has been identified for Lakes Victoria and Tanganyika. The dry season accompanies the onset of the southwest monsoon in the Indian Ocean off the coast of East Africa between May and June, characterised by persistent southerly winds that reach speeds exceeding 20 m/s offshore during June, declining to less than 20 m/s during July and August (Schott and Fernandez-Partagas 1981). Wind data for the lakes are not readily available in a form useful for comparison, but descriptions of the climate over Lakes Victoria, Tanganyika, Malawi and Turkana all mention a pattern of strong southerly winds during this period that can reach speeds in excess of 15 m/s, and that persist at lower speeds (5–11 m/s) for periods of several days. Superimposed on these seasonal winds are diurnal patterns that differ from lake to lake. In Lake Turkana strong winds occur at night and in the afternoon (Ferguson and Harbott 1982) while in Lakes Tanganyika, Malawi and Victoria onshore winds during the day alternate with land breezes at night in coastal areas (Coulter 1963, Eccles 1974, Fish 1957).

During the remaining months of the year (October through April) winds are generally lighter and more variable (although strong winds can accompany local storms), from a northeast quarter on Lakes Tanganyika and Malawi, while southeast and southerly directions remain predominant on Lakes Victoria and Turkana. The prevailing southerly wind direction over Lake Victoria is reflected in the decrease in evaporation from the southern to northern shores (Fig. 5), a result of increasing water vapour content of the atmospheric boundary layer as it develops over the lake from south to north. A similar effect on humidity has been noted by Coulter and Spigel (1991) for Lake Tanganyika during the south wind season.

Solar radiation (Fig. 7) does not show much seasonal variation over the East African lakes, the largest variation occurring for Malawi, presumably because of its greater distance from the equator. Taken together, Figs. 4–7 illustrate the great contrast in climatic influences on the North American and East African lakes. The wide range in solar radiation (Fig. 7) and air temperature (Fig. 6) dominate seasonal effects on the temperate lakes, leading to a much wider range in water surface temperatures (Fig. 6) than for the tropical lakes. With the exception of Lake Kivu, the range in water temperatures in the tropical lakes lies within the range in air temperatures throughout most of the year, while this is not the case for the temperate lakes. Of the tropical lakes, Turkana is notably warmer than the others, its surface temperatures exceeding 30°C and temperatures at depth (26–27°C) approaching or exceeding the peak surface temperatures reached in all the other lakes. Surface and bottom temperatures for Lake Kivu (Figs. 6 and 8) exceed air temperatures, presumably because of significant geothermal heat inputs to the lake (Degens et al. 1973, Newman 1976).

The importance of evaporative cooling during the dry season has been identified as an important factor in the heat balance and mixing regime, for example in Lake Victoria (Newell 1960, Talling 1966) and in Lake Tanganyika (Coulter and Spigel 1991), and this may be true for all the tropical African lakes. The annual wind cycle over the tropical lakes, with southerly winds prevailing for four months, is quite different from the meteorological regime of the temperate lakes. Over the North American lakes changes in wind accompany the passage of successive high pressure ridges and low pressure troughs moving from west to east with a periodicity of 5 to 10 days. Wind mixing and water movements are strongly influenced by large storms that can last several days and during which the wind vector rotates through a complete circle, in a direction that depends on the position of the track of the storm relative to the lake (Hamblin and Elder 1973, Hamblin 1987, Ivey and Patterson 1984). The dominant wind direction and path of the storm can vary considerably from event to event. This is in contrast to the more unidirectional nature of the large-scale wind field

R.H. SPIGEL and G.W. COULTER

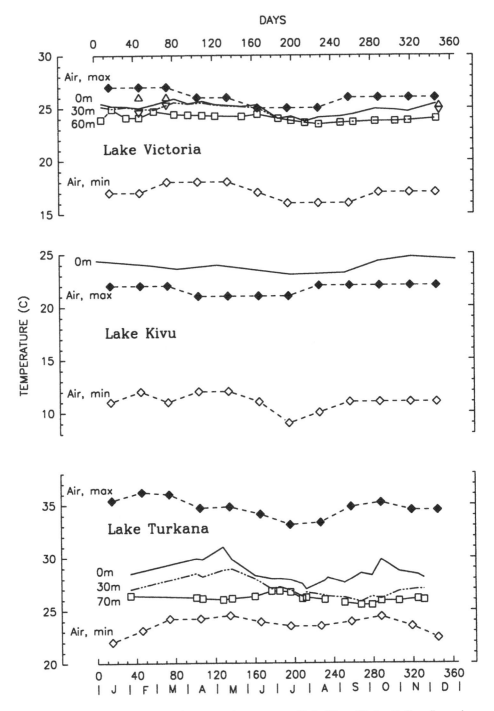

Figure 6. Maxima and minima of daily average air temperatures (dashed lines, filled and hollow diamonds, respectively), and surface and at-depth water temperatures (solid line, hollow squares, respectively). Intermediate-depth water temperatures are shown for Lakes Victoria and Turkana as double-dashed lines. Data for air temperatures for Lake Victoria are from Entebbe (Griffiths 1972a); Lake Kivu, Tshibinda, Bultot and Griffiths 1972); Lake Turkana, Lodwar (Griffiths, 1982a); Lake Tanganyika, Bujumbura (Griffiths 1972a); Lake Malawi, Karonga (Torrance 1972); Lake Ontario, IFYGL (Phillips 1974). Data for water temperatures for Lakes Victoria (1953), Kivu (1953), Tanganyika (1956) and Malawi (1958) are from

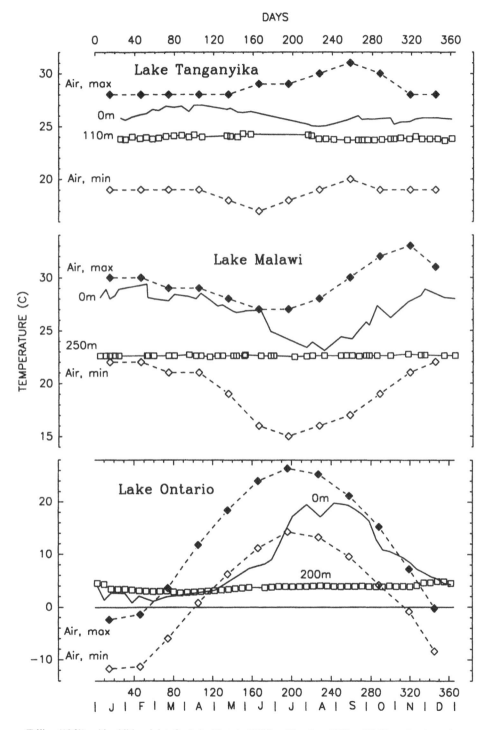

Talling (1969), with additional data for Lake Victoria (1953) at 30 m from Talling (1966), and at the surface and 30 m, for three days only (1989–90) from Hecky 1993. The recent data from Hecky (1993) for Lake Victoria (hollow triangles) show warmer surface temperatures in February and March. Data for Lake Ontario are from Boyce et al. 1977. All temperature scales for the East African lakes are the same, but these differ by a factor of two from that for Lake Ontario.

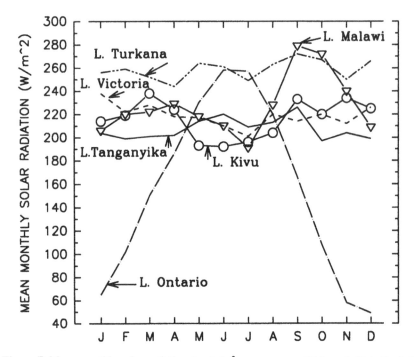

Figure 7. Mean monthly solar radiation (watts/m², average over 24 hours). Data for Lake Tanganyika are from Bujumbura (Griffiths 1972b); Lake Victoria, averages from Mwanza, Jijna and Entebbe, and correlations with sunshine hours for Entebbe and Kisumu (Griffiths 1972b); Lake Malawi, Lilongwe (Torrance 1972); Lake Turkana, Lodwar (Griffiths 1972a); Kivu, Lwiro (Bultot and Griffiths 1972); Lake Ontario, IFYGL (Phillips 1974).

over the tropical lakes. Diurnal sea and land breezes (on-shore winds by day, off-shore at night) occur in the temperate and tropical lakes alike.

GENERAL FEATURES OF STRATIFICATION

The strength of density stratification in all of the lakes — tropical and temperate — is sufficient to influence water movements and the distribution of oxygen and nutrients, over at least part of the year. In all of the lakes (with the exception of Kivu), density differences arise mainly from temperature gradients caused by the absorption of solar radiation. Dissolved salts are present in sufficient amounts to influence *absolute* water density, but generally do not exhibit sufficient vertical variation to significantly affect vertical stability due to density *differences* in the water column. (See Table 3, which summarises maximum temperature, conductivity and salinity ranges.) In the deep waters of Lakes Malawi and Tanganyika density changes due to temperature and salinity are very small and interactions between pressure, temperature, salinity and density must be considered when assessing water column stability. Wüest et al. (this volume) have shown that in Lake Malawi non-ionic compounds such as silicic acid and dissolved carbon dioxide are as important as ionic species for stability considerations. The effects of salinity are not so subtle in Lake Kivu.

Table 3. Relative Density Differences Related to Temperature and Salinity

	East Africa					North America	
	Victoria	Tanganyika	Malawi	Turkana	Kivu	Ontario	Superior
T_1, °C	26	27	28	31	22.4	22	18.6
T_2, °C	23.5	23.25	22.66	26.5	25.8	4	4.9
Z_1, m	0	0	0	0	50	0	0
Z_2, m	60	600	350	70	480	40	35
Pot. T_2, °C	23.49	23.15	22.60	26.49	25.71		
C_1, S/m (T_{ref}, °C)	0.0081 (20)	0.0640 (25)	0.0230 (20)	0.367 (25)	0.132 (20)		
C_2, S/m (T_{ref}, °C)	0.0088 (20)	0.0695 (25)	0.0240 (20)	0.374 (25)	0.400 (20)		
S_1 (ppt)	–	0.31	0.21	0.18	0.82		
S_2 (ppt)	–	0.34	0.23	0.18	3.57		
$\Delta\rho/\rho_0$	6.36×10^{-4}	1.01×10^{-3}	1.41×10^{-3}	1.32×10^{-3}	1.25×10^{-3}	2.20×10^{-3}	1.48×10^{-3}

T_1 and T_2 give maximum and minimum temperatures recorded at depths Z_1 and Z_2, respectively. Pot. T_2 is potential temperature corresponding to T_2 calculated from formula of Caldwall and Eide (1980). C_1 and C_2 give range of conductivities (siemens/metre); T_{ref} is the reference temperature for the conductivities. S_1 and S_2 are salinities as (g dissolved solids)/(kg solution); for Lake Malawi these include both charged (ionic) and uncharged species (Wüest et al. 1996); values for Lakes Tanganyika and Turkana have been estimated from conductivity and temperature using UNESCO formulas (Fofonoff and Millard 1983, Hill et al. 1986) and are therefore only approximations. Relative density difference $\Delta\rho/\rho_0$ was calculated from Chen and Millero's (1986) formulas using the temperatures (potential) and salinities in the table; $\Delta\rho$ is the density difference and ρ_0 is the average density. Data from Hecky 1993 (L. Victoria); Coulter and Spigel 1991, Ferro and Coulter 1974 (L. Tanganyika); Halfman 1993, Wüest et al. 1996 (L. Malawi); Ferguson and Harbott 1982 and Källqvist et al. 1988 (L. Turkana); Degens et al. 1973 and Newman 1976 (L. Kivu); Csanady 1972a,b (L. Ontario); Halfman and Johnson 1989 (L. Superior); Talling and Talling 1965 (L. Victoria, L. Kivu).

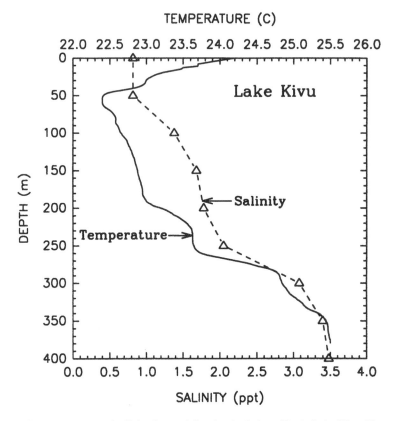

Figure 8. Temperature and salinity (as total dissolved salts) profiles in Lake Kivu. Temperature data measured by Newman (1976) with a microstructure profiler; salinity data from Degens but presented by Newman (1976).

Lake Kivu is unique among all the lakes described here in having an inverse or destabilising temperature gradient over much of its depth, with temperature increasing from 22.4°C at 50 m to 25.5°C at 375 m (1973 data, presented by Newman 1976 and shown here in Fig. 8; older data, also in Newman 1976, show temperatures increasing to 25.8°C at 450 m). Because of increasing salinity with depth below 50 m (Fig. 8), the overall density profile is stable. This feature of opposing temperature and salinity gradients, found also in solar ponds, in some ice-covered Antarctic lakes, in some regions of the ocean, and in many deep meromictic lakes, is thought to be caused in Lake Kivu by the input of hot, saline geothermal fluid at the bottom of the lake. The combination of warmer, saltier water underlying cooler, fresher water can lead to the formation of convecting layers if the stabilising effect of the salt is not too strong. This double-diffusive instability (so named because it is caused by the greatly different rates at which heat and salt diffuse on a molecular level; see Turner 1973) has been documented in Lake Kivu by Newman (1976), who used a temperature microstructure profiler to resolve centimetre-scale temperature variations. He found convecting layers that varied in thickness from 0.55 m to 1.40 m and within which temperature was isothermal; these were separated by layers approximately 10 cm thick containing sharp temperature

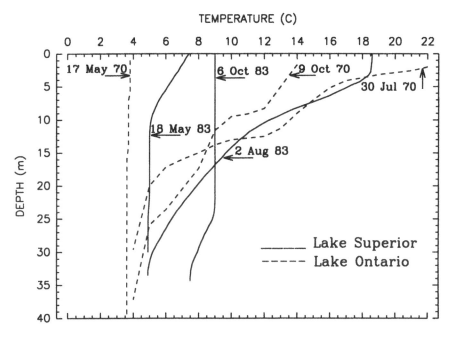

Figure 9. Temperature profiles from offshore locations in Lakes Ontario and Superior, illustrating the development of summer stratification and subsequent autumn cooling that leads to fall overturn and complete mixing.

gradients. Degens et al. (1973) found no dissolved oxygen below 50–60 m. The salinities, while high enough to stabilise the density profile and higher than those of any of the other lakes described here (Table 3), do not qualify Lake Kivu as a "saline" lake; its maximum salinity of 3.5 ppt is less than that of seawater by a factor of 10. Hecky and Kling (1987) show a mean Secchi-disk transparency of approximately 10 m for Lake Kivu, while H. Jannasch, quoted by Newman (1976), found no significant solar energy below 60 m. Hence the top 50–60 m appears to be the depth over which interaction with the atmosphere occurs in Lake Kivu, with the bottom 400 m or so of the water column dominated by geothermal effects. Degens et al. (1973) found that the gross temperature structure was uniform across the lake to within ±0.2°C, in agreement with the findings of previous investigators. This is quite different from the behaviour of the other lakes described here, all of which exhibit significant horizontal variations in temperature under the influence of wind. There is insufficient information available to determine whether the absence of horizontal temperature structure in Lake Kivu is due to a generally calmer wind regime, the lake's smaller size, lack of data or some combination of these factors.

Temperature stratification in the remaining four African lakes follows a more "normal" pattern, although still with important differences from that of most large temperate lakes. In Lakes Superior and Ontario, for example (Fig. 9), stratification begins to develop in May from an essentially isothermal condition of complete mixing near 4°C, the temperature of maximum density. Stratification does not develop uniformly over the entire lake. Differential heating and cooling rates between coastal and offshore regions, combined with the inverse density–temperature relation for water below 4°C, leads to formation of a "thermal

Table 4. Transparancy (Secchi-Disk Depth, Z_{SD}) and Light Extinction
(1% light level, $Z_{1\%}*$)

Lake	Transparancy and light extinction		Sources
Tanganyika	$5 \text{ m} \leq Z_{SD} \leq 17.5$ (1960–62; but $Z_{SD} > 22$ m recorded 1947)		Various sources, summarised by Hecky
	$10 \text{ m} \leq Z_{1\%} \leq 35$ m	(1975)	et al. 1991, pp. 92–93
Malawi	$12 \text{ m} \leq Z_{SD} \leq 23$ m	(1975–76)	Ferro 1977
	$15 \text{ m} \leq Z_{1\%} \leq 47$ m	(1940)	Beauchamp 1953
Victoria	$6.4 \text{ m} \leq Z_{SD} \leq 7.2$ m	(Historical)	Bugenyi and
	$1.5 \text{ m} \leq Z_{SD} \leq 2.7$ m	(Present)	Magumba 1993;
	$Z_{1\%} = 13.8$ m offshore,		Levring and Fish 1956
	11.0 m inshore	(1953)	
Turkana	$1 \text{ m} \leq Z_{SD} \leq 4.5$ m	(Central Lake, 1987–88)	Källqvist 1988
	$Z_{1\%}$	Distance from	
	(1972–75)	Omo R. mouth	
	0.5 m – 3.7 m	25 km	
	2.1 m – 9.8 m	105 km	Ferguson and
	4.0 m – 8.5 m	162 km	Harbott 1982
	10.2 m – 13.0 m	230 km	
Kivu	Mean $Z_{SD} = 10$ m		Hecky and Kling 1987
Ontario	$2.1 \text{ m} \leq Z_{SD} \leq 3.8$ m	(1965–82)	Dobson 1984
	$5.6 \text{ m} \leq Z_{1\%} \leq 16$ m	(1972–73)	Thomson et al. 1974 (using average vertical extinction coefficient)
Superior	$7.2 \leq Z_{1\%} \leq 26$	(1973)	Thomson et al. 1974

*Values as given in literature or computed from extinction coefficients or other light-depths assuming a single exponental decay for white light, $I(z)/I_0 = e^{-\eta z}$; hence for $I(Z_{1\%})/I_0 = 0.01$, $Z_{1\%} = -\ln 0.01/\eta = 4.605/\eta$.

bar" — a ring of warmer coastal water above 4°C separated by horizontal temperature gradients from offshore water below 4°C (Rogers 1965, Zilitinkevich et al. 1992). Stratification then spreads gradually to the deeper parts of the lake, reaching maximum strength in July and August. Surface temperatures can exceed 20°C in a relatively shallow mixed layer of approximately 10 m depth (or less) in Superior and Ontario, overlying deeper water still near 4°C (Table 4; Fig. 9). The mixed layer gradually deepens in late summer and its stability declines with autumn cooling, until complete overturn occurs. An inverse stratification develops in winter as surface water cools below 4°C.

No comparable cooling or mixing event occurs in the large tropical lakes. Solar radiation and air temperatures remain within a much narrower range in the tropics (Figs. 6 and 7); water temperatures also remain within a much narrower range (Fig. 6). Temperatures do not drop below 4°C, and no thermal bar develops. However, partly because of the higher thermal expansivity of water at warmer temperatures, the smaller temperature differences that do develop in the tropical lakes create sufficient stability to preserve stratification in some form throughout the year. In the shallower lakes complete vertical mixing may occur during periods of strong wind, but horizontal temperature variations normally remain and basin-wide isothermy is not reached. The reasons for this will be discussed later. Table 3 summarises maximum relative density differences for the lakes (as difference in density divided by mean density, $\Delta\rho/\rho_0$) due to observed temperature and conductivity differences. While the maximum relative density differences are slightly smaller for the African lakes, the contrast is not extreme.

The seasonal patterns of stratification in the East African lakes are illustrated by temperature profiles in Figs. 10–13 (except for Lake Kivu, for which seasonal data could not be found). Lakes Tanganyika and Malawi possess well defined mixed layers and seasonal thermoclines that strengthen during the wet season (Austral summer) and then weaken and deepen due to evaporative cooling and wind mixing during the south-wind season. Thermocline depths in the wet season and before the onset of the south winds (December and April profiles) are deeper than for the North American lakes, reflecting both the smaller density differences available from stratification as well as the dominant influence of the earth's rotation on water movements and stratification in the large temperate lakes. In large temperate lakes, Coriolis forces arising from the earth's rotation confine regions of deepest mixing, greatest thermocline tilting and upwelling, and highest current speeds to relatively narrow coastal boundary layers. These effects are discussed in more detail later. In the large East African lakes, the smallest density differences and the greatest mixing depths occur toward the end of the south-wind season (July, August profiles) but complete isothermy is not achieved. In Lake Tanganyika the thermocline disappears at the south end, where upwelling and mixing to depths of at least 250 m occurs (Coulter and Spigel 1991), but it sharpens and deepens at the north end (July profile). Similar horizontal variations, typical of the upwelling response to be expected under strong southerly winds, have been observed in Lake Malawi (Eccles 1974).

Both Lakes Tanganyika and Malawi are classified as meromictic, being of such great depth that complete vertical exchange does not occur, their hypolimnia (or monimolimnia) remaining as vast anoxic reservoirs of nutrients largely cut off from surface influences, probably exchanging with the upper layers only over limited periods and in limited areas, largely under the influence of upwelling at their southern ends. In these lower strata, extending below about 300 m in Tanganyika and about 350 m in Malawi, any significant general changes in temperature or other factors influencing stability occur on a much longer time scale than in the upper strata. Various hypotheses have been proposed to account for the apparent quasi-steady states of these deep hypolimnia. Profile-bound density currents down the sides of the basin, originating from cooler water at the lake margins and from the catchment, have long been postulated as a major factor (Beauchamp 1939, Capart 1952, Hutchinson 1957:476, Beadle 1974, Eccles 1974). (Evidence of such flow was demonstrated in Lake Albert by Talling [1963]). The influence of biogenically induced stratification from dissolved solids and gases, though minor, is significant in Lake Malawi (Wüest et al., this volume) and probably in Tanganyika too. Other factors contributing to deep convection are geothermal heat flow and emissions of hot water and gases from the basin (Tiercelin and Mondeguer 1991).

These factors may explain the high degree of uniformity in each hypolimnion (Coulter and Spigel 1991, Halfman 1993), but much more research is needed for a model to quantify their interactions. The problem of hypolimnion mixing is an important but difficult one, for which realistic but simple scaling estimates, like those to explain upwelling, are not yet available. Imberger and Patterson (1990) and Horsch and Stefan (1988) discuss some of the difficulties posed by free convection problems, such as may be involved if circulation in these hypolimnia is driven largely by cold water density currents from inshore. Isotopic

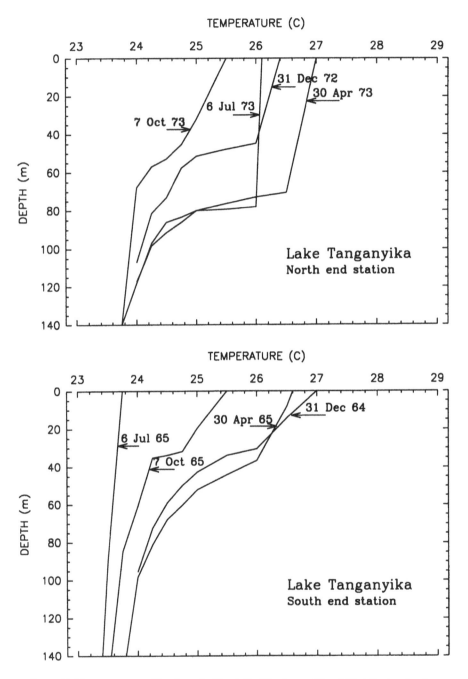

Figure 10. Temperature profiles from the North End Station and South End Station (stations about 10 km offshore from the major ports at either end of the lake), illustrating the seasonal pattern of stratification in Lake Tanganyika. Data are from Coulter and Spigel (1991). Note the deep mixed layer and sharp thermocline at the north end in July and nearly complete vertical mixing at the south end, characteristic of strong upwelling conditions.

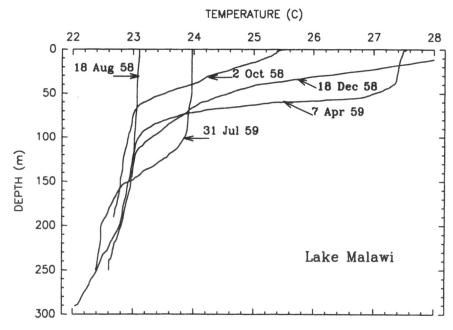

Figure 11. Temperature profiles from Lake Malawi, 5.5 km offshore from Nkhata Bay in water of 400 m depth. Data are from Eccles (1974).

tracers may be an indispensable tool, and tritium has been used in Tanganyika and Malawi to estimate transfer rates between density layers (Craig et al. 1974, Gonfiantini et al. 1977). Results by the latter authors indicate that more annual mixing between the hypolimnion and water above it takes place in Malawi than in Tanganyika. The hypolimnion in Tanganyika, at least, appears old enough to merit the term "relict." On the basis of enrichment of the stable isotopes deuterium and ^{18}O, Craig et al. (1974) suggested that the deep water is a relict hypolimnion relative to the present sources of water. This is supported by the helium isotope ratios they observed. The ^{14}C contents of dissolved CO_2 in near-bottom samples give a model age of about 800 years, and temperatures in the deep water during the past 50 years at least have been constant (Edmond et al. 1993).

Temperature profiles measured in Lakes Victoria and Turkana (Figs. 12 and 13) are notable for their lack of well-defined mixed layers and seasonal thermoclines, although this is truer for the modern Victoria profiles than the historic ones. Temperature gradients tend to be more diffuse, and horizontal variability greater (not shown in Figs. 12 and 13) than in Lakes Tanganyika and Malawi. These are characteristic of large, shallower lakes in which wind may cause complete vertical mixing (see Victoria profile for July) but is unable to achieve complete horizontal homogeneity. In Lake Turkana lake-length temperature transects during August and October (Ferguson and Harbott 1982) have revealed complete vertical mixing in the southern half of the lake coincident with stratified conditions in the north. This may be due to upwelling under southerly winds, although the structure does not appear to relax to one of horizontally uniform vertical stratification during calm conditions, and Ferguson and Harbott (1982) have invoked differential wind stress (higher winds in the south) to explain the persistence. Relaxation of tilted isotherms following the cessation of

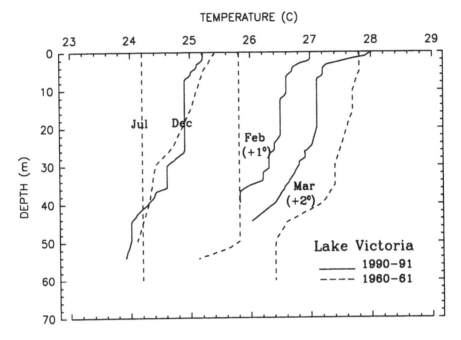

Figure 12. Temperature profiles from Lake Victoria (at the open water station of Fish 1957), approximately 32 km south of Jinja (historic data, dashed line, from Talling 1966; modern data, solid line, from Hecky 1993). Note that profiles for February and March have been offset by 1 and 2°C, respectively. Complete vertical mixing occurred in July 1961, but this is not meant to imply isothermy throughout the lake. Note the lack of well-defined mixed layers in the modern profiles, although these may be masked by diurnal thermoclines.

wind, in which colder upwelled water underflows warmer water that has been pushed by wind to the leeward end of the basin, plays an important role in the dynamics of Lake Tanganyika, probably setting in motion long internal waves that continue to oscillate throughout the wet season (Coulter and Spigel 1991). The longitudinal density difference present at the cessation of wind, as well as the damping characteristics of the basin, determine the speed and persistence of the relaxation; both weaker density differences and higher damping in Turkana would lead to a much slower and weaker response.

Conditions are further complicated in Lake Turkana by the importance of the Omo River inflow, which floods in response to heavy rainfall over its upland catchment from June through September, coincident with the period of strongest cooling and wind mixing. The river enters the northern end of the lake as a sediment-laden overflowing plume, which spreads as far south as the Central Island by September. Conductivity is much lower in the river water, 0.008 S/m as compared with values >0.350 S/m typical of the central basin (Ferguson and Harbott 1982). If the temperature of the inflow were warmer than the lakewater, this would further enhance stratification in the northern half of the lake.

Stratification in all of the lakes is influenced to an extent by the light attenuation properties of the water column. All other things being equal, and especially under conditions of weak to moderate winds, a lake in which light is attenuated rapidly will have a shallower and warmer mixed layer than one in which light penetrates to greater depths, until the mixed

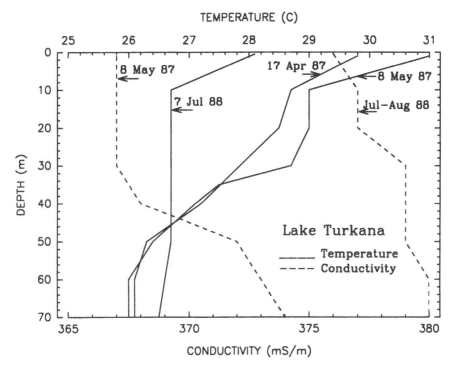

Figure 13. Temperature and conductivity profiles from the central basin station (H1; see Figure 3) of Ferguson and Harbott (1982) and Källqvist et al. (1988); data are from Källqvist et al. The lack of a well-defined mixed layer is a noticeable feature of the temperature profiles, although a 30 m deep mixed layer does appear in the May conductivity profile.

layer depth exceeds (very) approximately the 5% light level (Simpson and Dickey 1981 a,b). Then wind-mixing dominates optical effects. Measurements of Secchi disk depths (Z_{SD}) and 1% light levels ($Z_{1\%}$) are given in Table 4. No attempt has been made to convert from Secchi depth (a measure of transparency) to extinction depth (a measure of energy attenuation and absorption, and the more appropriate scale for considerations of mixing energetics). Lakes Tanganyika and Malawi, with the deepest summer thermoclines, also have the largest values of $Z_{1\%}$, while Victoria and Turkana have the smallest. The effect of the Omo River sediment plume on light attenuation in Lake Turkana is dramatic. Also of interest and concern is the significant decline of transparency in Lake Victoria from the 1950s and 1960s to the present as a consequence of increasing nutrient levels, productivity and eutrophication (Hecky 1993).

WIND MIXING AND WATER MOVEMENTS

In the previous section it was pointed out that in the absence of sufficient surface cooling, wind mixing alone does not normally achieve complete isothermy — although complete vertical mixing may occur, horizontal density gradients will remain, and when the wind ceases these will reversibly adjust under gravity to reestablish vertical stratification. The final stratification will, of course, be weaker than the initial stratification that existed at the onset of wind. The amount of mixing that occurs in any given wind event depends in large part on the strength of the applied wind stress relative to that of buoyancy forces that are associated with stable density stratification and that resist the wind induced vertical motions. Other factors are also involved, however.

Work done by wind stress on a water surface generates turbulence that causes mixing in the surface layer, leading to thermocline deepening by entrainment of metalimnion and hypolimnion water into the epilimnion. At the same time, wind stress generates horizontal currents and long waves that distort the thermocline, tilting it downward and bringing warmer surface water toward the leeward end of the lake. At the windward end of the lake the thermocline rises, bringing with it colder upwelling water from depth. Thermocline tilting thereby creates horizontal density gradients with surface layer temperatures increasing down-wind. These gradients generate buoyancy forces necessary to balance wind stress. If the wind is strong and the stratification weak enough, it is possible for complete overturn to occur before a balance can be reached. Once established, however, horizontal density gradients take much longer to overcome than the time required for vertical mixing alone, simply because of the relative distances over which mixing must occur (lake length versus lake depth). In large temperate lakes Coriolis forces arising from the earth's rotation generate large-scale horizontal motions that completely modify the wave and current response observed in smaller lakes or in large lakes at lower latitudes. The conditions under which Coriolis forces dominate motion will be discussed shortly.

For conditions under which Coriolis forces have a small influence, it is possible to quantify in a very approximate way the balance between vertical mixing, stratification, and thermocline tilting with upwelling. Imberger and Patterson (1990) have reviewed the theory of these interactions and show that the outcome of a wind event is largely governed by two dimensionless parameters, the Wedderburn (W) and LLake (L_N) numbers, that compare the work necessary to overcome stable stratification with the energy available from the wind for mixing. For a uniform wind stress (as force per unit area — see the list of symbols for a consistent set of units for all the parameters in the following equations), W and L_N can be expressed as

$$W = \Delta\rho g h^2/(\rho_0 u_*^2 L), \tag{3}$$

$$L_N = \frac{g\, S\, (1 - Z_T/H)}{\rho_0\, u_*^2\, A_s\, L(1 - Z_g/H)}, \tag{4}$$

where $\Delta\rho$ and ρ_0 are the density difference and average density corresponding to the mixed layer and deep waters (see Table 3), g is acceleration of gravity, h is mixed layer depth, $u_* = (\tau/\rho_s)^{1/2}$ is the shear velocity from wind, ρ_s is the density of water in the mixed layer, L is lake length, A_s is lake surface area, Z_T is the elevation of the mid-metalimnion above bottom, Z_g is the elevation above bottom of the centre of volume of the lake, and H is total depth. S is stability (Hutchinson, 1957), defined by

$$S = \int_0^H (Z_g - z)\, A(z)\, \rho(z)\, dz, \tag{5}$$

where $\rho(z)$ and $A(z)$ are density and horizontal area as functions of elevation z above bottom. The formulation for L_N in Eq. (4) assumes that the prevailing wind is along the main axis of the lake, a reasonable assumption for the African lakes during the dry season. Both W and

L_N incorporate the effects of water column stability ($\Delta \rho g h$, S), wind strength ($\rho_0 u_*^2$), wind fetch (L), and average slope of the thermocline at the onset of upwelling (proportional to h/L). Small values of W and L_N correspond to strong winds, weak stability, a high degree of mixing, and diffuse temperature gradients, while large values of W and L_N correspond to high stability, weak winds and little mixing. W applies only to the mixed layer however, and was originally derived with a two-layer stratification in mind, whereas L_N can incorporate complex basin shapes and arbitrary density profiles.

Experiments (Monismith 1986) have shown that "partial" upwelling — the surfacing of isotherms from the metalimnion alone, with the base of the metalimnion remaining mostly horizontal — occurs for W less than about 10. Total upwelling, in which the base of the metalimnion is deflected upwards and water from the hypolimnion may reach the surface, only occurs at much smaller values of W ($W < 1$) and only if L_N is also small enough (Imberger and Patterson 1990).

A rectangular basin containing a 3-layer stratification is perhaps the simplest configuration that captures the essential features of stratification needed to explain upwelling. Figure 14 illustrates the two types of periodic or wavelike modes that can occur for such a stratification in response to a disturbance, such as the onset or change of wind. The actual response that occurs in a given wind event can be described in terms of combinations of the two modes (Münnich et al. 1992). Eventually periodic motions may be damped by friction; but if the wind lasts long enough, aperiodic circulations (not shown in Fig. 14), and the isotherm deflections initiated by the modes, will remain until the wind ceases. Mode 2, in which the hypolimnion remains largely passive, is associated with partial upwelling, while mode 1, in which the entire metalimnion moves with the epilimnion, causes complete upwelling. The Wedderburn number can be calculated for the configuration of Fig. 14 from Eq. (3) with $h = h_1$ and $\Delta \rho = \rho_3 - \rho_1$; L_N can be evaluated from Eqs. (4) and (5) as

$$L_N = \frac{W}{h_1^2 H} \left(h_1 h_3 + \frac{h_2}{2} \left(h_1 + \frac{h_2}{2} + h_3 \right) \right) \left(h_1 + \frac{h_2}{2} \right), \qquad (6)$$

where $H = h_1 + h_2 + h_3$ is total depth. Values of W and L_N computed using Eq. (6) with density differences from Table 3 and layer depths based on temperature profiles in Figs. 10–13, are listed in Table 5. All values assume a shear velocity $u_* = 0.016$ m/s corresponding to a wind speed of approximately 10 m/s. For the examples and values chosen, both L_N and W are less than 1 for all the lakes, consistent with observations of hypolimnetic upwelling documented by Coulter (1963) in Lake Tanganyika, Eccles (1974) in Lake Malawi and Ferguson and Harbott (1982) in Lake Turkana. The major upwelling events reported for Tanganyika, Malawi and Turkana all occurred during the dry season during periods of sustained southerly wind in the southern part of the lakes. The Wedderburn and Lake number analyses are applicable only when the duration of wind is greater than 1/4 of the fundamental internal seiche (long internal standing wave) period — it takes at least this long to set up the thermocline. This condition is satisfied in Lakes Malawi, Tanganyika and probably Turkana during the dry season. In Lakes Tanganyika and Malawi, seasonal upwelling at the southern end of the basin is thought to play an important part in nutrient and production cycles (Coulter 1968, Coulter and Spigel 1991, Eccles 1974).

In Lake Victoria only Kitaka (1971) has so far published synoptic temperature data for 2 long north–south transects that cover most of the lake. During one of these a strong northerly wind blew for several hours and a localised upwelling event was observed in the south-central part of the lake but not along the shoreline. Kitaka has invoked pressure and wind

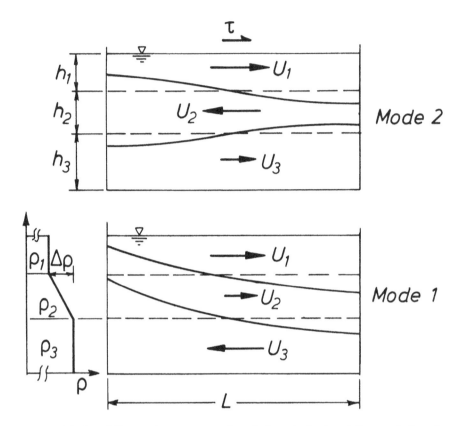

Figure 14. The difference between second vertical mode (top) and first vertical mode (bottom) responses for a 3-layer stratification. Second mode dominates for small W but large L_N. First mode is important for both W small and L_N small. Adapted from Imberger and Patterson (1990) and Münnich et al. (1992).

variations associated with the storm to explain the event. Strong tilting of isotherms and possible upwelling have also been hypothesised by Fish (1957) and Talling (1966) to explain temperature observations in the northern half of Lake Victoria, and localised upwelling and downwelling undoubtedly occurs in response to strong land-sea breezes as well as storms. Certainly the low values of W and L_N calculated for Lake Victoria (Table 5) imply that strong isotherm tilting and total upwelling could occur there on a lakewide scale if the wind duration were long enough.

In large lakes at higher latitudes the earth's rotation completely alters the periodic and aperiodic responses to wind, particularly for long internal waves or seiches. Instead of being aligned with the wind, or following paths dictated by a combination of wind stress, pressure gradients and bottom topography, currents in rotating basins are deflected by Coriolis force acting at right angles to the currents. As a result water particles can trace circular orbits in horizontal planes. By far the greatest amount of energy found in current meter records in large temperate lakes is due to rotation at the inertial period (Mortimer 1963, 1974, 1977; Saylor et al. 1981). The inertial period (also called a half pendulum day) is given by $T_r = 2\pi/f$, where $f = 2\Omega \sin \phi$ is the Coriolis parameter (two times the component of the earth's

Table 5. Mixing Parameters

	Victoria	Tanganyika	Malawi	Turkana
Wedderburn no., W	0.049	0.15	0.24	0.07
Lake number, L_N	0.059	0.20	0.31	0.11
h_1, m	25	50	50	20
h_2, m	20	20	20	25
h_3, m	47	1401	636	28

Calculations assume a 3-layer stratification with uniform densities in top and bottom layers and a linear gradient in the middle layer. Layer depths are based on profiles in Figs. 9–13. This is a different approximation to the density profiles than the step-profiles used in Table 6. The density differences used in the calculation are those in Table 3. The mixed layer depth used in the calculation for W (Eq. (3)) is h_1. Lake numbers (Eqs. (4) and (6)) are calculated assuming that the predominant wind fetch is along the main axes of the lakes, and is therefore based on the transect lengths in Table 1. Depths add up to H_{max}, Table 1.

angular velocity normal to the earth's surface), ϕ is latitude, and $\Omega = 7.29 \times 10^{-5}$ radians/sec is the angular velocity of the earth. Coriolis forces can only alter currents if the currents persist for a long enough time; motions with a time scale short compared with T_r will not feel the effects of rotation. T_r decreases from infinity at the equator ($f = 0$) to 12 hours at the poles; its ranges for the East African and North American Lakes are shown in Table 6. T_r is much longer for the African Lakes, except for Malawi because of its higher latitude. A further difference is the much larger *variation* in T_r and f for the African lakes, because of their north–south orientations and exceptional lengths. Water particles experience different rotational effects at different latitudes. The differences interact with large-scale variations in bottom topography and in the wind-stress field to generate restoring forces that can have significant dynamical effects on circulations and wave-like motions if north–south distances are large enough (Pedlosky 1979; Pond and Pickard 1983). It is uncertain whether these effects are important in the large African lakes.

The influence of Coriolis acceleration on thermocline tilting and internal seiching is determined by the ratio of the internal gravity wave speed, c, to a velocity scale associated with inertial motions, $2B/T_r$ (Csanady 1975), where B is lake width. Internal gravity wave speeds for the slowest mode have been estimated in Table 6 for the lakes (except Kivu) using the density differences in Table 3, a three-layer stratification (with uniform densities in all three layers) based on the profiles in Figs. 9–13, and equations given by Monismith (1985). The wave speeds do not differ significantly between the African and North American lakes; the speeds depend only on density stratification, being faster for stronger stratifications. However, the velocity scales $2B/T_r$, calculated using maximum lake widths, are much faster for the North American lakes, exceeding c by at least a factor of 10. One would therefore expect Coriolis effects to dominate motions associated with internal seiches and thermocline tilting in the North American lakes, while this would not be the case in the African lakes. In North American lakes upwelling has been observed to occur on the shore to the left of the wind (Mortimer 1974; corresponding to motion of surface water to the right of the wind in the northern hemisphere).

Table 6. Rotation, Seiching and Mixing Parameters

	Victoria	Tanganyika	Malawi	Turkana	Kivu	Ontario	Superior
Latitude, φ, degrees (positive north)	0.027	-3.36	-9.50	4.58	-1.58	46.44	49.00
	-2.50	-8.81	-14.50	2.38	-2.53	43.20	44.08
Coriolis parameter, f, rad/s	6.87×10^{-7}	-8.54×10^{-6}	-2.41×10^{-5}	1.16×10^{-5}	-4.02×10^{-6}	1.06×10^{-4}	1.10×10^{-4}
	-6.36×10^{-6}	-2.23×10^{-5}	-3.65×10^{-5}	6.06×10^{-6}	-6.44×10^{-6}	9.98×10^{-5}	1.01×10^{-4}
Inertial period, $T_r = 2\pi/f$, days	106	3.26	1.99	12.0	11.3	0.729	0.717
	11.4	8.51	3.02	6.25	18.1	0.688	0.661
Density difference, $\Delta\rho_{12}/\rho_0$	3.18×10^{-4}	9.12×10^{-4}	1.27×10^{-4}	8.78×10^{-4}	–	1.98×10^{-3}	1.33×10^{-3}
$\Delta\rho_{23}/\rho_0$	3.18×10^{-4}	1.01×10^{-4}	1.41×10^{-4}	4.39×10^{-4}	–	2.20×10^{-4}	1.48×10^{-4}
h_1, m	25	50	50	20	–	5	5
h_2, m	20	450	300	25	–	35	35
h_3, m	47	971	356	28	–	204	366
Wave speed, c, m/s	0.154	0.518	0.462	0.216	–	0.233	0.195
Seiche period, T_i, days	46.4	29.4	28.5	27.5	–		
Rossby radius, R, km	24.2	23.2	12.7	35.6	–	2.33	1.84
	224	60.7	19.2	18.5	–	2.30	1.77
$2B_{max}/T_r$, m/s	0.052	0.207	0.574	0.193	–	2.97	8.86
	0.486	0.540	0.872	0.100	–	2.92	8.51
B_{max}/R	1.1	1.3	3.9	2.8	–	40.1	143
	9.9	3.3	5.9	1.5	–	39.4	137
Damping time, T_d, days	65.4	>1 year	>1 year	25.5	–		

Northern and southern limits of latitude are given for each lake. The two values calculated for each rotational parameter correspond to the north–south limits of latitude. Internal wave speed c has been calculated for the slowest mode of a 3-layer stratification, uniform in each layer, using the method of Monismith (1985); $\Delta\rho_{12}/\rho_0$ and $\Delta\rho_{23}/\rho_0$ are relative density differences between top and middle layer and middle and bottom layer, respectively, and add up to $\Delta\rho/\rho_0$ in Table 3. Layer depths are based on profiles in Figs. 9–13. Depths add up to H_{max}, Table 1. Damping time calculated from (Coulter and Spigel 1991: 71; Spigel and Imberger 1980) $T_d = T_i \bar{H}/\delta$, where T_i is the seiche period, $\bar{H} = V/A_s$ is the average depth (Table 1; V is lake volume, A_s is lake surface area) and δ is the boundary layer thickness for the seiche. For a turbulent oscillating boundary layer on a rough surface, Kalkanis (1964) gives $\delta = u_{max} T_i^{1/2} e/(47 \nu^{1/2})$, where u_{max} is the maximum velocity due to oscillation $= (a/T_i)(2L/H_2)$, a is the amplitude of oscillation (taken as min $[30m, h_1]$, L is the lake length (Table 1), $H_2 = h_2 + h_3$, is the depth below epilimnion, e is the roughness of the surface adjacent to the boundary layer (taken as 6 cm as for a natural river channel), and ν is the kinematic viscosity of water (1.1×10^{-6} m^2/s).

Coriolis forces also restrict the fastest currents and the greatest isotherm deflections to coastal regions that extend a distance of approximately $R = c/f$ offshore (Csanady 1982, Mortimer 1974). R is known as the Rossby radius of deformation, and representative values are given in Table 6. The ratio of R to lake width can be used to gauge the importance of rotation. Mortimer (1974) suggests that lake width must be approximately $5R$ before rotational effects become significant, and greater than $20R$ before they become dominant. Ratios of B_{max}/R are estimated in Table 6, from which it can be seen that coastal boundary layers would be confined to relatively narrow strips in Lakes Ontario and Superior, but would cover almost the entire width in most of the African lakes. The appearance of width as the relevant length scale for rotation emphasises the importance for water movements of the narrow basin shapes of Lake Tanganyika, Malawi and Turkana.

The length of time that periodic or wavelike responses can be sustained during or after a wind depends on the frictional damping characteristics of a lake basin. Longest wave (seiche) periods, T_i, can be estimated from the slowest wave speeds, c, as $T_i = 2L/c$ (where L is lake length), for lakes in which rotation is not dominant. When rotation is dominant the longest waves are called Kelvin waves and travel in a cyclonic direction (in the sense of positive f, counterclockwise in the northern hemisphere and clockwise in the southern hemisphere) around the perimeter of a lake basin, again restricted to a coastal border whose width is of order R. When rotation is not important, a simple uninodal seiche, together with higher harmonics, dominate the response to wind. Damping times, T_d, for internal seiches have also been estimated in Table 6 using the method outlined by Spigel and Imberger (1980) and Coulter and Spigel (1991), which equates the fraction of kinetic energy lost due to friction during each wave period to work done against viscosity in the boundary layer along the lake bed. T_d is proportional to average lake depth, and is very long (more than a year) in Lakes Tanganyika and Malawi, but is of the same order as the seiche period T_i for Victoria and Turkana. Hence one would expect large-scale oscillations of the thermocline to be a prominent feature in Lakes Tanganyika and Malawi, while this would not be the case in Victoria and Turkana. Newell (1960) states the same conclusion but for a completely different reason. Using the stability analysis by Goldstein (1931) for two layers of water separated by a temperature discontinuity, Newell shows that such an idealised two-layer stratification is likely to be unstable under the circumstances of lake shape, temperature differences and wind speeds to be found in Lake Victoria. In fact this result has little bearing on whether periodic seiching could or could not occur, although it does explain, as does W and L_N scaling, why temperature profiles in Victoria tend to be diffuse rather than exhibiting a well-defined epilimnion–hypolimnion structure with a sharp thermocline.

In the absence of periodic motions, or after periodic motions have decayed, wind-driven circulations will exist so long as wind stress is applied. The time for these circulations to completely mix a lake containing horizontal density gradients can be estimated as T_m $L^2/(10u_*h)$, where h corresponds to the depth of vertical mixing (Imberger and Patterson 1990). This longitudinal mixing time is much longer than a year for all of the African lakes because of their great lengths. Hence one would not expect wind to achieve complete horizontal and vertical mixing throughout any of the lakes.

Long-term average circulations are important because of their role in transporting nutrients and pollutants. But because of their much smaller magnitudes than currents developed during wind events, long term circulations are difficult to identify and document. Coulter and Spigel (1991) discuss the presence of a persistent clockwise (cyclonic in the southern hemisphere) shorebound current in Lake Tanganyika, while a similar account is given by

Eccles (1974) for Lake Malawi. Newell (1960) gives evidence for the long-term circulation in Lake Victoria as being due to wind-drift from southeast to northwest in the surface waters. The possible existence of a clockwise circulation is supported by shorter travel times of lake steamers travelling clockwise around Lake Victoria as opposed to longer travel times for identical steamers travelling counterclockwise. In Lake Turkana wind-driven and river-influenced circulations appear to dominate any residual or long-term circulations (Ferguson and Harbott 1982). Emery and Csanady (1973) mapped patterns of surface circulation in more than 40 northern hemisphere lakes, lagoons, marginal seas and estuaries and found all except one to follow counterclockwise paths (cyclonic in the northern hemisphere). In Lake Ontario an average counterclockwise flow also occurs, although it is sometimes replaced by a double-cell reflecting windward (northeast) drift along southern and northern shores (Saylor et al. 1981). Various reasons have been advanced to explain the cyclonic circulation often observed in large lakes (Emery and Csanady 1973, Wunsch 1973, Bennett 1974). Csanady (1977) argued that greater mixing and heating in coastal regions would lead to deeper and warmer mixed layers near shore than in pelagic regions, resulting in a domed structure of the thermocline over the lake. A cyclonic circulation is then required to achieve geostrophic equilibrium, the balance between pressure gradients and Coriolis forces that one would expect to hold in the long-term. However, Csanady (1982) concluded that, at least in light of the more detailed studies that have been reported for Lake Ontario, observed circulation patterns are not adequately explained by any of the mechanisms proposed to date.

CONCLUSIONS

Although the East African lakes form a diverse group in terms of bathymetry, climate, stratification and water movements, it has been possible to identify some unifying themes, especially in comparison with large lakes in temperate regions. Solar radiation, which provides the greatest seasonal variation in temperate lakes, is fairly constant over the year in the tropics. Seasonality in the African lakes is dominated by the annual cycle of monsoon winds, with a cooler, drier period extending from May or June through August or September, and during which persistent southerly winds blow. Cooling is enhanced by evaporation, and upwelling at the south ends of Tanganyika, Malawi, Turkana (and possibly Victoria) occurs during this period. Although stratification is at its weakest in all of the lakes during this season, it never vanishes completely from any of the African lakes. Time scales, wave speeds, and dimensionless parameters (Wedderburn and Lake numbers) that describe the interaction between wind strength, water column stability, and thermocline motions have been estimated using fragmentary data from many sources. The actual numerical values that appear in the tables in this review are not to be seen as anything more than approximations. Nevertheless, it is felt that the orders of magnitudes of the terms provide useful information for comparison purposes and help to explain some of the prominent observations made by previous workers. These include the upwelling just mentioned, the general form of stratification and the presence of long internal waves in Tanganyika and Malawi (and their relative absence in Victoria and Turkana). The persistence of these long internal waves in Tanganyika and Malawi, which apparently keep pulsing for months after the wind that initiated them has ceased, is in marked contrast to the transient nature of seiches in the temperate zone. Parameters relating to the effects of the earth's rotation, so important in the North American Great Lakes, have also been estimated. The contrast between the tropical and temperate lakes is quite marked, and leads one to expect rotation to have only minor effects on thermocline motions in the African lakes. This influence is further diminished by the relative narrowness of the African basins furthest from the equator — Turkana, Tanganyika and Malawi. Lake Kivu is unique among the lakes in that its lower 400 m appear to be dominated by geothermal influences, with both temperature and salt concentration increasing markedly with depth.

It has not been possible to include all important physical processes in such a review. Chief among those neglected include the fate and influence of river inflows, and the dynamics of onshore–offshore, littoral–pelagic water exchanges. River inflows have been discussed in terms of their impact on sedimentation by Tiercelin and Mondeguer (1991) for Lake Tanganyika, by Johnson and Ng'ang'a for Lake Malawi and by Ferguson and Harbott (1982) for Lake Turkana. Onshore–offshore exchanges are subject to a much greater degree of speculation. These are but two of many unanswered questions for which further information is required if we are to make progress in understanding the dynamics of these remarkable lakes.

ACKNOWLEDGMENTS

It has been a privilege to be part of a group working on the East African lakes, and we acknowledge the leadership of Tom Johnson and the IDEAL committee for making this effort possible. We thank Alfred Wüest, Ray Weiss and Anthony Viner for constructive comments during the review process, and R. Spigel thanks the IDEAL committee for financial support to attend the Jinja meeting. Thanks also to Toh Seng Chee for help with data reduction, to Catherine Price for secretarial assistance, and to Val Grey for drafting.

LIST OF SYMBOLS

Units given in brackets are those which would result in consistent results for calculations shown by equations in the text.

A_s	=	lake surface area (m^2)
$A(z)$	=	area (m^2) within contour at elevation z above bottom
B	=	lake width (m)
B_{max}	=	maximum lake width (m)
c	=	wave speed corresponding to lowest mode internal seiche (m/s)
E	=	evaporation (m/s)
f	=	Coriolis parameter (= $2\Omega \sin \phi$; radians/sec)
g	=	acceleration of gravity (9.81 m/s^2)
H	=	$h_1 + h_2 + h_3$ = total depth (m)
H_{max}	=	maximum lake depth (m)
h	=	mixed layer depth; extent of vertical mixing (m)
h_1, h_2, h_3	=	depths of top, middle and bottom layers for an idealised 3-layer density stratification (m)
$\overline{H}$	=	average lake depth (= V/A_s; m)
L	=	lake length; also used to estimate wind fetch (m)
L_N	=	Lake number (dimensionless; see Eq. (4))
P	=	precipitation (rainfall; m/s)

Q_{in}	=	river inflow (m³/s)
Q_{out}	=	river outflow (m³/s)
Q_{gw}	=	net groundwater inflow (m³/s)
Q_{sro}	=	storm runoff from distributed sources (i.e., not accounted for by river inflows Q_{in}; m³/s)
R	=	Rossby radius of deformation (= c/f; m)
S	=	stability [(kg m) see Eq. (5); gS = increase in potential energy (joules) that would result from complete mixing of a stratified lake]
T_i	=	period of lowest mode internal seiche (= $2L/c$; seconds)
T_r	=	inertial period (= $2\pi/f$; seconds)
T_d	=	damping time for lowest mode internal seiche (see notes for Table 6; seconds)
T_m	=	longitudinal mixing time for a horizontally statified lake (seconds)
Δt	=	time interval (seconds) over which the difference in storage volume or some other parameter is calculated
u_*	=	shear velocity (in water) due to wind stress on water surface [= $(\tau/\rho_s)^{1/2}$; m/s]
ΔV	=	change in volume stored in a lake (m³)
W	=	Wedderburn number (dimensionless; see Eq. (3))
ΔZ_s	=	change in lake surface level (m)
Z_T	=	elevation above bottom of middle of metalimnion (m)
Z_g	=	elevation above bottom of centre of volume of lake (m)
z	=	elevation above bottom of any point in the water column (m)
$\Delta\rho$	=	density difference between mixed layer and deep waters (kg/m³)
ρ_0	=	average water density of mixed layer and deep waters (kg/m³)
ρ_s	=	density of water in mixed layer (kg/m³)
τ	=	shear stress applied by wind to lake surface (newtons/m², pascals)
Ω	=	angular velocity of the earth's rotation (= 2π radians/sidereal day length = 7.29×10^{-5} radians/sec)
ϕ	=	latitude (degrees)

REFERENCES

Balek, J. 1977. Hydrology and Water Resources in Tropical Africa. Elsevier, Amsterdam. 208 pages.

Beadle, L.C. 1974. The Inland Waters of Tropical Africa. Longman, London. 365 pages.

Beauchamp, R.S.A. 1939. Hydrology of Lake Tanganyika. Internationale Revue der gesamten Hydrobiologie und Hydrographie 39: 316–353.

Beauchamp, R.S.A. 1953. Hydrological data from Lake Nyasa. Journal of Ecology 41(2): 226–239.

Bennett, J.R. 1975. Another explanation of the cyclonic circulation in large lakes. Limnology and Oceanography 20: 108–110.

Bennett, E.B. 1978. Water budgets for Lake Superior and Whitefish Bay. Journal of Great Lakes Research 4(3–4): 331–342.

Boyce, F.M., Moody, W.J., and Killins, B.L. 1977. Heat content of Lake Ontario and estimates of average surface heat fluxes during IFYGL. Technical Bulletin No. 101. Inland Waters Directorate, Canada Centre for Inland Waters, Burlington, Ontario. 120 pages.

Bugenyi, F.W.B., and Magumba K.M. 1993. The present physicochemical ecology of Lake Victoria (Uganda). Presented at the International Symposium on the Limnology, Climatology and Palaeo-climatology of the East African Lakes; organised by the International Decade in the East African Lakes (IDEAL), 17–21 February 1993, Jinja, Uganda.

Bultot, F. 1972. Rwanda and Burundi. Chapt. 10 in: Griffiths, J.F. (Ed.), Climates of Africa, World Survey of Climatology, Vol. 10, Elsevier, Amsterdam. 604 pages.

Bultot, F., and Griffiths, J.F. 1972. The equatorial wet zone. Chapt. 8 in: Griffiths, J.F. (Ed.), Climates of Africa, World Survey of Climatology, Vol. 10, Elsevier, Amsterdam. 604 pages.

Caldwell, D.R., and Eide, S.A. 1980. Adiabatic temperature gradient and potential temperature correction in pure and saline water: an experimental determination. Deep-Sea Research 27A: 71–78.

Capart, A. 1952. Le milieu géographique et géophysique. Résultats scientifiques de l'exploration hydrobiologique du Lac Tanganika (1946–47). Institut Royal des Sciences Naturelles de Belgique 1: 3–27.

Chen, C.-T.A., and Millero, F.J. 1986. Precise thermodynamic properties for natural waters covering only the limnological range. Limnology and Oceanography 31(3): 657–662.

Coulter, G.W. 1968. Thermal stratification in the deep hypolimnion of Lake Tanganyika. Limnology and Oceanography 13(2): 385–387.

Coulter, G.W. 1963. Hydrological changes in relation to biological production in southern Lake Tanganyika. Limnology and Oceanography 8(14): 463–477.

Coulter, G.W., and Spigel, R.H. 1991. Hydrodynamics. Chapt. 3 in: Coulter, G.W. (Ed.), Lake Tanganyika and Its Life. Oxford, Oxford University Press. 354 pages.

Craig, H., Dixon, F., Craig, V., Edmond, J., and Coulter, G. 1974. Lake Tanganyika Geochemical and Hydrographic Study: 1973 expedition. Publication Scripps Institution of Oceanography, Series 75, 5: 1–83.

Csanady, G.T. 1977. On the cyclonic mean circulation of large lakes. Proceedings of the National Academy of Sciences USA 74(6): 2204–2208.

Csanady, G.T. 1982. Circulation in the Coastal Ocean. D. Reidel Publishing Co., Dordrecht, Holland. 279 pages.

Csanady, G.T. 1975. Hydrodynamics of large lakes. Annual Review of Fluid Mechanics 7: 357–386.

Csanady, G.T. 1972a. The coastal boundary layer in Lake Ontario. Part 1: The spring regime. Journal of Physical Oceanography 2: 41–53.

Csanady, G.T. 1972b. The coastal boundary layer in Lake Ontario. Part 2: The summer–fall regime. Journal of Physical Oceanography 2: 168–176.

DeCooke, B.G., and Witherspoon, D.F. 1981. Terrestrial water balance. Pages 199–219. In: Aubert, E.J., and Richards, J.L. (Eds.), IFYGL — The International Field Year for the Great Lakes. US National Oceanic and Atmospheric Administration, Great Lakes Environmental Research Laboratory, Ann Arbor, Michigan. 410 pages.

Degens, E.T., von Herzen, R.P., Wong, H.-K., Deuser, W.G., and Jannasch, H.W. 1973. Lake Kivu: structure, chemistry and biology of an East African rift lake. Sonderdruck aus der Geologischen Rundschau 62(1): 245–277.

Dobson, H.F.H. 1984. Lake Ontario water chemistry atlas. Inland Waters Directorate, National Water Research Institute, Canada Centre for Inland Waters, Scientific Series No. 139. Burlington, Ontario. 59 pages.

Eccles, D.H. 1974. An outline of the physical limnology of Lake Malawi (Lake Nyasa). Limnology and Oceanography 19(5): 730–742.

Eccles, D.H. 1984. On the recent high levels of Lake Malawi. Suid Afrikaanse Tydskrif vir Wetenskap 80: 461–468.

Edmond, J., Stallard, R., Craig, H., Craig, V., Weiss, R., and Coulter, G. 1993. Nutrient chemistry of the water column of Lake Tanganyika. Limnology and Oceanography 38(4): 725–738.

Emery, K.O., and Csanady, G.T. 1973. Surface circulation of lakes and nearly land-locked seas. Proceedings of the National Academy of Sciences USA 70: 93–97.

Ferguson, A.J.D., and Harbott, B.J. 1982. Geographical, physical and chemical aspects of Lake Turkana. Chapt. 1 in: Hopson, A.J. (Ed.), Lake Turkana. A report on the findings of the Lake Turkana Project 1972–1975. Government of Kenya and Ministry of Overseas Development, London. Overseas Development Administration, London.

Ferro, W., and Coulter, G.W. 1974. Limnological data from the north of Lake Tanganyika. United Nations Food and Agriculture Organisation Report, FI:DP/BDI/73/020/10: 1–19.

Ferro, W. 1977. A limnological baseline survey of the Chintheche area of Lake Malawi. Report for the Promotion of Integrated Fishery Development Project. Food and Agricultural Organisation of the United Nations (FAO), Rome. 23 pages.

Fish, G.R. 1957. A Seiche Movement and Its Effect on the Hydrology of Lake Victoria. Colonial Office Fishery Publications No. 10. London. 68 pages.

Fofonoff, N.P., and Millard, R.C. Jr. 1983. Algorithms for Computation of Fundamental Properties of Seawater. UNESCO technical papers in marine science 44. UNESCO, Paris. 53 pages.

Food and Agricultural Organisation of the United Nations (FAO). 1982. Fishery Expansion Project, Malawi. Biological Studies on the Pelagic Ecosystem of Lake Malawi. Technical Report 1. FAO for the United Nations Development Programme. Rome. 47 pages.

Goldstein, S. 1931. On the stability of superposed streams of fluid of different densities. Proceedings of the Royal Society of London, Series A, 132: 524–548.

Gonfiantini, G., Eccles, D., and Ferro, W. 1977. Isotope investigation of Lake Malawi. In Isotopes in Lake Studies: 195–207 Vienna: Publication of International Atomic Energy Agency, 1079.

Griffiths, J.F. 1972a. The Horn of Africa. Chapt. 4 in: Griffiths, J.F. (Ed.) Climates of Africa, World Survey of Climatology, Vol. 10, Elsevier, Amsterdam. 604 pages.

Griffiths, J.F. 1972b. Eastern Africa. Chapt. 9 in: Griffiths, J.F. (Ed.) Climates of Africa, World Survey of Climatology, Vol. 10, Elsevier, Amsterdam. 604 pages.

Halfman, J.D. 1993. Water column characteristics from modern CTD data, Lake Malawi, Africa. Journal of Great Lakes Research 19(3): 512–520.

Halfman, B.M., and Johnson, J.C. 1989. Surface and benthic nepheloid layers in the western arm of Lake Superior. Journal of Great Lakes Research 9(2): 190–200.

Hamblin, P.F., and Elder, F.C, 1973. A preliminary investigation of the wind stress field over Lake Ontario. Proceedings of the 16th Conference on Great Lakes Research, International Association for Great Lakes Research, pp. 723–734.

Hamblin, P.F. 1987. Meteorological forcing and water level fluctuations on Lake Erie. Journal of Great Lakes Research 13(4): 435–453.

Hecky, R.E. 1993. The eutrophication of Lake Victoria. Verhandlungen der Internationale Vereinigung fur theortische und angewandte Limnologie 25: 39–48.

Hecky, R.E., and Bugenyi, F.W.B. 1992. Hydrology and chemistry of the African Great Lakes and water quality issues: Problems and solutions. Mitteilungen der Internationale Vereinigung für theortische und angewandte Limnologie 23: 45–54.

Hecky, R.E., and Fee, E.J. 1981. Primary production and rates of algal growth in Lake Tanganyika. Limnology and Oceanography 26(3): 532–547.

Hecky, R.E., and Kling, H.J. 1987. Phytoplankton ecology of the great lakes in the rift valleys of Central Africa. Archiv für Hydrobiologie Ergebnisse der Limnologie 25: 197–228.

Hecky, R.E., Spigel, R.H., and Coulter, G.W. 1991. The nutrient regime. Chapt. 4 in: Coulter, G.W. (ed.), Lake Tanganyika and its Life. Oxford, Oxford Univ. Press. 354 pages.

Hill, K.D. Dauphinee, T.M., and Woods, D.J. 1986. The extension of the practical salinity scale 1978 to low salinities. IEEE Journal of Ocean Engineering OE-11: 109–112.

Horsch, G.M., and Stefan, H.G. 1988. Convective circulation in littoral water due to surface cooling. Limnology and Oceanography 33(5): 1068–1083.

Hutchinson, G.E. 1957. A Treatise on Limnology. Volume 1. Geography, Physics and Chemistry. John Wiley and Sons, New York. 1015 pages.

Imberger, J., and Patterson, J.C. 1990. Physical limnology. Advances in Applied Mechanics 27: 303–475.

Ivey, G.N., and Patterson, J.C. 1984. A model of the vertical mixing in Lake Erie in summer. Limnology and Oceanography 29(3): 553–563.

Johnson, T.C. 1980. Late glacial and postglacial sedimentation in Lake Superior based on seismic-reflection profiles. Quaternary Research 13(3): 380–391.

Johnson, T.C., Halfman, J.D., Rosendahl, B.R., and Lister, G.S. 1987. Climatic and tectonic effects on sedimentation in a rift-valley lake: Evidence from high-resolution seismic profiles, Lake Turkana, Kenya. Geological Society of America Bulletin 98: 439–447.

Johnson, T.C., Talbolt, M.R., Kelts, K., Cohen, A.S., Lehman, J.T., Livingstone, D.A., Odada, E.O., Tambala, A.F., McGill, J., Arguit, A., and Tiercelin, J.-J. 1990. IDEAL. An International Decade for the East African Great Lakes. Workshop Report 1 on the Palaeoclimatology of African Rift Lakes. Bern, Switzerland 29–31 March 1990. Duke University Marine Lab., Beaufort, N. Carolina. 39 pages.

Johnson, T.C., and Ng'ang'a, P. 1990. Reflections on a rift lake. Pages 113–135 in: Katz, B.J. (Ed.), Lacustrine Basin Exploration: Case Studies and Modern Analogs. American Association of Petroleum Geology, Memoir No. 50. Tulsa, Oklahoma.

Kalkanis, G. 1964. Transportation of Bed Material Due to Wave Action. United States Army Corps of Engineers, Coastal Engineering Research Centre, Technical Memorandum No. 2, Washington, DC.

Källqvist, T., Lien, L., and Liti, D. 1988. Lake Turkana limnological study 1985–1988. Norwegian Institute for Water Research (NIVA) and Kenya Marine and Fisheries Research Institute. NIVA, Oslo. 98 pages.

Kitaka, G.E.B. 1971. An instance of cyclonic upwelling in the southern offshore waters of Lake Victoria. African Journal of Tropical Hydrobiology and Fisheries 2(1): 85–92. Nairobi.

Levring, T., and Fish, G.R. 1956. The penetration of light into certain East African lake-waters. Oikos 7(1): 98–116.

Liti, L., Källqvist, T., and Lien, L. 1991. Limnological aspects of Lake Turkana, Kenya. Verhandlungen der Internationale Vereinigung für theoretische und angewandte Limnologie 24: 1108–1111.

Monismith, S.G. 1986. An experimental study of the upwelling response of stratified reservoirs to surface shear stress. Journal of Fluid Mechanics 171: 407–439.

Monismith, S.G. 1985. Wind-forced motions in stratified lakes and their effect on mixed-layer shear. Limnology and Oceanography 30(4): 771–783.

Mortimer, C.H. 1977. Internal waves observed in Lake Ontario during the International Field Year for the Great Lakes (IFYGL) 1972: part 1, descriptive survey and preliminary interpretation of near-inertial oscillations in terms of linear channel-wave models. Special Report No. 32, Centre for Great Lakes Studies, University of Wisconsin–Milwaukee, Milwaukee. 122 pages.

Mortimer, C.H. 1974. Lake hydrodynamics. Mitteilungen der Internationale Vereinigung für theoretische und angewandte Limnologie 20: 124–197.

Mortimer, C.H. 1963. Frontiers in physical limnology with particular reference to long waves in rotating basins. Proceedings of the 6th Conference on Great Lakes Research. Great Lakes Research Division Publ. No. 10: 9–42. University of Michigan Ann Arbor.

Münnich, M., Wüest, A., and Imboden, D.M. 1992. Observations of the second vertical mode of the internal seiche in an alpine lake. Limnology and Oceanography 37(8): 1705–1719.

Newell, B.S. 1960. The hydrology of Lake Victoria. Hydrobiologia 15: 363–383.

Newman, F.C. 1976. Temperature steps in Lake Kivu: A bottom heated saline lake. Journal of Physical Oceanography 6: 157–163.

Pedlosky, J. 1979. Geophysical Fluid Dynamics. Springer-Verlag, New York. 624 pages.

Phillips, D.W. 1974. IFYGL weather highlights. Proceedings of the 17th Conference on Great Lakes Research: 296–320. International Association for Great Lakes Research, Ann Arbor, Michigan.

Pond, S., and Pickard, G.L. 1983. Introductory Dynamical Oceanography (2nd ed.). Pergamon Press, Oxford. 329 pages.

Quinn, F.H. 1992. Hydraulic residence times for the Laurentian Great Lakes. Journal of Great Lakes Research 18(1): 22–28.

Quinn, F.H., and den Hartog, G. 1981. Evaporation synthesis. Pages 221–246 in: Aubert, E.J., and Richards, T.L. (Eds.), IFYGL — The International Field Year for the Great Lakes. US National Oceanic and Atmospheric Administration, Great Lakes Environmental Research Laboratory, Ann Arbor, Michigan. 410 pages.

Richards, T.L., and Aubert, E.J. 1981. The International Field Year for the Great Lakes — an introduction. Pages 1–14 in: Aubert, E.J., and Richards, T.L. (Eds.), IFYGL — The International Field Year for the Great Lakes. US National Oceanic and Atmospheric Administration, Great Lakes Environmental Research Laboratory, Ann Arbor, Michigan. 410 pages.

Rodgers, G.K. 1965. The thermal bar in the Laurentian Great Lakes. Proceedings of the 8th Conference on Great Lakes Research, University of Michigan Great Lakes Research Division Publication 13: 358–363.

Rzoska, J. 1976. Lake Victoria, physical features, general remarks on chemistry and biology. Chapt. 10 in: Rzoska, J. (Ed.), The Nile, Biology of an Ancient River. Dr. W. Junk b.v., The Hague. 415 pages.

Rzoska, J., and Wickens, G.E. 1976. Vegetation of the Nile Basin. Chapt. 5 in: Rzoska, J. (Ed.), The Nile, Biology of an Ancient River. Dr. W. Junk b.v., The Hague. 415 pages.

Saylor, J.H., Bennett, J.R., Boyce, F.M., Lin, P.C., Murthy, L.R., Pickett, R.L., and Simons T.J. 1981. Water movements. Pages 247–324 in: Aubert, E.J., and Richards, T.L. (Eds), IFYGL — The International Field Year for the Great Lakes. US National Oceanic and Atmospheric Administration, Great Lakes Environmental Research Laboratory, Ann Arbor, Michigan. 410 pages.

Schott, F., and Fernandez-Partagas, J. 1981. The onset of the summer monsoon during the FGGE 1979 experiment off the East African coast: a comparison of wind data collected by different means. Journal of Geophysical Research 86(C5): 4173–4180.

Simpson, J.J., and Dickey, T.D. 1981a. The relationship between downward irradiance and upper ocean structure. Journal of Physical Oceanography 11(3): 309–323.

Simpson, J.J., and Dickey, T.D. 1981b. Alternative parameterizations of downward irradiance and their significance. Journal of Physical Oceanography 11(6): 876–882.

Spigel, R.H., and Imberger, J. 1980. The classification of mixed layer dynamics in lakes of small to medium size. Journal of Physical Oceanography 10: 1104–1121.

Talling, J.F. 1957a. Some observations on the stratification of Lake Victoria. Limnology and Oceanography 11(3): 213–221.

Talling, J.F. 1957b. Diurnal changes of stratification and photosynthesis in some tropical African waters. Proceedings of the Royal Society of London, Series B 147: 57–83.

Talling, J.F. 1963. Origin of stratification in an African rift lake. Limnology and Oceanography 8(1): 68–78.

Talling, J.F. 1966. The annual cycle of stratification and phytoplankton growth in Lake Victoria. Internationale Revue der gesamten Hydrobiologie und Hydrographie 51(4): 545–621.

Talling, J.F. 1969. The incidence of vertical mixing, and some biological and chemical consequences, in tropical African lakes. Verhandlungen der Internationale Vereinigung für theoretische und angewandte Limnologie 17: 998–1012.

Talling, J.F., and Talling, I.B. 1965. The chemical composition of African lake waters. Internationale Revue der gesamten Hydrobiologie und Hydrographie 50(3): 421–463.

Thomson, K.P.B., Jerome, J., and McNeil, R. 1974. Optical properties of the Great Lakes (IFYGL). Proceedings of the 17th Conference on Great Lakes Research: 811–822. International Association for Great Lakes Research.

Tiercelin, J.-J., and Mondeguer, A. 1991. The geology of the Tanganyika trough. Chapt. 2 in: Coulter, G.W. (Ed.), Lake Tanganyika and Its Life. Oxford University Press, London. 354 pages.

Torrance, J.D. 1972. Malawi, Rhodesia and Zambia. Chapt. 13 in: Climates of Africa, World Survey of Climatology, Vol. 10, Elsevier, Amsterdam. 604 pages.

Turner, J.S. 1973. Buoyancy Effects in Fluids. Cambridge University Press, Cambridge. 367 pages.

Wüest, A., Piepke, G., and Halfman, J. 1996. The role of dissolved solids in density stratification of Lake Malawi. Pages 183–202 in: Johnson, T.C., and Odada, E.O. (Eds.), The Limnology, Climatology and Paleoclimatology of the East African Lakes. Gordon and Breach, Toronto, 1996.

Wunsch, L. 1973. On mean drift in large lakes. Limnology and Oceanography 18: 793–795.

Zilitinkevich, S.S., Kreiman, K.D., and Terzhevik, A.Yu. 1992. The thermal bar. Journal of Fluid Mechanics 236: 27–42.

The Present Physicochemical Ecology of Lake Victoria, Uganda

F.W.B. BUGENYI and K.M. MAGUMBA *Fisheries Research Institute, Jinja, Uganda*

Abstract — The ecosystem of Lake Victoria has undergone major changes during the past three decades. Among these are a significant increase in nutrient input from atmospheric and catchment sources which have enhanced eutrophication along with in-situ N-fixation by cyanobacteria. The average open water chlorophyll concentration has increased by 2× to 5×, the average light transparency has been reduced by 5-fold and dissolved Si has declined by 10-fold. There have been declines in hypolimnetic oxygen (anoxia below 40 m seasonally), and vertical mixing rates have changed profoundly; regular annual mixing is no longer sufficient to bring saturated oxygen levels from the surface to the bottom. In addition there has been an increase in water temperatures. These dramatic changes in the biological, chemical and physical characteristics of Lake Victoria are believed to be the result of several factors, among them the introduction of the Nile perch (*Lates niloticus*) which has greatly altered the food web, climate changes resulting in higher lake levels, and increased human activities in the watershed.

INTRODUCTION

The environmental degradation of Lake Victoria can be linked directly to burgeoning human population growth. This region is densely populated and is recording a 3-4% annual growth-rate (Table 1). Inhabitants depend on the lake fisheries (the cheapest animal protein source available in tropical Africa), subsistence agriculture (made possible by deforestation), and intense animal husbandry. Urbanized industrial centres, mainly Kampala, Jinja, Masaka and Entebbe have textile companies, mining and smelting businesses, food processing plants, beer breweries and soft drink manufacturers. Consequent waste effluents containing organic loads and other chemical contaminants are disposed of directly into the lake.

Primary productivity of the lake appears to have risen to about 2 to 3-fold over the past three decades (Fig. 1). This rise has likely been driven by the increased nutrient inputs and the altered fish community, and may have been aggravated by a warmer regional climate (Hastenrath and Kruss, 1992) which has produced a 0.5°C increase in average lake water temperature (Hecky, 1993) (Fig. 2).

The eutrophication (Hecky, 1993) has been accompanied by growth of blue-green algae which can now reach high biomasses inshore especially when they become buoyant. These can produce powerful phycotoxins and pose a potential threat to human inhabitants, livestock and others dependent on lake drinking water. A consequence of this increased phytobiomass production has been increased deoxygenation of the deep water, 35–55% of the bottom area of the lake now endures prolonged anoxia. The introduced water hyacinth, *Eichhornia crassipes*, which sustains anoxic conditions under its mat (Twongo et al., 1992) further deteriorates the lake waters. For most of the year now, fishes can live only in the upper-most 30 m of the lake, and away from the water hyacinth mat.

The "endless summer" (Kilham and Kilham, 1990a) of the tropical East African Great lakes, wherein biological processes govern internal mechanisms of lake ecosystems year

F.W.B. BUGENYI and K.M. MAGUMBA

Table 1. Increases in Human Population in Uganda Since 1921 to Today (including an estimation for 2000)

Year	Number of people (in millions)
1921	2.8
1931	3.5
1948	5.0
1959	6.5
1969	9.5
1980	12.6
1991	16.8
2000	24.6

Compiled from Hamilton (1984) and Uganda National Census Centre.

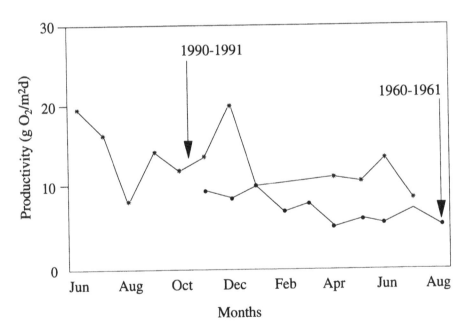

Figure 1. Primary productivity (g O_2 m^{-2} d^{-1}l) of the offshore site (Bugaia) 1960–61 compared to today (1990–91) (Mugidde 1993).

round, may predispose Lake Victoria to extreme sensitivity to eutrophication. This, together with warm deep water, the industrial development described above, plus a dominance of direct precipitation in the water budget (Hecky and Bugenyi, 1992) are considered as causative factors for the degradation of the lake waters.

Table 2. Composite Water Balance of Lake Victoria

Parameter	Input 10^9 (m^3)	Losses 10^9 (m^3)	Output 10^9 (m^3)
Rainfall on land	208.6		
Evapotranspiration and losses		189.8	
Runoff			18.8
Rainfall over lake	114.3		
Evaporation		99.6	
Rainfall–Evaporation			14.7
Initial storage	0.0		
Change in storage		4.7	
Output			4.7
Total	322.9	294.1	28.8

From UNDP/WMO, 1974.

HISTORICAL BACKGROUND

Lake Victoria presents a number of special features. It mixes throughout its depths at least once annually (Talling, 1966). It is shallow (80 m maximum depth), has a relatively short residence time and has a water budget dominated by precipitation (Table 2). Its annual thermal stratification and mixing regime, nutrient concentrations and productivity were recorded by Fish (1957) and Talling (1965, 1966) and showed pronounced seasonal changes, varying between inshore and offshore regions. Recent comparative studies, discussed below, have been done by FIRI/Freshwater Institute, Winnipeg, Canada (Bugenyi, 1992; Hecky and Bugenyi, 1992; Hecky, 1993; Mugidde, 1993).

Solar radiation, wind, rainfall and temperature interact within the water mass to determine the structure of the water column. The distribution of solar radiation over the land mass in tropical Africa (Fig. 3) is partly related to season, altitude and latitude. Lake water temperatures follow the magnitude of solar radiation input usually with some lag.

Two main wind systems affecting the climate of East Africa are the Northeast and Southeast Trade Winds. Conventional storms are received in March–May and September–November when wind systems become strongest (Fig. 4). These are periods when the trade winds meet over the equator and during equinoxes. The day-to-day diurnal winds also have localized effects. These processes interact to control lake water mixing and/or stratification.

Three phases of thermal stratification were distinguished in 1960–61: between September and December; January and May; and July and August (Talling, 1966). Complete mixing was in May–June and July. In 1961, Talling (1966) also reported complete mixing in the offshore area, Bugaia, during February–March. In the inshore areas (Buvuma Channel) Fish (1957) followed seasonal changes in the thermal stratification of temperature and dissolved oxygen (DO). The main cooling occurred in June/July as at the offshore station. However, in the subsequent period of warming (August to March) strong stratification with a marked discontinuity developed (and broke down) much earlier in the Buvuma Channel than in offshore sites. In other inshore areas, for example in shallow (<10 m) Pilkington Bay, thermal stratification was comparatively slight and largely diurnal. Changes of thermal

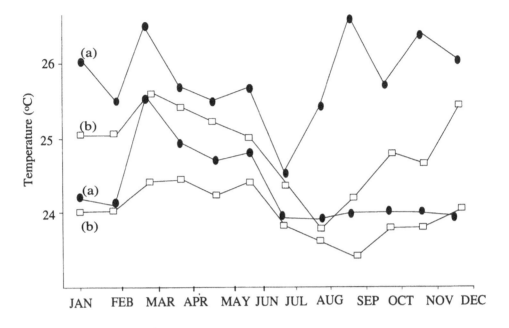

(a) Temperatures 1990/1992 of BUGAIA

(b) Temperatures 1960/1961 of BUGAIA

Figure 2. Monthly minimum and maximum water column temperatures of Bugaia site (a) in 1990/92 compared to (b) 1960/61, showing clearly that water temperatures are warmer today than three decades ago.

stratification were correlated with the annual variations of DO, pH, NO_3-N, PO_4-P and SiO_2 (Talling 1966).

Concentrations of DO below 10% saturation (approximately 0.7 mg l^{-1}) were within the lowermost 5 m of the water column from January to June 1961 (Talling 1966). Complete or almost complete deoxygenation (0.1 µg l^{-1}) was found much less frequently. The distributions of the three plant nutrients, NO_3-N, PO_4-P and SiO_2, had several features in common with DO. Lower concentrations were recorded in the photosynthetic zone (uppermost 20 m) with net consumption of nutrients, while higher concentrations of nutrients in the surface layers could be expected to depend upon the balance between biological consumption and replenishment by vertical mixing. In the tropics there seems to be an "endless summer" (Kilham and Kilham 1990a) where biological control (rather than physical control) dominates the cycles of essential elements in lakes all year, rather than for only a few summer months. Because the organisms control the cycles in the tropics, production will be maximum for the available resources (Kilham and Kilham, 1990b). Lake Victoria has generally been characterised in the past as having a great variety of phytoplankton (Thomasson, 1955) with 24 species: 5 diatoms; 4 blue-greens; 7 desmids; 7 chlorococcalean and other green algae and 1 dinoflagellate. Fossil diatom assemblages deposited in more than a dozen African lakes roughly 9500 years BP were dominated by a single planktonic species,

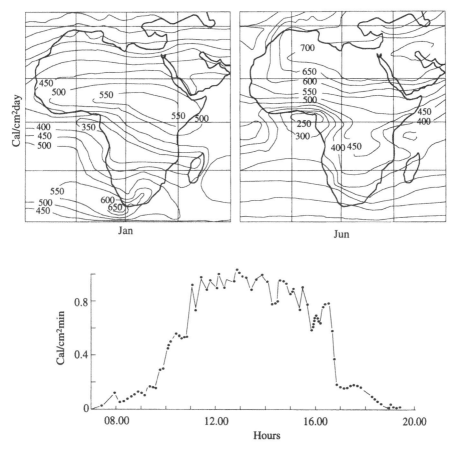

Figure 3. The distribution of incident solar radiation (January and June) over Africa, and the magnitude of daily sunshine. In January, the Lake Victoria basin has incident solar radiation between 500 and 550 cal cm^{-2} day^{-1}1 and slightly less in June. Intense sunshine is normally received between 11:00 a.m. and 4:30 p.m.

Stephanodiscus astraea (Ehrenb.) Grun. (Kilham and Kilham 1990a). These diatoms flourished when lake levels were maximal. The ecophysiology of *Stephanodiscus* is such that members of this genus are characterised as growing best at low Si:P ratios (Kilham, 1984; Kilham et al., 1986). The P-loading to such lakes must have been high enough to create such low ratios (Kilham 1971). The absolute amounts of Si and P producing a particular ratio can vary greatly leading to a complex relationship between the ratios and total biomass or productivity in a particular lake. However, the diatoms appear to be closely associated with a particular ratio and not with absolute amounts of Si or P. In Lake Victoria, the absolute numbers of diatom frustules were highest about 9000 years BP (Stager, 1984). *Stephanodiscus* has remained a dominant species in the sediment record although there has been a steady decrease in absolute abundance of total diatoms since ca. 5000 years BP (Kilham and Kilham, 1990a).

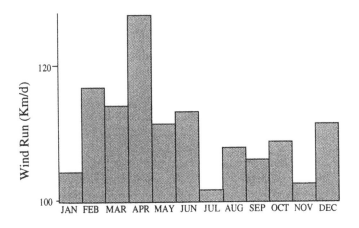

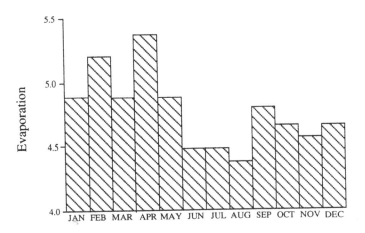

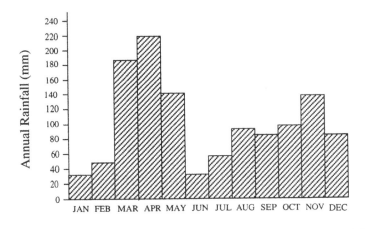

Figure 4. The two heavy rainy seasons (March–May and September–November) and the wind system within the region, with the strongest winds occurring in April, following the sunshine peaks. Periods of heavy rains correspond to periods of intense evaporation. (Wind, Evaporation and Sunshine data obtained from Hydromet, Entebbe.)

The nutrient dynamics of most tropical lakes are incompletely known, thus data for the biogeo-chemical factors which control the cycling of Si and P are unavailable. The molar biogenic Si:P ratio in sediments in Lake Victoria has been 100:1 throughout its recent history, which includes the rise in P deposition. However, P-loading from the catchment is affected by the marginal or fringing vegetation. This surrounding vegetation (usually papyrus) reduces P-loading either by direct uptake and storage (Gaudet, 1977; Chale, 1987; Bugenyi, 1993; Bugenyi et al., 1993) or by packaging the phosphorus in large particulates that are more easily buried. Howard-Williams and Gaudet (1985) concluded that papyrus swamps have a large capacity to take up nutrient elements such as N, P and S: from inflowing waters. Nutrient budgets for the tropical wetland swamps are still to be drawn for the tropical processes within the ecotone, and lake level rise could make these "sources" rather than "sinks."

THE CURRENT STATUS

The introduction of the piscivore, Nile perch, to Lake Victoria in the 1950s, overfishing, unregulated gill net mesh sizes and exploitive fishing techniques have led to the decline of nearly all endemic species, most notably the cichlid fish species. These practices led to the removal of the phytophagous haplochromines and native tilapines (Goldschmidt and Witte, 1992), greatly altering the biological trophic structure of the lake (Bugenyi, 1992), drastically reducing the grazing pressure and leaving an excess of phytobiomass.

The lake has been impacted by the increased human activities in the catchment (Bugenyi and Balirwa, 1989; Simons, 1989), mentioned earlier, urbanisation's sewage treatment and disposition problems and rampant deforestation (Hamilton, 1984). Additionally, the sudden rapid increase in lake level at the beginning of the 1960s (Fig. 5) covered large expanses of the fringing vegetation and possibly released nutrients hitherto locked up.

Lake Victoria is one of the few tropical lakes that have been characterised in terms of annual patterns of mixing, nutrient concentrations and primary productivity. Since the 1960s, the average open water chlorophyll concentrations have increased from about 2 μg l^{-1} to about 15 μg l^{-1} (Mugidde, 1993). As noted earlier, the lake now exhibits 2× higher phytoplankton photosynthesis rates and about 3× to 5× higher chlorophyll concentrations than before (Fig. 1). The increases in biomass have been accompanied by a 5-fold decrease in Secchi disc transparency. In the past, it varied from 6.4 to 8.4 m (Talling, 1966). Presently, it varies from 1.3 to 3.4 m. Maximum abundance of chl-a at the offshore station occurred during the months of May–June and September–November. The Daily Integral Productivity (IPD) shows the opposite trend possibly because of self shading. During periods of maximum chlorophyll, the SD figures are low (Fig. 6). The increases in phytobiomass have been accompanied by algal domination by blue-greens, especially by the filamentous N-fixing *Cylindrospermopsis* (Mugidde, 1993). Also because of elevated activities of the diatoms, possibly due in part to the increased temperature, the Si concentrations have declined 10-fold (Hecky, 1993; Hecky and Bugenyi, 1992; Bugenyi, 1992). The Si is likely to have sedimented with the diatom remains as it is not detected in either particulate form or in the deep waters. During the 1960s, anoxic conditions (concentrations of DO < 3.0 μg l^{-1}) were recorded below 55 m. Anoxia below 40–45 m is now encountered frequently during the stratified periods (Fig. 7) at the offshore station, Bugaia, and affects up to 50% of the lake's bottom area for prolonged periods of time. This greatly limits the habitable area for the fish (Bugenyi, 1992; Hecky et al., 1994). Massive fish kills have been frequently reported

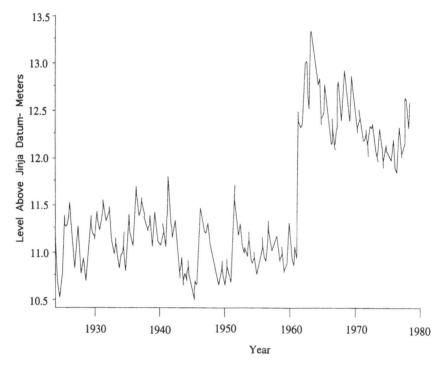

Figure 5. The sudden rise in lake level in early 1960s may have released the formerly locked up nutrients in the riparian wetland ecotone, and this could have contributed to the onset of the lake's eutrophication. Lake-level data obtained from Egyptian Embassy department, Jinja.

(Ochumba and Kibara, 1989) and are wholly or partly related to this widespread anoxia coupled with vertical mixing of the water. It is possible too that phycotoxins produced by the algal bloom are involved.

The strength and frequency of mixing govern the injection of nutrients from deeper waters to the euphotic zone. Figure 8 shows the variance of the nutrients, NH_4-N and PO_4-P during a calendar year. The stability of stratification regulates the diffusive flux upwards of nutrients from source areas in the hypolimnion or sediments, even as it affects the level of light available for photosynthesis through adjustments in mixed layer depth. Prevalent anoxic bottom conditions enhance rates of denitrification, resulting in low ratios of dissolved fixed N to P below the oxycline. This further reduces the TN/TP from 16:1 in the mixed layer to ca. 8:1 in the hypolimnion. A low N:P ratio in the offshore areas favours dominance by heterocystous cyanobacteria (Hecky, 1993). Inshore areas have higher TN and therefore higher N:P ratios. The apparent increasing eutrophication of the lake is thus occurring through an increased input of nutrients from the catchment drainage from rainfall (Bugenyi and Magumba, 1990) and from the in-situ N-fixation by cyanobacteria. The overall contribution of N-fixation to phytobiomass productivity is yet unknown. When Lehman and Branstrator (1993) calculated the loading of N from the catchment drainage and from rainfall, they found the figure from rainfall to be only slightly higher than that during the 1960s. Talling had rejected rainfall as the main contributor of nutrients. While investi-

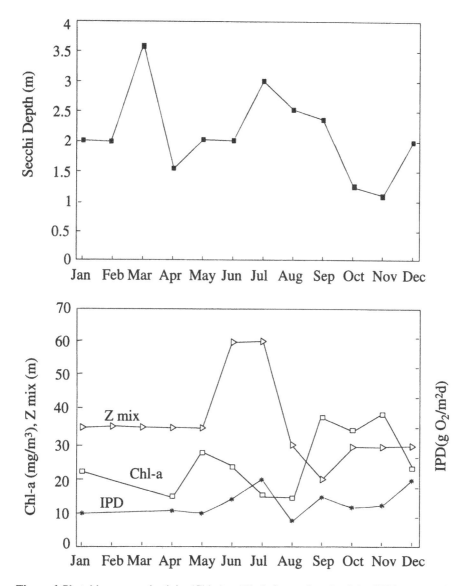

Figure 6. Phytobiomass productivity (Chl-a) and Daily Integral productivity (IPD) compared with Transparency (SD) and mixed layer thickness. The maximum mixing period corresponds to maximum IPD and least Chl-a. Transparency and Chl-a show an inverse relationship.

gating the effects of nutrients and grazing on the phytoplankton of Lake Victoria, Lehman and Branstrator (1993) concluded with Talling (1965), that N remains the most limiting nutrient element for the phytoplankton of Lake Victoria. Light is another limiting factor at present (Mugidde, 1993).

The depth of the mixed layer and the stability of the lake can change over diurnal, seasonal, yearly or even decadal time scales. In temperate lakes for example, solar insolation is most important in determining when lakes will mix or stratify. Wind patterns, evaporation

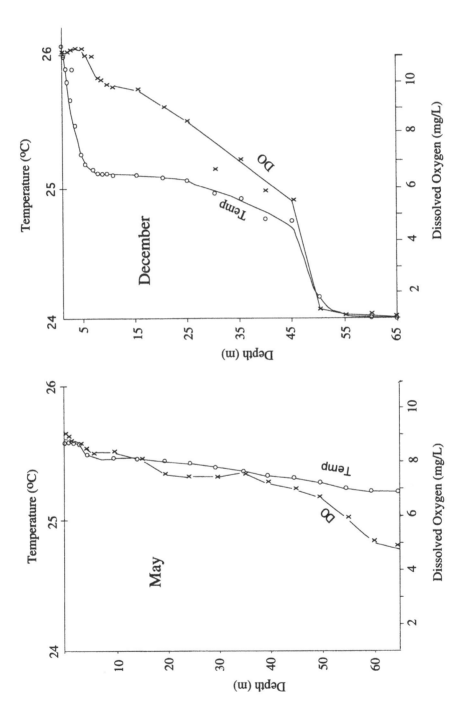

Figure 7. Temperature and DO profiles during the mixing period (May) and the stratification period (December) at the offshore station (Bugaia). DO at the surface is 9.0 mg l^{-1} and at the bottom, 7.0 mg l^{-1} during May. The lake is anoxic below 45 m during December.

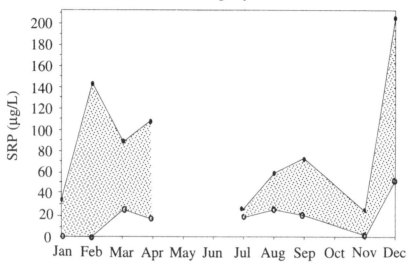

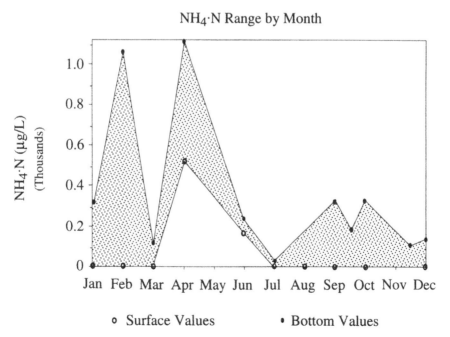

Figure 8. Concentration variations of NH₄-N and soluble reactive phosphate (SRP) over a calendar year. Bottom waters have the highest concentrations at any time. During stratification periods the surface concentrations of these nutrients are least.

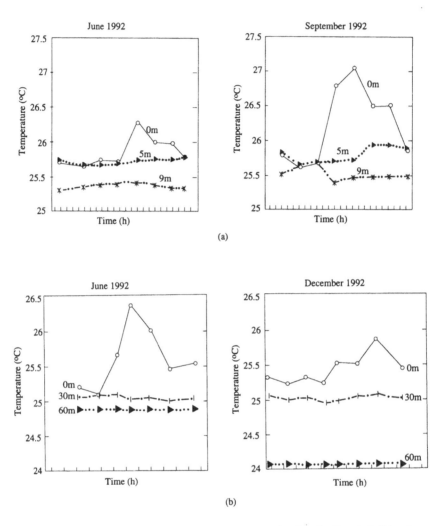

Figure 9. Diel temperature changes in Pilkington Bay (inshore) and at Bugaia (offshore).

and the onset of rainfall, all may interact to drive mixing and stratification of a lake. Talling (1966) determined that heat loss from evaporation and long wave radiation, and winds associated with the June through August monsoon drive the seasonal mixing patterns in Lake Victoria more so than do temperature effects.

Higher temperatures have already been recorded throughout the water column (Fig. 2) in most months (up to 26.7°C compared with 25.6°C in the 1960s. This is further evidence that the climate of East Africa has undergone a warming trend since the 1960s (Hastenrath and Kruss, 1992). Figures 9a and 9b show the diel temperature cycles in the shallow (Pilkington Bay) and offshore regions (Bugaia) of the lake.

DISCUSSION

Action must be taken now to improve water quality, species conservation, and control of the hyacinth weed (*Eichornnia crassipes*) in the whole lake ecosystem. We must identify, quantify and isolate causes. We should be asking ourselves the following questions: What activities in the watershed have led to the changed ecosystem of the lake? What nutrients control primary productivity? How do water column stability and mixed layer depth affect primary productivity? How important is herbivory in controlling phytoplankton species, biomass and production? And, how have historical changes in biogeochemistry, food webs, and climate shaped the contemporary status of the lake? The current research at the Fisheries Research Institute (formerly UFFRO), carried out in collaboration with scientists from North America, Europe and Africa, is trying to find answers to the above questions.

ACKNOWLEDGMENTS

Conduction of the above research has been due, mainly, to the International Development Research Centre, Ottawa, Canada, the National Science Foundation, USA and the Uganda Government's financial inputs for which we are very grateful. The Freshwater Institute, Winnipeg, Manitoba, Canada, The Water Environment Study Section of UFFRO, Mrs. F. Balirwa and Ms. Claire Mukobi are all greatly acknowledged for provision of analytical equipment and personnel for analysis of samples and for preparing this manuscript, respectively.

REFERENCES

Bugenyi, F.W.B., 1992: The effect of limnological factors and the changing environment on the fisheries of northern Lake Victoria: *in* Ssentongo, G. W., and Orach-Meza, F.L., Report on National Seminar on the development and management of the Ugandan Fisheries of Lake Victoria, Jinja, Uganda, 6–8 August 1991. UNDP/FAO, IFIP Regional Project Report RAF/8 7/099-TD/31/92 (EN). FAO/UNDP, Bujumbura, Burundi, pp. 36–58.

Bugenyi, F.W.B., 1993: Some considerations on the functioning of tropical riparian ecotones: Hydrobiologia: v. 251, pp. 33–38.

Bugenyi, F.W.B., and Balirwa, J.S., 1989: Human intervention in the natural processes of the lake ecosystem: The problem, *in* Salanki, J., and Herodek, S., Conservation and management of lakes: Symposium of Biology in Hungary, v. 38, Akademia; Kiado, Budapest, pp. 311–340.

Bugenyi, F.W.B., and Magumba, K.M., 1990: The physico-chemistry of the northern waters of Lake Victoria: Prevailing activities on the Lake Victoria basin with particular reference to the fisheries of the lake, Workshop, Mwanza, Tanzania, 8–9 March 1990, pp. 43–55.

Bugenyi, F.W.B., Magumba, K.M., Twinomujuni, E., and Wanda, F., 1993: Case studies on the functioning of a wetland ecotone on Lake Victoria (Uganda): Ecotones at the river basin scale — global land/water interactions: Proceedings of Ecotones Regional Workshop, Barmera, South Australia, 12–15 October 1992, pp. 213–220.

Chale, F.M.M., 1987: Plant biomass and nutrient levels of a tropical macrophyte (*Cyperus papyrus* L.) receiving domestic wastewater: Hydrological Bulletin, v. 21, pp. 167–170.

Fish, G.R., 1957: A seiche movement and its effect on the hydrology of Lake Victoria: Fisheries Publications, London, v. 10, pp. 1–68.

Gaudet, J.J., 1977: Uptake, accumulation, and loss of nutrients by papyrus in tropical swamps: Ecology, v. 58, pp. 415–422.

Goldschmidt, T., and Witte, F., 1992: Explosive speciation and adaptive radiation of haplochromine cichlids from Lake Victoria: an illustration of the scientific value of a lost species flock: Mitteilungen Internationale Vereinigung Limnologie, v. 23, pp. 101–107.

Hamilton, A.C., 1984: Deforestation in Uganda: Oxford University Press, Nairobi, pp.95.

Hastenrath, S., and Kruss, P.D., 1992: Greenhouse indicators in Kenya: Nature, v. 355, p. 503.

Hecky, R.E., 1993: The eutrophication of Lake Victoria: Verhandlungen der International Vereinigung fir Theoretische und Angewandte Limnologie, v. 25, pp. 9–48.

Hecky, R.E., and Bugenyi, F.W.B., 1992: Hydrology and chemistry of the African Great Lakes and water quality issues: Problems and solutions: Mitteilungen International Vereinigung Limnologie, v. 23, pp. 45–54.

Hecky, R.E., Bugenyi, F.W.B., Ochumba, P., Talling, J.F., Mugidde, R., Gophen, M., and Kaufman, L., 1994: Deoxygenations of the hypolimnion of Lake Victoria: Nature (in press).

Howard-Williams, C., and Gaudet, J.J., 1985: The structure and functioning of African swamps; in Denny, P. (ed.); The ecology and management of African wetland vegetation, Dr. W. Junk, Dodrdrecht, pp. 153–175.

Kilham, S.S., 1984: Silicon and phosphorus growth kinetics and competitive interactions between *Stephanodiscus minutus* and *Synedra* spp: Verhandlungen der Internationale Vereinigung fur Theoretische und Angewandte Limnologie, v. 22, pp. 435–439.

Kilham, P., 1971: A hypothesis concerning silica and the freshwater planktonic diatoms: Limnology and Oceanography, v. 16, pp. 10–18.

Kilham, P., and Kilham, S.S., 1990a: Endless summer: Internal loading processes dominate nutrient cycling in tropical lakes: Freshwater Biology, v. 23, pp. 379–389.

Kilham, P., and Kilham, S.S., 1990b: Do African lakes obey the "first law" of limnology?: Verhandlungen der Internationale Vereinigung fur Theoretische und Angewandte Limnologie, v. 24, pp. 68–72.

Kilham, P., Kilham, S.S., and Hecky, R.E., 1986: Hypothesised resource relationships among African planktonic diatoms: Limnology and Oceanography, v. 31, pp. 1169–1181.

Lehman, J.T., and Branstrator, D.K., 1993: Effects of nutrients and grazing on phytoplankton of Lake Victoria: Verhandlungen der Internationale Vereinigung fur Theoretische und Anagewandte Limnologie, v. 25, pp. 850–855.

Mugidde, R., 1993: The increase in Phytoplankton primary productivity and biomass in Lake Victoria (Uganda): Verhandlungen der Internationale Vereingung fur Theoretische und Angewandte Limnologie, v. 25, pp. 846–849.

Ochumba, P.B.O., and Kibara, D.I., 1989: Observations on blue-green algal blooms in the open waters of Lake Victoria, Kenya: African Journal of Ecology, v. 27, pp. 23–34.

Simons, M., 1989: High ozone and acid rain levels found over African rain forest: New York Times, v. 138, no. 47, p. 1.

Stager, J.C., 1984: The diatom record of Lake Victoria (East Africa): The last 17,000 years, in Mann, D.G. (ed.), Proceedings of 7th International Diatom Symposium, Philadelphia, pp. 455–621.

Talling, J.F., 1966: The annual cycle of stratification and phytoplankton growth in Lake Victoria (East Africa): Internationale Revue ges Hydrobiologie, v. 51, pp. 545–621.

Talling, J.F., 1965: The photosynthetic activity of phytoplankton in East African lakes: Internationale Revue ges Hydrobiologie, v. 50, pp. 1–32.

Twongo, T.K., Bugenyi, F.W.B., and Wanda, F., 1992: The potential for further proliferation of water hyacinth in Lakes; Victoria and Kyoga and some urgent aspects for research: Report of the sixth session of the sub-committee for the development and management of the fisheries of Lake Victoria, FAO Fisheries Report, No. 475, pp. 38.

Thomasson, K., 1955: A plankton sample from Lake Victoria: Svensk botanica Tkidskdsr, v. 49, pp. 2259–2274.

UNDP/WMO, 1974: Hydrometeorological survey of the catchment of Lakes; Victoria, Kyoga and Albert: RAF 66-025 Technical Report 1, v. 1, Meteorology and hydrology of the basin, Part II, pp. 925.

Measurement of Water Currents, Temperature, Dissolved Oxygen and Winds on the Kenyan Lake Victoria

P.B.O. OCHUMBA *Kenya Marine and Fisheries Research Institute, Kisumu Laboratory, Kisumu, Kenya*

Abstract — There have been distinctive changes in the physical structure of Lake Victoria over the last 40 years associated with climate change, eutrophication, introductions of exotic species, and poor watershed management. More than 100 stations were sampled between 1990 and 1992 to assess the temperature, dissolved oxygen, winds and currents structure on the Kenyan area of Lake Victoria. The lake was stratified between September and May and mixing occured between June and August. This mixing was not sufficient to bring saturated oxygen levels from the surface to the bottom. Oxygen levels were very low below a depth of 40 m and were associated with extensive fish kills.

Current speeds usually were between 0.3 cm/s and 56.5 cm/s with a maximum of 150 cm/s. The studied currents were vertically structured, weak and did not flow in any well developed pattern. Multiple inversions and reversals were observed that may be time dependent and related to local wind changes, internal seiches and solar radiation. Swift currents were observed at channel areas in the late afternoon. The dominant wind on the lake was towards the northeast in the afternoon and to the southwest in the mornings, and ranged in speed from 0.1 to 9.3 m/s.

INTRODUCTION

The circulation pattern of Lake Victoria responds to the large scale shift of the monsoon system and the cooling rainy season. The mixing season is characterized by isothermal structure and high oxygen levels throughout the water column. The stratification cycle over Lake Victoria has been described by Worthington (1930), Fish (1952 and 1957), Newell (1960), Talling (1957 and 1966), Kitaka (1972), Akiyama et al. (1977), Melack (1979), and Ochumba and Kibaara (1989). During the 1950s and 1960s, Lake Victoria lacked a permanent hypolimnion. Instead, the lake underwent two cycles of stratification and mixing that corresponded to the long and short rainy seasons, and the oxycline was found at 50 m beneath which oxygen levels dropped occasionally to as low as 1 mg/l. Fish life flourished in the lake's deeper water at all times of the year. In the 1980s and 1990s the situation for the June–August mixing seemed worse than that of the 1950s. Severe deoxygenation at shallow depths (Ochumba, 1990) indicated that a large volume of Lake Victoria was incapable of sustaining aerobic life.

Winds on Lake Victoria are mainly southwesterlies for every month of the year, with a maximum wind run of more than 6.7 m/s in February (Melack, 1976). Strong afternoon winds generate evaporative and nocturnal cooling on a daily basis (Hills, 1978) that influences the stability of the water column (MacIntyre and Melack, 1982). The daytime lake breeze and nighttime land breeze are sufficiently strong to influence the lake's circulation. Newell (1960) observed currents of 7 to 24 cm/s and assessed the impact of seiche and wind effects. Drift bottles released on Lake Victoria between Musoma and Mwanza indicated a

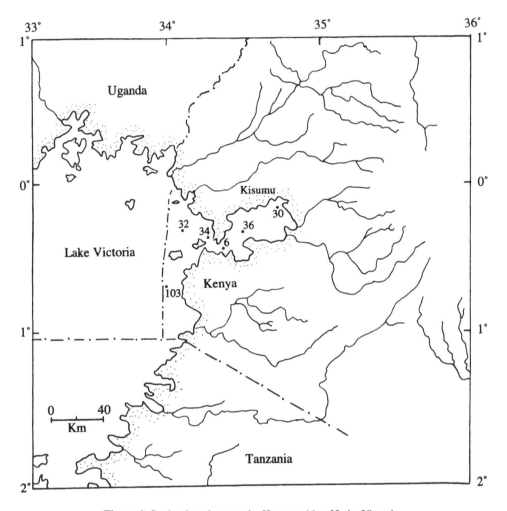

Figure 1. Station locations on the Kenyan side of Lake Victoria.

west–northwest drift of water from the centre and southeastern portion of the lake (Graham, 1929). Drift diverges near the west coast off Bukoba, with one current branching southward and the other northward to the Sesse Islands. Steamers on Lake Victoria have always reported that a clockwise voyage on Lake Victoria proceeds faster than an anticlockwise trip. A general northward flow of surface waters in response to the winds accompanied by a compensating flow of the lower layers southward has been noted by Newell (1960) and Kitaka (1972).

The purpose of this work is to present a record of the winds, currents, temperature and dissolved oxygen variations and distribution on the Kenyan side of Lake Victoria (Fig. 1). The knowledge of vertical and horizontal structure of these variables are important in that they may enable us to estimate mixing of anoxic water and rates of volume transport through various areas on Lake Victoria. The mixing season is especially important because it is the time of maximum exchange of surface and deep waters. Information about the lake's circulation is very important for proper management of the fisheries. It also sheds insight on the possible dispersion of contaminants that may be introduced into the lake.

STUDY AREA

Lake Victoria is about 400 km long and 280 km wide. It has a mean depth of 40 m and a maximum depth of 79 m. The lake volume is 2760 km^3, with a surface area of 68,800 km^2 and drainage basin area of 195,000 km^2. The shoreline length is 3440 km. Lake Victoria is a saucer-shaped lake lying between the the eastern and western arms of the East African Rift Valley. The present lake resulted from the amalgamation of many smaller lakes and westward flowing rivers due to tectonic activity (Kendall, 1969). The present lake is relatively young and has been controlled largely by the down-cutting of the Nile outlet (Scholz et al., 1990). Hydrologic balance is dominated by rainfall, river runoff and evaporation (Piper et al., 1986). Outflow via the Nile River at the northern end of the lake nearly equals river inflow estimated at 23.4 km^3/yr. Rainfall contributes most of the water input, 100 km3/yr, and is balanced by an equal amount of evaporation. Flushing and residence times are 140 and 23 years respectively. The mean monthly solar radiation is 240–270 W/m^2. The seasonal fluctuation in lake level is 0.4–0.7 m, and the internal lake level variability has been about 3 m (Kite, 1981).

The annual mean rainfall distribution for Lake Victoria varies between 812 and 2044 mm (Ogalo, 1981) with a double peak pattern, in which the long rainy season occurs between March and May and the short rainy season falls between October and December (Fig. 2). Daily fluctuations in lake level are common and are caused by wind-induced tilting towards the northwest corner of the lake in the late afternoon (Copley, 1953). Meteorological data on the northern shore at Mbita Point close to our stations offshore were supplied by the International Centre for Insect Physiology and Ecology (ICIPE). Since winds recorded on the shore may not reflect lake winds, our wind data were taken from the research vessel offshore.

METHODS

A Conductivity–Temperature–Depth (CTD) probe with additional sensors (Hydrolab Surveyor II) was used to measure temperature, pH, dissolved oxygen, conductivity and depth. Currents, winds and turbidity were measured at the same time as the CTD casts. The CTD sensors were calibrated frequently in the Kisumu Laboratory before going out in the field (Hydrolab, 1985) against known, freshly made standards.

Current speed and direction in the water column were determined by a four bladed impeller current meter with a surface readout (A.S. Sensordata, SD4 and SD2000), recently calibrated with known response characteristics. The 16 m research vessel, R/V UTAFITI, was anchored at each station, and current speed and direction were determined at 0, 1, 3, 5, 7, 10 m and every 5 m below 10 m within 1 min 40 sec of each other.

A 24-hour time series station was occupied near Rusinga Channel to estimate the effect of temporal variability at the sampled stations. The SD 2000 current meter had a recording interval of 24 minutes. Unfortunately, the instrument was lost in April 1992. Wind speed was measured with an Aanderaa Sensor Model 2750 and direction with an Aanderaa Sensor Model 3070. Sampling intervals varied from one month to four months. Data from the entire

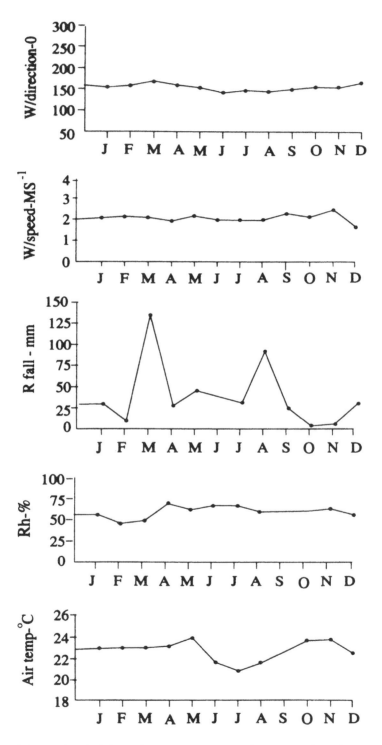

Figure 2. Seasonality of winds, rainfall, relative humidity and temperature in northeastern Lake Victoria.

study area were averaged to make a single year. Repeated measurements were made at the individual stations to assess seasonal variability.

RESULTS

Records of water temperature from stations 6, 32, 34, 99, 100 and 103 are plotted as isotherms in Fig. 3. The bottom water temperature ranged from a minimum of 24.1°C to a maximum of 26°C. The surface water temperatures were between 24.3 and 28°C. The observations indicate that thermal stratification occurred from September to June with a difference of between 2.5 and 3.9°C in the upper 40 m. Mixing occurred in July and August at Stations 34, 99, and 103, but was observed earlier in June and July at Station 32. At the end of June, temperatures were more uniform with depth and showed a drop of more than 3°C compared to data of February and March. A further cooling by about 2°C occurred between May and July at all stations. After the June, July and August isothermal cooling, water temperature began to rise and stratification set in. Unstable surface conditions with large temperature differences were observed in May at Stations 34 and 100. The lake frequently had two or more thermoclines at 5, 30 and 40 m. The greatest temperature changes were observed in the upper water column.

The oxygen profiles (Fig. 4) indicated the presence of a distinct anoxic water in the open Lake Victoria. Dissolved oxygen concentrations at the bottom ranged between 0 and 5.2 mg/l. Concentrations from 5.0 to 11.6 mg/l were recorded in the surface water during the period of study. Supersaturation at the surface was recorded mainly in the presence of algal blooms in November, December and February. Anoxic water below a depth of 35 m was observed at all stations during January to April and September to November. The oxycline occurred at 10–50 m depth at all stations, but was generally between 20 and 30 m most of the year. Between November and June, there were sporadic upsurges in the oxycline to depths as shallow as 10 m in the open lake. These events were associated with extensive fish kills as evidenced by floating carcasses of Nile perch (*Lates niloticus*), Nile tilapia (*Oreochromis niloticus*), the sardine (*Rastrineobola argentea*) and rafts of dead snails. The worst situation was in November when the oxycline was as shallow as 5 m at stations 34, 99 and 103. Fishermen reported mass catches of moribund fishes and migrations of fishes into nearshore areas. Bad oxygen conditions were associated with fish mortality at all stations. The oxygen concentration during the July mixing was between 7.6 mg/l at the surface and 1.9 mg/l at the bottom for most of the stations. This may indicate that complete mixing does not occur. Results from our 24 hr study show that anoxic water flowed into the Nyanza Gulf at 1700 hr and out at 2400 hr. Ngodhe Sill off Rusinga Channel at 20 m prevents anoxic water from flowing into the Nyanza Gulf.

Appreciable wind speeds were observed at the open lake stations 34, 99, 32, 103 and 33 compared to shallow Nyanza Gulf and bay stations 36, 6, 8, 7, 30 and 31 (Fig. 5). Negligible speeds were observed in February–May at the shallow stations, when there was a tendency of relatively calm morning periods followed by strong afternoon winds. The wind speeds varied from 1 m/s to 7 m/s with an overall mean of 3.9 m/s. The mean wind speed for the open lake stations was 4.6 m/s and for the Nyanza Gulf was 3.0 m/s. Afternoon mean wind speeds of 4.5 m/s exceeded the morning means of 3.0 m/s. The maximum speed was 9.3 m/s in March at Station 103. Out of 132 observations of wind directions, 30% indicated a southwesterly, 20% southeasterly, 17% northwesterly, 9% northeasterly and 25% registered no direction. The 24-hour monitoring showed that 40% of the wind direction was southeast-

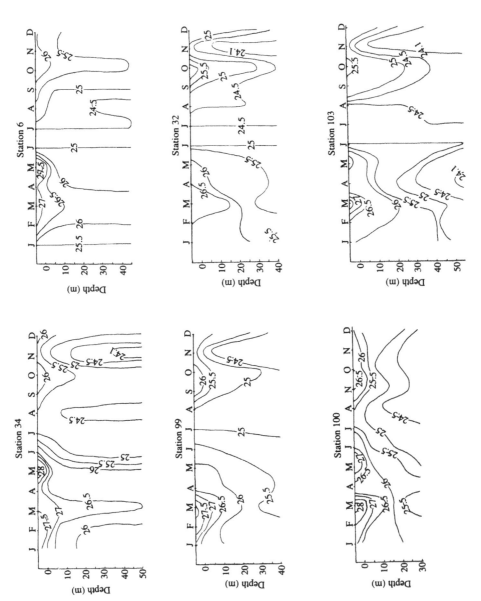

Figure 3. Isotherms at the Kenyan stations in 1992.

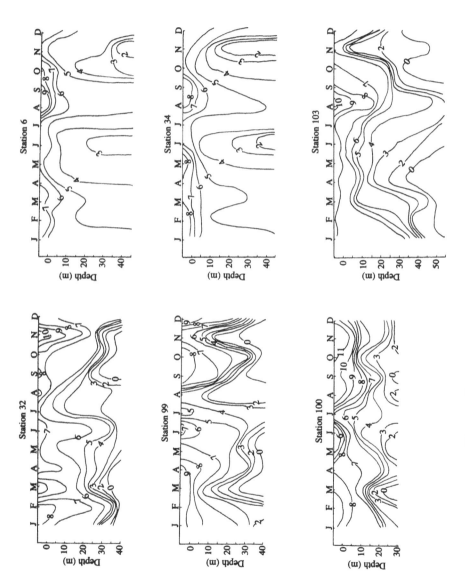

Figure 4. Oxygen profiles at the Kenyan stations in 1992.

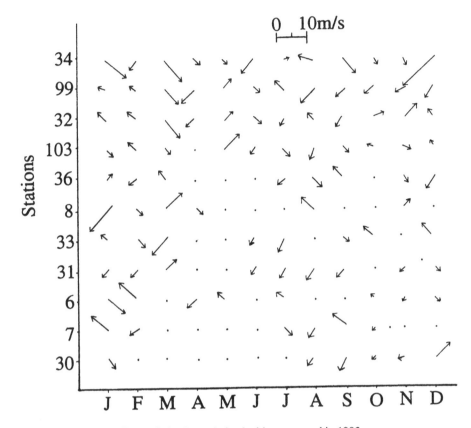

Figure 5. Surface wind velocities measured in 1992.

erly, 30% southwesterly, 20% northwesterly and 10% northeasterly. 24-hour monitoring results show that high wind speeds were observed in the late afternoon at 1400–1700 hrs and early morning between 0500 and 0600 hrs.

The currents are highly variable between stations, and the strongest currents measured were northwesterly (Fig. 6). The measurements show a relatively weak vertically structured flow, and measured between 1 and 42 cm/s. The near surface current speed off Rusinga Channel reached a maximum of 145 cm/s after a storm at 2100 hrs. Mean monthly patterns show that the highest velocities were observed at Station 34 in March and September. These coincide with the established rainy seasons for the region. Speeds were usually greater in the channel area (Stations 34 and 99) and in the rainy season. In July and August when mixing occurs, currents at different depths flow in the same direction, although deviations from this pattern could be observed at Station 100 and the Nyanza Gulf stations.

Periods of extremely low current speeds can persist for several days. This occurred at most stations in February and at Stations 100 and 103 in January and November. Very strong currents were observed in March, May, July and September at Station 34 in Rusinga Channel. Outflow was observed in the channel in May, July, August and September, whereas inflow prevailed in March and April.

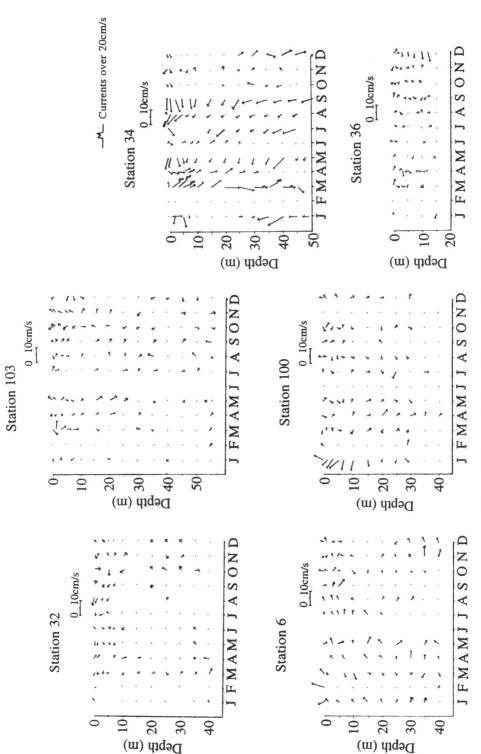

Figure 6. Current velocity profiles measured in 1992.

DISCUSSION AND CONCLUSIONS

Several studies on tropical lakes (Talling, 1969; Eccles, 1974; Melack, 1979a; Lewis, 1983 and 1987; Kling 1987 and 1988; Limon et al., 1989; and Coulter, 1988) show a consistent pattern of mixing during the southern hemisphere winter and the dry season. Isothermal conditions were observed between June and August with temperatures between 24.2 and 24.5°C. Maximum temperatures of 28°C were recorded in March and May. The mixed layer depth deepened from 5 m in May to 60 m in August.

Hypolimnion water temperatures as low as 23.5°C observed by Graham (1929), Worthington (1930), Fish (1957), Talling (1966) and Akiyama et al. (1977) were not seen during our studies on the lake. These data may suggest a response of Lake Victoria to a possible warming trend of the climate in East Africa (Hastenrath and Krus, 1992) that is likely to make the lake more stable today than in the previous years.

Oxygen conditions have deteriorated significantly offshore in Lake Victoria since the 1950s. Comparing data from Fish (1957), Talling (1966), Ochumba and Kibaara (1989) and this study reveals that there has been a considerably intensified oxygen deficiency in the lake. The unforeseen catastrophic oxygen depletion in November 1983 which affected most of the Nyanza Gulf (Ochumba 1987 and 1990) was first regarded to be an exception due to the extraordinary weather conditions. This study shows the origin of that water mass was from the open lake. Oxygen depletion occurred in the lake earlier (Graham 1929) when man began to modify the watershed for large-scale agriculture, but the conditions responsible for oxygen depletion today are worse than in the past. Changes in bathymetry, increased sedimentation rates and restricted water circulation may all have contributed to the lake's capacity to assimilate its oxygen demand which has severely affected the biota (Ogutu-Oh-wayo, 1990; Kaufman 1992). Hecky (1993) and Mugidde (1993) consider this development to be related to increased primary production associated with a significant increase in nutrient inputs from the watershed and air-shed.

The principal seasons in Lake Victoria consist of dry and rainy periods (Ogalo, 1981). Most cooling and mixing take place in the dry season and maximal stratification occurs during the rainy season (Talling, 1969; Lewis, 1973; MacIntyre and Melack, 1982; Coulter and Spigel 1991). The seasonality of rain affects biological events such as algal blooms, zooplankton production and fish spawning (Melack, 1979b; Talling, 1987; Ogari and Dadzie, 1988; Mavuti and Litterick, 1991; Ochumba and Manyala, 1992; Gophen et al., 1993). Winds also display a seasonal pattern. The wind speed was maximum during the dry season which maximized evaporation. The resulting heat losses and associated turbulence caused by the wind would enhance mixing in the water column.

Although our measurements covered only the Kenyan portion of the lake, the observed distribution of water masses is consistent with studies in Uganda (Hecky and Bugenyi 1992; Hecky 1993 and Mugidde 1993) and Tanzania (Witte et al., 1992). Throughout the water column there is a complex vertical structure with multiple reversals that is time and density dependent and not clearly related to local changes in the wind. Flow in one direction next to the coast often was associated with a flow reversal offshore. Our data shows that in all seasons the currents follow the bathymetry and generally flow southeast. During the mixing period, currents are uniform throughout the water column and can reach speeds of 30 cm/s. In the stratified period, strongest currents are confined to the surface layer and decrease as one moves down the water column and closer to shore. The seasonal differences in the currents are due to the strong density stratification during the dry and rainy seasons.

The major driving force for hypolimnetic currents in lakes are seiches. Internal waves exist in Lake Victoria during the stratified period (Fish, 1957 and Newell, 1960) for 8 months: September to April. Internal waves may be responsible for advecting cooler, nutrient-rich and less oxygenated water into the various bays and gulfs in the absence of winds in Lake Victoria (Coulter and Spigel 1992).

Marginal cooling may cause profile-bound density currents offshore (Talling 1963). Diurnal temperature differences of up to 3°C occur at Lake Victoria's margins (Ochumba 1990), and inflowing streams during the rains can be 6°C less than the surface temperatures (Ochumba and Manyala 1992). Such differences could give rise to coastal currents.

More detailed data on the wind stress, current fluctuations, temperature and dissolved oxygen as primary facing functions are still needed to qualitatively understand Lake Victoria's circulation and productivity. The major factors controlling the wind patterns involve the interaction between locally generated thermal air currents around island areas and the prevailing monsoonal trade winds. The prolonged stagnation of the anoxic water in Lake Victoria is damaging to the water quality and fisheries. The oxygen supply to the deep waters has been insufficient to prevent the occurrence of anoxia in deep hypolimnion since the 1960s.

ACKNOWLEDGMENTS

Financial support for equipment and field work came from the International Foundation for Science (IFS) Grant G/1083, United States Agency for International Development (USAID-CDR Grant DPE-5544-55-7075-00-7-080) and a grant to Les Kaufman by the National Science Foundation (NSF) (no. BSER-9016552). I am grateful to the Lake Victoria Research Team members Les Kaufman, Moshe Gophen, Bill Cooper, Bob Hecky, Paul Sackley, James Ogari and Ezekiel Okemwa. The assistance of S. Agembe, W. Oyieko, E. Odada, G. Odanga and D. Owage in data collection is acknowledged.

REFERENCES

Akiyama, T., Kajumulo, A. A., and Olsen, S., 1977, Seasonal variation of plankton and physicochemical parameters in Mwanza Gulf, Lake Victoria: Bulletin of the Freshwater Fish Resources Laboratory, v. 27, pp. 49–61.

Beadle, L. C., 1974, The inland Waters of Tropical Africa: An Introduction to Tropical Limnology: London, Longman (2nd. Edition 1981).

Copley, H., 1953, The Tilapia fishery of the Kavirondo Gulf: Journal of the East African Natural Historical Society, v. 22, pp. 57–61.

Coulter, G. W., 1988, Seasonal hydrodynamic cycles in Lake Tanganyika: Verh. International Verein. Limnologae, v. 23, pp. 86–89.

Coulter, G. W., and Spigel, R. H., 1991, Hydrodynamics, Ch. 3, *in* Lake Tanganyika and its Life: British Museum of Natural History and Oxford University Press.

Eccles, D. H., 1974, An outline of the physical limnology of Lake Malawi (Lake Nyasa): Limnology and Oceanography, v. 19, pp. 730–742.

Fish, G. R., 1952, Local hydrographical conditions in Lake Victoria: Nature, v. 169, p. 839.

Fish, G. R., 1957, A seiche movement and its effect on the hydrology of Lake Victoria: Fish. Publ. Lond., v. 10, pp. 1–68.

Ganf, G. G., 1974, Diurnal mixing and the vertical distribution of phytoplankton in a shallow equatorial lake (Lake George, Uganda): Journal of Ecology, v. 62, pp. 611–629.

Gophen, M., Pollinger, U., and Ochumba, P. B. O., 1993, Feeding Habits of Tilapias and Limnology of Lake Victoria: Final Report 1989/91 USAID-CDR Grant No. DPE 5544-55-7075-00-c7-080.

Graham, M., 1929, The Victoria Nyanza and its Fisheries: A Report on the Fishery Surveys of Lake Victoria: London, Crown Agents for the Colonies, 256 pp.

Hastenrath, S., and Kruss, P. D., 1992, Greenhouse indicators in Kenya: Nature, v. 355, p. 503.

Hecky, R. E., 1993, The eutrophication of Lake Victoria: Verh. International Verein. Limnologae.

Hecky, R. E., and Bugenyi, F. W. B., 1992, Hydrology and chemistry of the African Great Lakes and water quality issues: Problems and solutions: Mitt. Verein. International Limnologae, v. 23, pp. 45–54.

Hills, R. C., 1978, Meteorology for agriculturalists: East African Agriculture and Forestry Journal, v. 43, pp. 399–402.

Kaufman, L. S., 1992, Catastrophic changes in species-rich freshwater ecosystems: The lessons of Lake Victoria: Bioscience, v. 42, pp. 846–858.

Kendall, R. L., 1969, An ecological history of Lake Victoria basin: Ecological Monographs., v. 39, pp. 121–176.

Kilham, P., and Kilham, S. S., 1989, Endless summer: Internal loading processes dominate nutrient cycling in tropical lakes: Freshwater Biology, v. 23, pp. 379–389.

Kitaka, G. E. G., 1972, An instance of cyclonic upwelling in the Southern offshore waters of Lake Victoria: African Journal of Tropical Hydrobiology of Fish, v. 1, pp. 85–92.

Kite, G. W., 1981, Recent changes in the level of Lake Victoria: Hydrological Science Bulletin, v. 26, pp. 233–243.

Kling, G. W., 1987, Seasonal mixing and catastrophic degasslng in tropical lakes, Cameroon, West Africa: Science, v. 237, pp. 1022–1024.

Kling, G. W., 1988, Comparative transparency, depth of mixing, and stability of stratification in lakes of Cameroon, West Africa: Limnology and Oceanography, v. 33, pp. 27–40.

Lewis, W. M., 1973, The thermal regime of Lake Lanao (Philippines) and its theoretical implications for tropical lakes: Limnology and Oceanography, v. 18, pp. 200–217.

Lewis, W. M., 1983, Temperature, heat and mixing in Lake Valencia, Venezuela: Limnology and Oceanography, v. 28, pp. 273–286.

Lewis, W. M., 1987, Tropical limnology: Annual Revue of Ecological Systems, v. 18, pp. 159–184.

Mavuti, K. M., and Litterick, M. R., 1992, Composition, distribution and ecological role of zooplankton community in Lake Victoria, Kenya waters: Verh. International Verein. Limnology, v. 24, pp. 1117–1122.

Melack, J. M., 1976, Limnology and dynamics of phytoplankton in equatorial African Lakes [Ph. D Thesis]: Duke University, Durham.

Melack, J. M., 1979a, Photosynthetic rates in four tropical African freshwaters: Freshwater Biology, v. 9, pp. 555–571.

Melack, J. M., 1979b, Temporal variability of phytoplankton in tropical lakes: Oecologia, v. 44, pp. 1–7.

Mugidde, R., 1993, The increase in phytoplankton primary productivity and biomass in Lake Victoria (Uganda): Verh. International Verein. Limnology, v. 25 (in press).

Newell, B. S., 1960, The hydrology of Lake Victoria: Hydrobiologia, v. 15, pp. 363–383.

Ochumba, P. B. O., 1987, Periodic massive fish kills in the Kenyan part of Lake Victoria: Water Quality Bulletin, v. 12, pp. 119–121, 130.

Ochumba, P. B. O., 1990, Massive fish kills in the Nyanza Gulf of Lake Victoria, Kenya: Hydrobiologia, v. 208, pp. 93–99.

Ochumba, P. B. O., and Kibara, D. I., 1989, Observations on blue-green algal blooms in the open waters of Lake Victoria, Kenya: African Journal of Ecology, v. 27, pp. 23–34.

Ochumba, P. B. O., and Manyala, J. O., 1992, Distribution of fishes along the Sondu-Miriu River of Lake Victoria, Kenya with special reference to upstream migration, biology and yield: Journal of Aquac. Fish. Man., v. 23, pp. 701–719.

Ogalo, L., 1981, Trend of rainfall in East Africa: Kenya Journal of Science and Technology (A), v. 2, pp. 83–90.

Ogari, J., and Dadze, S., 1988, The food of Nile perch *Lates niloticus* after the disappearance of haplochromine cichlids in the Nyanza Gulf of Lake Victoria: Journal of fish. Biol., v. 32, pp. 571–577.

Ogutu-Ohwayo, R., 1990, The decline of the native fishes of Lakes Victoria and Kyoga (East Africa) and the impact of introduced species, especially the Nile perch, *Lates niloticus*, and the Nile tilapia, *Oreochromis niloticus*: Environment and Biology of Fish, v. 27, pp. 81–96.

Piper, B. S., Plinston, D. T., and Sutcliffe, J. V., 1986, The water balance of Lake Victoria: Hydrologic Science Journal, v. 31, pp. 25–38.

Scholz, C. A., Rosendahl, B. R., Verseth, J., and Rach, N., 1990, Results of high resolution echo sounding of Lake Victoria: Journal of African Earth Science, v. 11.

Serruya, C., and Pallingher, G., 1983, Lakes of the warm Belt: Cambridge University Press.

Smith, I. R., and Sinclar, I. J., 1972, Deep water waves in lakes: Freshwater Biology, v. 2, pp. 387–399.

Talling, J. F., 1957, Some observations on the stratification of Lake Victoria: Limnol Oceanogr., v. 2, pp. 213–221.

Talling, J. F., 1966, The annual cycle of stratification and phytoplankton growth in Lake Victoria (East Africa): International Rev. ges. Hydrobiol, v. 50, pp. 421–463.

Talling, J. F., 1969, The incidence of vertical mixing and some biological and chemical consequences in tropical African lakes: International Ver. Theor. Angew. Limnology Verh, v. 17, pp. 998–1012.

Talling, J. F., 1987, The phytoplankton of Lake Victoria (East Africa): Erqeb. Limnology, v. 25, pp. 229–256.

Witte, F., Goldschmidt, T., Goudswaard, P. C., Ligtvoet, W., Van Oijen, M. J. P., and Wannik, J. M., 1992b, Species extinction and concomitant ecological changes in Lake Victoria: Netherlands Journal of Zoology, v. 42, pp. 214–232.

Witte, F., Goldschmidt, T., Wannik, J. H., Van Oijen, M. J. P., Goudswaard, P. C., Witte-Mass, E. L. M., and Bouton, N., 1992a, The destruction of an endemic species Flock: quantitative data on the decline of the haplochromine species from the Mwanza Gulf of Lake Victoria: Environment and Biology of Fishes, v. 34, pp. 1–28.

Worthington, E. B., 1930, Observations on the temperature, hydrogen ion concentrations and other physical conditions of the Victoria and Albert Nyanzas: International Rev. ges. Hydrobiology, v. 25, pp. 328–357.

CTD-Transmissometer Profiles from Lakes Malawi and Turkana

J.D. HALFMAN   *Department of Geoscience, Hobart and William Smith Colleges, Geneva, New York, United States*

Abstract — Reconnaissance conductivity, temperature and total suspended solid profiles through the water column in Lakes Malawi and Turkana, two large end-member lakes in the East African Rift System, update water column measurements from the 1970s and earlier and investigate the relation between the physical limnology and sediment dispersal pathways at deltaic settings. In Lake Malawi, water conductivities, temperatures and dissolved oxygen concentrations were similar to those reported previously except that the 1992 data reveal a thermal inversion below 350 m that can be explained by adiabatic changes and an increase in conductivity between 150 and 250 m that contributes to the stratification of the lake. Fluvial inputs, biological production, resuspension events and biogeochemical processes at the chemocline introduce suspended sediments to the water column that are dispersed by hypopycnal (lake surface), intermediate (internal density contrasts) and hyperpycnal (lake floor) pathways. At Lake Turkana, temperature profiles are nearly isothermal and total suspended solid profiles reveal either a turbid layer at the surface or nearly homogeneous distributions through the water column. Omo River effluent is dispersed along the lake's surface even at low Omo flow and is probably maintained by salinity contrasts between the less saline effluent and the lake. Previously, frequent water column mixing was attributed to physical mixing by the strong nocturnal winds; however, surface water cooling and subsequent convective mixing induced by evaporitic cooling of the surface waters during the wind events, cooler nighttime air temperatures or a combination of the two are additional mechanisms to consider. The results highlight the need for additional investigations to determine the range and magnitude of intra- and inter-annual variability in limnologic conditions of the rift-valley lakes to ensure proper interpretation of climatic signals preserved in the sediments.

INTRODUCTION

A goal set forth by the International Decade for the East African Lakes (IDEAL) is to obtain and decipher long, high-resolution records of climate change in tropical east Africa that are preserved in 4 to 5 km of sediments within the large rift-valley lakes (Johnson et al., 1990). The typical climatic proxy however, is intertwined with biological, chemical, physical and geological processes within the basin. Unfortunately, our understanding of these processes in tropical basins is less understood than in basins at temperate latitudes.

The objective of this report is to present temperature and total suspended solid profiles through the water column from Lake Turkana and compare the results to modern conductivity, temperature and total suspended solid profiles from Lake Malawi. Turkana and Malawi are chosen because they represent two important end-members of the large, modern-day, tropical, rift lakes, i.e., hydrologically closed and well-mixed versus hydrologically open and meromictic systems. This objective also updates earlier results on the physical limnology which is based primarily on data collected in the 1970s and earlier, except for some continuing efforts, e.g., Lake Tanganyika (Coulter and Spigel, 1991), and investigates the interaction of the physical limnology with sediment source and dispersal mechanisms within these basins. These interactions are important to accurately decipher intra- and inter-annual variations in rift-lake sedimentology, that may ultimately be preserved in the sediment record, including sediment flux, sediment reworking, sediment focusing, and biogeochemical processes.

Lake Malawi (Lake Nyasa), the fifth largest lake in the world, is located at the southern end of the East African Rift System (see Johnson, this volume). It is approximately 570 km long by 75 km wide and has a mean and maximum depths of about 270 and 700 m, respectively. Variations in precipitation, wind velocity and air temperature define rainy, austral summer versus windy, austral winter seasons. The hydrological budget is balanced by precipitation, river runoff and evaporation — the Shire River presently contributes about 20% of the water loss (Beadle, 1981). The water column is thermally and chemically stratified with the upper oxygenated portion (above 200 to 250 m) experiencing distinct seasonal variability in temperature, 23 to 30°C, and the lower anoxic portion (below 300 m) experiencing slightly cooler but isothermal conditions (Eccles, 1974). The principle dissolved cations of calcium, magnesium, sodium are electrically balanced by bicarbonate and other minor anions with measured conductivities (κ), alkalinities and pH of approximately 250 μS/cm, 2.40 meq/l and 7.7, respectively (Gonfiantini et al., 1979). Recent seismic profiles and sediment coring reveal a complex history of sedimentation, i.e., widely disparate sedimentary facies occur in close juxtaposition (Crossley, 1984; Johnson and Ng'ang'a, 1990; Scholz et al., 1990).

Lake Turkana is the largest closed-basin lake in the East African Rift System. It is smaller than Malawi, only about 250 km long, averages 30 km wide with a mean and maximum depth of 30 and 100 m, respectively. The local climate is arid, precipitation <200 mm/yr, and hot, annual mean temperatures of 30°C. Hydrologic input is dominated by the Omo River, which provides about 90% of the total budget and drains the Ethiopian Plateau to the north. The lake is moderately saline, 2.5‰, alkaline, 20 meq/l with a pH of 9.2, and frequently well mixed by strong diurnal winds (Yuretich and Cerling, 1983). Principal dissolved ions are sodium, bicarbonate and chloride and conductivity gradients along vertical and horizontal planes, when they exist, are interpreted to reflect the influx of less saline Omo effluent over more saline lake water (Ferguson and Harbott, 1982). Recent seismic profiles and sediment coring reveal a sediment record dominated by detrital input from fluvial (Omo River) and aeolian sources since a late Pleistocene lowstand (Yuretich, 1979; 1986; Johnson et al., 1987; Halfman and Johnson, 1988; Halfman and Hearty, 1990).

METHODS

In Lake Malawi, water column profiles of conductivity, temperature, dissolved oxygen and total suspended solids (TSS) were measured at 28 stations (Fig. 1) during January, 1992, with an internally recording electronic conductivity, temperature and pressure (depth) profiler (CTD) interfaced with an aspirating pump and light transmissometer (SEACAT SBE-19; polarographic O_2 sensor, SBE 23Y; Sea Tech, 25-cm pathlength transmissometer). Deployment rates were about 1 m/s and data recorded every 0.5 seconds. CTD reproducibilities were: temperature 0.01°C, specific conductance 0.14 μS/cm, dissolved oxygen 0.17 mg/l and total suspended solids 0.1 mg/l, and determined by comparing upcast/downcast profiles, multiple cast data, and pre- and post-cruise calibrations (Halfman and Scholz, 1993; Halfman, 1993).

A transmissometer measures the percent light transmitted (T) over a fixed distance (z) and is related to beam attenuation (A, m^{-1} = $-\ln(T)/z$). The total attenuation is the sum of light

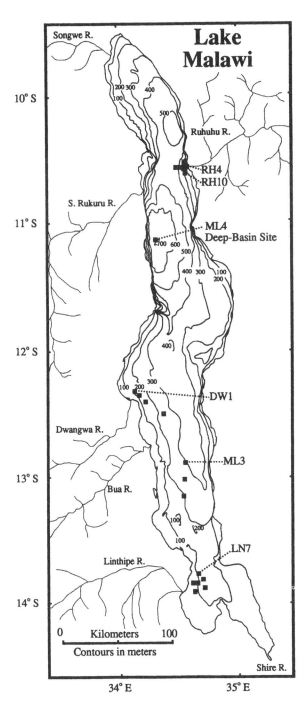

Figure 1. Bathymetric map of Lake Malawi with locations of offshore CTD-transmissometer stations. Nearshore stations with water depths less than 5 m are not plotted for clarity.

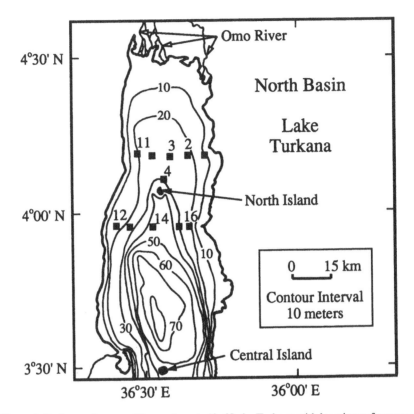

Figure 2. Bathymetric map of the northern half of Lake Turkana with locations of temperature/transmissometer stations.

absorption by water and dissolved substances in the water and light scattering by particles suspended in the water. The absorption of monochromatic light by water is constant over a fixed distance, and absorption by dissolved substances is minimized by using a wavelength of 660 nm. The degree of scattering by particles is affected by particle concentration, size, shape and refractive index. Fortunately, the last three parameters do not change much in a well defined region but variability between lakes may dictate independent TSS versus attenuation calibrations. The linear, least-squares regression of mean TSS values from duplicate filtered water samples and light attenuation at Lake Malawi is (Halfman and Scholz, 1993):

$$\text{TSS [mg/l]} = 1.02(A) - 0.362, \qquad r^2 = 0.98.$$

In Lake Turkana, water temperatures and total suspended solid concentrations were measured at 11 stations at 1-m increments (Fig. 2) during January, 1990, with an electronic temperature profiler (RSVP probe designed by Dr. T. Dillon, OSU) interfaced with a Sea Tech 25-cm pathlength transmissometer. Reproducibility for temperature was 0.1°C based on a laboratory calibration before the cruise and crosschecks between duplicate temperature sensors in the field. TSS concentrations were empirically related to light attenuation in a similar manner, i.e., attenuation data were calibrated to mean TSS concentrations from duplicate water samples filtered through precombusted, preweighed glass-fiber filters (Gel-

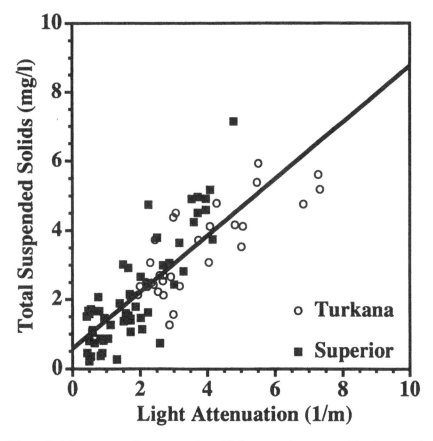

Figure 3. A least-squares linear regression of light attenuation measured by the transmissometer and total suspended solid concentrations from filtered water samples. Data from Lake Turkana (o) are supplemented with data from Lake Superior that used the identical transmissometer (■: Halfman and Johnson, 1989).

man A/E, 1.0 μm nominal pore diameter). The limited number of calibration data for Lake Turkana dictated augmentation with additional data for the identical transmissometer from Lake Superior (Halfman and Johnson, 1989). It assumed that the light scattering properties of Lake Superior clays are similar to Lake Turkana clays. The resultant linear, least-squares regression is (Fig. 3):

$$TSS[mg/l] = 0.818(A) + 0.576, \qquad r^2 = 0.70.$$

Even though comparison of upcast/downcast transmission data suggests data precision of 0.1 mg/l, the standard deviation of the least-squares fit limits the TSS accuracy to 0.8 mg/l. The inaccuracy may be partially related to different size, shape and refractive index of the clay populations between Lakes Turkana and Superior; however, the error is significantly larger than predicted by comparing TSS and attenuation calibrations among the Laurentian Great Lakes (Hawley and Zyren, 1990). Conductivity and dissolved oxygen concentrations were not measured on this cruise but 1970s data collected over a 2-1/2 year period are available (Ferguson and Harbott, 1982).

Specific Conductance (κ25, μS/cm)

Temperature (°C)

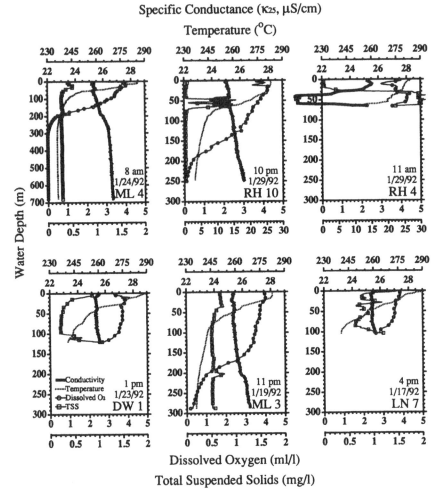

Figure 4. Representative downcast conductivity (κ25), temperature, dissolved oxygen and total suspended solid profiles from Lake Malawi. See Fig. 1 for locations. Note changes in depth (ML 4 site) and total suspended solids scales (RH 4 and RH 10 sites). See Fig. 1 for locations. Redrawn from Halfman (1993).

Station locations in both lakes are skewed to major fluvial systems, an obvious source of suspended sediment. Satellite navigation (GPS) provided site locations to within 15 m according to the manufacturer. Radar fixes were used on the rare occasions satellites were not available.

LAKE MALAWI: 1992 CRUISE

The shape of the temperature, dissolved oxygen and conductivity profiles in Lake Malawi are remarkably uniform throughout the open lake (Fig. 4: Halfman, 1993). Water temperatures generally decrease from above 28°C at the surface to below 23°C at the lake floor with the greatest variability in temperature within the mixolimnion (above the chemocline). The

Monimolimnion Temperatures

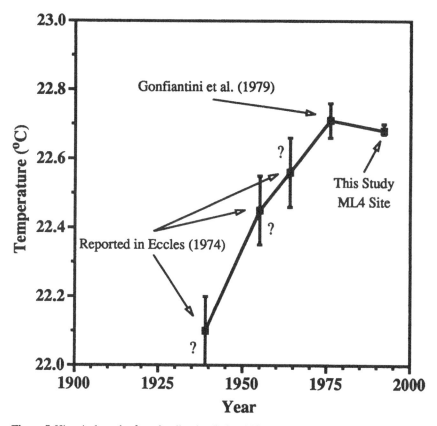

Figure 5. Historical trends of monimolimnion (below 300 m) water temperatures at the deep basin of Lake Malawi. The mean of all the available deep-water temperatures is plotted with error bars representing minimum and maximum temperatures below 300 m and estimated at 0.1°C for the earlier data due to a lack of information.

monimolimnion (below the chemocline) temperatures are approximately 22.7°C, and reveal a thermal inversion, i.e., temperatures cool to 22.66°C at 350 m but then warm to 22.70°C by 690 m.

A thermal inversion in the benthos of Lake Tanganyika is attributed to geothermal inputs (Coulter and Spigel, 1991). The inversion in Lake Malawi may reflect geothermal warming and is consistent with previous suggestions that a historic warming of monimolimnion waters reflects geothermal warming of the bottom waters and not instrument accuracy (Eccles, 1974). However, the 1992 monimolimnion temperatures are inconsistent with the historic trend (Fig. 5); and, the inversion can be explained by adiabatic temperature changes with water depth, i.e., calculated potential temperatures (T_q, adjusted for pressure) cool by 0.04°C from 350 to 690 m. Halfman (1993) suggests that geothermal inputs to the monimolimnion, if they exist, are offset by occasional introduction of cooler water from, e.g., turbidity events, fluvial discharge, nearshore cooling, or evaporitic cooling during austral winter.

Dissolved oxygen concentrations decrease with water depth from over 80% saturation to anoxic conditions below 200 to 300 m. The distributions are consistent with expected inputs from the atmosphere across the air/water interface and net-photosynthesis in the euphotic zone balanced by losses from net-respiration and geochemical–oxidation reactions lower in the water column (Fig. 4). The exact depth for the oxic/anoxic transition is difficult to pinpoint due to the slow response time and possible poisoning of polarographic sensors with other dissolved gasses. A persistent turbid layer just above 200 m, a 50% increase in conductivity between 150 and 250 m, and only sporadic reports of oxygen from depths below 250 m in the literature suggest that the oxic/anoxic transition is closer to 200 m.

Specific conductance (κ_{25}) of the water column typically increases from 258 μS/cm at the surface to just above 270 μS/cm at 690 m (Fig. 4). The conductivity data over the entire water column are slightly larger (5 to 10 μS/cm) and the surface to bottom water increase is slightly smaller (5 μS/cm) than previous reports (Gonfiantini et al., 1979). It suggests that the lake is getting more saline with time, but the trend may be an artifact of the season or instrumentation. The increase in conductivity (κ_{25}) with water depth is proportional to the increase in salinity defined by the relative distribution of major ions. Over 50% of the increase is between 150 and 250 m. Previously, temperature was postulated as the control on the relative density of the water column. Calculations of the relative increase in density (both $\rho_{0,T,0}$ and $\rho_{S,T,0}$) between 150 and 250 m using the conductivity, temperature and pressure data from the deep-basin station (site ML 4), the low-salinity equations of Hill et al. (1986), and freshwater PVT equations of Chen and Millero (1977, 1986) indicate that both temperature and salinity influence density stratification of the water column to the same order of magnitude, especially when dissolved silica (an uncharged species thus undetected by the CTD) is added to the salinity structure (260 to 266 μS/cm, 23.13 to 22.74°C, 20 to 270 μM SiO_2, respectively). If the salinity structure persists throughout the year, it may represent the only barrier to vertical mixing across the chemocline during austral winter (see Wüest et al., this volume).

Stations offshore of the Ruhuhu River reveal lower conductivities than typically detected elsewhere in the lake, as low as 213 μS/cm. The low values coexist with very turbid water (see below), are confined to water depths between 30 and 65 m, and gradually increase from the lowest values at the mouth of the Ruhuhu to values more typical of open-water conditions 1 km to the north, 5 km to the south and 10 km to the west of the river mouth.

Total suspended solid (TSS) profiles reveal less similarity across Lake Malawi (Fig. 4: Halfman and Scholz, 1993). TSS concentrations are typically 0.1 to 0.5 mg/l but are occasionally larger, i.e., to concentrations above the upper detection limit of the transmissometer (25 mg/l). Across the lake, site averaged TSS values segregate into two populations: nearshore sites with TSS concentrations >1 mg/l, and offshore sites with TSS concentrations typically <0.5 mg/l. At each delta, the largest TSS concentrations are detected near the respective river mouth. Offshore profiles typically reveal elevated TSS concentrations at or near the surface, at the lake floor, at or just above 200 m, and a mid-depth plume offshore of the Ruhuhu River. Probable compositions and/or sources include fluvial inputs, biological productivity, resuspension events and biogeochemical processes at the oxic/anoxic boundary but samples are lacking for confirmation. Curiously, benthic turbidity layers are absent at all of the sites with water depths below 200 m, i.e., deep enough to include monimolimnetic waters.

The largest TSS concentrations (>25 mg/l) were detected offshore of the Ruhuhu between 30 and 65 m. The layer is interpreted as a plume of fluvial sediments because it coexists with low-salinity water, TSS concentrations decrease with increasing distance from the river mouth, and the survey was just after a 2-day monsoon style rainstorm. A mid-depth injection suggests that conductivity is not as important as turbidity and possibly temperature to determine the relative density and relative importance of hypopycnal (lake surface), intermediate (along density contrasts within the water column) and hyperpycnal (lake floor) pathways for sediment dispersal. Estimates of the sediment burden to the delta fan due to this event range from 0.02 to 1.0 mg/cm^2/day and bracket the flux of sediments to the lake floor (0.05–0.1 mg/cm^2/day; Johnson and Ng'ang'a, 1990; Owen et al., 1990, Finney and Johnson, 1991; Halfman and Scholz, 1993).

LAKE TURKANA: 1990 CRUISE

Water temperatures at the 11 stations in the survey of Lake Turkana gradually decrease with water depth from just over 28°C at 1 m below the surface to just below 27°C at 40 m (Fig. 6). The largest change in temperature with depth is within the upper 1 to 3 m (0.2°C/m), and associated with profiles taken during relatively calm and warm afternoons. Temperatures are more uniform with depth along the western margin of the lake (Stations 11, 12 and 13H). Site averaged temperatures reveal a minor decrease in temperature from north to south (27.3 to 26.8°C); however, the trend may reflect increasing water depth to the south because the southern stations are deeper. Ferguson and Harbott (1982) report that seasonal changes in water temperature are less than 2°C with surface waters ranging from 27.2 to 29.4°C and bottom water temperatures from 25.5 to 26.4°C. Their data also reveal that dissolved oxygen concentrations are nearly uniform with water depth with near 80 to 100% saturation levels at depth in the lake. The results are consistent with their earlier conclusions that the entire water column mixes frequently through the year.

Total suspended solid concentrations at the 11 stations range from 2.0 to 6.8 mg/l (Fig. 6). TSS profiles reveal either a more turbid surface layer that overlies less turbid water with TSS concentrations decreasing by 1.5 to 3.2 mg/l at stations 2, 3 and 4 or homogeneous turbidities with water depth (max − min ≤ 1 mg/l) at the other stations. Stations 2, 3 and 4 are located in the northeastern quarter of the survey sites and were sampled 3 days before the other stations with a strong wind event between cruises. Site averaged turbidities reveal decreasing values from east to west, e.g., 3.1 to 6.8 mg/l along the northern transect, and decreasing values from north to south, e.g., from about 4.6 to 2.7 along the mid-lake transect. Additional turbidity data are not available in the literature for direct comparison. However, light extinction coefficients (ε, m^{-1}), an indirect indicator of water turbidity, decrease with increasing distance from the Omo River (over 20 m^{-1}) to clear-water levels (about 1 m^{-1}) approximately 80 km south of the delta (Ferguson and Harbott, 1982).

The spatial character of the TSS profiles suggests that suspended sediments from the Omo River are mixed with the rest of the lake along top to bottom, west to east and north to south gradients. Surface turbid layers at stations 2, 3 and 4 are more consistent with hypopycnal (surface) dispersal of fluvial sediments than resuspension of bottom sediments by wind-driven waves. Similar spatial trends are observed in the 1970s conductivity data (Ferguson and Harbott, 1982). Specific conductance (κ$_{25}$) increases from 80 μS/cm in the river to about 3500 μS/cm just to the south of North Island. Farther south, conductivities are remarkably similar along horizontal and vertical transects with offshore conductivities up to 3600

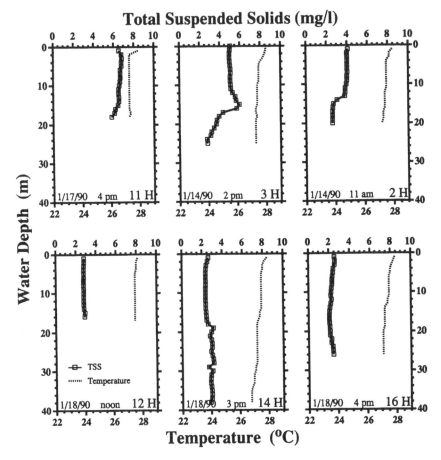

Figure 6. Representative temperature and total suspended solid profiles from Lake Turkana. See Fig. 2 for locations.

µS/cm. The freshwater plume is typically restricted to the upper and western half the lake during high discharge events in late August to early October but is more diffuse over the northern part of the lake during other times of the year. Satellite photos also reveal turbid water extending from the Omo to and beyond North Island at or near the surface of the lake and typically restricted to the western half of the lake (Yuretich, 1979). Analyses of suspended sediment and sediment cores reveal a spatially consistent Omo signature as well (Yuretich, 1979; 1986; Halfman and Johnson, 1988; Halfman and Hearty, 1990; Halfman et al., 1994). The temperature profiles do not differentiate the existence of Omo River effluent. Apparently, neither temperature nor turbidity control the distributional pathway of the Omo plume, rather the relative salinities, circulation patterns set by strong southeasterly winds, and possibly Coriolis acceleration are the primary factors.

DISCUSSION AND TESTABLE HYPOTHESES

Temperature profiles are strikingly different between Lakes Malawi and Turkana. A wide range of temperatures in the water column of Lake Malawi contrasts with a relatively small range in Lake Turkana. Previous reports suggested that the strong southeasterly trade winds at Lake Turkana completely mix the water column (Beadle, 1981; Ferguson and Harbott, 1982). Yet, averaged 24-hr wind runs are smaller at Lake Turkana (300 to 450 km/day: Ferguson and Harbott, 1982) than the Mweras (southeast trades) at Lake Malawi (up to 900 km/day: Eccles, 1974). The discrepancy is reduced when the diurnal nature of the winds at Lake Turkana are taken into account. Mixing depths in Malawi, defined by the depth of the thermocline, range from approximately 80 up to 200 m depending on the season. Similar depths are reported for Lake Tanganyika (Beadle, 1981; Coulter and Spigel, 1991). These depths are equal to or extend beyond the maximum depth of Lake Turkana and suggest that the available wind energy is probably sufficient to mix the lake. However, proximity of the lake floor in Lake Turkana should decrease the effectiveness of wind induced physical mixing.

The data do not exclude density induced, convective mixing by seasonal or nocturnal cooling of the surface waters (Talling, 1969; MacIntyre and Melack, 1982). Annual air temperatures rarely vary by more than 3°C at Lake Turkana but vary by 15°C at the southern end of Lake Malawi. These relative changes are reflected in the extent of the seasonal changes in water temperature at each lake with the thermal structure of the mixolimnion at Lake Malawi changing considerably during the year. Perhaps the small seasonal change in air temperature at Lake Turkana is are sufficient to augment the ability of the winds to mix the entire water column during the cooler season.

Average maximum and minimum daily air temperatures at Lake Turkana are 32.5 ± 0.2°C and 26.0 ± 0.2°C over the 2-1/2 year survey, that is primarily due to a daily temperature cycle of approximately 6°C through out the year (Ferguson and Harbott, 1982). Surface water temperatures (<1 m depth) also reveal a daily temperature cycle by as much as 2.5°C (Ferguson and Harbott, 1982). The coolest water temperatures coincided with the pre-dawn hours and occasionally isothermal conditions through out the water column. It suggests that the water column is mixed by diurnal changes in air temperature. The relative contribution of air temperature or wind induced mixing of the water column is difficult to ascertain with the available data because the winds may induce evaporitic cooling of the surface waters as well as physical mix the water column, and are strongest during late evening and early morning when air temperatures are lowest.

Distributional pathways of fluvial sediments are different between Lakes Malawi and Turkana, as well. First, the Omo River influences water turbidity significantly farther into the lake than the systems surveyed in Lake Malawi. The relationship is surprizing because profiles of the Ruhuhu plume were linked to a monsoon style rainfall event and the estimated sediment burden to the delta fan was a significant portion of the annual load; whereas, the Turkana survey was during low flow season of the Omo River. Second, hypopycnal dispersal is active in Lake Turkana; whereas, hypopycnal, intermediate and hyperpycnal pathways are active in Lake Malawi. Both observations can be related to the salinity contrasts between the river effluent and lake. The contrast in salinity between the Omo River (80 μS/cm) and Lake Turkana (3500 μS/cm) is significantly larger that the contrast between the Ruhuhu River (RH 4, 213 μS/cm) and Lake Malawi (258 μS/cm). At Lake Turkana, the significantly lower salinity of the Omo River compared to the lake retards the assimilation of Omo effluent into the lake and dictates hypopycnal dispersal irregardless of the contrast in sediment load or water temperature. At Lake Malawi, the intermediate dispersal of Ruhuhu effluent indicates that the sediment load and possibly cooler temperatures of the effluent counteracts the buoyant tendencies of the less saline effluent. Clearly, climate and geomorphology which dictates the temperature and salinity of the rivers and lakes also impacts sedimentation patterns.

The extent that limnological conditions impact the history of sedimentation is unclear at the present time. Seismic profiles and sediment cores suggest that hyperpycnal dispersal (turbidity currents) of fluvial inputs impact sedimentation in Lake Malawi, even in the deepest basin of the lake, but the total suspended solid profiles indicate intermediate dispersal is important as well. In contrast, turbidity events are not observed in seismic profiles, sediment cores or water column profiles from Lake Turkana. Is this a consequence of contrasting limnological characteristics as suggested above, or is drainage basin size, availability of sediment, geomorphology of the basin or other factors more important? Additional data are necessary to quantify these differences, the relationship between sediment discharge and climate change, and the extent that these changes are decipherable in sediment cores recovered from the lakes.

CONCLUSIONS

Preliminary CTD-transmissometer surveys from two end-member lakes in the East African Rift System reveal a range of limnological conditions from well-mixed but saline to stratified but freshwater systems. In Lake Malawi, a temperature inversion in the monimolimnion can be explained by adiabatic changes and salinity gradients across the chemocline contribute to the interannual stratification of the lake and anoxic waters below the chemocline. In Lake Turkana, daily cycles in wind strength and air temperature are probable mechanisms to promote mixing of the water column, which in turn maintains very small temperature and dissolved oxygen contrasts with water depth. Hypopycnal dispersal of fluvial effluents in Lake Turkana is controlled by salinity gradients but the variety of dispersal mechanism in Lake Malawi indicates that small changes in sediment load, water temperature or salinity may be sufficient to dictate the exact dispersal path. The reconnaissance surveys suggest that additional investigations are required to highlight the differences between the various limnological end-members to better discern hydrologic and other environmental conditions in the past.

ACKNOWLEDGMENTS

Thanks are extended to Thomas C. Johnson, Chris Scholz, Jim McGill and the captain and crew of the M/V Timba, operated by the Malawi Department of Surveys, for their invaluable assistance in the field at Lake Malawi; and, to the Governments of Malawi and Tanzania for research permission. Financial support for the 1992 Malawi research was provided by the Jesse H. Jones Faculty Research Program and Department of Civil Engineering and Geological Sciences at the University of Notre Dame. Thanks are extended to Bruce Finney and Paul Hearty for assistance in the field; and the Kenyan Government and Dr. E. Odada for their support in our continued research efforts at Lake Turkana. The 1990 Turkana research was funded by the National Science Foundation (ATM 91-05842). Finally, I thank Sally

MacIntyre and an anonymous reviewer for their constructive comments on an earlier draft of the manuscript.

REFERENCES

Beadle, L.C., 1981, The inland waters of tropical Africa: 2nd ed. New York, Longman Group, 475 pp.

Chen, C.T., and Millero, F.J., 1977, The use and misuse of pure water PVT properties for lake waters: Nature, v. 226, pp. 707–708.

Chen, C.T., and Millero, F.J., 1986, Precise thermodynamic properties for natural waters covering only the limnological range: Limnology and Oceanography, v. 31, pp. 657–662.

Coulter, G.W., and Spigel, R.H., 1991, Hydrodynamics, in Coulter, G.W., ed., Lake Tanganyika and Its Life: Oxford, Oxford University Press, pp. 49–75.

Crossley, R., 1984, Controls on sedimentation in Malawi rift valley, central Africa: Sedimentary Geology, v. 40, pp. 33–50.

Eccles, D.H., 1974, An outline of the physical limnology of Lake Malawi (Lake Nyasa): Limnology and Oceanography, v. 19, pp. 730–742.

Ferguson, A.D.J., and Harbott, B.J., 1982, Geographical, physical and chemical aspects of Lake Turkana, in Hopson, A.J., ed., Lake Turkana: A Report of on the Findings of the Lake Turkana Project, 1972–1975: London, Overseas Development Administration, pp. 1–107.

Finney, B.P., and Johnson, T.C., 1991, Sedimentation in Lake Malawi (east Africa) during the past 10,000 years: a continuous paleoclimatic record from the southern tropics: Palaeogeography, Palaeoclimatology, Palaeoecology, v. 85, pp. 351–366.

Gonfiantini, R., Zuppi, G.M., Eccles, D.H., and Ferro, W., 1979, Isotope investigations of Lake Malawi, in Isotopes in Lake Studies, International Atomic Energy Agency Panel Proceeding Series STI/PUB/511.

Halfman, B., and Johnson, T.C., 1989, Nepheloid layers in Lake Superior: Journal of Great Lakes Research, v. 15, pp. 15–25.

Halfman, J.D., 1993, Water column stability from modern CTD data, Lake Malawi, Africa: Journal of Great Lakes Research, v. 19, pp. 512–520.

Halfman, J.D., and Hearty, P.J., 1990, Cyclical sedimentation in Lake Turkana, Kenya, in Katz, B.J., ed., Lacustrine basin exploration: Case studies and modern analogs: Tulsa, OK, American Association of Petroleum Geologists Memoir #50, pp. 187–195.

Halfman, J.D., and Johnson, T.C., 1988, High-resolution record of cyclic climatic change during the past 4 ka from Lake Turkana, Kenya: Geology, v. 16, pp. 496–500.

Halfman, J.D., Johnson, T.C., and Finney, B.P., 1994, New AMS dates, stratigraphic correlations and decadal climatic cycles for the past 4 ka at Lake Turkana, Kenya: Palaeogeography, Palaeoclimatology, Palaeoecology, v. 111, pp. 83–98.

Halfman, J.D., and Scholz, C.A., 1993, Suspended sediments in Lake Malawi, Africa: A reconnaissance study: Journal of Great Lakes Research, v. 19, pp. 499–511.

Hawley, N., and J.E. Zyren, 1990, Transparency calibrations for Lake St. Clair and Lake Michigan: Journal of Great Lakes Research, v. 16, pp. 113–120.

Hill, K.D., Dauphinee, T.M., and Woods, D.J., 1986, The extension of the practical salinity scale 1978 to low salinities: Institute of Electrical and Electronic Engineers Journal of Oceanic Engineering, v. OE-11, pp. 109–112.

Johnson, T.C., 1996, Sedimentary processes and signals of past climatic change in the large lakes of the East African Rift Valley, in Johnson, T.C., and Odada, E.O., eds., The Limnology, Climatology and Paleoclimatology of the East African Lakes: Toronto, Gordon and Breach, pp. 367–412.

Johnson, T.C., Halfman, J.D., Rosendahl, B.R., and Lister, G.S., 1987, Climatic and tectonic effects on sedimentation in a rift-valley lake: Evidence from Lake Turkana, Kenya: Geological Society of America Bulletin, v. 98, pp. 439–447.

Johnson, T.C., and Ng'ang'a, P., 1990, Reflections on a rift lake, *in* Katz, B.J., ed., Lacustrine basin exploration: Case studies and modern analogs: Tulsa, OK, American Association of Petroleum Geologists Memoir #50, pp. 113–135.

Johnson, T.C., Talbot, M.R., Kelts, K., Cohen, A.S., Lehman, J.T., Livingstone, D.A., Odada, E.D., Tambala, A.F., McGill, J., Arquit, A., and Tiercelin, J.-J., 1990, IDEAL An International Decade for the East African Lakes: Technical Report, Duke University Marine Laboratory, Beaufort, NC, 40 pp.

MacIntyre, S., and Melack, J. M., Meromixis in an equatorial African soda lake: Limnology and Oceanography, v. 27, pp. 595–609.

Owen, R.B., Crossley, R., Johnson, T.C., Tweddle, D., Kornfield, I., Davison, S., Eccles, D. H., and Engstrom, D.E., 1990, Major low levels of Lake Malawi and implications for speciation rates in cichlid fishes: Proceedings of the Royal Society of London, v. 240, pp. 519.

Scholz, C.A., Rosendahl, B.R., and Scott, D.L., 1990, Development of coarse-grained facies in lacustrine rift basins: examples from east Africa: Geology, v. 18, pp. 140–144.

Talling, J.F., 1969, The incidence of vertical mixing and some biological and chemical consequences in tropical African Lakes: Verhandlungen der Internationalen Vereiningung Limnologie, v. 17, pp. 998–1012.

Wüest, A., Piepke, G., and Halfman, J. 1996, The role of dissolved solids in density stratification of Lake Malawi, *in* Johnson, T.C., and Odada, E.O., eds., The Limnology, Climatology and Paleoclimatology of the East African Lakes: Toronto, Gordon and Breach, pp. 183–202.

Yuretich, R.F., 1979, Modern sediments and sedimentary processes in Lake Rudolf (Lake Turkana) eastern rift valley, Kenya: Sedimentology, v. 26, pp. 313–331.

Yuretich, R.F., 1986, Controls on the composition of modern sediments, Lake Turkana, Kenya, *in* Frostick, L.E., et al., eds., Sedimentation in the African Rifts: London, Geological Society of London Special Publication #25, pp. 141–152.

Yuretich, R.F., and Cerling, T.E., 1983, Hydrogeochemistry of Lake Turkana, Kenya: Mass balance and mineral reactions in an alkaline lake: Geochemica et Cosmochemica Acta, v. 47, pp. 1099–1109.

Combined Effects of Dissolved Solids and Temperature on the Density Stratification of Lake Malawi

A. WÜEST and G. PIEPKE   *Swiss Federal Institute for Environmental Science and Technology (EAWAG) and Swiss Federal Institute of Technology (ETH), Environmental Physics, EAWAG, CH-8600 Dübendorf, Switzerland.*

J.D. HALFMAN   *Department of Geoscience, Hobart and William Smith Colleges, Geneva, New York, United States*

Abstract — The vertical density structure in a lake depends on temperature and chemical gradients in the water column. The relative importance of dissolved constituents for water column stability often increases in deep hypolimnia, such as those of the deep rift-valley lakes of East Africa, where thermal stability is usually low.

In this paper, we quantify the water column stability in the deep basin of Lake Malawi as a function of in-situ conductivity, temperature and silicic acid concentration. The salinity–conductivity relationship and the coefficient of haline contraction for Lake Malawi water are determined on the basis of its physicochemical properties. The analysis of water column stability shows that: 1) temperature gradients are nearly adiabatic, implying thermal stability to be low in the deep hypolimnion (below 350 m); 2) dissolved solids contribute up to 10% to the stability throughout most of the hypolimnion; and 3) the non-ionic constituents (silicic acid and dissolved gases such as CO_2, CH_4 and H_2S) cannot be neglected in stability calculations in the hypolimnion. During periods of low stratification in the austral winter, dissolved solids help maintain meromixis. We suggest that cool water occasionally introduced into the hypolimnion offsets geothermal warming and heat diffusion from above.

1. INTRODUCTION

Temperature, oxygen, [3]He and ^{3}H profiles from Lakes Malawi and Tanganyika suggest that both rift-valley lakes are meromictic; i.e., only the upper water column (the upper 100 to 250 m) mixes during periods of low overall stratification (Eccles, 1974; Coulter and Spigel, 1991; Craig, 1974). In addition to the thermal stratification, biogenically induced chemical gradients may contribute to the stability of the water column, inhibiting deep-water renewal (Hutchinson, 1975). Even Lake Victoria, which is much shallower than the other large East African lakes, has shown signs of incomplete vertical mixing in recent years (Hecky, 1993). The two most likely causes of this are changes in meteorological forcing (warmer or less windy conditions during the dry season, the period of deep mixing) and enhanced primary productivity. The eutrophication process, which has been suggested to operate in Lake Victoria (Hecky, 1993), increases the rate of mineralization and the release of dissolved solids to the deep water. The subsequent accumulation of dissolved solids may result in a biogenically induced density stratification, which will reduce the vertical exchange of hypolimnetic waters with the surface layer of the lake. Even though the biogeochemical boundary conditions are not well known, dissolved constituents such as solids or gases may play a significant role in the stability of the water column of the East African lakes, especially during the slightly cooler and/or windier dry season when thermal stratification is minimal.

Assuming the relative proportions of the major ions to be constant, the relationship between in-situ conductivity and salinity (Cox et al., 1967) and the dependence of density on temperature and salinity (UNESCO, 1981) have been empirically derived and are well established for seawater. In contrast, the relative proportions of the major ions present in lacustrine systems vary from lake to lake and sometimes even within a lake (especially vertically). A separate relationship between salinity and in-situ conductivity must therefore be established for each lake, and sometimes even for separate water masses (i.e., epi-, meta- and hypolimnion) within a lake.

The goal of this paper is to outline a general and easy-to-use procedure to calculate the salinity and static stability of low conductivity waters based on conventional CTD (Conductivity, Temperature and Depth) data and data on major ions, and to apply the derived relationships to a CTD profile (Seacat SBE-19) taken in the deepest basin of Lake Malawi in January 1992 (Halfman, 1993; Halfman and Scholz, 1993).

Based on the physicochemical properties of a specific freshwater lake, we derive the temperature-dependent salinity and conductivity relationship and determine the coefficient of haline contraction at 25°C and atmospheric pressure. This coefficient will be extrapolated over the entire temperature, pressure and salinity range in Lake Malawi, applying the conventional oceanographic equation of state for low salinity waters (Chen and Millero, 1986). We therefore use the term salinity to designate the mass of total dissolved solids per unit mass of water (g/kg = ‰) instead of per unit volume of water [g/m^3].

Lake Malawi, the world's fifth largest lake by volume, is approximately 570 km long and 75 km wide, covers a surface area of about 29,000 km^2 and attains a maximum depth of 700 m. The Shire River outflow currently accounts for only about 20% of the total water loss; at times this outflow is even dry (Beadle, 1981; Scholz and Rosendahl, 1988). Seasonal changes in precipitation, wind, and air temperature are responsible for variations in surface water temperatures ranging from about 23°C during July/August (austral winter) to about 29°C during February/March (austral summer) and for corresponding variations in thermal density stratification (Eccles, 1974; Beadle, 1981). Persistent anoxic conditions in the hypolimnion indicate that convective vertical mixing induced by surface cooling does not extend into deep hypolimnetic waters even during the austral winter.

2. DETERMINATION OF SALINITY

The following two assumptions are necessary to calculate salinity distributions based on conductivity data alone: 1) the relative proportions of the major ions within the lake are constant, and 2) the conductivity data represent all the dissolved solids; i.e., the major dissolved constituents are ionic species and will therefore be detected by the conductivity cell of the CTD. In Lake Malawi, the first assumption is reasonably well founded. The second assumption, however, is not valid as the vertical salinity gradient of the (non-ionic) silicic acid is of the same order of magnitude as the salinity gradient of the ionic dissolved species. Consequently, the total salinity, S, has to be considered as the sum

$$S = S_c + S_o \quad [‰] \tag{1}$$

of the ionic and non-ionic contributions to salinity, S_c and S_o, respectively (McManus et al., 1992). S_c is a function of all the ionic species concentrations, $c_1, ..., c_n$, and is related to the in-situ conductivity, $\kappa_T^p = \kappa(c_1, ..., c_n, p, T)$. S_c can consequently be measured by using a CTD, whereas S_o must be established by chemical analysis. An expression for the conductivity-related term, S_c, is presented below. The second term, S_o, will be discussed later at the end of §2. It is evident that both these parameters are functions of depth. For the sake of simplicity, this depth dependence will not usually be expressed explicitly in the formulas which follow; i.e., we will usually write "f" instead of "$f(z)$."

2.1. Conductivity as a Function of Temperature and Depth

In-situ conductivity is a function of the concentration of the dissolved ions, temperature and pressure. For dilute solutions, such as those contained in freshwater lakes, the in-situ conductivity can be approximated by the sum of the conductivities of the individual species:

$$\kappa_T^p = \kappa(c_1, ..., c_n, p, T) = \sum_{i=1}^{n} f_i \times \lambda_i^\infty \times Z_i \times c_i \quad [\mu S \; cm^{-1}], \tag{2}$$

where c_i is the concentration [mol/l], λ_i^∞ the conductance per unit charge at infinite dilution $[\mu S \; cm^{-1} \; (eq/l)^{-1}]$, Z_i the charge [–], and f_i the reduction coefficient [–] of the ion i (values for Lake Malawi are listed in Table A3). λ_i^∞ is a function of pressure, p, and temperature, T, i.e., $\lambda_i^\infty = \lambda_i^\infty(p,T)$ (Table A1), and f_i is a function of the ionic composition, i.e., $f_i = f_i(c_1, ..., c_n)$. A parameterization approximating f_i as a function of total ionic charge is provided in Appendix A2.

In order to establish a salinity-dependent quantity, it is common in lacustrine systems to convert the temperature-dependent in-situ conductivity, κ_T^p, to a temperature-independent reference conductivity, κ_{20}^0, by normalizing the in-situ conductivity from the in-situ temperature, T, and pressure, p, to a reference temperature $T = 20°C$ (Sorensen and Glass, 1987; McManus et al., 1992; Wüest et al., 1992) and to atmospheric pressure, which is set by convention to $p = 0$. For low-salinity water, this conversion can be approximated by:

$$\kappa_T^p = f_p \times f_{T,Comp} \times \kappa_{20}^0 \quad [\mu S \; cm^{-1}], \tag{3}$$

where $f_p = f_p(p,T)$ is the temperature-dependent pressure correction (relative to $p = 0$):

$$f_p = \kappa_T^p/\kappa_T^0 \quad [-], \tag{4a}$$

and $f_{T,Comp} = f_{T,Comp}(T)$ is the temperature correction (relative to the reference temperature $T = 20°C$) at atmospheric pressure ($p = 0$):

$$f_{T,Comp} = \kappa_T^0/\kappa_{20}^0 \quad [-]. \tag{4b}$$

Since the temperature dependence of the conductivity varies with the ionic species composition and the total number of charges, f_T depends on the relative proportions of the ions as well as on the absolute concentration. To stress this fact, f_T is labeled with the subscript "Comp."

The temperature-dependent pressure correction, f_p, accounts for the increase in conductivity due to the compression of water at pressure p. Since f_p is not known for specific ionic compositions (values of λ_i^∞ are available only for $p = 0$: Table A1), we use oceanic data from

Bradshaw and Schleicher (1965), extrapolated to zero salinity, and assume the following linear pressure dependence:

$$f_p = 1 + \gamma \times p \quad [-]. \tag{4c}$$

A second-order polynomial fit to Bradshaw and Schleicher's (1965) six temperature data points yields a temperature-dependent coefficient $\gamma(T)$ for the pressure range of 0 to 1700 dbar of

$$\gamma(T) = 10^{-5} \times (1.856 - 0.05601 \, [\text{dbar}^{-1} \, °\text{C}^{-1}] \times T$$
$$+ 0.0007 \, [\text{dbar}^{-1} \, °\text{C}^{-2}] \times T^2) \quad [\text{dbar}^{-1}]. \tag{4d}$$

γ changes by a factor of approximately 2 in the typical temperature range of 4 to 25°C found in lakes. In Lake Malawi, f_p varies from 1.0 at the surface to 1.0066 at the maximum depth of 700 m (with $T = 22.6°C$).

For a given water mass, the temperature correction, $f_{T,\text{Comp}}$, is determined from (2) and (4b). Numerator, κ_T^0, and denominator, κ_{20}^0, are calculated using (2) based on an ionic composition determined by chemical analysis and the temperature dependent equivalent conductance of the ions (compiled in Tables A1 and A2 from laboratory experiments). The quotient on the right hand side of (4b) is calculated for different temperatures and a third-order polynomial is fitted over the relevant temperature range of 20 to 30°C.

Since the ionic concentration in Lake Malawi increases slightly with depth (especially between 50 and 300 m; Table A3) we carried out this procedure for the water of the epilimnion, metalimnion and hypolimnion separately. For the major ion concentrations of the three water masses we used the 1976 values determined by Gonfiantini et al. (1979) (Table A3). The three expressions for $f_{T,\text{Comp}}$ resulting from this procedure are:

$$f_{T,\text{Epi}}(T) = 0.60448 + 0.016481 \, [°\text{C}^{-1}] \times T + 1.9942 \times 10^{-4} \, [°\text{C}^{-2}] \times T^2$$
$$- 1.7353 \times 10^{-6} \, [°\text{C}^{-3}] \times T^3,$$

$$f_{T,\text{Meta}}(T) = 0.60128 + 0.016964 \, [°\text{C}^{-1}] \times T + 1.7648 \times 10^{-4} \, [°\text{C}^{-2}] \times T^2$$
$$- 1.3955 \times 10^{-6} \, [°\text{C}^{-3}] \times T^3,$$

$$f_{T,\text{Hypo}}(T) = 0.59715 + 0.017501 \, [°\text{C}^{-1}] \times T + 1.5410 \times 10^{-4} \, [°\text{C}^{-2}] \times T^2$$
$$- 1.1023 \times 10^{-6} \, [°\text{C}^{-3}] \times T^3. \tag{5a}$$

In Fig. 1, the conductivity values calculated for epilimnetic, metalimnetic and hypolimnetic water masses using (2) are compared to in-situ conductivities measured by the CTD. The calculated conductivities (lines in Fig. 1) agree reasonably well with the continuous CTD data. Slight deviations from the lines representing average concentrations (Table A3) are caused by the gradual vertical changes in conductivity. The close agreement in the metalimnion and hypolimnion indicates that bottom-water salinities have not changed significantly since 1976. In-situ conductivities in the upper 50 m of the water column follow a line that parallels the calculated conductivity line for the 1976 epilimnion data. The parallel offset suggests that since 1976 the epilimnion has increased in salinity by about 3%. Even though 3% lies within the margin of error for water analyses, a small increase seems reasonable, as evaporation accounts for 80% of the water loss from the lake (Beadle, 1981). Due to this delicate hydrological balance, the epilimnetic salinity is subject to seasonal variations of a few percent (Eccles, 1974).

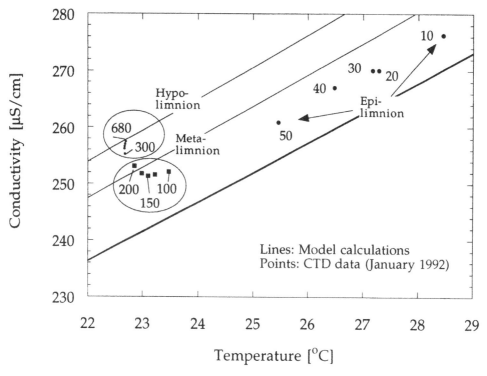

Figure 1. Conductivity as a function of temperature (eq. (2) for $p = 0$) based on dissolved ion concentrations in the hypolimnion (upper line), metalimnion (middle line) and epilimnion (lower line) of Lake Malawi as determined by Gonfiantini et al. (1979) by chemical analysis, and modern CTD data from Halfman (1993) (points). Both data sets are from the 700-m deep basin. Numbers indicate depth in m.

The $f_{T,Comp}$ relationships for each water mass are very similar: differences in $f_{T,Comp}$ are smaller than 1.2×10^{-4}. Based on the average composition of the major ions, a composite curve for Lake Malawi will therefore be used in this paper:

$$f_{T,Malawi}(T) = 0.60097 + 0.016982 \, [°C^{-1}] \times T + 1.7667 \times 10^{-4} \, [°C^{-2}] \times T^2$$
$$- 1.4110 \times 10^{-6} \, [°C^{-3}] \times T^3. \tag{5b}$$

Finally, the reference conductivity, κ_{20}^0, is calculated from in-situ conductivities by rearranging (3):

$$\kappa_{20}^0 = \frac{1}{f_p \times f_{T,Comp}} \times \kappa_T^p \quad [\mu S \, cm^{-1}]. \tag{6}$$

2.2. Salinity as a Function of Conductivity

The salinity due to the ionic species, S_c (eq. (1)), defined as the total mass of ionic dissolved solids per unit mass of water, is given by

$$S_c = \frac{1}{\rho} \sum_i M_i \times c_i \quad [g \, kg^{-1} = ‰], \tag{7}$$

where M_i is the molar mass [g/mol] and c_i the molar concentration [mol/m^3], respectively, of ionic species i, and ρ is the density of water [kg/m^3]. The reference conductivity, κ_{20}^0, is related to the salinity of the ionic dissolved solids, S_c, by simultaneously solving (2) and (7) at $T = 20°C$ for the ionic composition of the aquatic system. We employ the parameterization

$$S_c = g_{S,\text{Comp}} \times \kappa_{20}^0 \quad [\text{‰}], \tag{8}$$

where $g_{S,\text{Comp}}$ [‰ (μS/cm)$^{-1}$] is the factor relating the reference conductivity to the salinity of the ionic species. With the same argument as for f_T (eq. (4b)), g_S depends on the ionic composition of the water. We estimated $g_{S,\text{Malawi}}$ for average Lake Malawi water by fitting S_c/κ_{20}^0 ((7) and (2)) to a second-order polynomial in κ_{20}^0 (eq. (2)) over the present salinity range of the lake water:

$$g_{S,\text{Malawi}}(\kappa_{20}^0) = 0.873 \times 10^{-3} \ [\text{‰ } (\mu S/cm)^{-1}] + 1.861 \times 10^{-7} \ [\text{‰ } (\mu S/cm)^{-2}] \times \kappa_{20}^0$$
$$- 1.288 \times 10^{-10} \ [\text{‰ } (\mu S/cm)^{-3}] \times (\kappa_{20}^0)^2 \quad [\text{‰ } (\mu S/cm)^{-1}]. \tag{9}$$

Finally, combining (6) and (8), the conductivity-related salinity, S_c, can be expressed as a function of in-situ conductivity measurements, κ_T^p, as follows:

$$S_c = \frac{g_{S,\text{Comp}}}{f_p \times f_{T,\text{Comp}}} \times \kappa_T^p. \quad [\text{‰}] \tag{10}$$

As explained above, the functions $g_{S,\text{Malawi}}$ and $f_{T,\text{Malawi}}$ are specific to the ionic composition of Lake Malawi and are therefore unique to this lake. However, the recipe shown here provides a method determining g_S and f_T for any low-salinity lake.

2.3. Salinity Due to the Non-Ionic Species

The salinity due to the non-ionic species, S_o (eq. (1)), is given by the total mass of the non-ionic dissolved solids per unit mass of water,

$$S_o = \frac{1}{\rho} \sum_i M_i \times c_i,$$

where M_i, c_i and ρ have the same meaning as in (7).

3. COEFFICIENT OF HALINE CONTRACTION

The addition of small amounts of electrolytes i to water, with corresponding changes of concentration Δc_i, increases the volume of the solution from the initial volume V to

$$V \times (1 + \sum_i v_i \times \Delta c_i)$$

and modifies the mass from the initial $V \times \rho$ to

$$V \times (\rho + \sum_i \Delta c_i \times M_i).$$

To a first order approximation, the change of density from the initial ρ to $\rho(\Delta c_i)$ is given by:

$$\rho(\Delta c_i) = \frac{\text{Mass}}{\text{Volume}} = \frac{\rho + \sum_i \Delta c_i \times M_i}{1 + \sum_i v_i \times \Delta c_i} \quad [\text{kg m}^{-3}], \tag{11}$$

where v_i represents the partial molal volume of electrolyte i defined by

$$v_i = \frac{1}{V} \times \frac{\partial V}{\partial c_i},$$

and M_i is its molar weight. The former describes the relative change in volume of the solution with a change in concentration of electrolyte i. The v_i's are functions of the concentration of dissolved solids, temperature and pressure. For standard conditions ($p = 0$, $T = 25°C$), the partial molal volumes of various electrolytes have been determined in laboratory experiments (Millero et al., 1977). The values for those salts occurring in Lake Malawi are tabulated in the Appendix (Table A4). (Since we use partial molal volumes of electrolytes, the ions are stoichiometrically combined to salts in order to apply the procedure.) The coefficient of haline contraction for the ionic species is defined by:

$$\beta_c = \frac{1}{\rho} \frac{\partial \rho}{\partial S_c} \quad [\text{‰}^{-1}]. \tag{12a}$$

Using (7) and (11), it follows that

$$\beta_c = \lim_{\Delta c_i \to 0} \left(1 - \frac{\rho \sum_i v_i \Delta c_i}{\sum_i M_i \Delta c_i} \right).$$

If we assume that changes in salinity, S_c, occur in a "regular" manner, i.e., that the relative proportions of the various ions remain constant ($\Delta c_i = \eta c_i$, where η is a dimensionless constant), this expression reduces to:

$$\beta_c = 1 - \frac{\rho \sum_i v_i c_i}{\sum_i M_i c_i} \quad [\text{‰}^{-1}]. \tag{12b}$$

For the hypolimnetic water of Lake Malawi, (12) yields $\beta_c = 0.7759 \times 10^{-3}$ ‰$^{-1}$ for $T = 25°C$ and $p = 0$. The value is 0.0209×10^{-3} ‰$^{-1}$ or 1.028 (= 0.7759/0.7550) times greater than the corresponding value for β derived from the equation of state for seawater (Chen and Millero, 1986). This small difference is due to the fact that the ionic composition of Lake Malawi differs from that of seawater. (This method gives perfect agreement for seawater.)

We assume that for Lake Malawi, β_c can be explicitly calculated for any temperature, salinity and pressure by using this percentage offset to Chen and Millero's (1986) equations; i.e., we apply this correction wherever β_c is used explicitly:

$$\beta_c^{Malawi} = 1.028 \times \beta^{Chen\ \&\ Millero} \quad [\%o^{-1}]. \tag{13}$$

The coefficient of haline contraction for the non-ionic species (here silicic acid), β_0, is calculated by the same procedure using the partial molal volume of silicic acid (Table A4) and applying (12b) for the special case of only one component c_i. The coefficient of haline contraction for silicic acid is significantly smaller, $\beta_0 = 0.36 \times 10^{-3} \%o^{-1}$, than that for the ionic species β_c. Due to the large difference between the two coefficients of haline contraction we use, in contrast to McManus et al. (1992), β_0 instead of β_c for the calculation of the density contribution from silicic acid.

4. VERTICAL DENSITY STRUCTURE OF LAKE MALAWI

Both temperature and dissolved solids contribute to the vertical density structure and static stability of the water column. Estimation of the static stability, which is essential for studies of vertical exchange, is complicated by pressure effects. Pressure modifies in-situ density via the compressibility of water by: 1) decreasing the volume, and 2) increasing the in-situ temperature, T, due to adiabatic warming (no exchange of heat). Since both pressure-dependent effects on in-situ density are reversible during isentropic vertical displacement (no exchange of heat or salt), they have to be removed to estimate the static stability as shown below (eq. (16)). Therefore, the pressure-dependent in-situ temperature and in-situ density are commonly replaced by potential temperature (which compensates for the adiabatic warming) and potential density (which is a function of potential temperature at surface pressure). Another advantage is that potential temperature is the correct measure for heat budgets that allow quantification of mixing rates.

4.1. Potential Temperature

The adiabatic lapse rate, Γ, i.e., the rate of increase of in-situ temperature T with depth caused by isentropic compression of water, is given by the thermodynamic equation:

$$\Gamma(T,p,S) = \frac{g\alpha}{c_p}(T + 273.15\ K) \quad [K\ m^{-1}], \tag{14}$$

where $g = 9.81\ ms^{-2}$ is the acceleration of gravity, $\alpha = (-1/\rho)(\partial\rho/\partial T)\ [K^{-1}]$ the thermal expansivity, $c_p\ [J\ kg^{-1}\ K^{-1}]$ the specific heat at constant pressure and $T\ [°C] + 273.15\ K$ the absolute temperature. The two parameters α and c_p, which are functions of T, p and S, are calculated from Chen and Millero's (1986) equations. With respect to a reference depth of z_o (here $z_o = 0\ m$ = lake surface), the potential temperature, $\theta(z,z_o)$, of a water parcel at depth z (positive upwards), is defined as the temperature of that water parcel if it were to be displaced isentropically from z to z_o:

$$\theta(z,z_0) = T(z) - \int_z^{z_o} \Gamma(\theta(z,z'),p(z'),S(z))\ dz' \quad [°C]. \tag{15a}$$

Polynomial expressions for $\theta(T(z),p(z),S(z),z_o)$, provided by Bryden (1973) are valid only for the range of seawater salinities. Thus, for freshwater, we solve (15a) iteratively by the following procedure:

$$\theta^{(0)}(z,z_o) = T(z),$$

$$\theta^{(i+1)}(z,z_o) = T(z) - \int\limits_z^{z_o} \Gamma(\theta^{(i)}(z,z'),p(z'),S(z))\, dz' \quad [°C], \tag{15b}$$

which converges well enough after 2 iterations ($|\theta^{(2)} - \theta^{(3)}| < 10^{-4}$ °C). In warm waters, the adiabatic correction (integral term) is positive ($\Gamma > 0$) and, consequently, $\theta < T$. The differences between in-situ and potential temperature in Lake Malawi (up to 0.12°C; Fig. 2) are significantly larger than in temperate lakes. This is due to the high thermal expansivity α of warm water (eq. (14)). The minimum in-situ temperature occurs at a depth of approximately 400 m, below which the in-situ temperature gradually increases (Fig. 2), whereas the minimum potential temperature is about 40 m above the lake floor, below which the potential temperature increase is less than 0.001°C (Fig. 5). Consequently, the hypolimnetic water is thermally density-stratified up to approximately 40 m above the lake floor. We assume that the weak negative temperature gradient near the lake bottom reflects the flux of geothermal heat from the earth's interior.

4.2. Salinity

The conductivity-dependent salinity, S_c, is roughly constant at about 0.211‰ in the epilimnion (Fig. 3). From 50 to 250 m, S_c increases linearly with depth to about 0.218‰. The increase in S_c within the hypolimnion is extremely gradual; i.e., only about 0.0015‰ over the deepest 450 m of the water column. For the conductivity-independent salinity S_o (non-ionic species), only silicic acid is considered. We use a simplified profile of silicic acid data, measured down to a depth of 600 m in 1981 by Hecky (personal communication), which shows a constant epilimnetic concentration in the upper 100 m of 20 µmol/l, a linear increase to 180 µmol/l at 300 m depth and finally a linear decrease to 140 µmol/l at 300 to 700 m depth. In contrast to the salinity of the ionic species, S_c, the total salinity, $S = S_c + S_o$ (Fig. 3), decreases with depth in the deep water.

4.3. Static Stability

The static stability, expressing the strength of the density stratification, is defined as:

$$N^2 = g \times (\alpha \times [(dT/dz) + \Gamma] - [\beta_c \times (dS_c/dz) + \beta_o \times (dS_o/dz)]) \quad [s^{-2}] \tag{16}$$

(α, Γ, β_c and β_o depend on in-situ T, p and S). For a stably stratified water column, N^2 is required to be positive ($N^2 > 0$). As shown in Fig. 4, the stability N^2 decreases exponentially in the upper hypolimnion (from 50 to 350 m depth) and levels off in the lowest 200 m at a value of about 2×10^{-7} s^{-2}, which is quite low for freshwater lakes. In the deepest 40 m of the lake, the stability of the water column is indistinguishably close to neutral (Fig. 4).

In the present case, static stability can also be determined by using potential density, which is approximated here by

$$\rho_\theta(z) = \rho(\theta(z,0), p = 0, S_c(z)) + \beta_o \times \rho \times S_o(z) \quad [kg\ m^{-3}], \tag{17}$$

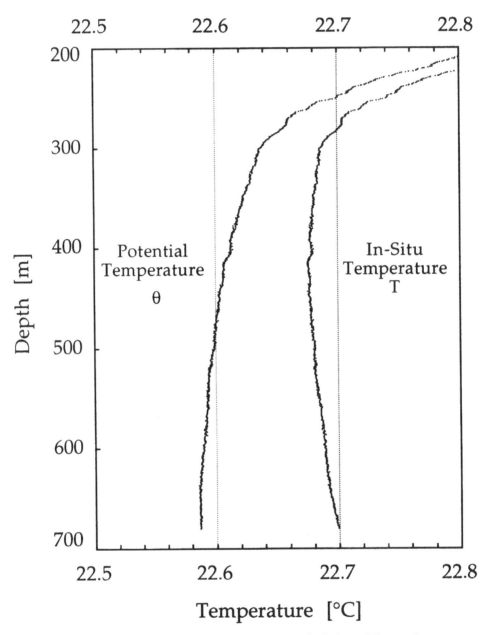

Figure 2. In-situ temperature, T, and potential temperature, θ. The large differences between in-situ and potential temperatures are related to the high thermal expansivity of warm water.

where $\rho(\theta, p, S_c)$ is given by Chen and Millero (1986) and corrected by $\beta_o \times \rho \times S_o$ for the non-ionic silicic acid (Fig. 4). For the warm waters of Lake Malawi, $N_\theta^2 = -(g/\rho_\theta)(d\rho_\theta/dz)$ is close to the exact definition of static stability (eq. (16)), i.e., $N_\theta^2 \approx N^2$.

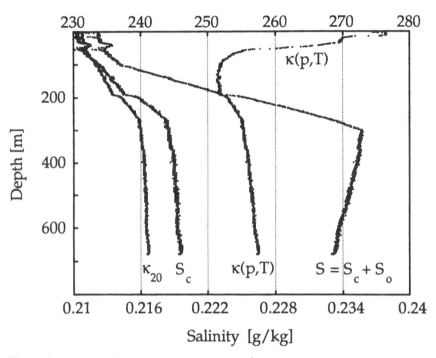

Figure 3. Vertical profiles of in-situ conductivity, κ_T^p; conductivity normalized to $T = 20°C$ and $p = 0$, κ_{20}^0; salinity due to ionic dissolved species, S_c (calculated from the in-situ conductivity); and total salinity, $S = S_c + S_o$, where S_o is salinity due to non-ionic species (e.g., silicic acid).

The relative contribution of temperature or salinity to the total stability in (16) is commonly expressed by the density ratio $R\rho$:

$$R_\rho = \frac{\beta_c \times \dfrac{dS_c}{dz} + \beta_o \times \dfrac{dS_o}{dz}}{\alpha \times \left[\dfrac{dT}{dz} + \Gamma\right]} \quad [-], \tag{18}$$

which is plotted in Fig. 6. This plot reveals that temperature dominates stability throughout the entire water column ($|R_\rho| < 1$), except in the lowest 50 m (Fig. 5). The conductivity-dependent salinity term, S_c, increases with depth and thus tends to increase the total stability (positive term in (16)), whereas S_o decreases with depth and has the opposite effect (negative term in (16)). The total salinity has an overall stabilizing effect on the upper hypolimnion, above 380 m depth ($R_\rho < 0$), and a destabilizing effect on the deep water below that depth ($R_\rho > 0$). The relative importance of dissolved solids for the total stability varies continuously with depth as shown in Fig. 6. A value of $R_\rho \approx -0.2$ at ~360 m depth for example (Fig. 6) indicates that the ions have a 20% stabilizing effect over the temperature gradient,

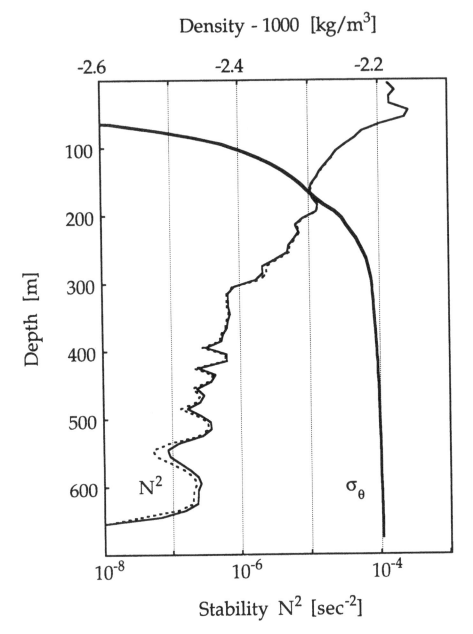

Figure 4. Potential density, σ_θ, and static stability, N^2 (solid line: based on S_c only; dashed line: based on total salinity), as a function of depth (from (16)). The CTD data are averaged over 10 m intervals (the original data were sampled with an average increment of 0.5 m). Both the uniform potential densities and the nearly vanishing N^2 indicate very low stability in the deepest layer.

whereas a value of $R_\rho \approx 0.5$ at ~550 m depth implies that the total salinity has a 50% destabilizing effect over the stabilizing effect of the temperature gradient.

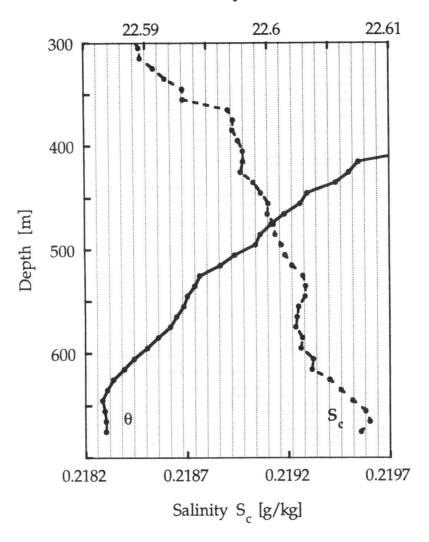

Figure 5. Profiles of potential temperature, θ, and salinity, S_c, below 300 m. The CTD data are averaged over 10 m intervals. The bulk of the data shows stabilizing gradients for both temperature and salt. The negative temperature gradient in the deepest 40 m is probably generated by the geothermal heat flux.

If there are other dissolved species gradients (e.g., dissolved gases) in the water column, additional terms must be added to (16) and (17). If the species is non-ionic, it will not be detected by the CTD. A rough estimate based on pH values (7.3 to 8.1) reported by Gonfiantini et al. (1979) indicates that not all of the dissolved inorganic carbon is present as HCO_3^-, and that the small proportion of CO_2 resulting does affect the density profile significantly. If other gases like CH_4 and H_2S also contribute significantly to the vertical density structure, precise concentration profiles of those gases will also be required to calculate the water column stability accurately.

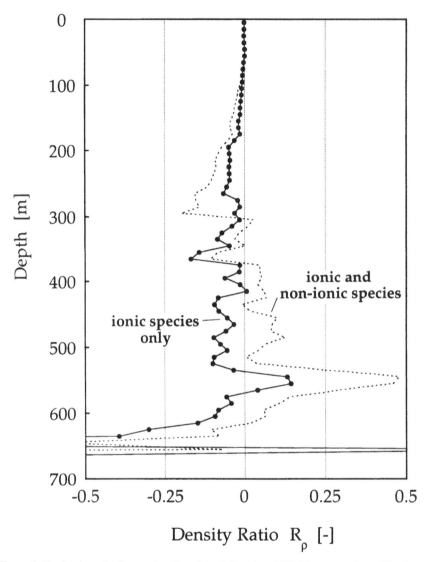

Figure 6. The density ratio, R_ρ, as a function of depth, based on CTD data averaged over 10 m intervals. This ratio defines the relative importance of heat and salt in stabilizing the water column of Lake Malawi. The contribution of ionic species to water column stability is less than 10% of the total stability even below 300 m (solid line). The inclusion of the non-ionic species (e.g., silicic acid) reduces the stabilizing contribution of the ionic species (dashed line). See also §4 for the interpretation of R_ρ.

5. STABILITY-RELATED GRADIENTS AND DEEP-WATER MIXING

Anoxic conditions, relatively stable temperatures over the past 30 years, and low vertical exchange rates inferred from tritium profiles, suggest that the deep hypolimnion of Lake Malawi is largely isolated from the oxic surface layer (Gonfiantini et al., 1979). For the sake of simplicity, let us first assume steady state conditions. In this case, the profiles of temperature and salinity in the deep hypolimnion reflect the balance between the upward geothermal heat and ion fluxes from the sediments and the turbulent vertical fluxes in the deep-water column (Wüest et al., 1992). (This assumption implies that the thin bottom layer of inverse temperature gradient is a result of continuous geothermal heating.) In equating the geothermal and turbulent heat fluxes, F_{geo} and $K_v \times \rho \times c_p \times \partial\theta/\partial z$, respectively, the vertical diffusivity is given by

$$K_v(z) = -\frac{F_{geo}}{\rho \times c_p \times \partial\theta/\partial z}(z) \quad [m^2\,s^{-1}]. \tag{19}$$

Assuming an average geothermal heat flux of 0.040 Wm^{-2} (Von Herzen and Vaquier, 1967) and 1.5×10^{-5} km^{-1} for $\partial\theta/\partial z$ (Fig. 5), the vertical diffusivity is ~6 $cm^2\,s^{-1}$ for the deepest part of the hypolimnion (eq. (19)). This value is high compared to other lakes and to the ocean; however, it is not inconsistent with the tritium profiles reported by Gonfiantini et al. (1979). It suggests that mixing near the lake floor (probably convectively driven by the geothermal heat flux) is intense, giving rise to nearly neutral stability (Fig. 4).

This result is confirmed by examining primary productivity estimates and hypolimnetic salinity or bicarbonate gradients. Given the vertical diffusivity (~6 $cm^2\,s^{-1}$) and the gradients of salinity S_c (Fig. 5) and bicarbonate (Gonfiantini et al., 1979), respectively, we can estimate primary productivity by assuming that 5% of the initially produced algal biomass is mineralized in the deep water. (This is a typical fraction for such a deep lake: Johnson et al., 1982.) In equating again the deep-water upward flux of salinity and bicarbonate with the algal mineralization flux (Wüest et al., 1992), primary productivity is estimated to be about 180 and 360 g C_{ass} m^{-2} yr^{-1}, respectively. In-situ estimates of primary productivity determined by Degnbol et al. (1982) and Bootsma (1993) fall between these two estimates. We can conclude that the deep profiles are consistent with the quasi-steady state assumption.

Why then is Lake Malawi meromictic in the long term? The biogenically induced density stratification in Lake Malawi is relatively small compared to that in other meromictic lakes. The salinity difference ΔS between surface and deep water can only compensate for a temperature difference of $\Delta T = \alpha^{-1} \times \beta \times \Delta S$ to maintain minimal stability (defined more precisely by (18) with $R_\rho = 1$); i.e., $\Delta S \approx 0.015‰$ (Fig. 3) leads to a potential excess cooling of only $\Delta T \approx 0.05°C$. Consequently, the surface water of Lake Malawi cannot cool below the hypolimnetic temperature less ΔT, i.e., $22.585 - 0.05 \approx 22.5°C$, without the onset of convective mixing. Even if global warming over the last few decades, which probably led to a warming of the upper hypolimnion (Eccles, 1974; Halfman, 1993), is taken into consideration, deep convection would occur at least sporadically during cold periods. We can, therefore, conclude that: 1) the surface temperature of the main water body is always warmer than the hypolimnetic temperature of ~22.5°C; and 2) a mechanism is active which sporadically cools the deep water and thereby compensates for geothermal warming and heat diffusion from above. Hypolimnetic cooling can be achieved either by the intrusion of cold river water or of cold lake water from nearshore shallow regions. A slight decrease in bottom water temperatures since 1976 (Halfman, 1993) and the lower silicic acid content in the deep water (Hecky, personal communication, Fig. 3) supports this assumption. We can conclude that the hypolimnion is not as stagnant as previously assumed, and that the quasi-steady state is maintained by non-local vertical exchange processes.

ACKNOWLEDGMENTS

We are grateful to Nicolas Gruber (1993) for compiling the ionic specific constants in the Appendix during his diploma thesis. We also thank Robert Hecky for providing the silicic acid profile, Harvey Bootsma for helpful hints and David Livingstone for carefully reading and constructively criticizing an earlier version of the manuscript.

APPENDIX

A1. Equivalent Conductance as a Function of Temperature

The temperature dependence of the equivalent conductance of the ion i is expressed by a third order polynomial approximation (except for CO_3^{2-} and Mg^{2+}: 2nd-order only) using the equation

$$\lambda_i^{\infty}(T) = A_i + B_i \times T + C_i \times T^2 + D_i \times T^3 \quad [\text{mS cm}^{-1} \, (\text{eq/l})^{-1}], \tag{A1}$$

where T is temperature [°C]. Table A1 lists the coefficients A_i to D_i for the major ions detected in Lake Malawi based on Robinson and Stokes (1970), MacInnes (1961), Franks (1973), Schwabe (1986) and Handbook of Chemistry and Physics (1991).

Table A1. Equivalent Conductance at Infinite Dilution of the Major Ions in Lake Malawi as a Function of Temperature at Atmospheric Pressure ($p = 0$)

Ion	A [mS cm^{-1} (eq/l)$^{-1}$] = [*]	B [* °C^{-1}]	C [* °C^{-2}]	D [* °C^{-3}]
Bicarbonate (HCO$_3^-$)	24.63	0.701	5.374×10^{-3}	-6.485×10^{-5}
Carbonate (CO$_3^{2-}$)	36.00	1.380	-1.073×10^{-3}	
Chloride (Cl$^-$)	41.08	1.242	8.597×10^{-3}	-6.999×10^{-5}
Nitrate (NO$_3^-$)	40.00	1.287	-4.870×10^{-3}	1.473×10^{-4}
Sulfate (SO$_4^{2-}$)	41.00	1.256	2.333×10^{-2}	-4.488×10^{-4}
Calcium (Ca^{2+})	31.20	0.898	1.117×10^{-2}	-7.249×10^{-5}
Sodium (Na$^+$)	26.46	0.758	9.090×10^{-3}	-6.179×10^{-5}
Potassium (K$^+$)	40.69	1.203	4.381×10^{-3}	3.433×10^{-6}
Magnesium (Mg^{2+})	28.90	0.662	1.214×10^{-2}	
Ammonium (NH$_4^+$)	40.20	1.298	-1.437×10^{-3}	1.130×10^{-4}

A2. Equivalent Conductance as a Function of Concentration

The concentration-dependent reduction coefficient of the equivalent conductance (see Table A2) of the ion i (eq. (2)) is defined by:

$$f_i(m) = \lambda_i(m)/\lambda_i^{\infty}(m) \quad [-].$$

It is approximated by a third-order polynomial of the square root of the total charge

$$m = \sum_i c_i Z_i \quad [\text{eq/l}]$$

Table A2. Reduction Coefficient of the Equivalent Conductance as a
Function of the Total Charge m for the Major Ions in Lake Malawi,
based on Data from MacInnes (1961) and Handbook of
Chemistry and Physics (1991)

Ion	A [-]	B [(eq/l)$^{-1/2}$]	C [(eq/l)$^{-1}$]	D [(eq/l)$^{-3/2}$]
Bicarbonate (HCO_3^-)	1.000	−0.887	1.221	−4.275
Carbonate (CO_3^{2-})[a]	1.002	−2.035	3.207	−2.043
Chloride (Cl^-)	1.000	−0.627	0.685	−0.270
Nitrate (NO_3^-)	0.999	−0.658	−0.146	0.777
Sulfate (SO_4^{2-})	1.002	−2.035	3.207	−2.043
Calcium (Ca^{2+})	1.001	−2.378	5.456	−5.633
Sodium (Na^+)	1.000	−0.825	1.059	−0.872
Potassium (K^+)	1.000	0.674	0.930	−0.658
Magnesium (Mg^{2+})	1.001	−2.516	5.863	−6.360
Ammonium (NH_4^+)	1.000	−0.627	0.666	−0.312

[a]The sulfate coefficients have been used for carbonate as these are not known.

using the equation:

$$f_i(m) = A_i + B_i \times m^{1/2} + C_i \times m + D_i \times m^{3/2}. \quad [-] \tag{A2}$$

A3. Conductivity in Lake Malawi

Table A3 gives the ionic concentration and conductivity of Lake Malawi water.

A4. Partial Molal Volume as a Function of Electrolyte Concentration

The concentration dependence of the partial molal volume of electrolyte i (see Table A4) is expressed by a second-order polynomial approximation of the square root of the ionic strength:

$$I = \frac{1}{2} \sum_i c_i Z_i^2 \quad [\text{mol/l}]$$

using the equation

$$v_i = v_i^o + 1.5 A_i \, I^{1/2} + 2 B_i I. \quad [\text{ml/mol}]. \tag{A3}$$

Table A3. Ionic Concentration and Conductivity of Lake Malawi Water

Depth [m]	HCO_3^- [e]	CO_3^- [e]	SO_4^{-2}	Cl^-	Ca^{2+}	Mg^{2+}	Na^+	K^+	κ_{20} in situ [a]	κ_{20} model [b]	Salinity S_c
	Concentration [10^{-3} eq/l]								[μScm^{-1}]	[μScm^{-1}]	[‰]
0–75	2.309	0.011	0.155	0.14	0.905	0.615	0.875	0.16	231 to 233	226.1	0.2058
150–275	2.408	0.008	0.130	0.20	0.968	0.627	0.903	0.17	235 to 240	236.8	0.2150
300–700	2.474	0.005	0.130	0.25	0.988	0.630	0.895	0.17	240 to 241	242.9	0.2209
Equivalent conductance λ_{20}^∞ at $T = 20°C$ and infinite dilution [c] [$mScm^{-1}$ (eq/l)$^{-1}$]											
	40.29	63.18	71.86	68.81	53.04	47.00	44.80	66.5			
Reduction coefficient f_i [-] [d]											
	0.957	0.897	0.907	0.970	0.893	0.887	0.962	0.968			

[a]Calculated from (6) (see text) based on the January 1992 CTD data (Fig. 2). [b]Calculated from (2) (with constants from Tables A1 and A2) based on data from Gonfiantini et al. (1979). The small differences between in-situ (1992) and modeled conductivity (1979) are due to 1) analytical errors, 2) changes in the evaporation to outflow ratio, and 3) neglecting the contributions of trace ions. [c]From data set compiled in Table A1. [d]Calculation based on Table A2. [e]HCO_3^- and CO_3^{-2} concentrations were calculated from alkalinity and pH [Gonfiantini et al., 1979].

Table A4. Partial Molal Volume as a Function of Ionic Strength at $T = 25°C$
(data from Millero et al. 1977)

Electrolyte	Molar mass [g mol^{-1}]	v_i^0 [ml mol^{-1}]	A_i [ml/mol (l/mol)$^{1/2}$]	B_i [ml/mol (l/mol)]
NaCl	58.443	16.613	1.811	0.094
Na$_2$CO$_3$	105.990	−6.195	7.71	−0.383
Na$_2$SO$_4$	142.050	11.559	6.75	0.0876
NaHCO$_3$	84.000	23.118	3.662	−0.115
Ca(HCO$_3$)$_2$	162.100	30.86	7.313	−0.430
Mg (HCO$_3$)$_2$	146.332	27.532	6.857	−0.312
KHCO$_3$	100.112	33.427	3.691	−0.024
KCl	74.555	26.85	1.839	0.087
MgCO$_3$	84.323	−24.652	5.002	0.303
CaCO$_3$	100.091	−21.316	5.457	0.08
Si(OH)$_4$	96.12	61.5[a]		

[a]From Duedall et al. (1976).

REFERENCES

Beadle, L.C., 1981, The inland waters of tropical Africa: 2nd ed. New York, Longman, 475 pp.

Bootsma, H.H., 1993, Algal dynamics in an African Great Lake and their relation to hydrographic and meteorological conditions. Ph.D. Thesis. University of Manitoba, XV+311 pp.

Bradshaw, A., and Schleicher, K.E., 1965, The effect of pressure on the electrical conductance of sea water: Deep-Sea Research, v. 12, pp. 151–162.

Bryden, H.L., 1973, New polynomials for thermal expansion, adiabatic temperature gradient and potential temperature of sea water: Deep-Sea Research, v. 20, pp. 401–408.

Chen, C.T., and Millero, F.J., 1986, Precise thermodynamic properties for natural waters covering only the limnological range: Limnology and Oceanography, v. 31, pp. 657–662.

Cox, R.A., Culkin, F., and Riley, J.P., 1967, The electrical conductivity/chlorinity relationship in natural sea water: Deep-Sea Research, v. 14, pp. 203–220.

Coulter, G.W., and Spigel, R.H., 1991, Hydrodynamics, in Coulter, G.W. ed., Lake Tanganyika and its Life: New York, Oxford University Press, pp. 49–75.

Craig, H., Dixon, F., Craig, V., Edmond, J., and Coulter, G., 1974, Lake Tanganyika geochemical and hydrographic study: 1973 expedition: Publication Scripps Institution of Oceanography, Series 75-5, pp. 1–83.

Degnbol, P., and Mapila, S., 1982, Limnological observations on the pelagic zone of Lake Malawi from 1978 to 1981: FAO MLW/75/019 Tech. Rep. 1.

Duedall, I.W., Dayal, R., and Willey, J.D., 1976, The partial molal volume of silicic acid in 0.725 m NaCl: Geochimica et Cosmochimica Acta, v. 40, pp. 1185–1189.

Eccles, D.H., 1974, An outline of the physical limnology of Lake Malawi (Lake Nyasa): Limnology and Oceanography, v. 19, pp. 730–742.

Franks, F., 1973, Waters: Vol. 2, New York and London, Plenum Press, 684 pp.

Gonfiantini, R. Zuppi, G.M., Eccles, D.H., and Ferro, W., 1979, Isotope investigations of Lake Malawi. In: Isotope in lake Studies. International Atomic Energy Agency Panel Proceedings Series STI/PUB/511.

Gruber, N., 1993, Kohlenstoff- und Sauerstoffkreislauf im Soppensee: Diploma thesis EAWAG/ETH, 119 pp.

Halfman, J.D., 1993, Water column characteristics from modern CTD data, Lake Malawi, East Africa: Journal of Great Lakes Research, v. 19, pp. 512–520.

Halfman, J.D., and Scholz, C.A., 1993, Suspended Sediments in Lake Malawi, Africa: A Reconnaissance Study: Journal of Great Lakes Research, v. 19. pp. 449–511.

Handbook of Chemistry and Physics (Ed. D.R. Linde), 1991, Boston, CRC Press.

Hecky, R.E., 1993, The eutrophication of Lake Victoria: Verh. Internat. Verein. Limnol., v. 25, pp. 39–48.

Hutchinson, G.E., 1975, A Treatise on Limnology: Vol. 1, New York, Wiley, 540 pp.

Johnson, T.C., Evans, J.E., and Eisenreich, S.J., 1982, Total organic carbon in Lake Superior sediments: Comparisons with hemipelagic and pelagic marine environments: Limnology and Oceanography, v. 27, pp. 481–491.

MacInnes, D.A., 1961, The Principles of Electrochemistry: New York, Dover.

McManus, J., Collier, R.W., Chen, C. A., and Dymond, J.,1992, Physical properties of Crater Lake, Oregon: A method for the determination of a conductivity- and temperature-dependent expression for salinity: Limnology and Oceanography, v. 37, pp. 41–53.

Millero, F.J., Laferriere, A., and Chetirkin, P.V., 1977, The partial molal volumes of electrolytes in 0.725 m sodium chloride solutions at 25°C: Journal of Physical Chemistry, v. 81, pp. 1737–1745.

Millero, F.J., and Poisson, A., 1981, International one-atmosphere equation of state of seawater: Deep-Sea Research, v. 28A, pp. 625–629.

Robinson, R.A., and Stokes, R.H., 1970, Electrolyte solutions, Butterworths, London, 571 pp.

Scholz, C.A., and Rosendahl, B.R., 1988, Low lake stands in Lakes Malawi and Tanganyika, East Africa, delineated with multi-fold seismic data: Science, v. 240, pp. 1645–1648.

Schwabe, 1986, Physikalische Chemie, Band 2: Elektrochemie Akademie-Verlag, Berlin, 443 pp.

Sorensen, J.A., and Glass, G.E., 1987, Ion and temperature dependence of electrical conductance for natural waters: Analytical Chemistry, v. 59, pp. 1594–1597.

UNESCO, 1981, Tenth Report of the Joint Panel on Oceanographic Tables and Standards: UNESCO Technical Papers in Marine Science, No. 36, 24 pp.

Von Herzen, R.P., and Vaquier V., 1967, Terrestrial heat flow in Lake Malawi, Africa: Journal of Geophysical Research, v. 72, pp. 4221–4226.

Wüest, A., Aeschbach-Hertig, W., Baur, H., Hofer, M., Kipfer, R., and Schurter, M., 1992, Density Structure and Tritium-Helium Age of Deep Hypolimnetic Water in the Northern Basin of Lake Lugano: Aquatic Sciences, v. 54, pp. 205–218.

Aquatic Chemistry

Phosphorus Pumps, Nitrogen Sinks, and Silicon Drains: Plumbing Nutrients in the African Great Lakes

R.E. HECKY   *Department of Fisheries and Oceans, Freshwater Institute, Winnipeg, Canada*

H.A. BOOTSMA   *Great Lakes Environmental Research Laboratory, Ann Arbor, Michigan, United States*

R.M. MUGIDDE and F.W.B. BUGENYI   *Fisheries Research Institute, Jinja, Uganda*

Both neolimnologists, who study modern lakes, and paleolimnologists, who study the history of lake ecosystems, require knowledge of the ecological determinants of algal community composition and productivity in the African Great Lakes. Algal productivity, in turn, sets general energetic limits on the productivity of higher trophic levels including fish (Melack 1979, Hecky et al. 1981, Hecky 1984). However, algal species differ in their nutritive value and their digestibility by primary aquatic consumers. Therefore, some of the differences in trophic efficiencies among the African lakes (Hecky 1984) may be attributed to marked differences in their phytoplankton communities and their utilization by other species in the food webs. In turn, the remains of both plants and animals are important contributors to sediments forming in lakes, and biogenic remnants are the most prominent sedimentary constituent in the African Great Lakes of Victoria, Tanganyika, Malawi, Kivu and Edward (Kendall 1969, Hecky and Degens 1973). These lakes have fossil-rich, highly organic sediments which have continuously accumulated for long periods of time. These deep water sediments are rich repositories of the geological and climatic history of tropical Africa since the Tertiary and are a major focus of the IDEAL program. Understanding of the processes leading to the formation of these biogenic sediments will set limits on what can be inferred about the lakes' histories. The most important question concerning sediment formation is what controls the primary productivity and the algal species composition of the phytoplankton communities of these great lakes.

Nitrogen and phosphorus are most often identified as the nutrients limiting algal biomass or productivity in aquatic ecosystems while silicon is an essential element for diatom growth (Hecky and Kilham 1988). The fluctuating availability of these three elements in lakes is often invoked to explain algal productivity and species succession on time scales from days to millenia (e.g., Sommer 1993, Haberyan and Hecky 1987). The organic sediments of the African Great Lakes and the aquatic microfossil record they contain owe their origins to the nutrient fluxes entering these lakes and to the internal cycling of nutrients in these ecosystems. Understanding and managing the productivity of these ecosystems and interpreting their fossil records requires quantification and modelling of the external and internal processes which determine nutrient availability for algal growth.

Despite the obvious utility of understanding nutrient cycling in the African Great Lakes for both modern management and historical understanding, our knowledge of these processes is still in its infancy especially compared to freshwater in general, to other large lakes of the world in particular and to the world's oceans. Good quality nutrient measurements are available for a few samples on a few lakes as early as the 1930s (e.g., Beauchamp 1939), the depth distribution of some nutrients in the deep lakes was recognized by the late 1940s and 1950s (Kufferath 1952, Verbeke 1957) and high quality seasonal data for Victoria are available for the late 1950s and 1960s (e.g., Talling 1966). More recently research by chemical oceanographers (Degens et al. 1971 and 1973, Craig 1975, Hecky et al. 1991, Edmond et al. 1993) has further advanced our knowledge of nutrient distributions in Lake Tanganyika. The regional comparative study of Talling and Talling (1965) especially impresses on any reader the rich chemical and nutrient diversity of the African lakes. Hecky et al. (1993) showed that the nutrient composition of particulate matter in the African Great Lakes is also quite diverse, especially in comparison to the Laurentian Great Lakes, indicating that nutrient loading and cycling was likely under quantitatively different controls among the African lakes.

There is growing concern that these great lakes are at risk from the activities of the rapidly increasing human populations in their watersheds as well as from climate change (Hecky 1993, Hecky and Bugenyi 1992, Bootsma and Hecky 1993, Hecky et al. 1994). The need for rational exploitation of their aquatic resources is currently driving intense scientific explorations of Lakes Victoria, Tanganyika and Malawi with multiyear programs being conducted by a variety of international investigators and donors including the IDEAL initiative. We are confident that the next decade will see an exponential growth of information including nutrient cycling on these lakes. The purpose of our effort here is not to rigorously review all information currently at hand, but to provide a conceptual synthesis which will structure previous information and highlight deficiencies which retard our understanding and limit scientific management of these Great Lakes. We conclude that what we don't know far outweighs what we do know; and, in particular, there are some quite obvious deficiencies in basic information which must be addressed before realistic models of nutrient cycling can be developed.

HYDROLOGY AND HYDROGRAPHY OF THREE GREAT LAKES

Victoria, Tanganyika and Malawi are nearly closed bodies of water hydrologically (Hecky and Bugenyi 1992) and complete closure has certainly been a long-term feature of their geological history (Kendall 1969, Hecky 1978, Haberyan and Hecky 1987). In fact, Malawi and Tanganyika may have been closed prior to the late 1800s and the lakes' chemistries may yet be adjusting to their geologically recent opening (Craig 1975, Hecky 1977, Owen et al. 1990, Hecky et al. 1991, Edmond et al. 1993). Substantial water level fluctuations have been historically recorded as well as a general synchrony of these fluctuations among the three systems under the common influence of the highly variable rainfall that characterizes East Africa (Nicholson, this volume). Extraordinarily high lake levels followed extremely rainy years in the early 1960s, and Victoria levels have not returned to their pre-1960 datum level since then. We must recognize that the hydrological water balance of these systems is quite dynamic in spite of their great size and that water exchange directly across their surfaces dominates their water budgets. Unfortunately open lake data on evaporation and precipitation are nonexistent, and lakeshore data probably are not representative of offshore surface

Table 1. Morphometric and Hydrological Data for Africa's Three
Largest Lakes (from Bootsma and Hecky 1993)

	Malawi	Tanganyika	Victoria
Catchment area (km^2)	100,500	220,000	195,000
Lake area (km^2)	28,800	32,600	68,800
Maximum depth (m)	785	1,470	79
Mean depth (m)	292	580	40
Volume (km^3)	8,400	18,900	2,760
Outflow (O) $(km^3\ y^{-1})$	11	2.7	20
Inflow (I) $(km^3\ y^{-1})$	29	14	20
Precipitation (P) $(km^3\ y^{-1})$	39	29	100
Evaporation $(km^3\ y^{-1})$	55	44	100
Flushing time (V/O) (years)	750	7,000	140
Residence time $(V/(P + I))$ (years)	140	440	23
Oxygenated sediment area $(\%)$[a]	50	25	100

[a]At time of deepest annual mixing in July–August (Hecky et al. 1994).

energy exchanges (Kite 1981). Better knowledge about input from precipitation and losses to evaporation will be essential for successful water quality and quantity modelling.

The responses of these three lakes to closure and lower precipitation:evaporation ratios will be quite different because of their different morphometries. All closed lakes with insignificant groundwater input or output must balance their water budget by adjusting their lake areas because the only water loss occurs across their surface. Because Victoria has quite gentle basin slopes, relatively minor declines in water level result in substantial changes in area compared to the steep sided basins of Tanganyika and Malawi. To effect a 10% change in surface area, Victoria must drop 12 m while Malawi would drop 32 m and Tanganika 56 m (Bootsma and Hecky 1993). However a 12 m drop in Victoria which has a 40 m mean depth would represent a greater than 30% loss of water volume with a resulting concentration of solutes including nutrients. The greater decline in level in the much deeper Tanganyika or Malawi (Table 1) in contrast would result in less than a 5% change in volume and negligible solute concentration directly attributable to volume change. The great mass of water and solutes contributes to the long residence time of solutes (Hecky and Bugenyi 1992; Table 1) which buffers the deep lakes against salinity changes on millenia and greater time scales. Haberyan and Hecky (1987) suggest that Tanganyika may never have been very saline in its long history despite geologically frequent closures. In contrast, Victoria was dry or nearly dry 12,000 y B.P. and has gone through stages in which calcite and gypsum, at least, have precipitated (Kendall 1969). Although definitive data are lacking, the effect of volume changes on nutrient concentrations in the deep lakes would likely be similar to the major ions, i.e., minor in the deep lakes and substantial in Victoria, assuming that nutrient inputs were relatively unaffected.

Although nutrient concentrations would not be very responsive to closure in the deep lakes, stratification changes should result in substantial alteration of rates of nutrient cycling and consequently concentrations. Within the period of scientific observation, the deep lakes have had permanently anoxic water below 250 m in Malawi and 100 to 200 m in the two major basins of Tanganyika. This permanent anoxia results from the thermal inertia of this deep water with a minor contribution of salts and dissolved silicon to the density gradient (Wüest et al., this volume). Therefore, even though the deep water temperatures are always in excess of 22–23°C in the deep lakes, they are stably stratified. There is a seasonal thermocline which sets up in all three lakes during September to April. Circulation in the dry season mixing period is restricted to the upper 200 m in the deep lakes while the shallower Victoria circulates to its maximum depth.

The persistence of the permanent deep thermocline during prolonged lake closure is less certain. Currently lake outflow removes water and has a diluting effect on the upper mixed layer in all seasons. Loss of outflow would lead to concentration of the mixed layer. If closure were of long duration, the resulting volume changes and increasing ionic concentration of the mixed layer could lead to destabilization and downward mixing which, in turn, could certainly weaken the density barrier to mixing and increase vertical circulation. The thermally imposed density gradients are weak, on the order of 1×10^{-7} g cm^{-1}; therefore, periods of full circulation might be expected from evaporative concentration when the deep lakes become closed. Low lake stands and loss of outlets occurred under cooler as well as drier climates in tropical Africa which were contemporaneous with full maximal glacial periods in the temperate regions (Livingstone 1975). Cooler surface temperatures would also weaken thermal stratification and increase the likelihood of full circulation at least seasonally. Weakened stratification would cause more rapid vertical mixing and tend to more uniform vertical nutrient distributions. It might also lead to increased oxygenation of the deep water if vertical mixing were rapid enough. However, high water temperatures limit the oxygen concentrations even at atmospheric saturation and also accelerate decomposition processes. Consequently anoxic or nearly anoxic deep waters have probably been a nearly continuous feature of the deep lakes when stratification prevents atmospheric exchange. This permanently warm and anoxic deep water is a feature unique to the African lakes (cf. the approximately 4°C temperatures and oxygenated deep waters of temperate Great Lakes and oceans) and has profound consequences for their nutrient cycles. Anoxic and hypoxic deep water has become a nearly continuous feature of Lake Victoria since 1960 as a consequence of its eutrophication and increased stability of thermal stratification (Hecky 1993, Hecky et al. 1994).

INPUTS, OUTPUTS AND STEADY STATE

Steady state conditions greatly simplify discussion of nutrient cycles because inputs of a nutrient equal outputs and the concentrations and fluxes within the ecosystem are invariant. Although a true steady-state condition may seem unlikely especially at short time scales, this assumption can become more tenable as one integrates over longer intervals of time (annual and beyond), and it will be assumed here in order to facilitate discussion. Elemental inputs are derived from rivers, wet and dry precipitation and ground water while outputs result from outflow, sedimentation, ground water and gas exchange. Unfortunately, we do not have a fully defined nutrient budget which characterizes all these fluxes for any of the great lakes. There is little or no geochemical evidence to indicate that groundwater inputs of outputs of nutrients are significant in any of the African Great Lakes except Kivu (Von Herzen et al. 1971, Degens 1973). If groundwater inputs are thus considered negligible, then the most

Table 2. Annual Inputs of S, N, P and Si to Lake Malawi and Loss at the Shire River Outflow of These Elements

	S	N	P	Si
		(megamoles)		
River inflow (I_R)	1,330	740	80	6,720
Wet and dry precipitation (I_P)	100	2,650	63	Trace
Biological N-fixation (I_B)		11,810		
Shire outflow (O)	140	120	4	240
Retention ($R = I_R + I_P + I_B - O$)	1,290	15,080	139	6,480
Renewal time (yr) (M/O)	1,770	1,120	3,030	2,860
Residence time (yr) (M/I)	170	9	90	100

Renewal and residence times are based on estimates of the total dissolved mass (M) of these elements in the lake (Hecky and Bugenyi 1992).

complete data set for constructing elemental budgets is for Malawi (Bootsma and Hecky 1993; Hecky and Bugenyi 1992). The input and output data for nutrient elements S, N, P and Si are summarized for Malawi in Table 2 which combines data from Hecky and Bugenyi (1992) and Bootsma and Hecky (1993).

The Great Lakes are nearly closed hydrologically (Table 1), and consequently outflow losses of the nutrients and other particle active substances are negligible. The high retention coefficients (= (input fluxes – outflow losses)/(input fluxes)) for S (90%), N (99%), P (97%) and Si (96%) confirm that sedimentation and/or loss to the atmosphere must be the dominant processes affecting these biologically active elements in Malawi. A similarly high nutrient retention would be expected in Tanganyika and Victoria as well. Although Bootsma (1993) measured much less atmospheric wet deposition of N and P at Malawi than did Ganf and Viner (1973) near Lake George, his inclusion of dry deposition (not reported by Ganf and Viner) would indicate that direct atmospheric deposition of N and P on the lake's surface may be the most important input for all the Great Lakes. The atmospheric input for Tanganyika should also be high relative to river inputs (although perhaps not as high as estimated from Ganf and Viner's data used by Hecky et al. 1991), and even higher for Lake Victoria because rainfall exceeds river inflow by an even greater ratio (Table 1) in these lakes.

All three lakes also have significant inputs of fixed nitrogen by biological fixation of atmospheric nitrogen. In Malawi, biological fixation of nitrogen may be 4.5 times greater than the atmospheric deposition (Bootsma and Hecky 1993). Heterocystous, photosynthetic cyanobacteria are prominent in the phytoplankton (Hecky and Kling 1987; Hecky 1993; Bootsma 1993) and benthic algal communities of the lakes (Bootsma 1993), and their nitrogen fixation represents an input of N which is not dependent on hydrological fluxes. Nitrogen fixation is also performed by some benthic Eubacteria. Estimates of the scale of N-fixation have been made for some of the lakes, and it seems that biological N-fixation is by far the most important input of nitrogen to these lakes (Bootsma and Hecky 1993).

When all inputs are accounted for and the mass of dissolved nutrients is known in Lake Malawi, then the residence time (=mass of element in the lake divided by rate of inputs) of

Table 3. Estimates of the Annual N, P and Si Loading to the Mixed Layer
of Lake Malawi (from Bootsma and Hecky 1993) and the Residence Time
of These Elements in the Annually Mixed Layer (upper 200 m)

	N	P	Si
		(moles m^{-2} y^{-1})	
Rivers	0.023	0.001	0.22
Atmospheric deposition	0.087	0.002	Trace
Vertical mixing (from below 250 m)	0.0	0.023	1.64
N deficit (biological fixation)	0.41	0.0	0.0
Total inputs	0.52	0.026	1.86
Mixed layer mass (moles m^{-2})	1	0.1	80
Residence time (yr)	2	4	40

the respective nutrients can be determined. The conservative element sodium which is not retained within the lake (Hecky and Bugenyi 1992) has a residence time of approximately 900 years which is 10 times longer than that of P and 100 times longer than N. The nitrogen mass in the whole lake is being replaced every 9 years and most of the replacement is due to biological fixation (the reason for this rapid cycling and high fixation will be discussed below), and substantial increases or decreases in N fixation would change the whole lake nitrogen mass in Malawi on a decadal time scale. In contrast there is a vast reservoir of Si and P in the deep water which would buffer their cycles against short term changes in input or output rates. The mixed layer with its smaller volume and lower nutrient concentrations has much lower residence times for the nutrient elements, and inputs are dominated by upward transfer from the deep water. The mixed layer can respond to changes in vertical mixing on less than decadal time scales (Table 3). Hecky et al. (1991) reached similar conclusions about these nutrient cycles in Tanganyika although the actual data are much less secure for that lake.

Losses of nutrients to burial in sediments are virtually unmeasured in the African great lakes. The problem is not trivial given the complexity of sedimentary regimes which can produce high variance in deposition rate estimates. Linear sediment accumulation rates can vary from zero in non-depositional areas, which can account for a significant portion of the lake's area (Johnson, this volume), to greater than a centimeter per thousand years in the quiescent, deep basins. Indirect estimates of sedimentary output can be generated by subtracting outflow losses from inputs. This is a reasonable first order estimate for phosphorus and silicon which do not have a gaseous phase in their biogeochemical cycle, but this approach would overestimate sediment burial for nitrogen which can be denitrified in hypoxic environments with attendant loss of nitrous oxide and nitrogen gas to the atmosphere.

Good measurements of nutrient burial rates require undisturbed, continuous sedimentation, interpretable radiochronology and appropriate nutrient analysis of the biogenic portion of the sediment matrix. Extrapolating site specific burial rates to the whole lake also requires knowledge of sedimentation patterns. Mean annual burial rates for Victoria prior to 1900 (Hecky 1993), and for Tanganyika and Kivu (Haberyan and Hecky 1987) are quite similar for P and N (approx. 1 mM P m^{-2} and 30 mM N m^{-2}). This surprising similarity might be explained if the regional atmosphere were the source of these nutrients, nitrogen processing were similar in the lakes and sediment focusing were similar at the coring sites. However, the available data for atmospheric deposition and sediment burial are too sparse to assess whether agreement or lack of agreement is simply chance. The estimated rates offered here are provided to give a sense of magnitude and to make the point that inputs must equal outputs at steady state. Therefore, reasonable agreement between inputs and outputs, after correcting burial rates for the sediment focusing, is expected. The IDEAL program should greatly improve our knowledge of nutrient burial rates.

VERTICAL MIXING

The equality of inputs and outputs is essential to the steady state condition, but these parameters alone do not predict the distribution of the nutrient elements in aquatic systems. At least four factors must be known to define fully the internal distribution of nutrients between surface water, deep water and sediments (Broecker and Peng (1982): 1) the rate of input, 2) the rate of loss at the outflow, 3) the rate of transfer of water and materials between surface and deep water, and 4) the fraction of these materials returned to solution in the deep water (or inversely, the fraction lost to permanent burial). Factors 1), 2), and 4) were discussed above. Factor 3) combines the sedimentation flux from the mixed layer and the return of deep water and its contained nutrients to the mixed layer. At steady state these downward and upward fluxes should be nearly equal, and offset only by the loss of material to permanent burial. However, a great range of sedimentary fluxes and vertical mixing coefficients can occur in aquatic ecosystems and still fulfill the steady state assumption.

The vertical mixing flux is the master variable for nutrient cycling because, in slowly flushed lakes, sedimentary flux cannot greatly exceed the return of nutrient for long without exhausting the mixed layer of nutrient. This is made apparent in Table 3 where the input fluxes for Malawi in Table 2 are recast and the supply of nutrient to the mixed layer from vertical mixing is estimated (Bootsma and Hecky 1993). The coefficient of vertical mixing estimated from "bomb" tritium invasion into Malawi is based on the data and interpretation of Gonfiantini et al. (1979). The vertical mixing flux exceeds the input of P and Si by nearly an order of magnitude. Therefore, substantial variation in surface inputs would easily be masked if the vertical mixing component were not changed. In contrast, any annual or longer reduction in intensity of vertical mixing would be quickly evident for P. The residence time for Si is much longer because there are substantial reserves of Si in the mixed layer. Little if any N is contributed by vertical mixing of the deep water because of consumption of dissolved nitrate and ammonia at the permanently anoxic interface (Fig. 1). This consumption will be discussed below. The N balance of the mixed layer is extremely dependent on biological fixation of N, but this N "deficit" results from the relatively "excess" supply of P from vertical mixing. Edmond et al. (1993) describes a similar "precarious" N balance for Tanganyika which must be met by N-fixation (Hecky et al. 1991).

Tanganyika has higher deep water concentrations of dissolved P and N than Malawi by approximately a factor of two while surface concentrations are similarly low or undetectable in both lakes. If the rates of vertical mixing in Malawi were a factor of two higher and input rates and output rates were comparable, then the rate of N and P supply to the surface layer of the two lakes might be quite similar. The two lakes have been found to have quite similar rates of annual primary productivity (Hecky 1984) which would be consistent with similar

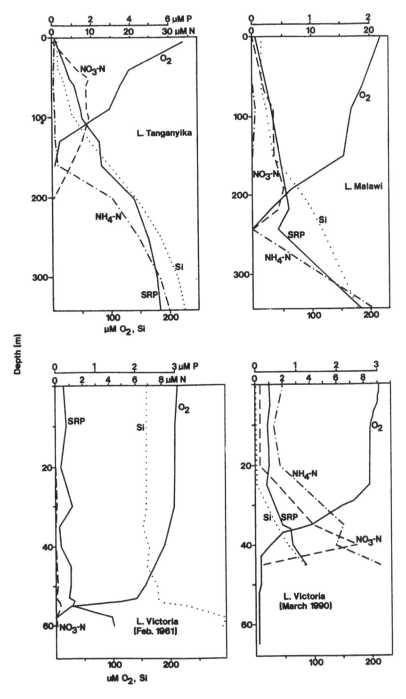

Figure 1. Profiles of inorganic nutrients and oxygen for Lake Tanganyika, Lake Malawi and Lake Victoria (1961 and 1990). SRP is soluble reactive phosphorus. From Bootsma and Hecky (1993).

rates of supply and two-fold higher rates of vertical mixing in Malawi. Hecky et al. (1991) postulated higher vertical mixing rates in Tanganyika and higher input rates of N and P than those in Malawi (Bootsma and Hecky 1993). The assumptions about the vertical mixing rates and the inputs for Tanganyika seem much too high given the results from the more complete Malawi data set. We need much better definition of vertical mixing rates in all the great lakes before we can successfully model or manage them.

Broecker and Peng (1982) illustrate that, under steady state conditions where inputs equal outputs, an increase in the vertical mixing can cause a transitory increase in burial rates and decline in deep water concentrations. The transitional state will persist until the downward sedimentary flux and the the burial rate have reached a new steady state with the higher mixing rate. Similar qualitative changes in burial rate could result from an increase in external inputs, but the response time of the increase in burial would be different. For example, in Lake Malawi response times to changes in vertical mixing would be relatively rapid, but its response to changes in riverine input would be much slower based on the proportional contributions of these nutrient sources to the mixed layer. The sediment record of Tanganyika, Kivu and Victoria offer several examples of fast and slow changes in sedimentation rates, nutrient burial and microfossil communities (Haberyan and Hecky 1987; Hecky 1993). Good mass-balance models, validated with appropriate values for modern inputs, outputs, vertical mixing coefficients and deep water regeneration will be necessary to help interpret the sediment records of these lakes and to forecast the effects of watershed and airshed changes on inputs or climate changes on rates of vertical mixing. The IDEAL program should strive to produce the necessary information.

NUTRIENT REGENERATION AND BURIAL

Chemical and physical weathering and atmospheric chemistry determine the rate of supply of nutrients to lakes, and solar energy and physical forces applied to the lakes determines their stratification behaviour and their rates of vertical and horizontal mixing. If these were the only major factors operating on the African Great Lakes, they would likely be much more similar than they are. What differentiates them biogeochemically from temperate Great Lakes and from each other are differences in the internal cycling of nutrients. Tropical lakes differ from temperate lakes in their endless summer of warm waters through all depths which allows biological processes to operate continuously and modify profoundly their chemical environments (Kilham and Kilham 1990). This endless summer results in continuously high levels of algal photosynthesis in surface waters, and in chronic hypoxia and anoxia in deep water which leads to selective regeneration of nutrients. In particular, we postulate that maintenance of a permanent or nearly permanent anoxic interface within the water column accelerates several microbial processes which have quite different effects on the cycling of N, P and Si. Effects on the internal cycling lead to unique stoichiometries of dissolved nutrients in the deep water among the lakes. The stoichiometries of particulate matter in surface waters differ among the lakes and also differ from the deep water stoichiometries indicating selective fluxes to the surface layers (Table 4).

THE PHOSPHORUS PUMP

The rate of phosphorus cycling sets the ecosystem tempo for all the nutrients in these great lakes as in most lakes. Phytoplankton take up nutrients in relatively narrow range of

Table 4. Stoichiometry (molar ratios) of Particulate Matter in Surface Waters and of Dissolved Inorganic Si, N and P in Deep Waters

	Surface particulates C:N:P	Deep water dissolved nutrients Si:N:P
Tanganyika	115:13:1[a]	38:5:1
Malawi	257:21:12[b]	100:12:1
Victoria	110:13:1[b]	57:66:1
Kivu	530:48:1[c]	22:81:1

[a]Edmond (1993). [b]Hecky et al. (1993). [c]Degens and Kulbicki (1973).

proportion (Hecky and Kilham 1988) from lake water to form the sedimentary particles, a fraction of which is then eventually lost to permanent burial. Although phytoplankton species and communities can be limited in space and time by local depletions of C, N, P or Si (the possibility of metal limitation is not treated here), atmospheric carbon and nitrogen are effectively infinitely available through gas exchange to the phytoplankton (Schindler et al. 1971, 1972, Schindler 1977, Hendzel et al. 1994); and Si is not required by all phytoplankton species. Light limitation occurs in the offshore of Lake Victoria (Mugidde 1993) where excess dissolved phosphate is continuously detected, but in shallow inshore waters phosphate becomes undetectable and P limits algal growth (Hecky 1993). The relative availability of the major nutrients and light in space and time will favor different species which, in turn, can differentially influence sedimentation losses of these nutrients. However, P will set the system rate of primary productivity, sedimentation and permanent burial because it is required by all species and is not readily available from the atmosphere.

The oxic–anoxic boundary in water or sediment columns has profound effects on rates of supply of phosphorus. The lowest concentrations of soluble reactive phosphorus (SRP) in lakes are set by the enzyme kinetics of phytoplankton which can take up SRP to subnano-molar concentrations. But, when light limits plant growth, e.g., in deep waters, the SRP concentrations are set by particle–SRP equilibria (Froehlich 1988). The redox state of Fe affects these equilibria markedly. Iron oxides coat inorganic particles under high redox conditions, and iron oxides have a high affinity for SRP. Under low redox conditions, Fe^{+3} compounds are reduced and the more soluble Fe^{+2} is released along with sorbed SRP. Under conditions of prolonged anoxia, nitrate, sulfate and eventually carbon dioxide are used as electron acceptors in dissimilatory metabolism. Reduction of sulfate to sulfide precipitates ferrous iron and limits iron concentrations in anoxic waters. The high retention of sulfur in Malawi (Table 2) can be equated with retention of Fe primarily as pyrite. The consequence of anoxia is loss of the iron oxide control on SRP concentrations in deep water. In contrast to the regeneration of P for eventual return to the mixed layer, Fe is retained in the deep water sediments. Return of SRP through the oxycline by vertical mixing has little affect on its concentration (Fig. 1). Although a slight "notch" is evident in the upwardly convex profile for SRP in Tanganyika and Malawi the general convexity of the curve is maintained indicating that the deep waters are continuously supplying P to the mixed layer. Further investigation is required to determine if the deeper "notch" in the Malawi profile which would prevent the upward transport of P is a permanent feature. In Kivu, there is evidence that P is consumed in the oxycline by a bacterial plate; consequently the oxycline short circuits the vertical transport of P and returns it to deep water without enriching the surface layers (Haberyan and Hecky 1987). This shortcircuiting must contribute to the extremely P deficient particulate matter in the mixed layer of Kivu (Table 4; Hecky et al. 1993). Vertical mixing of deep water P dominates the P budget of the the mixed layer in Malawi (Table 3), and it is the "pump" which maintains the high rates of biological activity in the surface layers. Decreased rates of vertical mixing would reduce algal production in the mixed layer, and increased mixing would have the opposite effect.

THE NITROGEN SINK

The presence of a continuous or nearly continuous oxic–anoxic interface in the water column of the African Great Lakes has an opposite effect on the vertical flux of N compared to its effect on P. This interface is a "sink" for dissolved inorganic nitrogen because of the activity of denitrifying bacteria which use nitrate as a terminal electron acceptor during the decomposition of organic compounds (Seitzinger 1988). The effect on dissolved inorganic nitrogen is dramatic and obvious in vertical concentration profiles (Fig. 1). As oxygen concentrations fall with depth to less than 5 μM denitrifying bacteria (Seitzinger 1988) become active. The activity of the denitrifiers reduces nitrate concentrations and a concave upward nitrate profile is observed. Below the oxic–anoxic interface, upward fluxing ammonium concentrations are also consumed to trace concentrations. In these hypoxic waters both nitrification and denitrification reactions occur contemporaneously with the net result that nitrate and ammonium are transformed to N_2 and N_2O which diffuse away as gases or to suspended N in microbial mass which then sediments. This effect is seen in all three lakes (Fig. 1). The consequence of the denitrifying layer to the mixed layer is to eliminate N loading from the deep water and to impose a very low N:P ratio on upward diffusing water (Fig. 2). The high N:P ratios just below the euphotic zone in Tanganyika for example (Fig. 2) must be maintained by biological N fixation which makes up the N deficit imposed by the vertical transport of deep water. The deep water is also affected by the mid-water sink. Sedimenting particles in the deep water do regenerate in a Redfield molar ratio of 16:1 (Edmond et al. 1993), but the N:P ratio of the dissolved inorganic nutrients never exceeds 10 in the deep water (Fig. 3) as a consequence of the midwater sink. Although the oxic–anoxic interface in the water column is not yet permanent in Victoria, it persists for adequate lengths of time to affect not only dissolved nitrate and ammonium, but also total N concentrations in the deep water (Fig. 3) resulting in low total N:total P ratios in the deep water. This profile clearly indicates that N is being input primarily from biological fixation of atmospheric nitrogen.

The oxic–anoxic interface is a feature of all lakes. If lakes mix fully and have oxygen continuously present to the surface of the sediments, then the interface occurs at some depth in the sediments. In the organic-rich sediments of the African lakes, which receive a continuous input of newly produced organic carbon from photosynthesis, this interface will always lie close to/or at the sediment surface, ready to emerge into the water column if thermal stratification reduces the downward flux of oxygen. As oxygen becomes unavailable, denitrification begins to be the dominant electron acceptor (Fig. 4). The movement of the oxycline to a shallower depth means that organic matter is in excess and more decomposition shifts to the denitrification pathway. If the interface lies within the sediment column, then flux rates for oxygen or nitrate will depend on molecular diffusion. When the

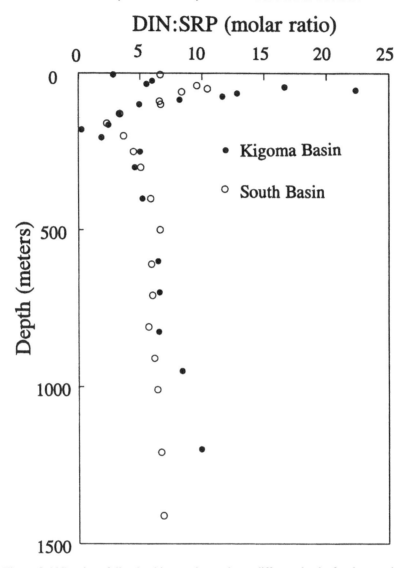

Figure 2. N:P ratios of dissolved inorganic species at different depths for the two deepest basins of Lake Tanganyika in April 1975 (data from Edmond et al. 1993).

interface is within the water column, however, supply rates are controlled by eddy diffusivity and can be orders of magnitude faster.

Much higher rates of nitrate consumption in the African Great Lakes can be calculated from the nitrate profiles in Fig. 1 with the application of reasonable coefficient of eddy diffusivity than have been measured in dimictic temperate lakes having oxic hypolimnions (Fig. 5). The lateral extent of the interface in the water column will depend on the depth of stratification, the morphometry of the basin and the input of new organic carbon to the deep water. As the water column interface spreads laterally when it becomes shallower, the area of oxygenated sediments where denitrification is controlled by molecular diffusion will consequently contract. Increasing the area of the water column interface will lead, in turn,

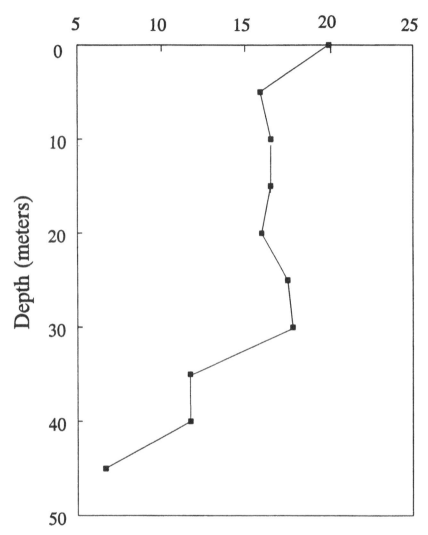

Figure 3. Total N:total P ratios at different depths in Lake Victoria (data from Hecky 1993). These data are for the same profile for which inorganic nutrients are given in Fig. 1.

to higher lakewide rates of denitrification and will create an increasingly effective sink for dissolved fixed nitrogen. Because P flux from the deep water remains unaffected or is enhanced by the spread of anoxia, the surface mixed layer will suffer an increasing N deficit which must be met on a system level by N fixation.

The extensive nitrogen sink in these three African great lakes poises them for dominance by nitrogen fixing, heterocystous cyanobacteria. The area of the permanent oxic–anoxic interface is a much higher proportion of Tanganyika's surface area than in the other two lakes, so shunting of nitrate to the nitrogen sink will dominate the nitrogen cycle to a greater

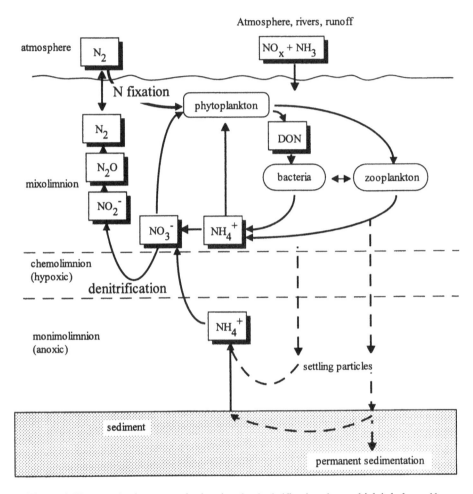

Figure 4. The aquatic nitrogen cycle showing the denitrification shunt which is balanced by biological nitrogen fixation in the African lakes.

extent than in Malawi. In fact, the deep water of Malawi has a higher N:P ratio than Tanganyika, and heterocystous cyanobacteria are a relatively minor component of the phytoplankton (Hecky and Kling 1987). Nitrogen fixation by benthic N-fixers may be adequate to satisfy the N-deficit because most of the lake bottom area is oxygenated and most N in settling organic material will be returned to algae (Fig. 4). In contrast, Lake Tanganyika has a spectacular bloom of *Anabaena* in October and November when the lake restratifies after its annual mixing period (Hecky and Kling 1981). The maximum biomass of the *Anabaena* bloom in Tanganyika in 1975 was 40 times greater than the maximum observed over the annual cycle in Malawi in 1980 (Hecky and Kling 1987). The eutrophication of Lake Victoria over the last three decades (Hecky 1993) has led to a deoxygenation of the seasonal hypolimnion (Hecky et al. 1994). This has led to higher losses of dissolved fixed nitrogen into the N sink (Figs. 1 and 5), and to nearly continuous dominance of heterocystous cyanobacteria over much of the lake (Kling and Mugidde, unpublished data). Similar changes in the lateral extent of the water column N sink and the denitrification flux

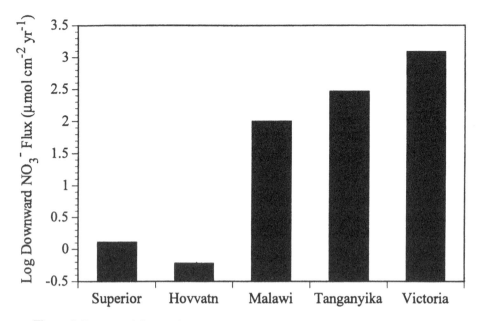

Figure 5. Downward fluxes of nitrate to the sites of denitrification which are within the sediments in Hovvatn and Superior but within the water column in the African lakes. Lake Superior and Hovvatn data are from Kelly et al. (1987). African lake data are estimated as described in the text.

have certainly been a recurrent phenomena in the history of these lakes, and understanding and quantifying these N fluxes will be crucial to interpreting the lake's history through paleolimnological analysis as well as predicting the lake's responses to anthropogenic impacts.

THE SILICON DRAIN

Silicon concentrations respond to increased P flux in a manner qualitatively similar to N resulting in falling concentrations in the mixed layer, but these changes occur by a quite different mechanism. Increased sedimentation rates and increased burial rates that result from increased primary productivity accelerate the sedimentary output of silicon contained in diatom frustules. In contrast to organic N and P compounds whose regeneration to their inorganic forms is accelerated by a variety of exo- and endo-enzymes generated by microbial decomposition, Si is dissolved from diatom frustules by a simple hydration reaction following first order kinetics (Lawson et al. 1978):

$$dC_{sol}/dt = K(C_{sat} - C_{sol})A,$$

where C_{sol} is the concentration of dissolved Si in the water, C_{sat} is the saturated concentration of Si, K is the rate constant, and A is the surface area of diatom silica per unit volume of solution. For a nominal saturation concentration of 1000 μM, dissolution will be relatively rapid in the surface waters of all three lakes where concentrations are quite low and will be significantly slower but still appreciable in the deep water where concentrations do not exceed 500 μM. Only in the sediments will interstitial water Si concentrations increase to saturation concentrations and frustule dissolution cease. Even there, the accumulating frustules must be buried to depth of saturation within the sediments which can mean exposure to continued dissolution for some years after reaching the sediment surface. The length of this exposure time will depend on the rate of sediment accumulation, the porosity of the sediments and the rates of vertical diffusivity within the upper sediment column.

Increased P flux into the mixed layer should lead to increased diatom productivity (until Si becomes limiting for diatom growth), increased downward flux of biogenic particulate Si and other biogenic constituents, increased sediment accumulation rates, and an increased burial rate of biogenic Si relative to P. The increased sedimentary output should strip the mixed layer of dissolved Si unless inputs of Si from the watershed increase in proportion. Because P regeneration is not dependent on the same reactions as Si, its burial rate need not increase in proportion to Si, and the hypolimnion Si concentrations may also fall over time under increased P fluxes. This sequence of events is well illustrated by the recent eutrophication of Lake Victoria. At the only coring site to have been studied in detail (Hecky 1993), sedimentary output of P has increased 3× since 1900, sediment accumulation rates have increased by 2.5× and biogenic Si output has increased 3.5×. Water column Si concentrations have fallen by a factor of 10 (Fig. 1). Water column P concentrations have increased somewhat, but not in proportion to the increase in sedimentary output illustrating how the Si and P cycles can be uncoupled.

Many pelagic diatom species are quite sensitive to Si concentrations and to the ratio of Si to N or P (Kilham et al. 1986). The diatom community of Lake Victoria has changed dramatically and rapidly (Hecky 1993) to its changing biogeochemical fluxes and the stoichiometry of these fluxes. Haberyan and Hecky (1987) invoked changing Si:P ratios in geological time to explain alterations observed in the diatom microfossil communities of Tanganyika and Kivu; in turn, they inferred varying positions of the thermocline and varying intensities of vertical mixing to explain the Si:P ratios. Because diatoms are the most abundant microfossils in sediments of the African Great Lakes, they will be a crucial indicator of past lake states. The hypothetical interactions proposed here need to be tested and quantified in the modern lakes in order to correctly interpret the rich histories of these lakes. Diatoms are often imputed to be the most desirable ration for zooplankton and other consumer organisms which are the basis of fish production (Officer and Ryther 1980). Improved modelling capacity for the interrelations of the P, N and Si cycles will be essential to the future maintenance of water quality and fish productivity. The IDEAL program and other ongoing studies should strive to attain this capacity.

CONCLUSION

The African Great Lakes exhibit a great range of morphometries and geological settings within a tropical climate which has been shown to vary on time scales from decades to eons. IDEAL proposes to recover long sediment cores in order to characterize and interpret the proxy data on the climatic history of continental eastern Africa available in these great lakes. The reconstructed histories of ecological and climatic change will be used to validate current global climate models which, in turn, will be used to forecast future climates. A reconstruction of the hydrological history of these lakes will be a critical component of that interpretive effort as their levels and internal circulation are dictated by climatic factors. Organic material and aquatic microfossils are abundant in the lakes' sediments and are rich in information content. To decode that information, however, we need greatly improved knowledge and validated models of how nutrients enter, circulate and leave the lakes. The current limited state of knowledge is unacceptable and precludes any rigorous scientific management of the lakes or detailed interpretation of their histories. Special attention must be given to improving our knowledge base for the following aspects:

1. Hydrological fluxes must be better defined, in particular open lake measurements of rainfall and evaporation are required.

2. The chemistry of the significant hydrologic inputs must be determined; little is known of the nutrient contributions from the rivers to the lakes, and atmospheric chemistry data are nearly nonexistent; what is known suggests that atmospheric chemistry varies widely and a well-defined network will be required to explain this variability.

3. The rates of vertical circulation are critical to any modelling effort, but they are essentially unknown; geochemical tracer techniques using anthropogenic compounds such as chlorofluorocarbons and bomb-tritium would provide this information and should be applied.

4. The nutrient chemistry of these lakes is defined on the short- to medium-time scale (years to centuries) by the internal cycles and activities of biological organisms; however, there are as yet no direct measurements of sulfate reduction, denitrification, methanogenesis, diatom dissolution, or nitrogen fixation on seasonal time scales in any of the lakes.

5. Burial in sediments is the eventual fate of the master nutrient variable, P, as well as Si, yet their loss rates on a lakewide basis are unknown for any of the lakes; a knowledge of the dependence of rates of burial on varying input rates and vertical mixing rates is an essential component of water quality modelling. These data are currently nonexistent, but within the grasp of the IDEAL program.

The African Great Lakes are an important water and biological resource for their riparian countries and the eastern African region as a whole. They are also a priceless genetic and historical resource for all the peoples of the world. The lakes deserve and demand the attention of scientists of many origins and disciplines to insure that these resources enrich everyone. That is the challenge to IDEAL.

REFERENCES

Beauchamp, R.S.A., 1939, Hydrology of Lake Tanganyika: Internationale Revue der Gesamten Hydrobiologie und Hydrographie, v. 39, pp. 316–353.

Bootsma, H.A., 1993, Algal dynamics in an African Great Lake, and their relation to hydrographic and meteorological conditions: Ph.D. Thesis, University of Manitoba, Winnipeg, Manitoba.

Bootsma, H.A., and Hecky, R.E., 1993, Conservation of the African Great Lakes: A limnological perspective: Conservation Biology, v. 7, pp. 644–656.

Broecker, W.S., and Peng, T.-H., 1982, What keeps the system in whack, Ch. 6, *in* Tracers in the sea: Eldigio Press, Columbia University, New York, pp. 275–316.

Craig, H., Dixon, F., Craig, V.K., Edmond, J., and Coulter, G.W., 1974, Lake Tanganyika geochemical and hydrographic study, 1973 expedition: Scripps Institute of Oceanography Reference Series 75–5, 83 pp.

Degens, E.T., and Kulbicki, G., 1973, Data file on metals in East African rift sediments: Woods Hole Oceanographic Institution Technical Report WHOI-73-15, 235 pp. + 20 figs.

Degens, E.T., von Herzen, R.P., and Wong, H.-K., 1971, Lake Tanganyika water chemistry, sediments, geological structure: Naturwissenschaften, v. 59, pp. 229–241.

Degens, E.T., von Herzen, R.P., Wong, H.-K., Deuser, W.G., and Jannasch, H.W., 1973, Lake Kivu: Structure, chemistry and biology of an East African rift lake: Geologische Rundschau, v. 62, pp. 245–277.

Edmond, J.M., Stallard, R.F., Craig, H., Craig, V., Weiss, R.F., and Coulter, G.W., 1993, Nutrient chemistry of the water column of Lake Tanganyika: Limnology and Oceanography, v. 38, pp. 725–738.

Froelich, P.N., 1988, Kinetic control of dissolved phosphate in natural rivers and estuaries: A primer on the phosphate buffer mechanism: Limnology and Oceanography, v. 33, pp. 649–668.

Ganf, G.G., and Viner, A.B., 1973, Ecological stability in a shallow equatorial lake (Lake George, Uganda): Proceedings of the Royal Society of London B, v. 184, pp. 321–346.

Haberyan, K., and Hecky, R.E., 1987, The Late Pleistocene and Holocene stratigraphy and paleolimnology of Lakes Kivu and Tanganyika: Palaeogeography, Palaeoclimatology, Palaeoecology, v. 61, pp. 169–197.

Hecky, R.E., 1978, The Kivu-Tanganyika basin: The last 14,000 years: Polish Archives of Hydrobiology, v. 25, pp. 159–165.

Hecky, R.E., 1984, African lakes and their trophic efficiencies: A temporal perspective, in Meyers, C., and Strickler, J.R., eds., Trophic interactions within aquatic ecosystems: AAAS Selected Symposium, v. 85, pp. 405–448.

Hecky, R.E., 1993, The eutrophication of Lake Victoria: Verhandlungen Internationale Vereingung fuer Theoretische und Angewandte Limnologie, v. 25, pp. 39–48.

Hecky, R.E., and Bugenyi, F.W.B., 1992, Hydrology and chemistry of the African Great Lakes and water quality issues: Problems and solutions: Mitteilungen Internationale Vereingung fuer Theoretische und Angewandte Limnologie, v. 23, pp. 45–54.

Hecky, R.E., Bugenyi, F.W.B., Ochumba, P., Gophen, M., Mugidde, R., and Kaufman, L., 1994, Deoxygenation of the deep water of Lake Victoria: Limnology and Oceanography, v. 39, pp. 1476–1480.

Hecky, R.E., Campbell, P., and Hendzel, L.L., 1993, The stoichiometry of carbon, nitrogen, and phosphorus in particulate matter of lakes and oceans: Limnology and Oceanography, v. 38, pp. 709–724.

Hecky, R.E., and Degens, E.T., 1973, Late Pleistocene–Holocene chemical stratigraphy and paleolimnology of the rift valley lakes of Central Africa: Woods Hole Oceanographic Institution Technical Report 73-28.

Hecky, R.E., Fee, E.J., Kling, H.J., and Rudd, J.M.W., 1981, Relationship between primary production and fish production in Lake Tanganyika: Transaction of the American Fisheries Society, v. 110, pp. 336–345.

Hecky, R.E., and Kilham, P., 1988, Nutrient limitation of phytoplankton in freshwater and marine environments: Limnology and Oceanography, v. 33, pp. 796–832.

Hecky, R.E., and Kling, H.J., 1987, Phytoplankton ecology of the great lakes in the rift valleys of Central Africa: Ergebnisse der Limnologie, v. 25, pp. 197–228.

Hecky, R.E., Kling, H.J., and Brunskill, G.J., 1986, Seasonality of phytoplankton in relation to silicon cycling and interstitial water circulation in large, shallow lakes of central Canada: Hydrobiologia, v. 138, pp. 117–126.

Hecky, R.E., Spigel, R.H., and Coulter, G.W., 1991, The nutrient regime, in Coulter, G.W., ed., Lake Tanganyika and its life: Oxford University Press, Oxford, England, pp. 76–89.

Hendzel, L.L., Hecky, R.E., and Findlay, D.L., 1994, Recent changes of N_2-fixation in Lake 227 in response to reduction of the N:P loading ratio: Canadian Journal of Fisheries and Aquatic Sciences (in press).

Johnson, T.C., 1996, Sedimentary processes and signals of past climatic change in the large lakes of the East African Rift Valley, in Johnson, T.C., and Odada, E., eds., The limnology, climatology and paleoclimatology of the East African lakes: Gordon and Breach, Toronto, pp. 367–412.

Kelly, C.A., et al., 1987, Prediction of biological acid neutralization in acid-sensitive lakes: Biogeochemistry, v. 3, pp. 129–140.

Kendall, R.L., 1969, An ecological history of the Lake Victoria basin: Ecological Monographs, v. 39, pp. 121–176.

Kilham, P., and Kilham, S.S., 1989, Endless summer: Internal loading processes dominate nutrient cycling in tropical lakes: Freshwater Biology, v. 23, pp. 379–389.

Kilham, P., Kilham, S.S., and Hecky, R.E., 1986, Hypothesized resource relationships among African planktonic diatoms: Limnology and Oceanography, v. 31, pp. 1169–1181.

Kite, G.W., 1981, Recent changes in level of Lake Victoria: Hydrological Sciences Bulletin, v. 26, pp. 233–243.

Kufferath, J., 1952, Le milieu biochimique. Résutats scientifiques de l'exploration hydrobiologique du Lac Tanganika (1946–47): Institut Royal des Sciences Naturelles de Belgique, v. 1, pp. 31–47.

Lawson, D.S., Hurd, D.C., and Pankratz, H.S., 1978, Silica dissolution rates of decomposing phytoplankton assemblages at various temperatures: American Journal of Science, v. 278, pp. 1373–1393.

Livingstone, D.A., 1975, Late Quaternary climate change in Africa: Annual Review of Ecology and Systematics, v. 6, pp. 249–280.

Melack, J.M., 1976, Primary productivity and fish yields in tropical lakes: Transactions of the American Fisheries Society, v. 105, pp. 575–580.

Mugidde, R., 1993, The increase in phytoplankton primary productivity and biomass in Lake Victoria: Verhandlungen Internationale Vereingung fuer Theoretische und Angewandte Limnologie, v. 25.

Nicholosn, S.E., 1996, A review of climate dynamics and climate variability in eastern africa, *in* Johnson, T.C., and Odada, E., eds., The limnology, climatology and paleoclimatology of the East African lakes: Gordon and Breach, Toronto, pp. 23–54.

Officer, C.B., and Ryther, J.H., 1980, The possible importance of silicon in marine eutrophication: Marine Ecology Progress Series, v. 3, pp. 83–91.

Owen, R.B., Crossley, R., Johnson, T.C., Tweddle, D., Kornfeld, I., Davison, S., Eccles, D.H., and Engstrom, D.E., 1990, Major low levels of Lake Malawi and their implications for speciation rates in cichlid fishes: Proceedings of the Royal Society of Lond., v. 240, pp. 519–553.

Schindler, D.W., 1977, The evolution of phosphorus evolution in lakes: Science, v. 195, pp. 260–262.

Schindler, D.W., Armstrong, F.A.J., Holmgren, S.K., and Brunskill, G.J., 1971, Eutrophication of Lake 227, Experimental Lakes Area, northwestern Ontario, by the addition of phosphate and nitrate: Journal of Fisheries Research Board of Canada, v. 28, pp. 1763–1782.

Schindler, D.W., 1972, Atmospheric carbon dioxide: Its role in maintaining phytoplankton standing crops: Science, v. 177, pp. 1192–1194.

Seitzinger, S.P., 1988, Denitrification in freshwater and coastal marine ecosystems : Ecological and geochemical significance: Limnology and Oceanography, v. 33, pp. 702–725.

Sommer, U., 1990, The role of competition for resources in phytoplankton succession, *in* Sommer, U., ed., Plankton ecology: Succession in plankton communities: Springer, pp. 57–106.

Stoffers, P., and Hecky, R.E., 1978, The Late Pleistocene–Holocene evolution of the Kivu-Tanganyika basin: International Association of Sedimentology, Special Publication, v. 2, pp. 43–55.

Talling, J.F., 1966, The annual cycle of stratification and phytoplankton growth in Lake Victoria (East Africa): Internationale Revue der Gesamten Hydrobiologie, v. 51, pp. 545–621.

Talling, J.F., and Talling, I.B., 1965, The chemical composition of African lake waters: Internationale Revue der Gesamten Hydrobiologie, v. 50, pp. 421–463.

Verbeke, J., 1957, Recherches ecologiques sur la faune des grands lacs de l'est Congo Belge, *in* Exploration hydrobiologiques des Lacs Kivu, Edouard et Albert, Institut des Sciences Naturelles de Belgique, Bruxelles, v. 3(1), 177 p.

Von Herzen, R.P., and Vacquier, V., 1967, Terrestrial heat flow in Lake Malawi, Africa: Journal of Geophysical Research, v. 72, pp. 4221–4222.

Wüest, A., Piepke, G., and Halfman, J., 1996, The role of dissolved solids in density stratification of Lake Malawi, *in* Johnson, T.C., and Odada, E., eds., The limnology, climatology and paleoclimatology of the East African lakes: Gordon and Breach, Toronto, pp. 183–202.

Pore Water Chemistry of an Alkaline Lake: Lake Turkana, Kenya

T.E. CERLING   *Department of Geology and Geophysics, University of Utah, Salt Lake City, Utah, United States*

Abstract — Pore fluids from two 12-meter piston cores in Lake Turkana, northern Kenya, show that methanogenesis following sulfate depletion is an important early diagenetic feature of this alkaline lake. CO_2 production during methanogenesis leads to calcite dissolution and dissolution of a magnesium silicate phase. The excess alkalinity/magnesium ratio is 2.8 in the southern core, close to the theoretical ratio of 2.25 for stevensite dissolution. Chloride concentrations in the southern basin increase downcore, but decrease in the northern basin core. This indicates that the two basins were separated in the late Pleistocene to early–middle Holocene, the southern basin being occupied by a saline lake while the northern basin was perhaps completely dry.

INTRODUCTION

Pore fluid chemistry has been important in understanding early diagenesis in marine systems (Berner, 1979; Broecker and Peng, 1982; Gieskes, 1974). Small changes in sediment composition resulting from mineral precipitation or dissolution often result in measurable changes in the dissolved ion concentration in the pore fluid. Marine systems are characterized by high dissolved sulfate concentrations so that diagenesis of organic matter is often dominated by sulfate reduction in the sediment column. Some alkaline lakes have lower sulfate concentrations than seawater so that complete depletion of sulfate is likely to occur near the sediment–water interface, and subsequent organic matter decomposition occurs by methanogenesis, which has a characteristic isotopic signature.

In this study I examine the early diagenesis of sediments in Lake Turkana, Kenya, an alkaline, closed-basin lake by studying the pore waters from 12-meter cores collected in the lake in 1984 by Duke University's Project PROBE.

REGIONAL BACKGROUND

Lake Turkana is a closed basin lake occupying a structural depression associated with the development of the East African Rift. In 1984 several cores were collected from the two major basins of Lake Turkana, Kenya (Fig. 1), including cores 1P, 2P and 2PG, 4P, 7P and 8P which are the subject of this study. Lake Turkana is fed principally by the Omo River, and secondarily by the combined flow of the Turkwel and Kerio Rivers (Fig. 1). The lake has a surface area of about 7560 km^2, and an average depth of 31 meters, the deepest part of the lake being about 114 meters in 1973. The lake level fluctuates by almost a meter per year; the average level has been dropping since the mid-1960s. The region has a mean annual temperature of about 29°C and receives about 280 mm annual precipitation, which comes principally in late-March to mid-May. The region has a considerable water deficit, estimated to be about 2.2 meters per year (Hopson, 1982; Cerling, 1986).

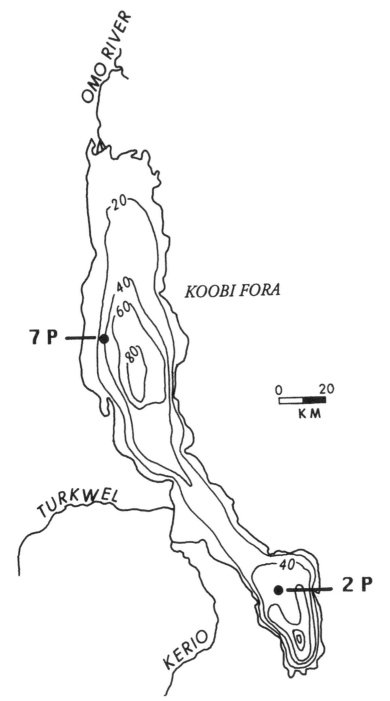

Figure 1. Location of cores 2P and 7P from Lake Turkana, Kenya. Core 2PG was a short gravity core collected at the 2P locality. Bathymetric contours (in meters) show that the lake is divided into two major basins, the North Basin and South Basin.

Lake Turkana is a brackish Na–HCO$_3$ lake with a total dissolved solids content of about 2500 mg/l (Table 1). Approximately 90% of the incoming ions are removed annually by mineral precipitation or buried as pore water (Yuretich and Cerling, 1983; Cerling, 1986).

Table 1. Chemistry of Modern Lake Turkana Waters Collected at Koobi Fora Spit from September 1984 to September 1985 (in mmol/l, except alkalinity, which is in meq/l)

Date	pH	Na	K	Ca	Mg	Alk	Cl	SO_4	F	Br	SiO_2
02 Sep 84	9.2	32.4	0.43	0.11	0.10	22.4	12.1	0.36	0.57	0.053	0.61
16 Sep 84	8.8	34.4	0.43	0.11	0.10	23.6	12.8	0.38	0.65	0.057	0.54
02 Oct 84	9.0	33.4	0.41	0.11	0.09	22.6	11.9	0.35	0.61	0.052	0.54
15 Oct 84	9.2	32.9	0.40	0.11	0.10	22.5	11.9	0.35	0.61	0.053	0.54
02 Nov 84	9.4	32.7	0.40	0.12	0.09	22.2	11.6	0.34	0.60	0.053	0.54
16 Nov 84	9.2	33.0	0.40	0.11	0.09	21.5	11.3	0.33	0.58	0.050	0.52
15 Dec 84	9.1	36.7	0.40	0.12	0.10	23.7	12.0	0.35	0.62	0.054	0.56
04 Jan 85	9.1	37.0	0.47	0.10	0.10	24.1	11.8	0.34	0.61	0.053	0.54
16 Jan 85	9.0	35.5	0.43	0.10	0.10	23.7	12.1	0.35	0.62	0.056	0.63
02 Feb 85	9.2	35.4	0.41	0.10	0.10	23.1	12.4	0.36	0.63	0.056	0.64
16 Feb 85	9.1	38.7	0.46	0.12	0.10	26.4	13.4	0.39	0.68	0.061	0.64
02 Mar 85	9.0	40.3	0.42	0.12	0.10	25.5	13.8	0.44	0.70	0.062	0.59
16 Mar 85	9.0	35.5	0.40	0.12	0.10	23.3	12.6	0.36	0.64	0.057	0.59
02 Apr 85	8.9	36.2	0.38	0.10	0.10	23.8	12.7	0.37	0.65	0.059	0.55
15 Apr 85	9.4	35.5	0.35	0.11	0.09	23.1	12.1	0.36	0.63	0.054	0.57
02 May 85	9.5	36.9	0.38	0.10	0.09	23.7	12.4	0.36	0.64	0.056	0.59
16 May 85	9.4	35.9	0.38	0.09	0.09	24.1	12.6	0.37	0.65	0.056	0.61
02 Jun 85	8.9	37.4	0.40	0.12	0.10	24.3	12.9	0.38	0.66	0.057	0.53
15 Jun 85	9.4	36.7	0.37	0.10	0.10	23.9	12.6	0.37	0.64	0.056	0.49
01 Jul 85	9.2	37.5	0.37	0.11	0.09	23.9	12.4	0.36	0.64	0.055	0.52
14 Sep 85	9.1	35.2	0.36	0.12	0.10	23.0	12.0	0.35	0.62	0.055	0.53
Average	9.1	35.7	0.40	0.11	0.10	23.5	12.4	0.36	0.63	0.055	0.56
±1σ	0.2	2.1	0.03	0.01	0.00	1.1	0.6	0.02	0.03	0.003	0.04

Lake Turkana, like all East African lakes, is deficient in magnesium, having only about 0.1 mmol/l. Mass balance calculations show that magnesium is most likely removed as a silicate phase, probably as a tri-octahedral smectite. Barton et al. (1987) showed that the chemistry of the modern lake is compatible with basin closure taking place sometime between about 4000 and 10,000 years ago.

METHODS

Piston and several gravity cores were collected in October 1984. They were sealed onboard ship and kept at approximately 25°C until February 1985 when they were opened. Pore fluids were centrifuged and filtered through 0.1 micron filters, and then analyzed within two weeks of filtration. It is not known what effect storage had on these cores, but they were kept at approximately the ambient temperature of Lake Turkana, which should minimize dissolution effects. Alkalinity was measured by titration, other anions by ion chromatography, cations except ammonium by atomic absorption, and silica and ammonium by colorimetric methods. Dissolved carbon isotopes were extracted by acidification with 100% H_3PO_4; CO_2 was equilibrated with the water for oxygen isotopic analysis. Cores 2P and 7P were sampled at approximately 30 to 50 cm intervals and are discussed in greatest detail. Cores 1P, 4P, and 7P were sampled at greater intervals than 2P and 7P. Halfman et al. (1992) found the sedimentation rate in Core 2P to be about 3 mm per year.

Water samples were collected each month from Koobi Fora spit on the northeastern end of the lake. Samples were stored in polypropylene bottles at Koobi Fora until they were shipped to the USA in August of each year. Waters were analyzed as described above.

RESULTS AND DISCUSSION

Chemistry of Lake Turkana

The chemistry of the surface lake waters was monitored at Koobi Fora spit, which is in the northern end of the lake, for about one year. As previously noted (e.g., Talling and Talling, 1965; Yuretich and Cerling, 1983) the lake is brackish and alkaline (pH 9.1). The salinity varies by about 10% over the course of a year, being the freshest in September through November, and the most saline in March through June. The salinity is controlled by the Omo River flood which peaks in August–October, and local rains in November and March–April. Conservative (sodium, chloride, bromide) and other major ions (alkalinity) show these trends, while the minor ions (sulfate, potassium, calcium, and magnesium) show little variation. Silica also varies little throughout the annual cycle. Of the conservative or major ions, chloride is most highly correlated with bromide ($r^2 = 0.9$), and less well with alkalinity ($r^2 = 0.7$) and with sodium ($r^2 = 0.5$). The lake is more saline at its southern end, as most of the inflow into the lake is at the extreme northern end of the lake (Hopson, 1982).

General Trends: Major Ions

The results of the gravity core 2PG from site 2P shows that piston core 2P from the same locality superpenetrated the sediment by about 2 meters (Fig. 2). The chemistry of the top of core 2PG is similar to the chemistry of the lake measured at Koobi Fora (Tables 1 and 2). In the following discussion and diagrams, the results from core 2P are shifted by 2 meters to account for the superpenetration. Core 7P is assumed to have superpenetrated about 1.5 meters based on Cl and Na profiles.

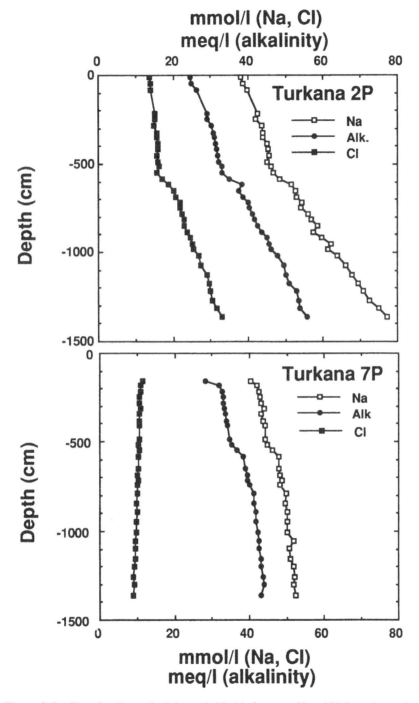

Figure 2. Profiles of sodium, alkalinity, and chloride for cores 2P and 7P from the southern and northern basins of Lake Turkana. Data from piston core 2P assume 2 meters of superpenetration into the sediment; data from core 7P assumes 1.5 meters of superpenetration. Total salinity, sodium, and alkalinity increase with increasing depth in both cores. However, the chloride concentration in core 2P from the southern basin increases with depth while that from core 7P in the northern basin decreases with depth.

Table 2. Chemistry of Pore Waters from Piston Core 2P and Gravity 2PG (in mmol/l, except alkalinity, which is in meq/l)

Sample	Depth interval	Average depth	Estim. depth	pH	Na	K	Ca	Mg	Alk	Cl	SO$_4$	F	Br	SiO$_2$
2PG-33	10–20	15.0	−15.0	8.7	38.0	0.60	0.14	0.13	24.7	13.7	0.02	0.61	0.06	0.98
2PG-34	44–55	49.5	−49.5	8.7	38.5	0.66	0.19	0.17	25.0	13.9	0.00	0.58	0.06	0.60
2PG-35	80–90	85.0	−85.0	8.6	39.7	0.73	0.14	0.24	26.4	14.1	0.00	0.54	0.06	0.65
2P-30	10–20	15.0	−215.0	8.6	42.7	0.80	0.13	0.43	29.1	15.1	0.01	0.50	0.07	0.90
2P-31	45–55	50.0	−250.0	8.7	42.0	0.86	0.19	0.50	29.2	15.2	0.01	0.54	0.08	0.68
2P-32	80–90	85.0	−285.0	8.7	43.5	0.89	0.19	0.55	30.2	14.9	0.00	0.46	0.07	0.75
2P-27	118–130	124.0	−324.0	8.6	43.8	0.94	0.15	0.64	30.9	15.6	0.01	0.48	0.07	0.80
2P-28	145–156	150.5	−350.5	8.6	43.9	0.86	0.15	0.67	31.1	15.7	0.01	0.47	0.07	0.73
2P-29	180–190	185.0	−385.0	8.6	44.9	0.93	0.17	0.72	31.3	15.8	0.01	0.49	0.07	0.73
2P-24	210–220	215.0	−415.0	8.5	45.4	0.87	0.16	0.75	31.6	16.1	0.00	0.46	0.07	0.68
2P-25	245–255	250.0	−450.0	8.6	45.6	0.90	0.21	0.78	32.0	15.7	0.00	0.44	0.07	0.73
2P-26	280–290	285.0	−485.0	8.5	45.1	0.86	0.19	0.92	32.1	16.1	0.00	0.44	0.07	0.80
2P-21	311–320	315.5	−515.0	8.6	46.2	0.87	0.20	1.02	32.9	16.1	0.00	0.42	0.07	0.73
2P-22	346–354	350.0	−550.0	8.6	46.8	0.87	0.21	1.01	33.1	15.6	0.01	0.44	0.07	0.75
2P-23	378–386	382.0	−582.0	8.6	48.3	1.07	0.22	1.26	34.9	17.0	0.01	0.46	0.07	0.70
2P-18	410–419	414.5	−614.5	8.5	51.4	1.36	0.26	2.02	38.4	18.7	0.01	0.52	0.07	0.67
2P-19	445–454	449.5	−649.5	8.6	52.6	1.05	0.31	1.85	37.6	20.2	0.01	0.47	0.09	0.60
2P-20	481–489	485.0	−685.0	8.5	53.0	0.95	0.38	2.04	38.5	20.6	0.01	0.43	0.09	0.93
2P-15	512–520	516.0	−716.0	8.6	54.3	1.06	0.39	2.32	40.0	21.7	0.01	0.44	0.09	0.65
2P-16	545–554	549.5	−749.5	8.7	53.9	0.98	0.46	2.46	40.4	21.8	0.00	0.44	0.09	0.82
2P-17	580–588	584.0	−784.0	8.7	55.9	1.01	0.42	2.52	41.1	22.3	0.00	0.45	0.09	0.78
2P-12	613–619	616.0	−816.0	8.7	56.7	1.10	0.42	2.64	41.8	22.9	0.00	0.45	0.09	0.67
2P-13	646–653	649.5	−849.5	8.7	58.5	1.27	0.44	2.79	42.7	23.0	0.01	0.49	0.09	0.63
2P-14	680–690	685.0	−885.0	8.7	57.5	0.98	0.51	3.09	43.6	23.8	0.01	0.46	0.10	0.70
2P-9	713–723	718.0	−918.0	8.6	59.5	1.02	0.62	3.27	44.9	24.5	0.00	0.41	0.10	0.55
2P-10	747–755	751.0	−951.0	8.7	62.1	1.09	0.61	3.35	45.6	25.1	0.00	0.49	0.10	0.85
2P-11	776–784	780.0	−980.0	8.6	61.2	1.04	0.58	3.49	46.2	25.5	0.00	0.44	0.10	1.02
2P-7	815–823	819.0	−1019.0	8.7	63.7	0.99	0.62	3.78	47.9	26.7	0.00	0.37	0.11	0.95
2P-8	869–877	873.0	−1073.0	8.6	65.9	1.05	0.66	3.98	49.4	27.5	0.01	0.37	0.11	0.98
2P-5	923–931	927.0	−1127.0	8.7	67.6	1.03	0.64	3.94	50.1	29.1	0.01	0.39	0.11	1.16
2P-6	968–976	972.0	−1172.0	8.7	69.4	1.08	0.72	4.10	50.9	29.5	0.00	0.39	0.12	1.22
2P-3	1014–1022	1018.0	−1218.0	8.8	70.7	1.08	0.77	4.32	52.8	30.0	0.01	0.38	0.12	0.88
2P-4	1069–1078	1073.5	−1273.0	8.6	72.6	1.07	0.84	4.18	53.5	30.5	0.01	0.37	0.12	1.40
2P-1	1108–1116	1112.0	−1312.0	8.8	74.9	1.08	0.83	4.20	53.8	31.6	0.01	0.40	0.11	0.88
2P-2	1159–1170	1164.5	−1364.0	8.8	77.2	1.08	0.82	4.23	55.6	33.7	0.01	0.38	0.12	0.83

Chloride increases downcore in south basin core 2P while it decreases downcore in northern basin core 7P. Barton et al. (1987) previously noted these trends and suggested that they resulted from a low stand of Lake Turkana that occurred below the sill depth separating the northern and southern basins in the early to middle Holocene. Johnson et al. (1987), using high resolution seismic profiles, noted a hard reflector and an erosional unconformity at about 60 meters depth in Lake Turkana indicting a previous low stand. This is well below the present sill depth of about 30 meters (Fig. 2). Chloride accumulation models (Barton et al., 1987) and measured sedimentation rates (Barton and Torgersen, 1988; Halfman et al., 1992) are compatible with a basin refilling time for the Turkana Basin of between 4000 and 10,000 years ago.

Sodium and alkalinity increase in all cores systematically down core. This probably is due to the in situ weathering of feldspar producing cations and alkalinity in equal molar proportions.

In the south basin, core 2P shows a distinct "kink" in the Cl, Na and alkalinity profiles at about 6 meters depth; the other southern basin core, 4P shows similar distinct "kinks" at about 9 meters depth. In the north basin, a distinct "kink" in the Na and alkalinity profiles occurs at about 5.5 meters depth. Finney (unpublished data) has shown that there is a subtle change in the chemistry of the sediments of core 2P at about 6 meters depth; the interval from 4.5 to 5.5 meters is characterized by high Mg/Al ratios (ca. ~0.45) compared with lower values both above and below this interval. There is no noticeable change in porosity in core 2P at this depth (Halfman, 1987).

General Trends: Minor Ions and Silica

Concentrations of the minor anions (Fig. 3) in cores 2P and 7P from Lake Turkana (sulfate, fluoride, and bromide) are all less than 1 mmol/l. Sulfate is at the detection limit (0.002 mmol/l) in waters below 15 cm depth (Table 2 and 3), far below the open lake concentration of 0.36 mmol/l. Fluoride decreases downcore in both the northern and southern basin. Calculations of mineral saturation shows that the waters are have an ion activity product $[Ca^{-2}][F^-]^2$ of about $10^{-11.5}$ throughout the cores, suggesting that the fluoride concentration is controlled by a calcium bearing phase such as fluorite ($K_{sp} = 10^{-10.6}$; Brown and Roberson, 1977). Bromide in core 2P increases from 0.06 mmol/l to 0.12 mmol/l at about 1400 cm depth, and although in 7P it does not have a discernable trend. The Cl/Br ratio is essentially constant (ca 225:1) throughout the core 2P, given analytical uncertainty. The Cl/Br ratio in core 7P is problematical but may decrease with depth; however, the apparent decrease is very close to analytical uncertainty. Dissolved silica in the southern basin core 2P at 0.7 to 1 mmol/l is much higher than in the northern basin core 7P (0.3 to 0.6 mmol/l); diatoms are much more abundant in the 2P sediments than in the 7P sediments (T. Johnson, pers. comm).

The minor cations K, Ca, and Mg, are present in the lake waters at levels of about 0.4, 0.1, and 0.1 mmol/l, respectively (Table 1). Potassium increases downcore to 1.0 mmol/l in core 2P, and to about 0.9 mmol/l in core 7P (Fig. 4), probably as a result of silicate dissolution. Calcium increases significantly in core 2P to 0.8 mmol/l (Fig. 4), probably as a result of calcite dissolution in core 2P. Calcium is essentially conservative in core 7P. Mg increases most dramatically, from 0.1 to over 4 mmol/l in core 2P and from 0.1 to 0.5 mmol/l in core 7P (Fig. 4), and may be related to the dissolution of a Mg–silicate phase (discussed below). A significant change in slope in Mg concentration is observed in core 2P, occurring at about 5.5 meters below the estimated sediment–water interface.

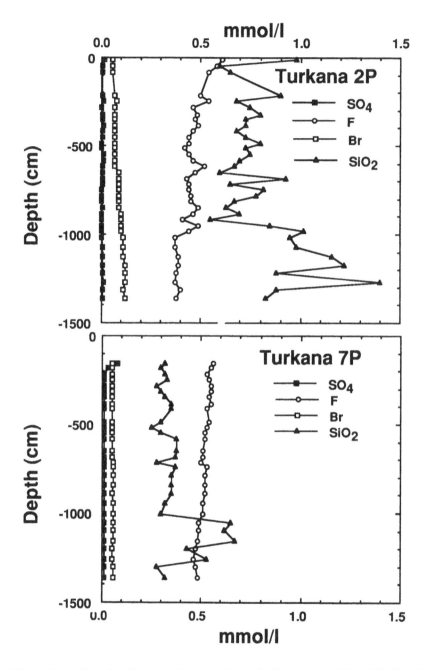

Figure 3. Profiles of sulfate, fluoride, bromide, and silica in cores 2P and 7P. Fluoride decreases downcore in both cores 2P and 7P, essentially maintaining a constant ion activity product with fluorite suggesting that fluorite solubility controls the concentration of fluoride in these waters. Bromide increases downcore in core 2P, maintaining a constant Cl/Br ratio (within error limits) down core. Bromide concentrations changes in core 7P were not detectable (Table 3) but were within the range expected from the decrease in chloride observed in the core.

Table 3. Chemistry of Pore Waters from Piston Core 7P (in mmol/l, except alkalinity, which is in meq/l)

Sample	Depth interval	Average depth	Estim. depth	pH	Na	K	Ca	Mg	Alk	Cl	SO4	F	Br	SiO2
7P-66	0–18	-9.0	-159.0	8.5	40.2	0.64	0.19	0.15	28.2	11.3	0.08	0.56	0.05	0.32
7P-65	26–33	-29.5	-179.5	8.5	42.1	0.82	0.18	0.21	32.0	10.9	0.03	0.55	0.05	0.30
7P-64	60–70	-65.0	-215.0	8.6	42.6	0.75	0.19	0.23	32.7	10.8	0.01	0.53	0.05	0.32
7P-63	95–105	-100.0	-250.0	8.5	42.7	1.05	0.24	0.28	32.9	10.7	0.01	0.54	0.05	0.33
7P-62	130–140	-135.0	-285.0	8.6	43.1	0.87	0.21	0.28	33.0	10.7	0.01	0.55	0.05	0.28
7P-61	160–170	-165.0	-315.0	8.7	44.0	0.88	0.18	0.29	33.4	10.7	0.01	0.55	0.05	0.30
7P-60	190–200	-195.0	-345.0	8.7	43.0	0.77	0.19	0.32	33.5	10.7	0.01	0.54	0.05	0.32
7P-59	228–238	-233.0	-383.0	8.5	43.6	0.79	0.19	0.33	33.9	10.7	0.01	0.55	0.05	0.35
7P-58	252–262	-257.0	-407.0	8.7	44.3	0.08	0.18	0.35	34.1	10.6	0.01	0.53	0.05	0.35
7P-56	330–340	-335.0	-485.0	8.6	44.3	0.80	0.20	0.34	34.7	10.6	0.01	0.54	0.05	0.30
7P-55	360–372	-366.0	-516.0	8.6	44.7	0.73	0.21	0.33	35.3	10.5	0.01	0.53	0.05	0.25
7P-54	390–401	-395.5	-545.5	8.7	46.1	0.75	0.21	0.30	36.7	10.5	0.01	0.52	0.05	0.30
7P-53	432–442	-437.0	-587.0	8.8	47.7	0.79	0.23	0.32	38.2	10.3	0.01	0.52	0.05	0.38
7P-51	494–505	-499.5	-649.5	8.8	47.9	0.76	0.22	0.32	38.9	10.2	0.01	0.51	0.05	0.38
7P-50	530–541	-535.5	-685.5	8.8	48.1	0.75	0.22	0.31	39.3	10.2	0.01	0.51	0.05	0.37
7P-49	560–570	-565.0	-715.0	8.7	48.6	0.83	0.21	0.31	39.6	10.3	0.01	0.50	0.06	0.28
7P-48	587–600	-593.5	-743.5	8.7	48.0	0.77	0.20	0.31	40.0	10.0	0.01	0.53	0.06	0.37
7P-47	635–645	-640.0	-790.0	8.9	49.8	0.78	0.22	0.33	41.0	10.1	0.01	0.52	0.06	0.35
7P-46	685–696	-690.5	-840.5	8.8	49.6	0.72	0.21	0.36	41.2	9.9	0.01	0.52	0.06	0.35
7P-45	737–747	-742.0	-892.0	8.8	50.0	0.77	0.19	0.37	41.8	10.0	0.01	0.52	0.05	0.35
7P-44	787–800	-793.5	-943.5	8.8	50.0	0.75	0.22	0.37	41.7	9.8	0.01	0.51	0.06	0.32
7P-43	850–862	-856.0	-1006.0	9.0	50.0	0.78	0.25	0.37	42.3	9.8	0.01	0.51	0.06	0.30
7P-42	900–912	-906.0	-1056.0	8.8	51.6	0.73	0.18	0.38	42.5	9.6	0.01	0.49	0.06	0.65
7P-41	940–951	-945.5	-1095.5	8.8	50.8	0.76	0.19	0.38	42.6	9.5	0.01	0.49	0.06	0.62
7P-40	1000–1012	-1006.0	-1156.0	8.9	50.9	0.79	0.20	0.39	43.0	9.5	0.01	0.48	0.06	0.67
7P-39	1043–1055	-1049.0	-1199.0	8.7	51.7	0.79	0.19	0.40	43.0	9.4	0.01	0.47	0.05	0.43
7P-38	1101–1114	-1107.5	-1257.5	8.7	52.1	0.76	0.22	0.44	43.6	9.1	0.01	0.46	0.05	0.53
7P-37	1147–1161	-1154.0	-1304.0	8.8	51.6	0.81	0.22	0.46	43.8	9.2	0.01	0.47	0.06	0.28
7P-36	1203–1218	-1210.5	-1360.5	8.9	52.2	0.83	0.27	0.49	43.2	9.0	0.01	0.48	0.06	0.32

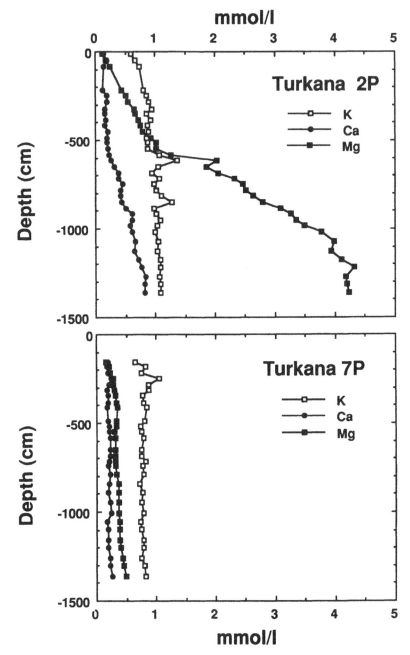

Figure 4. Profiles of potassium, calcium, and magnesium in cores 2P and 7P. Potassium and calcium show small but significant downcore increases in concentration, whereas magnesium exhibits a very significant increase in concentration in core 2P with a change in slope occurring at about 5.5 meters.

Core 2P has elevated Mg/Ca ratios reaching a molar ratio of 6, which is much higher than surface waters from East Africa (including Lake Turkana) which have ratios of about 1

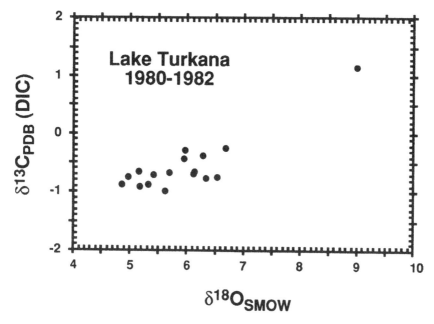

Figure 5. Carbon and oxygen isotopic composition of surface waters from Lake Turkana at Koobi Fora spit from September, 1980 to September, 1982.. Talbot and Kelts (1990) noted a correlation between carbon and oxygen for closed basin lakes. The higher values for both carbon and oxygen coincide with periods of high salinity and the low values coincide with periods of freshening due to river flooding.

(Cerling 1979). Dolomite is thought to require increased Mg/Ca ratios or low sulfate concentrations for precipitation, both of which are observed in core 2P. Dolomite in Plio–Pleistocene lacustrine sediments in East Africa are associated with minerals indicating high alkalinity and salinity (Cerling 1979). The high Mg/Ca ratios and the low sulfate values observed in core 2P would promote dolomite precipitation, although it has not yet been observed in the Lake Turkana cores.

General Trends: Stable Isotopes

The carbon and oxygen isotopic composition of surface waters collected at Koobi Fora spit from Lake Turkana are shown in Fig. 5. The single sample enriched in both ^{13}C and ^{18}O had abnormally high salinity probably represent an isolated pool of water collected at Koobi Fora spit. Both carbon and oxygen isotopic values are enriched in ^{13}C and ^{18}O during periods of evaporation but are depleted during periods of flooding due to local rains or to the more distant Ethiopian Highland rainy season which results in Omo flooding and the main annual lake level rise. The increase in $\delta^{13}C$ is probably related to isotopic exchange with the atmosphere: (Peng and Broecker, 1980) have found relatively rapid isotopic exchange in highly alkaline lakes compared to hard water lakes. The increase in $\delta^{18}O$ is related to enrichment of ^{18}O during evaporation. Talbot (1990) and Talbot and Kelt (1990) have noted a positive correlation between $\delta^{13}C$ and $\delta^{18}O$ for closed basin lake carbonates. This study confirms the positive correlation for lake waters.

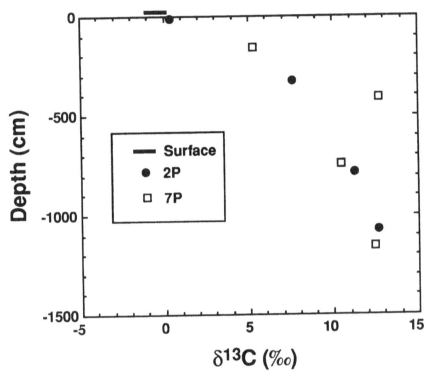

Figure 6. The carbon isotopic composition of dissolved inorganic carbon (DIC) from cores 2P and 7P versus depth.

The isotopic composition of dissolved inorganic carbon is cores 2P and 7P is shown in Fig. 6. $\delta^{13}C$ values for dissolved inorganic carbon may be in error (up to several tenths of ‰) because the sample bottles were not completely full and therefore some degassing of CO_2 could have taken place prior to measurements (Craig, 1976). $\delta^{13}C$ values increase with depth, from values near the surface $\delta^{13}C$ value of about 0‰ and reach +12‰ at several meters depth. Such high values indicate methanogenesis where CO_2 is enriched in ^{13}C relative to bulk organic carbon and methane is depleted:

$$2CH_2O = CH_4 + CO_2.$$

Methanogenesis is especially important in aqueous systems with low dissolved sulfate. Cerling et al. (1988) previously noted ^{13}C enriched diagenetic lacustrine carbonate nodules in Plio–Pleistocene sediments in the Turkana Basin.

The $\delta^{13}C$ values near the sediment–water interface are similar to open water values (ca. −0.5‰) and indicate that a significant injection of alkalinity in the sediment column due to sulfate reduction does not occur:

$$2CH_2O + SO_4^{-2} = 2HCO_3^- + H_2S.$$

This means that the alkalinity produced by sulfate reduction diffuses back into the lake.

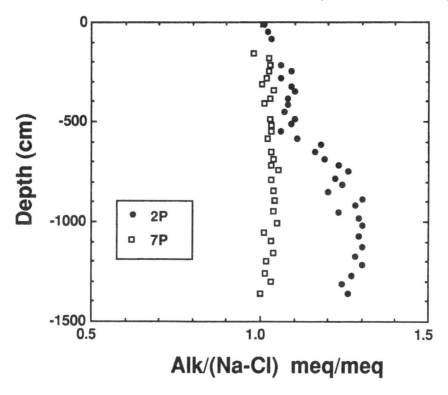

Figure 7. Ratio of alkalinity/(Na-Cl) in cores 2P, 2PG, and 7P in Lake Turkana. Ratios greater than one indicate excess alkalinity that must be derived from dissolution reactions.

Excess Alkalinity and Magnesium Anomaly

Figure 7 shows that the alkalinity produced in core 7P and in the upper part of core 2P is balanced by sodium. However, below about 6 meters depth in core 2P a considerable excess alkalinity (alkalinity − (Na-Cl)) is observed, that must relate from either production of alkalinity in the pore fluids, or from sodium consumption. In the following paragraphs we explore the relationships associated with this excess alkalinity.

The excess alkalinity appears to be associated with the high magnesium values observed in core 2P at depth. Figure 8 shows the excess alkalinity versus excess magnesium in core 2P, where both the magnesium and alkalinity have been corrected for calcite dissolution assuming calcite with 3 mole percent $MgCO_3$. Figure 8 shows that the excess alkalinity/Mg ratio averages 2.8, which is close to the theoretical value of 2.25 predicted from the dissolution of tri-octahedral stevensite:

$$0.66Ex + 2.67Mg^{+2} + 4\ SiO_2\ (aq) + 6HCO_3^-$$
$$= Ex_{0.66}Mg_{2.67}Si_4O_{10}(OH)_2 + 6CO_2 + 2H_2O$$

$$\text{stevensite} \qquad HCO_3^-/Mg^{+2} = 2.25$$

(where "Ex" is the exchangeable cation, in this case dominated by Na^+), which was predicted to be forming in modern lake by Yuretich and Cerling (1982) based on mass balanced considerations. These results indicate that it is highly likely that neoformed stevensite or a

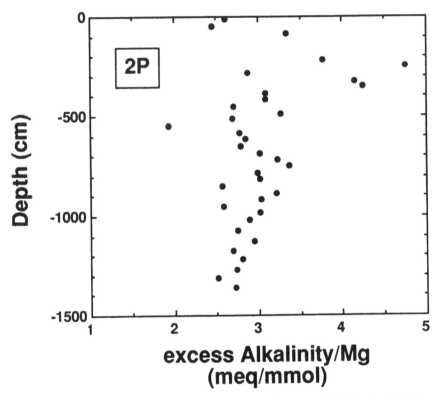

Figure 8. Excess alkalinity/Mg ratio in core 2P and 2PG. Excess alkalinity is: alkalinity −
(Na-Cl). The average ratio of 2.8 is close to the stoichiometric ratio of 2.25 for stevensite
dissolution.

similar phase which formed in Lake Turkana is undergoing dissolution in the south basin
sediments. An increase in $P(CO_2)$ would favor dissolution as the above equation shows. This
reaction should also release silica: Table 2 and Fig. 5 show that core 2P has high dissolved
silica, whereas core 7P has low levels of dissolved silica. Although the silica levels are
somewhat below amorphous silica saturation in core 2P indicating that further silica
reprecipitation may be taking place. Preliminary results on X-ray diffraction on the <0.1
micron clay fraction indicates a significant change in D-spacing in core 2P at the depth of
the "Mg-kink" (R. F. Yuretich, pers. comm). In addition, preliminary results from the bulk
chemistry of the cores shows that the ratio of Mg/Ti decreases at about the same level (B.
Finney, pers. comm.). Taken together, the dissolved water composition, bulk chemistry of
the sediments, and X-ray diffraction results indicate that a significant change in diagenesis
is taking place in the south basin cores which is probably related to dissolution of neoformed
tri-octahedral smectite (such as stevensite) or a similar magnesium silicate phase. The
dissolution may be driven by increased $P(CO_2)$ resulting from methanogenesis.

CONCLUSIONS

Pore fluids from Lake Turkana in East Africa show several important features that may be
in common with other alkaline rift valley lakes. Chloride profiles show that this lake was

isolated into two separate basins during lake level low stands, with a saline lake occupying the southern basin. Bromide profiles are compatible with this model. Total sodium and alkalinity increases in the pore fluid are mostly due to the production of alkalinity and sodium due to feldspar dissolution caused by high $P(CO_2)$ levels. Sulfate is reduced just below the sediment–water interface; below the zone of sulfate reduction, methanogenesis becomes important and is shown by the high $\delta^{13}C$ values for dissolved inorganic carbon at depth in the cores. Methanogenesis is accompanied by high $P(CO_2)$ which is generated along with CH_4. Slight calcite dissolution may occur in the cores, probably resulting from the high $P(CO_2)$ generated during methanogenesis. A significant increase in dissolved magnesium occurs in the south basin core; the ratio of excess alkalinity to magnesium is close to that expected from dissolution of a neoformed magnesium silicate phase such as stevensite, probably due to elevated $P(CO_2)$ levels.

ACKNOWLEDGEMENTS

These cores were collected by T. C. Johnson, G. S. Lister, and J. D. Halfman. I thank them for making these cores available, C. Pittlekow for laboratory assistance, and R. F Yuretich and B. Finney for access to unpublished data. Mathew Muteo collected samples from Lake Turkana at Koobi Fora spit. I thank Paul Baker and Berry Lyons for helpful reviews. Laboratory studies were supported by BLS.

REFERENCES

Barton, C. E., and T. Torgersen, 1988, Palaeomagnetic and ^{210}Pb estimates of sedimentation in Lake Turkana, East Africa. Palaeogeography, Palaeoclimatology, Palaeoecology 68: 53–60.

Barton, C. E., D. K. Solomon, J. R. Bowman, T. E. Cerling, and M. D. Sayer, 1987, Chloride budgets in transient lakes: Lakes Baringo, Naivasha, and Turkana, Kenya. Limnology and Oceanography 32: 745–751.

Beadle, L. C. 1974, The Inland Waters of Tropical Africa. Longmans, London.

Berner, R. A., 1980, Early Diagenesis: a Theoretical Approach. Princeton University Press. Princeton, NJ, 241 pp.

Broecker, W. S., and T.-H. Peng, 1982, Tracers in the Sea. Eldigio Press, Palisades, NY, 690 pp.

Brown, D. W., and C. E. Roberson, 1977, Solubility of natural fluorite at 25 C. Journal of Research, U.S. Geological Survey 5: 509–518.

Cerling, T. E., 1979, Paleochemistry of Plio–Pleistocene Lake Turkana, Kenya. Palaeogeography, Palaeoclimatology, Palaeoecology 27: 247–285.

Cerling, T. E., 1986, A mass balance approach to basin sedimentation: constraints on the recent history of the Turkana Basin. Palaeogeography, Palaeoclimatology, Palaeoecology 54: 63–86.

Cerling, T. E., J. R. Bowman, and J. R. O'Neil, 1988, An isotopic study of a fluvial–lacustrine sequence: The Plio–Pleistocene Koobi Fora Formation, East Africa. Palaeogeography, Palaeoclimatology, Palaeoecology 63: 335–356.

Craig, H., 1975, Lake Tanganyika geochemical and hydrographic study: 1973 expedition., SIO Reference Series 75-5. Scripps Institution of Oceanography, La Jolla, CA, 83 pp.

Gac, J. Y., A. Droubi, B. Fritz, and Y. Tardy, 1978, Geochemical behavior of silica and magnesium during the evaporation of waters in Chad. Chemical Geology 19: 219–228.

Gieskes, J. M., 1974, Interstitial water studies, Leg 25. Initial Reports of the Deep-Sea Drilling Project, Volume XXV, pp. 361–394.

Halfman, J. D., 1987, High-resolution sedimentology and paleoclimatology of Lake Turkana, Kenya. Ph.D. Dissertation, Duke University, Durham, NC, 188 pp.

Halfman, J. D., D. F. Jacobson, C. M. Cannella, K. A. Haberyan, and B. P. Finney, 1992, Fossil diatoms and the mid to late Holocene paleolimnology of Lake Turkana, Kenya: a reconnaissance study. Journal of Paleolimnology 7: 23–35.

Hopson, A. J. (Editor), 1982, Lake Turkana: a report on the findings of the Lake Turkana project, 1972–1975. Overseas Development Administration, London, 331 pp.

Johnson, T.C., J. D. Halfman, B. R. Rosendahl, and G. S. Lister, 1987, Climatic and tectonic effects on sedimentation in a rift valley lake: evidence from high resolution seismic profiles, Lake Turkana, Kenya. Geological Society of America Bulletin 98: 439–447.

Peng, T.-H., and W. S. Broecker, 1980, Gas exchange rates for three closed-basin lakes. Limnology and Oceanography 25: 789–796.

Talbot, M. R., 1990, A review of the palaeohydrological interpretation of carbon and oxygen isotope ratios in primary lacustrine carbonates. Chemical Geology (Isotope Geoscience Section) 80: 261–279.

Talbot, M. R., and K. Kelts, 1990, Paleolimnological signatures from carbon and oxygen isotopic ratios in carbonates from organic carbon-rich lacustrine sediments. In B. J. Katz (Ed.), Lacustrine Basin Exploration — Case Studies and Modern Analogues. American Association of Petroleum Geologists Memoir 50. American Association of Petroleum Geologists. Tulsa, Oklahoma, pp. 99–112.

Talling, J. F., and I. B. Talling, 1965, The chemical composition of African lake waters. Intern. Rev. Ges. Hydrobiol. 50: 421–463.

Yuretich, R. F., 1979, Modern sediments and sedimentary processes in Lake Rudolf (Lake Turkana) Eastern Rift Valley, Kenya. Sedimentology 26: 313–331.

Yuretich, R. F., and T. E. Cerling, 1983, Hydrogeochemistry of Lake Turkana, Kenya: mass balance and mineral reactions in an alkaline lake. Geochimica et Cosmochimica Acta 47: 1099–1109.

Seasonal Variation in the Nitrogen Isotopic Composition of Sediment Trap Materials Collected in Lake Malawi

R. FRANÇOIS *Department of Marine Chemistry and Geochemistry, Woods Hole Oceanographic Institution, Woods Hole, Massachusetts*

C.H. PILSKALN *Department of Oceanography, University of Maine at Orono*

M.A. ALTABET *Department of Chemistry, University of Massachusetts at North Dartmouth*

Abstract — We have measured the nitrogen isotopic composition of sediment trap materials collected in northern Lake Malawi between June 1987 and January 1991. Short-lived but prominent flux maxima, which occurred irregularly during the dry season, were accompanied by significant decreases in the $\delta^{15}N$ of the settling material. We interpret this observation as reflecting preferential uptake of $^{14}NO_3^-$ during phytoplankton blooms triggered by sporadic upwelling events which brought nitrate-rich water to the surface of the lake. We suggest that isotopic fractionation during nitrate uptake by phytoplankton might provide a means of identifying permanent or recurring upwelling centers from lacustrine sedimentary records, from which it might be possible to infer secular changes in regional wind patterns.

INTRODUCTION

In a recent study, Talbot and Johannessen (1992) have reported large excursions in the $\delta^{15}N$ record of a core taken in Lake Bosumtwi (Ghana). Their work suggests that lake sediment $\delta^{15}N$ might provide valuable information on the paleoclimatic evolution of tropical Africa. Unambiguous deciphering of sedimentary $\delta^{15}N$ record, however, requires a better understanding of the relative importance of the numerous processes (e.g., nitrate utilization in the euphotic zone, ammonia volatilization, denitrification, nitrogen fixation, early diagenesis, autochtonous vs allochtonous organic nitrogen input, etc.) which can affect the nitrogen isotopic composition of lake sediments. In a preliminary attempt to evaluate the importance of some of these processes, we have measured the nitrogen isotopic composition of settling particles collected monthly with a time-series sediment trap deployed at an average depth of 350 m depth (ca. 100 m above bottom), within the anoxic deep waters of northern Lake Malawi (10°14.4′S; 34°21.9′E), between June 1987 and January 1991 (C. Pilskaln; Fig. 1).

ANALYTICAL METHODS

Nitrogen isotopes were measured on all sediment trap samples for which sufficient material was available. The dried samples were combusted at 1000°C in a Europa Scientific Robo-Prep C–N analyzer on-line with a Finnigan MAT 251 mass spectrometer. The N_2 gas produced after combustion and reduction of NO_x by hot (600°C) copper was transported to the mass spectrometer in a He stream. After removing CO_2 and H_2O, a fraction of the

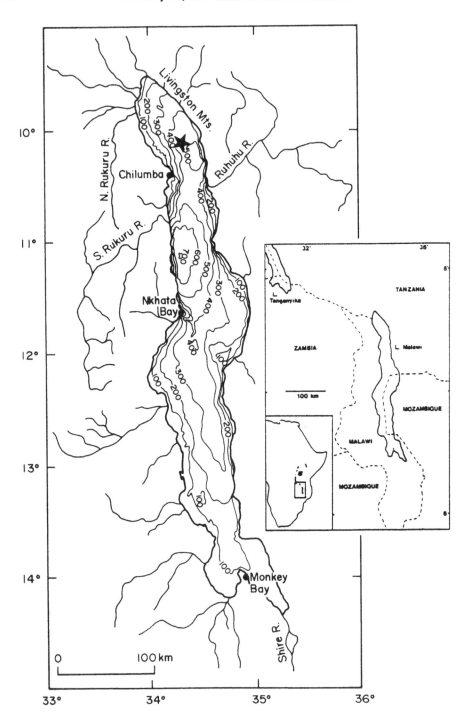

Figure 1. Northern Lake Malawi; contoured bathymetry in meters and position of the sediment trap mooring (*; 10°14.4′S; 34°21.9′E).

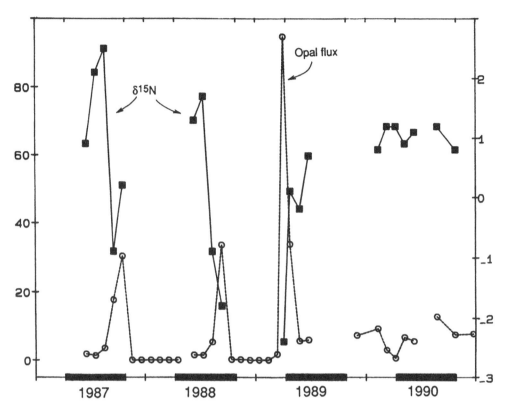

Figure 2. Seasonal variations in the $\delta^{15}N$ of settling particles (‰; filled squares) and opal flux (mg/m²·d; open circles). Bars on the year axis represent the approximate duration of the dry season.

N_2-containing He stream was diverted into the ion-source of the mass spectrometer by a splitter valve. Isotopic data are reported in the conventional δ notation against atmospheric nitrogen. Standard deviation on replicate measurements using glycine was 0.15‰.

SEASONAL VARIATIONS IN SETTLING FLUX

Large seasonal variations in sediment fluxes were recorded over the period studied (Pilskaln, 1989). Significant positive correlations were observed between total particulate, biogenic opal and organic carbon mass fluxes, indicating that particle flux was primarily controlled by variations in diatom productivity. The recorded flux variations did not follow a regular seasonal pattern. Short-lived but prominent flux maxima (Fig. 2) occurred either at the end (1987; 1988) or at the beginning (1989) of the dry season, and sometimes were totally missing (1990). During these episodic blooms, the diatom population was largely dominated by *Aulacoseira* sp. (formerly *Melosira* sp.) and *Stephanodiscus* sp. (Fig. 3). We suspect that the flux maxima reflect sporadic upwellings induced in the northern section of the lake by the strong southerly winds prevailing during the dry season between April and October (Jackson et al., 1963; Eccles, 1974). These events would have resulted in relatively brief but intense blooms followed by rapid depletion of the nutrients brought to the surface.

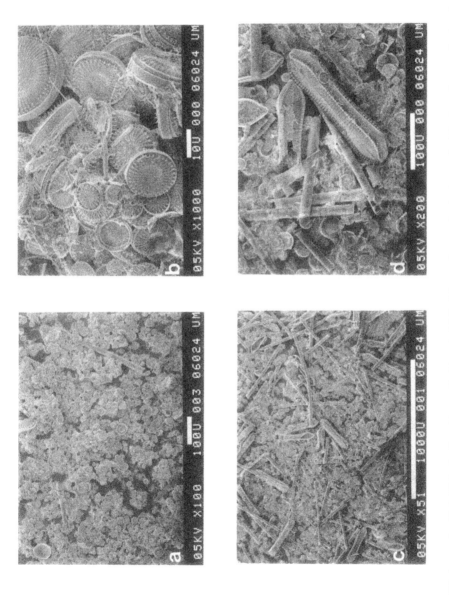

Figure 3. SEM micrographs of Lake Malawi sediment trap particulate material. Scale bars on bottom of each micrograph. a) *Stephanodiscus* sp.-dominated trap sample collected 10/5/87–11/4/87. b) High magnification view of *Stephanodiscus* valves within a particle aggregate from the above sample. c) Trap material obtained 8/27/88–9/26/88 consisting primarily of *Aulacoseira* sp. chains, aggregates of *Stephanodiscus* valves, and occasional *Surirella* sp. tests. d) Higher magnification view of diatomaceous material in c).

Table 1. $\delta^{15}N$ in Sediment Trap Materials Recovered from Lake Malawi Between June 1987 and November 1990

Period of deployment	$\delta^{15}N$ (‰)	N_{flux} (mg/m²·d)	Integrated $\delta^{15}N$*
1987			
6/4 – 7/4	+0.9	122	
7/5 – 8/4	+2.1	111	
8/5 – 9/4	+2.5	114	
9/5 – 10/4	−0.9	281	
10/5 – 11/4	+0.2	553	+0.4
1988			
5/26 – 6/26	+1.3	102	
6/27 – 7/26	+1.7	116	
7/27 – 8/26	−0.9	271	
8/27 – 9/26	−1.8	1302	−1.3
1989			
3/27 – 4/11	−2.4	2561	
4/12 – 5/12	+0.1	1440	
5/12 – 6/12	−0.2	160	
6/12 – 7/12	+0.7	172	−1.4
1990			
1/27 – 2/27	+0.8	500	
2/27 – 3/27	+1.2	160	
3/27 – 4/27	+1.2	49	
4/27 – 5/27	+0.9	252	
5/27 – 6/27	+1.1	29	
7/27 – 9/27	+1.2	316	
9/27 – 11/27	+0.8	218	+1.0

*$\Sigma(N_{flux} \times \delta^{15}N)/\Sigma N_{flux}$.

SEASONAL VARIATIONS IN THE $\delta^{15}N$ OF SEDIMENT TRAP MATERIAL

The three maxima in opal flux (associated with maxima in organic carbon flux; Pilskaln, in prep.) coincided with abrupt decreases in the $\delta^{15}N$ of the sediment trap materials (Fig. 2; Table 1). The data obtained from the settling particles collected in 1987 and 1988 document a 3 to 4‰ decrease in $\delta^{15}N$ during the inception of the diatom bloom, while those collected in 1989 reveal an increase of similar magnitude at the end of the bloom period. In contrast, during 1990, when the opal flux remained moderate to low without dramatic increases, the $\delta^{15}N$ of the material collected by the sediment trap was relatively constant and high.

There is a negative correlation between the $\delta^{15}N$ of settling particles and opal ($r = -0.74$; $n = 20$) or N ($r = -0.73$; $n = 20$) flux, reflecting the low $\delta^{15}N$ measured in periods of high fluxes. A similar correlation between organic carbon flux and $\delta^{15}N$ has been reported for settling particles collected with a deep sediment trap deployed in the Sargasso Sea (Altabet and Deuser, 1985). Altabet and Deuser attributed their observation to seasonal variations in the stratification of the upper water column, affecting the supply of nitrate to the euphotic zone. Deep winter convection brings nutrients to the surface, inducing a phytoplankton bloom and high settling flux in spring. Because phytoplankton discriminates against $^{15}NO_3^-$ during uptake (Wada and Hattori, 1978; Montoya, 1990), the biomass produced at the beginning of the bloom is isotopically light compared to the available nitrate (Altabet et al., 1991). As thermal stratification intensifies, reducing the rate of supply of new nitrate from intermediate depth to the euphotic zone, surface nitrate becomes gradually depleted and isotopically heavier, as dictated by Raleigh fractionation kinetics. Lower nitrate supply reduces the settling flux and increases the $\delta^{15}N$ of phytoplankton biomass and settling material. During summer, when the seasonal thermocline is fully developed, all the new nitrate entering the euphotic zone is immediately utilized. Under these conditions, isotope mass balance considerations require that the $\delta^{15}N$ of settling particles be equal to that of the new nitrate supplied (Altabet, 1988).

A similar scenario could explain the data from lake Malawi, although the forcing mechanism regulating the occurrence of phytoplankton blooms in this environment is clearly distinct from that observed in the Sargasso Sea. In Lake Malawi, the open-lake phytoplankton blooms are historically and presently dominated by *Aulacoseira* sp. and *Stephanodiscus* sp. (Fig. 3; Kilham et al., 1986; Hecky and Kling, 1987; Owen, 1989). We propose that the blooms recorded by the sediment trap result from sporadic wind-induced upwellings of nitrate-rich waters, promoting the rapid growth of diatom populations during brief periods of the dry season, when strong southerly winds predominate as a result of the northward migration of the ITCZ (Jackson et al., 1963; Eccles, 1974). Early during the blooms, as recorded during the 1987 and 1988 events, the rate of supply of nitrate to the euphotic zone exceeds the rate of utilization, promoting preferential uptake of ^{14}N, and resulting in low $\delta^{15}N$ in phytoplankton and sinking particles (Fig. 2). As the phytoplankton bloom evolves, surface nitrate becomes gradually depleted and isotopically heavier. This results in a gradual increase in the isotopic composition of sinking particles, as recorded during the 1989 event. In 1990, the observed opal (Fig. 2) and N (Table 1) fluxes were comparatively low but measurable, suggesting sustained primary production rates and a continuous but slow supply of new nitrate by mixing at the base of the euphotic zone. Under these circumstances, we would expect a rapid and total utilization of all the new nitrate supplied to the euphotic zone, resulting in settling particles with a nitrogen isotopic composition similar to that of new nitrate (Altabet, 1988). Integrating the $\delta^{15}N$ of particles settling in 1990 (Table 1) suggests that the $\delta^{15}N$ of new nitrate is approximately 1‰.

In a closed system, Raleigh fractionation dictates that nitrate becomes increasingly heavier with depletion (Fig. 4), and at the end of a bloom period, when the upwelled nitrate is nearly totally utilized, one could expect to observe a maximum in the $\delta^{15}N$ of settling particles to values greater than that of new nitrate (i.e., >1‰), depending on the rate of removal of particles from the euphotic zone (Altabet et al., 1991) and the sampling resolution of the sediment trap. With short particle residence times in the upper water column and high sampling resolution, variations in the $\delta^{15}N$ of sediment trap material as a function of surface nitrate depletion would approach the values predicted from the instantaneous product equation (Fig. 4). On the other hand, long particle residence time or coarse trapping resolution would integrate the $\delta^{15}N$ signal, damping temporal variations in the $\delta^{15}N$ of sediment trap materials which would then approach that predicted by the accumulative product equation, resulting in lesser $\delta^{15}N$ maxima at the end of the bloom. In a closed system, the accumulated product after total utilization of nitrate must have the same $\delta^{15}N$ as

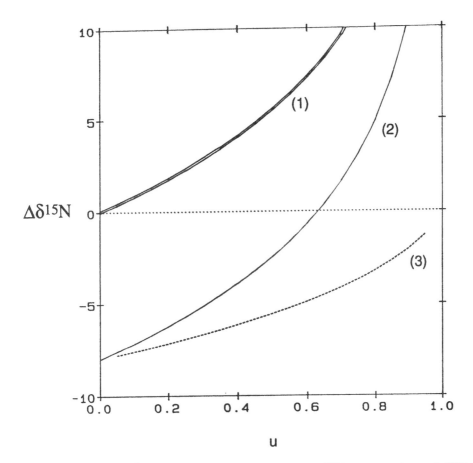

Figure 4. Variations in $\delta^{15}N$ as a function of nitrate utilization (u). $\Delta\delta^{15}N$ represents changes in $\delta^{15}N$, taking the $\delta^{15}N$ of nitrate at $u = 0$ as reference. (1) Changes in surface nitrate $\delta^{15}N$ ($\Delta\delta^{15}N = -\varepsilon_u \times \ln(1-u)$, where ε_u is the fractionation factor during nitrate utilization by phytoplankton, here taken as 8‰). (2) Changes in the $\delta^{15}N$ of particulate N according to Raleigh's instantaneous product equation ($\Delta\delta^{15}N = (-\varepsilon_u \times \ln(1-u)) - \varepsilon_u$). (3) Changes in the $\delta^{15}N$ of particulate N according to Raleigh's accumulated product equation ($\Delta\delta^{15}N = ((1-u)/u) \times \varepsilon_u \times \ln(1-u)$).

that of the initial nitrate. The lack of $\delta^{15}N$ maximum at the end of the 1989 event would thus suggest that the residence time of particles in the water column is large compared to the residence time of nitrate in the euphotic zone or that the monthly sampling resolution was too coarse to record the $\delta^{15}N$ maximum. However, when integrating the $\delta^{15}N$ of particles settling during each bloom (Table 1), we find cumulative $\delta^{15}N$ significantly lower than 1, particularly during the 1988 and 1989 events, indicating that the system here studied cannot be described as closed. Evidently, heavy (i.e., partially depleted) nitrate has been removed form the region sampled by the trap, probably by wind-induced surface currents. One would thus expect to find high $\delta^{15}N$ particles sinking in a region downwind of the sampling location. There was a $\delta^{15}N$ maximum in the settling particles collected at the beginning of the trap deployment, just prior to the 1987 maximum opal flux (Fig. 2). This could record

Table 2. δ^{15}N in Dark and Light Sediment Laminae from Cores Taken in the
Vicinity of the Sediment Trap Mooring Location
(see Pilskaln and Johnson 1991 for core description)

Core no.	Location	Depth in core (cm)		δ^{15}N (‰)	%N
30P	10°07′S; 34°10′E	420–424	dark	+1.2	0.27
			light	−0.2	0.17
31P	10°10′S; 34°11′E	763–767	dark	+2.2	0.17
			light	−0.5	0.26
33P	9°07′S; 34°09′E	302–304	light	+1.8	0.19
		305–308	dark	+1.5	0.25
			light	+1.7	0.16

the end of a bloom which occurred earlier during the 1987 dry season (which possibly started upwind from the sampling location), although, at the moment, we have no means to ascertain this interpretation.

Materials settling in land-locked basins such as lake Malawi are expected to contain significant amounts of terrigenous organic debris. The presence of such material has been recognized in the underlying sediments which were laminated during earlier periods of deposition and consisted of alternate light–dark laminae representing seasonal varves (Pilskaln and Johnson, 1991). The white laminae, which presumably record dry season phytoplankton blooms, are dominated by diatom frustules, and scanning electron microscope examination shows no evidence for the presence of significant amounts of terrigenous debris (Pilskaln and Johnson, 1991). In contrast, the dark laminae, which are thought to be deposited primarily by near-bottom density flows generated at the lake edge during the rainy seasons when maximum fluvial input occurs, contain a significant proportion of terrestrial plant debris. One would thus anticipate a significant terrigenous contribution to the δ^{15}N of settling particles during the rainy seasons, but not during the high opal flux periods. The rainy season sediment trap samples were too small for measuring their δ^{15}N. However, analysis of several dark laminae carefully removed after drying and foliation of a piece of laminated sediment varied between +1.2 and +2.2‰ (Table 2). If we take this value as resulting from a mixture of refractory terrigenous plant debris and traces of phytoplankton material with an isotopic composition equivalent to that of new nitrate (i.e., ca. 1‰), it follows that the terrigenous debris should have a similar or slightly higher isotopic ratio, which would be typical for organic nitrogen of terrestrial origin from nearby regions (Heaton, 1987). The presence of terrigenous material should thus have a minor effect on the δ^{15}N values reported in Fig. 2. We have also analyzed some of the light sedimentary laminae, dominated by opal, in three cores taken in the vicinity of the sediment trap location (Pilskaln and Johnson, 1991; Table 2). If the light laminae integrate an entire bloom, its δ^{15}N should be equal to that of deep nitrate which appears to be similar to the δ^{15}N of terrigenous material dominating the dark laminae. This seem to be the case for the laminae collected in core 33P (Table 2). On the other hand, the light laminae obtained from cores 30P and 31P appear to be depleted of [15]N. This could arise if these light laminae do not integrate the entire bloom, but only the onset of the bloom when phytoplankton is isotopically lighter and surface nitrate has only been partially depleted. The ^{15}N-enriched unutilized nitrate would have been laterally advected by wind-induced surface currents and utilized downwind, where we would expect to find a corresponding light laminae with $\delta^{15}N$ heavier than the dark laminae.

CONCLUSIONS

Our data suggest that nitrogen isotopic fractionation during nitrate uptake by phytoplankton is a major factor controlling the pattern of seasonal variations in the $\delta^{15}N$ of settling particles collected in northern lake Malawi. Since the underlying sediment will often integrate this seasonal signal, we might not be able to recover this information from the sedimentary record (unless sedimentation rates would be so high as to permit sampling at a sub-seasonal resolution). However, permanent or recurring centers of upwelling in lakes could be identified by relatively low $\delta^{15}N$ in the underlying sediments and gradual increase towards the limits of the region influenced by upwelled and laterally advected nutrients. Similar trends have already been observed in the equatorial Pacific Ocean and the Southern Ocean (François et al., 1992; Altabet and François, 1994a; 1994b). Lake Malawi is predominantly influenced by southerly winds during the dry season. We would thus expect that Ekmann transport would promote upwelling along the eastern shores of the lake, resulting in a east to west gradient in the $\delta^{15}N$ of its sediments. If that prediction can be confirmed, it would provide a means of evaluating past variations in wind patterns and possibly wind strength.

As mentioned in the introduction, there are many other processes which will affect the $\delta^{15}N$ of lacustrine sediments. Identification of upwelling centers in lakes, however, relies more on geographical patterns of distribution of sediment $\delta^{15}N$ than on their absolute values. Therefore, this application might not necessarily require a full understanding of the influence of these additional processes.

ACKNOWLEDGMENTS

Financial support for the sediment trap work was provided by NSF grant ATM88-11615, industry sponsors of Project PROBE (Duke University), and the David and Lucile Packard Foundation. We thank K. Murphy, J. B. Paduan and J. Donoghue for technical assistance and the government of Malawi and Malawi Hydrographic Survey Department for their cooperation in the study.

REFERENCES

Altabet, M. A., 1988, Variations in nitrogen isotopic composition between sinking and suspended particles: implications for nitrogen cycling and particle transformation in the open ocean: Deep-Sea Res., v. 35, pp. 535–554.

Altabet, M. A., and Deuser, W. G., 1985, Seasonal variations in seasonal abundance of ^{15}N in particles sinking into the deep Sargasso Sea: Nature, v. 315, pp. 218–219.

Altabet, M. A., Deuser, W. G., Honjo, S., and Stienen, C., 1991, Seasonal and depth-related changes in the source of sinking particles in the N. Atlantic: Nature, v. 354, pp. 136–139.

Altabet, M. A., and François, R., 1994a, Sedimentary nitrogen isotopic ratio as a recorder for surface ocean nitrate utilization: Global Biogeochemical Cycles, v. 8, pp. 103–116.

Altabet, M. A., and François, R., 1994b, The use of nitrogen isotopic ratio for reconstruction of past changes in surface ocean nutrient utilization. In: Carbon Cycling in the Glacial Ocean: Constraints on the Ocean's Role in Global Change. Edited by R. Zahn, M. Kaminski, L. D. Labeyrie, and T. F. Pedersen. NATO AST series. Springer Verlag.

Eccles, D. H., 1974, An outline of the physical limnology of Lake Malawi (Lake Nyasa): Limnol. Oceanogr., v. 19, pp. 730–742.

François, R., Altabet, M. A., and Burckle, L. H., 1992, Glacial to interglacial changes in surface nitrate utilization in the Indian sector of the southern ocean as recorded by sediment $\delta^{15}N$: Paleoceanography, v. 7, pp. 589–606.

Heaton, T. H. E., 1987, The $^{15}N/^{14}N$ ratios of plants in South Africa and Namibia: relationship to climate and coastal/saline environments: Oecologia, v. 74, pp. 236–246.

Hecky, R. E., and H. J. Kling, 1987, Phytoplankton ecology of the great lakes in the Rift Valley of Central Africa: Arch. Hydrobiol. Beih., v. 25, pp. 197–228.

Jackson, P. B. N., T. D. Iles, D. Harding, and G. Fryes, 1963, Report on the survey of northern Lake Nyasa 1954–1955. Joint Fisheries Research Organization Report, Malawi Government Printer, Zomba, Malawi.

Kilham, P., S. S. Kilham, and R. E. Hecky, 1986, Hypothesized resource relationships among African planktonic diatoms: Limnol. Oceanogr., v. 31, pp. 1169–1181.

Montoya, J. P., 1990, Natural abundance of ^{15}N in marine and estuarine plankton: studies of biological isotopic fractionation and plankton processes: Ph.D. thesis, Harvard University, Cambridge, MA.

Owen, R. B., 1989, Pelagic fisheries potential of Lake Malawi. ODA Project R4370 Rept., 300 pp.

Pilskaln, C. H., 1989, Seasonal particulate flux and sedimentation in lake Malawi, East Africa, Eos, v. 70, p. 1130.

Pilskaln, C. H., and Johnson, T. C., 1991, Seasonal signals in Lake Malawi sediments: Limnol. Oceanogr., v. 36, pp. 544–557.

Talbot, M. R., and Johannessen, T. A., 1992, A high resolution climatic record of the last 27,500 years in tropical West Africa from the carbon and nitrogen isotopic composition of lacustrine organic matter: Earth Planet. Sci. Lett., v. 110, pp. 23–27.

Wada, E., and Hattori, A., 1978, Nitrogen isotope effects in the assimilation of inorganic nitrogenous compounds by marine diatoms: Geomicrobiol. J., v. 1, pp. 85–101.

The Chemical Composition of Precipitation and Its Significance to the Nutrient Budget of Lake Malawi

H.A. BOOTSMA and M.J. BOOTSMA *Department of Botany, University of Manitoba, Winnipeg, Manitoba, Canada*

R.E. HECKY *Department of Fisheries and Oceans, Freshwater Institute, Winnipeg, Manitoba, Canada*

Abstract — Relative to other sites studied in Africa, precipitation on the southwestern shore of Lake Malawi is dilute with regard to most constituents, except Ca^{2+} and HCO_3^-, and has a chemical composition similar to that observed for remote, non-marine locations. Relatively high NH_4^+/cation and NO_3^-/anion ratios indicate a moderate influence of biomass burning on rain chemistry, while a high mean Ca^{2+} concentration suggests extensive soil deflation. Rain pH averaged ~6, and was controlled almost entirely by $Ca(HCO_3)_2$ concentration.

Phosphorus in rain appears to originate as aerosols, but the ultimate source is uncertain. It is suggested that soluble nitrogen and sulfate, which were correlated, are produced by biomass burning. Atmospheric deposition is estimated to account for 33% of new P and 72% of new N input into Lake Malawi (excluding N fixation). If the upward flux of nutrients from the monimolimnion is accounted for, atmospheric deposition accounts for 6% of P input to the surface mixed layer, while the atmospheric N contribution remains 72%. A comparison of nutrient input ratios with nutrient concentration ratios in the surface mixed layer suggests that nitrogen is recycled more efficiently than phosphorus within the surface waters of Lake Malawi.

INTRODUCTION

Recently, an increasing amount of attention has been given to the chemical composition of precipitation. This is largely due to the concern over acidic atmospheric deposition near industrialized areas and its impact on ecosystem functioning. However, much research has also resulted from the realization that knowledge of atmospheric deposition is fundamental to understanding the biogeochemical cycling of many elements. For example, atmospheric input may account for a significant proportion of the nutrient flux to lakes (Schindler et al. 1976; Manny and Owens 1983) and terrestrial ecosystems (Lindberg et al. 1986).

Because acid deposition is commonly considered an industrial problem, and because of the skewed global distribution of scientists, the majority of research on precipitation chemistry has been conducted in industrialized Europe and North America. However, it is now apparent that non-industrial human activities can also have a large impact on atmospheric chemistry in other regions of the world. In particular, vegetation burning, which is so prevalent in the tropics, has been shown to greatly increase flux rates of C, N, S and K through the atmosphere (Lewis 1981; Delmas 1982; Andreae et al. 1988). Although less well studied, there is also evidence that land exposure resulting from deforestation and intense agricultural activities increases atmospheric loads of soil-derived aerosols (Junge and Werby

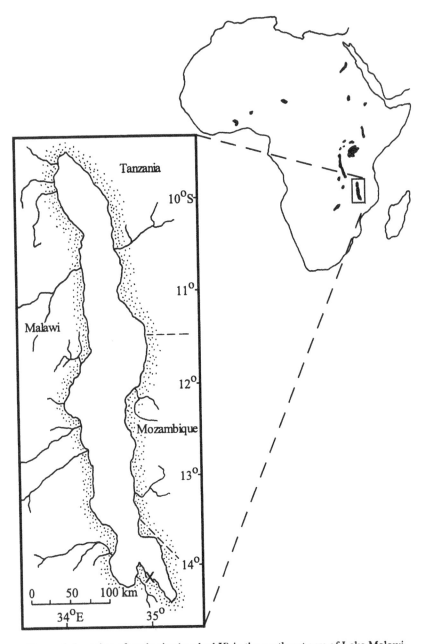

Figure 1. Location of study site (marked X) in the southeast arm of Lake Malawi.

1958; Munger 1982; Linsey et al. 1987), which may be an important source of Al, Si, Ca, Mg, Fe and P (Lawson and Winchester 1979; Munger 1982).

Data on precipitation chemistry in Africa is limited to a small number of studies (Visser 1961; Ganf and Viner 1973; Bromfield 1974; Bromfield et al. 1980; Gaudet and Melack 1981; Rodhe et al. 1981; Lacaux et al. 1992). In some regions, there have been apparent changes in the chemical composition of precipitation during the last few decades, possibly due to intense anthropogenic activities (Bootsma and Hecky 1993).

Here we present data on the chemical composition of rain near Lake Malawi, with the objectives of: 1) determining the potential influence of anthropogenic activities on precipitation chemistry, and 2) assessing the importance of atmospheric deposition as a nutrient source to Lake Malawi.

METHODS

Rain samples were collected during the 1990–91 rainy season (Nov. 25 through Feb. 26) at a station located on the western shore of the southeast arm of Lake Malawi (Fig. 1). Samples were collected in a polyethylene bucket with a top diameter of 24 cm. At the commencement of each precipitation event, the bucket was washed with deionized water and placed on top of an 8 metre tall tower. Rain water was usually removed from the bucket within 12 hours following a precipitation event, but on three occasions the time span between precipitation and collection was 12–24 hours, and on one occasion it was three days. Volume-weighted mean concentrations of all measured components in these "delayed" samples were not significantly different from the mean concentrations in immediately collected samples, and therefore the delayed samples were treated as wet-only samples and are included in the analyses below.

Out of a total of 29 rain events in the 1990–91 rainy season, 25 were sampled. Before analysis for dissolved components, each sample was filtered through a pre-ashed Whatman GF/C filter. Analyses for NO_3^-, NH_4^+, soluble reactive phosphorus (SRP) and soluble reactive silicon (SRSi) were usually done immediately after sample collection, but on several occasions filtered samples were stored frozen in polyethylene bottles for later analysis. Previous experiments had confirmed that this treatment had no significant effect on dissolved inorganic N or P. SRSi was also unaffected by freezing providing samples were left for more than 12 hours between thawing and analysis. When sample volume was sufficient, subsamples were placed in new, pre-rinsed polyethylene bottles and shipped to the Freshwater Institute in Winnipeg for the analyses of total dissolved nitrogen (TDN), total dissolved phosphorus (TDP), major ions (Na^+, K^+, Ca^{2+}, Mg^{2+}, Cl^-, SO_4^{2-}), dissolved organic carbon (DOC), conductivity and pH. In 6 samples alkalinity was also measured. No preservative was added to these samples, and therefore the partitioning of C, N and P between inorganic and organic forms may have been altered due to microbial activity between the time of sampling and the time of analysis. However, while there may have been some decompositional loss of organic C as CO_2, the effect of microbial activity on total N and P was likely minimal. Similarly, lack of preservation should have had little effect on the concentrations of major ions.

Filters were dried, packed with silica gel desiccant and sent to Winnipeg for the analyses of particulate carbon, nitrogen and phosphorus. All analyses were done using the methods of Stainton et al. (1974).

In order to estimate potential atmospheric nutrient deposition during the dry season, on 9 occasions between early August and late November approximately 1 L of deionized water was added to the bucket (to simulate a wet lake surface; Lewis 1983; Cole et al. 1990) and left on top of the tower for periods of 1 to 7 days, after which water volume was measured and samples were analyzed for NO_3^-, NH_4^+ and SRP. To prevent contamination by insects and spiders, the collection bucket was supported on bricks set in containers of oil. In one sample, two small flying insects were found after collection. These were removed before analysis. No measures were taken to prevent contamination by bird droppings, but birds were never observed roosting near the collection tower.

RESULTS

Volume-weighted mean (VWM) concentrations of solutes, particulates and conductivity are presented in Table 1. Because alkalinity was not routinely measured, the sum of measured anions was usually less than the sum of measured cations. For six samples in which alkalinity was measured, it ranged between 6 and 165 μeq l^{-1}, and total anions then balanced total cations within 10%. The most abundant cation was Ca^{2+}, which accounted for 67% of the cation equivalent concentration, followed by NH_4^+ which accounted for 18%. The dominance of Ca^{2+} is due primarily to three samples collected in early January (shortly after the commencement of the rainy season) in which the Ca^{2+} concentration was exceptionally high (66 to 134 μM). If these three values are excluded, the VWM Ca^{2+} concentration is reduced from 20.0 to 5.0 μM (Table 1) and accounts for only 33% of the cation equivalent concentration, while NH_4^+ accounts for 36%.

An inverse relationship between pH and Ca^{2+} was observed (Fig. 2), indicating that a large portion of alkalinity (as HCO_3^-) is associated with the Ca^{2+} ion. pH measurements were always between 5.9 and 7.7, with a mean of 6.25 (based on volume-weighted mean H^+ concentration).

Among measured anions, the concentrations of NO_3^- and SO_4^{2-} were similar in magnitude, followed by Cl^-. While these ions were nearly sufficient in some samples to balance cation concentration, on average the measured anions accounted for only 20% of the cations. If HCO_3^- makes up the anion deficit in each sample (an assumption that appears justified based on the samples in which alkalinity was measured), the volume-weighted mean HCO_3^- concentration is calculated as 38 μM if all Ca^{2+} measurements are used in the ion balance, and is lowered to 9 μM if the three high Ca^{2+} measurements are excluded.

The mean dissolved inorganic nitrogen (DIN) concentration was slightly greater than the mean measured TDN concentration. This discrepancy was also observed between SRP and TDP. DIN and SRP were measured in Malawi immediately after sample collection, while TDN and TDP were measured several weeks later in Winnipeg. The differences may be due to analytical error or loss due to microbial activity and volatilization. Considering that analyses were done in two different labs using different equipment, analytical error is probably responsible for at least part of the difference, especially with regard to dissolved P which was at concentrations approaching detection limits. The fact that concentrations of TDN and TDP were not greater than the dissolved inorganic concentrations of these elements indicates that concentrations of dissolved organic N and P were negligible.

In order to determine the importance of marine aerosols as a source of ions, ion/Cl^- ratios were calculated and compared to seawater ratios (Table 2). All ratios were higher in rain than in seawater, ranging from an enrichment of 2X for Na^+ to more than 2 orders of magnitude for Ca^{2+}.

To further determine potential origins of the various components, the data (volume-weighted) were treated with factor analysis, followed by varimax rotation. 71% of the total variation in the data was accounted for by the first three factors. A plot of variable loading scores on these three factors reveals four major rain component groups, within which variables were closely correlated (Fig. 3).

Table 1. Volume-Weighted Concentrations (μmol l^{-1}) of Solutes and Particulates in Rain from Lake Malawi, Other Parts of Africa, and Remote Non-Marine Sites

	Malawi				Remote non-marine[a]	Congo[b]	Kampala[c] (Uganda)	East Africa[d]
	25%	50%	75%	VWM				
Ca^{2+}	1.5	4.0	29.5	5.0	0.05–1.3	15	1.3	9.6
Mg^{2+}	0.4	1.0	2.5	1.1	0.1–1.0	–	–	3.3
Na^+	1.0	2.0	4.0	2.9	1.0–7.0	30	74	27
K^+	0.3	0.5	1.2	0.6	0.6–0.9	4.5	44	7.7
H^+	0.1	0.4	0.8	0.7	11–17	20	0.01	0.8
NH_4^+	2.5	5.4	24.6	5.4	1.1–2.3	18	37	0.0
NO_3^-	2.2	2.5	10.6	3.2	1.9–4.3	14	27	0.0
SO_4^{2-e}	1.6	2.2	4.4	2.5	1.4–3.5	12	19	7.8
Cl^-	0.3	0.8	4.0	1.5	2.5–11.8	33	25	8.5
Alk	6.6	11.4	67.6	9.0	–	–	–	23.5
SRP	0.07	0.13	0.54	0.18	–	–	–	–
TDN	6.4	7.5	14.3	8.4	–	–	–	–
TDP	0.07	0.13	0.32	0.10	–	–	–	–
DOC	50	60	60	56	–	–	–	–
PC	11	19	82	20	–	–	–	–
PN	0.6	0.9	5.5	1.0	–	–	–	–
PP	0.03	0.03	0.08	0.04	–	–	–	–

Malawi data are presented as quartiles and volume-weighted mean (VWM). Mean concentration of Ca^{2+} in Malawi rain excludes three exceptionally high measurements (see text). Alk = alkalinity (μeq l^{-1}); SRP = soluble reactive phosphorus; TDN = total dissolved N; TDP = total dissolved P; PC, PN and PP are particulate C, N and P, respectively; DOC = dissolved organic C.

[a]Galloway et al. 1982. [b]Lacaux et al. 1992. [c]Visser 1961 (median values). [d]Rodhe et al. 1981 (absence of NH_4^+ and NO_3^- from samples is likely due to long storage time). [e]SO_4^{2-} concentrations probably do not adequately reflect deposition rates, due to the use of a wet collector (Lewis 1983).

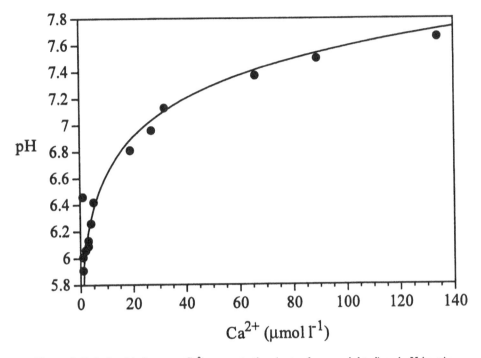

Figure 2. Relationship between Ca^{2+} concentration (not volume-weighted) and pH in rain. The curved line represents the pH predicted for distilled water containing HCO_3^- at a concentration that would balance the Ca^{2+} concentration.

Temporally, most components decreased in concentration over the course of the rainy season. However, this is likely due to the fact that rainfall per event tended to increase as the rainy season progressed, rather than reflecting a change in atmospheric concentrations. No seasonal trend was observed in the areal deposition rate of any chemical component.

Mean daily dry deposition rates of soluble inorganic N and P (70 and 3.0 $\mu mol\ m^{-2}\ d^{-1}$, respectively) were less than mean daily wet deposition rates of these solutes (241 and 4.6 $\mu mol\ m^{-2}\ d^{-1}$). It is possible that some dissolved inorganic N or P may be lost to microbial growth or volatilization when the collection period is longer than several days, as in the case of dry deposition samples. However, calculated daily areal deposition rates did not decrease with longer collection periods, and we concluded that for collection periods of up to one week there was no significant loss of these solutes.

DISCUSSION

pH Control

In North America and Europe, acidic precipitation is due primarily to H_2SO_4 and HNO_3 which are produced as a result of fossil fuel burning and other combustion processes. However, acidic precipitation can also occur in the less industrialized tropics (Lewis 1981; Crutzen and Andreae 1990). Possible causes of low pH in tropical rain include the production of nitric and sulphuric acids as a result of biomass burning (Lewis 1981; Crutzen and Andreae 1990; Lacaux et al. 1992) and the presence of organic acids resulting from either

Table 2. Ion ratios of Malawi Rain
(volume-weighted means), Seawater, and
Lake Malawi Surface Water, Based
on Equivalent Concentrations

	Ion/Cl$^-$	
	Seawater	Rain
Ca^{2+}	0.04	6.67
Mg^{2+}	0.19	1.47
Na^+	0.86	1.96
K^+	0.02	0.4

Σ Ions = Ca^{2+} + Mg^{2+} + Na^+ + K^+ + Cl^-. Adjusted ratios were calculated after correcting for apparent Ca^{2+} enrichment (see text). Three exceptionally high Ca^{2+} measurements in rain were not used in calculations.

direct gaseous emissions by vegetation (Keene and Galloway 1986; Lacaux et al. 1992) or biomass burning (Crutzen and Andreae 1990). None of these sources appears to have a large effect on the pH of rain near Lake Malawi. As in much of Africa, biomass burning is widespread in Malawi, Mozambique and Tanzania (Andreae 1993), the three countries surrounding Lake Malawi. While burning occurs throughout the year, much of it is done in the early part of the dry season, when fires are easier to control. This may explain why no large impact on rain chemistry was evident during this study, even at the beginning of the rainy season. However, the relatively high ratios of NO_3^- and NH_4^+ to other ions (Table 3) suggests that biomass burning does have some influence on rain chemistry (Lewis 1981; Crutzen and Andreae 1990).

The pH of distilled water at equilibrium with atmospheric CO_2 is 5.6, and therefore a rain pH higher than this must be due to the presence of alkali. Two other studies of precipitation chemistry in Africa have also reported high pH values. Visser (1961) observed a median pH of 7.9 near Kampala (Uganda), and Rodhe et al. (1981) reported a median pH of 6.1 for 9 East African sites (Table 1). In rain near Lake Malawi it appears that alkalinity can be attributed almost entirely to $Ca(HCO_3)_2$ (Fig. 2). Comparison of anion and cation concentrations measured by Visser reveals a large anion deficit. As in the present study, this may represent HCO_3^-. However, the median Ca^{2+} concentration reported by Visser was much lower than the mean concentration measured in Malawi precipitation (Table 1), and Na^+ and K^+ made up a much larger portion of total base cations. Ca^{2+} concentrations measured by Rodhe et al. (1981) are more similar to those in Malawi, but they also observed Na^+ and K^+ concentrations much higher than those measured during this study. Therefore, while precipitation in East Africa generally appears to have a relatively high pH, the relative importance of the various base cations and acid anions varies with location. In Malawi, SO_4^{2-} and NO_3^- concentrations are minimal, and rain pH can be higher than 5.6 due to the presence of $Ca(HCO_3)_2$. Further north (Tanzania, Kenya, Uganda), SO_4^{2-} and NO_3^- concentrations are greater, but rain pH remains above 5.6 due to the buffering capacity resulting from alkalinity associated with high Ca^{2+}, K^+ or Na^+ concentrations.

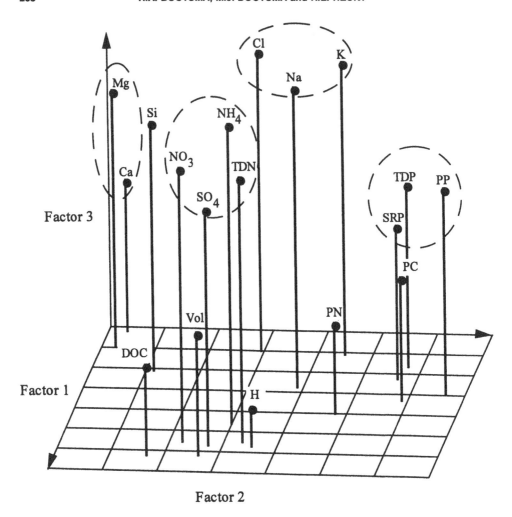

Figure 3. Plot of rain component loadings on the first three factors produced by factor analysis followed by varimax rotation. The first three factors represent 71% of the total data variance. Circled groups represent rain components for which deposition rates are controlled by similar mechanisms.

In contrast to the relatively high pH of rain observed in semi-arid East Africa, Lacaux et al. (1987, 1992) have found rain in Ivory Coast and Congo, which have a wetter climate, to have a low pH, particularly in the rainy season. Their measurements suggest this is due partly to biomass burning, which is particularly severe in the savannah near these locations (Crutzen and Andreae 1990), and partly to the emission of organic acids by the surrounding rain forest. In this area, although base cation concentrations are similar to those measured in East Africa (Table 1), alkali concentrations are insufficient to buffer high atmospheric fluxes of acids, except during the dry season when there is greater input of $CaCO_3$ in the form of dust particles from the Sahara desert (Lacaux et al. 1987, 1992).

Sources of Rain Components

Apart from Ca^{2+}, NH_4^+, H^+ and HCO_3^-, all measured solute concentrations in Malawi rain were similar to those reported for remote, non-marine sites by Galloway et al. (1982; Table 1). While this does not eliminate anthropogenic activity as an important factor influencing rain chemistry, it does indicate that during the rainy season the effects of anthropogenic activities are not particularly strong, and other possible solute sources, such as soil and vegetation, may be important. Clues to the relative importance of the various sources can be obtained from the factor analysis.

The circled groups in Fig. 3 represent rain components between which correlations were strong. Correlations between components imply that similar mechanisms are controlling the variance of those components. These mechanisms can be considered under three categories: 1) production mechanisms; 2) transport mechanisms; 3) scavenging mechanisms.

Ca^{2+} is often considered to be indicative of a soil source (Junge and Werby 1958; Lawson and Winchester 1979; Munger 1982; Linsey et al. 1987). Although Artaxo et al. (1988) and Lesack and Melack (1991) have indicated that a large portion of atmospheric Ca^{2+} over the Amazon Basin may be in the form of biogenically produced aerosols, the concentration of Ca^{2+} in Malawi rain (10 µeq l^{-1}) is several times greater than that reported for Amazon Basin rain (1.9–2.9 µeq l^{-1}; Lesack and Melack 1991), and it is probable that Ca^{2+} in Malawi rain originates as soil particles. This is further supported by the observation that the highest Ca^{2+} deposition rate occurred when winds were from the northwest (overland), which is uncommon for this region. Satellite images show that there is more cultivation and soil exposure to the immediate west of the lake. Our conclusion is that Ca^{2+}, Mg^{2+}, and probably Si, are controlled primarily by soil exposure and wind direction, i.e., production and transport mechanisms.

The correlation between Na^+ and Cl^- can be interpreted as reflecting a marine origin for these ions. Although the mean rain Na^+/Cl^- ratio was more than twice the seawater ratio (Table 2), Möller (1990) has shown that continental rain that has not been influenced by additions of Na^+ or Cl^- can be expected to have a Na^+/Cl^- ratio of 1.1 to 1.8, due to the selective loss of Cl^- from fine aerosols. The mean rain Na^+/Cl^- ratio of 1.96 indicates a slight enrichment of Na^+, probably from soil. Although K^+ is correlated with Na^+ and Cl^-, the high K^+/Cl^- ratio in rainwater indicates that K^+ does not have a strong marine source. Hence, simultaneous fluctuations in the deposition rates of these three constituents can be related to changes in transport pathways. This mechanism has been invoked by Andreae et al. (1990) to explain the close association between almost all chemical components of rain in the Central Amazon Basin.

The correlation between SRP and TDP is not surprising, since SRP makes up the majority of TDP. The grouping of dissolved and particulate P indicates that these two components likely have a common source. Soil is a potentially important source (Linsey et al. 1987), and soil exposure resulting from deforestation and agriculture has been suggested as a possible cause of increased P deposition in Uganda (Bootsma and Hecky 1993). However, the release by vegetation of low molecular weight molecules or aerosols may also be an important P source (Lewis et al. 1985). The association between soluble and particulate P suggests much P may be deposited in the form of soluble aerosols, but we cannot determine the relative importance of soil, vegetation and biomass burning.

Almost all previous studies of atmospheric P deposition have used bulk samples, and we could find no data on soluble or particulate P concentration in rain. Our measurement of wet

P deposition is similar to the lowest estimate of summer atmospheric P deposition in the literature summarized by Cole et al. (1990).

Dissolved nitrogen consisted almost entirely of NO_3^- and NH_4^+, explaining the correlation between TDN and each of these ions. But unlike P, the soluble forms of N do not appear to be related to particulate N. As mentioned above, the relatively high NH_4^+/cation and NO_3^-/anion ratios suggest biomass burning as a N source. This might explain the correlation with SO_4^{2-}, which is also a product of burning (Andreae et al. 1988; Lacaux et al. 1992). However, the propensity of NH_4^+ to react with nitric acid to produce ammonium nitrate (Harrison and Pio 1983) and with SO_4^{2-} to form ammonium sulfate (Pratt and Krupa 1985), suggests that correlations between these constituents may be due as much to a common scavenging mechanism as to a common origin.

Dry Deposition of N and P

There are few other studies reporting dry deposition rates of soluble inorganic N and P. Cole et al. (1990), using collection buckets filled with distilled water, calculated a soluble P deposition rate of about 5 μmol m^{-2} d^{-1} during summer on Mirror Lake (New Hampshire), which is within the range observed by Lewis et al. (1985) in Colorado. The mean dry SRP deposition rate measured in Malawi (3.0 μmol m^{-2} d^{-1}) is slightly lower than that measured in these two studies. Our measurements did not include organic P, but the similarity of SRP and TDP in rain suggests organic P was not a large component of soluble P in dry deposition. Cole et al. (1990) suggested that the source of dissolved P near Mirror Lake was primarily particles, which consisted largely of plant fragments, insects and spiders. This is supported by their observation that P deposition decreased exponentially with distance from shore. We rarely observed abundant visible particles in our samples, probably due to the fact that the sampling site was located on the leeward side of the lake, so that any particles deposited in the collector would have had to travel a minimum distance of 25 km across the lake. At the sampling site, gaseous or fine aerosol forms of P were likely more significant. It is probable that P deposition rates are greater on the east (windward) side of the lake, and our measurements of wet and dry deposition should be considered minimum estimates when applied to the whole lake.

The potential importance of dry deposition of soluble N in other regions can be determined based on the differences between bulk and wet deposition. Linsey et al. (1987) found NO_3^- concentrations to be similar in bulk and wet samples, implying that dry deposition is insignificant, and Scheider et al. (1979) found that the majority of NO_3^- and NH_4^+ in bulk deposition was usually due to wet deposition. Our results are in agreement with these findings. However, on an annual basis the large difference between daily wet and dry deposition is offset by the low number of rain days in the Lake Malawi rainy season. While our limited measurements allow only a crude estimate of dry N and P deposition, the differences between estimated wet and dry deposition rates are large enough to support the conclusion that annual dry deposition rates of N and P are much greater than wet deposition rates.

Contribution of Atmospheric N and P to Lake Nutrient Cycles

Nutrient sources for a lake include riverine input, atmospheric deposition, and N fixation. Groundwater inflow may be an additional source, but in Lake Malawi the contribution of groundwater to the water budget is likely negligible (Owen et al. 1990), and therefore groundwater inputs can also be assumed to be a small component of the nutrient budget.

Table 3. Regional Comparison of the NH_4^+/Cation and NO_3^-/Anion Ratios in Rain

Ion ratio	Malawi	Remote	Congo	Kampala
$NH_4^+/(Ca^{2+} + Na^+ + K^+ + NH_4^+)$	0.29	0.22	0.22	0.23
$NO_3^-/(Cl^- + SO_4^{2-} + NO_3^-)$	0.33	0.20	0.20	0.30

Only ions that were measured in all compared studies were used to determine ratios. Data sources as in Table 1.

A number of studies have found direct atmospheric nutrient inputs to be a significant component of total nutrient flux to lakes (Schindler et al. 1976; Scheider et al. 1979; Manny and Owens 1983; Cole et al. 1990). In order to determine the relative importance of atmospheric deposition to the Lake Malawi nutrient budget, measured deposition rates of N and P were compared with estimates of fluxes from other sources (Table 4). Based on the measured forms of N and P, atmospheric deposition directly on the lake surface represents 33% of P input and 72% of N input (excluding N fixation, which is unknown) to the lake. These can be considered minimum estimates, since particulate N and P were not measured in the dry deposition samples. However, if the ratio of particulate to soluble nutrients in dry deposition is similar to that in rain, the exclusion of particulate nutrients in dry deposition will not result in a large error, and the above estimates can be accepted as approximate.

Because nutrients play such an important role in controlling phytoplankton production, and because an understanding of factors controlling phytoplankton production and composition has important applications ranging from fisheries biology to paleolimnology, it is also of interest to examine the role of atmospheric deposition in the nutrient budget of the surface mixed layer (SML), where phytoplankton production occurs. In Lake Malawi, P flux to the SML is dominated by upward flux from the monimolimnion (Table 4), and only a small portion (~6%) is in the form of atmospheric deposition. In contrast, denitrification near the oxic–anoxic boundary prevents fixed N in the monimolimnion from reaching the SML (Bootsma and Hecky 1993), and as a result atmospheric deposition accounts for ~72% of N supply to the SML. This estimate does not account for any N-fixation that may occur, but N-fixing cyanobacteria are rarely abundant in the phytoplankton of Lake Malawi (Hecky and Kling 1987; Bootsma 1993a). Hence, the N:P supply ratio for the SML is determined to a large degree by monimolimnetic P flux and atmospheric N deposition. However, even though N may at times limit phytoplankton production in Lake Malawi (Bootsma and Hecky 1993), it is ultimately P supply, and therefore mixing between the monimolimnion and SML, that drives new phytoplankton production (Hecky et al., this volume).

The ratio of total N flux : total P flux to the SML of Lake Malawi, using the data in Table 4, is 4.4:1. This is much lower than the total N : total P concentration ratio of 63:1 in the SML (Bootsma 1993b). There are four possible explanations for this disparity. First, dry deposition of particulate N, which we did not measure, may be significant. Second, our estimate of upwelling P may be too high. This estimate is based on a vertical TDP concentration profile and the mixing rate calculated by Gonfiantini et al. (1979), which they concede may have a large degree of error. However, it is unlikely that an underestimate of N deposition or an overestimate of P upwelling could account for the 14-fold difference between flux and concentration ratios. Third, our estimates of N and P flux may be accurate for the period of study, but may not represent long term steady state conditions. Because of

Table 4. Fluxes of N and P to the Surface Mixed Layer of Lake Malawi

	Atmospheric		Rivers	Monimo-limnion	Total	Required by phyto-plankton	Deficit
	Wet	Dry					
TDP	0.13	1.55	1.51	23	26.2	–	–
PP	0.03	NM	1.91	0	1.94		
Total P	0.16	1.55+	3.42	23	28.1+	61	33
TDN	7.0	81	23	0	111		
PN	0.78	NM	12	0	12.8		
Total N	7.78	81+	35	0	123.8+	1,155	1,031

Annual wet deposition calculated as mean daily wet deposition × 29 rain days. Dry deposition calculated as mean daily dry deposition × 336 days. Riverine fluxes calculated as the average concentration of 13 inflowing rivers (Hecky unpubl.) times total annual river inflow. Monimolimnetic fluxes from Bootsma and Hecky (1993). Labels as in Table 1. NM = not measured. All fluxes in mmol m^{-2} y^{-1}. N and P deficits are assumed to be met by recycling within the surface mixed layer (see text).

its long flushing time (~750 years, Bootsma and Hecky 1993), N:P concentration ratios in Lake Malawi are the result of N and P inputs integrated over more than one year. We have no way of evaluating interannual variability of nutrient inputs, but they may vary several fold from year to year (eg. Linsey et al. 1987).

A fourth possible explanation for the difference between flux ratios and concentration ratios is a difference between internal recycling efficiencies of N and P within the mixolimnion. A crude estimate of internal recycling rates can be made by comparing phytoplankton N and P demand with N and P supply rates to the SML. Based on lake-wide measurements of phytoplankton photosynthesis (mean = 42 mmol m^{-2} d^{-1}; ^{14}C uptake method) and seston C:N:P atom ratios (mean = 251:19:1; Bootsma 1993b), phytoplankton N and P requirements can be estimated (Table 4). Comparison with N and P fluxes indicates that about 11% of phytoplankton N demand and 46% of P demand are met by new supply to the SML. If the assumption of negligible N-fixation is valid, this implies that 89% of phytoplankton photosynthesis is supported by N recycled within the SML and 54% is supported by recycled P.

CONCLUSION

While the relatively high Ca^{2+} concentration and high NH_4^+/cation and NO_3^-/anion ratios indicate that rain chemistry near the southern end of Lake Malawi is probably influenced by soil exposure and biomass burning, the impact of anthropogenic activities at present appears to be minimal in comparison to other regions of Africa. This is somewhat surprising in view of the high frequency of biomass burning in the region surrounding the lake (Andreae 1993). Low solute concentrations in rain might be attributed to atmospheric washout and reduced burning during the rainy season. However, in this case one might expect to observe high deposition rates of at least some rain constituents at the beginning of the rainy season (eg. Lewis 1981). This was not the case near Lake Malawi. Alternatively, the time span between the burning "season" and rainy season near Lake Malawi may be long enough to allow atmospheric chemicals produced by burning to be redeposited or transported away before the commencement of the rainy season.

There are presently no data on interannual variability of rain chemistry for any part of Africa, and we cannot determine to what degree the data presented here represent long term conditions. This lack of knowledge, along with the apparent recent change in rain chemistry in parts of Africa (Bootsma and Hecky 1993), and the important role of atmospheric deposition in lake nutrient cycles, are strong reasons to establish deposition monitoring programs on Lake Malawi and other African lakes.

ACKNOWLEDGMENTS

We are grateful to the Malawi Department of National Parks and Wildlife and the Malawi Fisheries Department for logistic support and permission to conduct research in Malawi. Financial support was provided by the International Development Research Centre (Ottawa) and the Department of Fisheries and Oceans, Canada. Analyses of major ions, TDN, TDP, DOC, and particulates were conducted by staff of the analytical chemistry lab at the Freshwater Institute. We thank M.O. Andreae, W.M. Lewis Jr., J. Willey and an anonymous reviewer for helpful comments on the manuscript.

REFERENCES

Andreae, M.O., E.V. Browell, M. Garstang, G.L. Gregory, R.C. Harriss, G.F. Hill, D.J. Jacob, M.C. Pereira, G.W. Sachse, A.W. Setzer, P.L. Silva Dias, R.W. Talbot, A.L. Torres, and S.C. Wofsy. 1988. Biomass burning emissions and associated haze layers over Amazonia. J. Geophys. Res. 93: 1509–1527.

Andreae, M.O., R.W. Talbot, H. Berresheim, and K.M. Beecher. 1990. Precipitation chemistry in Central Amazonia. J. Geophys. Res. 95: 16987–16999.

Andreae, M.O. 1993. Global distribution of fires seen from space. EOS 74: 129–135.

Artaxo, P., H. Storms, F. Bruynseels, R. Van Grieken, and W. Maenhaut. 1988. Composition and sources of aerosols from the Amazon Basin. J. Geophys. Res. 93: 1605–1615.

Bootsma, H.A. 1993a. Spatio-temporal variation of phytoplankton biomass in Lake Malawi, Central Africa. Verh. Internat. Verein. Limnol. 25:882–886.

Bootsma, H.A. 1993b. Algal dynamics in an African Great Lake, and their relation to hydrographic and meteorological conditions. Univ. of Manitoba, 311 pp.

Bootsma, H.A., and R.E. Hecky. 1993. Conservation of the African Great Lakes: A limnological prespective. Cons. Biol. 7: 644–656.

Bromfield, A.R. 1974. The deposition of sulphur in rainwater in Northern Nigeria. Tellus 26: 408–411.

Bromfield, A.R, D.F. Debenham, and I.R. Hancock. 1980. The deposition of sulphur in rainwater in central Kenya. J. Agric. Sci. 94: 299–303.

Cole, J.J., N.F. Caraco, and G.E. Likens. 1990. Short-range atmospheric transport: A significant source of phosphorus to an oligotrophic lake. Limnol. Oceanogr. 35: 1230–1237.

Crutzen, P.J., and M.O. Andreae. 1990. Biomass burning in the tropics: Impact on atmospheric chemistry and biogeochemical cycles. Science 250: 1669–1678.

Delmas, R. 1982. On the emission of carbon, nitrogen and sulfur in the atmosphere during bushfires in intertropical savannah zones. Geophys. Res. Lett. 9: 761–764.

Galloway, J.N., G.E. Likens, W.C. Keene, and J.M. Miller. 1982. The composition of precipitation in remote areas of the world. J. Geophys. Res. 87: 8771–8786.

Ganf, G.G., and A.B. Viner. 1973. Ecological stability in a shallow equatorial lake (Lake George, Uganda). Proc. R. Soc. Lond. 184: 321–346.

Gaudet, J.J., and J.M. Melack. 1981. Major ion chemistry in a tropical African lake basin. Freshwater Biol. 11: 309–333.

Gonfiantini, R., G.M. Zuppi, D.H. Eccles, and W. Ferro. 1979. Isotope investigation of Lake Malawi, pp. 192–205, in Isotopes in lake studies. Proceedings of a conference on application of nuclear techniques to study of lake dynamics. International Atomic Energy Agency, Vienna.

Harrison, R.M., and C.A. Pio. 1983. An investigation of the atmospheric HNO_3–NH_3–NH_4NO_3 equilibrium relationship in a cool, humid climate. Tellus 35: 155–159.

Hecky, R.E., Bootsma, H.A., Mugidde, R., and Bugenyi, F.W.B. 1996. Phosphorus pumps, nitrogen sinks, and silicon drains: plumbing nutrients in the African Great Lakes, pp. 205–224, in The limnology, climatology and paleoclimatology of the East African lakes, eds. T.C. Johnson and E.O. Odada, Gordon and Breach, Toronto.

Hecky, R.E., and H.J. Kling. 1987. Phytoplankton ecology of the great lakes in the rift valleys of Central Africa. Arch. Hydrobiol. 25: 197–228.

Junge, C.E., and R.T. Werby. 1958. The concentration of chloride, sodium, potassium, calcium, and sulfate in rain water over the United States. J. Meteor. 15: 417–425.

Keene, W.C., and J.N. Galloway. 1986. Considerations regarding sources for formic and acetic acids in the troposphere. J. Geophys. Res. 91: 14466–14474.

Lacaux, J.P., J. Servant, and J.G.R. Baudet. 1987. Acid rain in the tropical forests of the Ivory Coast. Atmos. Environ. 21: 2643–2647.

Lacaux, J.P., J. Loemba-Ndembi, B. Lefeivre, B. Cros, and R. Delmas. 1992. Biogenic emissions and biomass burning influences on the chemistry of fogwater and stratiform precipitations in the African equatorial forest. Atmos. Environ. 26A: 541–551.

Lawson, D.R., and J.W. Winchester. 1979. Sulfur, potassium and phosphorus associations in aerosols from South American tropical rain forests. J. Geophys. Res. 84: 3723–3727.

Lesack, L.F.W., and J.M Melack. 1991. The deposition, composition, and potential sources of major ionic solutes in rain of the Central Amazon Basin. Water Resour. Res. 27: 2953–2977.

Lewis, W.M. Jr. 1981. Precipitation chemistry and nutrient loading by precipitation in a tropical watershed. Water Resour. Res. 17: 169–181.

Lewis, W.M. Jr. 1983. Collection of airborne materials by a water surface. Limnol. Oceanogr. 28: 1242–1246.

Lewis, W.M. Jr., M.C. Grant, and S.K. Hamilton. 1985. Evidence that filterable phosphorus is a significant atmospheric link in the phosphorus cycle. Oikos 45: 428–432.

Lindberg, S.E., G.M. Lovett, D.D. Richter, and D.W. Johnson. 1986. Atmospheric deposition and canopy interactions of major ions in a forest. Science 231: 141–145.

Linsey, G.A., D.W. Schindler, and M.P. Stainton. 1987. Atmospheric deposition of nutrients and major ions at the Experimental Lakes Area in northwestern Ontario, 1970 to 1982. Can. J. Fish. Aquat. Sci. 44 (Suppl. 1): 206–214.

Manny, B.A., and R.W. Owens. 1983. Additions of nutrients and major ions by the atmosphere and tributaries to nearshore waters of northwestern Lake Huron. J. Great Lakes Res. 9: 403–420.

Möller, D. 1990. Na/Cl ratio in rainwater and the seasalt chloride cycle. Tellus 42: 254–262.

Munger, J.W. 1982. Chemistry of atmospheric precipitation in the North-Central United States: Influence of sulfate, nitrate, ammonia and calcareous soil particles. Atmos. Environ. 16: 1633–1645.

Owen, R.B., R. Crossley, T.C. Johnson, D. Tweddle, I. Kornfield, S. Davison, D.H. Eccles, and D.E. Engstrom. 1990. Major low levels of Lake Malawi and their implications for speciation rates in cichlid fishes. Proc. R. Soc. Lond 240: 519–553.

Pratt, G.C., and S.V. Krupa. 1985. Aerosol chemistry in Minnesota and Wisconsin and its relation to rainfall chemistry. Atmos. Environ. 19: 961–971.

Rodhe, H., E. Mukolwe, and R. Söderlund. 1981. Chemical composition of precipitation in East Africa. Kenya J. Sci. Technol. (Ser. A) 2: 3–11.

Scheider, W.A., W.R. Snyder, and B. Clark 1979. Deposition of nutrients and major ions by precipitation in South-Central Ontario. Water Air Soil Poll. 9: 309–314.

Schindler, D.W., R.W. Newbury, K.G. Beaty, and P. Campbell. 1976. Natural water and chemical budgets for a small precambrian lake basin in central Canada. J. Fish. Res. Board Can. 33: 2526–2543.

Stainton, M.P., M.J. Capel, and F.A.J. Armstrong. 1974. The chemical analysis of fresh water. Can. Fish. Mar. Serv. Misc. Spec. Publ. 25: 125 pp.

Visser, S. 1961. Chemical composition of rain water in Kampala, Uganda, and its relation to meteorological and topographical conditions. J. Geophys. Res. 66: 3759–3765.

Residence Times of Major Ions in Lake Naivasha, Kenya, and Their Relationship to Lake Hydrology

B.S. OJIAMBO and W.B. LYONS   *Hydrology/Hydrogeology Program, Mackay School of Mines, University of Nevada at Reno, United States*

Abstract — Lake Naivasha is "distinctive" among Rift Valley lakes of Ethiopia, Kenya and Tanzania in that it is fresh. Outflow of lake waters into the groundwater system is the prevailing explanation for the lake's freshness. Seepage outflow in the southern portion of the lake has been estimated to be as much as 20% of the total water loss (Gaudet and Melack, 1981). In order to establish a better hydrologic budget for the lake, residence times were calculated for the major elements in the lake water and fell in the range of 3–6 years. Chloride was assumed to be a conservative element. The "chloride age" of the lake was computed to be 5.4 years. This very young age can be compared to chloride ages for other lakes in the region such as 2500–5000 yrs for Lake Turkana (Yuretich and Cerling, 1983; Barton et al., 1987) and 100–200 yrs for Lake Baringo (Barton et al., 1987). The residence times of Na^+, K^+, Ca^{2+}, Mg^{2+}, and HCO_3^- are all within ± 1 yr of that of Cl^-, while those of SO_4^{2-} and H_4SiO_4 are even less. The loss of SO_4^{2-} through diffusion into the sediments and sulfate reduction and H_4SiO_4 through diatom uptake supports the earlier conclusions of Gaudet and Melack (1981). Ojiambo (1992) computed mean parameters of the hydrologic budget using data from four earlier reports and found evaporation loss to be 77%, irrigation 3% and seepage loss 20%. Using chemical mass approaches, our data suggest that about 55% of the annual inflow is evaporated, while another 15% is utilized for irrigation. If this is so, then about 30% of the total loss must be via seepage outflow. Therefore, the most recent estimates suggest that 20–30% of the annual water loss from Lake Naivasha is through groundwater outflow.

SURFACE WATER HYDROLOGY

General

The location of Lake Naivasha in both plan and topographic profile is shown in Figs. 1 and 2, respectively. The lake is the dominant surface water body in the study area apart from the Crater lake or Sonachi (0.2 km^2) to west of Lake Naivasha. All the gullies in the area are erosional features which carry water only during heavy storms.

On a regional scale, Lake Naivasha is mostly recharged by Gilgil and Malewa rivers, which are perennial and by an ephemeral Karati river (Figs. 2 and 3a). Other sources of recharge to the lake are from direct precipitation, surface runoff from the surrounding land and groundwater.

Hydrology of Lake Naivasha: Bottom Morphology of the Lake

Lake Naivasha is a closed basin, freshwater lake located at the apex of N–S Rift Valley floor dome (Fig. 2) with an average elevation of 1886 meters above sea level (masl). The total surface area of the lake, about 200 km^2 (1983 levels), consists of about 139 km^2 of open

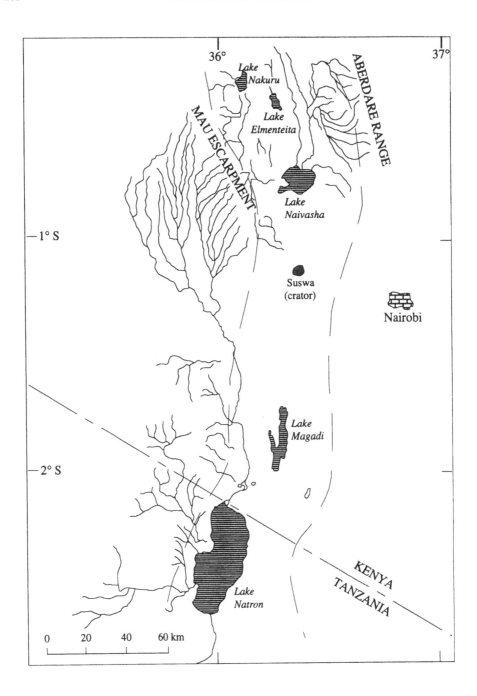

Figure 1. Location map of Lake Naivasha showing inflow rivers from the north. The two dashed lines are the approximate locations of the East and West scarps of the Rift Valley. From Jones et al. (1977).

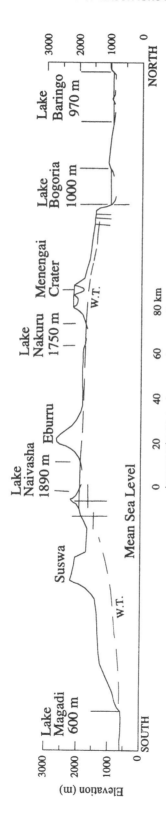

Figure 2. A North–South Topographical profile of the Rift Valley floor from Lake Baringo to Lake Magadi. Lake Naivasha is shown near the center. Postulated groundwater level is shown as dashed line. From Ojiambo (1992).

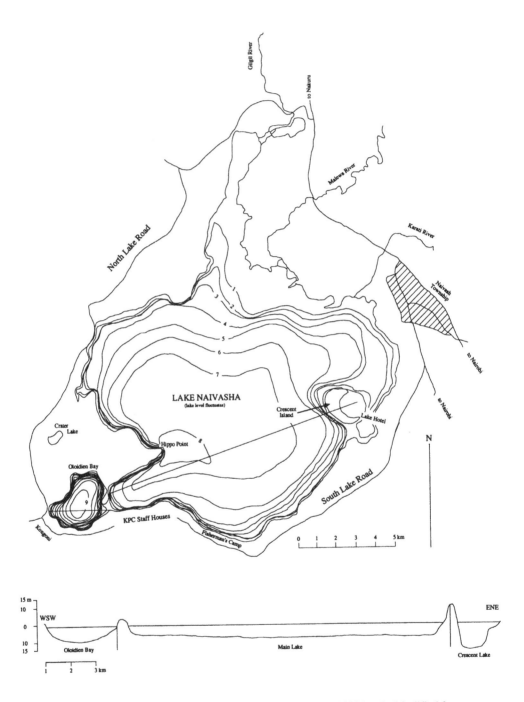

Figure 3. a) (top) Bathymetric map of Lake Naivasha based on 1983 levels. Modified from
Åse et al. (1986). b) (bottom) Bathymetric profile of Lake Naivasha from Oloidien Bay to
Crescent Lake, based on soundings 29, 26 and 22 in Fig. 3a. Modified from Åse et al. (1986).

water and 64 km^2 of swamp (McCann, 1974). The lake is shallow with an average depth of 4.7 meters and estimated volume of 900×10^6 m^3 based on 1983 levels (Åse et al., 1986). The deepest part of the main lake (8 meters) is located near Hippo Point (Fig. 3a,b). A WSW–ENE bathymetric profile of the lake bottom (Fig. 3b) shows the flatness of the central and main part of the lake and the crater-like morphology of the two deepest parts of the lake, the Oloidien bay (11.5 meters) to the WSW and Crescent lake (18 meters) to the ENE. Sonachi (Crater lake) has a maximum depth of about 4 meters according to Harper et al., 1990.

Åse et al. (1986) state that the flatness may be due to the fact that the basin has filled up with large quantities of sediments that has resulted in the development of even bottom topography. The two deepest parts of the lake have typical crater shaped morphology indicating volcanic origin of formation. The bathymetric map of the lake made in 1983 by Åse et al. (1986) shows similarities to the one drawn by the Public Works Department (PWD) in 1927 and reproduced by Thompson and Dodson (1963) except for the depth contours of Oloidien bay. Whereas the 1983 maps show the maximum depth of Oloidien bay as 11.5 meters, the PWD map gives a maximum depth of 14 ft or 4.3 meters despite the fact that the lake level in 1927 was nearly 3 meters higher than in 1983 (Fig. 4). Åse et al. (1986) contend that this large difference may be due to lack of sufficient depth data taken in 1927. This may be a plausible explanation but one may also ask why there is agreement in depth contours in other parts of the lake except in the Oloidien bay. Geologic maps of the area (Naylor, 1972; Clarke et al., 1990) show a number of young faults passing under the bay and, furthermore, the postulated subsurface outflow area from the lake is through this bay (Ojiambo, 1992). Some sediment cores and new bathymetric measurements in this area may help explain what is happening in the Oloidien bay.

A Note on the Lake Naivasha Water Level Elevations

Lake Naivasha water level measurements are shown in Fig. 4. Water level data have been kept by the Ministry of Water Development since 1908. While working on Lake Naivasha water level monitoring in 1972, Ojiambo observed that his surveyed water level elevations were lower than those kept by the Ministry of Water Development by about 3.6 meters. This matter was discussed with the Ministry authorities and, on checking their records, Ojiambo discovered that their Benchmark datum was changed by a similar factor of 3.6 meters in 1958. Cross checking was undertaken with the Survey of Kenya and a private survey company, Geosurvey Limited, and Ojiambo's figures were confirmed to be correct. Many researchers on the lake continued using the Ministry's water level drawing until Åse et al. (1986) noticed a similar discrepancy during their lake level measurements in October 1983. Åse et al. (1986), however, did not change their elevation values in their 1986 report but mentioned the existence of the differences. Åse carried out a further thorough investigation of the lake elevations discrepancies in 1986. His search through records kept on water level readings from Mr. Mennell of Korongo Farm, from Lake Naivasha Hotel at Crescent Island and from earlier workers resulted in Åse et al. (1986), drawing from which Fig. 4 has been developed to include the 3.6 meters correction (Åse, 1987). These records are the same values as those used by Ojiambo in 1972.

Lake Level Fluctuations

Since continuous water level measurements of Lake Naivasha began in late 1908 (Fig. 4) the largest water level drop has been 9.5 meters, which occurred between 1917 (1891 masl) and 1946 (1881.5 masl). Statistical analysis of the lake water level fluctuations since 1908

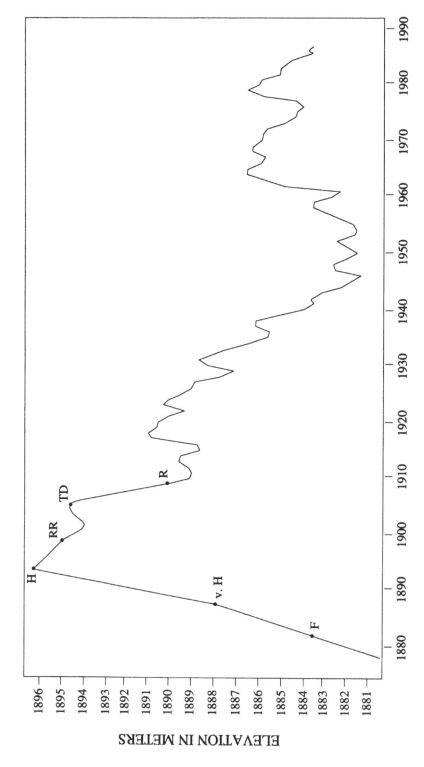

Figure 4. Lake Naivasha water level hydrograph from 1880 to 1988. Symbols represent source of data points: F = Fischer; v. H. = von Hohnel; H = Hobley; RR = railway records; R = recording start. This diagram incorporates the 3.6-m correction in text. Modified from Åse (1987).

by Vincent et al. (1979) shows a periodicity of around 7 years with an indication of an 11-year return period. Åse et al. (1986) attach some significance to the 11 years periodicity because it coincides with the lake level highs in 1905, 1917, 1937 and 1948. According to these authors, this return period corresponds to the sunspot cycle but they state that no 11-yr periodicity correlation is found after 1948. However, our closer analysis of the hydrograph (Fig. 4), shows that there is still some evidence for 11-yr periodicity after 1948 as shown by the coincidence of the hydrograph highs and 11-yr periodicity-predicted highs in 1957/58, 1968 and 1978/79 if the 1964 high is considered a statistical anomaly.

Lake Naivasha Hydrologic Water Budget

Kilham and Hecky (1973) and later Gaudet and Melack (1981), wrote that Lake Naivasha is "distinctive" among the Rift Valley lakes of Ethiopia, Kenya and Tanzania in that it is fresh. McCann (1974) states that the lake freshness indicates "that effluent seepage occurs." Local legend has it that the lake has a subsurface outflow (Åse et al., 1986). Early and more recent studies of the lake (Thompson and Dodson, 1963; McCann, 1972, 1974; Gaudet and Melack, 1981; Åse et al., 1986; Barton et al., 1987; Clarke et al., 1990) suggest that this is indeed the case. Gaudet and Melack (1981) show from their seepage meter measurements that there is seepage-out on the southwest and northeast sections of the lake, but that there is no explicit evidence for a subsurface river. Stable isotope measurements of δD and $\delta^{18}O$ indicate that the hot springs and Olkaria geothermal discharges to the south of the lake have isotopic compositions similar to those of the lake waters (Clarke et al., 1990; Darling et al., 1990; and Barton et al., 1987), suggesting to these investigators that these hot geothermal waters are derived from Lake Naivasha subsurface outflow. Gaudet and Melack (1981), using mass-balance techniques, calculated that water loss via seepage outflow in the southern portion of the lake could amount to as much as 20% of the total water loss. More recent research using stable isotopes (Darling et al., 1990) estimated that as much as 90% of the seepage loss from the lake is from the southern part of the basin.

Conservative estimates of this subsurface outflow range from 34 to 50×10^6 m^3 per year (McCann, 1974; Åse et al., 1986; Gaudet and Melack, 1981). A reevaluated mean value from this study is 38×10^6 m^3/yr (Table 1). This is in good agreement with the previous work especially given the input data uncertainties and the variations in the climatological factors used. Winter (1981) shows that errors that may be introduced into the equation for estimating water balances of lakes by the various parameters vary from 25% for precipitation to 53% for evaporation. Gaudet and Melack (1981) further estimate that 15×10^6 m^3/yr of the lake water is taken for irrigation. Environmental Services Australia (ESA, 1992), in their Environmental Assessment report on northeast Olkaria, used 1990 data of electric energy consumed for pumping and calculated that about 59×10^6 m^3/yr of water is pumped out from the lake for irrigation, although the licensed amount is only 30×10^6 m^3/yr. This amount is nearly the same as the outflow quantity computed by Åse et al. (1986) and Åse (1987), and about twice what was calculated by McCann (1974). Since most of the irrigated farms are adjacent to the lake itself, some of this water may still flow back into the lake as irrigation return after application. Further work is needed to establish the fate of the irrigation water since it forms substantial proportion of the lake water budget.

In an effort to make an improved water budget, a reanalysis of the data of McCann (1974), Åse et al. (1986) and of Gaudet and Melack (1981) was made. The Lake Naivasha water budget is difficult to calculate due to monthly and annual variations of the climatological

Table 1. Comparison of Lake Naivasha Hydrologic Water Balance Calculations
from Different Sources (last column shows weighted mean values using a range
of raw data; units in million cubic meters per year)

	McCann (1974) 1957–67	Gaudet and Melack (1981) 1973–75	Åse et al. (1986) 1972–74	Åse et al. (1986) 1978–80	Weighted mean
INPUT					
Precipitation	132	103	115	142	127
River discharge	248	185	187	254	230
Surface runoff	nd[a]	0.6	nd	nd	1
Seepage-in	nd	49	nd	nd	49
Total input	380	338	302	396	364
OUTPUT					
Evapotranspiration	346	313	308	301	329
Seepage-out	34	44	nd	nd	36
Irrigation + industrial	nd	12	nd	nd	12
Total output	380	369	308	301	356
Storage change	nd	–31	0.4	95	21

[a]nd = not determined.

factors and the postulated subsurface outflow. Use of mean annual values does not take into account these subtle changes (McCann, 1974). Both Gaudet and Melack (1981) and Åse et al. (1986) calculated their hydrologic balance based on monthly means and their data have been analyzed to arrive at the figures in Table 1. While McCann's (1974) data are from 1957 to 1967, Gaudet and Melack (1981) used 1973–75 and Åse et al. (1986) 1972–74 and 1978–80 data. McCann's (1974) data cover the period when the lake level was generally rising after a long low-level period. Åse et al. (1986) intentionally took data from both falling (dry) and rising (wet) periods of the lake level to better understand the lake budget variations (Fig. 4). The calculations in this study incorporate a longer data period (1957–80). The basic equation used in Table 1 may be written as:

$$INPUT = P + R + S_r + S_i$$
$$OUTPUT = E_o + ET + S_o + I_r \pm \Delta S$$
$$\pm \Delta S = INPUT - OUTPUT = P + R + S_r + S_i - E_o - ET - S_o - I_r,$$

where P is precipitation; R is inflow from rivers; S_r is surface runoff from surroundings (includes some of the yet unquantified return water from irrigation); S_i is seepage inflow, which also includes some of the irrigation water that reaches back to the lake via subsurface flow; E_o is open water evaporation from the lake; ET is evapotranspiration from the swamps and lake algae and other vegetation; S_o is seepage outflow, which may reach groundwater or recirculate back to the lake in shallow levels; I_r is irrigation and industrial water; $\pm \Delta S$ denotes changes in water storage.

Geochemistry and Ionic Residence Times

Residence times were calculated in order to gain some idea of the evolution of the present lake water chemistry. First, the mean annual water inflows from the various water sources that supply the lake from Table 1 were used. The mean chemical composition of the different water sources were taken from various studies: surface runoff and seepage (Gaudet and Melack, 1981); rainwater (Panichi and Tongiorgi, 1974; Gaudet and Melack, 1981); rivers and the lake (Glover, 1972; Gaudet and Melack, 1981). These were analyzed for consistency and ionic balance before they were selected and averaged to be used in the calculations. As shown in Table 2, all the waters have balanced ionic charges, except rain water which has an anion deficiency of 0.01 mequiv/L or 20%. This is probably because there were no data for fluoride or organic anions such as volatile organic acids. All the waters are sodium bicarbonate waters except surface runoff waters that have higher calcium and magnesium concentrations, probably derived from agricultural irrigation inputs.

Chloride was assumed to be a conservative element (i.e. it remains in solution once introduced). This is a reasonable assumption even for Lake Naivasha. Gaudet and Melack (1981) calculated sediment uptake and release of ionic solutes and found that of 4×10^6 tonnes/yr of chloride released by sediments in Lake Naivasha, only 0.7×10^6 tonnes/yr (17.5%) is involved in secondary chemical reactions in the sediments. Chloride was thus used to calculate the time taken to produce waters with chemical composition of the present day Lake Naivasha. This assumes all chloride was derived from the inflow waters. It was further assumed that the average chemical compositions of the inflow waters have been constant over the same time taken to produce the present day lake water. The "chloride age" was computed from the standard residence time equation (Yuretich and Cerling, 1983; Barton et al., 1987).

$$t_{Cl} = \frac{\text{chloride quantity in lake (moles)}}{\text{chloride annual flux (moles/yr)}},$$

where t_{Cl} is the chloride residence time = 5.4 years.

The budgets of the nine major ions were then estimated by multiplying their respective fluxes by the "chloride age" of the lake to get total inputs over the "chloride age" time span, and then subtracting these total inputs from the amounts in the lake to get residuals as shown in Table 3. The major result of these calculations is the very short turnover or residence time of the chloride and other ions in the lake (3–6 years). Yuretich and Cerling (1983) calculated chloride residence time of 2500 years for Lake Turkana whereas Burton et al. (1987) obtained 3000–5000 years for the same lake and 100–200 years for Lake Baringo. The calculated chloride residence time of about six years is extremely short compared to other lakes as mentioned above and is a strong proof of the hydraulic connection between the lake and the shallow groundwater system. From the calculations in Table 3, the annual ionic flow into the lake compared to what is in the lake varies, from 15% for sodium to 28% for sulfate. These low input ratios are cited by Gaudet and Melack (1981) as one of the reasons why the lake is fresh. Other reasons advanced by these authors for the freshness of the lake are loss of solutes via seepage and ionic exchange with sediments resulting in net accumulation of solutes at the lake bottom. It can be seen from Table 3 that the lake receives more Mg, SO_4, SiO_2 and F over the chloride residence time than is actually present, whereas there is enrichment of Na, K, Ca and HCO_3. We propose here that the net concentrations of Na, K and Ca in the lake waters are due in part to evaporative concentration. Magnesium uptake could occur by incorporation in plant cell walls, in hard parts of lake fish, and by ion-exchange. Gaudet and Melack (1981) calculated that most of the silica uptake is by diatoms in the lake. F and SO_4 are depleted via sediment loss and/or subsurface outflow.

Table 2. Chemistry of the Inflow and Lake Naivasha Waters

	Discharge 10^6 m^3/yr	Na ppm mol/m^3	K ppm mol/m^3	Ca ppm mol/m^3	Mg ppm mol/m^3	HCO$_3$ ppm mol/m^3	Cl ppm mol/m^3	SO$_4$ ppm mol/m^3	SiO$_2$ ppm mol/m^3	F ppm mol/m^3
Surface runoff	0.6	8 / 0.35	16 / 0.41	16 / 0.4	4 / 0.16	92 / 1.51	8 / 0.23	10 / 0.1	13 / 0.22	1.7 / 0.09
Seepage	49	44 / 1.91	31 / 0.79	23 / 0.57	10.2 / 0.42	231.5 / 3.79	28 / 0.79	11 / 0.11	47 / 0.78	1.7 / 0.09
Rainfall	127	0.54 / 0.02	0.31 / 0.01	0.18 / 0	0.23 / 0.01	1.2 / 0.02	0.41 / 0.01	0.72 / 0.01	nd / nd	nd / nd
Rivers	230	15 / 0.65	7 / 0.18	9 / 0.22	2.9 / 0.12	76 / 1.25	3.8 / 0.11	7 / 0.07	28 / 0.47	1.2 / 0.06
L. Naivasha	900 (vol)	40 / 1.74	22 / 0.56	20 / 0.5	5.8 / 0.24	190 / 3.11	14 / 0.39	8.4 / 0.09	34 / 0.57	1.5 / 0.08

Table 3. Major Ion Flux Budgets and the Residence Times in Years

	Na	K	Ca	Mg	HCO$_3$	Cl	SO$_4$	SiO$_2$	F
FLUXES (10^6 moles/yr)									
Surface runoff	0.2	0.25	0.24	0.1	0.9	0.14	0.06	0.13	0.05
Seepage	93.78	38.85	28.12	20.56	185.91	38.7	5.61	38.33	4.38
Rainfall	2.84	0.96	0.54	1.15	2.38	1.4	0.91	nd	nd
Rivers	138.32	37.96	47.6	25.3	264.06	22.72	15.45	98.79	13.39
Total flux	246	82	80	50	477	65	24	147	18
In lake (10^6 moles)	1566	504	450	216	2799	351	81	513	72
Ionic res. times, yrs	6.4	6.2	5.7	4.4	5.9	5.4	3.4	3.5	3.9
t_{Cl} × total flux (10^6 moles)	1320	439	427	266	2561	351	127	787	98
Residual in lake (10^6 moles)	246	65	23	−50	238	0	−46	−274	−26
Residual/in lake, %	16	13	5	−23	9	0	−57	−53	−36

REFERENCES

Åse, L. E., 1987, A note on the water budget of Lake Naivasha, Kenya — especially the role of Salvinia molesta Mitch and Cyperus papyrus L. Geograf. Annaler, v. 69A (3–4), pp. 415–429.

Åse, L. E., Sernbo, K and Syrén, P., 1986, Studies of Lake Naivasha, Kenya, and its drainage area. Stockholms Universitet Naturgeografiska Institutionen, Forskningsrapport 63, ISSN 0346–7406. 80 pp.

Barton, C. E., Solomon, D. K., Brown, J. R., Cerling, T. E., and Sayer, M. D., 1987, Chloride budgets in transient lakes: Lakes Baringo, Naivasha and Turkana. Limnol. Oceanogr., v. 32, no. 3, pp. 745–751.

Clarke, M. C. G., Woodhall, D. G., Allen, D., and Darling, G., with contributions by Kinyariro, J. K., Mwagongo, F. M., Ndogo, J. P., Burgess, W., Ball, K., Scott, S., Fortey, N. J., Beddoe-Stephens, B., Milodowski, A., Nancarrow, P.H. A., and Williams, L. A. J., 1990, Geological, volcanological and hydrogeological controls on the occurrence of geothermal activity in the area surrounding Lake Naivasha. Govt. of Kenya, Min. of Energy/British Geological Survey Report, 138 pp.

Darling, W. G., Allen, D. J., and Armannsson, H., 1990, Indirect detection of subsurface outflow from a Rift Valley lake. J. Hydr., v. 113, pp. 297–305.

Gaudet, J. J., and Melack, J. M., 1981, Major ion chemistry in a tropical African lake basin. Freshwater Biol., v. 11, pp. 309–333.

Glover, R. B., 1972, Chemical characteristics of water and steam discharges in the Rift Valley of Kenya, Unpublished UN Geothermal Expl. Proj. Report. 59 pp.

Harper, D. M., Mavuti, K. M., and Muchiri, S. M., 1990, Ecology and management of Lake Naivasha, Kenya, in relation to climatic change, alien species' introductions, and agricultural development. Environmental Conserv., v. 17, no. 4, pp. 328–336.

Jones, B.F., Eugster, H. P., and Rettig, S. L., 1977, Hydrochemistry of the Lake Magadi Basin, Kenya. Geochim. Cosmochim. Acta, v. 41, pp. 53–72.

Kilham, P., and Hecky, R. E., 1973, Fluoride: geochemical and ecological significance in East African waters and sediments. Limn. and Oceanogr., v. 18, pp. 932–945.

McCann, D. L., 1972, A preliminary hydrogeologic evaluation of the long-term yield of catchments related to geothermal prospect areas in the Rift Valley of Kenya. Unpublished UN Geoth. Proj. Report, 23 pp.

McCann, D. L., 1974, Hydrogeologic investigation of the Rift Valley catchments. Unpublished UN Geoth. Proj. Report, 56 pp.

Naylor, W. I., 1972, Geology of the Eburru and Olkaria geothermal prospects. Unpublished UNDP/EAPL Geothermal Project Report.

Ojiambo, B. S., 1992, Hydrogeologic, Hydrogeochemical and Environmental Isotopes Study of Possible Interactions Between Olkaria Geothermal, Shallow Subsurface and Lake Naivasha Waters, Central Rift Valley, Kenya. MS Thesis, University of Nevada at Reno.

Panichi, C., and E. Tongiorgi, 1974, Isotopic study of the hot water and steam samples of the Rift Valley, Kenya. Unpubished Report, UNDP/KPC Geothermal Project, 56 pp.

Thompson, A. O., and Dodson, R. G., 1963. Geology of the Naivasha Area., Geol. Sur. of Kenya, Report no. 55, 80 pp.

Vincent, C. E., Davies, T. D., and Beresford, A. K. C., 1979, Recent changes in the level of Lake Naivasha, Kenya, as an indicator of equitorial westerlies over east Africa. Climatic Change, v. 2, pp. 175–189.

Yuretich, R. F., and Cerling, T. E., 1983, Hydrogeochemistry of Lake Turkana, Kenya: Mass balance and mineral reactions in an alkaline lake. Geochim. Cosmochim. Acta, v. 47, pp. 1099–1109.

Food Webs
and Fisheries

Pelagic Food Webs of the East African Great Lakes

J.T. LEHMAN *Department of Biology and Center for Great Lakes and Aquatic Sciences, University of Michigan, Ann Arbor, Michigan, United States*

Abstract — The pelagic food webs of the East African Great Lakes exhibit features that suggest continuous, strong biological interactions as well as strong interactions between biota and their geochemical environment. Planktivory by both vertebrate and invertebrate predators appears to be a major structuring force affecting the animal communities. The phytoplankton appear to control or modify nutrient chemical conditions during an "endless summer" of strong interactions with resource levels. Several case studies illustrate the roles of these interactions in food web processes. The Dipteran larva *Chaoborus* appears to be an important predator whose distribution may be controlled by interactions between fish planktivory and the availability of oxygenated refugia. The strength of biological interactions is supported not only by the patterns of extant distributions among African Great Lakes, but also by the food web changes observed when efficient planktivorous fish are introduced into lakes. Lake Victoria is given special attention as a lake that has exhibited changes in limnological condition during the present century which seem to be driven primarily by changes in nutrient income and regional climate. The neolimnological changes in the lake are associated with conditions that leave interpretable signals in the stratigraphic record of the lake sediments. Thus Victoria represents a candidate lake for calibration of the tools and models that will be needed to infer past climate variations and ecosystem changes in other East African lakes.

The Great Lakes of East Africa are remarkable for their immense biological diversity. Some faunal elements, particularly fish and mollusks, have developed species richnesses that are of great scientific value for the study of speciation, extinction, and ecological interactions. One striking enigma of the African Great Lakes, however, is the contrast between levels of biodiversity in their littoral and benthic habitats with those of the lake pelagia. Most of the species richness in these lakes is confined to nearshore regions, and biological communities in offshore areas may be species-poor in comparison with some temperate Great Lakes ecosystems.

The purpose of this effort is to review existing information about the pelagic food webs of the East African Great Lakes, to identify generalizations, and to develop hypotheses about the open lake ecosystems. The analysis cannot be entirely satisfactory because the quantitative data necessary to compare and contrast the lakes are rarely available and are in no case complete. Nonetheless, these lakes present tantalizing patterns that are expressed in the form of ecological associations from basin to basin, and in the consistency of biological changes along spatial gradients. There is reason for optimism that careful study of the pelagic environments will reward us with rather precise information about the conditions that promote the success of alternative food webs, and thereby will give us a reliable way to interrogate the high resolution ecological and environmental stories recorded in the African lake sediments.

Present understanding of the African lakes points to a set of influential factors that affect pelagic food webs. First, even though the lakes are subject to seasonal variations in heat and water balance, and also to drainage basin and nutrient income effects, those external influences are less profound for these tropical great lakes than the external influences of seasonality that affect temperate great lakes. Kilham and Kilham (1990) have argued that the "endless summer" of tropical lakes permits biological control of elemental cycles to adjust to the nutrient regime fully, whereas temperate lakes are constrained by a physical environment that dominates seasonally and forcibly resets the biological system with hostile conditions of cold temperature and low light. Consequently, in the African Great Lakes the biota can interact with and alter their nutrient and chemical environment year-round, whereas for most of the year, temperate lake communities can only experience and adapt to their environment.

Prominence of biological forces in the tropical lakes does not mean that physical forces are unimportant or even of secondary importance. Physical influences promote important biological responses in African Great Lakes. Talling (1969) and Hecky and Kling (1987) have emphasized the importance of vertical stratification and mixing depth of the rift lakes as the major controlling variable for phytoplankton species composition, community changes, and perhaps biomass levels as well. Deep mixing and upwelling processes (Hecky et al. 1991) not only inject nutrient stores from hypolimnetic or monimolimnetic reservoirs into the surface water, but they subject algal cells to variable light regimes which favor taxa that can exploit the conditions. Changes in taxonomic representation among the algae cause feedbacks to the lake nutrient regimes, expressed through uptake of silica by diatoms, for example, or through active nitrogen fixation by diazotrophic cyanobacteria. The long-term consistency of most African Great Lakes phytoplankton communities, inferred from the presence of diatoms which leave stratigraphic records, argues that far from being haphazard assemblages of taxa, the lake communities are crafted by master controlling influences that are reproducible over sometimes thousands of years of record (Hecky 1984b).

The bulk of the phytoplankton biomass in the African Great Lakes is dominated by relatively few, widespread taxa mainly drawn from the green algae, diatoms, and cyanobacteria. This tendency toward parsimony is even greater in the offshore zooplankton communities of these lakes. The crustacean plankton is much simpler than in temperate lakes (Fernando 1980a, 1980b; Lehman 1988) and the pelagic is not rich in rotifers, either (see below). In Lake Tanganyika, ciliate protozoa account for a substantial fraction of the plankton biomass at times (Hecky and Kling 1981), but the dominant ciliate during stable stratification, *Strombidium*, contains endosymbiotic zoochlorellae, and is thus potentially an autotroph.

Quantitative comparisons of plankton biomass and relative composition among the lakes are extremely difficult owing first to the scarcity of relevant studies and second to the extreme differences in collection methods and mesh sizes of nets employed. The pioneering studies of Worthington (Worthington 1931; Worthington and Ricardo 1936), for instance, provide useful data on the vertical distributions of some planktonic species, but the nets used were most suitable for sampling oceanic calanoid copepods with such large mesh that most life stages of the abundant cyclopoid copepods that now dominate the lake plankton would have been missed. Similarly, biomass values for Lake Tanganyika based on collections with 300 μm and 400 μm mesh apertures (Burgis 1984) must be suspected of underestimating the zooplankton. In view of the probability that many herbivorous crustacean plankton were missed in the sampling, it is especially noteworthy that Burgis' estimates demonstrate that zooplankton biomass exceeds algal biomass in Lake Tanganyika (Hecky 1991), which in turn suggests the potential for grazer control of the algae through much of the year.

Offshore copepod assemblages of the African Great Lakes are typically represented by a single dominant diaptomid species (although *Diaptomus* has been reported to be missing from Lake Albert, see below), and one or more common cyclopoid species, usually including the predator *Mesocyclops*. The Cladocera are represented typically by two or more species, and the size structure of the community varies between inshore and offshore sites (Green 1967a, 1971). Based on historical reports, the Cladocera and Cyclopoida are usually more diverse inshore than offshore, and in the extreme case of Lake Tanganyika, the Cladocera are entirely absent. Meroplanktonic representatives include species of the planktivorous lake fly *Chaoborus* and atyid prawns (e.g., *Caridina*).

For reasons that will be explained in the following pages, the compositional features of the zooplankton communities of the African Great Lakes suggest that visual planktivory by fish is a dominant and persistent force in all the lakes. Moreover, it appears that in most of the lakes the intensity of the planktivory by fish is relaxed along an axis from inshore to offshore regions, to such an extent that invertebrate planktivory becomes a significant feature of the offshore environment. The exception to this pattern is Lake Tanganyika, where there is evidence that visual planktivory is great even in offshore waters. This exception is correlated with the presence of the endemic clupeid planktivores *Limnothrissa* and especially *Stolothrissa* in Lake Tanganyika. In lakes where transplants of one of these fish have occurred, striking and consistent changes in the resident zooplankton have followed. There is reason to believe that in the other lakes the indigenous zooplanktivorous fish, which are principally members of the families Cyprinidae and Cichlidae, are prevented by life history or physiology from exploiting the vast pelagic regions (Hecky 1984b; Turner 1982).

COMPARATIVE FOOD WEB COMPOSITION

Plankton and fish communities characterizing pelagic regions of the African Great Lakes are compared in Table 1. The categories are constructed to represent biological elements that recur consistently among basins. The family Centropomidae includes species of *Lates*, including the Nile perch, which are important piscivores in many of the lakes. Planktivorous fish include members of the Cyprinidae (*Neobola, Rastrineobola, Engraulicypris*) and Characidae (*Alestes*) which have been reported to be important pelagic zooplanktivores, although in many cases life history may make these taxa more common in inshore than offshore areas.

The category for prawns includes freshwater shrimp (Decapoda) from both the Atyidae (*Caridina, Limnocaridina*) and Palaemonidae (*Macrobrachium*). These taxa have been characterized as epibenthic detritivores, but some are known to be nocturnally planktonic. It is difficult to conceive why a strict detritivore would migrate upward from the detritus-rich benthos under cover of darkness. Rather, it seems possible that these animals may be facultative planktivores whose access to zooplankton prey is limited by risk of visual predation from fish. Additional study of their trophic interactions would be desirable.

The only freshwater Diptera present in the pelagic region are species of *Chaoborus*, which become meroplanktonic in the third and fourth instars (Macdonald 1956) and can be important invertebrate planktivores. The family Cyclopidae represents free-living cyclopoid copepods, including *Mesocyclops*, which is predatory in its subadult and adult stages, as well as smaller, herbivorous or omnivorous taxa: *Tropocyclops* and *Thermocyclops*. Calanoid

Table 1. Plankton and Fish Communities from Pelagic Regions of the
East African Great Lakes

	Lake Albert	Lake Edward
Centropomidae	*Lates niloticus*	(absent)
Planktivorous fish	*Rastrineobola argentae* *Alestes*	*Rastrineobola argentae*
Prawns	*Caridina nilotica*	
Planktivorous Diptera	*Chaoborus*	*Chaoborus*
Cyclopidae	*Mesocyclops* *Thermocyclops* *schuurmanae*	*Mesocyclops*
Diaptomidae	(absent)	(absent)
Cladocera	*Daphnia lumholtzi* *Diaphanosoma excisum* *Ceriodaphnia reticulata* *Ceriodaphnia cornuta* *Moina micrura* (inshore)	*Daphnia lumholtzi monacha* *Diaphanosoma excisum* *Ceriodaphnia dubia* *Ceriodaphnia cornuta* *Moina micrura* (inshore)
Rotifera	few	few
Phytoplankton	Bacillariophyta Chlorophyta Chrysophyta	Cyanophyta Chlorophyta
Sources	Rzoska 1975 Green 1971 Hecky and Kling 1987	Green 1971 Beadle 1981 Eccles 1985 Hecky and Kling 1987

	Lake Malawi	Lake Tanganyika
Centropomidae	(absent)	*Lates angustifrons* *Lates mariae* *Lates microlepis* *Lates stappersi*
Planktivorous fish	*Engraulicypris sardella*	*Limnothrissa miodon* *Stolothrissa tanganicae*
Prawns	*Caridina nilotica* (inshore)	*Limnocaridina* spp.
Planktivorous Diptera	*Chaoborus*	(absent)
Cyclopidae	*Mesocyclops* *Thermocyclops neglectus*	*Mesocyclops*
Diaptomidae	*Tropodiaptomus cunningtoni* *Thermodiaptomus mixtus*	(absent)
Cladocera	*Bosmina longirostris* *Diaphanosoma excisum*	(absent)
Rotifera	rare (<10 m^{-3})	few
Phytoplankton	Cyanophyta Chlorophyta Bacillariophyta	Cyanophyta Chlorophyta Bacillariophyta Chrysophyta Ciliata *Strombidium*
Sources	Twombly 1983 Hecky and Kling 1987 Fryer 1957	Coulter 1991 Beadle 1981 Hecky and Kling 1987

Table 1, cont'd

	Lake Turkana	Lake Victoria
Centropomidae	*Lates niloticus* *Lates longispinus*	*Lates niloticus*
Planktivorous fish	*Neobola stellae* *Alestes minutus*	*Rastrineobola argentae*
Prawns	*Caridina nilotica* *Macrobrachium niloticum*	*Caridina nilotica*
Planktivorous Diptera	(absent)	*Chaoborus*
Cyclopidae	*Mesocyclops* *Thermocyclops hyalinus*	*Mesocyclops* *Tropocyclops confinis* *Tropocyclops tenellus* *Thermocyclops neglectus* *Thermocyclops emini* *Thermocyclops incisus* *Thermocyclops oblongatus* *Th. schuurmanae*
Diaptomidae	*Tropodiaptomus banforanus*	*Thermodiaptomus galeboides* *Tropodiaptomus stuhlmanni*
Cladocera	*Moina brachiata* *Moina micrura* *Diaphanosoma excisum* *Ceriodaphnia cornuta* *Daphnia barbata*	*Daphnia lumholtzi* *Daphnia longispina* *Diaphanosoma excisum* *Ceriodaphnia dubia* *Ceriodaphnia cornuta* *Bosmina longirostris* *Moina micrura*
Rotifera		few
Phytoplankton	Cyanophyta	Cyanophyta
Sources	Beadle 1981 Eccles 1985 Ferguson 1978	Green 1971 Rzoska 1957 Beadle 1981 Mwebaza-Ndawula 1994a Mwebaza-Ndawula 1994b

	Lake Kivu (1950s)	Lake Kivu (1980s)
Centropomidae	(absent)	(absent)
Planktivorous fish	(absent)	*Limnothrissa miodon*
Prawns		
Planktivorous Diptera	(absent)	(absent)
Cyclopidae	*Mesocyclops* *Thermocyclops consimilis* *Tropocyclops confinus*	*Mesocyclops* *Thermocyclops consimilis*
Diaptomidae	(absent)	(absent)
Cladocera	*Daphnia curvirostris* *Diaphanosoma excisum* *Ceriodaphnia cornuta* *Moina micrura*	*Diaphanosoma excisum*
Rotifera	few	few
Phytoplankton		Cyanophyta Chlorophyta
Sources	Dumont 1986	Dumont 1986 Hecky and Kling 1987

copepods are represented in some lakes by the family Diaptomidae, including the genera *Tropodiaptomus* and *Thermodiaptomus*. Cladoceran zooplankton include the herbivorous genera *Daphnia, Diaphanosoma, Ceriodaphnia, Moina*, and *Bosmina*. No predatory Cladocera have been reported from these lakes.

The dominant phytoplankton in the African Great Lakes are principally members of the algal divisions Cyanophyta (bluegreen cyanobacteria), Chlorophyta (green algae), Bacillariophyta (diatoms), and in a few cases, Chrysophyta (golden-brown algae). Cryptophytes and dinoflagellates, although present, are not conspicuous in terms of biomass (Hecky and Kling 1987).

Missing from Table 1 is reference to the freshwater jellyfish (*Limnocnida*), of which species are known from both Tanganyika and Victoria (Coulter 1991; Beadle 1981). Although the medusae are capable of asexual reproduction by budding, and thus can develop populations independently of the littoral benthic polyp stage, the abundance and ecological significance of the taxa to the pelagic ecosystems of these lakes have not been well characterized, and so the potential importance of these jellyfish as invertebrate predators is not known at present.

Hecky (1984b) has attempted to summarize the relative magnitudes of primary production, zooplanktonic secondary production, and fish production from each of the lakes, and no substantial improvement over his effort is possible yet. Accurate, reproducible data are virtually nonexistent for zooplankton biomass and rates of production in the East African Great Lakes. In some cases where mesh sizes of plankton collections were specified (e.g., Worthington 1931 for Lake Victoria; Burgis 1984 for Lake Tanganyika) it is clear that quantitatively important elements of the fauna may have been missed. In other cases, the mesh sizes and construction of the sampling devices was not specified (e.g., Rzoska 1957, 1975; Green 1967a, 1971), and thus the observations cannot be repeated and compared.

CHAOBORUS DISTRIBUTION

One analysis that can be conducted with Table 1 is to observe taxonomic categories that are missing from some lakes, but which are present in others. A particularly intriguing element in this regard is the invertebrate planktivore *Chaoborus*. The predator is present in Albert, Edward, Malawi, and Victoria, but is absent from Kivu, Tanganyika, and Turkana. Together with members of the obligately benthic Chironomidae, *Chaoborus* is responsible for spectacular synchronized emergences of lake flies, sometimes at densities of thousands per square meter (Macdonald 1953, 1956). Possessing a winged adult stage, it has ample potential to disperse and to colonize suitable habitats. Overlapping cohorts of *Chaoborus* exist at African lake temperatures, with individual generation times of two months (Macdonald 1956). Eggs deposited at the water surface sink to the sediments and hatch within two to four days. Both first and second larval instars are holoplanktonic for about two weeks, but in the third and fourth instars, the larvae burrow in the mud during day and rise into the water column to forage only at night, similar to the behavior of most temperate chaoborids. At the end of two months, larvae in their fourth instar pupate and emerge from the water as adult midges. Daytime burial in benthic sediments seems to be a behavior evolved to evade visual planktivory, and is not obligatory for all chaoborids. In fishless temperate lakes, for instance, the large *Chaoborus americanus* remains holoplanktonic, but it is extirpated when fish are introduced (Northcote et al. 1978).

At Lake Malawi, the winged adults of *Chaoborus edulis* are collected as human food, and chemical analyses of congeners from Lake Victoria confirm their nutritional value (Okedi

Table 2. Presence or Absence of *Chaoborus* in East African
Great Lakes Ranked by Reported Conductivities, or
Ranges of Reported Conductivities of Their Surface Waters

Lake	Conductivity (µS)	*Chaoborus*
Victoria	97–187	+
Malawi	210	+
Tanganyika	610–649	–
Edward	456–925	+
Albert	691–735	+
Kivu	1110–1240	–
Turkana	3000	–

Data from Tallling and Talling 1965; Kilham 1971; Eccles 1985.

1992). Nonetheless, *Chaoborus* which pupate and emerge as winged, aerial adults have been regarded as lost potential productivity for the fishery of the lake (Turner 1982).

As an aquatic Dipteran, *Chaoborus* seems to have physiological constraints on its range of suitable habitats. Eccles (1985) suggested that its absence from Kivu, Tanganyika, and Turkana might be caused by the elevated salinities of those three lakes. As Table 2 demonstrates, however, Lakes Albert and Edward both possess higher conductivites than does Tanganyika, even though *Chaoborus* abounds in both of the former systems. There is precedent, however, for believing that plankton species distributions may be affected by total dissolved ion concentrations. LaBarbara and Kilham (1974) demonstrated chemical segregations of African copepod taxa by lake water conductivities. In temperate saline lakes, *Chaoborus* is restricted to salinities of 3–10 per mille (Topping 1971; Hammer 1986). Even more telling, however, is the fact that solute concentrations tolerated by aquatic Dipteran larvae are a function of water temperature (Hammer 1986, p. 429). At cold hypolimnetic temperatures in temperate lakes (5°C), the larvae can tolerate much higher salinities than they can bear at warm temperatures (20°C). It thus appears that experimental work is needed on the salinity tolerances of African chaborids at ambient lake temperatures (>22°C) in order to determine the role of lake chemistry in distribution patterns.

There is another factor related to the life history of *Chaoborus* that must be considered, moreover, to understand why it is excluded from some lake environments. The larvae are tactile, sit-and-wait predators with limited escape responses. Once observed by a mobile, visual predator, they have little chance of survival. Eggs deposited by adult females at the lake surface settle to the mud and hatch, and subsequently 75% of the larval life cycle (instars 3 and 4) is spent at depth or among the sediments during daylight hours. Lakes which have small proportions of their sediment surface available as suitable habitat for the larvae, or where steep submarine topography affords poor sediment refuge, may present insurmountable challenges to successful establishment of meroplanktonic chaoborid populations. Of the African Great Lakes, both Kivu and Tanganyika have the smallest proportions of their benthic areas exposed to aerobic waters, even during full circulation (Table 3). Larval *Chaoborus* have physiological mechanisms to withstand temporary anoxia, but prolonged anaerobic existence at elevated temperatures has not been reported, and the larvae may be

Table 3. Morphometry of East African Great Lakes (A_{ox} = maximum
area of bottom sediments oxygenated during full circulation;
z_{max} = maximum depth; z = mean depth)

Lake	A_0 (km^2)	z_{max}	z	A_{ox} (%)
Albert	6800	58	25	100
Edward	2325	117	40	95
Kivu	2370	485	240	20
Malawi	28800	704	290	45
Tanganyika	32600	1470	580	20
Victoria	68800	84	40	100
Turkana	7560	120		100

Data from Hecky and Kling 1987; Marshall 1991; Beadle 1981; Hecky and Bugenyi 1992; Talling 1965.

incapable of surviving those conditions. Consequently, planktivory by littoral fish communities alone could prove devastating in lakes where habitat is restricted to a littoral fringe. The extent to which deep, aphotic lake regions can serve as a refuge from visual planktivory is unknown at present.

ROTIFERA

Another striking feature of Table 1 is the fact that rotifers are rare in all African Great Lakes pelagic environments. Scarcity of these animals seems to be a particular feature of the large, deep lakes because shallow lakes like George and Kyoga have much higher abundances and species richness of rotifer taxa (Table 4). Individual body masses of rotifers are typically at least two orders of magnitude less than those of individual crustacean zooplankton (Bottrell et al. 1976). Consequently, by comparison with other plankton groups, the rotifers represent a trivial proportion of zooplankton biomass in these large lakes.

Rotifers tend to be more species rich and abundant in African rivers and lake littoral regions than in the open waters of East Africa (Green 1967b). Highest rotifer abundances may occur in saline, soda lakes which have an otherwise restricted planktonic fauna (Nogrady 1983). Rotifers are susceptible to invertebrate planktivory, especially from cyclopoid copepods (Stemberger and Gilbert 1987) and early instar *Chaoborus* (Neill and Peacock 1980). Their life history characteristics, including cyclic parthenogenesis and the interesting phenomenon of eutely, or cell constancy, permit them to exhibit short maturation times (hours) and very rapid rates of population growth under suitable environmental conditions (Bennett and Borass 1989). When these opportunistic organisms are consistently scarce, and the requisite predators are abundant, as in pelagic regions of the African Great Lakes, biological control of their distributions through invertebrate predation should be suspected. Unlike in temperate lakes, where predator populations wax and wane seasonally, rotifers have no seasonal refugium in the endless summer of East African lakes.

Table 4. Comparative Numerical Inventories of Planktonic Cladocera and Rotifera in Large East African Lakes

Lake	Station depth (m)	Cladocera		Rotifera	
		Indiv m^{-2}	Species	Indiv m^{-2}	Species
Albert (midlake)	47	160000	6	370	5
Edward (offshore)	24–30	60000–220000	5	620–1300	5
George	2–4	1400–10000	3	6800–25000	15
Kyoga	4–6	960–8600	13	2500–14000	24
Malawi	>50	50000–600000	2	<500	na
Tanganyika	>50	0	0	"rare"	na
Victoria	28	37000	9	3000	9

Data rounded to two significant digits from Green 1967a, 1967b, 1971; Twombly 1983; Coulter 1991.

CLADOCERA

The distribution of Cladoceran taxa in the African Great Lakes shows instructive complements to that of the Rotifera. Green (1967a, 1971) assembled an illuminating collection of data from the large lakes that drain to the White Nile. His data and interpretations argue convincingly that planktivory by fish is concentrated in the nearshore regions of the large lakes, especially Albert, Edward, and Victoria, and that the intensity of visual planktivory decreases toward offshore waters. He demonstrated the pattern with a series of species transitions. In both Lake Albert and Lake Edward, for instance, a large-bodied, conspicuous herbivore, *Daphnia lumholtzi* var. *monacha*, was present in highest abundance at midlake stations. The taxon was common in the stomachs of plantivorous fish, and its population abundance decreased rapidly near shore where planktivorous fish were present. Near shore in Lake Albert, the large bodied *Daphnia* was replaced by a smaller bodied, helmeted form, *Daphnia lumholtzi* (typical form), and by other small Cladocera. Similarly, within the genus *Ceriodaphnia*, the larger form *C. dubia* was present offshore, and the smaller bodied species *C. cornuta* was present inshore.

Green (1971) demonstrated that the larger Cladoceran taxa also had larger, more pigmented eyes, and thus by analogy with Zaret's (1969, 1972) evidence, they would be more visible to planktivorous fish. Abundances of Cladocera in the offshore waters of deep lakes compared with the shallow lakes George and Kyoga (Table 4) suggest that the Cladocera experience a refuge from fish planktivory offshore.

Green (1967a, 1971) sought to explain the success of large bodied Cladocera in offshore lake waters by the conventional theory of the time: the size–efficiency hypothesis of Brooks and Dodson (1965). He regarded the large animals to be competitively superior to the small ones, and thus better able to dominate when compensatory predation was removed. However, he offered no evidence about relative rates of fecundity, resource depletion, or growth efficiency that would be essential to establish a competitive interaction. Instead, the patterns he reports are tantalizingly consistent with the notion that whereas vertebrate planktivory wanes from inshore to offshore, the strength of invertebrate planktivory does the opposite.

Table 5. Lake Victoria Zooplankton: Comparison Between Inshore (Pilkington Bay, 10 m) and Offshore (NE open Lake Victoria, 67 m) Communities, 1956 (data from Rzoska 1957)

Taxon	Percent by count	
	Inshore	Offshore
Moina	27	5
Tropocyclops	8	0
Daphnia lumholtzi	0.4	0
Bosmina longirostris	2	1
Diaptomus	33	25
Thermocyclops	7	22
Mesocyclops	4	12
Diaphanosoma	3	5
Ceriodaphnia cornuta	2	8
Ceriodaphnia dubia	0	4
Daphnia longispina	0	5

Because invertebrate predators are typically smaller with respect to their prey than are fish, the invertebrates are subject to gape-limitation, sensu Zaret (1980). Their prey can find refuge in large body size such that capture and handling become ineffective (Kerfoot 1977a, 1977b). The larger bodied Cladocera present in the offshore waters of Albert, Edward, and Victoria would be less susceptible to predatory cyclopoids and diptera than would their smaller congeners. A particularly good example can be found in the genus *Ceriodaphnia*, involving the species *C. cornuta* and *C. dubia*.

C. dubia is an animal of larger body size than *C. cornuta*, and it is a characteristic species in the offshore waters of Lakes Albert, Edward, and Victoria (Green 1971; Rzoska 1957) (see Table 5). Specimens of *C. cornuta* from offshore regions tend to be of a morphological variety that has been called *C. rigaudi*, which is larger in body dimensions than is the typical form. Zaret (1969, 1972) studied these two forms, which also occurred in Gatun Lake, Panama, and demonstrated that visual predation by planktivorous fish was based mainly on the size of the conspicuous, pigmented eye. By this logic, more conspicuous animals would experience differential mortality and be excluded from habitats that possessed high concentrations of planktivorous fish. Green (1971) demonstrated that for the African lakes, eye-size is well correlated with body size. That explains the absence of larger forms from nearshore waters. But what prevents the smaller forms from being successful offshore? As with the rotifers, the likely answer is mortality imposed by abundant invertebrate predators, particularly *Mesocyclops*.

THE APPARENT INCONGRUITY OF LAKE TANGANYIKA

Based on the circumstantial evidence assembled, it appears that most of the Great Lakes of East Africa have food webs that are influenced most strongly by fish planktivory nearshore and by invertebrate planktivory offshore. Lake Tanganyika presents a contradiction to this pattern. No Cladocera are present in the offshore plankton of the lake, although some species of Chydoridae are found inshore and in adjoining waters (Coulter 1991). Moreover, there is evidence that the trend in vertebrate planktivory is opposite in Lake Tanganyika from that of the other lakes.

Mashiko et al. (1991) present demographic data contrasting lagoon and lake populations of the endemic shrimp *Caridina tanganyicae*. They demonstrate that lake animals are on average much smaller than those of the inshore lagoon. Lake populations mature at smaller body sizes than do lagoon populations, even though each population seems to invest similar fractions of total body mass in reproductive effort. Because there was no significant difference in reproductive effort scaled to body size between the populations, it seems less likely that the differences are caused by food limitation than that the lake populations are subject to more intense size-selective predation.

The critical, anomalous feature of Lake Tanganyika with respect to the other lakes is the presence of two endemic freshwater sardines, *Limnothrissa* and *Stolothrissa*, derived from marine stocks that successfully invaded rivers of West Africa and subsequently became established in Lake Tanganyika through the Zaire drainage (Coulter 1991). The fish are derived from effective oceanic planktivores, and one of them, *Stolothrissa*, is entirely pelagic, spawning and living its entire life away from shore. No comparable species exist in the other African Great Lakes, except now in Kivu where *Limnothrissa* was intentionally introduced. Before introduction of the planktivore, zooplankton biomass has been roughly 7.5 g DW m^{-2}, but subsequent to establishment of *Limnothrissa*, Dumont (1986) found only 0.15 g DW m^{-2} in a single, expeditionary sample. A similar experience was reported for Lake Kariba, an artificial impoundment, which was stocked with *Limnothrissa*. Both abundances and species richness of Cladocera were reduced, and *Chaoborus* was eliminated (Marshall 1991). A pattern of progressive reduction and elimination of zooplankton populations according to their body size in the face of visual planktivory was similarly documented by Gliwicz (1985) for Cahora Bassa, a reservoir downstream from Kariba that became colonized by *Limnothrissa*, as well. The case studies of species introduction to Lakes Kivu, Kariba, and Cahora Bassa constitute quasi-experiments at the whole ecosystem level that are consistent with intense offshore vertebrate planktivory by clupeoid fishes in Lake Tanganyika. There is no analog to these offshore planktivores in the other Great Lakes, where native freshwater Cyprinidae and Characidae exploit the plankton. The implications of the clupeoid stocks in Tanganyika for the anomalously high pelagic fishery potential of the lake have been discussed by many authors (Hecky et al. 1981; Hecky 1984b; Hecky 1991; Coulter 1991).

FOOD WEB DYNAMICS IN LAKE VICTORIA

Recent changes in the food web of Lake Victoria represent another case that deserves special attention, because the changes offer insight to trophic dynamics in large tropical pelagic ecosystems. Lake Victoria and its watershed have been transformed by events of the past century. The transformations have included loss of forests, lake eutrophication, deep water deoxygenation, and changing water levels, algal species, mixing dynamics and productivity. Understanding the processes that brought about these conditions is an essential first step in reconstructing the long histories of the African lakes, because the changes which occurred in Victoria may have pre-human analogs in the sedimentary stratigraphic records from the East African lakes. In fact, the most powerful way to test the validity of inferences about lake history is to examine systems that have been perturbed in a defined way, follow the perturbation signals through the ecosystem, and document their traces in the stratigraphic record.

The condition of Lake Victoria has changed profoundly during the last three decades. Talling (1966) reported a flora rich in large diatoms, particularly *Melosira* and *Stephanodiscus* in 1960 and 1961, with offshore Chl *a* ranging from 1.2 to 5.5 mg m^{-3} in the euphotic zone, SRSi generally between 67 and 75 μM (4.0–4.5 mg/l SiO$_2$), and O$_2$ extending to the deep sediments offshore. In contrast, the lake is now dominated by cyanobacteria, biomass is elevated, surface SRSi has declined, and deep waters are regularly anoxic (Hecky 1993).

Another conspicuous change to the Lake Victoria ecosystem has been the alteration of its fish community. During the 1980s, the fishery of the lake underwent radical transformation by the success of introduced species of the Nilotic fauna, in particular *Lates niloticus* (Nile perch, a piscivore) and *Oreochromis niloticus* (Nile tilapia, an herbivore). Vast endemic species flocks of haplochromine cichlids declined as the introduced species became established (Ogutu-Owayo 1990a, 1990b, 1992; Ligtvoet and Witte 1991; Witte et al. 1992a, 1992b). These species were introduced to the lake in the 1950s, but they did not become common or widespread for nearly two decades; by the late 1980s they dominated the fish catch to near exclusion of all native species except for *Rastrineobola argentea* (Oguto-Ohwayo 1990b).

Introduction and success of the nilotic species followed an episode of nearly unregulated reduction of gill net mesh sizes and collapse of the traditional tilapia fisheries, and was contemporaneous with initiation of commercial trawling for demersal haplochromines. The haplochromine stocks of Lake Victoria had been regarded as one of the great examples of vertebrate species radiation on the planet, with at least 300 species (van Oijen et al. 1984). Small clutch sizes and mouth breeding behaviors of the haplochromines are inconsistent with high rates of exploitation by either overfishing or predation. The only native species of Lake Victoria which has increased in yield during recent years is the small pelagic cyprinid, *Rastrineobola argentea*. This species had been common during the earliest surveys of the lake (Graham 1929, Worthington 1929), and it is not clear if present yields faithfully represent increased populations, or if they merely reflect higher fishing pressure on existing stocks.

The cause for the sweeping changes in physical, chemical, and biological properties of the lake is not known, because limnological observations were interrupted. As a result, the condition of Lake Victoria presents a forensic challenge to limnological theor.. and three main competing hypotheses have been advanced to explain the changes (Hecky 1993). Lake productivity mechanisms may have changed as a result of the introduced pisci'/ore and the ensuing food web changes. Alternatively, increased nutrients may be entering the lake from atmospheric precipitation, or watershed runoff, or both. Finally, there may also have been changes in the internal nutrient recycling mechanisms of the lake. Changes in regional climate, which caused changes in lake level, may have produced altered patterns of physical mixing and thermal stratification, which in turn influenced the development of diazotrophic cyanobacteria, hypolimnetic deoxygenation and altered sedimentary nutrient exchange.

Lake Victoria is the shallowest of the East African Great Lakes (80 m maximum depth), but it has the largest surface area of any freshwater lake other than Superior. Although its basin is ancient, the lake has endured marked changes in volume which have influenced the development of its fauna. During the late Pleistocene the lake was greatly reduced in area, more saline than today, and perhaps completely dry (Kendall 1969; Stager 1984; Stager et al. 1986). In very recent times the lake has responded to climate variation by significant increases in lake level, first in the early to mid 1960s and then again in the late 1970s (Kite 1981; Piper et al. 1986). Today its flushing time (volume/outflow) is 140 yr (Hecky and Bugenyi 1992), and its water budget is dominated by rainfall. The long flushing time means that nutrient retention is high, and it also means that the lake would recover slowly from chemical alterations. The annual cycle of nutrients in surface waters in Victoria, as in the other Great Lakes of East Africa, is believed to be controlled by vertical mixing and stratification (Hecky and Kling 1987). Victoria is one of the few lakes in Africa for which historical data are available about annual cycles of stratification, nutrient concentrations, and productivity (Talling 1965, 1966).

Over the period of the transformation of its fishery, silicon concentrations of Lake Victoria declined by an order of magnitude (Hecky and Bugenyi 1992). Sulfate concentrations appear to have declined as well, but Hecky and Bugenyi (1992) suspect that the apparent differences are possibly the result of more specific and more accurate modern methods (ion chromatography). The concentrations are probably the lowest of any large water body on earth. Moreover, residence time of SO_4 in the water column of Lake Victoria is less than 2 months based on rates of allochthonous income and water column inventories (Hecky and Bugenyi 1992), which supports Beauchamp's (1953) suggestion that it is behaving as a nutrient. Lehman and Branstrator (1994) have recently confirmed that turnover rates of sulfate are two weeks or less in the water column, and that the flux is dominated by biological uptake into algal particles larger than 1 μm. Despite the essential role of sulfate as a nutrient in Lake Victoria, however, the compound is not scarce enough to limit either uptake kinetics or biomass yields, and hence it may serve as a good biological tracer in the lake.

The decrease in Si in Lake Victoria is even more extreme than that observed in the Laurentian Great Lakes during their eutrophication (Schelske 1988). In the Laurentian lakes, increased phosphorus loading was believed to increase biomass and to cause increased algal demand for Si. In Lake Victoria, offshore TP concentrations have remained constant or slightly increased from 1961 to the present, although even in 1961 PO_4 concentrations were in excess of algal demand and unlikely to limit phytoplankton growth in the offshore (Talling 1966). Modern nutrient bioassay experiments have confirmed that P addition to lake water does not increase algal biomass (Lehman and Branstrator 1993), and that offshore concentrations of PO_4 are in excess of algal half saturation constants for uptake rates (Lehman and Branstrator 1994). Nearshore, concentrations of PO_4 are much reduced, and turnover times for the ambient dissolved pools which are as high as 5 days offshore decline to 5 min in the nearshore environment (Lehman and Branstrator 1994). The net result is that when water exits Lake Victoria as the White Nile, SRP concentrations are near zero, compared with 1 μM or more offshore, which means that on a lakewide, ecosystem basis virtually all inorganic P supplied to the lake is converted to organic matter (Hecky 1993). This is evidentally not a new situation for the lake, because Talling (1966) reported depressed concentrations of both SRSi and SRP at his northern inshore stations compared with offshore waters of the lake.

Nitrogen had been considered the macronutrient most likely to limit phytoplankton growth in Victoria (Talling and Talling 1965), based on low nitrate concentrations (<1 μM); modern bioassay experiments confirm that additions of inorganic nitrogen can increase algal biomass (Lehman and Branstrator 1993, 1994). There is also evidence that elevated algal biomass in recent years has reduced the vertical extent of light penetration, and that rates of photosynthesis show signs of light limitation (Mugidde 1993).

If the changes to Lake Victoria have been caused by nutrient addition (eutrophication), the observed increase in chlorophyll and algal biomass would have to be accompanied by increased supplies of nitrogen, because water column reserves of the element are small and incapable of supporting large biomass increases. Estimated export of N from catchment to lake is more than two orders of magnitude smaller than the flux required by measured rates of primary production (Lehman and Branstrator 1993). Internal sources of N are unlikely to make up the difference because sediments and overlying water are probably a sink rather than a source of inorganic N, owing to denitrification (Hecky 1993). Nitrate concentrations in rainwater are similarly insufficient (Talling 1966; Lehman and Branstrator 1993).

The other major source of N is elemental, atmospheric N_2. The algal community of Lake Victoria is now rich in cyanobacteria (Ochumba and Kibaara 1989; Ochumba 1990), including many diazotrophic forms (Hecky 1993). It is not yet known how much of the production is based on "new" nutrients rather than recycled ones, but clearly sedimentary fluxes, denitrification losses, and the increased algal biomass from historical levels must be balanced by accelerated N-fixation, because watershed inputs and precipitation account for a trivial fraction of algal demand for N. Hecky (1993) measured substantial rates of N-fixation and also showed that excess SRP offshore was converted to particulate, algal P inshore as N was added to the system through N-fixation. Stoichiometry of the particulate material inshore indicated possible P deficiency.

From observations about the P economy of the lake, it appears that despite the proximate roles of light and N limitation, the master controlling nutrient for Lake Victoria may be P. In the long term, and on the basin-scale, N-fixation during the "endless summer" causes all excess PO_4 to be used in biological processes.

An alternative hypothesis emphasizes the altered trophic structure of Lake Victoria after the introduction of nilotic fish species (Witte et al. 1992a; Baskin 1992). Many of the haplochromine species eliminated from the fauna had been primary and secondary consumers. If these food web changes, tied to success by an introduced piscivore, had reduced herbivory and detritivory by haplochromines, and had thereby increased sedimentation of organic matter, then some of the observed changes might result from a trophic cascade, sensu Carpenter et al. (1985). For instance, decreased herbivory by hapochromines directly on algae might permit increased algal biomass as well as increased rates of organic matter export from euphotic zone to sediments. The increased vertical particulate fluxes might lead to increased sediment oxygen demand, decreased hypolimnetic oxygen concentrations, and lower rates of SRSi regeneration, owing to higher burial rates of diatom frustules.

The hypothesis of lake ecosystem change by trophic cascade, or "top–down" control, depends critically on the ability of Lake Victoria haplochromines to suppress the algae of the lake well below carrying capacity. The presumption must be that historically, as at present, diazotrophic cyanobacteria would have been able to fix enough atmospheric N_2 to elevate algal biomass substantially, if only compensatory losses through herbivory were not so high. By this reasoning, the changes in nutrients and algal communities would have occurred contemporaneously with, or subsequently to, collapse of the haplochromine stocks, which is known to have occurred during the early 1980s (Ogutu-Owayo 1990a, 1990b, 1992; Kudhongania et al. 1992).

In contradiction to this second hypothesis, however, Hecky (1993) showed from paleolimnologic analyses that eutrophication by increased P loading and algal community change, including loss of *Melosira* (now named *Aulocosira*) preceded the rapid increase in Nile perch in the early 1980s. Hecky argued that eutrophication allowed increased survival of Nile perch larvae and immatures which then initiated the rapid population increase of the 1980s.

A third hypothesis explains some of the observed changes in Lake Victoria as consequences of changes in thermal stratification, mixing patterns, and regional climate. A region-wide pluvial episode caused a substantial elevation in lake level (2.5 m) from 1961 to 1964 (Kite 1981; Welcomme 1970), which was apparently caused by increased rainfall (Piper et al. 1986). The lake has not subsequently receded to previous levels and seems to possess a somewhat altered hydrodynamic state. The lake stratifies with greater physical stability than it did 30 years ago, and mean lake water temperatures, including temperatures at all depths, are about 0.5°C warmer than historical values (Hecky 1993). Victoria is known to respond biologically to seasonal mixing events, which reoxygenate its depths and redistribute nutrients through the water column. Particle-rich mixed layers are known to have altered heat balances and shallower mixed depths (Price et al. 1986; Kling 1988; Mazumder et al. 1990). Warmer surface temperatures are known to be one consequence of elevated algal biomass (Mazumder et al. 1990), but heat losses from increased surface temperatures and backscattering of light by the surface particulates decrease mean lake temperature (e.g., Hecky 1984a). Hence, the increase in water temperatures at all depths is strong evidence of regional climate variation.

Prolonged and intensified stratification, or restricted vertical mixing, encourages the development of more extreme hypolimnetic oxygen deficits, and decreased rates of vertical diffusive flux of SRSi to surface waters (Lehman and Branstrator 1993). Light limitation is important in optically deep lakes like Victoria, where diel mixing can suspend algae for substantial periods below the euphotic zone. In such lakes, more stable water columns permit the establishment of surface populations of diazotrophic cyanobacteria (Reynolds 1984), which rely on light energy for both N-fixation and C-fixation. Thus, increased thermal stability provides a mechanism for positive biotic feedback on the nitrogen cycle, and for the fact that present levels of algal biomass are from 2- to 10-fold higher offshore than 30 years ago (Talling 1966; Mugidde 1993).

By reason of either the first (eutrophication) or third (climate change) hypothesis, the changes at higher trophic levels could have been in part consequences of the nutrient and stability changes. Benthic trawl surveys (Bergstrand and Cordone 1971; Kudhongania 1973; Kudhongania and Cordone 1974) had documented the widespread distribution of haplochromines at all lake depths, although abundances were somewhat reduced in the deepest strata (Fig. 1), where Talling (1966) had reported seasonal O_2 depression. Nile perch, which had been introduced a decade earlier, were extremely rare and confined to littoral depths. If widespread deoxygenation of hypolimnetic regions during the 1970s had led to loss of haplochromine habitat and dispersal of demersal stocks, those displaced fishes may have provided an important resource to the Nile perch, and thereby stimulated their population explosion in the 1980s (Hecky 1993).

Changes in the recent paleostratigraphy of Lake Victoria sediments lend support to the view that lower food web events preceded the transformation of the fishery. Hecky (1993) demonstrated that *Melosira* species, which had formerly been common, declined precipitously shortly after 1960. *M. nyassensis*, in particular, is a heavily silicified species that grows best during periods of very deep, intense mixing, which delivers cells to the euphotic zone (Kilham et al. 1986). During 1960–61 (Talling 1966), this species developed popula-

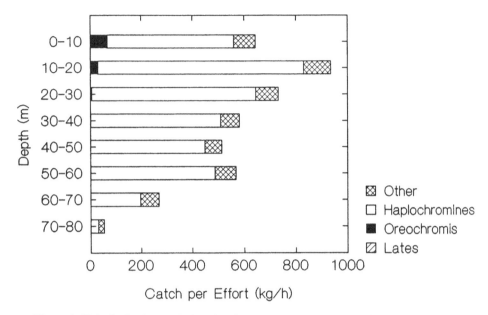

Figure 1. Fish distributions in Lake Victoria assessed by lakewide benthic trawl surveys, 1969–71, previous to major food web changes. Note the decreased abundances in the deepest strata. Data from Kudhongania (1973).

tions during strong mixing to 60 m. The species has been virtually absent in recent years. The contracted euphotic zone (Mugidde 1993, Hecky 1993) now contains many cyanobacteria, some of which are diazotrophic and thereby contributors to the nitrogen budget.

There is strong evidence that real changes have occurred to the mixing regime of Lake Victoria during this century, and the paleolimnological evidence indicates that the changes may have some temporal correlation with the changing lake levels and the conditions that produced them. The hypothesis that alteration of Lake Victoria was caused by a trophic cascade from piscivory triggered by introduction of the Nile perch appears untenable based on the temporal sequence of events. The increased algal biomass and lowered oxygen concentrations of Lake Victoria are a source of concern about the trajectory and future state of the lake. They imply there is an increased oxygen demand in the lake's deep water, or that oxygen levels are being recharged less effectively. Reduced oxygen levels lead to decreased hypolimnetic volume habitable by fish during seasonal stratification, and the loss of habitat most likely leads to loss of species and lower potential fishery yield.

Lake Victoria offers a compelling challenge to paleolimnology because the ecosystem changes of the last 30 years have been wholesale and dramatic (Hecky 1993). The record of these changes is now held in the sediments, and it is the study of recent lake history which may ultimately resolve among the competing hypotheses, as well as provide a modern calibration for interpreting more ancient records of change buried in the sediments. Many of the sedimentary changes, including loss of *Melosira* and rise of *Cyclostephanos* have analogs in the records of the other great lakes, such as Kivu and Tanganyika (Haberyan and Hecky 1987).

SUMMARY AND RESEARCH CHALLENGES

The Great Lakes of East Africa contain food webs that are characterized by strong interactions among organisms and between biota and geochemical processes. These biological interactions seem to be forced by absence of seasonal time lags and temporal refugia from interaction. In some cases, environmental conditions and history have produced enormous species richness, as among cichlid fishes and gastropods of littoral and benthic regions. In pelagic regions, however, the food webs are simple by comparison with temperate pelagia, and biological interactions, particularly predation, seem to have a dominating influence.

As tempting as first glimpses of food webs in these lakes may be, knowledge about them is tentative, and must be improved and placed on a firm, quantitative base. Comparative biomass inventories of plankton communities by standard methods in the lakes are needed, as are seasonal and interannual estimates of primary and secondary production. Only by collecting such data will it become possible to measure fluxes of carbon and essential nutrients through major trophic pathways and among ecosystem components. Time series observational and experimental data about fauna and flora, physiological rates, and biogeochemical fluxes are essential, as well, in order to characterize present conditions and variabilities among lakes. In lakes where some faunal elements are missing (e.g., *Chaoborus*, Cladocera in Tanganyika), improved sampling and experimentation is called for in order to determine if taxa are unable to colonize the habitats, or are instead suppressed to low abundances by strong biological forces.

Finally, these lakes offer excellent opportunities to establish the linkages between modern ecosystem processes and recent sedimentary records. Some of them have experienced significant perturbations in recent time (Hecky 1993), and effects of environmental change are accelerating (Cohen, this volume). It will be important to establish the ways that modern changes are recorded in lake sediments in these basins in order to make the long and often excellent paleostratigraphies of many basins interpretable as a history of ecological and environmental change.

ACKNOWLEDGMENTS

This review was facilitated by grants from the NOAA Undersea Research Program and from the University of Michigan Global Change Program. Both Dr. Robert Hecky and Dr. Daniel Livingstone provided helpful comments that improved the manuscript. Travel support to attend the symposium in Uganda was provided by the U.S. National Science Foundation. Much of the review material was assembled by D. A. Lehman.

REFERENCES

Baskin, Y. 1992. Africa's troubled waters. Bioscience. 42: 476–481.

Beadle, L. C. 1981. The inland waters of tropical Africa. Longman Group Ltd.

Beauchamp, R. S. A. 1953. Sulphates in African inland waters. Nature 171: 769–771.

Bennett, W. N., and M. E. Borass. 1989. A demographic profile of the fastest growing metazoan: a strain of *Brachionus calciflorus* (Rotifera). Oikos 55: 365–369.

Bergstrand, E., and A. J. Cordone. 1971. Exploratory bottom trawling in Lake Victoria. Afr. J. Trop. Hydrobiol. Fish. 1: 13–23.

Bottrell, H. H., et al. 1976. A review of some problems in zooplankton production studies. Norw. J. Zool. 24: 419–456.

Brooks, J., and S. I. Dodson. 1965. Predation, body size, and the composition of the plankton. Science 150: 28–35.

Burgis, M. J. 1984. An estimate of zooplankton biomass for Lake Tanganyika. Verh. Internat. Verein. Limnol. 22: 1199–1203.

Carpenter, S. R., J. F. Kitchell, and J. R. Hodgson. 1985. Cascading trophic interactions and lake productivity. Bioscience 35: 634–639.

Cohen, A. S., Kaufman L., and Ogutu-Ohwayo, R. 1996. Anthropogenic threats, impacts and conservation strategies in the african great lakes: a review. pp. 575–623. In T. C. Johnson and E. O. Odada [eds.], The limnology, climatology and paleoclimatology of the East African lakes. Gordon and Breach, Toronto.

Coulter, G. W. [ed.] 1991. Lake Tanganyika and its life. Oxford Univ. Press

Dumont, H. J. 1986. The Tanganyika sardine in Lake Kivu: another ecodisaster for Africa? Environ. Cons. 13: 143–148.

Eccles, D. H. 1985. Lake flies and sardines — a cautionary note. Biol. Conserv. 33: 309–333.

Ferguson, A. J. D. 1978. Studies on the zooplankton of Lake Turkana. Chapt. 3. In A.J. Hopson [ed.], Lake Turkana. A report on the findings of the Lake Turkana Project.

Fernando, C. H. 1980a. The freshwater zooplankton of Sri Lanka with a discussion of tropical freshwater zooplankton composition. Int. Rev. gesamten Hydrobiol. 65: 85–125.

Fernando, C. H. 1980b. The species and size composition of tropical freshwater zooplankton with special reference to the oriental region (Southeast Asia). Int. Rev. gesamten Hydrobiol. 65: 85–125.

Fryer, G. 1957. Freeliving freshwater Crustacea from Lake Nyasa and adjoining waters. Arch. für Hydrobiol. 53: 62–86.

Gliwicz, Z. M. 1985. Predation or food limitation: an ultimate reason for extinction of planktonic cladoceran species. Arch. Hydrobiol. Beih. Ergebn. Limnol. 21: 419–430.

Graham, M. 1929. The Victoria Nyanza and its fisheries. A report on the fish survey of Lake Victoria 1927–1928. Crown Agents for the Colonies, London 255 pp.

Green, J. 1967a. The distribution and variation of Daphnia lumholtzi (Crustacea: Cladocera) in relation to fish predation in Lake Albert, East Africa. J. Zool. Lond. 151: 181–197.

Green, J. 1967b. Associations of Rotifera in the zooplankton of the lake sources of the White Nile. J. Zool. Lond. 151: 343–378.

Green, J. 1971. Associations of Cladocera in the zooplankton of the lake sources of the White Nile. J. Zool. Lond. 165: 373–414.

Hammer, U. T. 1986. Saline lake ecosystems of the world. Junk. 616 pp.

Hecky, R. E. 1984a. Thermal and optical characteristics of Southern Indian Lake before, during, and after impoundment and Churchill River diversition. Can. J. Fish. Aquat. Sci. 41: 579–590.

Hecky, R. E. 1984b. African lakes and their trophic efficiencies: a temporal perspective. pp. 405–448. In D. G. Meyers and J. R. Strickler [eds.], Trophic interactions within aquatic ecosystems. AAAS.

Hecky, R. E. 1991. The pelagic ecosystem. pp. 90–110. In G. W. Coulter [ed.], Lake Tanganyika and its life. Oxford Univ. Press.

Hecky R. E. 1993. The eutrophication of Lake Victoria. Verh. Internat. Verein Limnol. 25: 39–48.

Hecky, R. E., and H. J. Kling. 1981. The phytoplankton and protozooplankton of the euphotic zone of Lake Tanganyika: species composition, biomass, chlorophyll content, and spatio-temporal distribution. Limnol. Oceanogr. 26: 548–564.

Hecky, R. E., E. J. Fee, H. J. Kling, and J. W. M. Rudd. 1981. Relationship between primary production and fish production in Lake Tanganyika. Trans. Am. Fish. Soc. 110: 336–345.

Hecky, R. E., and H. J. Kling. 1987. Phytoplankton ecology of the great lakes in the rift valleys of Central Africa. Arch. Hydrobiol. Beih. Ergebn. Limnol. 25: 197–228.

Hecky, R. E., R. H. Spigel, and G. W. Coulter. 1991. The nutrient regime. pp. 76–89. In G. W. Coulter [ed.], Lake Tanganyika and its life. Oxford Univ. Press.

Hecky, R. E., and F. W. B. Bugenyi. 1992. Hydrology and chemistry of the African Great Lakes and water quality issues: problems and solution. Mitt. Internat. Verein. Limnol. 23: 45–54.

Kendall, R. L. 1969. An ecological history of the Lake Victoria basin. Ecol. Monogr. 39: 121–176.

Kerfoot, W. C. 1977a. Implications of copepod predation. Limnol. Oceanogr. 22: 316–325.

Kerfoot, W. C. 1977b. Competition in cladoceran communities: the cost of evolving defenses against copepod predation. Ecology 58: 303–313.

Kilham, P. 1971. Biogeochemistry of African lakes and rivers. Ph.D. thesis, Duke University.

Kilham, P., S. S. Kilham, and R. E. Hecky. 1986. Hypothesized resource relationships among African planktonic diatoms. Limnol. Oceanogr. 31: 1169–1181.

Kilham, P., and S. S. Kilham. 1990. Endless summer: internal loading processes dominate nutrient cycling in tropical lakes. Freshwater Biol. 23: 379–389.

Kite, G. W. 1981. Recent changes in level of Lake Victoria. Hydrolog. Sci. Bull. 26: 233–243.

Kling, G. W. 1988. Comparative transparency, depth of mixing, and stability of stratification in lakes of Cameroon, West Africa. Limnol. Oceanogr. 33: 27–40.

Kudhongania, A. W. 1973. Past trends and recent research on the fisheries of Lake Victoria in relation to possible future developments. African J. Trop. Hydrobiol. Fish. Special Issue 2: 93–106.

Kudhongania, A. W., and A. J. Cordone 1974. Batho-spatial distribution patterns and biomass estimate of the major demersal fishes in Lake Victoria. African J. Trop. Hydrobiol. Fish. 3: 15–31.

Kudhongania, A. W., T. Twongo, and R. Ogutu-Ohwayo. 1992. Impact of the Nile perch on the fisheries of Lakes Victoria and Kyoga. Hydrobiologia 232: 1–10.

LaBarbera, M. C., and P. Kilham. 1974. The chemical ecology of copepod distribution in the lakes of East and Central Africa. Limnol. Oceanogr. 19: 459–465.

Lehman, J. T. 1988. Ecological principles affecting community structure and secondary production by zooplankton in marine and freshwater environments. Limnol. Oceanogr. 33: 931–945.

Lehman, J. T., and D. K. Branstrator. 1993. Effects of nutrients and grazing on the phytoplankton of Lake Victoria. Verh. Internat. Verein. Limnol. 25: 850–855.

Lehman, J. T., and D. K. Branstrator. 1994. Nutrient dynamics and turnover rate of phosphate and sulfate in Lake Victoria, East Africa. Limnol. Oceanogr. 39: 227–233.

Ligtvoet, W., and F. Witte. 1991. Perturbation through predator introduction: effects on the food web and fish yields in Lake Victoria (East Africa). pp. 263–275. In O. Ravera [ed.], Terrestrial and aquatic ecosystems perturbation and recovery. Ellis Harwood.

Macdonald, W. W. 1953. Lake-flies. Uganda Journal 17: 124–134.

Macdonald, W. W. 1956. Observations on the biology of chaoborids and chironomids in Lake Victoria and on the feeding habits of the "Elephant-Snout Fish" (Mormyrus kannume Forsk). J. Animal Ecol. 25: 36–53.

Marshall, B. E. 1991. The impact of the introduced sardine Limnothrissa miodon on the ecology of Lake Kariba. Biol. Conserv. 55: 151–165.

Mashiko, K., S. Kawabata, and T. Okino. 1991. Reproductive and populational characteristics of a few caridean shrimps collected from Lake Tanganyika, East Africa. Arch. Hydrobiol. 122: 69–78.

Mazumder, A., D. J. McQueen, W. D. Taylor, and D. R. S. Lean 1990. Effects of fish and plankton on lake temperature and minxing depth. Science 247: 312–315.

Mugidde, R. 1993. The increase in phytoplankton primary productivity and biomass in Lake Victoria (Uganda). Verh. Internat. Verein. Limnol. 25: 846–849.

Mwebaza-Ndawula, L. 1994a. Changes in relative abundance of zooplankton in northern Lake Victoria, East Africa. Hydrobiologia 272: 259–264.

Mwebaza-Ndawula, L. 1994b. Zooplankton studies. In Lake Victoria Biodiversity Project Technical Report: April 1992–April 1993. Fisheries Research Institute, Jinja, Uganda.

Neill, W. E., and A. Peacock. 1980. Breaking the bottleneck: Interactions of invertebrate predators and nutrients in oligotrophic lakes, pp. 715–724. *In* W. C. Kerfoot [ed.], Evolution and ecology of zooplankton communities. Univ. Press of New England.

Nogrady, T. 1983. Succession of planktonic rotifer populations in some lakes of the Eastern Rift Valley, Kenya. Hydrobiol. 98: 45–54.

Northcote, T. G., C. J. Walters, and J. M. B. Hume 1978. Initial impacts of experimental fish introductions on the macrozooplankton of small oligotrophic lakes. Verh. Internat. Verein. Limnol. 20: 2003–2012.

Ochumba, P. B. O. 1990. Massive fish kills within the Nyanza Gulf of Lake Victoria, Kenya. Hydrobiol. 208: 93–99.

Ochumba, P. B. O., and D. I. Kibaara. 1989. Observations on blue-green algal blooms in the open waters of Lake Victoria, Kenya. Afr. J. Ecol. 27: 23–34.

Ogutu-Ohwayo, R. 1990a. The reduction in fish species diversity in Lakes Victoria and Kyoga (East Africa) following human exploitation and introduction of non-native fishes. J. Fish Biol. 37: 207–208.

Ogutu-Ohwayo, R. 1990b. The decline of the native fishes of Lakes Victoria and Kyoga (East Africa) and the impact of introduced species, especially the Nile perch, *Lates niloticus*, and the Nile tilapia, *Oreochromis niloticus*. Environ. Biol. Fish. 27: 81–96.

Ogutu-Ohwayo, R. 1992. The purpose, costs and benefits of fish introductions: with special reference to the Great Lakes of Africa. Mitt. Internat. Verein. Limnol. 23: 37–44.

Okedi, J. 1992. Lake flies in Lake Victoria: Their biomass and potential for use in animal feeds. Insect Sci. Appl. 13: 137–144.

Piper, B. S., D. T. Plinston, and J. V. Sutcliffe. 1986. The water balance of Lake Victoria. Hydrolog. Sci. Journal 31: 25–37.

Price, J. F., R. A. Weller, and R. Pinkel. 1986. Diurnal cycling: observations and models of the upper ocean response to diurnal heating, cooling and wind mixing. J. Geophys. Res. 91: 8411–8427.

Reynolds, C. S. 1984. The ecology of freshwater phytoplankton. Cambridge.

Rzoska, J. 1957. Notes on the crustacean plankton of Lake Victoria. Proc. Linnean Soc. Lond. 168: 116–125.

Rzoska, J. 1975. Zooplankton of the nile system. pp. 333–344. *In* J. Rzoska [ed.], The Nile, biology of an ancient river. Junk.

Schelske, C. L. 1988. Historic trends in Lake Michigan silica concentrations. Int. Revue ges. Hydrobiol. 73: 559–591.

Stager, J. C. 1984. The diatom record of Lake Victoria (East Africa): the last 17,000 years, pp. 455–476. *In* D. G. Mann [ed.], Proc. 7th Diatom Symp.

Stager, J. C., P. N. Reinthal, and D. A. Livingstone. 1986. A 25,000-year history for Lake Victoria, East Africa, and some comments on its significance for the evolution of cichlid fishes. Freshwater Biol. 16: 15–19.

Stemberger, R. S., and J. J. Gilbert. 1987. Defenses of planktonic rotifers against predators, pp. 227–239. In W. C. Kerfoot and A. Sih [eds.], Predation. Univ. Press of New England.

Talling, J. F. 1965. The photosynthetic activity of phytoplankton in East African lakes. Int. Revue ges. Hydrobiol. 50: 1–32.

Talling, J. F. 1966. The annual cycle of stratification and phytoplankton growth in Lake Victoria (East Africa). Int. Revue ges. Hydrobiol. 51: 545–621.

Talling, J. F. 1969. The incidence of vertical mixing and some biological and chemical consequences in the tropical African lakes. Verh. Internat. Verein. Limnol. 17: 988–1012.

Talling, J. F., and I. B. Talling. 1965. The chemical composition of African lake waters. Int. Revue ges. Hydrobiol. 50: 421–463.

Topping, M. S. 1971. Ecology of larvae of *Chironomus tentans* (Diptera: Chironomidae) in saline lakes in central British Columbia. Can. Entomol. 103: 328–338.

Turner, J. L. 1982. Lake flies, water fleas and sardines, pp. 165–173. *In* Fishery expansion project, Malawi. Biological studies on the pelagic ecosystem. FI:DP/MLW/75/019. Tech. Rep. No. 1, Rome: UNDF/FAO.

Twombly, S. 1983. Seasonal and short term fluctuations in zooplankton abundance in tropical Lake Malawi. Limnol. Oceanogr. 28: 1214–1224.

Welcomme, R. L. 1970. Studies on the effects of abnormally high water levels on the ecology of fish in certain shallow regions of Lake Victoria. J. Zool. Lond. 160: 405–436.

Witte, F., T. Goldschmidt, P. C. Goudswaard, W. Ligtvoet, M. J. P. Van Oijen, and J. H. Wanink. 1992a. Species extinction and concomitant ecological changes in Lake Victoria. Neth. J. Zool. 42: 214–232.

Witte, F., T. Goldschmidt, J. Wanink, M. van Oijen, K. Goudswaard, E. Witte-Maas, and N. Bouton. 1992b. The destruction of an endemic species flock: quantitative data on the decline of the haplochromine cichlids of Lake Victoria. Environ. Biol. F. 34: 1–28.

Worthington, E. B. 1929. A report on the fishing survey of Lake Albert and Kyoga. Crown Agents for the Colonies: London, 136 pp.

Worthington, E. B. 1931. Vertical movements of freshwater zooplankton. Int. Revue gesamten Hydrobiol. 25: 394–436.

Worthington, E. B., and C. K. Ricardo. 1936. Scientific results of the Cambridge expedition to the East African lakes, 1930–1. No. 17. The vertical distribution and movements of the plankton in lakes Rudolf, Naivasha, Edward, and Bunyoni. J. Linn. Soc. (Zool.) 40: 33–69.

Van de Velde, I. 1984. Revision of the African species of the genus *Mesocyclops* Sars, 1914 (Copepoda: Cyclopidae). Hydrobiologia 109: 3–66.

van Oijen, M. J. P., F. Witte, and E. L. M. Witte-Maas. 1984. An introduction to ecological and taxonomic investigations on the haplochromine cichlids from the Mwanza Gulf of Lake Victoria. Neth. J. Zool. 31: 149–174.

Zaret, T. 1969. Predation-balanced polymorphism of *Ceriodaphnia cornuta* Sars. Limnol. Oceanogr. 14: 301–303.

Zaret, T. 1972. Predators, invisible prey, and the nature of polymorphism in the Cladocera (class Crustacea). Limnol. Oceanogr. 17: 171–184.

Zaret, T. 1980. Predation and freshwater communities. Yale Univ. Press, 187 pp.

Molecular Phylogenetic Inferences About the Evolutionary History of East African Cichlid Fish Radiations

A. MEYER *Department of Ecology and Evolution and Program in Genetics, State University of New York at Stony Brook, United States*

C.M. MONTERO *Department of Ecology and Evolution, State University of New York at Stony Brook, United States*

A. SPREINAT *Unterm Hagen 4, Göttingen, Germany*

Abstract — The species flocks of cichlid fishes from the Great East African Lakes, Victoria, Malawi and Tanganyika, are well-known among evolutionary biologists as extreme examples for adaptive radiation and explosive speciation. Of all radiations involving vertebrates, these species assemblages are the most species-rich and the most diverse, morphologically, ecologically and behaviorally. Traditionally, all knowledge about the evolution and phylogenetic relationships within and between these species flocks was derived from morphological analyses but recently molecular DNA-based data sets have provided new insights into the phylogenetic and biogeographic history of East African cichlid fishes. Phenotypic (e.g., morphological) and genotypic (e.g., DNA sequences of genes) data sets are expected to provide concordant phylogenetic information about these species assemblages, since both share identical evolutionary histories. Molecular data however have several advantages for phylogeny reconstruction over morphological data, e.g., DNA sequences tend to diverge with some regularity over time, which may or may not be true for morphological data. This "molecular clock" allows one to make time estimates of speciation events in the absence fossils. Our understanding of the phylogenetic relationships, history and evolutionary processes among East African cichlid fish species flocks has increased rapidly since the recent invention of the polymerase chain reaction (PCR) which dramatically facilitated the collection of DNA sequence data. Phylogenetic analyses of recent molecular data (mostly mitochondrial DNA sequences) in the context of the geological history of the Great East African Lakes helped to elucidate some aspects of the evolutionary historical patterns and evolutionary processes that might have led to the origin of these extraordinary fish faunas. Here, we summarize recent findings on the molecular phylogenetic relationships of endemic species of Lakes Malawi and Victoria and non-endemic, riverine species of haplochromine cichlids that phylogenetically connect these two species flocks. The DNA-sequence phylogeny confirms that the endemic species flock of Lake Tanganyika is by far the oldest and provided an evolutionary reservoire for the species diversity of East Africa. The Tropheini, among the tribes endemic to Lake Tanganyika, are found to be the closest living relatives to the haplochromine cichlids from Lake Tanganyika and outside of it. New mitochondrial control region DNA sequences, collected for this study, confirm that the vast majority of Lake Malawi cichlids can be assigned to two genetic and ecological groups, one that lives over sandy bottoms, and another one that lives over rocky substrate. Three other lineages are identified for Lake Malawi: *Rhamphochromis, Astatotilapia calliptera*, and possibly *Diplotaxodon. Copadichromis* which had been suggested to be another separate lineage based on mitochondrial restriction data (Moran, Reinthal and Kornfield, 1994) could, based on mitochondrial DNA sequences not be confirmed to be distinct from the "sand" group (called group "A" according to Meyer et al., 1990). Both Lake Victoria and Malawi species flocks are equally distantly related to the Tropheini. Non-endemic East African haplochromine cichlids (e.g., *Serranochromis, Astatoreochromis, Astatotilapia, Orthochromis*, and *Schwetzochromis*) are more closely related to these endemic flocks (and provide biogeographic links) than any Tanganyikan endemic cichlid species.

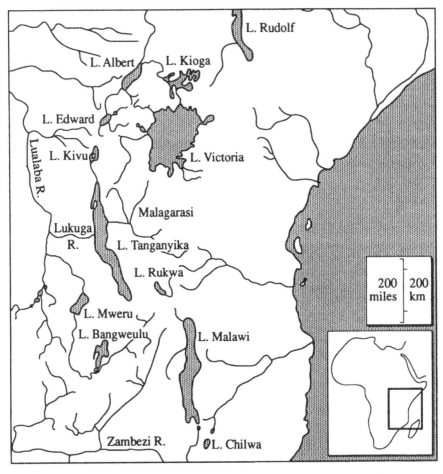

Figure 1. Map of East Africa. Figure redrawn after Fryer and Iles (1972).

INTRODUCTION

The cichlid fish faunas of the three Great East African Lakes, Victoria, Tanganyika, and Malawi, are enormously diverse and a testimony to the evolutionary success of cichlid fishes. Each of these lakes (Fig. 1) harbors a radiation of several hundred species (Fryer and Iles, 1972) almost all of which are endemic to their particular lake. These species flocks make the Cichlidae one of the most species-rich family of vertebrates (Fig. 2). The special history of cichlids is highlighted by the coexistence of other families of fishes in each of these three lakes, that have not undergone this kind of spectacular evolution. The evolutionary origin and ecological maintenance of the enormous cichlid species diversity has been much researched and debated (e.g., Mayr, 1942, 1984; Fryer and Iles, 1972; Coulter, 1991; Keenleyside 1991). Despite this long, still ongoing debate, the phylogenetic relationships among the endemic cichlid faunas have remained largely unresolved since no morphological feature could be found to be characteristic of all members of a particular radiation that might have unambiguously indicated that each of these radiations are, to some degree, independent of each other (e.g., Stiassny, 1981; Greenwood, 1983).

Figure 2. Body form variation found among East African cichlid fish from Lake Victoria (V), Lake Tanganyika (T) and Lake Malawi (M). Figure redrawn after Fryer and Iles (1972) and Greenwood (1984). The fish are (row by row) from left top to lower right: (V) generalized "*Haplochromis*," (T) *Telmatochromis vittatus*, (M) *Cyrtocara moori*, (M) *Rhamphochromis longiceps*, (M) *Rhamphochromis macrophthalmus*, (T) *Lobochilotes labiatus*, (V) *Pyxichromis parothostoma*, (T) *Xenotilapia sima*, (T) *Spathodus malieri*, (T) *Xenotilapia melanogenys*.

The species flocks (defined as monophyletic, i.e., containing a single ancestral species and all of its descendent species which inhabit one lake, Greenwood, 1984) of all three of the lakes contain a sweeping array of morphologically and behaviorally highly specialized cichlids (Fryer and Iles, 1972). An often mentioned reason for the evolutionary success and diversification of cichlids is a morphological novelty only they possess (Liem, 1973; Osse and Liem, 1975). Cichlids have a second set of jaws in back of their buccal cavity, the modified pharyngeal jaws, that are functionally de-coupled from their oral jaws. This key innovation is believed to have allowed cichlids to become highly specialized on particular types of prey and to have facilitated the evolution of fine ecological niche-partitioning. This second set of jaws might allow cichlids to out-compete other fish inhabiting the Great East African Lakes that do not possess them (Liem, 1973; Osse and Liem, 1975).

Some highly derived morphological and ecological specializations are similar between species endemic to different lakes, i.e., similar morphological solutions to the same ecological problems have been found in more than one cichlid species flock (Greenwood, 1983). The Lake Victoria endemic *Macropleurodus bicolor* and the Lake Malawi endemic *Chilotilapia rhoadesi* both have highly derived dentition, jaw structures and feeding behaviors — they prey on gastropods by crushing their shells with their oral jaws (Fryer and Iles, 1972; Greenwood, 1983). Unfortunately, the evolutionary relationships among the species assemblages remained unknown, the question of whether each of the assemblages is monophyletic, i.e., can be traced back to a single ancestral species, and consequently whether the above mentioned morphological similarities between members of different flocks evolved more than once independently, as parallelisms, remained unanswered. Alternatively, specializations could have arisen only once and would indicate polyphyletic origins (several ancestral species per lake) for the species flocks with each of several lineages having a geographic distribution that extends beyond the boundaries of a single lake (Stiassny, 1981; Greenwood, 1983). This interpretation would indicate that relationships of recent common ancestry exist among many of the members of the three species flocks (Fryer and Iles, 1972; Greenwood, 1983).

Much insight about the tempo and mode of evolution and about the origin of morphological solutions to ecological problems can be gained from an understanding of the evolutionary relationships among and between members of these species flocks. Estimates on rates of speciation in these flocks will hinge on basic knowledge (like monophyly versus polyphyly) and the age of the species assemblages. The identification of sister group relationships will help to pinpoint which characteristics in ancestors of these species flocks might have made them successful colonizers of these lakes and founders of species flocks.

TRACING EVOLUTIONARY HISTORIES WITH MORPHOLOGY AND MOLECULES

Each organism's phenotype and its underlying genotype have experienced the same evolutionary history, except for presumably rare cases of horizontal gene transfer. Hence, both general types of data sets should provide the same reliable estimates of phylogenetic relationships among species (Hillis, 1987; Patterson, 1987). Data derived from the phenotypes of organisms, which traditionally consist of morphological characters and various kinds of biochemical data reflecting the genotype are expected to share identical evolutionary histories. Molecular data sets are usually easier to obtain than morphological data sets. This is because often only experts of a particular group of organisms are able to identify meaningful morphological characters for a phylogenetic analysis which aims to reconstruct the phylogeny of the species under consideration. The number of molecular characters that can be identified in species is essentially without limits since each species' genome is made up of billions of DNA base pairs each of which potentially contains phylogenetic information. The number of characters that can be identified in the phenotype of organisms is limited by the morphologist's abilities working on the group to identify characters for a phylogenetic analysis and will tend to be orders of magnitude fewer.

Molecular data can have the added advantage over morphological data sets that they can be collected in objective metrics, e.g., DNA sequences of particular genes from several laboratories can be combined and applied to phylogenetic questions that were not intended in the original study. Such universal metrics are, e.g., small nuclear ribosomal RNA gene (18S) sequences that have been collected for a wide variety of organisms. This potential of some (but not all types of) molecular data sets to be "universal metrics" for the purpose of phylogeny reconstruction is not present in morphological data since each of these data sets must be newly established for every phenotype-based phylogenetic study and such data sets are only rarely transferable between studies. Still, one type of data set is not inherently better than another, both exhibit "phylogenetic noise" (e.g., homoplasy). Both provide useful phylogenetic information, while the signal-to-noise ratio is often similar in both kinds of data sets (Hillis, 1987; Sanderson and Donoghue, 1989; and also behavioral characters: DeQueiroz and Wimberger, 1993).

Since congruence in phylogenetic estimates is expected from both kinds of data sets, it has been argued that the combination of both morphological and molecular data sets should provide "total evidence" (Kluge, 1989). There are, however, numerous problems when both data sets are combined, and when different phylogenetic answers are obtained if these data sets are analyzed separately (reviewed in Swofford, 1991; Maddison and Maddison, 1992).

Several kinds of biochemical data are typically used to infer phylogenetic relationships among species. Allozyme, immunological and DNA–DNA hybridization data have been widely used but are now increasingly replaced by several types of DNA-based data (reviewed in Meyer, 1993b). Since the advent of the polymerase chain reaction (PCR) in 1985–86 (Saiki et al., 1985, 1988; Wrishnik et al., 1987), our knowledge about DNA and phylogeny of vertebrates has increased dramatically (reviewed in Kocher et al., 1989; Meyer, 1993a,b).

THE POLYMERASE CHAIN REACTION AND DIRECT SEQUENCING

The polymerase chain reaction (PCR) is an enzymatic cloning technique that allows the amplification of portions of genes (within size limits of maximally several thousand base pairs) that are defined by synthetic oligonucleotide "primers" (Saiki et al., 1985, 1988; reviews in, e.g., White et al., 1989; Arnheim et al. 1990; refs. in Erlich 1989; Innis et al., 1990) (Fig. 3). The primers are usually around 20 base pairs in length and define the beginning and the end of the double-stranded piece of DNA that is going to be amplified. The specificity of the amplification is accomplished through the need for an almost-perfect fit of the primers to the template DNA (Kwok et al., 1990). During each cycle of PCR, the number of copies of the DNA-fragment delineated by the primers at either end is doubled (Fig. 3). Usually 25–40 cycles are completed in a computer controlled heating block (thermal cycler) in about three hours creating millions and millions of identical copies of a piece of a DNA. PCR is much faster and cheaper than conventional cloning techniques. First, a double-stranded PCR product is produced that is then either sequenced (double stranded sequencing, or alternatively "cycle-sequenced"), or subcloned and then sequenced, or cut with restriction enzymes (RFLP data) or used as template DNA for a subsequent asymmetric amplification (Gyllensten and Erlich, 1988) or digested with an exonuclease to produce single-stranded DNA for direct sequencing of single-stranded DNA. Sequencing

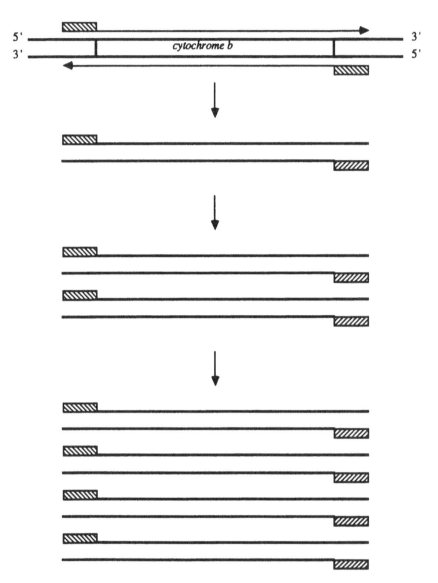

Figure 3. The principle of the polymerase chain reaction. Modified after Wrishnik et al., 1987. The hatched boxes represent "primers" small oligonucleotides (about 20 base pairs in length) whose DNA sequence is complementary to the stretch of DNA for which they are designed to attach, in this example the regions flanking the mitochondrial cytochrome *b* gene. During each cycle of the polymerase chain reaction the number of copies that are defined by the 5' ends of the two primers is doubled leading to an exponential increase in the number of identical pieces of DNA with each PCR-cycle. Primers are incorporated into the copied DNA strands as indicated. See text and cited references (e.g., Erlich, 1989; White et al., 1989) for more details.

gels of single-stranded DNA often allow one to read more base pairs than sequencing gels of double-stranded DNA. Single-stranded PCR amplified DNA can be as clean as sub-cloned DNA and routinely more than 300–400 bp can be unambiguously determined from a single sequencing reaction.

The determination of DNA sequences tends to be more time-consuming, costly and technically involved than several other molecular data sets that can be used for phylogenetic analysis, however DNA sequences of homologous mitochondrial and nuclear genes will allow direct comparisons and study of DNA from different species that have been determined in different laboratories — DNA sequences of the same genes are "universal metrics" that can be transferred between different studies and laboratories. DNA sequences are stored in data banks (e.g., EMBL, GenBank) and are universally usable, powerful data. The increased costs of DNA sequences compared to, e.g., RFLP data are far outweighed by their advantage as a universally retrievable, and applicable type of data, since homologous data from independent laboratories can be used in direct comparisons for several studies.

MOLECULAR DATA AND THE EVOLUTIONARY HISTORY OF EAST AFRICAN CICHLID RADIATIONS

Until lately, evolutionary studies on cichlids relied exclusively on morphological characters and were subject to the danger of circularity by interpreting the evolution of the same morphological specializations that were used to construct the phylogenetic relationships. Molecular approaches, specifically the study of the mitochondrial genome through restriction enzyme analysis and more recently through DNA sequences are providing many new insights and some surprising results (Kornfield, 1991; Meyer et al., 1990, 1991; Sturmbauer and Meyer, 1992, 1993; Kocher et al., 1993; Sturmbauer, Verheyen and Meyer, 1994). In these molecular studies, as in most other similar studies, evolutionary relationships among mitochondrial DNA haplotypes are used as proxy for the phylogenetic relationships among species (Avise and Ball, 1990; Meyer 1993b). Both morphological and molecular data are analyzed by identical phylogenetic methods (reviewed, e.g., in Swofford 1991, for fishes see Meyer, 1993b). All of these methods have weaknesses, strengths, and underlying assumptions; space does not allow the review of these methods here but excellent reviews are available (Felsenstein, 1988, Swofford and Olsen, 1990; Swofford, 1991). The paucity of cichlid fossils in Africa (VanCouvering, 1982) makes the molecular approach particularly valuable. DNA sequences of two sister species diverge with relative regularity ("molecular clock") over time from their ancestral species. If the approximate "ticking rate" of the molecular clock for a particular gene in a particular lineage is know, one can back-calculate how long ago a common ancestor of two species might have lived, based on the amount of DNA-sequence divergence observed. The regularity of the molecular clock is disputed and several simplifying assumptions enter into these calculations, which is why caution must be exercised in the interpretation of these data.

A SINGLE ANCESTOR FOR THE LAKE VICTORIA SUPER-FLOCK

Of the three large East Afrikan lakes, Lake Victoria is the youngest. Its origin is dated back to about about 250,000 to 750,000 years ago (Fryer and Iles, 1972), nonetheless it harbors a species flock of more than 300 endemic haplochromine cichlids, much of which is now threatened by extinction through the introduction of the Nile Perch: e.g., Witte et al., 1992). Lake Victoria originated from two westward flowing rivers, the proto-Kagera and the proto-Katonga, that were back-ponded in the Pleistocene by the uplifted western margin of the Victoria basin (Fryer and Iles, 1972). Geological data on the formation of Lake Victoria indicate that during the Pleistocene a connection existed between it and several smaller lakes to the west of it: Lakes Edward, George, and Kivu (Fig. 1). Hence, Lake Victoria species flock should be considered a super-species flock that goes beyond the current shores of Lake

Victoria (Greenwood, 1984). Greenwood christened the term super-species flock and laid out criteria for the use of "species flock" for species assemblages: (1) high levels of endemicity, (2) monophyly, (3) geographic circumscription (Greenwood, 1984).

Nearly all endemic cichlids of Lake Victoria had been assigned to the single genus "*Haplochromis*"; Greenwood later divided them into more than 20 different genera (Greenwood, 1980). It was not known for long whether more than one riverine ancestral species provided the initial "seed" to the present diversity in Lake Victoria. Among cichlid taxonomists, most believed that the Lake Victoria haplochromine cichlid assemblage had more than one ancestor (Fryer and Iles, 1972); Greenwood argued that neither the Victoria super-flock nor the Lake Malawi cichlid assemblage should be considered as single species monophyletic flocks (Greenwood, 1983). But, electrophoretic data demonstrated that the species in the Victoria cichlid super-species flock are extremely closely related (the mean genetic distance being only 0.006 substitutions per locus); this suggested that these species might have recently arisen from only a single ancestral species (Sage et al., 1984).

Mitochondrial DNA (mtDNA) sequences evolve faster than nuclear DNA (Brown et al., 1979; reviewed in Meyer, 1993a). Phylogenies based on mtDNA (particularly of the fast evolving part of the mitochondrial genome, the control region), therefore, can resolve evolutionary relationships among young, very closely related species (e.g., Meyer et al., 1990, 1991; Sturmbauer and Meyer, 1992). The amount of mtDNA variation among fishes of the Victoria flock was investigated and found to be extremely small (Meyer et al., 1990). In fact, no variation was detected in 363 base pairs (bp) of the cytochrome b gene, and only about 2–3 substitutions differentiate mitochondrial haplotypes and presumably species of Lake Victoria cichlids in 440 bp of the control region (Meyer et al., 1990). More variation had been found in the homologous portion of mtDNA genome in the single species, *Homo sapiens* than was found among the all 14 species of nine representative endemic genera of Lake Victoria haplochromine cichlids which had been studied (Vigilant et al., 1989). This high degree of mtDNA similarity and the earlier allozyme data suggested a very young age for this flock, it was estimated to be probably less than 200,000 years of age (Sage et al., 1984; Meyer et al., 1990) (Fig. 4). This age estimate for the Lake Victoria super species flock is younger than the lake itself, and supports the notion of intra-lacustrine speciation; i.e., the adaptive radiation of this species flock is likely to have occurred in the lake itself rather than being due to several founding species from different ancestral lineages. Phylogenetic relationships within the Victoria super-flock could not be established with certainty since too little phylogenetic information was contained even in the fastest evolving portion of the mitochondrial genome (Meyer et al., 1990). Comparisons of mtDNA sequences from Lake Victoria endemics with those from Lake Malawi, Lake Tanganyika, non-endemics and riverine cichlids of East and West Africa indicate that the Lake Victoria super-flock (which includes endemics from satellite lakes like Lake Edward) make it likely that the Victoria super-flock originated from a single ancestral species (Meyer et al., 1990, 1991) (Fig. 4). The mitochondrial-based suggestion of monophyly of the Lake Victoria super-flock still holds with the inclusion of more riverine East African cichlid species (see below, and Meyer and Montero, unpublished data).

For Lake Victoria, despite this extremely low level of mtDNA variation among morphologically very different species of cichlids (Fig. 2), there was only one case in which identical mtDNA haplotypes were detected among two morpho-types interpreted to be good biological species (Meyer et al., 1990). This might argue that lineage sorting of mtDNA haplotypes was fast and almost complete even among these very young species (Avise and

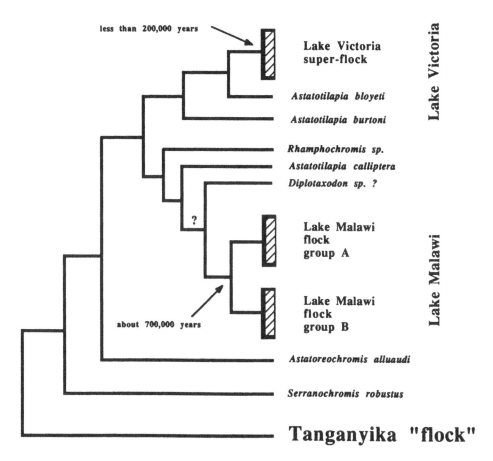

Figure 4. Phylogenetic tree relating the three endemic species flocks of Lake Victoria, Lake Malawi and some riverine species of haplochromine cichlids from East Africa to part of the Lake Tanganyika flock (the tribe Tropheini is the sister group to the haplochromine cichlids that are found in East African rivers and whose ancestors colonized Lakes Victoria and Malawi to form their species flocks, see Fig. 6) based on (Meyer et al., 1990, 1991; Moran, Reinthal and Kornfield, 1994; Sturmbauer and Meyer, 1993; Sturmbauer, Verheyen and Meyer, 1994). Presumed monophyletic assemblages are indicated with shaded boxes. Branches are not drawn to scale with time since divergence. *Astatoreochromis alluaudi* and *Serranochromis robustus* are widespread East African species. *A. alluaudi* also lives in Lake Victoria. The "?" is meant to indicate that we have not tested the finding that *Diplotaxodon* is a separate lineage from the other four found in Lake Malawi as suggested by a RFLP mitochondrial DNA study (Moran et al., 1994). *Astatotilapia bloyeti* is a generalized haplochromine (Fig. 2) that is found throughout much or East Africa and probably resembles the ancestral species of the Lake Victoria super-flock. *Astatotilapia burtoni* is found in Lake Tanganyika and surrounding areas. *Astatotilapia calliptera* is not strictly endemic to Lake Malawi.

Ball, 1990). These data might also argue that the estimated number of species in Lake Victoria which were solely based on, at times, slight morphological differences, is supported by genetic differences. The data might cautiously be interpreted to suggest that the different mitochondrial DNA haplotypes might indeed represent reproductively isolated biological species (Avise and Ball, 1990). A more detailed characterization, with larger intraspecific sample sizes and the inclusion of nuclear DNA markers will provide more insights into the question of the validity of the species ranks and the dynamics of speciation in Lake Victoria haplochromine cichlids. Preliminary data with larger sample sizes than the original study (Meyer et al., 1990), confirm that intraspecific levels of variation in endemic Lake Victoria cichlids are extremely low (Meyer et al., in prep.). However, the level of variation in the nuclear major histocompatibility complex was found to be extensive among Lake Malawi cichlids (Klein et al., 1993; Ono et al., 1993).

PHYLOGENY OF THE LAKE MALAWI FLOCK AND ITS RELATIONSHIP TO THE VICTORIA SUPER-FLOCK

Preliminary electrophoretic data determined that the endemic cichlids of Lake Malawi are extremely closely related (Kornfield, 1978) but suggested that the Lake Malawi and Tanganyika flocks are not monophyletic but share at least one lineage (Kornfield et al., 1985). In contrast, data from mtDNA sequences, suggested that the Lake Malawi species flock appears to be monophyletic (Meyer et al., 1990; Kocher et al., 1993). MtDNA of the highly derived *Macropleurodus-Chilotilapia* species pair from Lake Victoria and Malawi respectively (which had been used to argue for a polyphyletic origin of both flocks [Greenwood, 1983]) were compared, and demonstrated that these two species are not sister taxa but rather are more closely related to the rest of their monophyletic Victoria and Malawi assemblages (Meyer et al., 1990). Species from the Lake Victoria and Lake Malawi species flocks differ by at least 54 substitutions (in the 803 base pairs compared from two mitochondrial genes) from each other; therefore any morphological or behavioral similarity that appears to link particular species from these lakes must now be interpreted as parallelism or homoplasy rather than as an indicator of common descent (Meyer et al., 1990). Kocher et al. (1993) compared other species pairs from these species flocks that show striking morphological similarities and also concluded that in all cases these morphological similarities are merely parallelism and do not represent evidence for common descent. The Malawi and the Victoria flocks, despite being genetically distinct, are still very closely related: in the mitochondrial cytochrome *b* gene, they differ by only 5% sequence divergence whereas congeneric cichlid species of the Neotropical genus *Cichlasoma* differ by up to 11% (Meyer et al., 1990; Kocher et al., 1993).

Among the members of the Lake Malawi species flock two genetically distinct groups can be identified — each is composed of about 200 species (Eccles and Trewavas, 1989). Based on mtDNA sequences, these groups differ from each other by at least 24 substitutions (Meyer et al., 1990; Meyer and Montero, unpublished data) (Fig. 4). Based on mtDNA sequence divergence, the age of the this flock has been preliminarily estimated at around 700,000 years, suggesting that this radiation took place in the 1–2 million years old Lake Malawi basin (Fryer and Iles, 1972). One group of species is largely confined to rocky habitats (the mbuna), and the second lives over sandy habitats, and is composed of species that were until recently (Eccles and Trewavas, 1989) largely assigned to the genus *Haplochromis*. We suggested that both groups can be traced back to a common ancestral species for probably almost the whole Lake Malawi flock with the exception of the *Astatotilapia calliptera* lineage (Meyer et al., 1990, 1991). *Astatotilapia calliptera*, which is not strictly endemic to Lake Malawi, is, based on mitochondrial DNA sequences (Meyer et al., 1991), distinct from the two major lineages and might be representative of the ancestral stock which colonized the early Lake Malawi from rivers in East Africa (Meyer et al., 1991) (Fig. 4); this had been previously suggested by morphological data (Trewavas, 1949).

The origin of the Lake Malawi flock is probably due to a very small number of ancestral lineages (Meyer et al., 1990, 1991; Moran and Kornfield 1993; Moran, Reinthal, and Kornfield 1994; reviewed in Meyer, 1993b; but see Klein et al., 1993; and Ono et al., 1993 on *MHC* variation in Lake Malawi cichlids). Moran et al. (1994) based on mitochondrial RFLP data, suggest that there are six independent lineages in Lake Malawi: *Serranochromis robustus* is clearly a basal lineage; this species had not been included in our original studies (Meyer et al. 1990, 1991). Moran et al.'s (1994) recent restriction analyses of mtDNA further suggests that aside from the two major groups, the mbuna and non-mbuna (Fig. 4) also *Rhamphochromis, Diplotaxodon, Astatotilapia calliptera* and *Copadichromis* may represent other discrete endemic lineages for a total of six (not considering *Serranochromis robustus*) (Fig. 4). Further, these data tentatively indicated that the *Rhamphochromis* lineage may be more basal than *Astatotilapia calliptera* (Moran et al., 1994) (Fig. 4). Figure 4 is a composite of Moran et al.'s (1994) and our (Meyer et al., 1990, 1991; Sturmbauer and Meyer 1992, 1993) work.

We extended our mtDNA sequence analysis of East African cichlid species to include several other Malawian and riverine haplochromine taxa that had not been studied previously. The mtDNA sequences confirm some of Moran et al.'s (1994) findings, but differ somewhat in others (Fig. 5). *Serranochromis robustus* is a distant relative of both the Malawi and the Victoria haplochromine cichlids, we used it as an outgroup in our analysis of the relationships among the Victoria and Malawi species flocks plus some of the East African non-endemic haplochromines. We can confirm Moran et al.'s (1994) finding that *Rhamphochromis* represents another independent lineage of the Lake Malawi species flock, bringing to four the lineages represented in the lake (not counting *Serranochromis*) (Figs. 4 and 5). We disagree with Moran et al. (1994) in that we find that *Copadichromis* does not represent an independent lineage but appears to be a member of the non-mbuna group. We have not sequenced *Diplotaxodon*, and cannot comment as to whether it is another independent lineage (indicated with a "?" in Fig. 4). Whether *Rhamphochromis* is the most basal lineage in Lake Malawi, even more basal than *Astatotilapia calliptera* (Figs. 4 and 5) as has been suggested by Moran et al. (1994), was not clear from our data. Our 50% majority rule bootstrap tree based on a parsimony analysis of the control region sequences (Fig. 5) does not resolve the branching order among the four Malawian lineages, but our most parsimonious trees agree with Moran et al. (1994) in placing *Rhamphochromis* most basal. Within the non-mbuna group we find that *Cyrtocara* and *Tyranochromis* represent the most basal groups; within the mbuna group *Melanochromis labrosus* appears to be the most basal member (Fig. 5), these findings must remain tentative until a more complete representation of Lake Malawi species is accomplished.

MtDNA sequences identified the non-endemic *Astatotilapia burtoni*, a generalist species found in Lake Tanganyika and surrounding waters, to be the closest living relative of the Lake Victoria flock (Meyer et al., 1991). However, the bootstrap values supporting this branching are rather low (Fig. 5) making this finding tentative. More recently, several other non-endemic East African riverine cichlids from the Malagarasi river, the Ruahu river, Lake Rukwa and Lake Kitangiri (e.g., *Astatotilapia bloyeti*) have been characterized mitochondrially and are found to be even more closely related to the Victoria flock than *Astatotilapia burtoni* (Meyer and Montero, unpublished data) (Figs. 4 and 5). The Victoria super-flock, mitochondrially speaking, appears to include the endemics of Lake Victoria and its satellite lakes plus some riverine haplochromine cichlids of East Africa, e.g., a species of *Astatotilapia* from the Manago river from Tanzania (collected by L. Seegers) and has very close

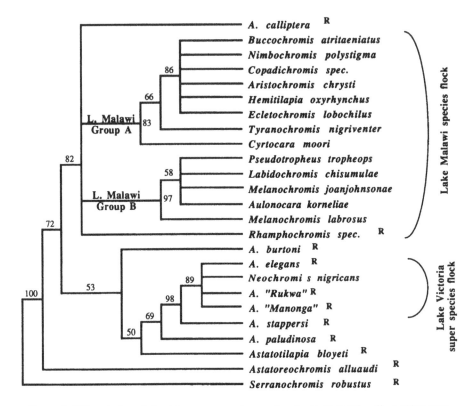

Figure 5. 50% majority rule bootstrap consensus tree analyzed with PAUP (Swofford 1991) (200 replications) of previously unpublished mitochondrial control region sequences of Lake Malawi, Lake Victoria and East African non-endemic haplochromine cichlids (Meyer and Montero, unpublished data). "A" stands for *Astatotilapia* indicating tentative assignment of these riverine, non-endemic haplochromine species (and "R" stands for riverine species) to this mitochondrial not-monophyletic genus.

affinities to other riverine Tanzanian *Astatotilapia* species of uncertain species status (collected and provided by Lothar Seegers and Tuur Vos). Some members of the widespread mainly riverine genus *Astatotilapia* (mitochondrially, an unnatural group, Figs. 4 and 5; see Meyer et al., 1991) is likely to represent the body plan and lifestyle of the ancestors of the Victoria and possibly the Malawi species flocks (Meyer et al., 1991). We are currently expanding our phylogenetic mitochondrial DNA analysis to include more East African riverine haplochromines (Meyer, Voos, Seegers, Montero unpublished data). Greenwood previously recognized that *Astatotilapia* is not monophyletic, when he assigned species into this genus, that this genus is a "depository"-genus for generalized non-endemic haplochromine cichlids.

LAKE TANGANYIKA, THE EVOLUTIONARY RESERVOIR FOR EAST AFRICAN CICHLIDS

Lake Tanganyika is estimated to have an age of 9–12 million years (Cohen et al., 1993, or even 26 mya Johnson, this volume), making it by far the oldest of the Great East African lakes. Being the oldest, it may not be surprising that it harbors the morphologically and behaviorally most diverse cichlid fauna, consisting of about 171 species (however, by comparison with the Victoria and Malawi flocks this is a relatively small number) in 49 endemic genera that are assigned to twelve tribes (Fryer and Iles, 1972; Poll, 1986; Coulter, 1991). These tribes are assemblages of genera that seem to represent ecologically, morphologically, as well as behaviorally well defined phylogenetic lineages (Poll 1986). Morphological and electrophoretic data suggest that the lineages of cichlids from Lake Tanganyika are old and can be traced back to at least seven distinct ancestral lineages (Poll, 1986; Nishida, 1991). Phenotypic differences between the tribes are pronounced, and mtDNA data turned out to be generally in good agreement with Poll's classification (Poll, 1986) and assignments of genera into tribes (Sturmbauer and Meyer, 1993; Sturmbauer, Verheyen and Meyer, 1994). Comparisons of electrophoretic and mtDNA data demonstrated that several Tanganyikan lineages are much older than the lineages of Lakes Victoria and Malawi (Nishida, 1991; Sturmbauer and Meyer, 1992, 1993; Kocher et al., 1993; Sturmbauer, Verheyen and Meyer, 1994).

The age of some endemic Tanganyikan cichlid lineages was recently estimated for two tribes of cichlids, the Tropheini and the Ectodini (Sturmbauer and Meyer 1992, 1993). The genus *Tropheus* was estimated to be about twice as old as the entire cichlid species flock from Lake Malawi and six times older than the entire flock of endemic haplochromines from Lake Victoria (Sturmbauer and Meyer, 1992). The Ectodini, represented by twelve endemic genera, seem to be approximately twice as old as *Tropheus* and five times older than the Malawi flock (Sturmbauer and Meyer, 1993). Estimates based on mtDNA sequences suggest that the Ectodini, a large variable tribe of endemics, are probably about 3.5 to 4 million years old, and some other lineages (e.g., Bathibatini and Lamprologini) might be even older than 5 million years (Nishida, 1991; Sturmbauer and Meyer, 1993; Kocher et al., 1993; Sturmbauer, Verheyen and Meyer, 1994.). The "short branches" at the base of the phylogenetic tree relating the major lineags of Lake Tanganyika cichlids suggests that the formation of lineages proceeded rapidly and that the tempo and mode of speciation and morphological diversification at the early stage of the Tanganyika radiation was dramatic (Sturmbauer, Verheyen and Meyer, 1994.).

The existence of several old lineages of cichlids might be evidence for the polyphyletic origin of the Lake Tanganyika species flock if it could be shown that more than one of these lineages is older than the lake itself or if basal members of more than one of those lineages are found outside the lake. Based on an earlier lower age estimate for Lake Tanganyika (2–4 million years) it had been assumed that an age of 5 million years for the old lineages implied a polyphyletic origin for this species flock (Nishida, 1991). A reevaluation of the geological age of Lake Tanganyika indicated that the age of the lake is likely to be greater than those of the tribes (Cohen et al., 1993) which might argue that virtually the whole Tanganyika flock could have evolved within the lake basin from a single ancestral lineage. This remains to be tested. Lake Tanganyika cichlids, probably unlike those of the other two species flocks, apparently have been able to leave the confines of the lake — several species of the *Lamprologus* group occur in the Zaire river. They appear be closely related to derived endemic lamprologine cichlids and are not basal lamprologine lineages (Sturmbauer, Verheyen and Meyer, 1994).

Both electrophoretic and mtDNA sequences suggest that the Victoria and Malawi flocks are closely related to a particular Tanganyikan tribe, the Tropheini (Nishida, 1991; Sturmbauer and Meyer, 1993; Sturmbauer, Verheyen and Meyer, 1994) (Fig. 6). This may

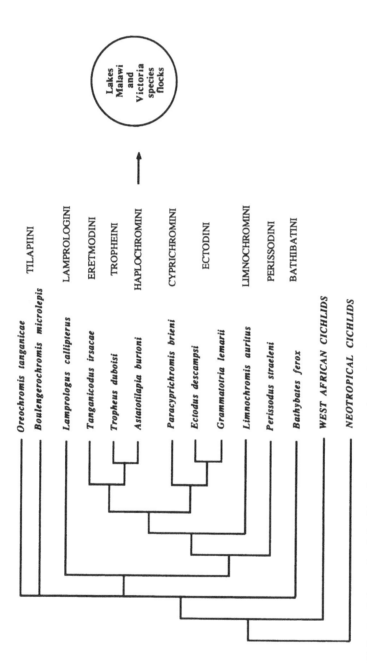

Figure 6. Molecular phylogenetic relationships of some representative species of ten of the eleven tribes (names of tribes in capitals following Poll (1986) endemic to Lake Tanganyika, figure based on (Sturmbauer and Meyer, 1993; Sturmbauer, Verheyen and Meyer, 1994). Some of the members of these tribes are shown in Fig. 2. e.g., *Lobochilotes* belongs to the same tribe as *Tropheus*, the Tropheini; *Spathodus* belongs to the Eretmodini; *Xenotilapia* to the Ectodini. The sister group relationship of the endemic Lake Tanganyika tribe Tropheini to the Lake Tanganyika haplochromine cichlids and the haplochromines of the two species flocks of Lakes Malawi and Victoria is indicated with the arrow and the bubble on the right and based on Nishida (1991) and Sturmbauer and Meyer (1993). Some of the relationships of Tanganyikan tribes remain somewhat tentative, they are indicated here as polytomies (and are being tested further, Sturmbauer and Meyer, in preparation).

not be surprising since considerable similarities between *Tropheus* and *Pseudotropheus* from Lakes Tanganyika and Malawi respectively, had been interpreted to argue for a polyphyletic origin of Lake Malawi cichlids (Fryer and Iles, 1972). However, the molecular phylogeny strongly suggests that these similarities are merely homoplasies due to convergent evolution since *Pseudotropheus* is genetically more closely related to all other species of Lake Malawi even including Malawian morphological generalists that have no resemblance to *Tropheus* from Lake Tanganyika (Kocher et al., 1993; Figs. 4–6).

Unlike the monophyletic species-flocks of the lakes Victoria and Malawi (Meyer et al., 1990, 1991), the cichlid fauna of Lake Tanganyika is believed to be of polyphyletic origin with affinities to the cichlid faunas of other African regions (Fryer and Iles, 1972; Poll, 1986; Nishida, 1991). The species of the tribe Lamprologini comprise species from Lake Tanganyika, as well as from the Zaïre river. Some "haplochromine" cichlids from Lake Tanganyika (i.e., *Tropheus* and *Astatotilapia burtoni*) appear to be the sister groups of the endemic species flocks of Lakes Malawi and Victoria (Meyer et al. 1990, 1991; Sturmbauer and Meyer 1993; Meyer 1993). The single endemic Tanganyikan species of the Tylochromini even has its closest relatives in central and western Africa (Stiassny, 1990). The Tanganyikan cichlid fauna hence can be viewed as an evolutionary reservoir of ancient African cichlid fishes (Nishida, 1991), the understanding of which might allow to eludicate the origin of the modern African cichlid fauna, as well as the interrelationships between the endemic flocks of the large Eastern African lakes. The Tanganyika flock has been viewed as a reservoir of old phylogenetic lineages that gave rise to the ancestors of the Victoria and Malawi flocks (Nishida, 1991). Lake Tanganyika does not harbor all the members and descendants of some of its endemic lineages since some lamprologine cichlids, which are endemic to the Zaire river were recently found not to be basal in the tribe Lamprologini and must hence be interpreted to have left the confines of the Lake Tanganyika (Sturmbauer, Verheyen and Meyer, 1994). More accurate calibrations of the age estimates for the lineages, and more work on riverine cichlids, particularly from West Africa, will be required to test the presumed polyphyly of the Lake Tanganyika species flock.

SPECIATION WITHIN THE LAKE BASINS (INTRA-LACUSTRINE SPECIATION)

The current Lake Victoria, with a an area of 68,000 km^2 about the size of Ireland, appears to have experienced a period of almost complete desiccation as recently as 14,000 years b.p. (Stager et al., 1986; Roche, 1991). Numerous ponds and rivers around the margins of the lake shore might have persisted and to have provided refugia for fish. Over evolutionary time spans there was probably ample opportunity for spatial isolation within the larger lake basin, providing the necessary preconditions for geographic speciation. Amalgamation of separate faunas of several smaller bodies of water after the lake level rose again is likely to have occurred (Worthington, 1937; Brooks, 1950; Fryer and Iles, 1972; Mayr, 1984; and more reference in Echelle and Kornfield, 1984). Geographic isolation of formerly interbreeding populations brought about by lake level fluctuations which split up larger bodies of water into smaller ones, followed by the acquisition of reproductive isolation before the geographically separated populations reunited, is a likely scenario for speciation in Lake Victoria. Hence, allopatric speciation (inter-lacustrine between separated bodies of water) might have been an important mechanism of speciation (Worthington, 1937; Brooks, 1950; Fryer and Iles, 1972). This possibility should not diminish the likely importance of microallopatric speciation for Lake Victoria cichlids.

Periods of aridity that led to drastic lake-level fluctuations (drops in water level of up to 600 m), and splits of the single lake are also documented for Lakes Tanganyika and a lesser extent Lake Malawi (Stager et al., 1986; Scholz and Rosendahl, 1988; Gasse et al., 1989; Roche, 1991; Tiercelin and Mondeguer, 1991). The lake topography consist of two (Lake Malawi) or three (Lake Tanganyika) extremely deep basins (up to 704 and 1470 m, respectively). These lake level changes will have separated populations that once exchanged genes and will have brought into contact populations that previously did not. In species of the *Tropheus* species complex from Lake Tanganyika, mtDNA sequences suggest that these large lake level fluctuations might have influenced the distribution of genetic variation and probably speciation (Sturmbauer and Meyer, 1992). Genetic distances and geographic patterning of genetic variation mirror the topology of the presumed paleo-lake shores during periods of low water levels (Scholz and Rosendahl, 1988; Gasse et al., 1989).

Intra-lacustrine speciation, probably by micro-allopatric speciation, through isolation by distance or appropriate habitat type, would appear to be the most important mode of speciation for all three species flocks. *Tropheus*, which only live over rocky habitats, from different sites in Lake Tanganyika appear to be effectively prevented from exchanging genes by discontinuities in the available habitat (Sturmbauer and Meyer, 1992). For example, long beaches or estuaries are evidently effective barriers to gene flow since large genetic differences were found between populations separated by only a few kilometers of shoreline (Sturmbauer and Meyer, 1992). Particularly for Lakes Tanganyika and Malawi, but probably also for Lake Victoria, it seems that speciation clearly can take place in single bodies of water. It would therefore appear that physical separation of water masses is not a necessary precondition for the establishment of genetic discontinuities and speciation.

Intra-lacustrine speciation should however, not be equated with sympatric speciation and should not be interpreted as refutation of allopatric speciation models (Kondrashov and Mina, 1986). It is often not appreciated that these lakes are vast and have extremely long varied coast lines, and that almost all endemic species have much restricted geographic distributions (Fryer and Iles, 1972). Only very few species, that live along the shores, occur throughout a whole lake (Fryer and Iles, 1972) and species often are restricted to single localities in which population size can be as small as a few hundred individuals (Ribbink et al., 1983). Most cichlids are poor dispersers, they are philopatric, they show homing, and males tend to defend feeding and breeding territories for several years (Hert, 1992; but see Turner, 1994). All of this points toward restricted gene flow. In addition, brood sizes are small and both factors are ingredients for fast diversification by micro-allopatric speciation (Cohen and Johnston, 1987). It seems important to point out, however, that many species of cichlids live in open water, roam freely and seem to be distributed widely or even occur in the whole lake (Coulter, 1991). In these species the models of restricted gene flow etc. do not seem to hold and the processes responsible for speciation might differ from the ones responsible for species restricted to small areas along the shore (George Coulter, pers. comm.). If the open water species return to restricted spawning areas on the shore for reproduction, some of the conditions potentially responsible for reproductive isolation among shore species might be met as well in these open water species.

RATES OF SPECIATION AND MORPHOLOGICAL DIVERSIFICATION, AND THE ROLE OF SEXUAL SELECTION

It is not clear how many species of the current flock of 300+ species of Lake Victoria survived the episode of drying 14,000 years ago. They may have survived in smaller marginal lakes, springs, or headwaters of rivers and recolonized the lake again after it filled up again. It may or may not appear likely (but not unthinkable, see below) that most of the 300+ species of Lake Victoria arose in less than 14,000 years. It seems possible also, that the Victoria flock is derived from East African riverine haplochromines that recolonized Lake Victoria after this period of aridity (Figs. 4 and 5).

Rates of speciation in cichlids can be astonishingly fast; this has been known since the discovery of five endemic species of cichlids in Lake Nabugabo (Greenwood, 1965), a small lake that is less than 4,000 years old and separated from Lake Victoria only by a sand bar. These five species are believed to have close relatives in Lake Victoria that chiefly differ in the male's breeding coloration, pointing to the potential importance of sexual selection for the fast rates of speciation in cichlids (Dominey, 1984).

Still faster rates of speciation were suggested by the finding that the southern end of Lake Malawi was dry only two centuries ago and is now inhabited by numerous endemic species and "color morphs" that are only found there and are believed to have originated during the last 200 hundred years! (Owen et al., 1990). This in situ speciation hypothesis seems supported by the fact that almost all endemic cichlids in all lakes have restricted geographic distributions (see also Turner, 1994). Ancestral genetic polymorphisms are retained across some species boundaries among some (but not all, Reinthal and Meyer, unpublished data) closely related species of mbuna consistent with the extreme rates of speciation observed in Lake Malawi cichlids (Moran and Kornfield, 1993; Moran, Reinthal and Kornfield, 1994).

Coloration appears to evolve quickly since there are several cases in which genetically closely related species of *Tropheus* and mbuna have dramatically different colorations (Ribbink et al., 1983; Sturmbauer and Meyer, 1992; Moran and Kornfield, 1993). Interestingly, despite pronounced variation in coloration among populations of *Tropheus*, this group of species has remained otherwise virtually unchanged for probably more than one million years (Sturmbauer and Meyer, 1992). Concurrently, the explosive speciation and morphological radiation of the Lake Victoria and Malawi flocks occurred, underscoring that morphological evolution can experience periods of rapid change and long periods of stasis (Avise, 1977).

The potential importance of coloration and sexual selection in the speciation of cichlid fish has been debated for some time (Dominey, 1984; Mayr, 1984; Turner, 1994). Sexually and socially selected traits might undergo more rapid diversification than traits under survival selection and might facilitate or fuel the explosive pace of speciation in cichlids. Coloration is of importance in intraspecific interactions and is a trait that might be more strongly shaped by sexual rather than by survival selection; coloration might act as a reproductive barrier without concordant morphological diversification. In *Tropheus*, coloration can vary tremendously among genetically closely related populations, alternatively it can also be very similar among genetically distant populations (or species) (Sturmbauer and Meyer, 1992). The importance of sexual selection in the formation of the species flocks is hotly debated — the verdict is still out.

ACKNOWLEDGMENTS

Andrew Cohen, George Coulter, Irv Kornfield, Ernst Mayr, Melanie Stiassny, and Paul Wilson commented on an earlier version of this paper. We thank Thomas Johnson, Dennis Tweddle and one anonymous reviewer for suggestions on this manuscript. Colleen Davis kindly provided technical assistance and Catherine Sexton drew the map and the fishes in Fig. 2. Lothar Seegers, Tuur Vos and Ole Seehausen collected riverine haplochromine cichlids in Burundi and Tanzania. This paper was prepared with partial support through grants from the U.S. National Science Foundation (BSR-910738) and a NATO collaboration grant with Erik Verheyen, Brussels (CRG-910911).

REFERENCES

Arnheim, N., White, T., and Rainey, W.E., 1990, Application of PCR: organismal and population biology: BioScience, v. 40, pp. 174–182.

Avise, J.C., 1977, Is evolution gradual or rectangular? Evidence from living fishes: Proceedings of the National Academy of Sciences U.S.A., v. 74, pp. 5083–5087.

Avise, J.C., and Ball, R.M., Jr., 1990, Principles of genealogical concordance in species concept and biological taxonomy, in Futuyma, D., and Antonovics, J., eds., Oxford Surveys in Evolutionary Biology Vol. 7: Oxford, University of Oxford Press, pp. 45–67.

Brooks, J.L, 1950, Speciation in ancient lakes: Quartl. Rev. Biol., v. 25, pp. 30–60.

Brown, W.M., George, M. Jr., and Wilson, A.C., 1979, Rapid evolution of mitochondrial DNA: Proceedings of the National Academy of Sciences U.S.A., v. 76, pp. 1967–1971.

Cohen, A., and Johnston, M.R., 1987, Speciation in brooding and poorly dispersing lacustrine organisms: Palaios, v. 2, pp. 426–435.

Cohen, A., Soreghan, M.J., and Scholz, C.A., 1993, Estimating the age of ancient lakes: an example from lake Tanganyika, East African rift system: Geology, v. 21, pp. 511–514.

Coulter, G.W. , ed., 1991, Lake Tanganyika and its life: Oxford, Oxford University Press.

DeQueiroz A., and Wimberger, P., 1993, The usefulness of behavior for phylogeny estimation: levels of homoplasy in behavioral and morphological characters: Evolution, v. 47, pp. 46–60.

Dominey, W.J., 1984, Effects of sexual selection and life history on speciation: species flocks in African cichlids and Hawaiian Drosophila, in Echelle, A.A., and Kornfield, I., eds., Evolution of fish species flocks: Orono, University of Maine at Orono Press, pp. 231–254.

Echelle, A.A., and Kornfield, I., eds., 1984, Evolution of fish species flocks: Orono, University of Maine at Orono Press.

Eccles, D.H., and Trewavas, E., 1989, Malawian cichlid fishes, The classification of some haplochromine genera: Herten, Lake Fish Movies.

Erlich, H. A., ed., 1989, PCR technology: principle and applications for DNA amplification: New York, Stockton Press.

Felsenstein, J., 1988, Phylogenies from molecular sequences: inference and reliability: Annual Reviews in Genetics, v. 22, pp. 521–565.

Fryer, G., and Iles, T.D., 1972, The cichlid fishes of the great lakes of Africa: their biology and evolution: Edingburgh, Oliver and Boyd.

Gasse, F., Ledee, V., Massault, M., and Fontes, J.C., 1989, Water-level fluctuations of Lake Tanganyika in phase with oceanic changes during the last deglaciation: Nature, v. 342, pp. 57–59.

Greenwood, P.H., 1965, The cichlid fishes of Lake Nabugabo, Uganda: Bulletin of the British Museum of Natural History (Zoology), v. 12, pp. 315–357.

Greenwood, P.H., 1980, Towards a phyletic classification of the "genus" Haplochromis (Pisces, Cichlidae) and related taxa. Part II; The species from Lakes Victoria, Nabugabo, Edward, George and Kivu: Bulletin of the British Museum of Natural History (Zoology), v. 39, pp. 1–101.

Greenwood, P.H., 1983, On Macropleurodus, Chilotilapia (Teleostei, Cichlidae) and the interrelationships of African cichlid species flocks: Bulletin of the British Museum of Natural History (Zoology), v. 45, pp. 209–231.

Greenwood, P.H., 1984, What is a species flock? in Echelle, A.A., and Kornfield, I., eds., Evolution of fish species flocks: Orono, University of Maine at Orono Press, pp. 13–19.

Gyllensten, U.B., and Erlich, H.A., 1988, Generation of single-stranded DNA by the polymerase chain reaction and its application to direct sequencing of the HLA–DQA locus: Proceedings of the National Academy of Sciences U.S.A., v. 85, pp. 7652–7655.

Hert, E., 1992, Homing and home-site fidelity in rock-dwelling cichlids (Pisces: Teleostei) of Lake Malawi, Africa: Environmental Biology of Fishes, v. 33, pp. 229–237.

Hillis, D.M., 1987, Molecular versus morphological approaches to systematics: Annual Reviews of Ecology and Systematics, v. 18, pp. 23–42.

Innis, M.A, Gelfand, D.H., Sninsky, J.J., and White, T.J., eds., 1990, PCR protocols: a guide to methods and applications: San Diego, Academic Press.

Johnson, T.C., 1996, Sedimentary processes and signals of past climatic change in the large lakes of the East African Rift valley, in Johnson, T.C., and Odada, E.O., eds., The limnology, climatology and paleoclimatology of the East African lakes: Toronto, Gordon and Breach, pp. 367–412.

Keenleyside, M.H.A., ed., 1991, Cichlid fishes: behavior, ecology and evolution: London, Chapman and Hall.

Klein, D., Ono, H., O'hUigin, C., Vincek, V., Goldschmidt, T., and Klein, J., 1993, Extensive *Mhc* variability in cichlid fishes of Lake Malawi: Nature, v. 364, pp. 330–334.

Kluge, A.G., 1989, A concern for evidence and a phylogenetic hypothesis of relationships among *Epicrates* (Boidae, Serpentes): Systematic Zoology, v. 38, pp. 7–25.

Kocher, T.D., Conroy, J.A., McKaye, K.R., and Stauffer, J.R., 1993, Similar morphologies of cichlid fish in Lakes Tanganyika and Malawi are due to convergence: Molecular Phylogenetics and Evolution, v. 2, pp. 158–165.

Kondrashov, A.S., and Mina, M.V., 1986, Sympatric speciation: when is it possible?: Biological Journal of the Linnean Society, v. 27 pp. 201–233.

Kornfield, I., 1978, Evidence for rapid speciation in African cichlid fishes: Experientia, v. 34, pp. 335–336.

Kornfield, I., 1991, Genetics, in Keenleyside, M.H.A., ed., Cichlid fishes: behavior, ecology and evolution: London, Chapman and Hall, pp. 103–128.

Kornfield, I., McKaye, K.R., and Kocher, T., 1985, Evidence for the immigration hypothesis in the endemic cichlid fauna of Lake Tanganyika: Iisozyme Bulletin, v. 18, p. 76.

Kwok, S., Kellogg, D.E., McKinney, N., Spasic, D., Goda, L., Levenson, C., and Sninsky, L., 1990, Effects of primer-template mismatches on the polymerase chain reaction: human immunodeficieny virus type 1 model studies: Nucleic Acids Research, v. 18, pp. 999–1005.

Liem, K.F., 1973, Evolutionary strategies and morphological innovations: cichlid pharyngeal jaws: Systematic Zoology, v. 22, pp. 425–441.

Liem, K.F., and Osse, J.W.M., 1975, Biological versatility, evolution and food resource exploitation in African cichlid fishes: American Zoologist, v. 15, pp. 427–454.

Maddison, W.P., and Maddison, D.R., 1992, MacClade: analysis of phylogeny and character evolution. Version 3.0: Sunderland, Sinauer.

Mayr, E., 1942, Systematics and the origin of species: New York, Columbia University Press.

Mayr, E., 1984, Evolution of fish species flocks: a commentary, in Echelle, A.A., and Kornfield, I., eds., Evolution of fish species flocks: Orono, University of Maine at Orono Press, pp. 3–11

Meyer, A., 1993a, Evolution of mitochondrial DNA in fishes, in Hochachka, P.W., and Mommsen, T.P., eds., The biochemistry and molecular biology of fishes Vol. 2: London, Elsevier Press, pp. 1–38.

Meyer, A., 1993b, Phylogenetic relationships and evolutionary processes in East African cichlid fishes: Trends in Ecology and Evolution, v. 8, pp. 279–284.

Meyer, A., 1994, Molecular phylogenetic studies of fish, in Beaumont, A.R. ed., Genetics and evolution of aquatic organisms: London, Chapman and Hall (in press).

Meyer, A., Kocher, T.D., Basasibwaki, P., and Wilson, A.C., 1990, Monophyletic origin of Lake Victoria cichlid fishes suggested by mitochondrial DNA sequences: Nature, v. 347, pp. 550–553.

Meyer, A., Kocher, T.D., and Wilson, A.C., 1991, African fishes: Nature, v. 350, pp. 467–468.

Moran, P., and Kornfield, I., 1993, Retention of an ancestral polymorphism in the mbuna species flock (Pisces: Cichlidae) of Lake Malawi: Molecular Biology and Evolution, v. 10, pp. 1015–1029.

Moran, P., Kornfield, I., and Reinthal, P., 1994, Molecular systematics and radiation of the haplo-chromine cichlids (Teleostei: Perciformes) of Lake Malawi: Copeia, v. 1994 (in press).

Nishida, M., 1991, Lake Tanganyika as evolutionary reservoir of old lineages of East African cichlid fishes: inferences from allozyme data: Experientia, v. 47, pp. 974–979.

Ono, H., O'hUigin, H. Tichy, and Klein, J., 1993, Major-histocompatibility complex variation in two species of cichlid fishes from Lake Malawi: Molecular Biology and Evolution, v. 10, pp. 1060–1072.

Owen, R.B., Crossley, R., Johnson, T.C., Tweddle, D., Kornfield, I., Davidson, S., Eccles, D.H., and Engstrom, D.E., 1990, Major low levels of Lake Malawi and their implications for speciation rates in cichlid fishes: Proceeding of the Royal Society Series B, v. 240, pp. 519–553.

Patterson, C., ed., 1987, Molecules and morphology in evolution: conflict or compromise?: Cambridge University Press.

Poll, M., 1986, Classification des cichlidae du lac Tanganyika: tribus, genres es aspeces. F ac. Royale De belgique Memories se la classe des sciences: Collection in 8°-2ᵉ serie. T.XIV, fasc., v. 2, pp. 1–163.

Ribbink, R.J., Marsh, B.A., Marsh, A.C., Ribbink, A.C., and Sharp, B.J., 1983, A preliminary survey of the cichlid fishes of rocky habitats in Lake Malawi: South African Journal of Zoology, v. 18., pp. 149–310.

Roche, E., 1991, Evolution des paleonenvironnements en afrique centrale et orientale au pleistocene superieur et a l'holocene, influences climatiques et anthropiques: Bulletin Societe Geographique Liege, v. 27, pp. 187–208.

Sage, R.D., Loiselle, P.V., Basasibwaki, P., and Wilson, A.C., 1984, Molecular versus morphological change among cichlid fishes of Lake Victoria, in Echelle, A.A., and Kornfield, I., eds., Evolution of fish species flocks: Orono, University of Maine at Orono Press, pp. 185–197.

Saiki, R.K., Scharf, S., Faloona, F., Mullis, K.B., Horn, G.T., Erlich, H.A., and Arnheim, N., 1985, Enzymatic amplification of beta-globin genomic sequences and restriction site analysis for diagnosis of sickle cell anemia: Science, v. 230, pp. 1350–1354.

Saiki, R.K., Gelfand, R.H., Stoffel, S., Scharf, S., Higuchi, R., Horn, G.T., Mullis, K.B., and Erlich, H.A., 1988, Primer-directed enzymatic amplification of DNA with a thermostable DNA polymerase: Science, v. 239, pp. 487–491.

Sanderson, M.J., and Donoghue, M.J., 1989, Patterns of variation and levels of homoplasy.: Evolution, v. 43, pp. 1781–1795.

Schluter, D., and McPhail, J.D., 1993, Character displacement and replicate adaptive radiation: Trends in Ecology and Evolution, v. 8, pp. 197–200.

Scholz, C.A., and Rosendahl, B.R., 1988, Low lake stands in Lakes Malawi and Tanganyika, East Africa, delineated with multifold seismic data: Science, v. 240, pp. 1645–1648.

Stager, J.C., Reinthal, P.N., and Livingstone, D.A., 1986, A 25000-year history for Lake Victoria, East Africa, and some comments on its significance for the evolution of cichlid fishes: Freshwater Biology, v. 16, pp. 15–19.

Stiassny, M.L.J., 1981, Phylogenetic versus convergent relationship between piscivorous cichlid fishes from Lake Malawi and Tanganyika: Bulletin of the British Museum of Natural History (Zoology), v. 40, pp. 67–101.

Sturmbauer, C., and Meyer, A., 1992, Genetic divergence, speciation and morphological stasis in a lineage of African cichlid fishes: Nature, v. 358, pp. 578–581.

Sturmbauer, C., and Meyer, A., 1993, Mitochondrial phylogeny of the endemic mouthbrooding lineages of cichlid fishes from Lake Tanganyika, East Africa: Molecular Biology and Evolution, v. 10, pp. 751–768.

Sturmbauer, C., Verheyen, E., and Meyer, A., 1994, Mitochondrial phylogeny of the Lamprologini, the major substrate spawning lineage of cichlid fishes from Lake Tanganyika, Eastern Africa: Molecular Biology and Evolution, v. 11 (in press).

Swofford, DL., 1991, When are phylogeny estimates from molecular and morphological data incongruent? in Miyamoto, M.M., and Cracraft, J., eds., Phylogenetic analysis of DNA sequences: Oxford, Oxford University Press, pp. 295–333.

Swofford, D.L., and Olsen, G.J., 1990, Phylogeny reconstruction, in Hillis, D.M., and Moritz, C., eds., Molecular systematics: Sunderland, Sinauer, pp. 411–501.

Tiercelin, J.-J., and Mondeguer, A., 1991, The geology of the Tanganyika through. in Coulter, G.W., ed., Lake Tanganyika and its life, Oxford, Oxford University Press, pp. 7–48,

Trewavas, E., 1949, The origin and evolution of the cichlid fishes of the Great African Lakes, with special reference to Lake Nyasa: 13th international Congress in Zoology, v. 1948, pp. 365–368.

Turner, G.F., 1994, Speciation mechanism in Lake Malawi cichlids: a critical review, in Martens, K., Goodeeris, B., and Coulter, G.W., eds. Speciation in ancient lakes: (in press).

VanCouvering, J.A.H., 1982, Fossil cichlid fish of Africa: Special Papers in Palaenotology, v. 29, pp. 1–103.

Vigilant, L., Pennington, R., Harpeding, H., Kocher, T.D., and Wilson, A.C., 1989, Mitochondrial DNA sequences in single hairs from a southern African population: Proceedings of the National Academy of Sciences U.S.A. v. 86, pp. 9350–9354.

White, T.J., Arnheim, N., and Erlich, H.A., 1989, The polymerase chain reaction: Trends in Genetics, v. 5, pp. 185–189.

Witte, F., Goldschmidt, T., Wanink, J., Von Oijen, M., Goudswaard, K., Witte-Maas, E., and Bouton, N., 1992, The destruction of an endemic species flock: quantitative data on the decline of the haplochromine cichlids of Lake Victoria: Environmental Biology of Fishes, v. 34, pp. 1–28.

Worthington, E.B., 1934, On the evolution of fish in the great lakes of Africa: Int. Rev. Ges. Hydrobiol. Hydrogr., v. 35, pp. 304–317.

Wrishnik, L.A., Higuchi, R.G., Stoneking, M., and Wilson, A.C., 1987, Length mutations in human mitochondrial DNA: direct sequencing of enzymatically amplified DNA: Nucleic Acids Research, v. 15, pp. 529–542.

Anthropogenic Impact on Fisheries Resources of Lake Naivasha

P.A. ALOO   *Department of Zoology, Kenyatta University, Nairobi, Kenya*

Abstract — Since the 1920s the Lake Naivasha community has seen the introduction of 8 or 9 new species of fish. These species established themselves and supported a very productive fishery during the 1950s and 1960s. However, in recent years reports have revealed that the species composition and yield in the lake have changed considerably (Okorie, 1972; Malvestuto, 1974; Siddiqui, 1979).

Lake Naivasha has been fished commercially for the past 40 years primarily for two species of tilapia, *Oreochromis leucostictus* (Trewavas) and *Tilapia zillii* (Gervais), and for large mouth bass, *Micropterus salmoides* (Lacepede). The red swamp Louisiana crayfish, *Procambarus clarkii*, is also commercially harvested. Also present is the cyprinid, *Barbus amphigrama*, which is of riverine origin but has extended its non-breeding phase to the lake. *Gambusia* sp., *Poecilia* sp. and *Lebistes reticulata*, were introduced in the swamp areas to control mosquitoes but have not been recorded since 1977. The only endemic species, the small tooth carp, *Aplocheilichthyes antinorii*, was last recorded in 1962 (Siddiqui, 1979).

INTRODUCTION

Lake Naivasha is a freshwater lake situated 100 km north of Nairobi, Kenya, in the eastern arm of the East African Rift Valley. The lake is a small closed basin with a surface area of 150 km^2 and a mean depth of 6 m. The basin is roughly circular and is made up of four waterbodies: the main lake, Crescent Lake, Oloidien Bay and Sonachi Crater Lake (Fig. 1). The basin is unique in being the only freshwater body in the eastern arm of the Rift Valley in Kenya (300 μs cm^{-1}); all other lakes in the Rift Valley, except Lake Baringo, are saline to some extent. The mechanisms which maintain the lake's low salinity include dilute inflows, biochemical and geochemical sedimentation and seepage loss (Gaudet and Melack, 1981).

The Naivasha basin receives water from the highest part of the Rift Valley floor and the flanking escarpments. Three major rivers, Malewa, Gilgil and Karati, enter the lake to the north (Fig. 1). The Malewa River (1730 km^2 watershed) originates from the Nyandarua Range and Kinangop Plateau and contributes about 90% of the discharge into Lake Naivasha. Most of the remainder is provided by the Gilgil River which drains the Bahati Highlands.

In addition to river inflow, the lake receives water through underground seepage. Thompson and Dodson (1963) described Lake Naivasha as a "hydrographic window" because water passes freely through the extremely porous volcanic rocks which form 80% of the lake basin. Water input by seepage has been shown to occur in the northeastern and northwestern sections, while water loss by seepage occurs to the south and southeastern region of the lake (Gaudet and Melack, 1981).

The Naivasha basin has a very rich and diversified plant and animal life (Litterick et al., 1979). Emergent macrophytes are mainly dominated by sedges of which *Cyperus papyrus* (L.) is the most important. Papyrus forms a fringe zone separating the lake from surrounding

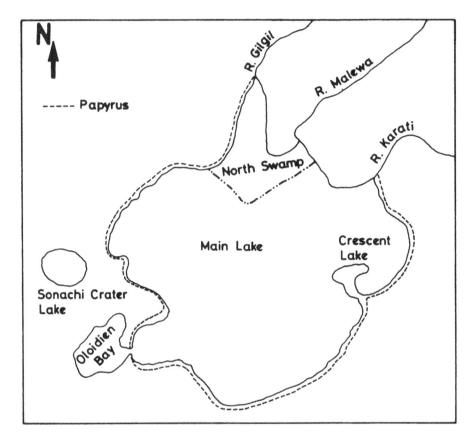

Figure 1. Map of lake basin showing the three water bodies.

farmland. In the open water, the floating macrophyte is dominated by the fern, *Salvinia molesta*, and the water hyacinth, *Eichhornia crassipes*.

Lake Naivasha is well known as a bird sanctuary and is routinely included on bird watchers' itineraries. About 350 species of birds have been recorded either as a resident or as migrant species from the northern temperate regions. At least ten of these species are piscivorous of which the most important are the African fish eagle, the white pelican, the pink backed pelican, cormorants and the king fishers. Herons, coots and ducks are also present.

Lake Naivasha attracts other wildlife to its shores, including herds of zebra, kongoni, gazelle, impala, and dikdik. Monkeys are plentiful, as are hippoppotami, which are seen all over the lake and augment the tourist attraction.

The lake has always been of economic value to the nearby human inhabitants. Agriculture, which relies heavily on irrigation, has replaced the more traditional use of the lake as a watering lake for cattle (Litterick et al., 1979). The lake basin has become an important area for vegetable production and supports a large vegetable drying factory. Cut flowers are grown on large plantations along the southern shore for export, and lucerne (alfalfa) is grown under sprinkler irrigation to support an important dairy industry. Tropical climate, good

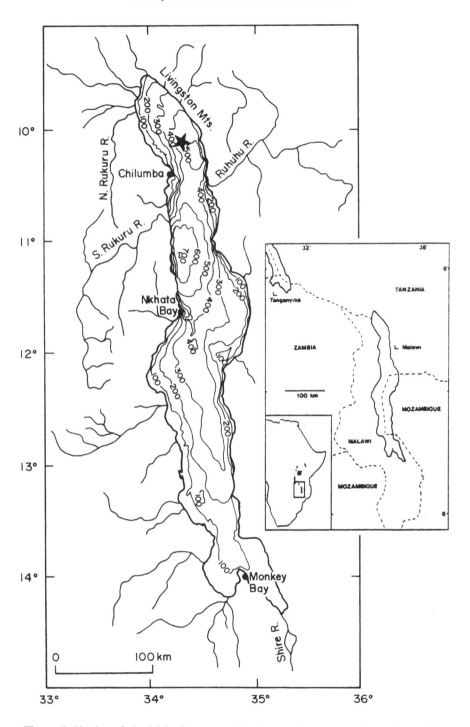

Figure 1. Northern Lake Malawi; contoured bathymetry in meters and position of the sediment trap mooring (*; 10°14.4'S; 34°21.9'E).

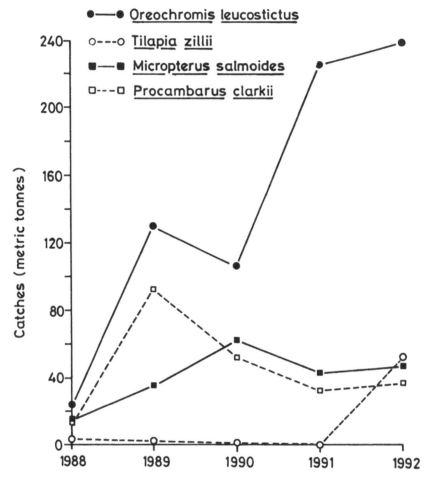

Figure 2. Lake Naivasha fish yields between 1988 and 1992 (Government of Kenya, Fisheries Department, annual statistics, 1992).

introduction was extremely successful, and by 1975 the species was being commercially exploited.

Several other species of fish have been introduced, but records of dates and numbers are not available. The cyprinodonts, *Gambusia* sp. and *Poecilia* sp., were introduced in the swamp areas for mosquito control, but have not been recorded since 1977. Another cyprinodont, *Lebistes reticulata*, which was assumed to have disappeared, is now present in the lake (Aloo, unpublished). Rainbow trout, *Salmo gairdneri* (Richardson), are caught occasionally, probably as strays from the affluent rivers into which they were introduced as game fish (Elder et al., 1971). The cyprinid, *Barbus amphigrama* (Boulenger), has extended its non-breeding phase into the lake from the Malewa River. This fish has been recorded mainly at the North Swamp where the Malewa River enters the lake and occasionally at Crescent Lake and in the Oloidien Bay, where it is now caught in large numbers (Aloo, unpublished).

THE LAKE NAIVASHA FISHERY

Presently, four species of fish are exploited commercially in the lake: *Oreochromis leucostictus*, *T. zillii*, *M. Salmoides* (large mouth bass) and *P. clarkii* (Louisiana crayfish) (Fig. 2). All three species of fish are caught in gill-nets, but only the bass contribute to the sports fishery.

The canoe-based commercial gill net fishery began by exploiting mainly *O. niger* and a small number of *T. zillii* (Garrod and Elder, 1960). Following the development of the fishery, *O. niger* was gradually replaced by hybrids of *O. niger* and *O. leucostictus*. These hybrids accounted for 57% of the total catch in experimental gill nets in 1962 (Elder et al., 1971). Lake levels rose considerably during 1962–64 leading to the formation of lagoons that were favourable habitats for *O. leucostictus* and *T. zillii*. By 1971, all *O. niger* and most of the hybrids had disappeared due to the loss of their favourable weed-free breeding and nursery grounds, giving way for the expansion of *O. leucostictus* populations (Siddiqui, 1977).

The fishery began with 13 cm and 14 cm stretched mesh gill nets which were in general use by 1961. By 1968 gill nets of small mesh size (10 cm) were introduced illegally by the local fishermen (Mann and Ssentongo, 1969). The 10 cm mesh nets were legalised in 1970 and the number of fishermen increased, leading to a considerable rise in fish catches and a highest recorded yield of 1,150 metric tons. Over-fishing led to a sharp decline in catches in the following years to which the fishermen responded by using even smaller meshes culminating in the near collapse of the fishery.

According to Okorie (1972) and Malvestuto (1974), the 10 cm mesh size gill net caught immature fish and/or first time breeders (18 cm total length for *O. leucostictus*) preventing population replenishment. Malvestuto (1974) suggested that the decline in fish yields during this time was also attributable to a decline in lake level, while Siddiqui (1977) attributed the decline to overfishing and Salvinia infestation (Table 2).

FISH SPECIES COMPOSITION

Today there are six species present in Lake Naivasha, five of which were introduced — *O. leucostictus*, *T. zillii*, *M. salmoides*, *L. reticulata* and *P. Clarkii* — and the riverine *B. amphigrama*. Data obtained from the commercial fishery from the year 1988 indicate that *O. leucostictus* is the most abundant species of fish in Lake Naivasha. The *O. leucostictus* catch has been rising steadily since 1988 with a brief decline in 1990. This abundance of *O. leucostictus* is due to its mouth brooding behaviour since it is not affected by the fluctuating water levels. *M. salmoides* have also increased, except for 1991 which recorded its lowest yield. This fish, having been introduced from a temperate climate into the tropical Lake Naivasha, has therefore taken time to stabilise in its new environment. This decline was also partly caused by fishermen who were illegally practicing beach seining, a method unsuitable for the bass, instead of using the gill net. Since 1988, *T. zillii* has contributed to an insignificant proportion of the commercial fishery although there was a sudden rise in 1992. According to Muchiri (1992), *T. zillii* distribution is highly related to aquatic macrophytes, since this vegetation cover fluctuates with changes in water level. *T. zillii* population has remained low over the years due to periodic loss of the macrophytes. Flooding also destroys their nests and this interferes with their breeding pattern. The crayfish *P. clarkii* has fluctuated over the five years with a maximum yield of 93.8 metric tons in 1989 (Fig. 2).

Table 2. Summary of Some Biologically Significant Factors Relating to the Commercially Exploited Fish in Lake Naivasha (from Malvestuto, 1974; Siddiqui, 1979; Aloo, 1989; Dadzie and Aloo, 1990; Muchiri, 1990; Aloo, unpublished)

Fish species	Habitat	Feeding	Maturity	Growth rate (mm/month)	Abundance
Oreochromis leucostictus	Mainly in papyrus lagoon zones and shallow macrophyte, very few in deep water.	Feeds mainly on detritus, planktonic algae, chrinomid larvae, oligochaete worms and insects such as Micronecta.	Reach maturity at about 1700 mm fork length. Breeding maxima in Sep–Nov and minima Jun–Aug.	5–7	Most abundant
Tilapia zillii	Large numbers in littoral zones, some in papyrus lagoons and a few in deep waters.	Detritus is the most important food as in *O. leucostictus*. Macrophytes and insects are also consumed. Best described as omnivorous browses.	Mature at 1300 mm fork length. Breed mainly in Sep–Feb and minima in Jul–Aug.	2.5	Not many
Micropterus salmoides	Predominant in shallow rocky shores with reasonable numbers of juveniles in open deep waters.	Feeding habits change with size; up to about 260 mm fork length, there is an almost total dependence on invertebrates, mainly *Micronecta scutellaris* (Stal.). Adults feed on the crayfish *Procambarus clarkii*, followed by fish fingerlings. Frogs are occasionally taken in.	Length at maturity is 250 mm. Breeding maxima in Aug–Nov and minima in Mar–May.	7	Many

All the three species have an essentially littoral distribution with particular preferences for papyrus fringes and lagoon areas; the conservation of these areas is therefore of paramount importance in fisheries management. There is little or no feeding overlap between these species, so that competition for food is minimal. *Micropterus salmoides* adults, though usually piscivorous (Robbins and Maccrimond, 1974) consumes crayfish in Lake Naivasha along with its own fry (Aloo, 1989). Bass predation on tilapia fry since the crayfish became abundant in the early 1970s is slight. All the three species breed continuously throughout the year with maximum intensity — September–November for the tilapias and August–November for the bass (Aloo, 1989; Dadzie and Aloo, 1990).

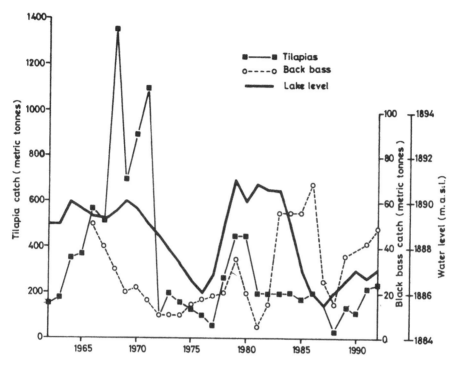

Figure 3. Lake Naivasha fish catches (1962–92) fluctuating with changes in the water level ((Litterick et al., 1979; Government of Kenya, Fisheries Department, annual statistics; Muchiri 1990).

LAKE LEVEL FLUCTUATION AND FISH POPULATION

Lake Naivasha is a highly unstable waterbody whose levels fluctuate rapidly (see Verschuren, this volume). The fluctuations that are seen today are a reflection of wide scale fluctuations that were witnessed in the past (Richardson and Richardson, 1972). Fish catches have responded positively to the hydrological forcing; rising lake levels are followed by increased catches and vice versa (Fig. 3).

Lake-level fluctuations influence fish numbers through effects on food, breeding grounds and predator-prey relationships. Food has been shown not to be a limiting factor for the fishes in Lake Naivasha (Aloo, 1989; Muchiri, 1990; Hickley et al., 1994). A more probable effect is that on the breeding behaviour. The tilapiines, especially *Tilapia zillii*, are very sensitive to fluctuating water levels since they breed in shallow waters. Lowe-McConnell (1987) listed factors such as habitat drying and flooding as controllers of tilapia numbers in fish communities. Fryer and Iles (1972) described how predation and fishing pressure on the cichlids of Lake Victoria were minimised by flooding of marginal terrestrial vegetation. It appears likely that fluctuating water levels of Lake Naivasha have similar effects leading to varying levels of predation and fishing pressure.

It is unlikely that lake levels will stabilise in the near future, and human interference is bound to add to the already existing problem. Lakeside farms are irrigated by water from the lake. The amount of water extracted for irrigation has long exceeded the amount permitted by the Ministry of Water Development (Muchiri, 1990). This and water loss through evaporation are higher than water input to the lake, so the decline in Lake Naivasha levels will continue at the expense of the fisheries resources of the lake (Aloo, unpublished). Interviews with fishermen who have operated on the lake for the past 20–30 years revealed a commonly held opinion that the lake is dying and no amount of rain can bring it back to its original condition. The present extraction of water from the rivers and the proposed damming of the River Malewa assures the steady decline of Lake Naivasha water level. Also the clearing of the vast areas of the river catchment and the papyrus of the North Swamp will lead to rapid siltation of the lake.

DISCUSSION AND CONCLUSIONS

As previously described, Lake Naivasha is unique among the East African Rift Valley lakes in its salinity which in turn creates a large and sustaining habitat for a wide assortment of terrestrial and lacustrine fauna. The lake is equally unique for its extremely low endemic fish count, primarily thought to be the result of fluctuating lake levels. Attempts began during the 1920s to introduce 8 or 9 new species of fish, with varying degrees of success. Currently five introduced and one riverine species survive and each depends on a relatively unstable lake level.

The Lake Naivasha Fishery, established during the 1950s, struggles to meet the ever increasing demands of the Naivasha human population, while trying to maintain a balanced ecosystem. A significant percentage of the Naivasha community relies on the fishery for its daily subsistence. In 1992, for example, there were 72 licensed fishermen operating 45 boats, each boat having an average of 3 crew members. In addition, there are transporters who carry fish from the main landing beach to the selling beach, watchmen who are hired to guard the boats when not in use, and fish traders who operate both in and outside of the town. Besides all these, there are poachers who also depend on the lake for their livelihood. Several enterprises rely indirectly on the fishery, including the Elsamere Conservation Camp and Fishermen's Camp who operate sports fishing facilities, as well as lakeside hotels which cater to tourists and offer primarily fish on their menus. One of Lake Naivasha's attractions is the diversity of birds that can be found there. Since many of the birds are piscivorous, a decline in fish population could also affect the bird community.

As the Fisheries Department seeks ways to maintain itself as a resource, overfishing threatens to undermine its efforts. Malvestuto (1974) suggested that improvement to the Lake Naivasha fisheries be obtained by cautious introductions of new fish species, specifically those preferring the somewhat more stable open water habitats. Several facts tempt one to support additional suggestions by Siddiqui (1977) and the Fisheries Department for such introductions. As mentioned, the Lake Naivasha Fishery is currently based on fish adversely affected by changing lake levels. The fish population of Lake Naivasha at present does not fully exploit available food resources or habitats: The mud feeders category which represent fish that feed on silt, decaying matter and associated microorganisms is filled up by the tilapia *T. zillii* and *O. leucostictus*; the niche of micro-herbivore taking free algae, diatoms and zooplanktons is partly filled up by *O. leucostictus*; large plant material is partly eaten by *T. zillii*, but *P. clarkii* is a much more dedicated macro herbivore; *B. amphigrama* and *L. reticulata* fall in the category of omnivores that feed on insects, insect larvae and zooplanktons; the large mouth bass has a mixed diet of macro invertebrates, fish fry and frogs thus fills the position of a generalized macropredator (Muchiri and Hickley, 1991). Therefore the benthic fauna, comprised of oligochaete worms and chironomid larvae, flourishes due to the

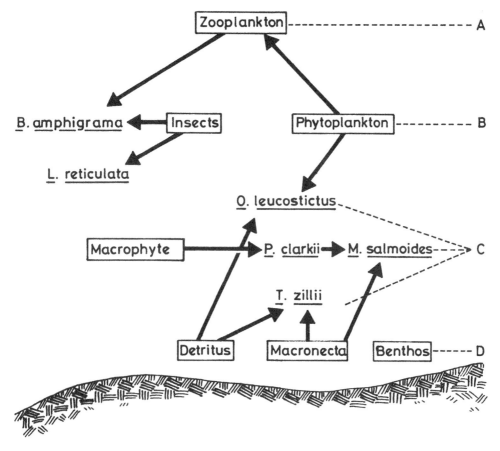

Figure 4. Lake Naivasha food web showing vacant niches (Muchiri et al., 1992).

absence of bottom feeding fishes; phytoplankton, zooplankton, and small fish are also abundant reflecting a lack of specialist feeders (Siddiqui, 1977; Fisheries Dept.) (Fig. 4). Littorial and shallow water zones are the preferred habitats, leaving the open water areas largely unoccupied.

Further introductions to exploit vacant niches could potentially diversify the ecosystem, provide continued essential protein for the local people, and ensure socioeconomic stability. Possible species under consideration are: *Limnothrissa miodon*, *Stolothrissa tanganyikae*, or *Alestes* sp., which occupy open water and feed on zooplankton, and *Mormyrus* spp. or *Haplochromis angustifrons*, which feed on the benthos (Muchiri and Hickley, 1991). The history of various introductions, however, shows repeated instances of inappropriate selections with often disastrous and irreversible results (Fryer and Iles, 1972), hence careful study of feeding requirements and breeding behaviour of potential introductive fish are crucial. Muchiri and Hickley (1991) and Muchiri et al. (1992) strongly suggest such proposals comply with strict recommendations of the European Inland Fisheries Advisory Commission (EIFAC 1988).

Additional recommendations include strict enforcement by the Fisheries Department of current policies concerning net size, number of nets per boat and number of licensed fishermen. Attempts should also be made to educate present and future fishermen on the effects their fishing methods have on the fish population. The Fisheries Department should also take tougher measures on poachers who use undersize nets and fish on prohibited breeding grounds.

ACKNOWLEDGMENTS

Supported by the German Academic Exchange Center (DAAD). I am grateful to the Government of Kenya Fisheries Department at Naivasha for providing fish yield data. To Mr. Ayugu, Kahareri, Kundu and Okoth Owino, I am very grateful. This paper has been developed from the author's ongoing Ph.D. research.

REFERENCES

Aloo, P., 1989, Studies of the ecology of the black bass, *Micropterus salmoides* (Lacepede) (Pisces: Cenrachidae) in Lake Naivasha [M.Sc. thesis]: Nairobi, Kenya, University of Nairobi.

Aloo, P.A., Parasites of fishes of Lake Naivasha, Kenya (ongoing Ph.D.)

Dadzie, S., and Aloo, P., 1990, Reproduction of the North American black bass, *Micropterus salmoides* (Lacepede), in an equatorial lake, Lake Naivasha, Kenya: Aquaculture and Fisheries Management, v. 21, pp. 49–458.

Elder, H.Y., Garrod, D.J., and Whitehead, I., 1971, A natural hybrid of *T. nigra* and *T. leucosticta* from Lake Naivasha, Kenya. Biol. J. Linn., Soc. 3, pp. 103–106.

Fryer, G., and Iles, T.D., 1972, The cichlid fishes of the Great Lakes of Africa: their biology and evolution: Edinburgh, Oliver and Boyd, 641 pp.

Garrod, D.J., and Elder, H.Y., 1960, The fishery of Lake Naivasha. Rep. East Africa Freshwater Research Organization, pp. 24–25.

Gaudet, J.J., and Melack, J.M., 1981, Major ion chemistry in a tropical African Lake basin: Freshwater Biology, v. 11, pp. 309–333.

Litterick, M.R., Kalff, J.J., and Melack, J.M., 1979, The limnology of an African lake, Lake Naivasha, Kenya Workshop on African limnology: SIL-UNEP, 73 pp.

Lowe-McConnell, R.H., 1987, Ecological Studies in tropical fish communities. Cambridge tropical biology series, 382 pp.

Lowery, R.S., and Mendes, A.J., 1977, *Procambanus clarkii* in Lake Naivasha, Kenya, and its effects on the established and potential fishery: Aquaculture, v. 11, pp. 111–121.

Malvestuto, S., 1974, The fishes of Lake Naivasha, their biology, exploitation and management: Report to Fisheries Department, Government of Kenya, 37 pp.

Mann, M.J., and Ssentongo, S.W., 1969, A first report on a survey of the fish and fisheries of Lake Naivasha, Kenya: East Africa Freshwater Fisheries Research Organization Report, pp. 28–33.

Muchiri, S.M., 1990, The feeding ecology of *Tilapia* and the fishery of Lake Naivasha, Kenya [Ph.D. thesis]: Leicester, England, University of Leicester.

Muchiri, S.M., and Hickley, P., 1911, Catch effort sampling strategies: their application in freshwater fisheries management: In Cowx, I.G., ed., 1991, Catch effort sampling strategies: their application in freshwater management: Oxford, Blackwell Scientific Publications, Fishing News Books.

Muchiri, S.M., Hickley, P., Harper, D.M., and North, R., 1992, The potential for enhancing the fishery of Lake Naivasha: International Symposium on Rehabilitation of Inland Fisheries Report, University of Hill.

Okorie, O.O., 1972, On the management of the Lake Naivasha Fishery: East Africa Freshwater Fisheries Research Organization Annual Report, pp. 18–23.

Parker, I.S.C., 1974, The status of the Louisiana red swamp crayfish, *P. Clarkii* (Girard) in Lake Naivasha: Nairobi, 45 pp.

Richardson, J.L., and Richardson, A.E., 1972, History of an African Rift Lake and its climatic implications: Ecol. Monogr., v. 42, pp. 499–534.

Robbins, W., and Maccrimmon, H.R., 1974, The black bass in America and overseas: Ontario, Canada. Biomanagement and Research enterprise, 196 pp.

Siddiqui, A.Q., 1977, Lake Naivasha Fishery, Kenya, and its management together with a note on the food habits of fishes: Biol. Cons. v. 12, pp. 217–227.

Siddiqui, A.Q., 1979, Changes in the fish species composition of Lake Naivasha–Kenya. Hydrobiol. 64: 131–138.

Thompson, A.O., and Dodson, R.G., 1963, Geology of the Naivasha area: Government of Kenya, Geological Survey of Kenya, Report no.55.

Verschuren, D., 1996, Comparative paleolimnology in a system of four shallow tropical lake basins: In Johnson, T.C., and Odada, E.O., eds., The limnology, climatology and paleoclimatology of the East African lakes: Toronto, Gordon and Breach.

Watson, C.E.P., 1969, East Africa Freshwater Fisheries Research Organization, Lake Naivasha File Reference DRL, v. 12, pp. 242, 247, 262.

Williams, R., 1972, Relationship between the water levels and the fish catches in Lakes Mweru and Mweru wa Ntipa, Zambia: African Journal of Tropical Hydrobiology and Fisheries, v. 2, pp. 21–32.

Zooplankton Dynamics in Lake Victoria

D.K. BRANSTRATOR and J.T. LEHMAN   *Department of Biology and Center for Great Lakes and Aquatic Sciences, University of Michigan, Ann Arbor, Michigan, United States*

L.M. NDAWULA   *Fisheries Research Institute, Jinja, Uganda*

INTRODUCTION

The limnology of Lake Victoria, East Africa, has changed in remarkable ways during the past three decades (Hecky and Bugenyi 1992; Hecky 1993). Since the early 1960s, phytoplankton biomass in the lake has increased nearly 5-fold, concentrations of mixed layer soluble reactive silica have declined 10-fold, and oxygen concentrations in the hypolimnion have declined markedly over much of the lake (Talling 1965, 1966; Talling and Talling 1965; Hecky and Bugenyi 1992; Hecky 1993; Mugidde 1993). The phytoplankton community has also shifted in its species composition. In 1960 to 1961, *Melosira* was the dominant taxon during the mixing period (July–August) followed by *Anabaena* during the first few months of stratification (October–December) (Talling 1966). The current phytoplankton community is dominated by the cyanobacteria *Cylindrospermopsis* and *Planktolyngbya*, and the diatom *Nitzschia* (Komarek and Kling 1991; Hecky 1993).

Contemporaneous with these changes has been a transformation of the lake's fish community. In the early 1960s the piscivorous Nile perch, *Lates niloticus*, was introduced to Lake Victoria (Ogutu-Ohwayo 1988). At that time the lake contained a diverse assemblage of haplochromine cichlids with an estimated 300+ endemic species (Witte et al. 1992). In the early 1980s, approximately 20 years after Nile perch was introduced, it suddenly became the dominant species of the commercial catch biomass (Ogutu-Ohwayo 1990). Many native fish species, particularly the haplochromines, drastically declined in abundance at that time (Ogutu-Ohwayo 1990). One estimate suggests that 200 species of haplochromines either disappeared or became threatened with extinction (Witte et al. 1992). By the mid-1980s, three species including the introduced Nile perch, one introduced tilapian cichlid (*Oreochromis niloticus*), and one native cyprinid (*Rastrineobola argentea*) dominated Lake Victoria's fishery (Ogutu-Ohwayo 1990).

Understanding the mechanisms of cause and effect that underlie these physical, chemical, and biological changes in Lake Victoria is a major challenge to limnologists. Proper tests of alternative hypotheses will rely in part upon comparisons between the historic and modern conditions and this will require a firm knowledge of the lake's past and present food webs. To this end we sought to characterize an important component of the food web, the crustacean plankton. In this chapter we report on a 10-month time series of zooplankton collections (1992 to 1993) from nearshore and offshore regions in the Ugandan waters of Lake Victoria. We analyze the taxonomic composition, abundance, and temporal variability of copepods and cladocerans from these two regions of the lake. We report on the grazing

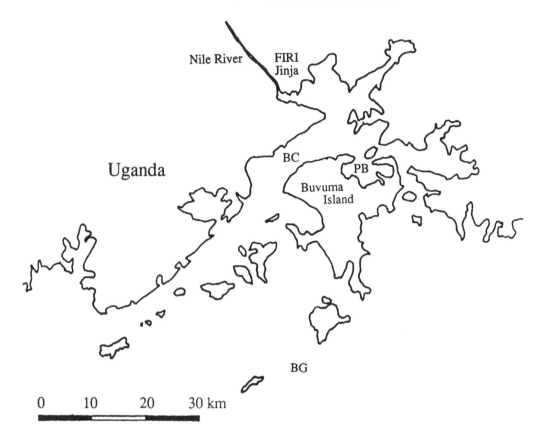

Figure 1. Northern shore of Lake Victoria showing locations of the three reference stations sampled: Pilkington Bay (PB), Buvuma Channel (BC), and Bugaia Island (BG). The Nile River, Buvuma Island, and the Fisheries Research Institute in Jinja are identified for reference.

potential of copepods and cladocerans as determined with shipboard enclosure experiments. Finally, we compare our results of zooplankton community composition with historic records. Of note, we discuss the significance of a large-bodied zooplankter, *Daphnia lumholtzi* var. *monacha*, previously unreported from Lake Victoria. We suggest that its new presence may reflect recent changes in the lake's planktivorous fish community and that fossilized remains of this species in the sediments may be an important marker for paleoecological studies.

METHODS

Zooplankton Collections

Zooplankton were collected at two stations designated as a nearshore station in Pilkington Bay (Station PB: 0°17.97′N, 33°19.56′E; depth = 10 m), and an offshore station near Bugaia Island (Station BG: 0°3.16′S, 33°16.59′E; depth = 62 m) (Fig. 1). All collections were made by conical nets of 0.5-m diameter opening and 1.5-m length fitted with Nitex mesh of either 60- or 100-μm aperture (Research Nets, Inc.). They were equipped with harnesses to permit attachment of weights for vertical orientation and were retrieved vertically from a few meters above the sediment while the ship lay at anchor. The content of each tow was preserved in 10% sugar-Formalin and later split into fractions with a plankton splitter (Aquatic Research Instruments). A wide-bore injection pipette was used to subsample representative fractions and the subsamples were counted until at least 100 specimens of the most abundant taxa were encountered. Copepods were assigned as either calanoid or cyclopoid, and cladocerans were identified to species except for *Ceriodaphnia* which were identified to genus. Zooplankton collections from April 1992 were later subsampled by the same methods described above. From these collections, specimens of cyclopoid copepods (C1–C6), *Bosmina longirostris*, and *Ceriodaphnia* sp. were measured for body length and these lengths were converted to dry weights from existing weight at length regressions (Dorazio et al. 1987; Culver et al. 1985). For larger taxa including calanoid copepods (C1–C6), *Diaphanosoma excisum*, and *Daphnia lumholtzi* dry weights were obtained by pooling animals drawn as encountered and drying them at 60°C to a constant weight (Cahn 29 electrobalance). An original weight at length regression was developed for pooled specimens of *Daphnia longispina* and *Daphnia lumholtzi* var. *monacha* which have quite similar body shapes. Mean individual dry weights, estimated from the April 1992 collections only, were multiplied by population abundances on April and on all subsequent dates to obtain population dry weights for dates shown here from 1992 to 1993 (Fig. 2; see Table 2).

Grazing Experiments

Lehman and Branstrator (1993) demonstrated that the epilimnetic zooplankton communities at the nearshore PB and offshore BG stations were unable to significantly suppress algal biomass in 1-day and 2-day incubation experiments conducted during April 1991. They also noted that those experiments were initiated during daylight when a portion of the zooplankton was absent from the epilimnion. We thus conducted a similar grazing experiment at the offshore BG station during October 1992 that was initiated at night. Experimental enclosures were 2-L polyethylene bags ("ziplock bags") that were filled at 2100 h with epilimnetic water collected by a hand pump with a 5-cm diameter intake hose (Forestry Suppliers, Inc.). In order to reduce the variance among initial plankton densities in the bags, all water needed to set up the experiment was delivered first to a 100-L, plastic-lined cooler (Coleman) on deck and gently mixed. Treatments consisted of ambient concentrations of lake water with zooplankton (1X); controls consisted of water sieved through 100-μm aperture Nitex mesh to exclude the crustacean zooplankton (0X). Nitex mesh of 100-μm aperture was used consistently throughout our study to fractionate zooplankton and will hereafter be referred to as 100-μm mesh. Initial concentrations of total Chl. *a* in the bags were estimated with water in the 100-L reservoir that was filtered (GF/C) either raw or after being sieved (100-μm mesh) (90% acetone, fluorescence). Bags were incubated for 48 h in a water bath at ambient lake water temperatures, and illuminated by natural sunlight reduced to 10% of surface intensity with neutral density filters. Size-fractioned particulate Chl. *a*, in total and <50-μm size fractions, were determined by filtering bag water (GF/C) at the end of incubation.

We conducted a second experiment during the same week with the purpose of quantifying the community grazing rate. Phytoplankton for this experiment were collected from offshore

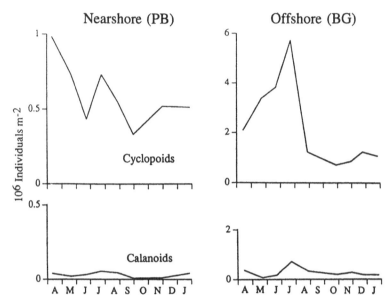

Figure 2. Abundances of copepods (C1–C6) and cladocerans reported as individuals m^{-2} at nearshore PB and offshore BG stations estimated by a conical net with a 0.5-m diameter opening and 100-μm mesh. Note that the Y-axis scales are different for copepods and cladocerans. Dates of sample collections are nearshore PB: 10 APR 92 (2330 h), 19 MAY 92 (1900 h), 21 JUN 92 (1100 h), 21 JUL 92 (2100), 24 AUG 92 (2230 h), 26 SEP 92 (2030 h), 25 NOV 92 (1945 h), 21 JAN 93 (2100 h), and offshore BG: 13 APR 92 (2200 h), 21 MAY 92 (2015 h), 19 JUN 92 (2300 h), 20 JUL 92 (2030 h), 24 AUG 92 (0130 h), 24 OCT 92 (2115 h), 24 NOV 92 (2000 h), 19 DEC 92 (1130 h), 20 JAN 93 (1130 h).

BG station but due to logistical constraints, zooplankton for the experiment were collected at a station in the Buvuma Channel (Station BC: 0°20.40′N, 33°15.35′E; depth = 20 m) (Fig. 1). To begin the experiment, sieved epilimnetic water (100-μm mesh) from offshore BG station was dispensed to 2-L polycarbonate bottles amended by addition of either carrier-free ^{35}S as Na$_2$^{35}SO$_4$, or carrier-free ^{33}P as H$_3$^{33}PO$_4$. Bottles were incubated in the same water bath used for the other experiment under simulated in situ conditions for approximately 48 h at approximately 25°C. Phosphorus pools in algae can turn over within minutes to hours (Lean 1973; Lean and Nalewajko 1976), and doubling times of light-saturated algae can range from 1 to 2 d^{-1} (Reynolds 1989). Therefore, 48 h should have been sufficiently long for algae to have come into isotopic equilibrium with the water. After incubation, the water was dispensed to 4 separate 1-L polyethylene bottles (2 for ^{35}S, 2 for ^{33}P). At channel BC station, zooplankton were collected from 10 m depth by 4 successive casts of a 10-L Schindler trap. This depth was well below the photic zone as current secchi disc readings in Lake Victoria are 1 to 3 m (Mugidde 1993). The content of each cast was concentrated with 100-μm mesh and added to one 1-L bottle containing radiolabelled particles; thus zooplankton densities were elevated 10× above ambient levels during the experiment. Bottles were incubated in the dark for 15 minutes which should have been short enough to prevent loss of the radioisotope label in fecal pellets at our experimental temperature of approximately

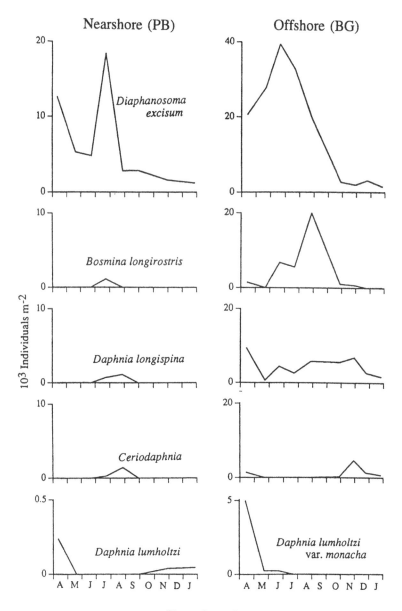

Figure 2, cont'd

25°C (Dam and Peterson 1988). The zooplankton were removed by sieving (100-μm mesh) and immediately rinsed and preserved in 10% sugar-Formalin. From each bottle, 20-ml subsamples were filtered onto replicate 0.2-μm and replicate 1.0-μm Millipore filters. Approximately 2 weeks later, zooplankton were concentrated from the preservative on 1.0-μm Millipore filters and aliquots of the zooplankton preservative solution were collected as well to recover isotope that leaked from the animals into solution during their preservation

Table 1. Coefficients of Variation (CV = SD/mean) of Zooplankton Abundance for 3 Sets of Replicate Net Tows (*n* = 4 tows per set) from the Offshore BG Station

| Taxon | Coefficient of variation | | | |
	Set 1	Set 2	Set 3	**Mean**
Daphnia longispina	0.29	0.27	0.27	0.28
D. lumholtzi var. *monacha*	0.46	0.39	0.31	0.39
Diaphanosoma excisum	0.32	0.10	0.25	0.22
Bosmina longirostris	0.86	0.36	–	0.61
Ceriodaphnia	0.46	0.70	0.78	0.65
Calanoid copepods C1–C6	0.23	0.15	0.10	0.16
Cyclopoid copepods C1–C6	0.21	0.11	0.15	0.16

All collections were made by conical nets with 0.5-m diameter openings towed vertically from 50 to 0 m. The coefficients of variation are reported by taxon for each set and the means for the three sets are given. The sets are identified by **number**, date, time (mesh aperture); **1**, 13 April 1992, 2300 h (60 µm); **2**, 13 April 1992, 2200 h (100 µm); **3**, 16 April 1992, 1030 h (100 µm). (–) indicates no specimens were found in collections.

(Holtby and Knoechel 1981). The filters and aliquots of preservative were counted individually in 10 ml Bio-Safe II counting cocktail (Research Products International) on a Beckman LS 6800 counter. Zooplankton grazing rates were computed as

$$\text{Grazing rate} = (\text{DPM in zooplankton/DPM } L^{-1})/(10 \times t), \tag{1}$$

where t is incubation time and DPM is disintegrations per minute of the radioisotope.

Most of the particles that became radioactively labeled during the initial 48 h of incubation were probably larger than 1.0 µm diameter. Lehman and Branstrator (1994) reported nutrient mass uptake rates by Lake Victoria plankton less than 100 µm diameter from offshore BG station with experiments conducted during the same week as this grazing experiment. They found that 91% of sulfate uptake by particles larger than 0.2 µm diameter could be ascribed to particles that were larger than 1.0 µm diameter. Similarly, 76% of phosphate uptake by particles larger than 0.2 µm diameter could be ascribed to particles that were also larger than 1.0 µm diameter. Less than 10% of the total uptake of either isotope could be ascribed to killed particles (1% Formalin). Therefore, we believe our experiment measured grazing on living particles that were primarily between 1 to 100 µm diameter.

RESULTS AND DISCUSSION

Species Composition and Seasonality

The mean coefficient of variation (SD/mean) for replicate tows from one station ranged from 0.16 for copepods up to 0.65 for *Ceriodaphnia* (Table 1). This large disparity in the coefficients of variation could be interpreted to suggest that different levels of patchiness exist among different zooplankton taxa in Lake Victoria. In a recent review on the interreplicate variance associated with zooplankton collections, Downing et al. (1987) showed that two variables, mean organism density and sample collection volume, explain a majority of

Table 2. Zooplankton Mean Abundance (individuals m^{-2})
During 10-Month Time Series

Taxon	Abundance (individuals m^{-2})				Dry wt (mg m^{-2})	
	PB		BG		PB	BG
Cladocera						
Daphnia longispina	230		4,370		4	85
Daphnia lumholtzi	40		0		<1	0
Daphnia lumholtzi var. *monacha*	0		610		0	13
Diaphanosoma excisum	6,180		16,560		15	39
Bosmina longirostris	140		3,960		<1	8
Ceriodaphnia	210		890		1	4
Total Cladocera*	6,800	1%	26,390	1%	≈20	149
Total Calanoids	31,110	5%	277,740	11%	172	1,533
Total Cyclopoids	598,640	94%	2,236,050	88%	1,161	2,862

Percent composition by number for total categories, and mean dry weight (mg m^{-2}) at nearshore PB and offshore BG stations for cladocerans and C1–C6 stages of calanoid and cyclopoid copepods. Means are based on samples collected by a conical net with a 0.5-m diameter opening and 100-μm mesh from April 1992 to January 1993 as reported in Fig. 2.

Moina sp. appeared occasionally at nearshore PB station but never reached an abundance greater than 1,000 individuals m^{-2}.

observed variance. They used a large data set drawn from the literature to develop the equation

$$s^{2*} = 0.745m^{1.622} V^{-0.267} \qquad (2)$$

that describes a relationship between interreplicate variance (s^{2*}) and the two variables, mean organism density (m) and sample collection volume (V). In order to assess the potential source of the interreplicate variance in our zooplankton collections from Lake Victoria we selected a set of mean abundance estimates and compared the observed variance (s^2, $n - 1$ weighting, Neter et al. 1985) to the predicted variance (s^{2*}) from Eq. 2. For this analysis we used the same three sets of replicate zooplankton collections reported in Table 1. *Bosmina longirostris* was absent from one set of net tows and thus we had 20 pairs of values for the observed and predicted variances. A correlation analysis that compared log$_{10}$ transformed observed vs predicted variances was highly significant ($r = 0.99$, $p < 0.005$, $n = 20$; SYSTAT). This close 1:1 relationship between the observed and predicted variances suggests that the two variables, mean organism density (m) and sample collection volume (V), may largely account for the range in observed variance associated with our estimates of zooplankton abundance (Downing et al. 1987) (Table 1).

It is important to point out that our calculations of zooplankton abundance assumed that nets sampled with 100% efficiency. This is unlikely, especially considering the large biomass of phytoplankton in Lake Victoria that could have clogged the mesh (Mugidde

1993). Zooplankton abundance data in this chapter should thus be regarded as conservative estimates.

Although the taxonomic identification of crustacean zooplankton in Lake Victoria is generally agreed upon, there appears to be some questions regarding the copepods. In a detailed monograph on the copepods of Lake Victoria, Sars (1909) reported twenty species of cyclopoids all of the genus *Cyclops*. A review by Rzoska (1957) of the early literature, including Sars (1909) study, reports only five species of cyclopoids representing the three genera *Thermocyclops*, *Tropocyclops*, and *Mesocyclops*. The most recent published study on Lake Victoria zooplankton (Mavuti and Litterick 1991) used material collected from 1984 to 1987 and reports *Thermocyclops*, *Mesocyclops*, and *Microcyclops* from the Winam Gulf and the main water of Lake Victoria bordering Kenya. Mavuti and Litterick (1991) also report that *Thermocyclops* was the dominant genus of cyclopoids in that study. Regarding the calanoid copepods, Sars (1909) reports five species of *Diaptomus*. In Rzoska's (1957) review, *Diaptomus* is the only genus of calanoid copepod listed. Mavuti and Litterick (1991) report three calanoid species representing the two genera *Thermodiaptomus* and *Tropodiaptomus* from the Kenyan waters. Again, in our study we distinguished only between the cyclopoid and calanoid copepods, and lumped all potential species within these two groups. It is agreed that the cladocerans *Bosmina longirostris* and *Diaphanosoma excisum* are the only species of these two genera in the lake (Rzoska 1957; Green 1971; Mavuti and Litterick 1991). Green (1971) found two species of *Ceriodaphnia*, *C. dubia* and *C. cornuta*, but we did not distinguish between different species of *Ceriodaphnia* in our study. Previous reports consistently identified only two daphnid species, *Daphnia longispina* and *D. lumholtzi* (Rzoska 1957; Green 1971; Mavuti and Litterick 1991). The non-helmeted and large-bodied daphnid, *D. lumholtzi* var. *monacha*, that we found occasionally at offshore BG station had not been previously reported from Lake Victoria. In a later section we discuss its potential significance as a new member of the plankton.

Abundances of cladocerans and copepods changed markedly during 1992 to 1993 (Fig. 2). With few exceptions, the patterns of change were similar as most taxa peaked during the spring and summer months, primarily between April and August. The relative proportions of major taxa remained remarkably constant across dates and always ranked as cyclopoids > calanoids > cladocerans in terms of mean water column abundances (individuals m^{-2}) (Fig. 2). Whether or not this seasonality in abundance is an annual feature of the crustacean zooplankton is not yet possible to assess since the data set covers only approximately one year. In contrast to temperate lakes that experience a relatively large annual range in water temperatures, Lake Victoria fluctuates only 1 to 2°C during the year (Hecky 1993). There-fore, the effect of water temperature on metabolic rates and secondary production probably plays a much smaller role in the annual variability of zooplankton abundance in Lake Victoria compared to temperate lakes (Hutchinson 1967). Seasonal variability in the abundance of tropical crustacean zooplankton has been documented in several other large lakes including Lake Chad (Robinson and Robinson 1971), Lake Lanao (Lewis 1978), Lake Valencia (Infante 1982), and Lake Malawi (Twombly 1983). In many instances the seasonal zooplankton maxima have been observed to coincide with lake mixing and increased primary productivity but not necessarily with maximum phytoplankton biomass. For exam-ple, in Lake Valencia, cyanobacteria dominate the phytoplankton year round but the crusta-cean zooplankton tend to wax and wane seasonally in response to peaks in the biomass of diatom and green algae which are not as abundant as the cyanobacteria (Infante 1982). Judging from the single year of data presented here from Lake Victoria, the crustacean zooplankton appear to peak during spring and summer months when the lake mixes. This period was historically (1960 to 1961) associated with the annual peak biomass of phytoplankton in the water column (Talling 1966). Estimates of the modern (1990 to 1991) Chl. *a* concentrations indicate that phytoplankton biomass now peaks from about September to January during lake stratification suggesting that zooplankton and phytoplankton community biomass no longer peak at the same time (Mugidde 1993). There is evidence that the modern phytoplankton community is largely composed of cyanobacteria (Hecky 1993), particularly *Cylindrospermopsis* and *Planktolyngbya* (Komarek and Kling 1987) which may be poor quality food for the zooplankton (Haney 1987; Lampert 1987). The relative contribution of cyanobacteria versus diatoms (primarily *Nitzschia*) during different seasons is unknown. Based on our understanding of plankton succession in lakes we might expect there to exist seasonal variability in the quantity and quality of the phytoplankton resource, and that zooplankton productivity might track the changes in phytoplankton (Demott 1989). A possible relationship between zooplankton seasonality and the phytoplankton resource deserves further study, particularly since previous workers have demonstrated such relationships in other tropical lakes (Lewis 1978; Infante 1982).

Seasonal variability in the intensity of fish planktivory is a second potential factor controlling the fluctuations in the abundance of crustacean zooplankton. Witte (1981) noted that spawning activity of zooplanktivorous haplochromines peaks during the period of lake turnover which was again historically the period when phytoplankton biomass peaked. Witte (1981) suggested that haplochromine breeding cycles were set to coincide with this period of relative high food abundance because such timing favors the growth of juveniles. Our data indicate that lake turnover is also the period when crustacean zooplankton abundance peaked and declined during 1992 (Fig. 2). Whether or not fish predation pressure is currently driving this seasonal cycle in the abundance of zooplankton remains difficult to assess because the fish stocks of Lake Victoria have undergone extreme changes during the past few decades. The once dominant zooplanktivores, the haplochromines, have been replaced by the cyprinid, *Rastrineobola argentea* (Ogutu-Ohwayo 1990). Haplochromine stocks have declined so severely as to cast doubt on whether they are currently an important factor driving zooplankton dynamics. *R. argentea* has progressively become a dominant component of the commercial fishery yield, yet how predation pressure by *R. argentia* is affecting the seasonal dynamics of the cladocerans and copepods cannot be determined without a better understanding of this fish's reproductive cycle and distribution (Ogutu-Ohwayo 1990).

Grazing

Results of the grazing experiment conducted at the offshore BG station in the 2-L bags indicated that initial concentrations of Chl. *a* (we measured the total fraction only in initials) was not significantly different between the control (0X) and grazer (1X) treatments (ANOVA; $p = 0.08$; SYSTAT) (Fig. 3). Therefore, the final concentrations of Chl. *a* were directly compared between controls and treatments. There were no significant differences between the final concentrations of Chl. *a* in the controls (0X) versus treatments (1X) for either the total or <50-µm fractions (ANOVA; total: $p = 0.67$, <50-µm: $p = 0.06$; SYSTAT) (Fig. 3). In regards to the relatively small p-value associated with the <50-µm fraction, the trend in the data suggests that final Chl. *a* was actually higher where zooplankton were present (1X) (Fig. 3). This does not suggest that zooplankton grazing had a negative effect on phytoplankton biomass. These results are consistent with grazing experiments of a

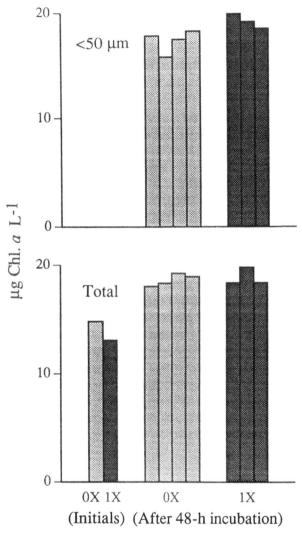

Figure 3. Final concentrations after 48-h incubation of Chl. *a* in particles that passed through a 50-μm mesh (upper panel) and total Chl. *a* (lower panel) in a grazing experiment conducted at offshore BG station during October 1992. 0X is the control (zooplankton excluded by 100-μm mesh) and 1X is the treatment with ambient concentrations of zooplankton. Each bar represents the mean value of replicate Chl. *a* measurements for one enclosure. The lower panel also shows estimates of initial total Chl. *a* in the control and treatment enclosures where each bar represents the mean of triplicate Chl. *a* measurements from a reservoir of lake water used to set up the experiment.

similar design conducted during April 1992 at the nearshore PB and offshore BG stations in which the zooplankton community also had no significant effect on phytoplankton biomass (Lehman and Branstrator 1993).

Table 3. Results of the Radiotracer Grazing Experiments
Conducted in 1-L Carboys at Channel BC Station
During October 1992

Isotope replicate	Removal rates on particles	
	>0.2 μm	>1.0 μm
P:1	3.9	4.4
P:2	8.0	9.1
S:1	7.6	8.3
S:2	7.4	8.1
Mean	**6.7**	**7.5**

Particles were pre-labelled with either ^{33}P or [35]S. The experiment was replicated for each isotope. Particle removal rates are expressed as the % of
standing crop that would be eaten over 24 h by the zooplankton assemblage
retained on a 100-μm mesh. Removal rates were estimated by assuming that
the zooplankton grazed either all particles of spherical diameter >0.2 μm
(bacteria and phytoplankton) or only partlicles of spherical diameter >1.0
μm (phytoplankton). The mean removal rates for the different size classes of
particles, as estimated by both replicated experiments, are in bold type.

Results of the radioisotope grazing experiment indicated that the crustacean zooplankton
community at the channel BC station could remove from 6.7% (>0.2-μm diameter particles)
to 7.5% (>1.0-μm diameter particles) of the standing crop of particles per day (Table 3).

Peters and Downing (1984) reviewed a large literature on cladoceran and copepod feeding
in which they report a general, median feeding rate of 40.0 μg wet wt individual^{-1} day^{-1}. In
order to compare our results we had to estimate zooplankton density in the 1-L bottles as
well as the total μg wet weight of particles eaten by the zooplankton during the experiment.
Zooplankton density in the bottles was estimated as 30 individuals L^{-1} based on zooplankton
abundance data from the nearshore PB station. We assumed that the concentration of Chl. *a*
in the 1-L bottles could be adequately estimated from the final concentration of total Chl. *a*
in the 2-L bag experiment controls (Fig. 3). The water for the bag experiment had been
treated in virtually the same manner as water for the radioisotope experiment in that both
batches of water had been collected from the offshore BG station, screened (100-μm mesh),
and permitted to incubate for 48 h in the shipboard water bath. The experiments were
conducted within a week of each other. The only difference was that water for the radioisotope experiment had received some carrier-free ^{33}P and ^{35}S which should not have affected
phytoplankton growth since neither P nor S limit phytoplankton growth in Lake Victoria
(Lehman and Branstrator 1994). Therefore, based on the final concentration of total Chl. *a*
in the 2-L bag experiment (Fig. 3, 0X Total), we estimated the starting concentration of Chl.
a in the radioisotope experiment as 18.6 μg L^{-1}. Wet mass was estimated as 1,431 μg given
the conversions Chl. (μg):vol. (10^6 μm^3) = 0.013, and vol. (10^6 μm^3):wet mass (μg) = 1
(Peters and Downing 1984). Our experimental data show that zooplankton removed ap-

proximately 7.5% of chlorophyll-containing particles per day, those >1.0 μm diameter (Table 3). Therefore, the estimated total mass of particles removed by 30 zooplankters was 107 μg wet wt., or approximately 3.6 μg wet wt. animal^{-1} day^{-1}. This feeding rate is an order of magnitude lower than the general median feeding rate of 40 μg wet wt. animal^{-1} day^{-1} reported by Peters and Downing (1984).

Taken together, the grazing experiments conducted here and by Lehman and Branstrator (1993) demonstrate that the cladocerans and copepods (>100 μm) exercise little grazer control over phytoplankton biomass in Lake Victoria. This result is not surprising considering the makeup of the grazer assemblage that we manipulated in these experiments. The assemblage was dominated by small-bodied cyclopoid copepods. Their mean individual length in the April 1992 nearshore and offshore samples combined was 0.50 mm ($n = 354$) and their mean individual dry weight was 1.9 μg. Cyclopoid copepods are clearly diverse in their food preferences and are known to feed on a variety of phytoplankton and zooplankton prey of a broad size range (Fryer 1957). The dominant crustaceans in Lake Victoria may thus be largely omnivorous and therefore unlikely to control phytoplankton biomass. The low grazing rates may also be attributable to the composition of the phytoplankton community. As mentioned above, the phytoplankton is partly dominated by cyanobacteria of which potentially inedible or toxic forms could reduce grazing rates of the zooplankton (Haney 1987; Lampert 1987). At times Lake Victoria also experiences prolonged offshore blooms of *Microcystis* (Ochumba and Kibaara 1989) that may contribute temporarily to reduced grazing rates. Lehman and Branstrator (1993) found that most of the total Chl. *a* in Lake Victoria during April 1991 and 1992 was in the <50-μm fraction. Results of our 2-L bag experiment also indicate that most of the Chl. *a* during October 1992 was in the <50-μm fraction (Fig. 3). Therefore, the phytoplankton probably do not interfere with the filtering apparatus of the cladocerans since the cells appear to be primarily within the size range preferred by zooplankton (Sterner 1989). Carney and Elser (1990) have argued that the importance of zooplankton grazing and related nutrient regeneration by zooplankton should diminish along a trophic gradient from mesotrophic to eutrophic systems. Our experimental results on zooplankton grazing in Lake Victoria are consistent with that hypothesis and suggest that grazing pressure by the crustacean zooplankton (>100 μm) is not strong in this eutrophic lake (Hecky 1993).

Because we used 100-μm mesh to isolate zooplankton in the experiments, neither rotifers nor copepod nauplii were excluded from the control enclosures and thus our results underestimate the full grazing potential of the metazoan, zooplankton community. The exclusion of rotifers, at least, may be of minor importance since rotifers are consistently scarce in pelagic regions of African Great Lakes including Lake Victoria (Green 1967b; Mavuti and Litterick 1991). Our 60-μm mesh nets rarely caught rotifers from nearshore PB and offshore BG stations, even though they would have been adequately retained. We acknowledge that the grazer impact of copepod nauplii remains unknown and deserves careful attention in future studies considering that copepods dominate the crustacean zooplankton (Table 2). Mazumder et al. (1992) found that small-bodied zooplankton, particularly rotifers, dominate the metazoan herbivores in Lake Ontario of the Laurentian Great Lakes. Their results provide an example of the principal role in trophic transfer carried out by small-bodied grazers in large lake food webs.

Nearshore vs. Offshore

The crustacean zooplankton of the nearshore PB and offshore BG stations are similar in that cyclopoids are always most abundant with fewer calanoids and still fewer cladocerans throughout the year (Fig. 2). The seasonal peaks in zooplankton abundance also occur during the same period, spring to summer, in both regions. Notable differences between the two regions can be seen among the distributions of cladoceran taxa. We found *Daphnia longispina* at the nearshore PB station during July and August only but at the offshore BG station it maintained a relative constant density throughout the year. This pattern is consistent with Mavuti and Litterick (1991) who report that daphnids are also more successful in open water of the Kenyan region of Lake Victoria in comparison to nearshore regions in the Winam Gulf. In similar fashion, *Bosmina longirostris* was present at the nearshore PB station during July only but was present at the offshore BG station throughout most of the year. There was a striking contrast in the spatial distribution of the two forms of *Daphnia lumholtzi* as well. The typical form of *D. lumholtzi* which bears a long pointed helmet, long tailspine, and long pointed fornices (Green 1967a) was found exclusively at the nearshore station. The other form, termed *D. lumholtzi* var. *monacha*, has much shorter spine ornamentation. It was found exclusively at the offshore station. Green (1967a) compared the two forms of *D. lumholtzi* with specimens from Lake Albert and found that the *monacha* form attained a much greater mean carapace length than the typical form. In Lake Victoria, specimens of the *monacha* form exceed 2.5 mm length from the eye to the base of the tailspine. After analyzing the samples our impression is that the *monacha* form of *D. lumholtzi* is also a much larger and robust daphnid in comparison to the typical form in Lake Victoria.

These results suggest that the large-bodied cladocerans, *Daphnia longispina* and *D. lumholtzi* var. *monacha*, are more successful at the offshore BG station in comparison to the nearshore PB station. In a comprehensive study on the distribution of cladocerans in East African lakes, Green (1971) previously described a nearshore-offshore gradient in species distributions, particularly for *D. lumholtzi* in Lakes Albert and Edward. He found that the larger daphnids were more successful in offshore regions and that the smaller forms tended to dominate nearshore regions of the large lakes. Results here indicate a similar pattern for Lake Victoria (Fig. 2). Green (1971) also found that large-bodied species of *Ceriodaphnia*, including *C. reticulata* and *C. dubia*, were most abundant in the offshore regions of Lakes Albert, Edward, and Victoria. He used these patterns of plankton body size and associated data on the distribution of planktivorous fishes in Lake Albert to conclude that fish planktivory is greatest in the nearshore regions of these large African lakes and decreases along a gradient to offshore waters. The distributional pattern of daphnids that we observed in Lake Victoria is consistent with a gradient in fish planktivory that declines from nearshore to offshore. Lehman (this volume) reviewed Green's (1967a, 1971) work on the nearshore-offshore distribution of cladocerans in the African lakes. In that chapter Lehman suggests that a reverse gradient in the intensity of invertebrate predation may exist from the nearshore to offshore regions of these large lakes as this would account for the paucity of small-bodied zooplankton prey, particularly rotifers, observed in the offshore regions.

Modern vs. Historical

Given the dramatic changes in the limnology and fishery of Lake Victoria in recent years, we were interested in comparing the modern and historical zooplankton communities. There are several published reports on the historical zooplankton community of Lake Victoria but most provide only presence and absence data (reviewed by Rzoska 1957). Studies by Worthington (1931) and Rzoska (1957) provide the only quantitative estimates of zooplankton in the lake prior to changes in the phytoplankton community, nutrient chemistry, and fishery (Hecky 1993; Ogutu-Ohwayo 1990). The only quantitative estimates of the modern zooplankton community are by Mavuti and Litterick (1991) who collected samples from 1984 to 1987, and this study.

Our data indicate that cyclopoid copepods dominate the modern community of cladoceran and copepod zooplankton (Table 2). The average percent composition by number at the nearshore PB station is cyclopoid copepods (94%), calanoid copepods (5%), and cladocerans (1%), and at the offshore BG station is cyclopoid copepods (88%), calanoid copepods (11%), and cladocerans (1%). Owing to the small relative body size of the cyclopoids, their contribution to total zooplankton biomass is less extreme than their numerical dominance might suggest (Table 2). Mavuti and Litterick (1991) did not distinguish between cyclopoid and calanoid copepods in their study but they found that the copepods always comprised >80%, and the cladocerans always comprised <2% by number of the zooplankton in the Kenyan region of the lake. Their numbers are basically consistent with the numbers reported here for zooplankton composition in the lake.

In general, the relative contribution of these main groups to the zooplankton community does not appear to have changed substantially since the 1950s but because so few historical samples exist and the collection dates are so sporadic it is not possible to draw firm conclusions (Rzoska 1957). There is some indication based on Worthington's (1931) study that the cladocerans and calanoid copepods historically comprised a larger percentage of the plankton than they do at the present time. Worthington (1931) collected samples at 3-hour intervals during a 24-hour period in 1927 from a deep (69 m) area of the lake. In summary statistics he reports that the percent composition of the three main groups was cladocerans (39%), calanoid copepods (50.1%), and cyclopoid copepods (5.4%). He collected specimens from other taxa as well that made up the remaining few percent. His results would at first appear to be strikingly different from the modern zooplankton community. Yet Worthington (1931) used a net with mesh of approximately 640-µm aperture which would have surely excluded many cladocerans and copepods, particularly the now abundant small-bodied cyclopoid copepods. Therefore, his data do not provide a fair comparison with modern data given the different methods of collection and potential bias.

Lehman (this volume) reviewed the literature on pelagic food webs of the East African lakes. He points out that the composition of tropical crustacean zooplankton is typically much simpler and the dominant species are smaller bodied in comparison to temperate crustacean zooplankton (Fernando 1980a, 1980b; Lehman 1988). The zooplankton of Lake Victoria are characteristic of tropical lakes insofar as the community is dominated by small-bodied taxa, the cyclopoids, and the species richness of cladocerans is much lower than temperate lakes, particularly in the numbers of daphnid species (Fernando 1980a). The number of copepod species in Lake Victoria appears to be uncertain given the discrepancies between the modern reports and that of Sars (1909). Why the average body size of plankton is smaller in tropical lakes remains unanswered. Fernando (1980a, 1980b) discussed this topic with reference to tropical lakes of Sri Lanka. He argued that annual warm temperatures in the tropics render large-bodied taxa such as *Daphnia* metabolically less efficient than small-bodied species. He also noted that size-selective predation by planktivorous fishes can be an important structuring force in tropical lakes that reduces the average body size of plankton. Both warm temperatures and fish predation may be largely responsible for the predominance of small-bodied cyclopoid copepods in Lake Victoria. Nonetheless, a raptorial feeding behavior and broad diet breadth of cyclopoid copepods may also permit them to forage more efficiently than the cladocerans in an environment with poor food quality and thereby contribute to their success as well (Richman and Dodson 1983). In tropical Lake George, *Thermocyclops hyalinus* uses *Microcystis* as food (Burgis 1971; Burgis et al. 1973; Moriarty et al. 1973). This trophic link with cyanobacteria may permit *T. hyalinus* to dominate the crustacean zooplankton of that lake where cyanobacteria maintains a high annual biomass (Burgis 1971). A direct trophic link between cyclopoids and cyanobacteria in Lake Victoria has not been demonstrated but could potentially also account for the success of cyclopoids there.

Finally, a striking discovery in our net collections was *Daphnia lumholtzi* var. *monacha*, a species previously unreported from Lake Victoria (Rzoska 1957; Green 1971; Mavuti and Litterick 1991). We first found it in April 1992 at the offshore BG station. Some specimens measured >2.5 mm body length making them clearly the largest cladocerans in the community. The *monacha* form may have been a historical member of the plankton but it seems unlikely that previous workers would have overlooked this species given its conspicuous size. Worthington (1931) and Green (1971) found the typical form of *D. lumholtzi* but did not find any specimens of the *monacha* form in the lake. As recently as June 1984 and March 1987, no specimens of the *monacha* form were found by Mavuti and Litterick (1991) who collected large numbers of samples from the Kenyan region of the lake. It thus appears that the *monacha* form of *D. lumholtzi* is a recent addition to the plankton of Lake Victoria

Although the annual contribution of *Daphnia lumholtzi* var. *monacha* to total zooplankton biomass is negligible, its large-bodied, conspicuous nature raises questions regarding recent changes in the lake's fish community. As described above, in Lake Albert the abundance of *D. lumholtzi* var. *monacha* is inversely related to the intensity of fish predation (Green 1967a). With the tremendous collapse in the haplochromine stocks of Lake Victoria in the early 1980s, it seems feasible that fish planktivory declined and opened a refuge for the persistence of large-bodied zooplankton such as *D. lumholtzi* var. *monacha*. Before the haplochromines collapsed approximately 11% to 53% were zooplankton feeders (Witte 1981), an estimate that may be conservative considering that species belonging to other trophic guilds also forage on zooplankton as juveniles (van Oijen et al. 1981). It is believed that the zooplankton feeders once maintained a lakewide distribution utilizing the offshore pelagic zone as well as shallow bays, the latter primarily during the breeding season (van Oijen et al. 1981; Witte et al. 1992). A considerable fraction of them also occupied areas of the lake deeper than 30 m (Kudhongania and Cordone 1974). We suspect that the appearance of the large-bodied daphnid is related to the dramatic reductions in the fish stocks. Moreover, we believe that paleostratigraphy of *D. lumholtzi* var. *monacha* in lake sediments could provide a useful marker regarding the spatial distribution of fish planktivory between nearshore and offshore regions, and temporal variability in the intensity of fish planktivory during the past several decades in Lake Victoria (Uutala et al. 1993). An analysis of this nature would require that fossils of the *monacha* form be distinguishable from other daphnids in the lake. Green (1967a) described several morphological differences between the two forms of *D. lumholtzi*. He noted that the relative length of the helmet, tailspine, and fornices, and the length and spacing of spines lining the ventral edge of the carapace differ substantially between the two forms. Therefore, fossils of the carapace and helmet from the two forms of *D. lumholtzi* should be distinguishable. We performed microdissections on several specimens of both the *monacha* and typical forms of *D. lumholtzi* and on *D. longispina* and found that the postabdomen differ in their number and arrangement of anal spines and size of pectin on the claw (unpublished data). The postabdomen should thus provide a means of distinguishing among the three daphnid species in the lake.

SUMMARY

The crustacean zooplankton of Lake Victoria are clearly dominated in number and biomass by small-bodied, cyclopoid copepods. We did not identify them to species but Mavuti and Litterick (1991) note that the dominant cyclopoids in the Kenyan region of the lake are of the genus *Thermocyclops*. Our results show that cyclopoid copepods dominated on all dates during the 10 months that were studied from April 1992 to January 1993. The abundance of cladocerans and copepods varied markedly during the 10-month time series. Copepods were most abundant during lake turnover from April to August and declined thereafter. The dominant cladoceran was *Diaphanosoma excisum* and it followed a pattern similar to the copepods. Other cladoceran taxa were more sporadic in their appearance but with few exceptions also tended to be most abundant during the period of lake turnover.

The small-bodied nature of the crustacean zooplankton suggests that fish planktivory may be generally an important factor affecting the community. There was spatial variability in the distribution of the large daphnids, *Daphnia longispina* and *Daphnia lumholtzi* var. *monacha*, which were both most successful at the offshore BG station. In an attempt to explain the greater success of the larger cladocerans offshore, we suggest that fish planktivory may be most intense nearshore but decline at the deeper, offshore regions of the lake. As discussed, Green (1967a) found a striking inverse correlation between the spatial distribution of fish planktivory and zooplankton body size in neighboring Lake Albert.

In general, grazing by crustacean zooplankton (>100 μm) does not appear to control phytoplankton biomass in Lake Victoria. A 48-h enclosure experiment conducted at the offshore BG station did not detect a significant effect of zooplankton grazing on algal biomass. This result was consistent with the results of a previous grazing experiment of similar design conducted at the offshore BG station in April 1992 (Lehman and Branstrator 1993). A radioisotope grazing experiment of 15 minutes duration indicated that ambient concentrations of zooplankton at channel BC station could potentially remove from 6.7% to 7.5% of the ambient biomass of particles per day. When the biomass of particles was converted to wet mass, results suggested that grazing rates were low in comparison to average grazing rates typical of crustacean zooplankton. We note that our experiments did not measure grazing by microzooplankton (<100 μm) such as copepod nauplii and we consider this a topic for further study.

Although few quantitative descriptions of the zooplankton in Lake Victoria are available, the data from this study, Mavuti and Litterick (1991), and Rzoska (1957) suggest that the relative composition of cladocerans, calanoid copepods, and cyclopoid copepods in the modern community is largely unchanged from historical conditions. One important difference is the presence of the large-bodied daphnid, *Daphnia lumholtzi* var. *monacha*, which has not been previously described in the zooplankton of Lake Victoria. The success of this species in offshore regions of neighboring Lake Albert has been associated with low levels of fish planktivory (Green 1967a). We discuss how recent changes in the fish community of Lake Victoria may have led to the establishment of *D. lumholtzi* var. *monacha* in the modern zooplankton and how the distribution of its fossils in the sediments could provide an important paleoecological marker of the distribution of fish planktivory and the timing of the fishery transformation in Lake Victoria.

ACKNOWLEDGMENTS

We gratefully thank members of the Fisheries Research Institute in Jinja, Uganda, for assistance with field sampling. A. Mazumder, J. White, and one anonymous reviewer provided many useful comments on the manuscript. This work was supported by the Undersea Research Program of NOAA and by a grant from the University of Michigan Global Change Program.

REFERENCES

Burgis, M. J. 1971. The ecology and production of copepods, particularly *Thermocyclops hyalinus*, in the tropical Lake George, Uganda. Freshwater Biol. 1: 169–192.

Burgis, M. J., J. P. E. C. Darlington, I. G. Dunn, G. G. Ganf, J. J. Gwahaba, and L. M. McGowan. 1973. The biomass and distribution of organisms in Lake George, Uganda. Proc. R. Soc. Lond. 184: 271–298.

Carney, H. J., and J. J. Elser. 1990. Strength of zooplankton–phytoplankton coupling in relation to lake trophic status, pp. 615–631. *In* M. M. Tilzer and C. Serruya [eds.], Large Lakes, Ecological Structure and Function. Springer-Verlag.

Culver, D. A., M. M. Boucherle, D. J. Bean, and J. W. Fletcher. 1985. Biomass of freshwater crustacean zooplankton from length-weight regressions. Can. J. Fish. Aquat. Sci. 42: 1380–1390.

Dam, H. G., and W. T. Peterson. 1988. The effect of temperature on the gut clearance rate constant of planktonic copepods. J. Exp. Mar. Biol. Ecol. 123: 1–14.

Demott, W. R. 1989. The role of competition in zooplankton succession, pp. 195–252. *In* U. Sommer [ed.], Plankton Ecology, Succession in Plankton Communities. Springer-Verlag.

Dorazio, R. M., J. A. Bowers, and J. T. Lehman. 1987. Food-web manipulations influence grazer control of phytoplankton growth rates in Lake Michigan. J. Plankton Res. 9: 891–899.

Downing, J. A., M. Perusse, and Y. Frenette. 1987. Effect of interreplicate variance on zooplankton sampling design and data analysis. Limnol. Oceanogr. 32: 673–680.

Fernando, C. H. 1980a. The freshwater zooplankton of Sri Lanka, with a discussion of tropical freshwater zooplankton composition. Int. Revue ges. Hydrobiol. 65: 85–125.

Fernando, C. H. 1980b. The species and size composition of tropical freshwater zooplankton with special reference to the oriental region (South East Asia). Int. Revue ges. Hydrobiol. 65: 411–426.

Fryer, G. 1957. The food of some freshwater cyclopoid copepods and its ecological significance. J. Animal Ecol. 26: 263–286.

Green, J. 1967a. The distribution and variation of *Daphnia lumholtzi* (Crustacea: Cladocera) in relation to fish predation in Lake Albert, East Africa. J. Zool. Lond. 151: 181–197.

Green, J. 1967b. Associations of Rotifera in the zooplankton of the lake sources of the White Nile. J. Zool. Lond. 151: 343–378.

Green, J. 1971. Associations of Cladocera in the zooplankton of the lake sources of the White Nile. J. Zool. Lond. 165: 373–414.

Haney, J. F. 1987. Field studies on zooplankton–cyanobacteria interactions. N. Zeal. J. Mar. Freshwat. Res. 21: 467–475.

Hecky, R. E. 1993. The eutrophication of Lake Victoria. Verh. Internat. Verein. Limnol. 25: 39–48.

Hecky, R. E., and F. W. B. Bugenyi. 1992. Hydrology and chemistry of the African Great Lakes and water quality issues: problems and solutions. Mitt. Verein. Internat. Limnol. 23: 45–54.

Holtby, L. B., and R. Knoechel. 1981. Zooplankton filtering rates: error due to loss of radioisotopic label in chemically preserved samples. Limnol. Oceanogr. 26: 774–780.

Hutchinson, G. E. 1967. A Treatise on Limnology, Volume 2. Wiley.

Infante, A. de. 1982. Annual variations in abundance of zooplankton in Lake Valencia (Venezuela). Arch. Hydrobiol. 93: 194–208.

Komarek, J., and H. Kling. 1991. Variation in six plantonic cyanophyte genera in Lake Victoria (East Africa). Archiv. Hydrobiol. Suppl. 61: 21–45.

Kudhongania, A. W., and A. J. Cordone. 1974. Batho-spatial distribution pattern and biomass estimate of the major demersal fishes in Lake Victoria. Afr. J. Trop. Hydrobiol. Fish. 3: 15–31.

Lampert, W. 1987. Laboratory studies on zooplankton–cyanobacteria interactions. N. Zeal. J. Mar. Freshwat. Res. 21: 483–490.

Lean, D. R. S. 1973. Phosphorus dynamics in lakes. Science 179: 678–680.

Lean, D. R. S., and C. Nalewajko. 1976. Phosphate exchange and organic phosphorus excretion by freshwater algae. J. Fish. Res. Board Can. 33: 1312–1323.

Lehman, J. T. 1988. Ecological principles affecting community structure and secondary production ⱶ zooplankton in marine and freshwater environments. Limnol. Oceanogr. 33: 931–945.

Lehman, J. T. 1996. Pelagic food webs of the East African Great Lakes, pp. 281–301. In T. C. Johnson and E. O. Odada [eds.], The Limnology, Climatology and Paleoclimatology of the East African Lakes, Gordon and Breach.Toronto.

Lehman, J. T., and D. K. Branstrator. 1993. Effects of nutrients and grazing on the phytoplankton of Lake Victoria. Verh. Internat. Verein. Limnol. 25: 850–855.

Lehman, J. T., and D. K. Branstrator. 1994. Nutrient dynamics and turnover rates of phosphate and sulfate in Lake Victoria, East Africa. Limnol. Oceanogr. 39: 227–233.

Lewis, W. M. 1978. Dynamics and succession of the phytoplankton in a tropical lake: Lake Lanao, Philippines. J. of Ecol. 66: 849–880.

Mavuti, K. M., and M. R. Litterick. 1991. Composition, distribution and ecological role of zooplankton community in Lake Victoria, Kenya waters. Verh. Internat. Verein. Limnol. 24: 1117–1122.

Mazumder, A., D. R. S. Lean, and W. D. Taylor. 1992. Dominance of small filter feeding zooplankton in the Lake Ontario foodweb. J. Great Lakes Res. 18: 456–466.

Moriarty, D. J. W., J. P. E. C. Darlington, I. G. Dunn, C. M. Moriarty, and M. P. Tevlin. 1973. Feeding and grazing in Lake George, Uganda. Proc. R. Soc. Lond. 184: 299–319.

Mugidde, R. 1993. The increase in phytoplankton primary productivity and biomass in Lake Victoria (Uganda). Verh. Internat. Verein. Limnol. 25: 846–849.

Neter, J., W. Wasserman, and M. H. Kutner. 1985. Applied Linear Statistical Models. Irwin.

Ochumba, P. B. O., and D. I. Kibaara. 1989. Observations on blue-green algal blooms in the open waters of Lake Victoria, Kenya. Afr. J. Ecol. 27: 23–34.

Ogutu-Ohwayo, R. 1988. Reproductive potential of the Nile perch, Lates niloticus L., and the establishment of the species in Lakes Kyoga and Victoria (East Africa). Hydrobiologia 162: 193–200.

Ogutu-Ohwayo, R. 1990. The decline of the native fishes of lakes Victoria and Kyoga (East Africa) and the impact of introduced species, especially the Nile perch, Lates niloticus, and the Nile tilapia, Oreochromis niloticus. Environ. Biol. Fish. 27: 81–96.

Oijen, M. J. P. van, F. Witte, and E. L. M. Witte-Maas. 1981. An introduction to ecological and taxonomic investigations on the haplochromine cichlids from the Mwanza Gulf of Lake Victoria. Neth. J. Zool. 31: 149–174.

Peters, R. H., and J. A. Downing. 1984. Empirical analysis of zooplankton filtering and feeding rates. Limnol. Oceanogr. 29: 763–784.

Reynolds, C. S. 1989. Physical determinants of phytoplankton succession, pp. 9–56. In U. Sommer [ed.], Plankton Ecology, Succession in Plankton Commuities. Springer-Verlag.

Richman, S., and S. I. Dodson. 1983. The effect of food quality on feeding and respiration by Daphnia and Diaptomus. Limnol. Oceanogr. 28: 948–956.

Robinson, A. H., and P. K. Robinson. 1971. Seasonal distribution of zooplankton in the northern basin of Lake Chad. J. Zool. Lond. 163: 25–61.

Rzoska, J. 1957. Notes on the crustacean plankton of Lake Victoria. Proc. Linn. Soc. Lond. 168: 116–125.

Sars, G. O. 1909. Zoological results of the Third Tanganyika Expedition, conducted by Dr. W. A. Cunnington, F. Z. S., 1904–1905: report on the copepoda. Proc. Zool. Soc. Lond.

Sterner, R. W. 1989. The role of grazers in phytoplankton succession, pp. 107–170. *In* U. Sommer [ed.], Plankton Ecology, Succession in Plankton Commuities. Springer-Verlag.

Talling, J. F. 1965. The photosynthetic activity of phytoplankton in East African lakes. Int. Revue ges. Hydrobiol. 50: 1–32.

Talling, J. F. 1966. The annual cycle of stratification and phytoplankton growth in Lake Victoria (East Africa). Int. Revue ges. Hydrobiol. 51: 545–621.

Talling, J. F., and I. B. Talling. 1965. The chemical composition of African lake waters. Int. Revue ges. Hydrobiol. 50: 421–463.

Twombly, S. 1983. Seasonal and short term fluctuations in zooplankton abundance in tropical Lake Malawi. Limnol. Oceanogr. 28: 1214–1224.

Uutala, A.J., N. D. Yan, A. S. Dixit, S. S. Dixit, and J. P. Smol. 1993. Paleolimnological assessment of damage to fish communities in three acidic, Canadian Shield lakes. Fish. Res. 19: 157–177.

Witte, F. 1981. Initial results of the ecological survey of the haplochromine cichlid fishes from the Mwanza Gulf of Lake Victoria (Tanzania): breeding patterns, trophic and species distribution. Neth. J. Zool. 31: 175–202.

Witte, F., T. Goldschmidt, J. Wannik, M. van Oijen, K. Goudswaard, E. Witte-Maas, and N. Bouton. 1992. The destruction of an endemic species flock: quantitative data on the decline of the haplochromine cichlids of Lake Victoria. Environ. Biol. Fish. 34: 1–28.

Worthington, E. B. 1931. Vertical movements of freshwater zooplankton. Int. Revue ges. Hydrobiol. 25: 394–436.

A Review of Lake Victoria Fisheries with Recommendations for Management and Conservation

E.F.B. KATUNZI   *Senior Research Officer, Mwanza Fisheries Research Centre, Mwanza, Tanzania*

Abstract — The Lake Victoria fisheries have undergone significant change in recent years along with a shift in the lake's ecosystem. The multispecies fishery, based on a variety of exploitation techniques, has been reduced to mainly a trispecies fishery exploited by gill-nets, beach seines, and hooks. *Lates niloticus*, *Rastrineobola argentea* and *Oreochromis niloticus* form the basis of this activity, although there are signs of recovery of the inshore and rocky haplochromine cichlids. This paper reviews the lake fisheries before and after the introduction of the Nile perch. It also analyzes the existing management issues and discusses proposed management and conservation strategies leading to the rational use of the resources.

INTRODUCTION

Lake Victoria is the largest lake in Africa, with a surface area of about 69,000 km². Small scale artisanal fishermen exploit the fish stocks in the lake. Trawling is not popular because of costs involved in the fishing operations. The fisheries of Lake Victoria evolved considerably after the onset of this century. From 1905 to 1916, gill-nets were introduced into the lake to exploit the tilapiine cichlids, *Oreochromis esculentus* and *Oreochromis variabilis*. Catches declined with time as the fishing activity increased (Graham, 1929). At the time, *Protopterus aethiopicus*, *Bagrus docmac* and *Clarias gariepinus* were among the species targeted for exploitation, and they soon became overfished. The fisheries then moved to other species which also subsequently declined.

Lack of scientific knowledge on the exploitable species, and fishing without strict management, contributed greatly to the decline of *Labeo victorianus* in the late 1950s. This species was normally caught by blocking rivers during spawning. In an attempt to increase Tilapia catches, *T. zilii*, *T. melanoopleura*, *Oreochromis niloticus*, and *O. leucostictus* were introduced into the lake. The outcome, however, never showed positive returns (Welcome, 1967; Fryer, 1972; Kudhongania and Cordone, 1974; Lowe-McConnel, 1987). The attitude of intensifying fishing on the large targeted species and extensive use of beach seines starting in the late 1960s made the situation worse for the tilapiines. The use of beach seines in the inshore waters disturbed the breeding nests and eggs of cichlids that tended to breed on sandy inshore waters.

In the 1970s, beach seines and a few trawlers near Mwanza were used to exploit the haplochromines. These caught large numbers of spawners and young Tilapia. At the same time minor fisheries based on *Schilbe*, *Alestes*, *Barbus*, *Synodontis* and *Momyrus* were being carried out using gill nets. As a result of the decline of target species, interest to exploit the *Rastrineobola argentea* grew, and a light fishery was developed for this purpose in Kenyan and Tanzanian waters (Marten, 1979; Okedi, 1981). Interest in the haplochromines also grew despite the unpopularity of their bony nature.

This period became difficult for artisanal fishermen whose income greatly declined due to low catches. Further decline on the targeted species in the early 1970s forced the fishermen that remained in the industry to concentrate more on the haplochromines, as these were the only abundant species (80% of the dermesal ichthyomass) (Kudhongania and Cordone, 1974). A small scale trawl fishery was therefore developed on the Tanzanian side of Lake Victoria. It was on this basis that a fish meal plant was constructed in the Mwanza area to convert the bony haplochromine into chicken feed. The capacity of the factory was to convert 60 tons in a day, but the target was never reached. With the demand for both human and chicken feed, there was overfishing of the haplochromines both in the Mwanza Gulf and Nyanza Gulf (Marten, 1979; Witte, 1981; Witte and Goudswaard, 1985). Despite the declines brought about by uncontrolled fishing, the Lake Victoria ecosystem remained diverse with a complex fish fauna of ecological importance.

In order to boost the fisheries, *Lates niloticus* was introduced to the lake in 1954 (see Kudhongania et al., this volume). The effects of introduction were not apparent in Tanzanian waters until towards the end of the 1979. The introduced *Lates* spread very fast and became very successful, at the expense of a vast number of haplochromines. As the establishment of the Nile perch continued, stocks of most of the native species declined rapidly and some completely disappeared, with the exception of *R. argentea*. Many haplochromine species disappeared or are presently nearing extinction. The diverse forage specialities developed by the cichlids in the lake disappeared, hence causing new pathways in the food-chain. Other environmental effects such as frequent blue-green algal blooms and mass mortalities of fish from anoxia became common (Ochumba and Kibara, 1989).

The changes brought about by the predation pressure on haplochromines and other native fish species have resulted in the significant changes in the fishery. Perhaps surprisingly, there has been a fourfold increase in the fishery output in the three riparian states (CIFA, 1989). The Nile perch is the most important species caught in each of the three countries. The increase in fishery output is also due to *Oreochromis niloticus* and *Rastrineobola argentea*.

THE CURRENT FISHERIES REGIMES

There is immediate need to identify the resource base for a sustainable level of exploitation of Lake Victoria fish. The economic base of the small scale fishermen greatly depends on the sustainability of the fishery. The original multispecies fisheries has now been reduced to one based on *L. niloticus*, *R. argentea*, and *O. niloticus*. The total fish production in the lake has greatly increased as a result of the Nile perch, *R. argentea* and *O. niloticus*. The biggest challenge to the riparian countries is to devise a concerted management strategy to avoid possible consequences of over exploitation and loss of socioeconomic benefits from the new fisheries regimes.

Lack of information on the size of the Nile perch stock has strongly limited the planning and development of the fishery. The initial approach on a Nile perch study is to investigate the forage strategies and population parameters (Ogari, 1985; Ocere, 1985; Asilla and Ogari, 1987). Information on the structure of the Nile perch artisanal fishery is described by Ligtvoet and Mkumbo (1991). The description and interpretation of events taking place in the lake should be interpreted with great caution, as changes can occur within a short period of time.

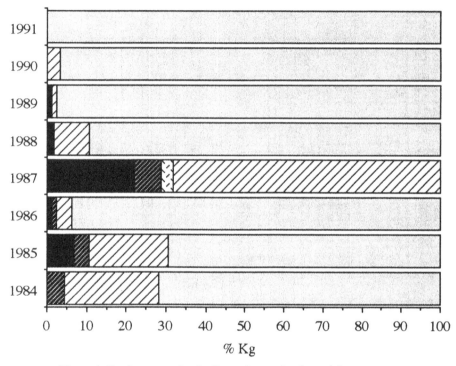

■ Clarias Sp.
▨ Bagrus docmac
▨ Oreochromis niloticus
▨ Protopterus arthiopicus
☐ Lates Sp.

Figure 1. Catch per year for the five major species, by weight percent.

The research vessel M/V Kiboko, based on the southern part of the lake, has been making surveys since 1984. As observed in Figs. 1–3, the fish species composition has been changing very rapidly since that time. *Lates* increased, reaching a climax in 1986 (Fig. 2), and then declined. At the same time other target species have been decreasing and are approaching extinction.

A recent survey made in the lake revealed information on the changes in trends of the fishery over the past three years (Bwathondi et. al., 1992). The gill-net fishery on the Tanzanian side is still dominant except for the eastern side (Mara region) where beach seines contribute about 50% of the landings. Earlier studies indicated a progressive decrease in mesh size of the gill-nets in the Nile perch population from 7–8" in 1987–88 (Ligtvoet and Mkumbo, 1991) to 5–8" in 1992, with 7" being common to all the regions. This continual decrease in mesh size increases the likelihood of over exploitation. The average size of fish caught in the nets has decreased over the years. The current modal length falls between 50 and 60 cm compared with the earlier recorded range of 60–80 cm.

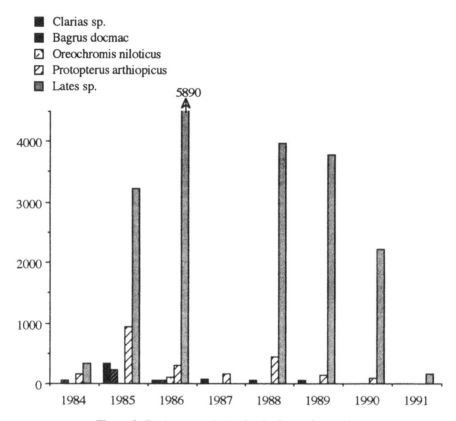

■ Clarias sp.
■ Bagrus docmac
▣ Oreochromis niloticus
◨ Protopterus arthiopicus
▥ Lates sp.

Figure 2. Catch per year by kg for the five major species.

Gill-nets are still in use despite a proposal to prohibit them. This is the most common fishing technique employed on the eastern side of the lake (Mara region). The cod-end meshes for the seines in use ranges between 1 and 3". In the recent survey of Bwathondi et al. (1992), more than 80% of the fish caught were shorter than 50 cm. This is less than the modal length typical of the other regions. The majority of fish caught with hooks are longer than 50 cm, although few fishermen use hooks. The use of hooks is encouraged since the required investment is low and the gear is highly selective.

The *Rastrineobola* fishery is also contributing greatly to the catch landings. Much of the catch data are from inshore and coastal areas (Greenwood, 1966). In the Mwanza Gulf, studies indicate an increase in biomass, both in terms of population density and mean size, with depth up to 10–20 m, and decrease below this depth range (Wanink, 1989). Apart from contributing to the fishery, *R. argentea* is known to play an ecological role as a link in the food chain between the herbivores and predators higher up in the ecosystem (Katunzi, in prep). The artisanal fishery on *R. argentea* is based on beach seines, scoop nets and lift-nets. There are basic characteristics of each gear type but the following features are common to all of them:

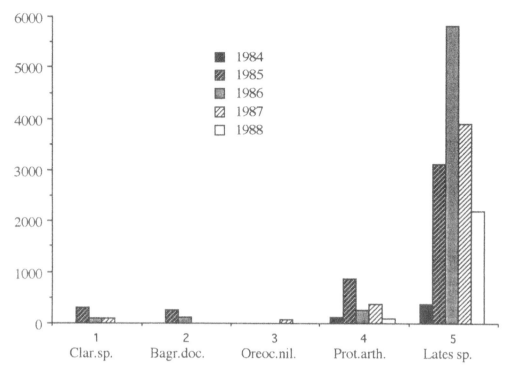

Figure 3. Species catch, in kg, 1984–90.

— All gears operate better at night and depend on external artificial sources of light to concentrate the fish.

— Since light concentration is a prerequisite to the fishing operations, fishing stops during full moon.

— Many of the fish are sold dry. Only 20% are sold fresh. Sundrying is carried out during the day on the beaches.

— Fishermen involved in the operations are normally hired or employed and will change fishing grounds depending on the catches.

The beach seine is the most common technique for exploitation of *Rastrineobola*. These have mesh sizes ranging between 8 and 10 mm. A light source is attached to a raft and pulled along with the net along the beach. In scoop net fishery, the lamps are hauled close to a canoe after the fish have been attracted so that they can be scooped into the canoe. Estimating the catch of *Rastrineobola* is difficult because recorders have to stay out over night and landing beaches change from time to time depending on the catch.

There are no records of *Oreochromis* fishery, but observations confirm their increase in the market. Like *Lates*, *Oreochromis* is exploited by gill-nets, beach seines and hooks. Continual use of beach seines in the shallow, sandy beaches catch a lot of juvenile *Oreochromis*. Breeding sites and eggs are destroyed by the dragging net. The fishery is seriously threatened if measures are not put into effect to correct or officially prohibit the use of gill-nets.

CONSERVATION AND MANAGEMENT ISSUES

The present status suggests that there is no hope of restoring the past species diversity in the lake. Species of *Haplochromis* that occupy inshore and rocky habitat show signs of recovery but for many species, their numbers are greatly reduced or at the verge of extinction. In lakes where *L. niloticus* is native, the haplochromine species are very few (i.e., Lake Albert and Turkana) (Greenwood, 1974).

The possible causes of the decline in fish diversity have not been studied fully apart from the effects brought about by the introduction of the Nile perch. We know very little about what it takes to keep most species functional. We have to study the details of natural processes maintaining the ecosystem and evaluate their relative importance in affecting the fish community. Throughout the development of the Lake Victoria fisheries, overfishing has always been the main cause of decline of quite a number of species (Graham, 1929; Marten, 1979; Witte, 1981). The continued reduction of mesh size of the gill-nets must be stopped to reduce further exploitation of immature fish. The existing fisheries regulation concerning mesh sizes needs review because they are outdated and are no longer binding.

Stricter enforcement of existing regulations must be accomplished. Although trawling in bays, gulfs and inlets at depths of 20 meters or less is prohibited, most of the trawling and beach seining operations are made in these places without permission. It has been proposed that in order to save both the Nile perch and *Oreochromis* populations, the minimum mesh of the gill-nets in use should be 5". No beach seining should be allowed on the lake. This proposal was met with success in Uganda and Kenya, but is yet to be implemented in the Tanzanian waters.

Another possible threat to the biodiversity of species in the lake is alteration of the habitats. Stocks of *Lates* greatly influenced the trophic efficiency of the system. The interaction of species and conditions upon which the overall diversity and stability of the lake depended no longer exist (Ogutu-Ohwayo, 1990). Environmental change brought about by loss of species and trophic diversity, and the accompanied alterations in food web, are linked with algal blooms and the deoxygenation of the hypolimnion (Ochumba and Kibara, 1989).

Although pollution is not yet a big threat to many parts of the lake, precaution must be taken with the rapid expansion of both agriculture and industry in the drainage basin. The move of populations from rural to urban areas has greatly increased the demand for water resources. The urban centres in Bukoba, Musoma and Mwanza are engaged in industrial activities that could impact the lake's ecosystem. The coffee curing plant in Bukoba, the textile factory and fish processing plant in Musoma, and the vegetable, oil and soap factories in Mwanza are a few examples where industrial effluent is draining into the lake. Agricultural activities along the lake have grown and in certain localities forests have been cleared in order to get more land for agriculture. This has resulted in the use of pesticides for crops which in many cases drain into the lake. Selective use of chemicals in the agricultural sector should be encouraged to avoid possible consequences of pollution. The Tanzania Bureau of Standards should set measures to control industrial discharge into the lake.

Apart from the environmental perturbation caused by the alien *Lates niloticus*, the socioeconomic impact of the species in the neighbouring areas has been positive. The fisheries of the lake, which were almost collapsing, were revitalized with the Nile perch. Similarly, catches of *Rastrineobola* and *Oreochromis niloticus* supplemented the increase.

However the rapid expansion of the fish processing factories on the Tanzanian side of the lake call for alarm because they have been established with high expectations for production, without regard for keeping the fishery on a sustainable level for future rational exploitation.

REFERENCES

Acere, T.O., 1985, Observations on the biology of the Nile perch *L. niloticus* and growth of its fishery in the northern waters of Lake Victoria: Food and Agricultural Organization Fishery Report, v. 335, pp. 42–61.

Asila, A.A., and Ogari, J., 1987, Growth parameters and mortality rates of Nile perch *Lates niloticus* estimated from length data in the Nyanza Gulf (Lake Victoria), *in* Contributions to Tropical Fisheries Biology, Venema, S., Christesten, J. Moller, and Pauly, D., eds. Food and Agricultural Organization Fishery Report, v. 389, pp. 272–287.

Bwathondi, P.O.J., Katunzi, E.F.B., Budeba, Y.L., Temu, M.M., Kashindye, J.J., and Mkumbo, O.C., 1992, The Artisanal Fishery of the Nile perch in Lake Victoria (Tanzanian Waters): A report submitted to the European Commission under Project No. 5100.36.94.401

CIFA, 1988, Report of the 4th session of the sub-committee for the development and management of the fisheries in Lake Victoria, Kisumu, Kenya, 6–10 April, 1987: Food and Agricultural Organization Fishery Report, v. 388, pp. 1–112.

Fryer, 1972, Conservation of the Great Lakes of East Africa: A lesson and a warning: Biological Conservation, v. 5, pp. 222–223.

Graham, M., 1929, Fishing Survey of Lake Victoria (1927–1928): London: His Majesty's Stationery Office, 255 pp.

Greenwood, 1966, The Fishes of Uganda, Second Revised edition, Kampala: The Uganda Society, 131 pp.

Greenwood, 1974, The cichlid fishes of Lake Victoria, East Africa. The biology and evolution of a species flock: Bulletin of the British Museum Natural History (Zool.), suppl. 6, pp. 1–134.

Kudhongania, A.W., Cordone, A.W., and Cordone, A.J., 1974, Past trends, present stocks and possible future state of the fisheries of the Tanzania part of the Lake Victoria: Hydrobiological Fishery, v. 3, pp. 167–181.

Kudhongania, A.W., Ocenodongo, D.L., and Okaronon, J.O., 1996, Anthropogenic perturbations on the Lake Victoria ecosystem, *in* The Limnology, Climatology and Paleoclimatology of the East African Lakes, T.C. Johnson and E.O. Odada, eds. Toronto, Gordon and Breach, pp. 625–632.

Ligtvoet, W., and Mkumbo, O.C., 1991, A pilot sampling survey for monitoring the artisanal Nile perch *Lates niloticus* fishery in the southern Lake Victoria (East Africa), *in* catch Effort Sampling Strategies. Their Application in Fresh Water Fisheries Management, I.G. Cowx, ed. Fishing News Books, pp. 349–360.

Lowe-McConnel, R.H, 1987, Ecological studies in tropical fish communities: Cambridge, University Press, 382 pp.

Marten, G.G., 1979, Impact of fishing on the inshore fishery of Lake Victoria (East Africa): Journal of Fisheries Research Board of Canada, v. 36, pp. 891–900.

Ochumba, P.B.O., and Kibara, D.I., 1989, Observations on blue green algal blooms in the open waters of Lake Victoria, Kenya: African Journal of Ecology, v. 27, pp. 23–34.

Ogutu-Ohwayo, R., 1990, The reduction in fish species diversity in Lakes Victoria and Kyoga (East Africa) following human exploitation and introduction of non-native fishes: Journal of Fish Biology, Supplement A., pp. 207–208.

Okedi, J., 1981, Integrated management for Dagaa fishery of Lake Victoria, *in* Proceedings of the workshop of the Kenya Marine and Fisheries Research Institute on Aquatic Resources of Kenya, July 13–19, 1981, pp. 440–444.

Wanink, J.H., 1989, The ecology and fishery of Dagaa *Rastrineobola argentea* (Pellagrin 1904), *in* Fish Stock and Fisheries in Lake Victoria. A handbook to the HEST/TAFIRT/FAO/DANIDA regional seminar, Mwanza, Tanzania, January 1989. Report of the Haplochromis Ecology Survey Team (HEST) and Tanzania Fisheries Research Institute (TAFIRI) No. 53. Leiden, The Netherlands.

Welcomme, R.L., 1967, Observations on the biology of the introduced species of Tilapia in lake Victoria: Revue of Zoology and Botany of Africa, v. 76, pp. 249–276.

Witte, F., 1981, Initial results of the ecological survey of the Haplochromine Cichlids from the Mwanza gulf of Lake Victoria, Tanzania: breeding patterns. Trophic and species distribution: Netherlands Journal of Zoology (In HEST-BUNDEL 1981), v. 31, pp. 175–202.

Witte, F., and Goudswaard, P.C., 1985, Aspects of the haplochromine fishery in southern Lake Victoria: FAO Fishery Report, v. 335, pp. 81–88.

Witte, F., Goldschmidt, Wanink, J., Van-Oijen, M.V., Goudswaard, P.C., Els, W.M., and Bouton, N., 1992, The destruction of an endemic species flock: quantitative data on the decline of haplochromine cichlids of Lake Victoria: Environmental Biology Of Fishes, v. 34, pp. 1–28.

Sedimentary Processes and Deciphering the Past in the Large Lakes

Sedimentary Processes and Signals of Past Climatic Change in the Large Lakes of the East African Rift Valley

T.C. JOHNSON   *Large Lakes Observatory, University of Minnesota, Duluth*

Abstract — The large lakes of the East African Rift Valley contain a unique record of past climatic change in the tropics, extending back several million years in time. Thus far sediment sampling has been restricted to piston coring or less effective methods for recovering long records, so our knowledge of past climate variability has been restricted to the last 20,000 years in most lakes, with glimpses of conditions as far back as 40,000 years in Lake Tanganyika and Lake Malawi. Patterns of sedimentation in the large lakes are complex, and acoustic remote sensing techniques must be used to identify promising coring targets that have not been disturbed by bottom currents, turbidity currents or mass wasting processes. All of the large lakes in the Rift Valley north of about 9° South Latitude exhibit a coherent pattern of lake level fluctuations over the past 20,000 years, with lowstands from 20 ka (thousand of years before present) to about 12 ka, followed by a rise to outlet levels between 12 and 10 ka, and highstands maintained throughout most of the Holocene. Where records extend further back in time, e.g., Lakes Tanganyika, Mobutu and Victoria, the onset of the lowstand appeared to have occured around 25 ka. Lake Malawi, by contrast, has a lake level history that appears to be exactly out of phase with the lakes to the north, and probably indicates the southern hemisphere response to orbital forcing on insolation and the millenial scale variability in the position of the Intertropical Convergence Zone. The paleoclimate records of the Rift Valley lakes frequently contain intriguing signals of high temporal resolution, including evidence for annual and decadal scale periodicities.

INTRODUCTION

The large lakes of the East African Rift Valley (Fig. 1) contain an unsurpassable record of past climatic change in the tropics. Sedimentation rates are fast, averaging about 1 mm/y, and bioturbation is minimal in the deep basins, often due to anoxia (Johnson, 1984; Pilskaln and Johnson, 1991). As a result, temporal resolution of the paleoclimate record is resolvable easily to decades, if not to individual years (e.g., Owen and Crossley, 1992). Lake levels have changed dramatically in response to Holocene climatic change (Beadle, 1981). All of the large lakes have been in a closed-basin configuration at certain times in the past 20,000 years, so their water chemistry and biota have undergone significant change (Kendall, 1969; Haberyan and Hecky, 1984; Gasse et al., 1989; Finney and Johnson, 1991). These changes are recorded as strong signals in the microfossil assemblages, mineralogy and chemical composition of the sediments. Because of the great depth of some of the lakes (1500 m for Tanganyika and 700 m for Malawi), the deep basins have remained submerged even during the driest of climatic episodes (Scholz and Rosendahl, 1988), so sedimentation has been continuous at times when the sedimentary record of smaller lakes and bogs in East Africa were not. More than four kilometers of sediment underlie some of the rift lakes (Rosendahl, 1987), representing on the order of 10–15 million years of sediment accumulation (Cohen et al., 1993). All of these factors contribute to a paleoclimate record that may have the

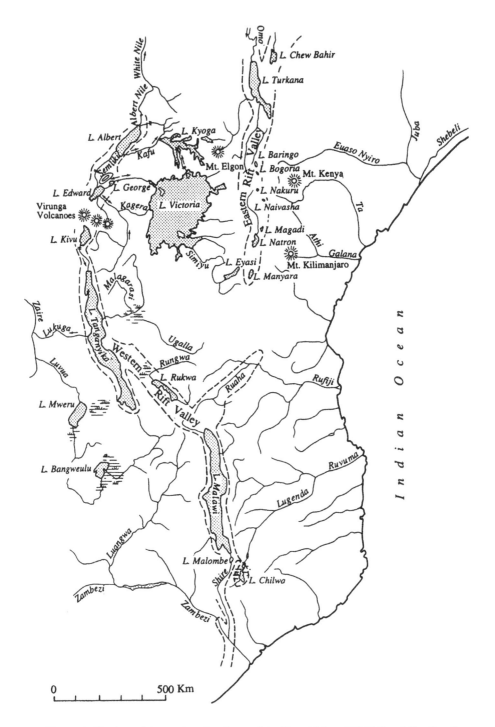

Figure 1. The large lakes and major rivers of the East African Rift Valley (after Beadle, 1981).

temporal resolution of glacial ice cores, and the longevity and continuity of deep-ocean cores.

Unfortunately, recovery of this unique record of past climatic change is plagued with difficulty. Sedimentation patterns in large lakes are complex because of several dynamic processes that affect the composition of the sediment rain in time and space. The flux of biogenic material to the lake floor varies with seasonal and interannual changes in biological productivity which depends on the intensity of upwelling and other factors affecting the supply of nutrients to the photic zone. Water chemistry varies significantly with depth in the large lakes, and affects the dissolution or precipitation of certain mineral phases (see, for examples, Ricketts and Johnson, this volume, and Owen et al., this volume). Sediment redistribution by waves, bottom currents, gravity flows and turbidity currents are common. Tectonic processes can also imprint on the sedimentary record, and distinguishing the tectonic signals from the climatic ones is not always simple. Standard coring techniques (e.g., Kullenberg piston corer, box corer, Mackereth corer) seldom recover more than 12 m of sediment from the deep basins, representing in most cases just 12,000 or so years because of very fast sedimentation rates. Finally, our ability to date rift lake sediments is limited at the present time, especially outside the range of the radiocarbon time scale (ca. 40,000 years).

This paper reviews the evidence for various sedimentary processes that have been inferred from seismic reflection and side scan sonar profiles from the east African lakes, the signals of past climatic change that have been derived from their sediments, and the problems we currently face with regard to geochronology. The paleoclimate records deduced from several of the large East African lakes are compared to demonstrate their potential as proxies of past climatic change, and to illustrate some of their similarities and differences.

SEDIMENTARY PROCESSES DETERMINED FROM ACOUSTIC REMOTE SENSING

The extent to which sediments are reworked by physical processes depends on the size of the lake basin, the depth and steepness of the basin morphology, and the rate of sediment input. Superimposed on these factors are the frequency and magnitude of past lake level changes that can shift the boundaries of erosion and deposition tens of kilometers landward or lakeward. The two largest lakes in the East African Rift Valley, Tanganyika and Malawi, each exhibit complex patterns of sedimentation, with sediment erosion and redeposition occurring even in deep basins as a result of many processes (e.g., Tiercelin and Mondeguer, 1991; Johnson and Ng'ang'a, 1990). Lake Turkana, on the other hand, is much smaller with a gently sloping lake floor in most regions, and a very high sediment influx; the pattern of sedimentation is much simpler in this lake than in the larger, deeper lakes to the south (Johnson et al., 1987).

Calculations by standard oceanographic methods indicate that surface waves generated by winds can impact sedimentation easily to a depth of 100 m in lakes the size of Malawi and Tanganyika (Johnson, 1980). High-resolution and side-scan sonar records from Lake Malawi (Johnson and Ng'ang'a, 1990) and from Lake Tanganyika (Tiercelin et al., 1992) reveal a highly reflective sandy facies that usually extends to a depth of about 100 m, except close to river deltas (Fig. 2). Side-scan sonar records from the wave-swept regions often reveal a mosaic of high and low acoustic reflectivity that indicates the presence of a thin, patchy mud veneer that accumulates ephemerally and is swept into deeper water during

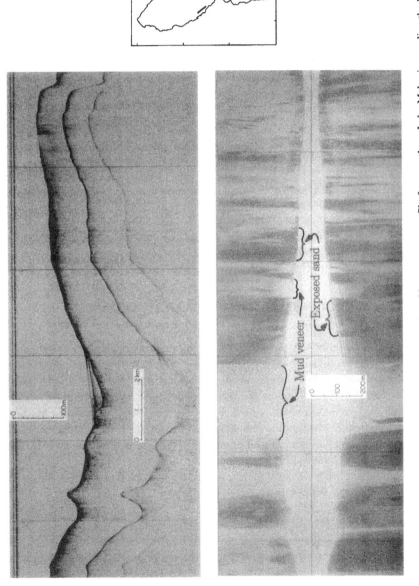

Figure 2. A high-resolution seismic reflection profile and accompanying side scan sonar profile from northern Lake Malawi, revealing the lack of a permanent sediment cover in water shallower than 100 m, due to erosion by oscillatory currents associated with surface waves (from Johnson and Ng'ang'a, 1990).

storms (Fig. 2). In Lake Turkana, the boundary between the sandy, wave affected environment and the mud accumulation zone farther offshore is much shallower than in the larger lakes because of its much higher sedimentation rate and smaller fetch.

Winds blowing over the rift lakes set up longshore currents in the nearshore zone that can generate significant lateral transport of sediment. Cohen et al. (1986) provide evidence for this in the mineral distribution of sands off some small rivers in Lake Turkana. Side scan sonar records from Lake Malawi show well developed sand waves aligned perpendicular to the shoreline, particularly along the western shore of the lake in water depths shallower than 20 m, where strong, northward flowing currents are generated by the southeast trade winds and are funneled up the rift valley during the dry, windy season of the austral winter (Johnson et al., in press) (Fig. 3).

There is also abundant evidence for strong, deep-water currents that erode and redeposit sediments in the large lakes. Erosional moats have been scoured out from around bathymetric highs and fields of mud waves have been observed in water 100 to 200 m deep in Lake Malawi (Johnson and Ng'ang'a, 1990; Scott et al., 1991) (Fig. 3). Large sediment waves form at 400 m depth in the southern basin of Lake Tanganyika (Tiercelin et al., 1988; Tiercelin and Mondeguer, 1991) (Fig. 3). Deep currents have not been measured directly in any of the rift lakes, so their origins are not known, but they undoubtedly include the presence and funneling of seiches or density currents (see Spigel and Coulter this volume).

Evidence for downslope transport of sediment by creep, debris flows or turbidity currents is also observed frequently on seismic reflection profiles from Lakes Tanganyika and Malawi. Near major border faults, catastrophic debris flows are triggered by periods of heavy rain or earthquakes and create a line of small fan deltas with poorly sorted coarse-grained deposits, including boulders, near their apices, that become progressively finer grained, better stratified and bedded with distance from the border fault (Tiercelin et al., 1992). Seismic reflection profiles from these regions show a chaotic lake floor with high reflectivity and numerous hyperbolic echoes near the fault, gradually changing to flatter lake floor morphology, better stratification, and better penetration of acoustic energy with distance towards the basin center (Fig. 4). Sediment diapirs are widely observed in the stratified deposits at the distal portions of the fan deltas. These result from the rapid sediment loading of coarse-grained deposits onto the interbedded muds and sands offshore, causing sufficient hydrodynamic force to drive the diapiric flow (Tiercelin et al., 1992) (Fig. 4). Diapirs such as these have been observed off a border fault in Lake Malawi as well (Johnson and Ng'ang'a, 1990) and off major river deltas in northern Lake Tanganyika (Bouroullec et al., 1991) and northern Lake Malawi (Johnson et al., in press). Evidence for slower down-slope movement of sediment is seen in the formation of compressional ridges formed in sandy mud on the slopes off some river deltas (Scott et al., 1991; N'gang'a, 1993; Johnson et al., in press) (Fig. 4). These ridges probably form by the process of creep, and are erased by the more dynamic action of debris flows and turbidity currents when a certain threshold of instability is achieved.

The lake floor off the large river deltas in Lake Malawi often shows a chaotic or hummocky relief on high-resolution seismic profiles, along with prolonged and diffuse acoustic reflectivity (Johnson and Ng'ang'a, 1990; Ng'ang'a, 1993; Johnson et al., in press). Sediment cores from such areas contain a substantial amount of gas, presumably methane derived from early diagenesis of organic matter. The importance of methanogenesis as a physical agent of sediment disturbance in the rift lakes is not known. It is conceivable that

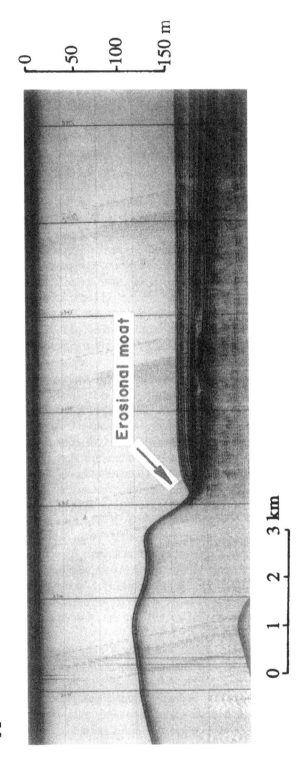

Figure 3. A) An erosional moat and mud waves from 155 m water depth in southern Lake Malawi, indicating the presence of strong bottom currents, probably associated with seiches. B) Large sand waves on the outer shelf of Lake Malawi, just south of the Dwangwa River, caused by longshore currents associated with wind-driven circulation. These sand waves are about 50–100 m long (see Johnson et al., in press). C) Mud waves from about 400 m depth in southern Lake Tanganyika. These have wavelengths of about 500 m and are believed to be formed by density currents (from Tiercelin et al., 1992).

B

0 m

100

200

C

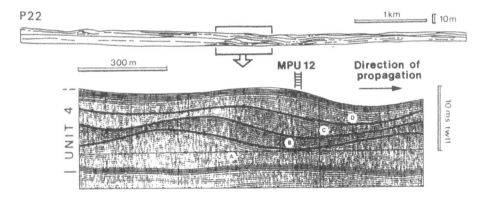

Figure 3, cont'd

it contributes to the formation of diapirs, hummocky relief and downslope transport of sediment.

Evidence for turbidity currents is widespread in Lakes Tanganyika and Malawi. Turbidite channels are well developed off major river deltas (Johnson and Ng'ang'a, 1990; Bouroullec et al., 1991; Tiercelin and Mondeguer, 1992; Ng'ang'a, 1993; Scholz et al., 1993; Johnson et al., in press). Turbidite channels several hundred meters wide are found in many of the deep basins, with well developed levees that rise 10 or more meters above the channel floor (Fig. 5). Attempts to core the turbidite channels usually fail because of the difficulty in coring sand and gravel. Johnson and Ng'ang'a (1990) report the recovery of gravel approximately 1 cm in diameter in a core catcher from a turbidite channel in northern Lake Malawi at a water depth of 400 m.

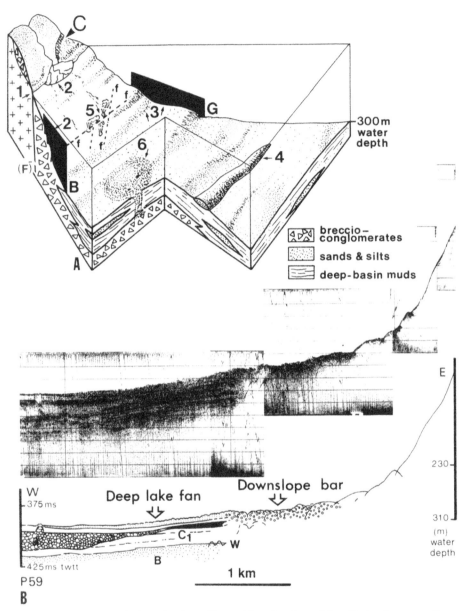

Figure 4. High-resolution seismic profiles and line drawings of the effects of mass wasting along a border fault in Lake Tanganyika (from Tiercelin et al., 1992).

The evidence for hydrothermal input to most of the large rift valley lakes is not extensive, with the exception of Lake Kivu, which is impacted more by volcanism than any of the other large rift lakes. Both temperature and salinity rise with depth in the water column, indicating input of warm, saline water at depth (Degens et al., 1973). A hyperbolic echo observed in the water column in Lake Kivu has been attributed by Degens et al. (1973) and Wong and Von Herzen (1974) to a sub-lacustrine hydrothermal spring (Fig. 6A). Hydrothermal vents

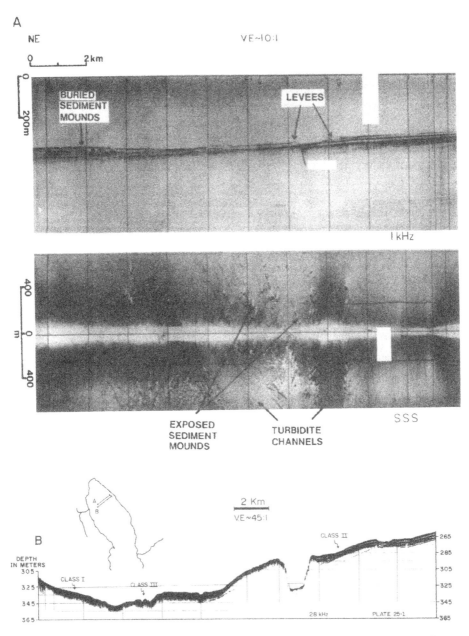

Figure 5. High-resolution seismic profiles, side scan sonar records, and echo sounding record across a major, fault-controlled turbidite channel in northern Lake Malawi (from Scott et al., 1991).

in northern Lake Tanganyika have been examined by Tiercelin and co-workers, who have found localized deposits of carbonates and metal sulfides, as well as chemosynthetic communities of bacteria, in the vicinity of the vents (Fig. 6B) (Tiercelin et al., 1993). More recently, De Batist et al. (this volume) suggest a hydrothermal source for some of the diapirs

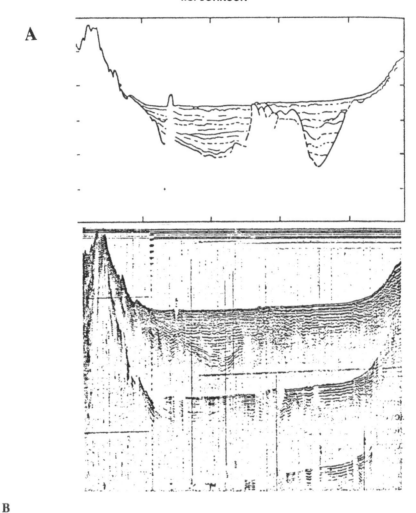

Figure 6. Evidence for hydrothermal input to rift lakes. A) A seismic reflection profile from Lake Kivu, showing a hyperbolic reflector near the left side of the sequence that is interpreted to be a jet of hot water injected from the underlying lake floor (from Degens et al., 1973). B) Photographs of a region of hydrothermal activity in northern Lake Tanganyika: Left, hydrothermal chimneys of aragonite at 6 m depth; Right, gas escaping from sandy bottom sediment at 10 m depth. (Photos provided by J.-J. Tiercelin.)

in deep water sediments of the Livingstone Basin in northern Lake Malawi, although this has not yet been confirmed by geochemical evidence.

Despite the complexity of sedimentation that arises from the various physical processes described, there are extensive regions within the large lake basins where sediment appears to accumulate quite uniformly and unaffected by physical mixing processes, making it most suitable for paleoclimatic study. These "hemipelagic regions" contain sequences of sediments that often are laminated in response to seasonal or interannual variability in the composition of the sediment flux (e.g., Pilskaln and Johnson, 1991; François et al., this volume) and, if situated in sufficiently deep water, have relatively steady sedimentation rates without erosional hiatuses. These areas typically are found on gentle slopes in regions that are isolated from strong bottom currents or downslope sediment transport. Seismic reflection profiles from these regions show highly stratified, acoustically transparent sediment, with no evidence of erosion for long periods (Fig. 7). Even these areas, however, can be affected by recent faulting, so care must be taken in the location of coring sites for paleoclimatic analysis.

GEOCHRONOLOGY

One of the most difficult problems associated with paleoclimatic studies of the African rift lakes is generating a reliable chronology for the sediments. Radiocarbon dates using both conventional methods and accelerator mass spectrometry (AMS) are easily obtained on organic matter or carbonates in the rift lake sediments, but they are not always reliable. Replicate dates on various fractions from nearly the same levels in sediment cores can vary considerably. Two examples are presented that illustrate this problem, and demonstrate why chronologies based on just a few radiocarbon dates should be suspect.

A reliable chronology was obtained in Core LT84-8P from Lake Turkana only after realizing that the dates of bulk carbonates were too old because of an aeolian influx of micrite and small ostracodes from east of the lake. Three fractions of carbonates were dated in this core: bulk carbonate, ostracodes in the 63–150 µm size range, and ostracodes coarser than 150 µm. Halfman et al. (1994) determined that AMS dates on the coarsest ostracodes provided the most consistent age vs. depth relationship, and the youngest dates for any given horizon (Fig. 8A). The resultant chronology in this core has been verified by stratigraphic correlations of the core's lithology and magnetic properties to other cores in the lake basin that have also been dated by the AMS method applied to the coarse carbonate fraction. Core LT84-2P, from the southern basin, did not exhibit such large age discrepancies among the various carbonate fractions, presumably because carbonate-bearing lacustrine deposits are not common on shore, in an upwind direction from this core site (Halfman et al., 1994).

In Lake Malawi, radiocarbon dates of organic and carbonate fractions of the sediments have often compared favorably in many of the cores examined. However there are outlier dates in some of the Malawi cores that are not readily explained. Core M86-18P, for example, yielded a radiocarbon date of $22,050 \pm 510$ ybp on a carbonate horizon from 660 cm depth, sandwiched between other samples of carbonate that dated between $28,110 \pm 420$ ybp and greater than 27,100 ybp. Organic carbon from near the same horizon yielded a date of $33,250 \pm 790$ ybp (Fig. 8B) (Finney et al., this volume). Radiocarbon dates on carbonates and organic carbon from another core nearby, coupled with stratigraphic correlations of diatom assemblages, lithology and acoustic profiles between the two core sites, indicate that the age on the organic fraction is the correct one (Finney et al., this volume).

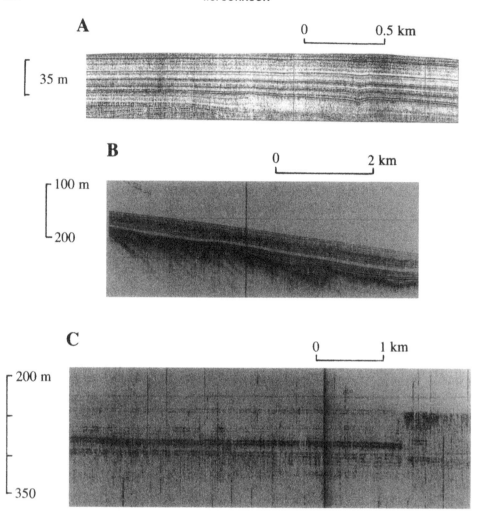

Figure 7. High-resolution seismic reflection profiles from: A) Lake Tanganyika. Approximately 600 m deep in the Mpulungu Basin in the southern part of the lake (from Tiercelin et al., 1992), B) northern Lake Malawi (from Johnson and Ng'ang'a, 1990), and C) southern Lake Malawi (from Johnson and Davis, 1989). All show evidence for long periods of deposition, relatively undisturbed by physical processes, and so would make promising targets for paleoclimatic study. The major unconformity in the middle profile defines the base of the "Songwe Sequence", estimated to be about 75–80,000 years old by Scholz and Finney (in press).

Dating sediments beyond the range of radiocarbon techniques, i.e., older than about 40,000 years, poses even more difficult problems. Volcanic ash layers may be found in the rift lake sediments that can be dated directly or correlated to ashes on land nearby that have been dated by K–Ar or other techniques (e.g., Walter and Aronson, 1982; Sarna-Wojcicki et al., 1985; Haileab and Brown, 1992). Volcanic ash layers have been reported in all of the rift lakes, and should be sufficiently common to provide a reasonable geochronology by Ar/Ar

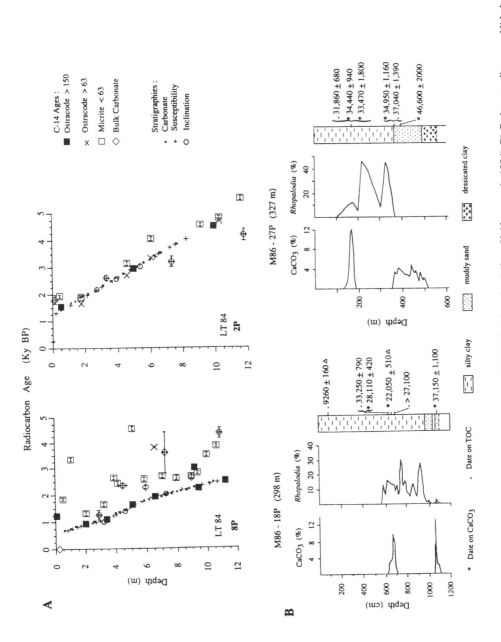

Figure 8. A) Radiocarbon dates from two cores in Lake Turkana (see Fig. 12A for their locations) (from Halfman et al., 1994). B) Carbonate, diatom and lithological stratigraphy and radiocarbon dates for two cores from Lake Malawi (see Fig. 18 for their locations) (from Finney et al., this volume).

or K/Ar dating in sediments older than a few hundred thousand years. In sediments younger than this, K/Ar dating is problematic because of its long half life (Dalrymple and Lanphere, 1969). However recent developments in $^{238}U-^{230}Th$ and $^{235}U-^{231}Pa$ disequilibria dating of volcanic glass using mass spectrometry have yielded dates on mid-oceanic ridge basaltic glass in the range of 0–130 ka, with an accuracy of ±1–7 ka (Goldstein et al., 1993). These techniques may apply in certain instances to the dating of ash layers in the rift lake sediments.

Th/U dating of fossil stromatolites exposed on the margins of some East African paleo-lakes has yielded dates back to about 90 ka, and there is reasonable agreement with radiocarbon dates for the youngest ^{230}Th age of about 26 ka (Hillaire-Marcel et al., 1986; Casanova and Hillaire-Marcel, 1992). The biogeochemical cycling of uranium-series iso-topes need to be studied in much greater detail in the African rift lake systems before their full potential is realized, but the initial results from Hillaire-Marcel and colleagues are very promising.

PALEOCLIMATIC RECORDS FROM THE LARGE EAST AFRICAN LAKES

The sediments of the large East African lakes carry strong and diverse signals of past climatic change. Much has been learned of East Africa's climatic history for the past 20,000 years from the analysis of diatom and pollen assemblages and of the inorganic composition of the sediment such as the abundance, mineralogy and isotopic composition of the carbon-ate fraction. The organic compounds stored in the sediments of the large East African lakes have not been investigated in any detail, but may well provide one of the richest records for future investigation. Talbot and Livingstone (1989), for example, have shown that the Hydrogen Index and the carbon isotopic composition of the organic fraction of sediments from Lake Victoria and Lake Rukwa vary with past changes in lake level. Lipiatou et al. (this volume) provide preliminary results of organic molecular analysis of Lake Victoria sedi-ments that shed light on the origin of the organic matter, past bacterial activity and anthropogenic impact. The isotopic composition of bulk organic matter from the west African Lake Bosumtwi has been interpreted to reflect changes in past algal communities (Talbot and Johannessen, 1992). Isotopic analysis of individual organic compounds found in the sediment will provide even better information as the analytical methods are improved.

The following summary of past investigations and results from paleoclimate studies of the large East African lakes demonstrates what approaches and results have been obtained to date. Some of the results were summarized previously by Livingstone and Van der Hammen (1978). They recognized the very dynamic history of vegetation change in tropical Africa, even under Holocene conditions that were relatively stable, the value of the paleo-limnological record to be recovered from the deep rift-valley lakes, and the relatively uniform signal of climatic change in tropical Africa that marked the transition from the last ice age into the Holocene. Here we expand on their review of the record from the large lakes, and provide an update of what has been learned since 1978.

Lake Tanganyika

Most of our knowledge on past variability of lake level and climate for Lake Tanganyika is derived from three cores from the Mpulungu Basin in the southern part of the lake (Fig. 9). These are core T2 recovered by Livingstone and Kendall in 1961 from 440 m depth, and cores MPU12 and MPU3 recovered by Project Georift of Elf-Aquitaine in 1985 from

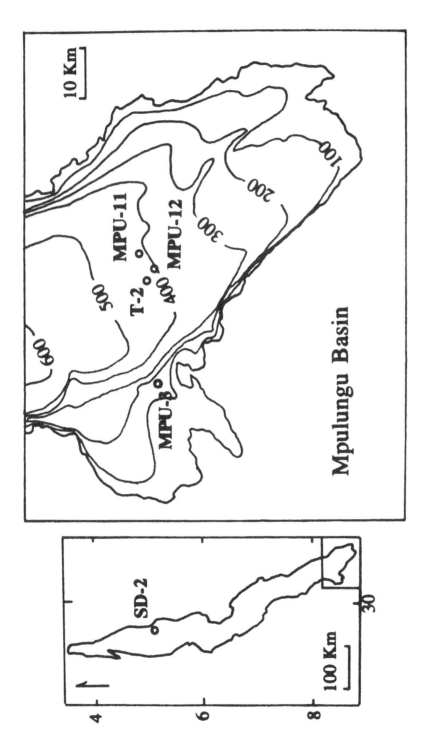

Figure 9. Core locations in Lake Tanganyika (after Gasse et al., 1989).

422 m and 130 m depth, respectively. Data from these cores are consistent with and expand on the paleolimnological interpretations that had previously been made by Degens, Hecky and co-workers from the analysis of 7 cores taken along a north–south transect in the lake in 1970 by the Woods Hole Expedition (Degens et al., 1971; Stoffers and Hecky, 1978; Hecky, 1978). There are also three cores recovered by Project Georift from the northern part of Lake Tanganyika that were analyzed for their pollen content (Vincens, 1989 and 1993). These extend back 12, 13 and 32 ka in time, respectively, and are consistent with the information published on the cores from the Mpulungu basin.

Core T2 appears to have reliable radiocarbon dates that extend from the present back to about 15.9 ka over the 1060 cm long core. The sedimentation rate is fairly constant at 0.4–0.5 mm/y between 11.5 ka and the present, and was about twice that prior to 11.5 ka. Haberyan and Hecky (1987) divided the roughly 16,000-year history into 8 zones based on the diatom assemblages in the core. Major aspects of this history include a closed-basin phase between 15.9 and 9.8 ka, and open-basin conditions between 9.8 and about 3 ka. The water column mixed more completely between 9.8 and about 5 ka than after 5 ka, particularly at 3–4 ka when stratification was strongest. In sediments younger than 3000 years, $CaCO_3$ is relatively abundant, suggesting that the lake was closed for much of that time. Haberyan and Hecky (1987) estimate that lake level may have been no more than 75 m below present during the past three millenia if closure was due only to a low stand of Lake Kivu, cutting off the supply of the Ruzizi River to Lake Tanganyika.

Radiocarbon dates on Core MPU12 range from 12.7 ka at about 230 cm depth to 25.6 ka at the bottom of the core (1013 cm) (Gasse et al., 1989). Extrapolation of the radiocarbon curve to the core top, as well as correlation of the diatom stratigraphy in MPU12 to T2 indicates a core-top age for Core MPU12 of about 5000 years (Fig. 10). The reason for the long hiatus at the top of the core could be due either to over-penetration of the piston corer or to erosion by bottom currents. The core site is positioned in a field of migrating sediment waves (Fig. 3C) (Tiercelin et al., 1988). Given its close proximity, Core T2 may also have been recovered from this dynamic sedimentary environment. Despite the implication this may have for the validity of the paleoclimate record from either of these cores, the coherence of their diatom records, along with the reasonable progression in radiocarbon dates down each core, suggest that, taken together, they accurately depict the history of the lake for the past ca. 26,000 years.

Gasse et al. (1989) interpret the diatom record in MPU12 as indicating closed-basin status but only a slight lowstand of the lake between 25.9 and 21 ka; a major lowstand, perhaps 400 m below present, between 21 and 14.5 ka, with the lowest lake level centered on 18 ka; a sporadic rise beginning at about 14.5 ka; then deep-lake status from 12.7 ka to the top of the core (ca. 5 ka) (Fig. 10). Note that Gasse et al. (1989) place the transition from closed- to open-basin status about 3000 years earlier than Haberyan and Hecky's (1987) estimate. The evidence for a slight lowstand prior to 21 ka is relatively weak, and is based on the abundance of *Cyclotella ocellata*, which is regarded as a littoral species. Gasse et al. (1989) report that there is no analog for this assemblage in modern African lakes, but that such assemblages existed in late Pleistocene Lake Abhé when it was 100–160 m deep. Whether the abundance of *C. ocellata* reflects a slight lowstand, resuspended diatoms transported down slope, or a change of some other limnological factor such as nutrient variability or temperature is subject to further inquiry. Given the presence of mud waves at the core site, reworking by density currents may well be the cause of the no-analog assemblage.

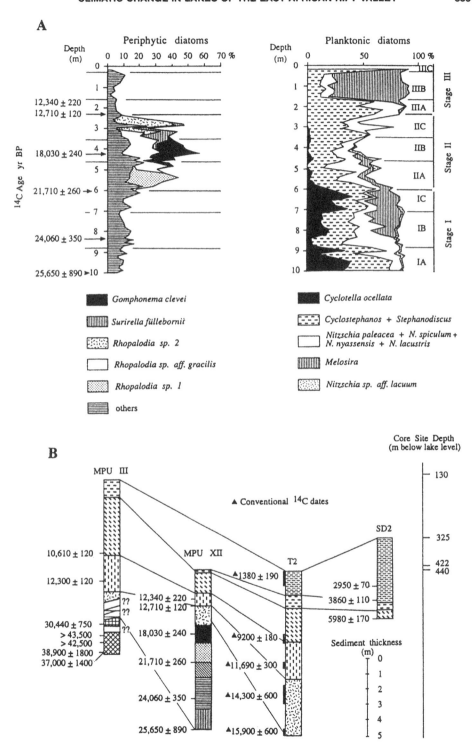

Figure 10. Diatom stratigraphies of several cores from Lake Tanganyika (based on Gasse et al., 1989 and Haberyan and Hecky, 1987). A) Core MPU XII. B) Stratigraphic correlations based on the diatom profiles. C) A summary of the diatom zones and their limnological significance (from Gasse et al., 1989).

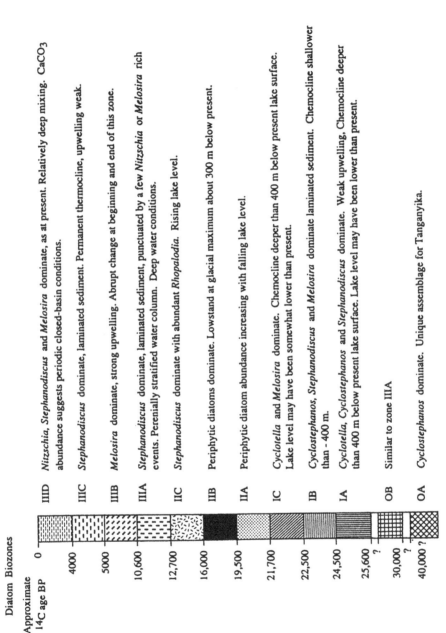

Figure 10, cont'd

Core MPU3, recovered from 130 m water depth to the west of T2 and MPU12 (Fig. 9) can be correlated to the latter two cores on the basis of diatom stratigraphy to a depth in core of about 8 m, where the age of the sediment is about 13,000 years old (Fig. 10). This is underlain by an interval of sediment more than 1 m thick that has abundant planktonic diatoms and is about 30,000 y old and, at the bottom of the core, an interval of sediment with abundant planktonic diatoms somewhat different in assemblage composition, that yields radiocarbon dates that are greater than or equal to 40 ka. The record of this core indicates that the lowstand of Lake Tanganyika at 21.7–13 ka probably was lower than 130 m below present lake level, causing the hiatus just above the 30 ka interval. The lake was at a higher stand and perhaps in an open basin configuration around 30 ka, then again around 40(?) ka.

Although the hiatus in MPU3 constrains the lowstand of the last ice age (22–14 ka) to at least 130 m below present lake level, there is no sedimentological evidence to suggest that the lowstand was 400 m below present, as was suggested by Gasse et al. (1989) but had been disputed by Livingstone (1965) and subsequent reports on the basis of the sediment in T2. There is no significant change in sediment texture reported for either MPU12 or T2, nor do the seismic reflection profiles show a change in acoustic character that surely would have accompanied an increase in sand content in the sediments had they been deposited within 20–40 m of the lake surface (Tiercelin et al., 1988). On this basis, we suspect that the lowstand was no more than 300 m below present lake level. This estimate was recently arrived at independently by Vincens et al. (1993), based on pollen-derived rainfall and precipitation estimates between 25 and 9 ka.

The pollen stratigraphy in cores MPU3, MPU12 and MPU11, taken near MPU12, and in cores from the northern part of the lake, provides insight into the history of temperature and humidity of the catchment area that complements nicely the lake-level history inferred from the diatoms (Vincens, 1991; Vincens et al., 1993). A high abundance of Afro-montane herb pollen in sediments of 15–25 ka indicate a cool, dry period that was estimated to be about 4°C cooler than present (Vincens et al., 1993). A transition to warmer, moister conditions is reflected in an increased representation of Zambezian arboreal taxa between 15 and 12 ka. More humid pollen types are found in increasing abundance until 2600 ybp, after which time there is evidence for a return to distinctly drier conditions similar to today (Vincens et al., 1993).

Lake Kivu

Lake Kivu is the youngest of the deep Rift Valley lakes. It is underlain by at most 500 m of sediment of low seismic velocity (Degens et al., 1971, 1973) that probably is no older than early Pleistocene (Haberyan and Hecky, 1987). There is some evidence for about another kilometer of sediment of unknown age underlying the unconsolidated sediment, having an acoustic velocity of about 2.8 km/sec (Wong and Von Herzen, 1974).

Lake Kivu is unusual among the deep rift lakes because it is affected significantly by hydrothermal activity. Warm, saline waters injected at the lake floor cause the bottom waters to be warmer than the overlying surface waters (see Spigel and Coulter, this volume). Stability of the water column is weakly maintained by the salinity gradient. Salinity and temperature profiles show a step-like structure that suggests the existence of individual convection cells within the water column, each well mixed, that perhaps reflect individual hydrothermal events (Degens et al., 1971). The water chemistry results from the mixing of two water masses: the intensely evaporated surface waters that are characterized by rela-

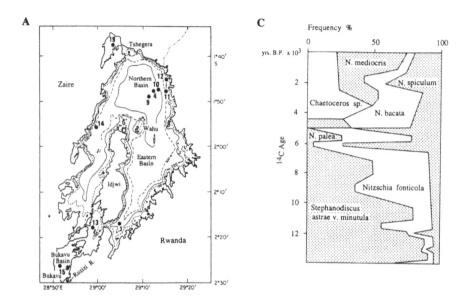

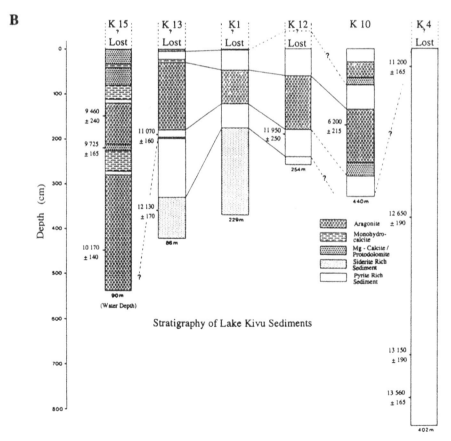

Stratigraphy of Lake Kivu Sediments

Figure 11. A) Core locations in Lake Kivu. B) Lithostratigraphy of six cores from Lake Kivu. C) Diatom stratigraphy based on the Lake Kivu cores (from Stoffers and Hecky, 1978).

tively low salinity and high $\delta^{18}O$, and the more saline deep water derived from hydrothermal input with isotopically lighter composition (Degens et al., 1971, 1974).

Sediments in Lake Kivu cores display a wide range of sediment lithology, from beach gravel to tuff, siderite-rich soil, pyrite-rich horizons, laminated diatomites and carbonate-rich layers. Stoffers and Hecky (1978) worked out a history of the lake spanning the past 14 ka based on lithology and diatom assemblages in 7 cores (Fig. 11). Only 11 radiocarbon dates were obtained on this suite of cores, with the two youngest dates being $6,200 \pm 215$ ybp and $9,460 \pm 240$ ybp. Sufficient overlap exists in the stratigraphy of the cores to assemble a reasonably coherent and continuous record from 13.5 ka to the present (Stoffers and Hecky, 1978; Haberyan and Hecky, 1987). Given just the single date younger than about 9,500 y in any of the cores, however, the timing of events after this date should be considered tentative at best.

A gravel deposit in one of the Kivu cores, K14, recovered from 330 m water depth, was interpreted as a beach deposit by Stoffers and Hecky (1978). If so, the lake was at least 300 m lower than present during the last deglaciation. (Actually, the age of this gravel is not known, but was estimated by Stoffers and Hecky (1978) to be about 13 ka based on stratigraphic correlations with K-4.) A change in lithology from siderite-rich sediment (soil?) to pyrite-rich lacustrine sediment is dated at about 12 ka in Core K-13 recovered from 86 m water depth (Fig. 11B), indicating that the lake had risen significantly by that time. The lake level was high between about 10 and 5 ka, with widespread carbonate deposition and a shift in diatom assemblage from dominance by *Stephanodiscus astrea* to *Nitzschia fonticola*, then *Nitzschia spiculum*, reflecting an increase in the Si:P ratio in the surface waters (Fig. 11B,C) (Haberyan and Hecky, 1987). Hydrothermal input established the salinity-controlled water column stability (crenogenic meromixis) at about 5 ka, with an abrupt change to diatom assemblages with abundant *Chaetoceras*, indicating high salinity waters, and species of *Nitzschia* that reflect stratification of the water column (Fig. 11C).

The abundance of $CaCO_3$ is high in the sediments dating between about 9 and 5 ka, suggesting significant evaporation of surface waters and overturn of the water column to keep its pH favorable for the preservation of carbonates. Carbonates are also abundant in the sediments at about 1–3 ka, suggesting that the meromixis incurred by hydrothermal input was occasionally interrupted when the lake was closed and surface water salinity rose to the point where the water column could overturn and allow carbonates to be preserved (Stoffers and Hecky, 1978; Haberyan and Hecky, 1987).

In summary, the paleoclimatology and paleolimnology of Lake Kivu is not well con-strained by the available radiocarbon dates, especially after 9.5 ka. However, there are strong signals in the sediments that show a major rise in lake level from the last glacial maximum to high-stand conditions by about 10 ka, in agreement with the records from other lakes in the region. The diatom assemblage suggests relatively deep mixing in the water column between about 10 and 5 ka, then an abrupt change in conditions in the lake at about 5 ka with the onset of significant hydrothermal input and stratification of the water column. The carbonate record since 5 ka suggests that at times the lake has been only weakly stratified or non-stratified.

Lake Turkana

Lake Turkana is the largest closed-basin rift lake in East Africa today, although both Lakes Malawi and Tanganyika have been closed in historical times. High stand terraces within the

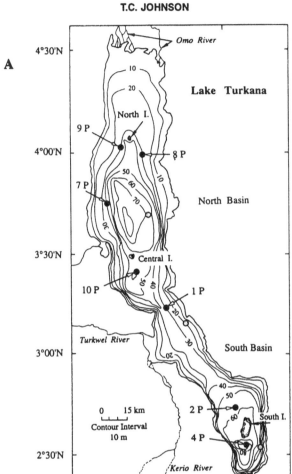

A

Lake Turkana

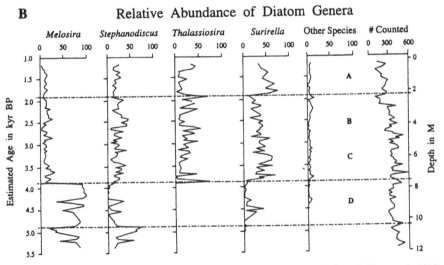

B   Relative Abundance of Diatom Genera

Figure 12. A) Core locations in Lake Turkana. B) Diatom stratigraphy of LT84-2P (from Halfman et al., 1992).

lake basin have been extensively dated, with most dates falling between 4000 and 10,000 years ago (Owen et al., 1982). Butzer et al. (1972) reports a level 60–70 m above the 1971 level of the lake "just prior" to 35,000 ybp, but this is based on three radiocarbon dates that are older than 35,000 years. How much older? No dates have been reported for high stand deposits that fall between the >35,000 y and 10,320 ± 150 ybp, suggesting that the lake was low during that period. Johnson et al. (1987) report evidence for lake level as much as 60 m below its 1984 datum, based on high-resolution seismic profiles. They estimate the age of the lowstand to fall between 10 and 35 ka. Many dates from terraces 60–80 m above the 1971 level fall within the range of 4–10 ka, with some evidence for a brief lowstand at about 6–8 ka. Deposits younger than 4 ka are, for the most part, significantly lower in elevation, at about 20 m above the 1971 datum. The lake level record is better delineated from sediment cores for this time period.

Because of the very high sedimentation rate in Lake Turkana, piston cores (Halfman and Johnson, 1988) and Mackereth cores (Barton and Torgesen, 1988) have recovered sediments no older than 5500 years. Although the paleoclimate record from Lake Turkana sediment cores is of short duration, it is the highest-resolution record available for any of the East African rift lakes. The record presented here is based primarily on the analyses of two piston cores, one from the north basin of the lake, LT84-8P, and one from the southern basin, LT84-2P, collected by Duke University's Project PROBE in 1984 (Fig. 12). The cores are each about 12 m long and register rather steady and fast sedimentation rates of about 3 mm/y, based on an extensive array of radiocarbon dates and stratigraphic correlations with six other dated cores from the basin (Halfman et al., 1994). Unfortunately, the piston corer over-penetrated the very fluid sediment-water interface, resulting in core-top ages of about 600 ybp in 8P and about 1250 ybp in 2P.

The sedimentary parameters used as proxies of paleoclimate in these cores vary; %CaCO$_3$ profiles are available for both (Halfman et al., 1994), but the profiles in the north basin are probably affected as much by dilution by detrital material introduced from the Omo River as by chemical changes in the water column. The isotopic composition of authigenic calcite in the sediments has been related to past changes in lake level by Johnson et al. (1991), and more recently in much greater detail by Muchane (1994) (Fig. 13A) and Muchane (this volume). Pollen concentrations in LT84-8P are relatively sparse, but they provide a signal that can be related to the isotopic composition of the authigenic calcite in the same core (Mohammed et al., submitted), covering the interval between 600 and 2600 ybp. The diatom record of 2P from the south basin provides a strong record of limnological change for the period between 5.4 and 1.2 ka (Fig. 12B) (Halfman et al., 1992). Diatoms are not preserved in the northern basin of the lake (Yuretich, 1979).

CaCO$_3$ abundance increases up Core LT84-2P from about 10% at the base to about 20% at its top (Fig. 13B). The carbonate consists of both authigenic calcite and ostracodes, and in the south basin of the lake the increased abundance of carbonate through time is attributed to rising salinity of the lake (Halfman and Johnson, 1988; Halfman et al., 1992). There is a major rise in carbonate content at about 9.5 m depth in the core (4.5 ka) that Halfman et al. (1992) interpret as the first lowering of lake level in this time interval to closed-basin status (Fig. 13B). Closed-basin status was achieved permanently at 3.9 ka, when there was a dramatic shift in diatom assemblage in which *Melosira* ceased to be the dominant genus, and was replaced by an assemblage with relatively abundant *Stephanodiscus, Thalassiosira,* and *Surirella*, reflecting a higher salinity than before. These conditions persisted until 1.9 ka, when the abundance of diatoms dropped significantly, perhaps from a rise in the pH of

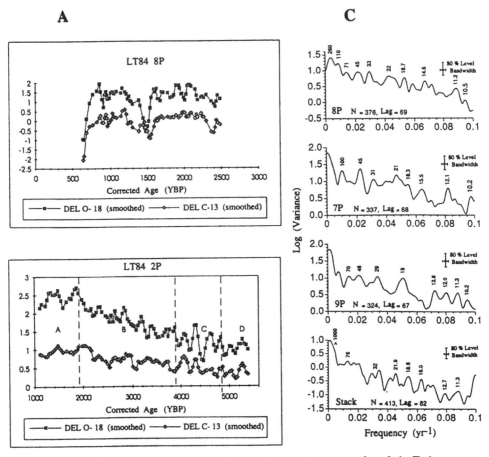

Figure 13. A) Isotope stratigraphy of authigenic carbonate in two cores from Lake Turkana. B) Carbonate stratigraphy in Lake Turkana cores. C) Time-series analysis of carbonate profiles from lake Turkana (from Halfman et al., 1994).

the lake waters (Halfman et al., 1992), and the abundance of *Surirella* greatly surpassed *Stephanodiscus* and *Thalassiosira*, suggesting a further drop in lake level (Fig. 12B). The diatom assemblage at the top of the core is similar to that of 1.9–3.9 ka, and is evidence for a modest rise in lake level at about 1.2 ka (Halfman et al., 1992).

The $\delta^{18}O$ record in LT84-2P agrees quite well with the diatom record (Fig. 13A). The sharp rise in $CaCO_3$ at 4.5 ka is accompanied by an equally dramatic increase in $\delta^{18}O$. The shift to permanently closed-basin conditions at 3.9 ka is also reflected by an abrupt rise in $\delta^{18}O$. Diatom Zone B is characterized by increases both in carbonate abundance and its $\delta^{18}O$ through time, punctuated by higher frequency variability. This is succeeded by Diatom Zone A, in which the Surirella-dominated assemblage, the high carbonate abundance and its heaviest isotopic values, all attest to maximum salinity.

The $\delta^{18}O$ profile from core 8P in the north basin was interpreted by Johnson et al. (1991) to reflect lake level change for the past 4000 years. Two recent developments mandate a revised interpretation. First, the geochronology of the core has been improved with many

Northern Cores : Carbonate (wt. % CaCO3)

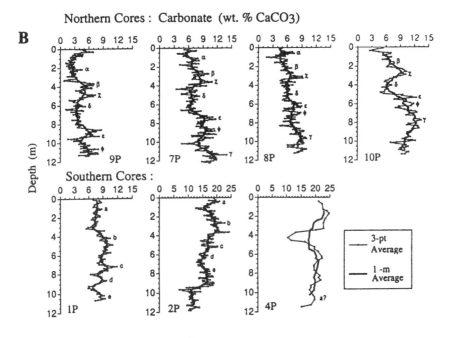

Figure 13, cont'd

new radiocarbon dates (Fig. 8) (Halfman et al., 1994), and second, the implication for lake level change of the $\delta^{18}O$ curve must be revised according to the model presented by Ricketts and Johnson (this volume). Application of the Ricketts and Johnson model to the $\delta^{18}O$ profile yields a lake level that is relatively low and dropping from about 2500 to 1600 ybp, a significant rise in lake level at 1600 ybp that is maintained until about 1000 ybp, a progressive drop from 1000 bp to about 700 ybp, then perhaps another rise just before 600 ybp.

The isotopic profiles from 2P and 8P show similarities and differences during the time of their overlap. The $\delta^{18}O$ values are somewhat higher in 2P than 8P, reflecting their respective distances from the Omo River, the major source of isotopically light water to the lake. While both profiles show rising mean values from 2.6 to 1.9 ka, the rise is steeper in 2P than 8P. Both cores show the light event centered on about 1.5 ka, and about 500 years duration. It is more pronounced in 8P than 2P, again most likely due to 8P's closer proximity to the Omo River. These results show that, while correlations can be made between the two cores, there is significant spatial variability in isotopic records from a lake as big as Turkana. This variability must be accounted for in any quantitative reconstruction of past climate based on isotopes alone.

The profiles of $\delta^{13}C$ track the oxygen isotope profiles quite closely, as has been observed in many other closed-basin lakes. This correlation is attributed to the relatively strong influence of carbon isotopic composition of rivers and carbon isotope fractionation associated with lake-atmosphere exchange of CO_2 compared to the impact of primary productivity on the carbon isotopic composition of dissolved inorganic carbon (DIC) in the surface waters (Talbot, 1990; Talbot and Kelts, 1990). However, there are deviations from the correlation between $\delta^{18}O$ and $\delta^{13}C$, and these may well reflect the effect of primary productivity, in which light carbon is preferentially photosynthesized and the remaining DIC becomes enriched in ^{13}C, as does the carbonate that precipitates from the surface waters (Stuiver, 1970; McKenzie, 1984).

The sediments of Lake Turkana are laminated despite the fact that the water column is oxygenated and can support a benthic community. Laminations are preserved because the sedimentation rate is too fast to allow complete reworking by the sparse benthic community that is limited by food-resource availability (Cohen, 1984). The laminations do not represent annual varves. Halfman and Johnson (1988) reported that each light–dark couplet represented, on average, 4–5 years, and mentioned a possible connection to the El Niño–Southern Oscillation (ENSO) phenomenon. Subsequent modification of the radiocarbon geochronology in the cores, however, reveals that each light–dark couplet represents an average of about 2 years (Halfman et al., in press). This could reflect the quasi-biennial oscillation that frequently shows up in time-series analysis of rainfall anomalies in East Africa (see Nicholson, this volume).

On a still longer time scale, individual and stacked carbonate profiles from the Lake Turkana cores have been subjected to time-series analysis to investigate possible cyclicity in the paleoclimate record. A number of spectral peaks recur in the records, exceeding the 90% confidence level. These correspond to periods of 32, 22, 18, 16 and 11 years (Halfman et al., 1994) (Fig. 13A). Time-series analysis of down-core variability in lamination thickness in the sediments also reveals periods of 11, 18.6 and 23 years, in close agreement with the carbonate results. These suggest cyclic variability in the climate of the Ethiopian Plateau on a time scale of decades to centuries, perhaps associated with the northern hemisphere monsoon. Halfman et al. (1994) mention other occurences of periodic variation in the climate of Africa on the same time scale: Hameed (1984) reports an 18.4 year periodicity in the discharge of the Nile River, and Campbell et al. (1983) report an 18.5-year periodicity in the Indian monsoon. Faure and Gac (1981) report a 31-year periodicity in the discharge of the Senegal River. Possible links between these observed frequencies and the lunar nodal period of 18.6 years or the sunspot and double sunspot cycles of 11 and 22 years, respectively (Crowley, 1983; Morner and Karlen, 1984; Currie and Fairbridge, 1985) need further study. S. Nicholson (pers. comm.) believes the 11- and 22-year cycles are more likely to be multiples of the 5–6 year recurrence of the El Niño–Southern Oscillation (ENSO) than due to sunspot activity. West (in press) has just concluded from Bayesian statistical analysis of the Lake Turkana proxy records that periodicities shorter than a century cannot be confirmed, given our present uncertainty in the calibration of the radiocarbon time scale.

Lake Victoria

Only five piston cores have been recovered from Lake Victoria and analyzed for past climatic change (Fig. 14) (Kendall, 1969; Stager, 1984; Stager et al., 1986; Talbot and Livingstone, 1989). All of these cores were obtained from the northern-most periphery of the lake, so they may not reflect conditions representative of the basin as a whole, although their paleoclimatic records are consistent on a millenial time scale. Only Core P-2, recovered from 9 m depth in Pilkington Bay on Bivuma Island, is well dated with 28 radiocarbon dates, showing a remarkably steady sediment accumulation rate (Kendall, 1969), ranging from 860 ± 120 ybp at 72 cm depth to 14,730 ± 200 ybp at 1756 cm depth (Fig. 15A). The lowest dated sample comes from below a discontinuity, and may be older than the C-14 date

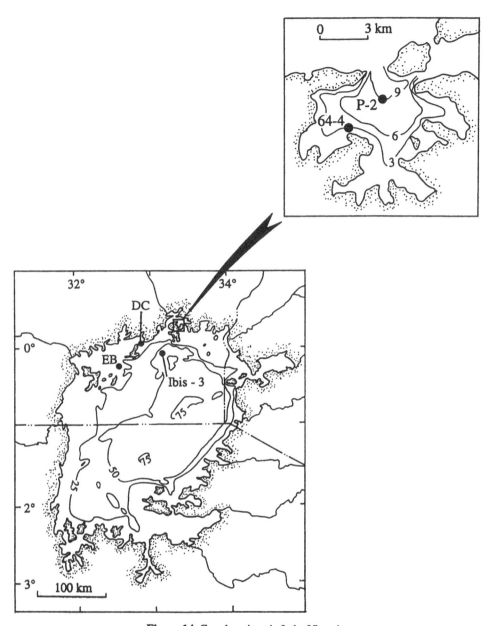

Figure 14. Core locations in Lake Victoria.

because of an admixture of root and soil organic matter (Talbot, pers. comm.). This is the best dated sequence of climatic change over the entire Holocene available for any of the large east African lakes. The other four cores have been correlated with Core P-2 by pollen (Kendall, 1969) and diatom (Stager, 1984; Stager et al., 1986) assemblages.

Kendall (1969) found evidence for climatic and lacustrine change based on several parameters, most notably water content, carbonate abundance and changes in assemblages

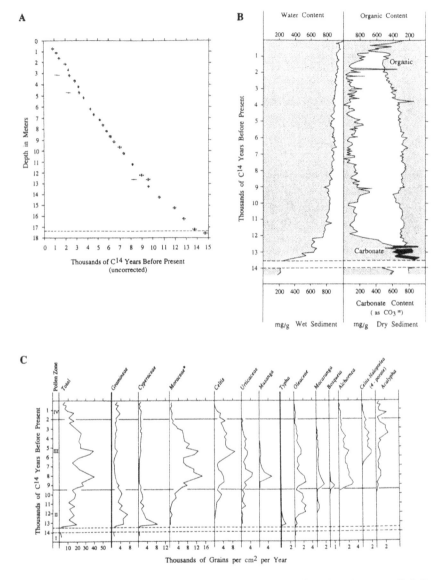

Figure 15. A) Radiocarbon ages (conventional) on bulk organic carbon from core P-2. B) Water content, organic carbon abundance and carbonate abundance in Core P-2. C) Pollen stratigraphy in core P-2. (all from Kendall, 1969).

of diatoms, green algal remains and pollen. Carbonate content is relatively high between the bottom of the core, dated at 14.5 ka and the horizon corresponding to 12 ka (Fig. 15B), and is interpreted to reflect a low stand of the lake (Kendall, 1969). A sandy discontinuity is found in sediments dating between 13.5 and 14 ka, indicating that lake level dropped below the elevation of this site (–25m, relative to 1960 level) prior to 13.5 ka. The greatest change in diatom assemblage composition occurred just prior to 12 ka, with a dramatic drop in the abundance of *Stephanodiscus astrea* and a concurrent rise in *Nitzschia* and *Melosira* spp..

This coincides with the shift towards fresher water suggested by the drop in carbonate abundance at this time.

Talbot and Livingstone (1989) analyzed sediments from cores DC and IBIS-3 for total organic carbon (TOC), hydrogen index (HI) and $\delta^{13}C$ of the TOC. Core DC, from 32 m depth, was first referred to as the "Domba Channel core" by Stager et al. (1986). Talbot and Livingstone (1989) refer to it as IBIS-1. The core contains a major erosional surface at about 8.8 m depth in core, with evidence of subaerial exposure, desiccation structures and root molds. Core IBIS-3, from 66 m water depth, has a similar discontinuity at about 9.4 m down core. Organic carbon near these horizons is in relatively low abundance, with low values of HI and $\delta^{13}C$, characteristic of relatively high proportions of woody and herbaceous material and C-4 pathway plant residue in the organic matter. Mottling and subvertical prismatic cracks in the core suggest brief subaerial exposure (Talbot and Livingstone, 1989). These horizons date between 15.1 and 17.3 ka, and represent the Late Pleistocene minimum level for Lake Victoria (Talbot and Livingstone, 1989). Lacustrine sediment at about 10.2 m depth in core IBIS-3 was dated at 23.6 ka, and represent a somewhat higher lake level at that time.

The oldest lacustrine sediment in Core 64-4, taken near P-2 in Pilkington Bay (Fig. 14), is 10,270 radiocarbon years old, so either the lake did not achieve open-basin status until 10 ka or, if it had, it dropped briefly again just prior to 10 ka to a level approximately 12 m below the 1960 level. The green algae record from Core P-2 indicates a brief and minor drop in lake level at about 10 ka, lasting about 500 years. The pollen record also indicates a brief period of drier and cooler conditions at this time (Fig. 15C) (Kendall, 1969). This is the best evidence for a Younger Dryas Event to be found in any of the large east African lakes, but it comes from an embayment and there is no apparent signal of the Younger Dryas Event in either Core DC or IBIS-3 from farther offshore (Stager, 1984; Talbot and Livingstone, 1989). Other evidence for the Younger Dryas Event is reported from one site each in the Sahara and Sahel (Gasse et al., 1990), from Lake Magadi (Roberts et al., 1993), and a highland bog in Burundi (Bonnefille et al., submitted).

A high abundance of *Melosira* spp. between 7 and 9 ka corresponds to a time of relatively high dissolved silica concentration in the water column, and also the highest lake levels in Holocene East Africa, at least north of the equator. The rise in *Nitzschia* at 7 ka, followed by a rise in other pennate diatoms at 5 ka, could reflect a shift to somewhat drier and more seasonal conditions (Stager, 1984). The pollen record in Core P-2 obtained by Kendall (1969) is consistent with the diatom and carbonate records (Fig. 15C). It indicates that the vegetation near the core site was predominantly savanna before 12.5 ka. From 12.2 to 7 ka a major forest developed, with the brief arid period around 10 ka. After 6 ka, the pollen indicate a reduced and more seasonal annual rainfall compared to the preceeding 4000 years. A significant reduction in forest pollen occured after 3 ka that is attributed to the expansion of human activity in the lake basin (Kendall, 1969).

Lake Albert (Mobutu)

Lake Albert is located just 2° north of the equator and is fed primarily by the Victoria Nile, draining Lake Victoria, and the Semliki River, draining Lake Edward (Fig. 16). Victoria Nile inflow accounts for about 65% of the lake's freshwater input (Hurst et al., 1966). The lake has a maximum depth of 58 m.

The record of past climate change on Lake Albert is inferred from various analyses on four piston cores taken from the lake in the early 1970s: cores 2 PC and 3 PC of the 1972

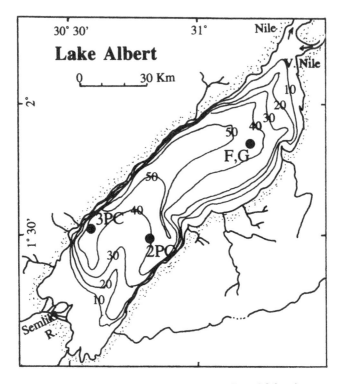

Figure 16. Core locations in Lake Albert (Mobutu).

Woods Hole Expedition (Hecky and Degens, 1973) and cores F and G obtained by D. A. Livingstone and T. J. Harvey of Duke University in 1971 (Fig. 16).

Age assignments in the cores are based primarily on 7 radiocarbon dates on core G, ranging from $1,850 \pm 90$ years in the 10–60 cm depth interval to $28,180 \pm 860$ y at the base of the core (1010–1060 cm). Extrapolation of the age-vs.-depth curve to the top of the core yields an age of about 100 years (Harvey, 1976). Core F, taken very close to core G, has 4 radiocarbon dates, ranging from $8,750 \pm 400$ years at 140–150 cm depth, to $29,900 \pm 750$ y at 902–907 cm depth at the bottom of the core (Sowumni, 1991). Neither of these cores are well constrained in their chronology in sediments younger than 9 ka (Fig. 17). Core 2 PC has no radiocarbon dates and Core 3 PC has only one date of $2,660 \pm 90$ ybp at 140–180 cm depth (Hecky and Degens, 1973). All other ages in core 3 PC have been deduced from unspecified stratigraphic correlations with cores F and G (Cohen, 1987; Ssemmanda and Vincens, 1993).

Harvey (1976) established 4 diatom zones in core G, spanning the last 28,000 years (Fig. 17). The upper 90 cm of the core, representing roughly the last 5,000 years, is dominated by a *Stephanodiscus-Nitzschia*–mixed assemblage, and is assumed to represent modern conditions in the lake. From 90 to 645 cm depth in the core, representing 5–12.5 ka, a *Stephanodiscus-Melosira* assemblage is present that Harvey (1976) interpreted as representing conditions of lower alkalinity and higher dissoved silica content than today. At 650–930 cm depth in core, diatoms are rare and poorly preserved. The upper 80 cm of this interval has a calcareous silty sand with abundant ostracodes. The 650–930 cm interval, ranging from

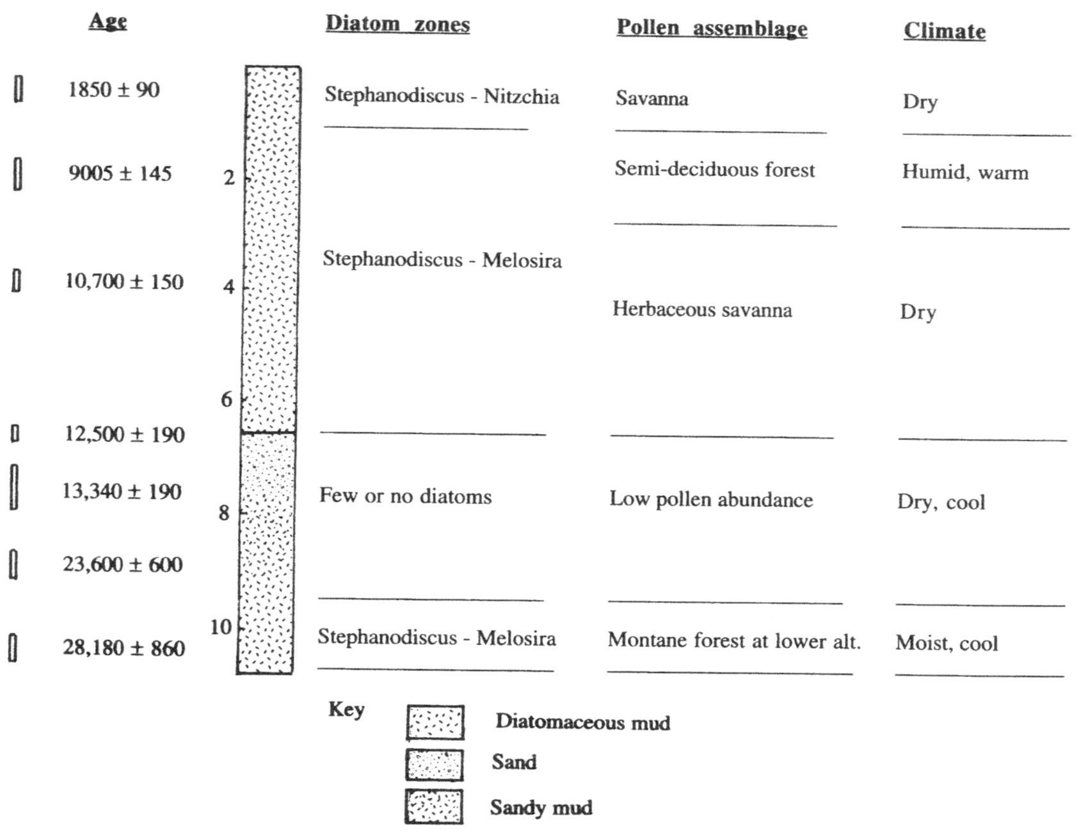

Figure 17. Diatom and pollen records from Lake Albert, with radiocarbon dates from Core G.

12.5–25 ka, is inferred to represent a lowstand of the lake, with highly alkaline waters causing the poor preservaton of diatoms. Within the interval, however, is a 30-cm section representing approximately 14–18 ka, in which diatoms are abundant and dominated again by species of *Stephanodiscus* and *Melosira*. The bottom 130 cm of the core, representing 25–28 ka, has a *Stephanodiscus-Melosira* assemblage that indicates that low alkalinity waters enriched in dissolved silica characterized that time as well (Harvey, 1976).

Results of the pollen analyses of Ssemmanda and Vincens (1993) supplement the diatom results of Harvey (1976) for the last 13,000 years, based on the analyses of Core 3 PC, although it must be remembered that age control in this core is tenuous. Ssemmanda and Vincens (1993) divide the past 13 ka into 4 pollen zones: 13–9.5 ka is characterized by widespread herbaceous savannas, indicating conditions around the lake that are cooler and drier than today. From 9.5 to 5 ka there was significant expansion of humid, semi-deciduous forests, but with savanna still retained near the lake, indicating warmer and moister conditions than before. From 5 to 3 ka the vegetation reflects more semi-deciduous character (drier) than the previous 4 millenia. Finally the last 3 ka shows a dramatic increase in savannas, with no accompanying evidence for human impact. Ssemmanda and Vincens (1993) interpret this to reflect still drier conditions than those of the early Holocene. Dates on all but the youngest pollen zone should be considered as rough estimates.

The pollen record in Core F provides more insight into vegetation zones in the Lake Mobutu basin prior to 12.5 ka. Sowumni (1991) identifies four pollen zones: the oldest, 30–25 ka, indicates conditions much cooler than today, with montane forest vegetation spreading to lower altitudes. From 25 to 15 ka, pollen abundance is very low in Lake Mobutu sediments, interpreted by Sowumni (1991) to indicate a lowstand of the lake and arid conditions in the basin. Silts and sands are more abundant in this interval than elsewhere in the core, supporting Sowumni's interpretation of low lake level. From 14.5 to 12.5 ka the pollen assemblage is enriched and indicates the establishment of grasses and sedges, followed by shrubs and trees.

The unusual aspect of this interpretation of the pollen record vis a vis all the other paleoclimate data from East Africa is the onset of moist, post-Pleistocene conditions at 14.5 ka, or at least 2,000 years earlier than has been reported elsewhere. However this age is based on a radiocarbon date on bulk organic matter in sandy sediment at 658–673 cm depth in core. It is quite likely that this interval contains reworked, older organic matter that yields an older date than it should.

Cohen (1987) analyzed the ostracode assemblages in Cores 2 PC and 3 PC. Three assemblages were identified: a nearshore assemblage associated with sands, suggesting deposition in water less than 10 m deep, a profundal assemblage indicating deep-water conditions, and a mixed assemblage of corroded ostracode carapaces that suggests sediment redeposition. The shallow water assemblage was found in the basal sand of Core 3 PC and is estimated to be 13,000 years old. This sand may correlate with the ostracode-rich sand at 650 cm depth in Core G. Since Core 3 PC was recovered from a water depth of 47 m, it suggests that Lake Albert was at least 30 m lower than present at the end of the Pleistocene.

Harvey (1976) modeled the hydrological budgets of Lakes Albert, Edward and Victoria to determine the circumstances under which Lake Albert could have been closed. Assuming a mean annual temperature that was 5°C cooler than present, he determined that a 29% decrease in rainfall in the region would have resulted in Lake Victoria becoming closed, Lake Albert thereby closing, but Lake Edward remaining open and providing outflow to Lake Albert. An open-basin configuration for Lake Albert between 25 and 28 ka requires an

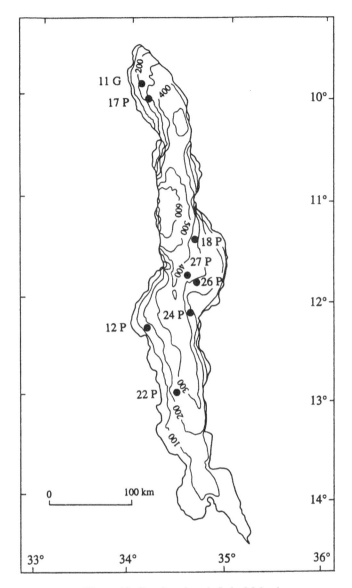

Figure 18. Core locations in Lake Malawi.

open Lake Victoria as well, to provide inflow from the Victoria Nile. This agrees reasonably well with the results of Talbot and Livingstone (1989), who inferred a relatively high level for Lake Victoria at 24 ka based on the isotopic composition of organic matter in the sediments of that age.

In summary, Lake Albert appears to have been open between about 25 and 30 ka, with temperatures about 5°C cooler than present, a montane forest relatively near the lake and deep seasonal mixing providing silica-rich waters to the euphotic zone. Conditions became significantly drier around 25 ka, with annual precipitation dropping by perhaps 30%, resulting in closed basin conditions for both Lake Albert and Lake Victoria. At about 12 ka,

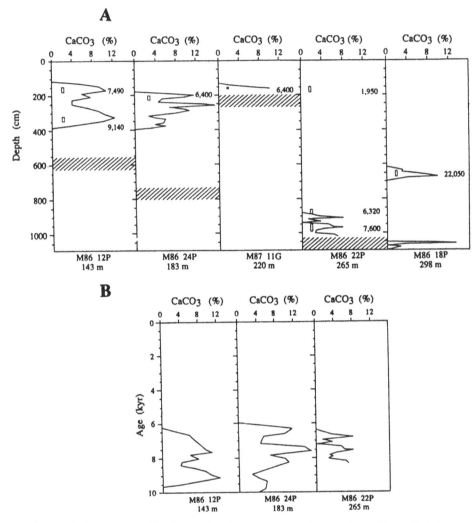

Figure 19. Carbonate profiles from Lake Malawi plotted against: A) depth, B) age. Relative abundances of *Stephanodiscus* spp. and *Melosira* spp. in Lake Malawi cores plotted against: C) depth and D) age (from Finney and Johnson, 1991).

humid, warmer conditions ensued, and savanna was replaced by a semi-deciduous forest and silica-rich deep waters once again supported a *Melosira*-rich community of diatoms. This lasted until about 5 ka when slightly drier conditions developed, leading up to the still drier of recent times. The diatom assemblage shifted from being dominated by *Stephanodiscus-Melosira* to one dominated by *Stephanodiscus-Nitzschia*, suggesting less intense vertical circulation in the lake than before. The vegetation around the lake shifted to more wide-spread savanna at the same time.

Lake Malawi

The climatic history of Lake Malawi is unique among the rift lakes, because the variations in lake level during the past 40 ka are out of phase with the large lakes to the north. These

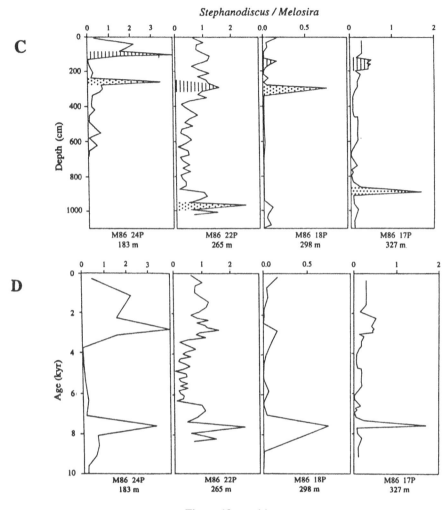

Figure 19, cont'd

results are based on analysis of seven cores taken from a suite of piston cores collected by Duke University's project PROBE in 1986 (Johnson et al., 1988) (Fig. 18).

Primary evidence for lake level fluctuations in Lake Malawi is the deposition of carbonate-bearing sediment during lowstands. At present, the surface waters of Lake Malawi are slightly over-saturated with respect to calcite, but the deeper waters are undersaturated, so calcite does not accumulate in the bottom sediments (Ricketts and Johnson, this volume). During times when the lake level was 100 m or more below present, for which there is abundant evidence in high-resolution seismic profiles (e.g., Johnson and Ng'ang'a, 1990; Scholz and Finney, in press) and deep-penetration, low frequency seismic lines (Scholz and Rosendahl, 1988), the water column would have been saturated with respect to calcite to 200 m below the paleo-lake surface (Ricketts and Johnson, this volume). Several cores from Lake Malawi contain carbonate-bearing horizons, interpreted as low stand deposits, that date at about 6–10 ka (Finney and Johnson, 1991) (Fig. 19A,B) and at about 28–37 ka (Finney et al., this volume) (Fig. 8B). Diatom assemblages within or adjacent to the younger carbonate horizons support the stratigraphic correlations and provide evidence of varying past changes in the relative supply of dissolved phosphorus and silica to the euphotic zone (Fig. 19C,D) (Finney and Johnson, 1991). Diatom assemblages adjacent to the older carbonate horizons provide additional evidence for lowstands in the relative abundance of benthic diatoms, especially *Rhopalodia* spp. (Fig. 8B) (Finney et al., this volume). The timing and duration of the 6–10 ka low stand is reasonably well constrained by the radiocarbon dates in three cores (Fig. 19A,B). The older low stand at about 28–37 ka, however, is not as well constrained, as discussed earlier in the paper. Eight of the ten dates from this horizon fall in the range of 28–37 ka, but there is so much scatter in the depth–age relationships of cores M86-18P and 27P that the precise timing of the fall and rise of lake level is not known to better than about ±1000 y (Fig. 8B).

The existence of low stands on Lake Malawi at 6–10 ka and 28–37 ka coincides with the timing of relatively low summer insolation in the southern hemisphere and is consistent with climate model simulations that predict weaker monsoonal circulation in the southern hemisphere at these times (Finney et al., this volume). Lake Malawi appears to be the only rift valley lake to behave like a "southern hemisphere lake." Lake Rukwa, just to the north of Lake Malawi, lying at a latitude of about 9°S, was at a high stand during the early Holocene (Haberyan, 1987), as was Lake Tanganyika, which extends from 3–9°S (Haberyan and Hecky, 1987; Gasse, et al., 1989) and Lake Chelshi, Zambia, at 9°S (Stager, 1988). While there is much evidence for drier-than-modern conditions during the early Holocene farther to the south in Africa, including the Namib (Vogel, 1989), the Kalahari (Partridge, 1992), and much of eastern South Africa (Scott, 1989), the pattern does not appear to be as spatially and temporally homogeneous as the coincidental wet period in equatorial and northern Africa (Partridge, 1993; Scott, 1993).

Finney and Johnson (1991) analyzed the Malawi sediment cores for their Fe/Al ratio to determine past depth of the chemocline, or boundary between the oxygenated epilimnion and the anoxic hypolimnion. The rationale for this investigation was that iron oxide, or oxyhydroxide, precipitation should occur near the chemocline in Lake Malawi as a result of reduced soluble iron in the anoxic deep waters becoming oxidized and precipitating when it comes into contact with the overlying oxygenated water (see Owen et al., this volume). If this occurs within the water column in the middle of the lake basin, the precipitated iron oxyhydroxides would very likely re-dissolve as they sink back into the hypolimnion. However, if precipitation occurs where the chemocline intersects the lake floor, the iron oxyhydroxides could be preserved in the bottom sediments, thereby increasing the Fe/Al ratio in the sediments. Profiles of this ratio in a suite of cores from a range of water depths shows peak values that Finney and Johnson (1991) interpreted to be tracking the chemocline as it travered the depths of the respective core sites through time (Fig. 20).

There are climatic signals in the cores from Lake Malawi that are of higher frequency than the millenial variations cited above. Ricketts and Johnson (this volume) determined that the carbonates deposited during the last major low stand of the lake at 5–9 ka contain an isotopic signal that reflects four "drying events," each of a few hundred years duration. Owen et al. (1990) examined the tops of many gravity cores from the southern embayments of Lake Malawi, and found evidence for a depressed lake level between 1500 and 1850 AD in the form of an erosional hiatus and an abrupt change in diatom assemblages in water depths to 108 m. The presence of a low stand at this time is substantiated by the oral history of some of the indigenous people of the Lake Malawi basin, as well as by the distribution of radiocarbon dates of archeological sites exposed along the present shoreline of the lake

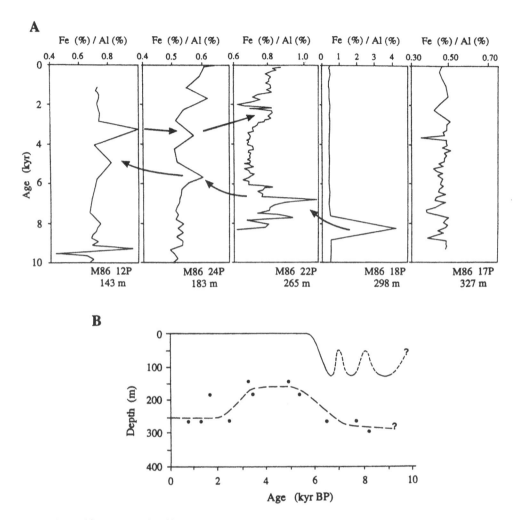

Figure 20. A) The ratio of iron to aluminum in the bulk chemical composition of cores from Lake Malawi, plotted against age. Note the peaks in the Fe/Al ratio, and the water depth of each of the cores, designated below their respective profiles. The arrows mark the transgression and regression of the chemocline through time and depth. B) The resultant curves of lake level (upper light line) and chemocline (lower dark line) vs. time, based on the profiles of carbonate (Fig. 15B) and Fe/Al ratio (Fig. 16A) (from Finney and Johnson, 1991).

(Owen et al., 1990). The timing of this brief low stand nearly coincides with the Little Ice Age. The significance of this observation remains to be seen. On an even higher frequency time scale, we have measured the thickness of about 1300 laminations in a core from northern Lake Malawi, employing computerized image analysis of sediment X-radiographs. The laminations are annual layers (varves) (Pilskaln and Johnson, 1991), averaging about a millimeter in thickness. Time-series analysis of the lamination record shows periodicities at 2.6 and 3.5 years, very similar to periodicities in rainfall anomalies for this part of Africa (Fig. 21) (see Nicholson, this volume).

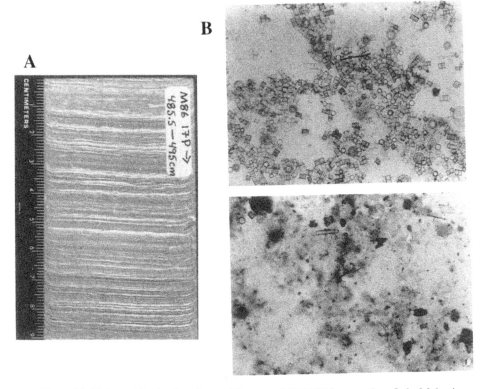

Figure 21. A) Annual laminations (varves) from core M86-17P from northern Lake Malawi. B) Photomicrographs of light layers (above) and dark layers (below), showing the seasonal variation in sediment flux to the lake floor (from Pilskaln and Johnson, 1991). C) Computerized image scan of an X-radiograph of M86-17P, showing the relative intensity of reflected light off the X-radiograph. This is converted to a time series of lamination thickness. D) Time-series analysis of lamination thickness, showing spectral peaks primarily at 2.7 and 3.6 years (Johnson et al., in prep.).

CONCLUSIONS

Patterns of sedimentation in the large lakes of the East African Rift Valley are sufficiently complex to require careful surveys of prospective coring sites for paleoclimate analysis. Surface waves generated by strong winds can sort and redistribute sediments in water depths as great as 100 m. Deep-water currents, perhaps associated with seiches, can scour channels and create erosional unconformities in water several hundred meters deep. Gravitational mass wasting and turbidity currents are commonplace in certain regions of large lakes, particularly off border faults and the deltas of large rivers, where such processes create hiati in the sedimentary record as well as wildly varying rates of sediment accumulation. Despite these complicating factors, regions of all the lakes can be found where the sedimentary record is well suited for paleoclimatic investigation.

In such regions, strong signals of past climatic and limnological change are archived in the sediments, on time scales ranging from millenia to individual years. The most reliable signals to be extracted thus far are biogenic, namely, in the relative abundances of diatom

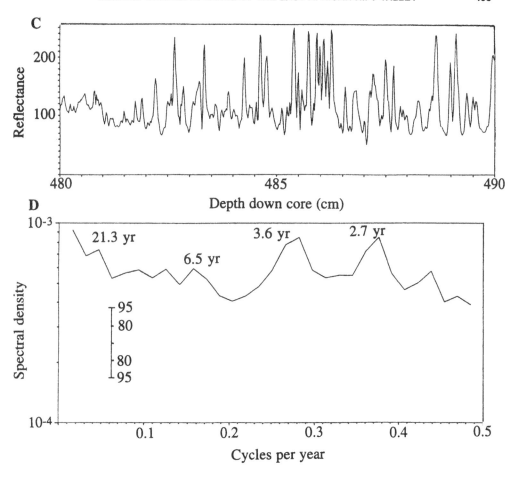

Figure 21, cont'd

species and pollen. Geochemical records of past climatic change are somewhat more difficult to interpret, but show promise for further development. Variation in the abundance of $CaCO_3$, for example, seems to provide a reliable signal of previous lowstand in Lakes Malawi and Victoria, but may not in Lakes Tanganyika and Kivu. The isotopic composition of the carbonates is just beginning to be examined in any detail in the rift valley lakes, and appears to provide an independent signal of past variations in lake water chemistry. Unlike carbonate abundance, the isotopic signal should not be affected by dilution by the non-carbonate fraction. Other geochemical indicators of past lacustrine conditions, such as the Fe/Al ratio in bulk sediment, the isotopic composition of total organic matter or of specific organic compounds, and the trace element composition of biogenic carbonates, all need further investigation.

One of the greatest hindrances to progress in East African paleoclimatology at present is the difficulty in obtaining reliable dates on sediments. There are numerous examples of inconsistent radiocarbon dates from sediment cores and, beyond the range of radiocarbon dating, the techniques for determining sediment age are limited. Perhaps the most promising approach will involve radiometric dating of volcanic ashes coupled with the establishment

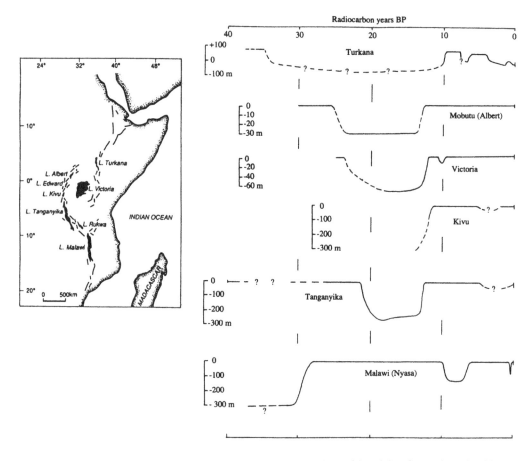

Figure 22. A summary of lake level curves for the large East African lakes for the last 10–40 ka.

of a high-resolution magnetic stratigraphy based on secular variation in inclination and declination.

All of the large East African lakes except Malawi exhibit nearly the same history of lake level change for the past 20,000 years, at least on a millenial time scale (Fig. 22). All of the large lakes north of Malawi were low during the last glacial maximum, and began to rise to high levels around 10–12 ka. Lake Victoria's level reverted briefly to low stand conditions around Younger Dryas time, but came back up in the subsequently warmer early Holocene. There is no firm evidence for the Younger Dryas in any of the other large lakes, although this requires further investigation.

Change within the Holocene is recognized in all of the lakes, particularly in the succession of diatom assemblages. The nature and timing of these changes varies among the lakes, perhaps indicating variations in the intensity of upwelling or steepness of nutrient gradients within the water column, that are unique to each lake. Equally dramatic changes in geochemical parameters are also recognized. The exact timing of these changes need to be worked out for each lake basin, and their climatic significance must be determined. On shorter time scales (interannual to centuries), the search for cyclic variability and abrupt climatic events must continue. Teleconnections between global atmospheric and oceanic circulation and east African climate are beginning to be realized on various time scales, but this work is in its infancy. There is much to be done.

ACKNOWLEDGMENTS

This manuscript was written while on a Fulbright Fellowship at the Laboratoire de Geologie du Quaternaire, CNRS–Luminy, in Marseille, France. I thank my French colleagues for their hospitality, financial support, and exchange of ideas on the paleoclimate of Africa. I am especially grateful to Raymonde Bonnefille, Yves Lancelot and Edith Vincent for all that they did to make my stay in France a memorable and enjoyable one. This manuscript benefitted significantly from thoughtful review by Andy Cohen, Dan Livingstone and Mike Talbot. Financial support was provided by CNRS-France, NSF Grant EAR-9421566, AID Grant HRN-5600-G00-2056-00, and NOAA Grant NA36PO317. I thank Christine Vanbesien and Michel Decobert for their excellent drafting of many of the figures.

REFERENCES

Barton, C. E., and T. Torgesen, 1988, Palaeomagnetic and ^{210}Pb estimates of sedimentation in Lake Turkana, East Africa: Palaeogeography, Palaeoclimatology, Palaeoecology, v. 68, pp. 53–60.

Beadle, L. C., 1981. The Inland Waters of Tropical Africa, An Introduction to Tropical Limnology. Longman, London, 475 p.

Bonnefille, R., Riollet, G., Buchet, G., Arnold, M., Icole, M., and Lafont, R., submitted, High resolution pollen and carbon records for the glacial/interglacial in equatorial Africa: Quaternary Science Reviews.

Bouroullec, J.-L., Rehault, J.-P., Rolet, J., Tiercelin, J.-J., and Mondeguer, A., 1991, Quaternary sedimentary processes and dynamics in the northern part of the Lake Tanganyika trough, east African rift system. Evidence of lacustrine eustatism? Bulletin Centres Recherches d'Exploration et Production d'Elf-Aquitane, v. 15, pp. 343–368.

Butzer, K. W., Isaac, G. L., Richardson, J. L., and Washbourn-Kamau, C., 1972, Radiocarbon dating of East African lake levels: Science, v. 175, pp. 1069–1076.

Campbell, W. H., Blechman, J. B., and Bryson, R. A., 1983, Long-period tidal forcing of Indian monsoon rainfall: a hypothesis: Journal of Climatology and Applied Meteorology, v. 22, pp. 289–296.

Casanova, J., and Hillaire-Marcel, C., 1992, Chronology and paleohydrology of Late Quaternary high lake levels in the Manyara Basin (Tanzania) from isotopic data (^{18}O, ^{13}C, y^{14}C, Th/U) on fossil stromatolites: Quaternary Research, v. 38, pp. 205–226.

Cohen, A. S., 1984, Effects of zoobenthic standing crop on laminae preservation in tropical lake sediments, Lake Turkana, East Africa: Journal of Paleontology, v. 58, pp. 499–510.

Cohen, A. S., 1987, Fossil ostracodes from Lake Mobutu (Lake Albert): paleoecologic and taphonomic implications: Paleoecology of Africa, v. 18, pp. 271–281.

Cohen, A. S., Ferguson, D. S., Gram, P. M., Hubler, S. L., and Sims, K. W., 1986, The distribution of coarse grained sediments in modern Lake Turkana, Kenya: implications for clastic sedimentation models of rift lakes, *in* Frostick, L., et al., eds., Sedimentation in the African rifts: Geological Society of London Special Publication 23, pp. 121–133.

Cohen, A. S., Soreghan, M. J., and Scholz, C. A., 1993, Estimating the age of ancient lakes: an example from Lake Tanganyika, East African Rift system: Geology, v. 21, pp. 511–514.

Crowley, T. J., 1983, The geologic record of climatic change: Reviews of Geophysics and Space Physics, v. 21, pp. 828–877.

Currie, R. G., and Fairbridge, R. W., 1985, Periodic 18.6-year and cyclic 11-year induced drought and flood in northwestern China and some global implications: Quaternary Science Reviews, v. 4, pp. 109–134.

Dalrymple, G. B., and Lanphere, M. A., 1969, Potassium–Argon Dating: San Francisco, W. H. Freeman, 258 p.

De Batist, M., Van Rensbergen, P., Back, S., and Klerkx, J., 1996, Structural framework, sequence stratigraphy and lake level variations in the Livingstone Basin (Northern Lake Malawi): first results of a high-resolution reflection seismic study, *in* Johnson, T. C., and Odada, E. O., eds., The Limnology, Climatology, and Paleoclimatology of the East African Lakes: Gordon and Breach, Toronto, pp. 509–521.

Degens, E. T., Von Herzen, R. P., and Wong, H.-K., 1971, Lake Tanganyika: water chemistry, sediments, geological structure: Naturwissenschaften, v. 58, pp. 229–240.

Degens, E. T., Von Herzen, R. P., Wong, H.-K., Deuser, W. G., and Jannasch, H. W.,1973, Lake Kivu: structure, chemistry and biology of an east African rift lake: Geologisches Rundschau, v. 62, pp. 245–277.

Faure, H., and Gac, J. Y., 1981, Will the Sahelian drought end in 1985? Nature, v. 291, pp. 475–478.

Finney, B. P., and Johnson, T. C., 1991, Sedimentation in Lake Malawi (East Africa) during the past 10,000 years: a continuous paleoclimate record from the southern tropics: Palaeogeography, Palaeoclimatolgy, Palaeoecology, v. 85, pp. 351–366.

Finney, B. P., Scholz, C. A., Johnson, T. C., Trumbore, S., and Southon, J., 1996, Late Quaternary lake-level changes of Lake Malawi, *in* Johnson, T. C., and Odada, E. O., eds., The Limnology, Climatology, and Paleoclimatology of the East African Lakes: Gordon and Breach, Toronto, pp. 495–508.

François, R., Pilskaln, C. H., and Altabet, M. A., 1996, Seasonal variation in the nitrogen isotopic composition of sediment trap materials collected in Lake Malawi, *in* Johnson, T. C., and Odada, E. O., eds., The Limnology, Climatology, and Paleoclimatology of the East African Lakes: Gordon and Breach, Toronto, pp. 241–250.

Gasse, F., Ledee, V., Massault, M., and Fontes, J.-C.,1989, Water-level fluctuations of Lake Tanganyika in phase with oceanic changes during the last glaciation and deglaciation: Nature, v. 342, pp. 57–59.

Gasse, F., Tehet, R., Durand, A., Gilbert, E., and J.-C. Fontes, 1990, The arid–humid transition in the Sahara and the Sahel during the last deglaciation: Nature, v., 346, pp. 141–146.

Goldstein, S. J., Murrell, M. T., and Williams, R. W., 1993, ^{231}Pa and ^{230}Th chronology of mid-ocean ridge basalts: Earth and Planetary Science Letters, v. 115, pp. 151–160.

Haberyan, K. A., 1987, Fossil diatoms and the Paleolimnology of Lake Rukwa, Tanzania: Freshwater Biology, v. 17, pp. 429–436.

Haberyan, K. A., and Hecky, R. E., 1987, The late Pleistocene and Holocene stratigraphy and paleolimnology of Lakes Kivu and Tanganyika: Palaeogeography, Palaeoclimatology, Palaeoecology, v. 61, pp. 169–197.

Harvey, T. J., 1976, The Paleolimnology of Lake Mobutu Sese Seku, Uganda–Zaire: the Last 28,000 Years. Ph.D. Dissertation, Duke University Department of Zoology, Durham, NC, 104 pp.

Haileab, B., and Brown, F. H., 1992, Turkana Basin–Middle Awash valley correlations and the age of thee Sagantole and Hadar Formations: Journal of Human Evoloution, v. 22, pp. 453–468.

Halfman, J. D., and Johnson, T. C., 1988, High-resolution record of cyclic climatic change during the past 4 ka from Lake Turkana, Kenya: Geology, v. 16, pp. 496–500.

Halfman, J. D., Jacobson, D. F., Cannella, C. M., Haberyan, K. A., and Finney, B. P., 1992, Fossil diatoms and the mid to late Holocene paleolimnology of Lake Turkana, Kenya: a reconaissance study: Journal of Paleolimnology, v. 7, pp. 23–35.

Halfman, J. D., Johnson, T. C., and Finney, B. P., 1994, New AMS dates, stratigraphic correlations and decadal climatic cycles for the past 4 ka at Lake Turkana, Kenya: Palaeogeography, Palaeoclimatology, Palaeoecology, v. 111, pp. 83–98.

Hameed, S., 1984, Fourier analysis of Nile flood level: Geophysical Research Letters, v. 11, pp. 843–845.

Hecky, R. E., 1978, The Kivu–Tanganyika basin: the last 14,000 years: Polskie Archiwum Hydrobiologii, v. 25, pp. 159–165.

Hillaire-Marcel, C., Carro, O., and Casanova, J., 1986, ^{14}C and Th/U dating of Pleistocene and Holocene stromatolites from East African paleolakes: Quaternary Research, V. 25, pp. 312–329.

Johnson, T. C., 1980, Sediment redistribution by waves in lakes, reservoirs and embayments, *in* Stefan, H., ed., Proceedings of the Symposium on Surface Water Impoundments: New York, American Society of Civil Engineering, pp. 1307–1317.

Johnson, T. C., 1984, Sedimentation in large lakes: Annual Reviews of Earth and Planetary Science, v. 12, pp. 179–204.

Johnson, T. C., Halfman, J. D., Rosendahl, B. R., and Lister, G. S., 1987, Climatic and tectonic effects on sedimentation in a rift valley lake: Evidence from high-resolution seismic profiles, Lake Turkana, Kenya: Geological Society of America Bulletin, v. 98, pp. 439–447.

Johnson, T. C., Davis, T. W., Halfman, B. M., and Vaughan, N. D., 1988, Sediment Core Descriptions: Malawi 86, Lake Malawi, East Africa: Beaufort, N. C., Duke University Technical Report, 101 p.

Johnson, T. C., and Davis, T. W., 1989, High-resolution seismic profiles from Lake Malawi, Africa: Journal of African Earth Sciences, v. 8, pp. 383–392.

Johnson, T. C., Howd, P. A., and Ofsanko, R., in prep. Time series analysis of varve thickness in Lake Malawi, East Africa, shows rainfall anomaly spectral peaks.

Johnson, T. C., and Ng'ang'a, P., 1990, Reflections on a rift lake, *in* Katz, B. J., ed., Lacustrine Basin Exploration: Case Studies and Modern Analogs: Tulsa, Oklahoma, American Association of Petroleum Geologists Memoir No. 50, pp. 113–135.

Johnson, T. C., Halfman, J. D., and Showers, W. J., 1991, Paleoclimate of the past 4000 years at Lake Turkana, Kenya based on isotopic composition of authigenic calcite: Palaeogeography, Palaeoclimatology, Palaeoecology, v. 85, pp. 189–198.

Johnson, T. C., Wells, J. T., and Scholz, C. A., submitted, Deltaic sedimentation in a modern rift lake: Lake Malawi, East Africa: Geological Society of America Bulletin.

Kendall, R. L., 1969, An ecological history of the Lake Victoria basin. Ecological Monographs, v. 39, pp. 121–176.

Lipiatou, E., Hecky, R. E., Eisenreich, S. J., Lockhart, L., Muir, D., and Wilkinson, P., 1996, Recent ecosystem changes in Lake Victoria reflected in sedimentary natural and anthropogenic organic compounds in sediments, *in* Johnson, T. C., and Odada, E. O., eds., The Limnology, Climatology and Paleoclimatology of the East African Lakes: Gordon and Breach, Toronto, pp. 523–541.

Livingstone, D. A., 1965, Sedimentation and the history of water level change in Lake Tanganyika: Limnology and Oceanography, v. 10, pp. 607–610.

Livingstone, D. A., and Van der Hammen, T., 1978, Palaeogeography and palaeoclimatology, in: Tropical Forest Ecosystems, UNESCO/UNEP/FAO, Paris, pp. 61–90.

McKenzie, J. A., 1984, Carbon isotopes and productivity in the lacustrine and marine environment, *in* Stumm, W., ed., Chemical Processes in Lakes: Wiley, New York, pp. 99–118.

Mohammed, U. M., Bonnefille, R., and Johnson, T. C., submitted.

Morner, N. A., and Karlen, W. (eds.), 1984, Climate Changes on a Yearly to Millenial Basis. Reidel, 667 p.

Muchane, M. W., 1994, Stable Isotope Analyses of Authigenic Calcite from Lake Turkana, Kenya: High Resolution Paleoclimatic Implications for the Past 5000 Years. Ph.D. dissertation, Duke University Department of Geology, Durham, NC, 190 pp.

Muchane, M. W., 1996, Comparison of the isotope record in micrite, Lake Turkana, with the historical weather record over the last century, *in* Johnson, T. C., and Odada, E. O., eds., The Limnology, Climatology and Paleoclimatology of the East African Lakes: Gordon and Breach, Toronto, pp. 431–441.

Ng'ang'a, P., 1993, Deltaic sedimentation in a lacustrine environment, Lake Malawi, Africa: Journal of African Earth Sciences, v. 16, pp. 253–264.

Nicholson, S. E., 1996, A review of climate dynamics and climate variability in Eastern Africa, *in* Johnson, T. C., and Odada, E. O., eds, The Limnology, Climatology and Paleoclimatology of the East African Lakes: Gordon and Breach, Toronto, pp. 25–56.

Owen, R. B., Barthelme, J. W., Renaut, R. W., and Vincens, A., 1982, Palaeolimnology and archeology of Holocene deposits north-east of Lake Turkana, Kenya: Nature, v. 298, pp. 523–528.

Owen, R. B., and Crossley, R. 1992, Spatial and temporal distribution of diatoms in sediments of Lake Malawi, Central Africa, and ecological implications: Journal of Paleolimnology, v. 7, pp. 55–71.

Owen, R. B., Crossley, R., Johnson, T. C., Tweddle, D., Kornfield, I., Davidson, S., Eccles, D. H., and Engstrom, D. E., 1990, Major low levels of Lake Malawi and implications for speciation rates in cichlid fishes: Proceedings of the Royal Society of London. B. v. 240, pp. 519–553.

Owen, R. B., Renaut, R. W., and Williams, T. M., 1996, Characteristics and origins of laminated ferromanganese nodules from Lake Malawi, Central Africa, *in* Johnson, T. C., and Odada, E. O., eds., The Limnology, Climatology and Paleoclimatology of the East African Lakes: Gordon and Breach, Toronto, pp. 461–474.

Partridge, T. C., 1992, The evidence for Cainozoic aridification in southern Africa: Quaternary International, v. 17, pp. 105–110.

Partridge, T. C., 1993, Warming phases in southern Africa during the last 150,000 years: an overview: Palaeogeography, Palaeoclimatology, Palaeoecology, v. 101, pp. 237–244.

Pilskaln, C., and Johnson, T. C., 1991, Seasonal signals in Lake Malawi sediments: Limnology and Oceanography, v. 36, pp. 544–557

Ricketts, R. D., and Johnson, T. C., 1996, Early Holocene changes in lake level and productivity in Lake Malawi as interpreted from oxygen and carbon isotopic measurements of authigenic carbonates, *in* Johnson, T. C., and Odada, E. O., eds., The Limnology, Climatology and Paleoclimatology of the East African lakes: Gordon and Breach, Toronto, pp. 475–493.

Roberts, N., Taieb, M., Barker, P., Damnati, B., Icole, M., and Williamson, D., 1993, Timing of the Younger Dryas event in East Africa from lake-level changes: Nature, v. 366, pp. 146–148.

Rosendahl, B. R., 1987, Architecture of continental rifts with special reference to East Africa: Annual Reviews of Earth and Planetary Science, v. 15, pp. 445–504.

Sarna-Wojcicki, A. M., Meyer, C. E., Roth, P. H., and Brown, F. H., 1985, Age of tuff beds at East African early hominid sites and sediments in the Gulf of Aden: Nature, v. 313, pp. 306–308.

Scholz, C. A., and Finney, B. P., in press, Late-Quaternary sequence stratigraphy of Lake Malawi (Nyasa), Africa: Sedimentology.

Scholz, C. A., Johnson, T. C., and McGill, J. W., 1993, Deltaic sedimentation in a rift valley lake: new seismic reflection data from Lake Malawi (Nyasa), East Africa: Geology, v. 21, pp. 395–398.

Scholz, C. A., and Rosendahl, B. R., 1988, Low lake stands in Lakes Malawi and Tanganyika, East Africa, delineated with multifold seismic data: Science, v. 240, pp. 1645–1648.

Scholz, C. A., Rosendahl, B. R., Versfelt, J. W., and Rach, N., 1990, Results of high-resolution echo sounding of Lake Victoria: Journal of African Earth Sciences, v. 11, pp. 25–32.

Scott, D. L., Ng'ang'a, P., Johnson, T. C., and Rosendahl, B. R., 1991, High-resolution character of Lake Malawi (Nyasa), East Africa, and its relationship to sedimentary processes: International Association of Sedimentologists, Special Publication 13, pp. 129–145.

Scott, L., 1989, Climatic conditions in southern Africa since the last glacial maximum, inferred from pollen analysis: Palaeogeography, Palaeoclimatology, Palaeoecology, v. 70. pp. 345–353.

Scott, L., 1993, Palynological evidence for late Quaternary warming episodes in southern Africa: Palaeogeography, Palaeoclimatology, Palaeoecology, v. 101, pp. 229–235.

Sowunmi, M. A., 1991, Late Quaternary environments in equatorial Africa: palynological evidence: Paleoecology of Africa, v. 22, pp. 213–238.

Spigel, R. H., and Coulter, G. W., 1996, Comparison of hydrology and physical limnology of the East African Great Lakes: Tanganyika, Malawi, Victoria, Kivu and Turkana (with reference to some North American Great Lakes), in Johnson, T. C., and Odada, E. O., eds., The Limnology, Climatology and Paleoclimatology of the East African Lakes: Gordon and Breach, Toronto, pp. 103–139.

Ssemmanda, I., and A. Vincens, 1993, Vegetation et climat dans le bassin du lac Albert (Ouganda, Zaire) depuis 13000 ans B.P.: apport de la palynologie: Comptes Rendues de l'Academie des Sciences Paris, t. 316, Serie II, pp. 561–567.

Stager, J. C., 1984, The diatom record of Lake Victoria (East Africa): the last 17,000 years: Proceedings of the 7th Diatom Symposium, pp. 455–476.

Stager, J. C., 1988, Environmental changes at Lake Chelshi, Zambia since 40,000 years B.P.: Quaternary Research, v. 29, pp. 54–65.

Stager, J. C., Reinthal, P. N., and Livingstone, D. A., 1986, A 25,000 year history for Lake Victoria, East Africa and some comments on its significance for the evolution of cichlid fishes: Freshwater Biology, v. 16, pp. 15–19.

Stoffers, P., and Hecky, R. E., 1978, Late-Pleistocene and Holocene evolution of the Kivu–Tanganyika basin: International Association of Sedimentologists Special Publication 2, pp. 43–55.

Street, F. A., and A. T. Grove, 1979, Global maps of lake-level fluctuations since 30,000 yr BP. Quat. Res., v. 12, pp. 83–118.

Stuiver, M., 1970, Oxygen and carbon isotope ratios of fresh-water carbonates as climatic indicators: Journal of Geophysical Research, v. 75, pp. 5247–5257.

Talbot, M. R., 1990, A review of the paleohydrological interpretation of carbon and oxygen isotopic ratios in primary lacustrine carbonates: Chemical Geology, v. 80, pp. 261–279.

Talbot, M. R., and Kelts, K., 1990, Palaeolimnological signatures from carbon and oxygen isotope ratios in carbonates from organic-rich lacustrine sediments, in Katz, B., ed., Lacustrine Basin Exploration: Case Studies and Modern Analogs: American Association of Petroleum Geologists Memoir No. 50, pp. 99–112.

Talbot, M. R., and Livingstone, D. A., 1989, Hydrogen index and carbon isotopes of lacustrine organic matter as lake level indicators: Palaeogeography, Palaeoclimatology, Palaeoecology, v. 70, pp. 121–137.

Talbot, M. R., and Johannessen, T., 1992, A high resolution palaeoclimatic record for the last 27,500 years in tropical West Africa from the carbon and nitrogen isotopic composition of lacustrine organic matter: Earth and Planetary Science Letters, v. 110, pp. 23–37.

Tiercelin, J.-J., and Mondeguer, A., 1991, The geology of the Tanganyika trough, in Coulter, G. W., ed., Lake Tanganyika and its Life: London, Oxford Univeristy Press, pp. 7–48.

Tiercelin, J.-J., Mondeguer, A., Gasse, F., Hillaire-Marcel, C., Hoffert, M., Larque, P., Ledee, V., Marestang, P. Ravenne, C., Raynaud, J. F., Thouveny, N., Vincens, A., and Williamson, D., 1988, 25,000 ans d'histoire hydrologique et sedimentaire de lac Tanganyika, Rift Est-africain: Paris, Academie des Sciences, Comptes Rendus, v. 307, pp. 1375–1382.

Tiercelin, J.-J., Soreghan, M., Cohen, A. S., Lezzar, K.-E., and Bouroullec, J.-L., 1992, Sedimentation in large rift lakes: example from the middle Pleistocene–Modern deposits of the Tanganyika trough, East African Rift System: Centres de Recherches Exploration-Production Elf-Aquitane, Bulletin, v. 16, pp. 83–111.

Tiercelin, J.-J., et al., 1993, Hydrothermal vents in Lake Tanganyika, East African Rift System: Geology, v. 21, pp. 499–502.

Vincens, A., 1989, Paleoenvironnements du bassin Nord-Tanganyika (Zaire, Burundi, Tanzanie) au cours des 13 derniers mille ans: apport de la palynologie: Review of Palaeobotany and palynology, v. 61, pp. 69–88.

Vincens, A., 1991, Late Quaternary vegetation history of the south-Tanganyika basin. Climatic implications in south central Africa: Palaeogeography, Palaeoclimatology, Palaeoecology, v. 86, pp. 207–226.

Vincens, A., 1993, Nouvelle sequence pollinique du Lac Tanganyika: 30,000 and d'histoire botanique et climatique du bassin nord: Review of Palaeobotany and Palynology, v. 78, pp. 381–394.

Vincens, A., Chalie, F., Bonnefille, R., Guiot, J., and Tiercelin, J.-J., 1993, Pollen-derived rainfall and temperature estimates from Lake Tanganyika and their implication for Late Pleistocene water levels: Quaternary Research, v. 40, pp. 343–350.

Vogel, J. C., 1989, Evidence for past climatic change in the Namib desert: Palaeogeography, Palaeoclimatology, Palaeoecology, v. 70, pp. 355–366.

Walter, R. C., and Aronson, J. L., 1982, Revisions of K/Ar ages for the Hadar hominid site, Ethiopia, Nature, v. 296, pp. 122–127.

West, M., and Johnson, T. C., in press, Some statistical issues in palaeoclimatology, Proc. of the Fifth Valencia International Meeting on Bayesian Statistics, Alicante, Spain, 1994.

Wong, H.-K., and Von Herzen, R. P., 1974, A geophysical study of Lake Kivu, East Africa: Geophysical Journal of the Royal Astronomical Society, v. 37, pp. 371–389.

Yuretich, R. F., 1979, Modern sediments and sedimentary processes in Lake Rudolf (Lake Turkana) eastern rift valley, Kenya: Sedimentology, v. 26, pp. 313–331.

Some Aspects of the Physical and Chemical Dynamics of a Large Rift Lake: The Lake Turkana North Basin, Northwest Kenya

D.O. OLAGO   *Tropical Palaeoenvironments Research Group, School of Geography, University of Oxford, Oxford, United Kingdom*

E.O. ODADA   *Department of Geology, University of Nairobi, Nairobi, Kenya*

Abstract — The Omo River accounts for about 90% of the lakes water budget, and thus contributes significantly to its physical and chemical identity. The seasonal Kerio and Turkwel rivers contribute most of the remaining fluvial input. The lake is moderately saline, alkaline, and is well mixed by strong, diurnal, southeasterly winds. The dissolved salt composition is characterized by high sodium and bicarbonate concentrations. Wind stress generated on the lake surface, coupled with the basin morphometry results in a closed-gyre circulation pattern centred along the basins north–south trending axis. The Omo River plume seasonally augments the subsurface currents and effects reductions in salinity. The relative difference in the concentration of solvated ions with large hydration sizes between the lake and river water determine the rate limiting step during mixing of the two waters. Other processes modulating dissolved salts concentrations include authigenic mineral precipitation, adsorption/exchange with suspended clay particles, biogenic uptake and bottom sediment resuspension.

INTRODUCTION

Lake Turkana is located in the arid north of Kenya at about 3°N, 36°E (Fig. 1). It lies within the broad Turkana depression, between the Kenya and Ethiopia domes in the East African Rift. The Turkana Depression has generally been regarded as a diffuse zone of faulting, linking the rift segments to the north and to the south (Dunkelman et al., 1988). The mean annual temperature is 30°C, mean annual rainfall is below 255 mm/yr, and the evaporation rate (piche evaporimeter data) is about 3.2 m per year, and relatively high evaporation rates persist throughout the year, with no apparent seasonal changes (Ferguson and Harbott, 1982). The rainfall is low, normally less than 255 mm/yr. The annual mean maximum temperature range is 30 to 34°C, while the annual mean minimum temperature is 23.7°C (Survey of Kenya, 1977).

The palaeochemistry of Lake Turkana has been estimated using the relative concentrations of the exchangeable clay mineral cations sodium and calcium, as well as from variations in the mineralogical assemblages of bottom sediments (Cerling, 1979). Since 1.8 Ma, the lake has varied from brackish (Cerling, 1979) to its presently alkaline and moderately saline state (Yuretich and Cerling, 1983). Recent research work on the water body has centred mainly around its aquatic ecosystem (e.g., Hopson, 1982). The importance of cation exchange reaction between suspended clay sediments and the lake water has also been investigated (Yuretich, 1986).

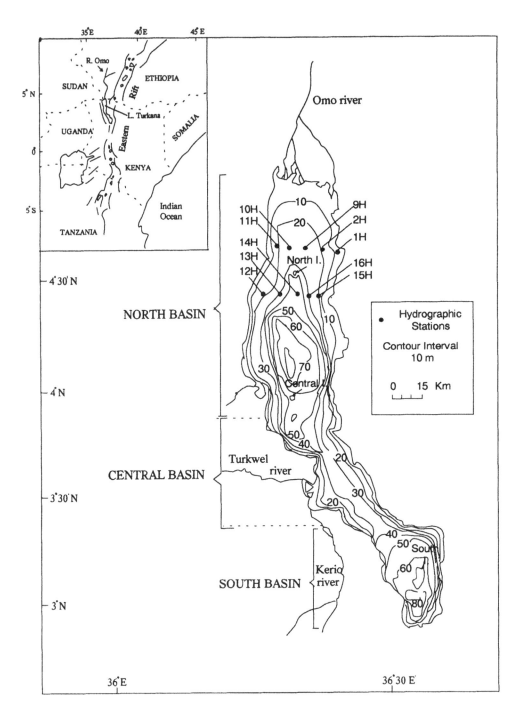

Figure 1. Location map of Lake Turkana and the hydrographic stations in the North Basin.

Although fluvial activity is generally infrequent (only the Omo River is perennial), the sediment load is high, in common with other arid environments, and delta construction is rapid (Frostick and Reid, 1986). Shoreline features of the lake include major spits of the western lake shore and are associated with high-energy coastlines (Ferguson and Harbott, 1982). Primary spits of the eastern shore such as Mvite and Koobi Fora are subject to relatively little wave action but are maintained by currents running along both the river and lake margins, creating extensive submerged and often steep-sided sand bars (op. cit.). During episodes of infrequent and short-lived flash-flooding along the eastern and western margins of the North Basin, significant amounts of sediment, derived from the thick Plio–Pleistocene deposits such as the Galana Boi Formation in the eastern shore, are probably supplied to the lake. These would probably significantly influence the lake water chemistry on the short term. However, the changes they may effect in water chemistry need to be monitored as they occur, prior to, during, and after flooding episodes. This has to date not been done.

In this paper we elaborate upon the processes and controls on mixing of the Omo River and lake water in the North Basin. We examine changes in the long-term steady state of the lake's chemical regime as effected by the perennial Omo River. Due to our sampling design, the effects of short-term, and possibly long-term influences on lake water chemistry due to the intermittent sediment and water supply of ephemeral streams (arising from the thick Plio–Pleistocene fluvio-lacustrine and tuffaceous sediments along the eastern and western shores) draining into the North Basin of the lake could not be evaluated.

METHODS

The North Basin water samples were collected during a cruise in January/February 1990, headed by Prof. T. C. Johnson of the Duke University Marine Laboratory. To establish our positions on the lake (to within 500 m^2), we used a standard model computerized Furuno radar system, which was mounted on top of one of the boats. As a backup navigation system we had a Magellan satellite GPS portable receiver.

Water samples were taken using standard water-samplers at various stations covering two east–west transects (Fig. 1). The water samples were filtered on site to remove organic matter and suspended sediments, and then stored in labelled plastic bottles. Conductivity was measured on a standard conductivity meter, and pH was likewise measured on a standard pH meter in the laboratory. The elements Na, Mg, Li, Al, Ca, and Fe were analyzed using the Inductively Coupled Plasma (ICP) Mass Spectrometer, model 7310. K was determined using a Coning Flame Photometer. SiO_2, Cl and PO_4 were determined by automated colorimetry. Chemical equilibria calculations were performed using the U.S.G.S PC version of the computer program WATEQF. Alkalinity values for Lake Turkana were computed from the anion–cation charge balance equation since it was not measured in the field. Dissolved oxygen concentration values were also not determined in the field and hence the results of Hopson A.J (1982) are used.

THE GEOLOGICAL SETTING

To the north, the rocks consist of basalts with rhyolites, basalt, trachyte and phonolite in the Ethiopian highlands with outcroppings of granitic gneiss basement in the lower regions to the south, an area covered largely by alluvium deposited by the Omo River (Fig. 2).

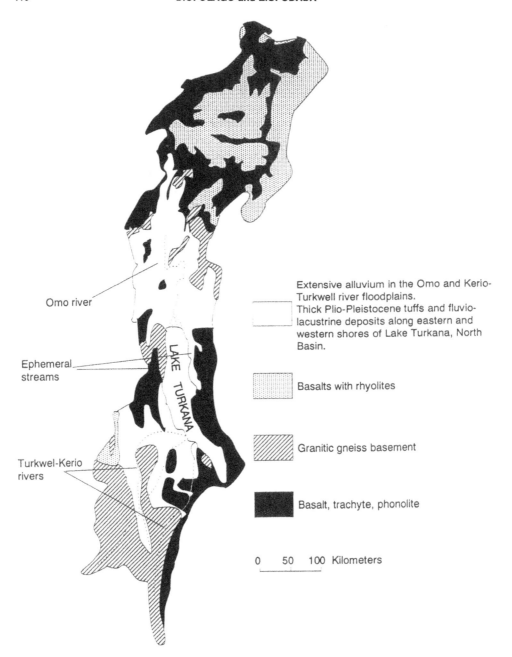

Figure 2. Geology of the Lake Turkana drainage basin (partly modified from Yuretich, 1979).

Precambrian rocks (mainly quartzite and amphibolite schists) are present in only 7% of the Omo River catchment area, which accounts for 58% of the total catchment area of the Lake Turkana basin (Ferguson and Harbott, 1982). On the western side of the lake, most of the uplands consists of soda-rich basalts, trachytes and phonolites, remnants of Tertiary volcanism, with few and scattered outcrops of Precambrian Basement rocks (Fig. 2). The alluvium cover is thin and runs in a narrow belt parallel to the lake margin. However, in the southwest part there occur extensive outcrops of Precambrian Basement rocks consisting of biotite gneisses, hornblende gneisses, migmatites and plagioclase amphibolites of the Upper Proterozoic Turbo-Kitale Group (Fig. 2) (Pallister, 1971). Closer to the lakeshore the alluvium cover is wide and expansive, deposited by the Turkwel and Kerio Rivers as they wind their way to the lake. The southern part of the lake basin consists largely of basalts, trachytes and phonolites, with small and interspersed outcrops of granitic gneiss Basement rocks (Fig. 2).

THE MODERN LAKE SEDIMENTS

The modern sediments of Lake Turkana are primarily detrital silicates and are dominantly fine grained. The relative composition of the silicate minerals is primarily controlled by the source rock composition and weathering intensity (Yuretich, 1979, 1986). The North Basin is rich in kaolinite and resistates that reflect the weathering of mafic volcanics in the humid tropical setting of the Omo Basin. In contrast, the sediments adjacent to the Turkwel–Kerio drainage system are coarser; they contain quartz, feldspar, illite and smectite, and reflect the arid climate and exposures of Pre-Cambrian gneiss and schists (Halfman et al., 1989). Carbonate is the next most abundant component to the detrital silicate fraction in the sediments of Lake Turkana (Yuretich, 1979; Halfman, 1987). Carbonate concentration of the sediment increases with increasing distance from the Omo River, i.e., from rare and poorly preserved diatom clays in the North Basin to well-preserved ostracod-diatom silty clays in the South Basin; this trend resulting from decreased dilution by detrital silicates with increasing distance from the Omo River delta (Halfman, 1987). The carbonate fraction in the offshore sediments has two main components; ostracod carapaces, and micron-sized crystals of carbonate (Halfman, 1987).

THE LAKE

Lake Turkana is the largest closed-basin lake in the East African Rift. It can be considered as the "arid region end-member" of large rift valley lakes and an important modern analogue for ancient rift environments in Africa and elsewhere (Halfman et al., 1989). It is 250 km long and has a mean width of 30 km, with a surface area of about 7500 square kilometers (Fig. 1). The average depth is 35 m while the maximum depth is 115 m (Fig. 1). Lake Turkana receives runoff and sediment from a wide geographical area. The Omo River provides about 90% of the water that flows into the basin (Cerling, 1986), draining southward from the Ethiopian plateau where mid-year monsoonal rainfall exceeds 1500 mm (Halfman and Johnson, 1988). The seasonal Turkwel and Kerio Rivers contribute most of the remaining fluvial input. Other streams, direct rainfall and subterranean flow are considered insignificant in the water budget (Ferguson and Harbott, 1982; Yuretich and Cerling, 1983). The Omo River thus contributes significantly to the physical and chemical identity of the lake, particularly in the North Basin. All water input is approximately balanced by evaporation, the surface level lying at ca. 375 m a.s.l.

The water column has a relatively uniform temperature of 25 to 26°C and dissolved oxygen concentrations that range from about 70% to 100% saturation (Hopson, 1982). The lake is moderately saline (2.5‰), alkaline (surface water pH = 9.2), and is well mixed by strong diurnal winds (Yuretich and Cerling, 1983). Its high alkalinity promotes rapid equilibration of CO_2 with the atmosphere (Peng and Broecker, 1980). The dissolved salt composition is characterized by high sodium and bicarbonate concentrations and relatively

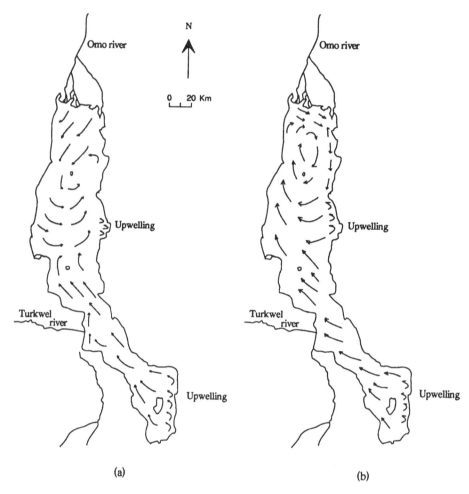

(a)   (b)

Figure 3. Circulation patterns within Lake Turkana. a) Principal currents during flood season (May to November). b) Current patterns during low discharge (December to April) (From Yuretich, 1979).

low abundances of magnesium, calcium and sulphate ions (Yuretich and Cerling, 1983; Halfman et al., 1989). The lake undergoes far greater changes in salinity than in temperature in the low latitude desert setting (Johnson, 1990). In the north, salinity is seasonally reduced through mixing with dilute Omo River flood-waters. Slight variations in the water analyses for Lake Turkana, as recorded by various workers are accounted for mainly by the changes which take place during the influx of fresher water from the Omo River (Hopson, 1982). Lake-wide conductivities range between 3.5–3.7 mS/cm South and Central Basins), to 3.2–3.5 mS/cm for the southern part of the North Basin, about 3.0 mS/cm for the middle part of the North Basin, to 1.0–3.0 mS/cm for the northern part of the North Basin (Hopson, 1982). Hopson's (1982) conductivity values have an almost constant conductivity depth profile in the South and Central Basins for measurements taken at any given time during the year. The average value of this profile may shift depending on the time of the year, generally within the ranges mentioned above. Since the lake lies in an enclosed basin with no surface outlet, a relatively constant lake level is maintained by a balance between inflow and evaporation, and, as a result, the concentration of solutes increases continually (Hopson, 1982). Conductivity values indicate that the concentration of solutes is increasing at a much slower rate than expected; over longer time periods, relatively low conductivities in the lake are maintained by sediment–water interactions (Hopson, 1982).

For most of the year an easterly wind, whose strength varies diurnally, prevails and is particularly strong near the lake shore. It drives the surface currents of the lake, leading to closed-gyre circulations in the North Basin, which are periodically reversed during the Omo River flood seasons (Fig. 3).

LAKE DYNAMICS: PHYSICAL

Lower concentrations of most ions in the northern, as compared to the southern transect reflect the decreasing influence of the Omo River plume with increasing distance south of the Omo River delta. The Omo River plume is strongest close to the western shoreline, with lowest conductivities and ion concentrations being registered at station 11H (Table 1; Fig. 4). The two major distributaries of the Omo River, the Dielerhiele arm (which extends 15 km into the lake terminating in a delta fan), and the Erdete branch, which enters at the northwest corner of the lake (Ferguson and Harbott, 1982), explain why the plume is centred in the western half of the lake. The Murdizi branch in the northeast corner of the lake became extinct in recent years (Ferguson and Harbott, 1982).

The low variations in conductivity above 15 m depth (Table 1) indicate that this part of the lake water is very well mixed, and can be attributed to wind-induced surface wave currents. The northward decrease in conductivity reflects the importance of the Omo River as a lake water diluting factor. This is especially significant in the western side of the lake where the Omo River plume is most concentrated. Above 20 m depth in the northern transect (Fig. 4), conductivity values are lowest on the western side (2.62 mS/cm), rising up towards the eastern side (2.86 mS/cm) of the lake. Towards the south in the southern transect (Fig. 4), there is a rise in the conductivities recorded on the west as compared to the northern transect, and the lowest conductivity region moves away from the western shore region towards the centre of the lake. This represents an eastward deflection of the Omo River plume. It is due to the effects of a lateral easterly directed component of velocity generated by plume pressure on the hard shoulder of the western coastline, as well as on the density differences between the river plume and the lake water and slope direction (easterly) of the lake bottom. A possible eastward-directed component arising from the geostrophic effects of the Coriolis force may be small or insignificant in this low-latitude setting. Although the river plume has a general effect on the conductivity levels by depressing the values throughout the water column in the western side of the lake, the lower conductivities at depths of 10–15 m at hydrographic stations 11H and 13H (13H is directly south of 11H) (Fig. 4) as compared to the conductivity levels of the surface and bottom waters, indicate that this is the region of the mainstream of the Omo River plume.

Wind energy is a most important physical, year-round control on circulation in Lake Turkana, and exerts a major control on temperature and dissolved gases as well, while the influence of the Omo River influx is more seasonal, being enhanced twice yearly for periods of about three months each. There is a three month time lag between the onset of seasonal rain in the Ethiopian Highlands catchment area and the resultant increase in lake level (Ferguson and Harbott, 1982). Rain falls principally between April and September, so the

Table 1. Lake Turkana North Basin Water Chemistry (alkalinity computed using charge balance equation)

Hydro-graphic station	Depth (m)	Conduc-tivity (mS/cm)	pH	Mg (mg/l)	Al (mg/l)	Ca (mg/l)	Na (mg/l)	Fe (mg/l)	K (mg/l)	SiO$_2$ (mg/l)	Cl (mg/l)	PO$_4$ (mg/l)	Alk. (meq/l)
1H	0	2.86	8.7	2.294	BDL	5.593	837.0	BDL	1.75	7.2	425.86	3.57	24.67578
	12	2.85	8.6	2.326	0.23	5.842	805.6	0.212	1.15	7.78	297.94	2.29	26.88463
2H	0	2.67	8.7	2.335	BDL	5.982	764.1	BDL	1.15	7.88	258.98	1.89	26.18647
	20	2.79	8.7	2.309	BDL	5.595	828	BDL	0.8	7.12	147.06	1.2	32.11041
9H	0	2.67	8.75	2.326	0.394	5.612	805.8	0.473	1	6	172.4	1.2	30.43626
	15	2.67	8.7	2.336	BDL	5.775	804	BDL	0.8	4.42	257.27	1.7	27.95819
10H	0	2.63	8.7	2.252	BDL	5.995	799.8	BDL	1.5	7.14	353.53	1.54	25.08197
	16	2.72	8.65	2.222	BDL	5.857	835.3	BDL	0.9	6.52	403.61	2.63	25.18206
	26	2.72	8.7	2.272	BDL	5.576	824.3	BDL	0.9	6.22	130.43	1.42	32.41679
11H	0	2.62	8.6	2.176	BDL	6.212	779.5	BDL	1	5.74	217.38	1.21	28.03224
	10	2.51	8.6	2.282	BDL	6.035	754.2	BDL	0.8	5.4	179.89	0.92	27.98709
	16	2.65	8.6	2.362	BDL	6.152	761.4	BDL	1.3	6.16	215.17	0.88	27.32458
12H	0	2.85	8.7	2.284	BDL	5.632	834.3	0.027	0.55	5.08	198.26	0.84	30.93757
	14	2.85	8.6	2.336	BDL	5.698	834.5	0.021	1.3	6.32	467.55	3.25	23.34818
13H	0	2.76	8.7	2.25	BDL	6.08	862.5	0.023	1.15	5.74	228.73	1.08	31.32736
	15	2.7	8.5	2.132	BDL	5.746	836.1	BDL*	0.8	6.12	287.03	1.28	28.51034
	30	2.83	8.6	2.243	BDL	5.307	844.5	BDL*	0.8	6.12	253.62	1.2	29.81255
14H	0	2.74	8.7	2.218	BDL	5.622	829.6	BDL	0.9	4.08	144.89	1.65	32.23597
	15	2.74	8.7	2.3	BDL	5.695	830.5	BDL	0.8	5.62	152.8	1.08	32.06064
	35	2.82	8.7	2.278	BDL	5.146	859.5	BDL	0.8	5.66	159.49	0.87	33.12098
15H	0	2.76	8.6	2.174	BDL	5.442	826.8	BDL	1	6.08	211.42	0.91	30.24165
	10	2.86	8.7	2.319	BDL	5.427	795.6	0.01	1	4.5	155.93	0.63	30.45824
	25	2.85	8.6	2.315	BDL	5.255	854.1	0.054	0.9	6.8	116.42	0.34	34.11332
16H	0	2.88	8.7	2.348	BDL	5.578	866.7	BDL	1	5.8	302.44	1.44	29.41484
	10	2.89	8.7	2.315	BDL	5.359	854.8	BDL	1.75	4.26	305.32	1.32	28.82961
	25	2.95	8.65	2.283	BDL	5.242	858	BDL	1.4	0.04	265.37	1.24	30.08330

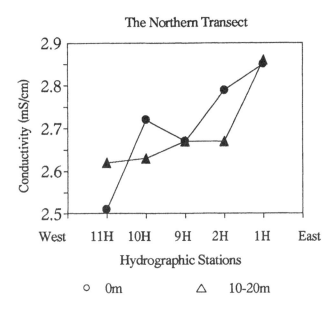

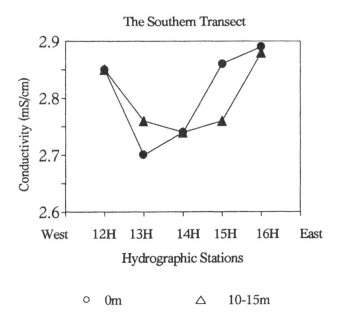

Figure 4. Conductivities in the northern and southern transects of the hydrographic stations in the North Basin.

samples collected in January 1990 reflect a period of diminishing Omo River influence on the composition of the North Basin waters.

The wind stress causes a closed-gyre circulation pattern, with circulation centred on the north–south trending axis of the lake basin (Fig. 3). Wright and Nydegger (1980) note that lake circulation patterns are strongly influenced by inflowing river currents and geostrophic effects. Thus, removed from the Omo River influence, a northwesterly flowing surface current dominates, driven by the prevailing, strong southeasterly winds (Yuretich and Cerling, 1983). The wind-induced surface currents are restricted to the top 6 or 7 meters but are strong enough to create subsurface countercurrents which in turn cause mixing through-out the entire water column (Ferguson and Harbott, 1982).

Circulation patterns are further modified by the seasonally deep incursions of the Omo River plume into the North Basin, which probably result in decreased strengths of the northwesterly surface currents and enhanced subsurface currents. In deeper parts of the basin during the Omo River excursions, physical energy is therefore supplied principally by the Omo River plume. Geostrophic effects (Coriolis force) are superimposed on these circula-tion patterns, causing a deflection to the right. During periods of low Omo River discharge (December to April), the gyre within the North Basin may be reversed and Omo River water is carried to the northeastern corner of the lake (Fig. 3) (Yuretich and Cerling, 1983). Hopson (1982) observed upwelling of deeper water at Loiyengalani in the southeastern part of the lake (Fig. 3). Upwelling may occur in numerous other areas along the shore (Yuretich and Cerling, 1983).

LAKE DYNAMICS: CHEMICAL

It is evident that the dissolved salt composition of the middle to northern parts of the North Basin are characterised by high Na and Cl concentrations, and much lower concentrations of Mg, Ca, K and PO_4, with very low quantities of Al and Fe and traces Of SO_4 (Table 1). This is generally true for the rest of the lake with the exception of PO_4. The alkalinity range of ca. 24–33 meq/l indicates that it is also high in bicarbonate concentrations. It is thus a Na–HCO_3 lake (Cerling, 1979; Hopson, 1982).

Chemical equilibria data indicate that Fe exists principally as $Fe(OH)_4^-$ and silica as H_4SiO_4(aq). Ca occurs as Ca^{2+}(aq) and $CaCO_3$(aq) in approximately equal concentrations, with the latter being slightly higher in concentration. Mg is present as Mg^{2+}(aq) and $MgCO_3$(aq), with slightly higher concentrations of Mg^{2+}(aq). Na occurs mainly as Na^+(aq), with minor quantities of $NaHCO_3$(aq). K and Cl are present as free ions. HCO_3^- is present mainly as free HCO_3^- (approximately one third of the total molality), with minor CO_3^{2-} and complexes of Na, Mg and Ca. PO_4 occurs as HPO_4^{2-}, and is complexed with Mg, Ca, Na and K in order of decreasing concentrations.

Mg and K concentrations in the lake and the Omo River are similar. They reflect Omo River influence on the composition of the North Basin waters. These two ions are most probably heavily adsorbed by the suspended sediment load associated with the river plume. There is little free aqueous Al or Fe as reflected by their concentrations. Ca has a concentra-tion factor of <1 between the River Omo and the lake, indicating active uptake in the lake. The high pH of the lake results in the precipitation of $CaCO_3$ and hence the loss of Ca from the water column. Na is by far the most abundant cation (Table 1). It seems to have a very high rate of diffusion in the lake, showing relatively equivalent concentration values as compared to the other ions. Cation exchange is of considerable importance for Na and Ca abundance in the lake: Na replaces Ca and some Mg as the principal exchangeable cation when fluvial clays enter the lake (Yuretich, 1986). K concentrations are lower in the southern transect than in the northern transect implying active K uptake by sediments in regions where the effect of the Omo River plume is becoming increasingly diminished, and the suspended sediment load consists of increasingly finer particles. Cl is the second most abundant anion after HCO_3 and has widely and randomly varying concentrations. The Cl ion has a very high concentration factor of about 300 (Table 2). PO_4 is generally supplied by the Omo River, although at stations 1H and 12H it may partly be released from the bottom sediments.

The concentrations of Cl^- fluctuate by quite large amounts as compared to those of the other elements (Table 1). It has been suggested that this may be due to large error bars incurred during the analysis of the water (T.C. Johnson, pers. comm.). Rather, these variations reflect the chemical identity of the river water entering the lake (based on a comparison with data obtained by other workers for the Omo River) (Table 2). K^+ and PO_4^{2-} register large and correlative variations with Cl^- (Tables 1 and 3). A possible explanation for this is that, due to their larger hydration sizes as compared to the other aqueous ions, K^+, Cl^- and PO_4^{2-} are relatively slow in diffusing across the Omo River plume-lake water boundary as the two chemically dissimilar waters tend towards an equilibrium state. Allen and Collinson (1986) note that river water entering a lake may maintain its chemical identity for some time before it eventually mixes with the lake water, and thus there may be significant vertical and lateral variation of water chemistry within the lake, the pattern depending on how the water circulates. These variations are therefore locationally specific for each of these ions depending on the circulation pattern, although the other ions display an overall tendency to smoothen this "noise," and hence an overall pattern of the behaviour of these two different water bodies can be observed by examining the variations in conductivity. One consequence of this is that, although there is an overall tendency for the two waters to achieve an equilibrium state, the rate at which it can move to this state is partly limited by the fact that each of the water bodies was initially in its own state of equilibrium, and has a tendency to maintain its initial respective equilibrium state. A fair amount of energy is therefore expended in bringing the two waters to a new equilibrium state. Principal components analysis of the water chemistry data set (Table 3) indicates that the first principal score, which accounts for 41% of the total variance, has a strong positive loading for alkalinity, and relatively high negative loadings for PO_4^{3-}, Cl^- and K^+, and smaller negative loadings for Ca^{2+} and SiO_2. This first principal component can therefore be viewed as reflecting the controls on the *rate of mixing* of the two different waters, this being dependent mainly on the initial independent compositions of the two waters, with the limiting factor being the product of the *difference in* concentration of the ions with the largest hydration sizes, principally, the Cl^-, PO_4^{3-} and K^+ ions in the two water bodies.

Chemical equilibria data suggest that these waters are saturated with respect to calcite $(CaCO_3)$, huntite $(CaMg(CO_3)_4)$, hydroxyapatite $(Ca_4(CaOH)(PO_4)_3)$, magnesite $(MgCO_3)$, the iron hydroxides goethite and $Fe(OH)_3(aq)$, and the aluminium hydroxide diaspore. Na behaves as a conservative ion (Fig. 5a). Active Ca and possibly Mg uptake is important in the lake (Fig. 6b,c). In the surface waters there is active uptake of Ca, PO_4 and SiO_2 (Fig. 6a–c). The carbonate phase is important in the regulation of Ca and PO_4, and, although only trace amounts of hydroxyapatite has been found in the sediments (Olago, 1992), the alkalinity has a strong influence on the migration Of PO_4, K and Ca in the lake water (Fig. 7a–c). The concentrations of the aqueous ions were also examined for trends in their behaviour in surface water, intermediate water, and bottom water. Ca uptake is ubiquitous in the water column, while Mg uptake appears to be significant only in the hypolimnion. Silica uptake occurs in the surface and bottom water. In the surface water this can be attributed to uptake by plankton, while in the bottom sediments it may involve the transfor-

Table 2. Temporal Variations in Chemical Composition of Lake Turkana North Basin and Omo River Waters

Year	Source	No. of samples	Site	Conductivity (mS/cm)	pH	Mg (mg/l)	Ca (mg/l)	Na (mg/l)	K (mg/l)	SiO$_2$ (mg/l)	Cl (mg/l)	PO$_4$ (mg/l)	Alk. (meq/l)
1954	Fish (1954)	1	North Basin	3190	9.5	2.4	2.45	–	–	72	245	2	23
1954	Fish (1954)	1	Omo	80	–	1.05	8	–	–	26	trace	0.05	8.4
1956–57	Dodson (1963)	1	Omo	–	–	0.66	1.1	–	–	–	2.8	–	2.5
1975	Yuretich (1976)	3	Omo	–	7	3.5	9.5	9.7	1.8	–	2.7	–	–
1973–75	Hopson (1982)	4	Omo	1156	–	2.15	4.68	462	12.49	–	265	–	10.85 (2)
1973–75	Hopson (1982)	3	North Basin	3140	–	2.68 (2)	3.93	820	24.3	–	454 (2)	–	22.65 (2)
1990	Olago (1992)	27	North Basin	2763.1	8.7 (lab)	2.28	5.671	822.6	1.05	5.762	242.6	1.445	29.184

Table 3. Principal Component Analysis of Lake Turkana North Basin Water Chemistry

	PC1	PC2	PC3	PC4	PC5	PC6	PC7
Mg	-0.076	-0.042	-0.888	-0.282	-0.258	-0.224	0.089
Ca	-0.279	0.539	0.094	0.230	0.022	-0.744	-0.121
Na	0.154	-0.618	0.244	-0.278	0.052	-0.575	0.102
K	-0.344	-0.304	-0.291	0.291	0.769	0.011	-0.168
SiO$_2$	-0.211	0.332	0.127	-0.801	0.410	0.083	0.106
Cl	-0.485	-0.277	0.165	0.009	-0.163	-0.077	0.464
PO$_4$	-0.463	-0.209	0.115	-0.208	-0.333	0.085	-0.753
Alk.	0.529	-0.044	-0.035	-0.143	0.181	-0.213	-0.382
Eigenvalue	3.2916	1.8198	1.0949	0.8615	0.4585	0.2798	0.194
Proportion	0.411	0.227	0.137	0.108	0.057	0.035	0.02
Cumulative	0.411	0.639	0.776	0.883	0.941	0.976	1

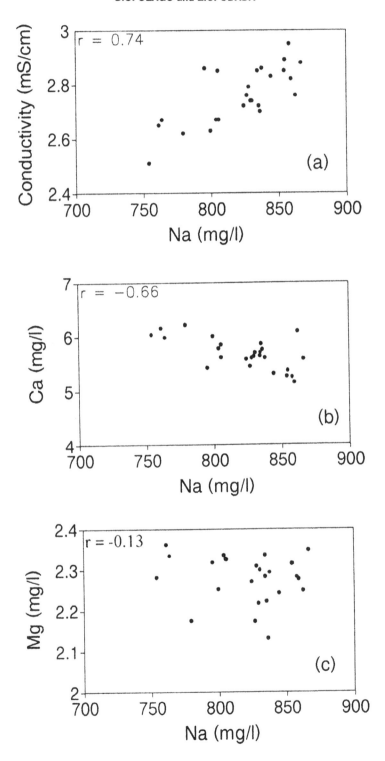

Figure 5. Aqueous ions interactions in the water column: the whole water body.

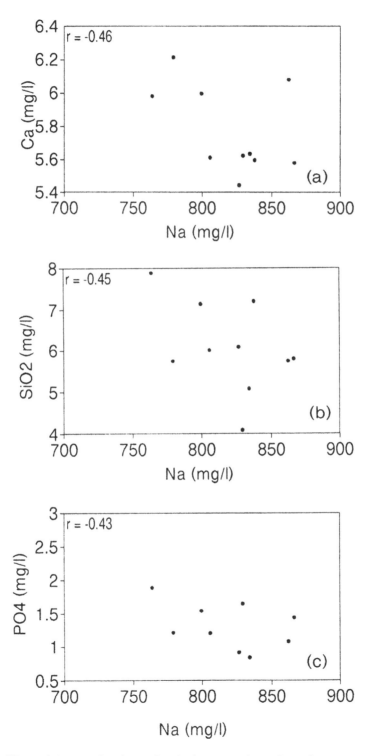

Figure 6. Aqueous ions interactions in the water column: the surface water.

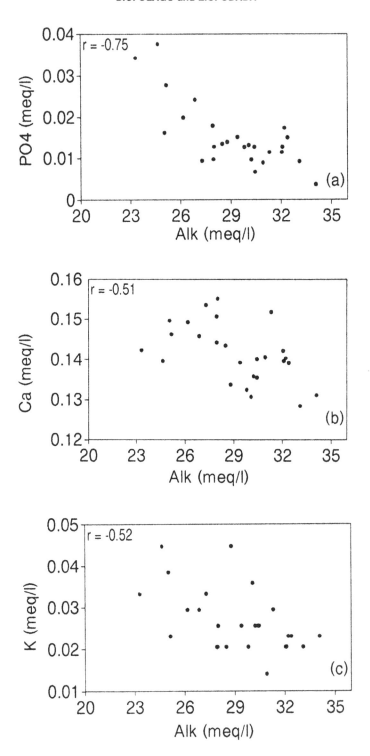

Figure 7. Aqueous ions interactions in the water column: the influence of alkalinity.

mation of dissolved silica into amorphous colloids as is suggested by the chemical equilibria data.

CONCLUSION

The water column is particularly well mixed above 15 m depth as a result of wind induced surface waves. The subsurface countercurrents generated further mix the rest of the lake water at depth, and are augmented seasonally by incursions of the Omo River plume. The mainstream of the plume occurs at 10 to 15 m depth and is centred close to the western shoreline. Controls on the rate of mixing of the Omo River and North Basin waters is effected mainly by the difference in concentrations of the largest hydrated ions Cl^-, PO_4^{3-} and K^+ between the two water bodies. The Omo River is important in constantly modifying the water chemistry balance of the North Basin. Its seasonal incursions effect large changes in salinity in the lake, and cation exchange with Omo River fluvial clays enrich Ca and Mg in the lake, while resulting in Na depletion. Ca and Mg are removed from the water column as authigenic low Mg-calcite precipitates. The carbonate phase is thus important in the migration of Ca, Mg and minor Na in the lake. It restricts the mobility Of PO_4^{3-}, which tends to be precipitated as hydroxyapatite. Silica uptake in the surface waters is regulated by aquatic organisms, while in the bottom waters amorphous silica colloids may undergo neoformation to clay silicates. Sediment resuspension is important in releasing exchangeable Fe ions to the water column.

ACKNOWLEDGMENTS

Thanks to Prof. T.C. Johnson of the Duke Marine Laboratory, U.S.A. for useful comments on the manuscript; and Mr. C. Jackson, Mr. P. Hayward and Miss E. Allen of the School of Geography, Oxford University, Oxford for their help in laboratory analyses and diagram reproduction, respectively.

REFERENCES

Allen, P.A., and Collinson, J.D., 1986. Lakes. In: Sedimentary Environments and Facies (Edited by Reading, H.G), pp. 63–94.

Cerling, T.E., 1979. Palaeochemistry of Plio–Pleistocene Lake Turkana, Kenya. Palaeogeography, Palaeoclimatology and Palaeoecology 27, 247–285.

Cerling, T.E., 1986. A mass balance approach to basin sedimentation: constraints on the recent history of the Turkana basin. Palaeogeography, Palaeoclimatology and Palaeoecology 43, 129–151.

Dixey, F., 1948. Geology of Northern Kenya. Rep. Geol. Surv. Kenya, No. 15, 43 pp.

Dunkelman, T.J., Karson, J.A., and Rosendahl, B.R., 1988. Structural style of the Turkana Rift, Kenya. Geology 16, 258–261.

Ferguson, A.J.D., and Harbott, B.J., 1982. Geographical, physical and chemical aspects of Lake Turkana, In: Lake Turkana: A Report on the Findings of the Lake Turkana Project, 1972–1975 (Edited by Hopson, A.J.), pp. 1–107. Overseas Development Administration, London.

Frostick, L.E., and Reid, I., 1986. Evolution and sedimentary character of lake deltas fed by ephemeral rivers in the Turkana basin, northern Kenya. In: Sedimentation in the African Rifts (Edited by Frostick, L.E., et al.), pp. 113–125. Geol. Soc. Spec. Publ. No. 25.

Halfman J.D., 1987. High-resolution sedimentology and palaeoclimatology of Lake Turkana, Kenya. Ph.D. thesis, Duke University, Durham, North Carolina, 180 pp.

Halfman, J.D., and Johnson, T.C., 1988. High-resolution record of cyclic climatic change during the past 4 ka from Lake Turkana, Kenya. Geology 16, 496–500.

Halfman, J.D., Johnson, T.C., Showers, W.J., and Lister, G., 1989. Authigenic low Mg-calcite in Lake Turkana, Kenya. Journal of African Earth Sciences 8, 533–540.

Hopson, A.J. (ed.), 1982. Lake Turkana: A Report on the Findings of the Lake Turkana Project, 1972–1975. 6 Vols., 1900 pp. Overseas Development Administration, London.

Johnson, T.C., 1990. Text to NSF Proposal. Unpubl.

Olago, D.O., 1992. The Mineralogy and Sedimentology of Recent Subaqueous Sediments in Lake Turkana, Kenya. M.Sc. thesis, University of Nairobi, Nairobi.

Pallister, J.W., 1971. The Tectonics of East Africa, In: Tectonique de l'Afrique (Tectonics of Africa), pp. 511–542, UNESCO.

Peng, T., and Broecker, W.S., 1980. Gas exchange rates for three closed basin lakes. Limnology and Oceanography. 25, 789–796.

Survey of Kenya, 1977. National Atlas of Kenya. 3rd edition, Nairobi.

Wright, R.F., and Nydegger, P., 1980. Sedimentation of detrital particulate matter in lakes: influence of currents produced by inflowing rivers. Water Resources Res., 16, 597–601.

Yuretich, R.F., 1979. Modern sediments and sedimentary processes in Lake Rudolf (Lake Turkana), Eastern Rift Valley, Kenya. Sedimentology, 26, 313–331.

Yuretich, R.F., 1986. Controls on the composition of modern sediments, Lake Turkana, Kenya. In: Sedimentation in the African Rifts (Edited by Frostick, L.E., et al.), pp. 141–152. Geol. Soc. Spec. Publ. No. 25.

Yuretich, R.F., and Cerling, T.E., 1983. Hydrogeochemistry of Lake Turkana, Kenya: mass balance and mineral reactions in an alkaline lake. Geochim. Cosmochim. Acta 47, 1099–1109.

Comparison of the Isotope Record in Micrite, Lake Turkana, with the Historical Weather Record Over the Last Century

M.W. MUCHANE   *Geological Sciences Department,*
University of Tennessee, Knoxville, United States

INTRODUCTION

Climate changes that are most consequential to society take place on time scales of decades to centuries. The record of the past 1000 years which overlaps instrumental records of temperature and precipitation offers opportunities for the study of natural climatic variations during a period when the extent of ice sheets and ocean surface temperatures were presumably fairly close to their present conditions (Hecht, 1979).

This study assesses the reliability of the isotopic signature of Lake Turkana carbonates as a measure of past climatic conditions by comparing isotopic records from three cores in the lake with historical records of rainfall on the Ethiopian Plateau.

Lake Turkana is a closed-basin lake in the eastern branch of the East African Rift Valley (Fig. 1). It is approximately 260 km long and 30 km wide with an average and maximum depth of 35 m and 115 m respectively. Lake Turkana is located in semi-arid northern Kenya. Very little rainfall falls around the lake so it is fed primarily by the Omo River in the north, draining the moist Ethiopian Plateau.

The level of Lake Turkana has fluctuated over a range of 20 meters within the last century. The peak level was in 1895 (+15 m) relative to that in 1968 (Butzer, 1971). It had dropped to +5 m by 1917 and to −5 m between 1940 and 1955. The level rose again in the 1960s, then dropped in the early 1970s (Butzer, 1971; Flohn, 1987) rising again in the late 1970s. Since the early 1980s, the lake has dropped considerably, reaching the low levels of the 1940s and 1950s (Kolding, 1993).

Most rain in East Africa occurs during the passage of the intertropical convergence zone (ITCZ) (Hamilton, 1982), the ill-defined low pressure zone where the northeast and southeast trades converge and cause convectional rainfall. The ITCZ migrates north and south with the sun and there is a tendency for there to be two well-marked wet seasons near the equator, but only one farther to the north or to the south. The Ethiopian highlands derive their moisture from the southeast trades coming off the Indian Ocean as well as the moist equatorial westerlies originating in the South Atlantic coming across the Congo basin. The dry season occurs when the rainfall belt shifts south and the area is influenced by dry northeastern trades originating in Arabia and northeast Africa.

Johnson et al. (1991) measured the $\delta^{18}O$ and $\delta^{13}C$ of the authigenic calcite in a Lake Turkana piston core, LT84 8P, and hypothesized that the $\delta^{18}O$ record was a proxy for rainfall and lake level fluctuations spanning the past 4000 years. One of the objectives of this study is to test the hypothesis of Johnson et al. (1991) by examining several cores from northern Lake Turkana in great detail for a period spanning the past century. Sedimentation in the

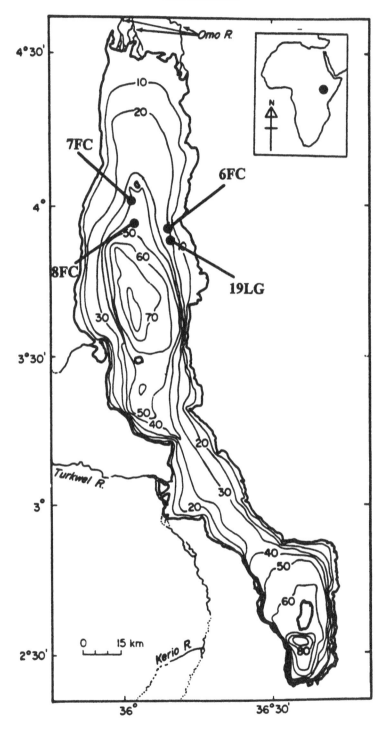

Figure 1. Map of Lake Turkana showing bathymetry and position of freeze cores collected in 1990.

northern part of Lake Turkana is faster than farther south due to proximity to the Omo delta. The aim is to examine the response of the lake sediments to lake level fluctuations over the last century to provide a basis for longer record interpretations.

METHODS

Three freeze cores of about one and a half meters in length were collected from the northern part of Lake Turkana in 1990 in water depths of between 30 and 40 meters (Fig. 1). Freeze coring was used to recover the fluid sediment and to enable the laminations in the sediment to be seen and photographed. The freeze cores were then subsampled at 5-cm intervals in the field. Unfortunately, the uppermost sediments were not well preserved using this procedure perhaps because efficient heat transport in the benthic boundary layer, where the temperature is 26°C, prevented the formation of ice.

A 7-cm diameter "Ligi" core also was taken in 1990 that provided the best sample of the sediment–water interface. The "Ligi" corer, designed by Dr. Guy Lister and Mr. Kurt Ghilardi of ETH-Zurich, has no core catcher to disturb inflowing sediment, and is closed by spring-loaded valves at its top and bottom when activated from the surface by a messenger. While the "Ligi" corer obtained a good sample of the sediment–water interface, it had to be extruded in short segments in the field and therefore could not be used for lamination analysis. The core was subsampled at 1-cm intervals for the upper 7- and 2-cm intervals from 7 to 63 cm. All the samples were sealed in plastic bags and sent by air freight back to the United States.

After porosity measurements by freeze-drying, the samples were washed with a 0.25% Calgon solution through nested 150 and 63 μm sieves to isolate ostracod carapaces (>63 μm) from the authigenic calcite (<63 μm). The <63 μm fraction was washed three times with deionized water and once with ultra-pure deionized water and then placed in an oven to dry. The samples were then ground and roasted for an hour under vacuum at 320°C to remove any organic matter. They were then reacted on a carbonate extraction line with 100% orthophosphoric acid at 90°C to release carbon dioxide. The gas was purified using standard cryogenic techniques. The extracted carbon dioxide was analyzed for $\delta^{18}O$ and $\delta^{13}C$ on a VG Sira Series II isotope ratio mass spectrometer at Duke University's Botany Department with a precision of 0.007 for $\delta^{18}O$ and 0.006 for $\delta^{13}C$.

GEOCHRONOLOGY

Age assignment in Lake Turkana cores is difficult. Halfman et al. (1994) acquired ^{14}C dates on different carbonate fractions on Turkana cores that often were not consistent. AMS dating of bulk carbonate (all of the carbonate in a core subsample) yielded old core top ages in LT90 6FC and LT90 8FC. The initial ^{14}C content of calcium carbonate deposited in water depends primarily on the source of the bicarbonate ions in solution. The old ages observed in the Turkana carbonate are not due to "hard-water" effect as the geology of the basin is mainly volcanics and granitic gneiss with no carbonate rocks. The old ages are probably due to contamination of the sediments by old detrital carbonate, primarily by wind, from lacustrine deposits east of the lake (Halfman et al., 1994).

The amount of contamination was calculated to be about 10% from the following equation (Geyh and Schleicher, 1990):

$$\Delta t = t_A - t = -\frac{\tau}{\ln 2} \ln\left(1 - \frac{k}{100}\right)$$

where Δt is the difference between the apparent and actual age, t_A is the apparent age (915 ± 65 BP for 6FC, 1075 ± 55 BP for 8FC), t is the actual age, assumed to equal 0 years, τ is the conventional half-life of ^{14}C, 5568 years, and k is the amount of contamination (%). AMS dating of ostracod carapaces from 56–58 cm in LT90 19LG gave a post-1950 age. This implies that all 63 cm of 19LG were deposited within a space of about 40 years or less, giving a sedimentation rate of 1.5 cm/yr or more. Unfortunately, the freeze cores did not contain enough ostracod carapaces to obtain ^{14}C dates so other dating techniques had to be tried. These included lamination measurements and ^{210}Pb activity.

Lamination Measurements

Modern sediments of Lake Turkana are well laminated throughout much of the lake (Fig. 2). Black and white photographs of the freeze cores taken in the field show thin beds or laminations of varying clarity. They frequently appear as light–dark couplets. Due to the strong seasonal signal of rainfall in the Ethiopian highlands, the light–dark couplets in the Lake Turkana sediment are interpreted to be annual with the dark, silty layer laid down during the flood season (May–November) and the light, clay-rich layer laid down during the period of low discharge (December–April). This contradicts Halfman et al., who concluded that each light–dark couplet represents about two years.

If the laminations in LT90 6FC, 7FC and 8FC are annual, the resultant sedimentation rates (average) are 1.65 cm/year for 6FC, 2.24 cm/year for 7FC and 2.09 cm/year for LT90 8FC (Fig. 3). They appear to be fairly constant over time. The sedimentation rate for LT90 6FC is close to the sedimentation rate of ≥1.5 cm/year estimated from radiocarbon dating of LT90 19LG. LT90 6FC and LT90 19LG were collected just a few hundred meters apart.

^{210}Pb

^{210}Pb activity was measured in twelve samples from LT90 8FC. Unfortunately, the ^{210}Pb activity was very low, so the estimated level of supported ^{210}Pb in the sediments is not well constrained. The unweighted least-squares fit to all of the data gives an average accumulation rate of 2.8 cm/year (Fig. 4). While this varies considerably from the sedimentation rate of 2.08 cm/year obtained from the lamination analysis of LT90 8FC, the age of the sediment at 125 cm depth is not much different using the two estimates: 53 ybp (^{210}Pb) or 59 ybp (lamination count). This lends credibility to the use of annual light–dark couplets as a dating tool.

RESULTS

Isotopic Results

The sediment–water interface was missing in freeze cores LT90 6FC and 7FC, so the isotope record covers from 1977 to 1895 and 1985 to 1920, respectively. The isotope record of the "Ligi" core, LT90 19LG, covers 1990 to 1949. The complete isotopic results are tabulated in Muchane (1994).

Oxygen isotope values range between about –0.6 per mil and +2.0 per mil PDB, and carbon isotope values range between about –1.6 per mil and +1.7 per mil PDB. From 2 to 4 analyses were run on all the samples in LT90 7FC, and showed variations of about 0.3 per mil for $\delta^{18}O$ and 0.1 per mil for $\delta^{13}C$. Replicate analyses on thirteen samples in LT90 6FC

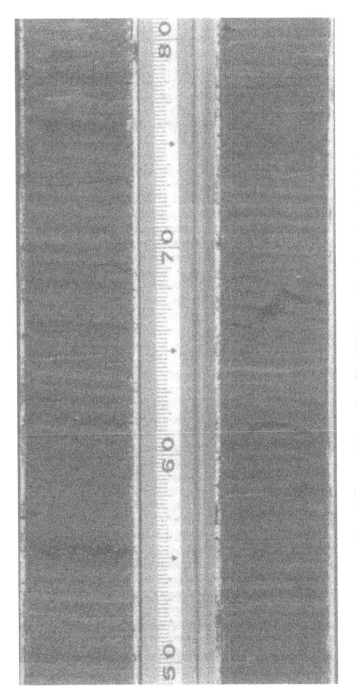

Figure 2. Representative section of Lake Turkana sediment core showing laminations.

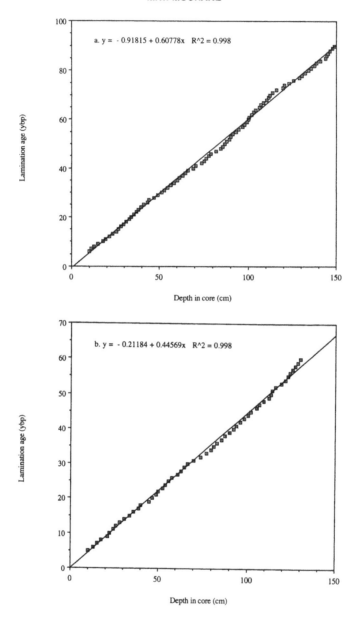

Figure 3. Age (years before 1990) inferred from lamination counting plotted against depth downcore in cores LT90 6FC (a), LT90 7FC (b), and LT90 8FC (c).

yielded variations in $\delta^{18}O$ of about 0.1 per mil and in $\delta^{13}C$ of about 0.020 per mil (Muchane, 1994).

Isotopic values generally are lighter in LT90 7FC than in the other two cores because of its location. The sites of 19LG and 6FC are on the eastern part of the lake, while 7FC is more to the west (Fig. 1). Landsat images of northern Lake Turkana show a tendency of the Omo River plume to flow down the western side of the lake. Since Omo River water ($\delta^{18}O \cong 5$–7

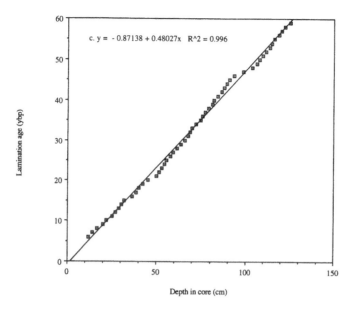

Figure 3, cont'd

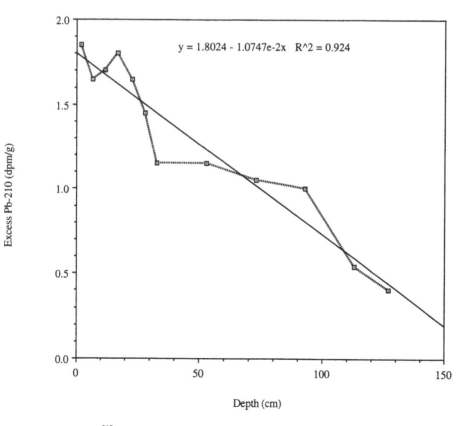

Figure 4. Plot of ^{210}Pb activity vs. depth in core in LT90 8FC. The average accumulation rate from the least squares fit is 2.8 cm/year.

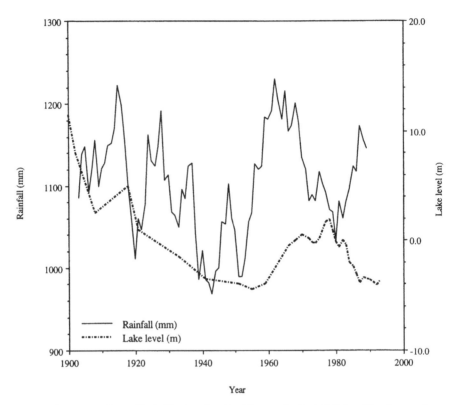

Figure 5. A comparison of Lake Turkana levels with smoothed rainfall data (5-point running average) from the Ethiopian highlands. The lake level data between 1900 and 1970 have intervals ranging from 2 to 10 years (from Butzer, 1971). After 1970, the data are yearly (from Kolding, 1993). The zero datum is taken to be the level in 1968.

per mil SMOW) is isotopically light (Cerling et al., 1988), calcite precipitated directly in the Omo River plume will have lighter oxygen and carbon isotopic composition than that precipitated in the surrounding lake waters.

Comparison with Historical Records

Continuous records of interannual variability in rainfall on the Ethiopian highlands are documented ranging from 1901 to present (Nicholson, 1988; University of Addis Ababa). Lake level data for Lake Turkana since 1888 are based on geomorphological and photographic evidence (Butzer, 1971) as well as monthly gauge readings (Ferguson and Harbott, 1982; Kolding, 1993). Gaps in the record exist from 1920 to 1930 and 1940 to 1950. When the lake level curve is compared with rainfall in the Ethiopian highlands (Fig. 5), it is in these time periods that corresponding rises in lake level are not observed for increased rainfall. Also noteworthy is the lag in the lake level response to changing rainfall amounts.

The $\delta^{18}O$ curves of the freeze cores correspond quite well with historical records of rainfall (Fig. 6). The $\delta^{18}O$ curves have been turned upside down so that light values are up and heavier values are down. This allows for a more direct comparison with the rainfall: a

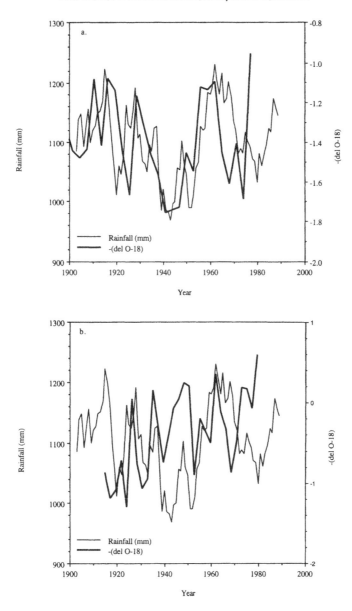

Figure 6. Comparison of rainfall record (5-point running average) in the Ethiopian highlands with $\delta^{18}O$ curves (per mil PDB) of LT90 6FC (0 lag) (a), LT90 7FC (5-year lag) (b), and LT90 19LG (0 lag) (c).

peak in the rainfall (more dilute lake waters) should correspond to a peak in the oxygen isotope values (lighter). Oxygen isotopes in 6FC show a fairly good correlation with rainfall in the Ethiopian highlands at zero lag and relatively good correspondence emerges in 7FC upon increasing the age of each horizon by five years. Since the laminations at the tops of the cores may not have been recovered, a five-year error in the ages inferred from laminations is not unreasonable.

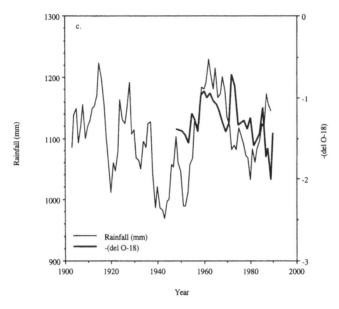

Figure 6, cont'd

The oxygen isotope record of 19LG also shows some correspondence with the rainfall record (Fig. 6c).

DISCUSSION AND CONCLUSIONS

The isotope record sequestered in authigenic calcite in the sediments of Lake Turkana has been shown by this study to reflect to some degree the rainfall record in the Omo River catchment and lake level history over the past century. One reason for the lack of a stronger correlation between the isotopic signature in the micrite and the historical weather record could be the position of the coring sites within or close to the point of entry of the Omo River, where the river plume can strongly affect the isotopic composition of calcite as it precipitates. Isotope records from farther south than the core sites of this study may be more representative of the lake's hydrological budget at any time because the short-term variations in Omo River discharge should be buffered. A 5000-year isotopic record from the southern basin of Lake Turkana (Muchane et al., in prep.) shows long-term changes in the lake level, on the order of centuries to millennia, supported by other proxy records.

Some allochthonous calcium carbonate may be blown in from the adjacent lake beds on the northeastern side of Lake Turkana and this could interfere with the paleoclimate signal. Estimates of contamination levels of about 10% presented earlier based on radiocarbon dates suggest that the authigenic calcite record should strongly dominate the signal.

ACKNOWLEDGMENTS

This research was supported by NSF and NOAA grants to T.C. Johnson. K. Sturgeon provided technical support and R. Anderson ran the ^{210}Pb analyses.

REFERENCES

Butzer, K.W., 1971, Recent history of an Ethiopian delta, University of Chicago, Department of Geography, Research paper no. 136, Chicago.

Cerling, T.E., Bowman, J.R., and O'Neil, J.R., 1988, An isotopic study of a fluvial–lacustrine sequence; the Plio–Pleistocene Koobi Fora sequence, East Africa. Palaeogeography, Palaeoclimatology, Palaeoecology, v. 63, pp. 335–356.

Flohn, H., 1987, East African rains of 1961/62 and the abrupt change of the White Nile discharge, *in* Coetzee, J.A., ed., Paleoecology of Africa, v. 18, pp. 3–18, A.A. Balkema, Rotterdam.

Geyh, M.A., and Schleicher, H., 1990, Absolute age determination. Physical and chemical dating methods and their application, Springer-Verlag, p. 174.

Halfman, J.D., Johnson, T.C., and Finney, B.P., 1994, New AMS dates, stratigraphic correlations and decadal climatic cycles for the past 4 ka at Lake Turkana, Kenya, Palaeogeography, Palaeoclimatology, Palaeoecology, v. 111, pp. 83–98.

Hamilton, A.C., 1982, Environmental history of East Africa, Academic Press.

Hecht, A.D., ed., 1979, Paleoclimatic research: status and opportunities. Quaternary Research, v. 12, pp. 6–17.

Johnson, T.C., Halfman, J.D., and Showers, W.J., 1991, Paleoclimate of the past 4000 years at Lake Turkana, Kenya, based on isotopic composition of authigenic calcite: Palaeogeography, Palaeoclimatology, Palaeoecology, v. 85, pp. 189–198.

Kolding, J., 1993, Population dynamics and life-history styles of Nile tilapia, *Oreochromis niloticus*, in Ferguson's Gulf, Lake Turkana, Kenya. Environmental Biology of Fishes, v. 37, pp. 25–46.

Muchane, M.W., 1994, Stable isotope analyses of authigenic calcite from Lake Turkana, Kenya: High-resolution paleoclimatic implications for the past 5000 years. Ph.D. dissertation, Duke University, Durham, NC.

Muchane, M.W., Ng'ang'a, P., and Johnson, T.C., in prep., Evidence of past climatic change in the inorganic and biogenic calcite record of Lake Turkana, Kenya.

Nicholson, S.E., 1989, African drought: characteristics, causal theories and global teleconnections, *in* Verger, A., Dickinson, R.E., and Kidson, J.W., eds., Understanding climate change. Geophysical Monograph 52, pp., 79–100, American Geophysical Union, Washington, DC.

Nicholson, S.E., Kim, J., and Hoopingarner, J., 1988, Atlas of African rainfall and its interannual variability, Department of Meteorology, Florida State University, Tallahassee.

Oldfield, F., and Appleby, P.G., 1984, Empirical testing of ^{210}Pb-dating models for lake sediments, *in* Haworth, E.Y., and Lund, J.W.G., eds., Lake sediments and environmental history, University of Minnesota Press, Minneapolis, pp. 93–124.

Distribution and Origin of Clay Minerals in the Sediments of Lake Malawi

L.S.N. KALINDEKAFE *Geological Survey Department, Zomba, Malawi*

M.B. DOLOZI *University of Malawi, Chancellor College, Zomba, Malawi*

R. YURETICH *Department of Geology and Geography, University of Massachusetts, Amherst, Massachusetts, United States*

Abstract — Lake Malawi occupies the southernmost part of the East African Rift Valley. Piston cores and gravity cores collected along the whole length of the lake were analyzed for clay minerals by X-ray diffraction and scanning electron micrography. The clays consist of smectite (8 to 96%), kaolinite (<1 to 76%) and illite (0 to 29%) with minor amounts of halloysite. Smectite is dominant in the finer-grained sediment facies whereas illite and kaolinite are more prevalent in the coarser-grained facies. Smectite also increases towards the southern parts of the lake. Scanning electron micrographs show broken and rounded crystal faces which, together with the shallow depth of deposition under low pressure and temperatures, suggest a detrital origin. Relative abundance of the three main clay groups show vertical fluctuations in the cores. Some authigenic nontronite is present in the southern part of the lake.

The spatial distribution and association with particular sedimentary facies suggest that the principal controls on clay mineral types is related to the source area (provenance) and sorting during deposition. Vertical fluctuations are caused by shifts in these variables, although influence by climatic change cannot be excluded. Further investigations of the clay mineral changes through time are needed in order to resolve these influences.

INTRODUCTION

Geologic Setting

Lake Malawi has a depth of over 700 m and occupies most of the Malawi Rift which lies at the southernmost extent of the East African Rift system (Fig. 1). The lake basin is surrounded by rocks belonging primarily to the Precambrian Basement Complex, comprising various gneisses and schists which have been metamorphosed to granulite and amphibolite facies. Granitic intrusions occur in several locations and sedimentary sequences occur as two major groups: Cretaceous-Karoo rocks in the northeastern parts of the rift, primarily the Ruhuhu River valley; and Cenozoic sediments, such as the Chiwondo Beds, which were deposited during the earlier stages of rift formation. Recent alluvial and beach sediments are found along most parts of the shoreline. Pyroclastic rocks of the Cenozoic Rungwe volcanic field in Tanzania, which is in the northern part of the basin, occur along the Songwe River valley (Harkin, 1982).

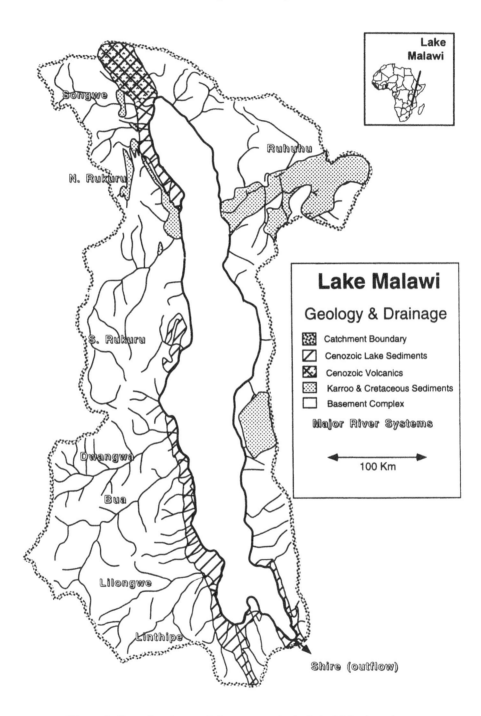

Figure 1. General geology and river systems in the Lake Malawi catchment.

The modern sediments of Lake Malawi contain a wide variety of facies. Deep-water accumulations consist of homogeneous diatomaceous clays and silts, varved and homogeneous diatomite and localized micaceous, quartzofeldspathic turbidites. The laminated sediments are found below the permanent chemocline, which occurs at a depth of about 250 m. In shallow water, local concentrations of ferromanganese nodules, sheet sands, sandy fish bone concentrates and phosphatic sands and muds are found (Johnson et al., 1988; Owen, 1989; Kalindekafe, 1991). Pelagic sedimentation rates in Lake Malawi are between 1 and 3 mm per year (Johnson and Ng'ang'a, 1990). Total sediment thickness as determined by seismic profiling exceeds 4km in some parts of the lake (Scholz and Rosendahl, 1988).

Climate and Hydrology

The Malawi Rift is located in a region of tropical climate. Mean annual temperature in the uplands around the lake is about 22°C, whereas at the rift floor it is closer to 25°C. The southern part of the rift is semi-arid tropical with annual rainfall averaging approximately 700 mm. Extensive areas of high elevation, especially in the northern part of the catchment, have a humid tropical climate with average annual rainfall ranging from 1500 mm to 2500 mm (Agnew and Stubbs, 1972; Berry, 1972). Most of the precipitation falls during the single wet season from November to April which results in a seasonality in clastic and organic inputs to the lake (Crossley, 1984). Climatic changes altering the amount of rainfall and runoff have been proposed as the major cause of lake-level fluctuations (Scholz and Rosendahl, 1988; Owen et al., 1990; Finney and Johnson, 1991).

Most of the major rivers drain the western side of the rift valley (Fig. 1); the Ruhuhu River is the only major fluvial input on the eastern side. Accordingly, the geographic extent of the drainage basin is greater along the western shore.

Lake Malawi is essentially a fresh-water lake, with total dissolved solids (TDS) of about 200 ppm. The pH of the water ranges from 7.3 to 8.5 with alkalinity between 2.31–2.51 meq/l (Jackson et al., 1963; Talling and Talling, 1965; Gonfiantini et al., 1979; Beadle, 1981). The lake is now permanently stratified. The upper part (mixoliminion) is oxic and the lower monimolimnion is anoxic, with the boundary separating these two layers located at a depth of approximately 250 m (Eccles, 1974).

METHODS

Sediment Sampling

Most sediment samples were collected by coring. The University of Malawi retrieved cores up to ~1 m in length in 1989 using a modified gravity corer. This device consisted of a 50 mm diameter plastic pipe strengthened by a metal casing. A 50 kg mass was attached part way down the core so that its broad surface helped prevent overpenetration and loss of the sediment–water interface. In some cases, a grab sampler was used where rapid recovery of large volumes of sediment was necessary, or where the sediments were too hard or soft and unconsolidated for coring. Duke University expeditions in 1986 and 1987 collected piston cores and gravity cores throughout the lake (Johnson et al., 1988). These cores are as long as 12 m.

X-Ray Diffraction

The University of Malawi cores were sampled at intervals down the cores; the <2 μm size fraction was separated by centrifugation and oriented mounts were prepared by smearing

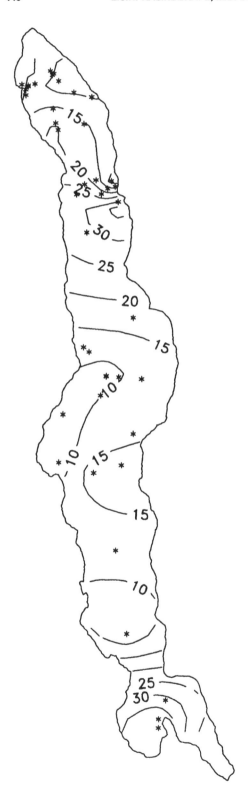

Figure 2. Distribution of fine-grained sediments (<2 mm) in the sediments of Lake Malawi. * = core location (Duke University/University of Massachusetts samples).

the separated clay paste onto glass slides. In addition, whole samples were powdered to <10 μm and these were used for unoriented mounts. Grain size of the samples was measured by sieving the sand and separating the silt and clay fractions by centrifuge. Subsamples were taken from the upper 5 cm of the Duke University cores. These were processed further at the University of Massachusetts by suspension in distilled water and separation of the <2 μm size by centrifugation. Oriented mounts were prepared by smearing a clay paste on glass slides (Gibbs, 1971). Grain-size distribution of several samples of the Duke University cores was determined on a Sedigraph particle size analyzer. The sand and silt fractions of the University of Malawi samples were examined under a petrographic microscope.

X-ray diffraction of the samples was undertaken using either a Rigaku X-ray diffractometer (University of Malawi cores) or a Siemens XRD updated with a Databox digital data processing system (Duke University/University of Massachusetts samples). Samples were scanned untreated and after solvation with ethylene glycol. Selected samples were also heated to 550°C or Mg-saturated and solvated with glycerol to further assist in the identification of the clay-mineral phases (Brindley and Brown, 1980; Hardy and Tucker, 1988). Relative abundance of clay minerals was determined by ratios of diffractogram peak height of the 001 diffraction line of each clay (Griffin, 1971). Although this does not provide a true estimate of the actual amount of each clay mineral, it is very reliable for comparing relative changes among the samples.

Scanning Electron Microscopy

Selected samples were analyzed with an SEM/microprobe (Model JXA-8600). This enabled a more detailed examination of the physical characteristics of the clay minerals to look for a detrital vs. authigenic origin and also permitted precise chemical analyses of small areas. In general, the SEM/microprobe methods employed are similar to those outlined by Trewin (1988) and Fairchild et al. (1988).

RESULTS

The clay-size fraction forms between 5 and 50% of the sediments in Lake Malawi. In the uppermost sediments, the finest grain size occurs in the vicinity of the Ruhuhu Delta, which drains Karroo sedimentary rocks, and in the extreme south of the lake where direct detrital influx is at a minimum (Fig. 2). In the sand and silt, micas make up about 30% of the minerals while microcrystalline quartz and feldspars take up the remaining volume. Other minerals and particles that were identified include magnetite, rutile, ilmenite, hornblende and fish-bone debris. Some of the muds, sands and diatomites reacted with dilute hydrochloric acid, indicating the presence of carbonates. However, these were not identified petrographically because of their small size and low abundance. X-ray diffraction of the <2 μm fraction also showed no traces of carbonates, although some crystals were observed under SEM (Ricketts and Johnson, this volume).

In general, the clay fraction ranges from dark yellowish brown (Munsell 10YR 5/4) to light olive gray (5Y 6/1) in colour. Most clay particles are subangular in shape as viewed on SEM and the associated quartz is typically subrounded. X-ray diffractograms indicate the presence of three major clay minerals in Lake Malawi sediments: smectite, kaolinite and illite (Fig. 3). The proportions of these various clays change depending upon the geographic position in the lake, but in general, smectite is the most abundant clay mineral followed by

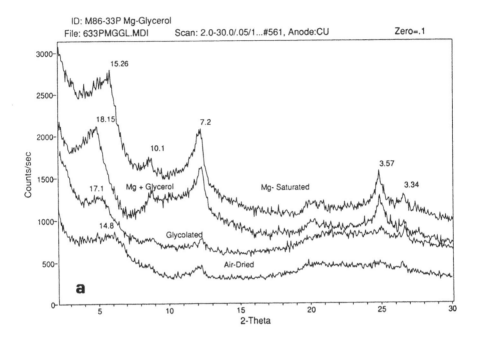

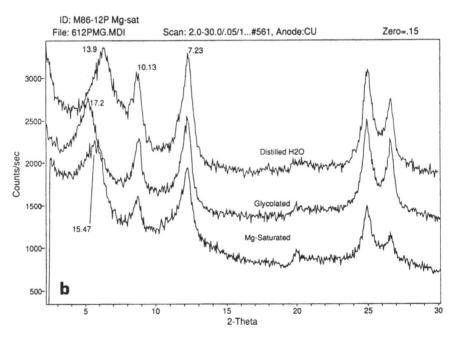

Figure 3. X-ray diffraction patterns illustrating the principal varieties of clay minerals in the sediments of Lake Malawi (<2 mm fraction); smectite identified by expansion from 14 to 17 Å upon glycolation; illite at 10 Å and kaolinite at 7 Å. a) extreme northern end of the lake; b) central part of lake near western shoreline; c) extreme south end of the lake; d) authigenic nontronite from southern part of lake.

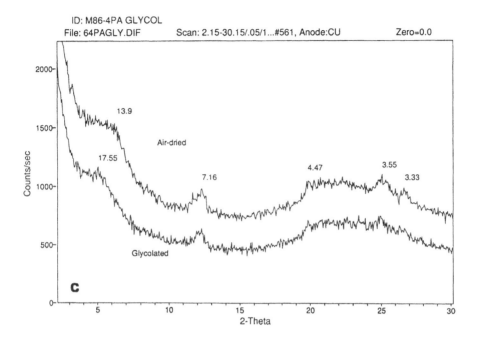

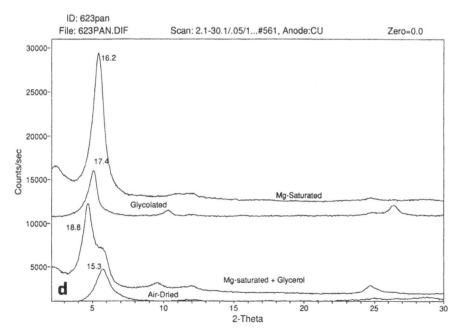

Figure 3, cont'd

kaolinite and then illite (Table 1). The relative proportion of smectite is generally lower in the samples closest to the sediment–water interface.

Table 1. Relative Clay Mineral Abundances in Lake Malawi Samples as Determined From Peak-Height Analysis of Glycolated Samples

Core no.	University of Malawi data			
	Depth	% smectite	% illite	% kaolinite
122	2	68	10	23
122	3	76	11	13
122	5	68	15	16
122	7	43	9	18
122	9	46	17	37
122	13	52	8	40
122	15	75	9	16
192	2	52	15	33
192	5	49	17	34
192	10	46	24	31
326	0	55	33	13
326	15	28	65	7
326	25	44	37	19
326	40	54	26	19
326	60	44	48	8
129	0	25	39	36
129	15	35	28	37
129	30	27	22	52
129	45	33	33	34
129	60	62	16	22
129	75	81	8	11
129	90	63	14	23
181	0	51	17	32
181	15	37	27	36
181	30	44	29	27
181	50	51	25	24
181	65	20	32	48
181	80	58	13	29
181	87	41	27	32
165	60	22	32	46
165	80	24	34	43
165	105	8	34	58
123	7	64	15	22
178	25	9	35	56
115	6	14	10	76
81	3	94	4	3
79	0	81	6	13
75	5	96	3	1
87	2	11	69	20

Table 1, cont'd

Core no.	Duke University/University of Massachusetts Data			
	Depth	% smectite	% illite	% kaolinite
86-1P	2	50	17	33
86-3P	2	56	14	30
86-4P	1	47	0	53
86-5P	4	46	15	39
86-6P	3	51	18	31
86-8P	3	18	21	60
86-9P	4	36	23	41
86-10P	3	46	33	20
86-11P	2	23	34	40
86-12P	3	34	24	42
86-13P	2	22	23	55
86-14P	2	20	32	48
86-15P	2	30	26	44
86-17P	2	32	22	46
86-18P	7	26	22	53
86-19P	4	29	24	47
86-20P	2	24	16	60
86-21P	3	26	14	60
86-22P	5	44	22	34
86-23P	2	94	2	3
86-24P	4	36	16	48
86-27P	2	31	26	43
86-28P	2	25	15	60
86-31P	4	33	19	48
86-32P	2	28	14	58
86-33P	2	36	19	45
87-3G	2	83	7	10
87-4G	2	56	21	23
87-5G	2	29	17	54
87-6G	4	52	24	23
87-7G	6	46	23	31
87-8G	3	38	18	44
87-9G	4	38	21	40
87-10G	3	24	18	58
87-11G	2	21	23	56
87-12G	2	22	26	52
87-13G	4	26	15	59
87-14G	3	37	12	51
87-15G	3	27	22	51
87-16G	2	23	8	69
87-17G	4	33	12	55
87-18G	1	51	18	30
87-19G	2	30	18	52
87-20G	3	34	10	56
87-21G	3	25	18	57
87-22G	3	24	19	57
87-23G	3	30	21	49

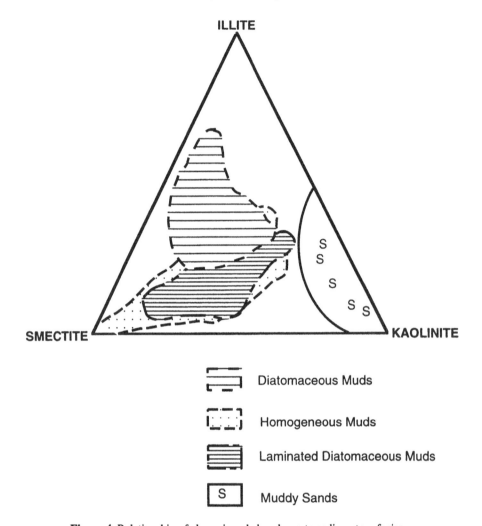

Figure 4. Relationship of clay-mineral abundance to sedimentary facies.

There is a pronounced clay mineral preference in different sedimentary facies as determined from the University of Malawi cores (Fig. 4). Smectite is the dominant clay in homogeneous and laminated muds and least common in muddy sands. Conversely, kaolinite has its greatest abundance in muddy sands whereas illite is the preferred mineral in diatomaceous muds. This suggests a relationship between the grain size of the clay minerals and that of the sediment facies in which they predominate. In the predominantly fine-grained sediments measured on the Sedigraph, kaolinite abundance increases slightly in siltier (coarser) samples (Fig. 5). Although the correlation is very slight, it is statistically significant according to the F-test.

The highest concentration of kaolinite in the uppermost sediments occurs in the middle and northern sections of the lake, especially in a region extending from the Ruhuhu Delta across the lake and along the eastern shoreline. Some localized kaolinite concentrations are

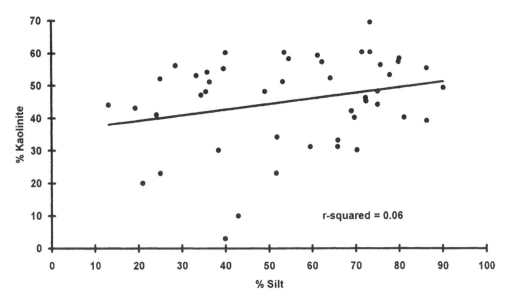

Figure 5. Scatter plot of kaolinite abundance versus silt content, data from Sedigraph analysis of Duke University/University of Massachusetts samples even though correlation is slight, trend is significant at the 0.01 level.

also found in the southern parts of the lake, although these values do not approach those for the northern sites (Fig. 6). Illite abundances are much more equally distributed throughout the lake, averaging 20 to 25% although there is a region of very high concentration (over 50%) along the western shoreline in the central region of the lake (Fig. 6). Illite is less prevalent in the southern part of the lake.

Smectite shows a distinctive pattern of abundance, with a general increase in the southern part of the lake (Fig. 6). The X-ray diffractograms of clay from this area suggest poorer crystallinity of the minerals. However, in a few localities, the sediments consist almost exclusively of very crystalline smectite which has a high iron content (Fig. 3). This is presumably nontronite which has been reported previously from the sediments of Lake Malawi (Müller and Förstner, 1973).

The abundance of the three main clay mineral types shows appreciable variability with depth (Fig. 7). These fluctuations occur both in cores containing a uniform sedimentary sequence as well as in cores where lithologic changes are more obvious. The relative changes from site-to-site, as influenced by the sedimentary facies, are of greater magnitude than these vertical fluctuations.

DISCUSSION

Origin of Clay Minerals

Three possible origins of clay minerals are commonly recognized: 1) detrital; 2) endogenesis or neoformation, usually in soil environments; and 3) transformation or diagenesis (Hardy and Tucker, 1988; Jones and Bowser, 1978; Parke, 1970; Chamley,1989). In Lake Malawi, the SEM results show clay mineral particles with subangular shapes. This suggests that these

% Smectite   % Kaolinite   % Illite

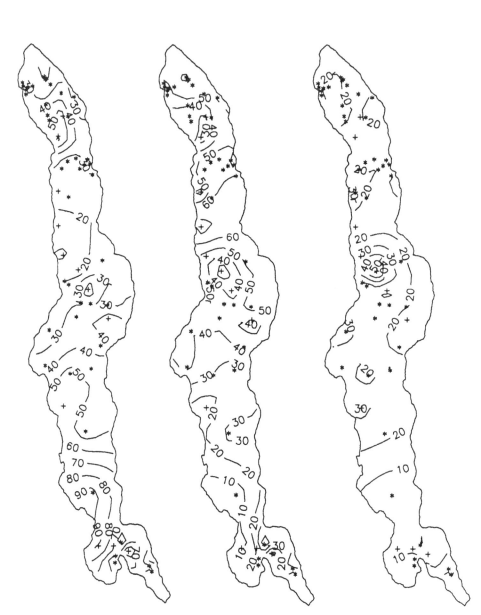

Figure 6. Distribution of clay minerals in Lake Malawi, based upon relative peak heights derived from glycolated diffractograms of uppermost sample at each site. + = University of Malawi cores; * = Duke University/University of Massachusetts cores.

minerals have been transported for a considerable distance and hence are detrital in origin. Diagenetic clays have been identified in some lakes. For example, Jones and Weir (1983) reported the formation of trioctahedral smectite and subsequent illite from a disordered

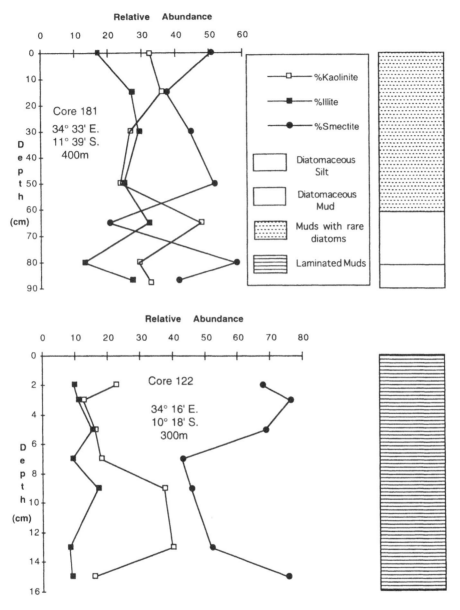

Figure 7. Relationship between clay type, lithology and sediment depth in representative cores from Lake Malawi (University of Malawi data); geographic coordinates and water depth of samples are listed under core number.

detrital smectite in Lake Abert, Oregon, USA. Yuretich and Cerling (1983) also proposed an uptake of Mg to form smectite in Lake Turkana, Kenya, but this has not been proven conclusively. In any case, the lakes where such processes occur are usually alkaline and at least moderately saline. In contrast, Lake Malawi is a fresh-water lake characterized by relatively low alkali levels and neutral pH (Gonfiantini et al., 1979; Beadle, 1981). Very limited authigenesis occurs under these conditions. The crystalline nontronite found in isolated localities (Fig. 3) is one candidate, although the mechanism is obscure. Müller and Förstner (1973) thought that this might result from the dissolution of Fe- and Mn-rich sediments at depth (under low pH and reducing conditions) and the re-precipitation of these at the oxic sediment–water interface. This will produce the iron oolites commonly found in these areas (Williams and Owen, 1990), but some additional reducing influence would seem to be required to incorporate Fe^{2+} into the smectite lattice.

Other diagenetic processes that can form clay minerals are known to occur under higher temperatures and pressures during burial of sedimentary sequences; specifically the trans-formation of smectite to illite (Hower et al., 1976; Boles and Franks, 1979; Chang et al., 1986). This requires burial depths of 1.8 km and temperatures between 90 and 110°C. Since the sediments in this study were taken at depths within 1 m of the sediment–water interface, such an origin for these clays is beyond reason.

Kaolinite forms as a product of hydrothermal alteration or weathering of feldspars, feldspathoids and other silicates. Hurst and Kunkle (1985) have observed that progressive leaching of alkalies and alkaline earths, under oxidizing conditions, transforms smectites and other silicates to kaolinite. This would occur largely within the soil environments since the water chemistry of Lake Malawi should minimize such leaching post-depositionally. Illite will form from muscovite at temperatures above 25°C (McDowell and Elders, 1980; Yates and Rosenberg, 1987; Rosenberg and Kittrick, 1990). Lake Malawi waters have temperatures which range from 22.7°C at a depth of 640 m to 25.6°C near the surface (Gonfiantini et al., 1979; Wüest et al., this volume). Although micas form up to 30% of clastic particles, their degradation to illite in Lake Malawi would proceed slowly at the prevailing water temperatures. All of these conditions favor a detrital origin for the mass of the clays.

Crystal Size and Facies Associations

Diffractogram peaks of kaolinite and illite from Lake Malawi are usually sharper than those of smectite (Fig. 3). One explanation is that the weathering reactions which led to kaolinite and illite formation proceeded more slowly than those that produced smectite, which resulted in better crystallinity. Another factor might be the generally larger grain size of kaolinite and illite. Hardy and Tucker (1988) mention that kaolinite grains tend to be large, up to 5 mm in diameter, whereas illite is smaller (0.1 to 0.3 mm) and smectite is even finer grained. This grain-size association is reasonable in Lake Malawi, since the finer-grained sediments are enriched in smectite while kaolinite dominates coarser sediments (Fig. 4). The Sedigraph analyses were run only on samples with a high proportion of fine-grained sediments; this may explain the lack of correlation with clay-mineral composition (Fig. 5). Additional data are needed.

Spatial Distribution of the Clays

The spatial distribution of the three clay mineral types shows that illite and kaolinite are prevalent in the northern part of the lake while smectite dominates the southern sediments (Fig. 6). The most important factors controlling this distribution are: 1) the geology of the catchment area; and 2) climate. Part of the catchment area of northern Lake Malawi consists of Karroo sedimentary rocks (Fig. 1). The oldest of these sedimentary rocks were deposited under cold climatic conditions and consist of mudstones, calcareous marls, limestones, sandstones, shales, coal seams and conglomerates (Carter and Bennett, 1973; Ray, 1975;

Thatcher, 1974). Illite is a major mineralogical component of the clay sequences in some of the northern Malawi Karoo deposits (Yemane, 1992), and these sedimentary rocks could serve as the source of detrital illite to the modern lake, particularly near the discharge of the South Rukuru River. In addition, the highland areas of central Malawi have lower temperatures which may favor the formation of illite (Engelhardt, 1977; Chamley, 1989).

The abundance of kaolinite off the mouth of the Ruhuhu River may be related to more intense leaching of the uplands in this part of the catchment. Soils in and around the extreme northern end of the lake encompassing the Songwe and Ruhuhu drainage basins are red, thoroughly leached, sesquioxide associations (Berry, 1972). In addition, there is appreciably greater runoff from the northern part of the lake watershed (Agnew and Stubbs, 1972; Berry, 1972). Longer transport distances in the Ruhuhu River basin may also cause some of the Karoo sedimentary source materials, especially feldspathic sandstones, to decompose to kaolinite enroute. The deltas of the rivers in the northern lake are also very steep and relatively coarse-grained (Scholz and Rosendahl, 1990), so that the finer-grained illite and smectite would be winnowed from the sediment influx and carried out to the distal and deeper parts of the lake basin. Johnson and N'gang'a (1990) have interpreted a moderately hard acoustic layer extending across the lake from the Ruhuhu delta region and a turbidite facies. The sands in these deposits should contain proportionally greater amounts of kaolinite.

The smectite distribution reflects climatic influences, grain-size sorting and perhaps some diagenesis. The increasingly arid climate towards the southern end of the lake should favor the formation of smectites in the less effectively leached soil horizons (Chamley, 1989). Consequently, after transport from the uplands, only poorly crystalline varieties of smectite are found in the southernmost part of the lake, usually associated with diatom-rich sediments. A secondary maximum of smectite occurs in the northern part of the lake. This area receives sediment input from the Cenozoic volcanics at the northern end of the lake catchment (Fig. 1). Smectites are the usual product of the breakdown of these rocks (Chamley, 1989), and some soils along the lower reaches of the Songwe River have been classified as smectite-bearing ("montmorillonoid"; Berry, 1972). The diagenetic origin of nontronite has been mentioned previously, but there is also a small chance that some of the poorly crystalline smectite in the diatom oozes may also be diagenetic, formed by the reaction of the warm pore-water solutions with the amorphous silica of the diatoms.

Fluctuations in Clay Abundance with Depth

The fluctuations of clay mineral abundance in vertical profile may be due to: 1) sedimentary facies changes and 2) changes in the types of clays supplied to the lake possibly induced by climatic changes. Since these fluctuations are evident even in relatively homogeneous sediments (Fig. 7), the climatic factor may be significant. Scholz and Rosendahl (1988), Crossley at al (1984) and Owen et al. (1990) have documented evidence for both historical (1500–1850 AD) and older declines in the water level of Lake Malawi. Climatic changes related to amount of rainfall have been regarded as the major cause of these recessions. High lake levels presume abundant rainfall which would cause greater leaching of alkalies from soil materials. Eventually, acidic conditions develop which could favor the development of kaolinite (Hardy and Tucker, 1988; Chamley, 1989). More arid conditions generally means lower kaolinite whereas subtropical conditions will promote the formation of smectite. In a similar fashion, cooler climate would also favor illite over smectite or kaolinite. Such climatic interpretations are quite speculative at this point, and will need to be scrutinized in conjunction with other proxy climatic information from the long piston cores.

CONCLUSIONS

1. Clay minerals in the modern sediments of Lake Malawi consist of smectite, kaolinite and illite.

2. The clays are predominantly detrital as viewed on SEM, although limited formation of trioctahedral smectite (nontronite) occurs in the southern part of the lake.

3. Kaolinite is most abundant in the north-central part of the lake where it is derived largely from extensive weathering of Karoo-age sedimentary rocks.

4. Illite is reasonably uniform in the modern sediments, although there is a localized concentration on the western side of the lake in the north-central region. Illite may originate directly from the uplands along the western margin and become concentrated in the siltier deposits closest to shore.

5. The greatest relative amounts of smectite, but lowest overall "crystallinity" of clays is found in the southern end of the lake farthest removed from detrital sources.

6. Fluctuations in clay minerals with depth in the cores may conceivably be related to changes in weathering and transport into the lake as affected by climatic factors.

In summary, although the clay minerals in this large rift-valley lake have not undergone the same kind of chemical transformation that is often apparent in the clays from the more saline, alkaline lakes of the eastern branch of the African Rift Valley, they contain some key information about the weathering conditions in the lake catchment. Once the geological constraints are evaluated, then the influence of climate on the clay-mineral assemblage becomes clear and this can be used in conjunction with other proxy data for paleoclimatic reconstruction.

ACKNOWLEDGMENTS

Part of this work led to an MSc thesis by the senior author. Special thanks to the thesis supervisors R.B. Owen, formerly of the University of Malawi, and Robin Renaut of the University of Saskatchewan. Analytical equipment at the Department of Geology of the University of Saskatchewan was used for this project. Sample preparation and analysis at the University of Massachusetts was ably assisted by Lori Barg and Constance Hayden Scott. Financial support came from IDRC (for LSNK and MBD) and the U.S. National Science Foundation (EAR90-17512 to RY).

REFERENCES

Agnew, S., and Stubbs, M., 1972, Malawi in Maps: London, University of London Press, 143 pp.

Beadle, L.C., 1981, The Inland Waters of Tropical Africa., 2nd edition: New York, Longman Inc., 475 pp.

Berry, L., 1972, Tanzania in Maps: Graphic Perspectives of a Developing Country: New York, Africana Publishing Corp., 172 pp.

Boles, J.R., and Franks, S.G., 1979, Clay diagenesis in Wilcox sandstones of southwest Texas: implications of smectite diagenesis on sandstone cementation: Journal of Sedimentary Petrology, v. 49, pp. 55–70.

Carter, G.S., and Bennett, J.D., 1973, The Geology and Mineral Resources of Malawi: Zomba, Bulletin No. 6, Government Printer.

Chamley, H., 1989, Clay Sedimentology: Berlin, Springer Verlag, 623 pp.

Chang, H.K., Mackenzie, F.T., and Schoonmaker, J., 1986, Comparisons between the diagenesis of dioctahedral and trioctahedral smectite, Brazilian off-shore basins: Clays and Clay Minerals, v. 34, pp. 407–423.

Crossley, R., 1984, Controls of sedimentation in the Malawi Rift Valley, Central Africa: Sedimentary Geology, v. 40, pp. 33–50.

Crossley, R., Davison-Hirschmann, S., Owen, R.B., and Shaw, P., 1984, Lake level fluctuations during the last 2000 years in Malawi, in Vogel, J., ed., Late Cenozoic Palaeoclimates of the Southern Hemisphere: Rotterdam, Balkema, pp. 305–316.

Eccles, D.H., 1974, An outline of the physical limnology of Lake Malawi (Lake Nyasa): Limnology and Oceanography, v. 19, pp. 730–742.

Engelhardt, W.V., 1977, The Origin of Sediments and Sedimentary Rocks: New York, John Wiley and Sons, 359 pp.

Fairchild, I., Hendry, G., Quest, M., and Tucker, M., 1988, Chemical analysis of sedimentary rocks, in Tucker, M., ed., Techniques in Sedimentology: London. London, Blackwell Scientific Publications, pp. 274–354.

Finney, B.P., and Johnson, T.C., 1991, Sedimentation in Lake Malawi (East Africa) during the past 10,000 years: a continuous paleoclimatic record from the southern tropics: Palaeogeography, Palaeoclimatology, Palaeoecology, v. 85, pp. 351–366.

Gibbs, R.J., 1971, X-ray diffraction mounts, in Carver, R.J., ed., Procedures in Sedimentary Petrology: New York, Wiley, pp. 531–539.

Gonfiantini, R., Zuppi, G.M., Eccles, D.H., and Ferro., W., 1979, Isotope investigations of Lake Malawi, in Proceedings on the Application of Nuclear Techniques to the Study of Lake Dynamics: Vienna, International Atomic Energy Agency Symposium, pp. 195–205.

Griffin, G.M., 1971, Interpretation of X-ray diffraction data, in Carver, R.J., ed., Procedures in Sedimentary Petrology: New York, Wiley, pp. 541–570

Hardy, R., and Tucker, M., 1988, X-ray powder diffraction of sediments, in Tucker, M., ed., Techniques in Sedimentology: London, Blackwell Scientific Publications, pp. 191–228.

Harkin, D.A., 1982, The Rungwe volcanics at the northern end of Lake Nyasa, in Quennell, A.M., ed., Rift Valleys Afro-Arabian: Stroudsburg, Pennsylvania Hutchinson Ross Publishing Co.

Hower, J.,M., Eslinger, E.V., Hower, M.E., and Perry, E.A., 1976, Mechanisms of burial metamorphism of argillaceous sediments: mineralogical and chemical evidence: Geological Society of America Bulletin, v. 87, pp. 725–737.

Hurst, V.J., and Kunkle, A.C., 1985, Dehydroxylation, rehydroxylation and stability of kaolinite: Clays and Clay Minerals, v. 33, pp. 1–14.

Jackson, P.B.N., Iles, T.D., Harding, D., and Fryer, G., 1963, Report on a Survey of Northern Lake Nyasa by the Joint Fisheries Research Organization, 1953–55: Zomba, Malawi, Government Printer.

Johnson, T.C., Davis, T.W., Halfman, B.M., and Vaughan, N.D., 1988, Sediment core descriptions: MALAWI 86, Lake Malawi, East Africa: Project PROBE Technical Report, Beaufort, North Carolina, Duke University Marine Laboratory.

Johnson, T.C., and Ng'ang'a, P., 1990, Reflections on a rift lake, in Katz, B.J., ed., Lacustrine Basin Exploration: Case Studies and Modern Analogs: Tulsa, Oklahoma, American Association of Petroleum Geologists Memoir #50, pp. 113–135.

Jones, B.F., and Bowser, C.J., 1978, The mineralogy and related chemistry of lake sediments, *in* Lerman, A., ed., Lakes: Chemistry, Geology and Physics: New York, Springer-Verlag, pp. 179–235.

Jones, B.F., and Weir, A.H., 1983, Clay minerals of Lake Abert, an alkaline, saline lake: Clays and Clay Minerals, v. 31, pp. 161–172.

Kalindekafe, L.S.N., 1991, Terrigenous and Authigenic Mineralogy of the Lake Malawi Sediments [MSc. thesis]: Zomba, Malawi, Chancellor College, University of Malawi.

McDowell, D., and Elders, W.A., 1980, Authigenic layer silicates in borehole Elmore 1, Salton Sea Geothermal Field, California, USA:Contributions to Mineralogy and Petrology, v. 74, pp. 293–310.

Müller, G., and Förstner, U., 1973, Recent iron ore formation in Lake Malawi, Africa: Mineralium Deposita, v. 8, pp. 278–290.

Owen, R.B., 1989, Pelagic Fisheries Potential of Lake Malawi: United Kingdom, Overseas Development Administration Project R4370.

Owen, R.B., Crossley, R., Johnson, T.C., Tweddle, D., Kornfield, I., Davison, S., Eccles, D., and Engstrom, D.E., 1990, Major low levels of Lake Malawi and implications for speciation rates in cichlid fishes: Proceedings, Royal Society of London, B, v. 240, pp. 519–553.

Parke, K.G., 1970, Montmorillonite, Bentonite and Fuller's Earth Deposits in Nevada: Reno, Nevada, Bulletin 76, Mackay School of Mines, University of Nevada.

Ray, G.E., 1975, The Geology of the Chitipa-Karonga area: Zomba, Malawi, Bulletin No. 42, Government Printer.

Ricketts, R.D., and Johnson, T.C., 1996, Early Holocene changes in lake level and productivity in Lake Malawi as interpreted from oxygen and carbon isotopic measurements of authigenic carbonates, *in* Johnson, T.C., and Odada, E.O., eds., The Limnology, Climatology and Paleoclimatology of the East African Lakes: Toronto, Gordon and Breach, pp. 475–493.

Rosenberg, P.E., and Kittrick, J.A., 1990, Muscovite dissolution at 25°C: implications for illite/smectite–kaolinite relations: Clays and Clay Minerals, v. 38, pp. 445–447.

Scholz, C.A., and Rosendahl, B.R., 1988, Low lake stands in Lakes Malawi and Tanganyika, East Africa, delineated with multifold seismic data: Science, v. 240, pp. 1645–1648.

Scholz, C.A., and Rosendahl, B.R., 1990, Coarse-clastic facies and stratigraphic sequence models from Lakes Malawi and Tanganyika, East Africa, *in* Katz, B.J., ed., Lacustrine Basin Exploration: Case Studies and Modern Analogs: Tulsa, Oklahoma, American Association of Petroleum Geologists Memoir #50, pp. 151–168.

Talling, J.F., and Talling, I.B., 1965, The chemical composition of African lake waters: Int. Rev. Ges. Hydrobiol., v. 50, pp. 421–463.

Thatcher, E.C., 1974, The Geology of the Nyika Area: Zomba, Malawi, Bulletin No. 40, Government Printer.

Trewin, N., 1988, Use of the scanning electron microscope in sedimentology, *in* Tucker, M., ed., Techniques in Sedimentology: London, Blackwell Scientific Publications, pp. 229–273.

Williams, T. M., and Owen, R. B., 1990, Authigenesis of ferric oolites in superficial sediments from Lake Malawi, Central Africa: Chemical Geology, v. 89, pp. 179–188.

Wüest, A., Piepke, G., and Halfman, J.D., 1996, Combined effects of dissolved solids and temperature on the density stratification of Lake Malawi, *in* Johnson, T.C., and Odada, E.O., eds., The Limnology, Climatology and Paleoclimatology of the East African Lakes: Toronto, Gordon and Breach, pp. 183–202.

Yates, D.M., and Rosenberg, P.E., 1987, Muscovite stability in solutions between 100°C and 250°C: Geological Society of America, Abstracts with Programs, v. 19, pp. 901–902.

Yemane, K., 1992, Late Permian climate imprints in the clay mineral assemblages of Gondwana lake deposits, northern Malawi: 1992 Agronomy Abstracts, Madison, Wisconsin, American Society of Agronomy, p. 383.

Yuretich, R.F., and Cerling, T.E., 1983, Hydrogeochemistry of Lake Turkana, Kenya: Mass balance and mineral reactions in an alkaline lake: Geochimica et Cosmochimica Acta, v. 47, pp. 1099–1109.

Characteristics and Origins of Laminated Ferromanganese Nodules from Lake Malawi, Central Africa

R.B. OWEN   *Department of Geography, Hong Kong Baptist University, Kowloon, Hong Kong*

R.W. RENAUT   *Department of Geological Sciences, University of Saskatchewan, Saskatoon, Saskatchewan, Canada*

T.M. WILLIAMS   *British Geological Survey, Applied Geochemistry Unit, Keyworth, Nottingham, United Kingdom*

Abstract — Fields of laminated ferromanganese nodules up to several decimeters thick are found in waters 80–160 m deep in Lake Malawi, commonly near the boundary of littoral sand and deeper-water mud facies. The nodules (<1–8 mm diameter) have a laminated cortex variably composed of goethite, manganite and other oxides. Compositionally, they range from ferric nodules with <2% Mn to examples in which Mn ≥ Fe. Total Fe+Mn in most nodules exceeds 50 weight percent. A genetic model is proposed whereby Fe^{2+} and Mn^{2+} mobilized from anoxic sediments are oxidized close to the sediment–water interface. The precipitation of Fe and Mn oxides and development of lamination are controlled by many factors, including seasonal and longer term variations in the level of the oxic–anoxic boundary in the lake, and the oxygen status and pH of the lake waters.

INTRODUCTION

The presence of ferruginous nodules on the floor of Lake Malawi was first reported by Müller and Förstner (1973), who described examples composed mainly of nontronite and limonite. A more comprehensive coring and lake-sediment sampling program (Owen, 1989; Owen and Crossley, 1989) has revealed the presence of several other varieties of nodule. We now recognize four nodule facies in Lake Malawi sediments: (1) laminated ferromanganese oxide and ferric oxide nodules; (2) nontronite nodules and peloids; (3) limonite nodules containing opaline silica (nontronite); and (4) vivianite nodules and concretions. Facies 2 and 3 correspond to those originally described by Müller and Förstner (1973).

As part of a broader study of Lake Malawi sediments, we are attempting to understand the origins of each nodule facies and the different physicochemical conditions that control their genesis. In this paper we present a preliminary account of the distribution, sedimentological characteristics, geochemistry and possible origins of the ferromanganese nodules of facies 1, which are widely distributed (Owen, 1989; Owen et al., 1991; Owen and Renaut, 1993). Aspects of their geochemistry have been discussed previously by Williams and Owen (1990, 1992). These nodules have concentric laminae composed of ferric oxides or inter-laminated ferric and manganese oxides. Although morphologically similar, they range from examples dominated by ferric iron with little Mn (<2%), through to examples in which Mn ≥ Fe.

GEOLOGICAL SETTING AND LIMNOLOGY

Lake Malawi, the fifth largest water body in the world by volume (6,140 km^3), lies at ~500 m above sea level, between 9°30'S and 14°30'S (Fig. 1). The lake occupies an elongate basin, approximately 570 km long by 50–60 km wide, and is ~700 m deep.

Lake Malawi lies at the southern end of the western branch of the East African Rift System (Fig. 1A). The lake is surrounded by basement rocks composed largely of granulites and amphibolites (Bloomfield, 1966, 1968). Sequences of Permo-Triassic Karoo sediments intersect the lake axis at two locations (Fig. 1C). Additionally, several small faulted Karoo troughs occur in northern Malawi. Jurassic to Cretaceous sediments occur northwest of the lake (Dixey, 1928), and are overlain by the Plio–Pleistocene Chiwondo Formation (lake marginal sediments) and the Chitimwe Beds (subaerial deposits). Volcanic rocks only occur north of the lake at the Rungwe volcanic center.

The Lake Malawi basin consists of three major opposing half-grabens (Fig. 1B), with a full graben in the extreme south. Much of the lake is bounded by steep escarpments, but gently sloping littoral plains are also present. The sedimentary fill is >4 km thick near the major border faults. Scholz and Rosendahl (1988) suggest an average sedimentation rate of ~1 mm y^{-1}. A wide range of lacustrine facies is present that shows a close relationship to the tectonic setting (Crossley, 1984; Crossley and Owen, 1989; Johnson and Ng'ang'a, 1990; Owen et al., 1990, 1991; Scholz et al., 1990, 1993). These include turbidites, diatomites, varved pelagic muds, homogeneous clays, and quartzo-feldspathic sands.

The lake is permanently stratified into oxic and anoxic regions, with the boundary between the layers at about 200 to 250 m depth (Eccles, 1974). Lake Malawi is fresh with a salinity of about 200 mg l^{-1} TDS, a surface pH of ~8.0–8.5, and a surface conductivity of ~250 µmhos cm^{-1} (Gonfiantini, 1979; Owen and Renaut, 1993). The dominant ions are Ca^{2+}, Mg^{2+}, Na^+ and HCO_3^-. Evaporation from the lake surface accounts for 80% of the total water loss.

DISTRIBUTION AND FIELD CHARACTERISTICS OF FERROMANGANESE NODULES

Extensive fields of ferromanganese nodules up to 50 cm thick, and comprising up to 95 weight percent of the sediment, occur off the Nkhotakota shoreline and on the slopes of tilt-block structures at Mbenje and Likoma Islands (insets D, A and B, respectively, in Fig. 2). These fields occur at the present sediment/water interface and are confined to water depths of 80–160 m; isolated nodules are disseminated in muds at greater depths. Known regions with high concentrations of ferromanganese nodules always occur at the boundary between littoral quartzo-feldspathic sands and detrital muds. They are rare or absent in regions of the lake bounded by major border faults, where steep slopes restrict littoral sand development and favor formation of turbidites. Nodules also occur in the extreme southeast arm of Lake Malawi, but these are predominantly of the nontronite and limonite-opal facies (Fig. 2C). Nontronite nodules are most common near diatomite deposits.

In places, nodule layers are buried by sediments. One example occurs ~40 km northwest of Likoma Island (Core 184 in Owen, 1989), where both limonite and ferromanganese

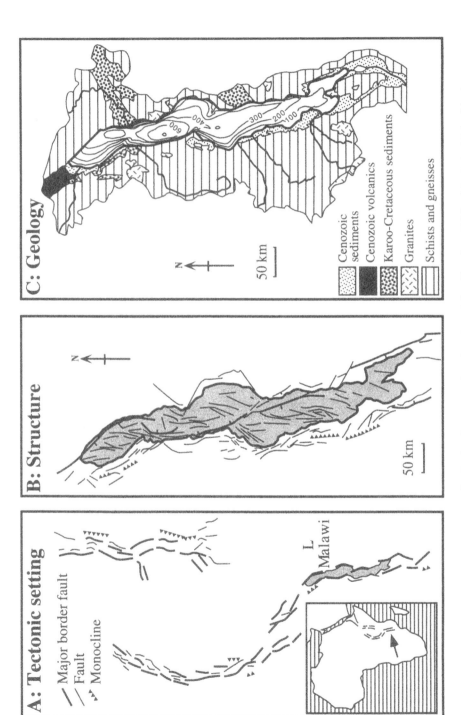

Figure 1. Tectonic setting, structure (after Project Probe maps) and geology (after Crossley, 1984) of the Lake Malawi basin.

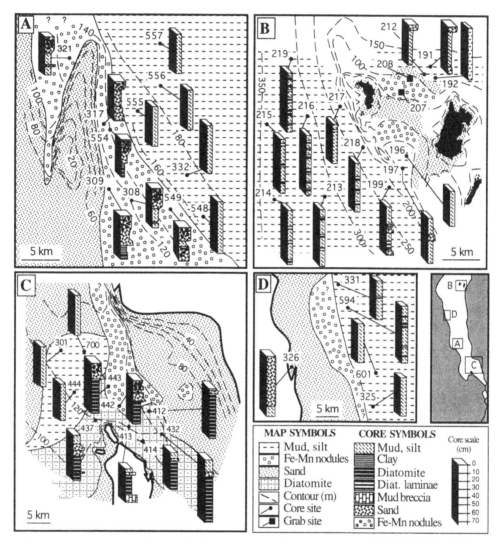

Figure 2. Local facies maps showing distribution patterns of ferromanganese nodules in Lake Malawi. A) Mbenje nodule field; B) Likoma nodule field; C) Southern Malawi field (nontronite nodules and peloids are also abundant in this area); D) Nkhotakota field. Maps were compiled using both cores and grab samples. Only grab-sample sites 207 and 208 are shown; other sites are given in Owen (1989).

nodules occur at a (dry) core depth of 67–70 cm in water that today is 265 m deep. The overlying micaceous silts (52–67 cm) contain carbonate laminae that elsewhere Finney and Johnson (1991) have related to lake levels 100–150 m lower than today during the period between 10,000 and 6,000 yr BP. This implies a very tentative, latest Pleistocene age for these particular nodules, assuming no prolonged hiatus in sedimentation. Scholz (pers. comm.) and Scholz and Finney (1994) have recorded extensive reflectors in shallow seismic profiles near Likoma Island that may also represent zones of Fe-enrichment and buried nodule fields. Waters at these sites are >280 m deep.

Near Mbenje Island, Core 549 (Fig. 2A; see Owen, 1989, for details) has several distinct nodule horizons, with limonite-opal types toward the base and ferromanganese types near the top. The various nodule layers are separated by siliciclastic muds and silts. This suggests discrete periods of nodule formation, the variations in type presumably relating to significant changes in the lake environment. However, the possible effects of shallow burial diagenesis need investigation before conclusions can be drawn. Discrete levels of Mn enrichment are also apparent in deep-water muds. Harper (unpublished) described a six-fold increase in Mn concentration over background levels at 110–120 cm depth in Müller and Förstner's (1973) Core no. 7, which came from a few kilometers south of the Mbenje nodule field.

Wherever they occur, the ferromanganese nodules are dark brown, range in diameter from 0.1 to 8 mm, and have a dull opaque or weakly metallic luster. Most are coated grains with well-defined concentric laminae surrounding a nucleus and can, therefore, be termed ooids (<2 mm) or pisoids (>2 mm). Individual nodules are well rounded with variable sphericity, ranging from spheres to rollers (Owen et al., 1991). Nodules exceeding 6 mm in diameter are scarce, and are subangular to subrounded. The larger nodules are commonly chipped or abraded. In places, ferromanganese oxide laminae also encrust pebble-size clasts.

NODULE FABRIC AND MINERALOGY

The nuclei of the nodules (from Likoma) are variously composed of quartz and feldspar grains, clay pellets and aggregates (including nontronite), smaller Fe–Mn nodules (i.e., composite forms), vivianite concretions, fish bone debris or rare shell fragments (Fig. 3). Some nodules are hollow, implying dissolution of the nucleus or plucking during preparation. Nontronite aggregates in some nuclei show transitional contacts with the cortex, possibly reflecting diagenetic replacement of iron oxide with the addition of silica, as described by Pedro et al. (1978) from Lake Chad.

The mineralogy of the cortex, as determined by X-ray diffraction (XRD) of 32 samples, consists both of crystalline and poorly crystalline or amorphous phases. Of the crystalline oxides, goethite dominates the Fe-rich laminae in most nodules; nodules with weak lamination tend to be poorly crystalline. The Mn-rich laminae are mostly manganite (γ-MnOOH) with variable crystallinity. Electron microprobe analyses also revealed a Ca-bearing phase (several percent CaO) on EDS spectra, with a composition similar to todorokite, but this was not confirmed by XRD. Clay minerals, mica and an unidentified Si-poor, Al-rich phase (oxide?) also occur in the cortices of some nodules. Rhodochrosite ($MnCO_3$) was confirmed by XRD in a Fe–Mn nodule (5 mm) from Core 184.

Back-scattered electron imagery (BEI) reveals well-defined cortical laminae that typically range in thickness from 2 to 10 μm (Fig. 3). Although Fe (26) has a higher atomic number than Mn (25), the brighter laminae commonly have a higher Mn concentration. The explanation is unclear, but may reflect the higher capacity of manganese oxides for sorption of transition metals and Ba, or a difference in surface charge. The succession of the light and dark laminae is highly variable, but in several examples examined the lighter laminae become more abundant toward the outer margin of a nodule. Single laminae also show a tendency for Mn-enrichment in their uppermost parts. Some individual laminae are dense throughout; others display irregular pores, many of which are subradial. Subradial microfractures a few microns across are also common; some cut the entire cortex; others terminate at a particular lamina. There are several possible origins, including grain-to-grain

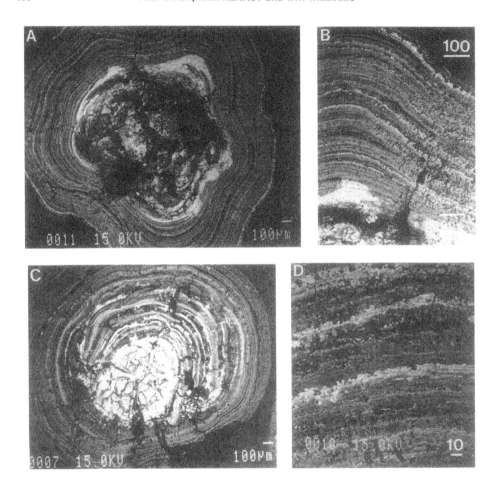

Figure 3. Back-scattered electron images of Lake Malawi ferromanganese nodules from Likoma field. A) Typical nodule showing well-defined concentric lamination. Nucleus is partially dissolved or plucked, and is composed of aggregates of porous nontronitic clays. Note common microfractures. Discontinuous laminae visible bottom left. B) Detail of microfractures and porous zones within the lamination. C) Nodule showing well-defined disconformable lamination. Growth initially occurred only on the upper surface of the nucleus, possibly while it was partially buried. D) Detail of lamination in C. Polished sections were examined using a JEOL JXA-8600 electron microprobe.

impact, but some microfractures and subradial pores may have resulted from contraction due to a volume change — perhaps when amorphous or poorly crystalline hydrous phases recrystallized as goethite or manganite. Some microfractures are infilled by Fe–Mn oxide cements, supporting an early origin for the fractures. The microfractures provide pathways for fluid movement toward the nucleus, possibly explaining hollow nodules where the nucleus may have dissolved, and, by addition of dissolved silica, the apparent replacement of the interior of some ferric nodules by nontronite. Several nodules examined show clear microdisconformities (Fig. 3C) that demonstrate periodic particle movement during growth.

Electron microprobe analyses of the cortex of a single nodule from the Mbenje area (Figs. 2 and 4C) showed a trend of increasing Mn enrichment toward the exterior surface, accompanied by a decrease in Fe. CaO, MgO and BaO (and to a lesser extent, P_2O_5 and K_2O)

Table 1. Nodule (~1 cm diameter) from the Upper Part of Core 549

	Nucleus (clay peloid)	Fe-lam.	Mn-lam.	Fe-lam.	Mn-lam.	Fe-lam.	Mn-lam.	Fe-lam. (outer lamina)
Fe_2O_3	4.58	68.42	11.57	55.48	2.15	56.47	5.44	44.52
MnO_2	0.08	8.64	71.25	18.30	84.17	24.08	86.20	37.14
CaO	1.47	0.25	1.25	0.56	1.01	0.47	0.66	0.57
MgO	2.09	0.02	1.02	0.04	1.87	0.18	1.18	1.01
P_2O_5	0.21	0.59	0.61	0.44	0.48	0.07	0.67	1.18
K_2O	5.34	0.06	1.01	0.84	0.87	0.27	0.08	1.02
Na_2O	0.32	0.00	0.01	0.02	0.01	0.00	0.00	0.08
SiO_2	59.77	6.87	5.21	4.25	2.59	3.24	1.11	2.15
Al_2O_3	17.66	4.57	2.64	14.50	4.09	5.17	0.58	4.17
TiO_2	0.02	0.11	0.22	0.08	0.35	0.48	0.10	0.22
BaO	0.01	0.00	2.14	0.00	0.06	0.04	0.08	0.03
Total	91.55	89.53	85.36	94.51	97.65	90.47	96.10	92.09

All iron is expressed as Fe_2O_3. Fe-lam = iron-rich lamina; Mn-lam = Mn-rich lamina. Laminae were sampled at 1 mm intervals from the nucleus to the outer margin. Analyses were made using a JEOL JXA-8600 electron microprobe (WDS) on polished thin sections.

tend to be more concentrated in the Mn-rich laminae (Table 1). Al_2O_3 is slightly elevated in the Fe-rich laminae and in the detrital nucleus. TiO_2 is higher in Mn-rich laminae of the inner cortex, but the pattern reverses in the outer layers. SiO_2 is higher in the nucleus but poorly correlated with either Fe or Mn in the cortex, suggesting that it is present mainly in detrital components. Na_2O is consistently low.

BULK CHEMISTRY OF THE NODULES

Williams and Owen (1992) analyzed bulk samples from the Likoma field (stations 207 and 208 in Fig. 2B) both for major and trace elements. The total Fe+Mn exceeds 50% in most samples, which exceeds that in comparable nodules from marine settings and most lake basins. The chemical composition of the ferromanganese nodules is highly variable even within individual nodule fields. Thus, analyses of individual nodules may not represent the total population. In the analyzed samples from near Likoma, Mn exceeds Fe in some small nodules (up to 70 weight percent of the oxides), whereas Fe dominates in others.

The average Mn-enrichment factor relative to mean upper-crustal values (166: Fig. 4A) is similar to the global average for nodules from deep ocean (170) and marine shelf (159) environments, but much higher than in most lacustrine nodules. The corresponding Fe value (6.5) is high compared to most other depositional settings. Subcrustal values are recorded for Mg, Al, Si, K, Ca and Ti, although the latter element locally shows high concentrations. The abundance of these major elements mainly reflects the amount and mineralogy of detrital impurities.

Of 26 analyzed trace elements, Zn, Co, Pb, Ba, Y, La, Zr, Ag, Be and Nb are present at supracrustal levels. Subcrustal concentrations occur for Sr, Ni, Cu, Cr, Sc, Rb, Ga, Li, B,

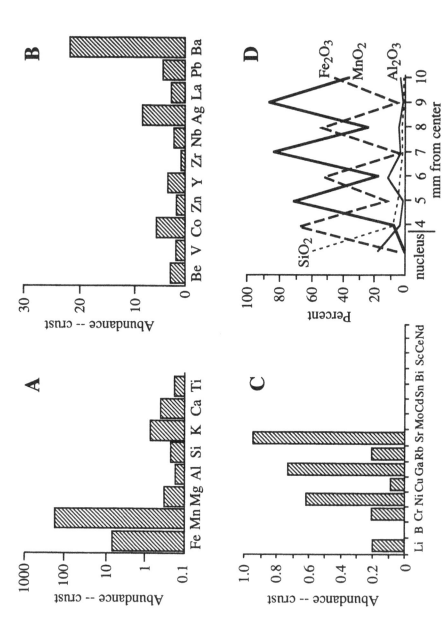

Figure 4. Geochemical data for Lake Malawi nodules. Average crust-normalized abundance of A) 8 major elements; B) enriched trace elements; C) elements at subcrustal concentrations. D) Plot of the distribution of Fe2O3, MnO2, SiO2 and Al2O3 in the laminae of a single nodule (see Table 1 for analyses).

Mo, Cd, Bi, Sn, Ce and Nd. The high enrichment of Co (5.5) and Zn (2), and the depletion of Ni (0.6) and Cu (0.09), characterize most lacustrine ferromanganese nodules. For further details, see Williams and Owen (1992) and Owen and Renaut (1993).

FORMATION OF THE FERROMANGANESE NODULES

The Fe–Mn nodules form both by physical and chemical processes. Although spherical coated grains can form with minimal movement (e.g., cave pearls), the presence of continuous laminae, disconformable laminae with uneven thickness, spheroidal and roller shapes, abrasion features, and moderately sorted shoal-like accumulations, all suggest recurrent nodule movement during growth by bottom currents. These features favor accretion on the lake floor, rather than displacive or replacive growth within shallow soft sediments.

The variable compositions of the laminae imply that the water chemistry near the sediment/water interface changed frequently, but perhaps subtly, during nodule formation. Chemically, formation of the ferric iron nodules and laminated ferromanganese nodules can be explained by variations in redox conditions and the pH of the lake and pore waters near the sediment/water interface. The stability fields for Fe and Mn oxides as a function of Eh and pH are well established (e.g., Krauskopf, 1957; Stumm and Morgan, 1981) and are shown in Fig. 5B, with two theoretical pathways for fluids that could have precipitated the nodules.

Path 1 shows the conditions when Fe^{2+} and Mn^{2+} in anoxic pore waters released from the underlying sediments or in anoxic lake-bottom waters, gradually mix with an overlying oxygenated water mass at a near neutral pH. The first phase to precipitate is iron, as $Fe(OH)_3$, probably as amorphous iron hydroxide or goethite (Point 2 in Fig. 5B). During this stage, most Mn^{2+} remains in solution and ferric iron nodules or Fe-rich laminae would form, containing relatively little Mn. With increased mixing and oxygenation the MnO_2 field is reached, and Mn-rich nodules or laminae should precipitate. If the pH and redox conditions remain constant for a period within either stability field (i.e., that of $Fe(OH)_3$ or MnO_2) the nodules should be predominantly Fe-rich or Mn-rich. If, however, the Eh fluctuates periodically (e.g., Point 3 in Fig. 5B), then alternating Fe-rich and Mn-rich laminae might form. Such fluctuations could be seasonal or of irregular occurrence.

Path 2 (Fig. 5B) shows that it is theoretically possible to produce compositional variations in the laminae by periodically changing the pH, while maintaining a constant Eh and a steady delivery of Fe^{2+} and Mn^{2+}. This condition could result from seasonal changes in the chemistry of the overlying water mass (discussed below) or from processes that modify pH at or just below the sediment/water interface, such as the supply rate of organic matter and its subsequent decay, including bacterial sulfate reduction (e.g., Berner et al., 1970) and methanogenesis. Many other paths, where both Eh and pH change, could be constructed to explain the compositionally variable laminae.

Such theoretical changes in pH and Eh require validation before acceptance in any genetic model. Regrettably, measurements of pH and Eh in Lake Malawi waters are sparse. However, Jackson et al. (1963) reported that the pH of bottom waters 150 m deep, offshore from Salima (near Mbenje nodule field: Fig. 2A), varied from 7.5 in January 1955 to 8.3 in August 1954. This is within the range that should favor accretion of alternating Fe- and Mn-rich laminae (Fig. 5B). Jackson et al. (1963) also reported dissolved oxygen levels of 0.5 mg l^{-1} in January 1955 and 6.3 mg l^{-1} in August 1954, at the same site and depth. Thus, the lower pH and low oxygenation levels in January should favor precipitation of $Fe(OH)_3$,

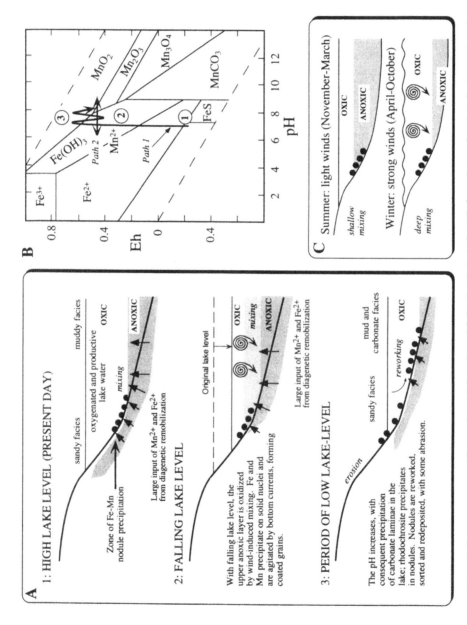

Figure 5. Inferred conditions for ferromanganese nodule genesis. A) Schematic model showing the loci of nodule formation as a function of oxic–anoxic environments within the lake and sediments. Shaded zone represents anoxic conditions. No scale is implied. B) Eh–pH diagram showing the possible pathways to produce the nodules. For explanation see text. Plot is modified from Bühn et al. (1992) and other sources. C) The effect of seasonal winds on the level of the oxic–anoxic boundary layer in the lake (no scale implied).

whereas the more oxygenated waters with higher pH in August should favor MnO_2 precipitation. Theoretically, alternating laminae could result.

These seasonal variations in oxygen status and pH may relate to the variable depth of mixing of the water column (Fig. 5C) and associated changes in productivity and organic loading. Mixing is driven by strong winds between April and October that oxygenate the waters significantly to at least 150 m depth, based on the data of Jackson et al. (1963). Wind-induced mixing decreases in the rainy season from November until March (Eccles, 1974). Mixing may also affect phytoplankton productivity. Owen and Crossley (1992), for example, reported periodic short-term diatom blooms in areas near nodule fields. They suggested that these are linked to release of nutrients after exceptional, deeper water, mixing events.

While helpful, the thermodynamic approach fails to account for reaction kinetics (e.g., Hem, 1981; Stumm and Morgan, 1981), other dissolved and solid species in the system, microbial influences (Roy, 1981), and the dynamics of the site of precipitation. The surface charge conditions of nodules may vary with time, depending on the surface composition of the major and trace metals. Consequently, there may be recurrent swings between FeOOH (and similar Fe-phases) and MnO_2 phases as the dominant adsorbed species.

We tentatively propose the following genetic model (Fig. 5A). Overall, the location of the nodule fields requires a two layer oxidizing–reducing system (Williams and Owen, 1990). A lower anoxic zone is needed to reduce and mobilize Fe and Mn in the detrital sediments. Fe^{2+} and Mn^{2+} then migrate to the oxic zone, the boundary of which may occur either at or just below the sediment/water interface, or within the stratified water column itself. Precipitation will occur on suitable nuclei at or near the site of mixing and oxidation of the waters. Within the lake, the position of the reducing–oxidizing boundary and degree of oxidation will vary with the strength and depth of wind-induced mixing (Fig. 5C), lake-level fluctuations and productivity, all of which are interrelated and vary over a range of timescales. Long-term changes in productivity (e.g., related to nutrient supply) and organic loading, for example, might modify the depth of the permanent oxic–anoxic boundary layer and the degree of oxygenation in shallower waters.

Although the buried nodules are clearly fossil features, from present evidence we cannot determine whether nodules at the sediment surface are actively forming, relic features, or both. The depth range of formation is thus also unclear, but their present situation is not incompatible with our model (Fig. 5A,1). The common association of nodule fields with former littoral sands may indicate (but does not prove) that many nodules are fossil features that formed either during or following a period of relatively low lake-levels, in which case they could have formed at shallower water depths than today. If fossil, they may have been kept free from burial by younger sediments and sorted by strong bottom currents, which are known to occur in Lake Malawi (Owen and Crossley, 1989). Many nodule fields lie at sites that currently receive little siliciclastic sediment influx.

Nodule fields lie in oxygenated waters 80–160 m deep, significantly above the permanently anoxic layer (below ~200–250 m: Eccles, 1974). If actively forming, most Fe^{2+} and Mn^{2+} derive from pore-waters in the underlying sediments, rather than an anoxic water column (Fig. 5A,1; Fig. 4 in Williams and Owen, 1990). Had they formed below a lower lake level, the permanently anoxic layer would probably have been displaced deeper than today. However, accompanying falling lake level, wind mixing would erode the upper part of the anoxic layer, which could have promoted Fe–Mn sedimentation in shallow oxygenated waters (Fig. 5A,2; Johnson and Ng'ang'a, 1990). The buried nodule fields in waters now >260 m deep (i.e., below the modern oxic–anoxic boundary) must have formed under lower lake levels than today, unless the oxic–anoxic boundary was itself formerly lower.

Lower lake levels occurred at several stages during the last few thousand years (Owen et al., 1990), between 6,000 and 10,000 yr BP (Finney and Johnson, 1991), and during the late Pleistocene (Scholz and Finney, 1994). During low lake levels, increased current and wave energy can be expected as nodules become near, or even within, the littoral zone (Fig. 5A,3). The carbonate laminae (Johnson and Finney, 1991) suggest that the pH may have increased accompanying the lower lake levels, increased evaporation and closed-basin conditions. With falling lake level photosynthesis may have increased temporarily as formerly deeper, anoxic bottom waters were disturbed by wind mixing and bottom currents, releasing nutrients. Rhodochrosite implies a pH of >8.5 (Fig. 5B), but possibly in pore-waters, rather than open lake waters.

The common association of nodule fields with sands could indicate a secondary control. Coarse sands may act as a permeable membrane for mixing of oxic lake waters and anoxic interstitial waters, as well as providing abundant favorable substrates for nucleation.

Although major fluctuations in the oxic–anoxic boundary layer may account for the occurrence of buried nodule horizons and changes in location of nodule fields, alone they cannot explain compositional variations in laminae. Fluctuations between Fe and Mn in cortical laminae may require shifts in redox conditions and/or pH. Seasonal changes in lake level in Lake Malawi are small (up to 1 m), with decadal variations of a few meters. These are unlikely to affect nodules in waters 80–160 m deep. In contrast, the seasonal and longer term wind-induced mixing of the water column described causes shifts in pH and dissolved oxygen content to this depth range and is, therefore, a more plausible mechanism for forming laminae of different compositions.

CONCLUSIONS

Favorable conditions for nodule formation have probably existed several times in Lake Malawi's recent history. Ferromanganese nodules form above the anoxic-oxic boundary, where Fe^{2+} and Mn^{2+} mix with oxygenated lake waters. The Fe^{2+} and Mn^{2+} derive from anoxic pore-fluids in the sediment or from wind mixing of the uppermost anoxic zone during falling lake level. The Eh and pH of the lake bottom waters vary at different times in the year, mainly in response to wind-induced mixing and related changes in organic productivity and loading. During periods of relatively low oxygenation and neutral pH, only Fe laminae are likely to precipitate. During periods of wind-induced mixing, bottom waters become more oxygenated and the pH rises, both of which favor precipitation of Mn-rich laminae. This may explain the alternation of Fe-rich and Mn-rich laminae in some nodules.

ACKNOWLEDGMENTS

Research at Lake Malawi has been supported by the Overseas Development Administration (U.K.), the Natural Sciences and Engineering Research Council (Canada), the International Development Research Centre (Canada), and Amoco Petroleum Corporation. We thank the Malawi Government Department of Fisheries, Dennis Tweddle, and Robert Crossley for their assistance, and Dr. Dan Engstrom and an anonymous referee for their helpful comments on the original manuscript.

REFERENCES

Berner, R.A., Scott, M.R., and Thomlinson, C., 1979, Carbonate alkalinity in the pore waters of anoxic marine sediments: Limnology and Oceanography, v. 15, pp. 544–549.

Bloomfield, K., 1966, Geological map of Malawi: 1:1,000,000 scale: Zomba, Malawi, Geological Survey of Malawi.

Bloomfield, K., 1968, The pre-Karroo geology of Malawi: Geological Survey of Malawi Memoir 15.

Crossley, R., 1984, Controls of sedimentation in the Malawi Rift Valley, Central Africa: Sedimentary Geology, v. 40, pp. 33–50.

Bühn, B., Stanistreet, I.G., and Okrusch, M., 1992, Late Proterozoic outer shelf manganese and iron deposits at Otjusondu (Namibia) related to the Damarian oceanic opening: Economic Geology, v. 87, pp. 1393–1411.

Crossley, R., and Owen, R.B., 1988, Sand turbidites and organic-rich diatomaceous muds from Lake Malawi, Central Africa, *in* Fleet, A.J., Kelts, K., and Talbot, M.R., eds., Lacustrine petroleum source rocks: Geological Society of London Special Publication 40, pp. 369–374.

Dixey, F., 1928, The Dinosaur Beds of Lake Nyasa: Transactions of the Royal Society of South Africa, v. 16, pp. 55–56.

Eccles, D.H., 1974, An outline of the physical limnology of Lake Malawi (Lake Nyasa): Limnology and Oceanography, v. 19, pp. 730–742.

Finney, B.P., and Johnson, T.C., 1991, Sedimentation in Lake Malawi (East Africa) during the past 10,000 years: A continuous paleoclimatic record from the Southern Tropics: Palæogeography, Palæoclimatology, Palæoecology, v. 85, pp. 351–366.

Gonfiantini, R., Zuppi, G.M., Eccles, D.H., and Ferro, W., 1979, Isotope investigations of Lake Malawi, *in* Proceedings on the applications of nuclear techniques to the study of lake dynamics: International Atomic Energy Agency Symposium, Vienna, pp. 195–205.

Harper, M.A., unpublished, A note on diatoms in two sediment cores from Lake Malawi: Cambridge, Culture Centre for Algae and Protozoa.

Hem, J.D., 1981, Rates of manganese oxidation in aqueous systems: Geochimica et Cosmochimica Acta, v. 45, pp. 1369–1374.

Jackson, P.B.N., Iles, T.D., Harding, D., and Fryer, G., 1963, Report on the survey of northern Lake Nyasa: Zomba, Nyasaland, Government Printer, 171 pp.

Johnson, T.C., and Davis, T.W., 1989, High-resolution seismic profiles from Lake Malawi, Africa: Journal of African Earth Sciences, v. 8, pp. 383–392.

Johnson, T.C., and Ng'ang'a, P., 1990, Reflections on a rift lake, *in* Katz, B., ed., Lacustrine basin exploration: Case studies and modern analogs: American Association of Petroleum Geologists Memoir 50, pp. 113–135.

Krauskopf, K.B., 1957, Separation of manganese from iron in sedimentary processes: Geochimica et Cosmochimica Acta, v. 12, pp. 61–84.

Müller, G., and Förstner, U., 1973, Recent iron ore formation in Lake Malawi, Africa: Mineralium Deposita, v. 8, pp. 278–290.

Owen, R.B., 1989, Pelagic fisheries potential of Lake Malawi — Palæolimnology and palæoproductivity of Lake Malawi and its implications for fisheries management: London, Overseas Development Administration Project R4370, 300 pp.

Owen, R.B., and Crossley, R.B., 1989, Rift structures and facies distributions in Lake Malawi, *in* Rosendahl, B.R., Rodgers, J.J.W., and Rach, N.M., eds., Rifting in Africa — Karoo to Recent: Journal of African Earth Sciences, v. 8, pp. 415–427.

Owen, R.B., and Crossley, R.B., 1992, Spatial and temporal diatom variability in Lake Malawi and ecological implications: Journal of Paleolimnology, v. 7, pp. 55–71.

Owen, R.B., Crossley, R.B., Johnson, T.C., Tweddle, D., Kornfield, I., Davison, S., Eccles, D.H., and Engstrom, D.E., 1990, Major low levels of Lake Malawi and their implications for speciation rates in cichlid fishes: Proceedings of the Royal Society of London, Series B, v. 240, pp. 519–553.

Owen, R.B., Crossley, R.B., Williams, T.M., and Sefe, F., 1991, Facies distributions associated with a submerged fault-controlled platform in Lake Malawi, Central Africa: Journal of African Earth Sciences, v. 13, pp. 449–456.

Owen, R.B., and Renaut, R.W., 1993, Resource potential of Lake Malawi: Ottawa, International Development Research Centre Project 3-P-86-1028-02, 203 pp.

Pedro, G., Carmouze, J.P., and Velde, B., 1978, Peloidal nontronite formation in Recent sediments of Lake Chad: Chemical Geology, v. 23, pp. 139–149.

Roy, S., 1981, Manganese deposits: London, Academic Press, 458 pp.

Scholz, C.A., and Finney, B.P., 1994, Late Quaternary sequence stratigraphy of Lake Malawi (Nyasa), Africa: Sedimentology, v. 41, pp. 163–179.

Scholz, C.A., Johnson, T.C., and McGill, J.W., 1993, Deltaic sedimentation in a rift valley lake: New seismic reflection data from Lake Malawi (Nyasa), East Africa: Geology, v. 21, pp. 395–398.

Scholz, C.A., and Rosendahl, B.R., 1988, Low lake stands in Lakes Malawi and Tanganyika, delineated by multifold seismic data: Science, v. 240, pp. 1645–1648.

Scholz, C.A., Rosendahl, B.R., and Scott, D.L., 1990, Development of coarse grained facies in lacustrine rift basins: Geology, v. 18, pp. 140–144.

Stumm, W., and Morgan, J.J., 1981, Aquatic chemistry, 2nd Edition: New York, Wiley, 780 pp.

Williams, T.M., and Owen, R.B., 1990, Authigenesis of ferric oolites in superficial sediments from Lake Malawi, Central Africa: Chemical Geology, v. 89, pp. 179–188.

Williams, T.M., and Owen, R.B., 1992, Geochemistry and origins of lacustrine ferromanganese nodules from the Malawi Rift, Central Africa: Geochimica et Cosmochimica Acta, v. 56, pp. 2703–2712.

Early Holocene Changes in Lake Level and Productivity in Lake Malawi as Interpreted from Oxygen and Carbon Isotopic Measurements of Authigenic Carbonates

R.D. RICKETTS and T.C. JOHNSON   *Large Lakes Observatory, University of Minnesota, Duluth, United States*

Abstract — Analyses of stable carbon and oxygen isotopes from authigenic lacustrine carbonates were undertaken in order to determine fluctuations in lake level and epilimnetic productivity during the early Holocene. The carbonates form a horizon in three cores taken from the bottom of Lake Malawi. The horizon has been interpreted as having been deposited during a lake lowstand dated to have existed between 5 and 10 kyr B.P. by ^{14}C dating. Modeling of the response of the oxygen isotopic composition of the lake waters to climate forcing indicates that a direct comparison between oxygen isotopic composition and lake level is difficult. Carbon isotope data from authigenic carbonates in two of the cores may indicate that primary productivity in the lake fluctuated widely from 7.5 to 9 kyr B.P. Between 5 and 7.5 kyr B.P. carbon isotopic values are less variable and probably indicate that primary productivity was relatively low during that time. Carbon isotopic data from the third core indicate that a portion of its carbonate formed diagenetically in a methanogenic environment.

INTRODUCTION

Analysis of carbon and oxygen stable isotopes in primary lacustrine carbonates has become an accepted method of investigating past climates. Work with $\delta^{18}O$ has ranged from atmospheric paleo-temperature reconstructions (Stuiver, 1970; Lister, 1989) to studies of fluctuations in precipitation and evaporation balances (McKenzie and Eberli, 1987; Kelts and Talbot; 1989, Talbot, 1990; Lister et al., 1991). $\delta^{13}C$ variations in primary lacustrine carbonates have been used to quantify changes in primary productivity or as an indication of bacterial processes in lake waters and sediment (McKenzie, 1982; McKenzie, 1985; Lister, 1988; Kelts and Talbot, 1989; Talbot and Kelts, 1990). Lake Malawi offers an opportunity to study primary lacustrine carbonates deposited in carbonate-rich intervals presumably during lake lowstands. Stable oxygen and carbon isotope compositions of the most recent of these intervals, deposited from 6 to 10 kyr B.P., were investigated to establish whether or not the isotopic signals of the carbonates are consistent between cores and if so, are they a high-resolution proxy for past lake levels and their climatic controls.

Lake Malawi is located in the western branch of the East African Rift system (Fig. 1). The lake is the fifth largest lake in the world by volume and has a maximum depth of nearly 700 meters. It is permanently anoxic below 250 meters water depth. Sixty-two percent of the water input is from rainfall onto the lake surface during the monsoon season, and 82% of water output is by evaporation (Owen et al., 1990). The remaining output is through the

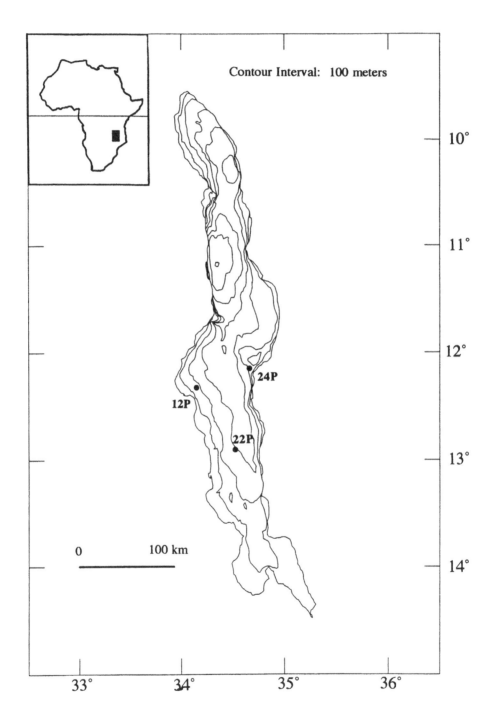

Figure 1. Map of the location of Lake Malawi and core sites.

Table 1. Core Descriptions

	M86-12P	M86-22P	M86-24P
Latitude	12°19.5'S	12°54.2'S	12°09.0'S
Longitude	34°08.8'E	34°31.2'E	34°39.4'E
Water depth	143 meters	265 meters	183 meters
Core length	555 cm	952 cm	733 cm
Depth interval of CaCO$_3$	80–410 cm	810–940 cm	200–410 cm

Shire River at the southern end of the basin. A six meter drop in present lake level would turn the lake into a closed basin (Beadle, 1981). The surface waters range in temperature from 27°C in the summer to 23.5°C in the winter. Below the chemocline the water temperature ranges between 22.66 ± 0.01°C and 22.70 ± 0.01°C during all seasons (Eccles, 1974; Halfman, this volume; Wüest et al., this volume).

High-resolution seismic reflection profiles have revealed evidence for several previous lowstands of the lake at water depths between 150 and 250 meters below present lake level (Johnson and Ng'ang'a, 1990). These previous lowstands are probably manifested in the sediments by calcite-rich intervals. The latest of these intervals is dated as being deposited during the early Holocene (Finney and Johnson, 1991).

METHODS

The sediments in three piston cores from Lake Malawi (Table 1, Fig. 1) containing carbonate-rich intervals corresponding to the early Holocene (Finney and Johnson, 1991) were subsampled within these intervals at a 5 cm spacing. Samples were freeze-dried and analyzed for calcium carbonate content using a vacuum-gasometric method (as described by Jones and Kaiteris, 1983). The standard deviation of percent calcium carbonate for replicate samples was approximately 0.5%.

For isotopic analyses, 20–40 mg of powdered bulk sediment were roasted under vacuum for two hours at 325°C to remove organic matter. The sediment was then reacted with 100% ortho-phosphoric acid at 90°C, and the evolved CO_2 gas collected using standard cryogenic techniques. The stable-isotope ratios for oxygen and carbon were determined using a VG Isotech mass spectrometer model Sira Series II located in the Duke University Department of Botany and are reported in the standard δ notation using the PDB reference standard (Craig, 1957). The analytical precision of this technique, based on replicate analyses of the same sample, is 0.10‰ for ^{18}O and 0.05‰ for ^{13}C. The standard deviations of at least four replicate samples from the same subsample for δ^{18}O are 0.10‰ for 12P, 0.34‰ for 22P, and 0.24‰ for 24P. The standard deviation of at least four replicate samples from the same subsample for δ^{13}C is 0.04‰ for core 12P, 0.11‰ for 22P, and 0.04‰ for 24P.

X-ray diffractometry analyses of mineral composition were done using a Philips XRG diffractometer with Cu–K$_\alpha$ radiation. Sediment samples were ground by mortar and pestle and mounted on glass slides. Minerals were identified from the XRD peaks using standard tables (Chao, 1969).

Table 2. ^{14}C Dating Results

Lab I.D.	Core	Depth (cm)	Material dated	Age (yr BP)
Beta 29140	M86-12P	120–141	CaCO$_3$	7490 ± 260
CAMS 2114		194–196	CaCO$_3$	7980 ± 210
CAMS 2116		194–196	TOC	7660 ± 70
Beta 29141		292–312	CaCO$_3$	9140 ± 170
CAMS 2115		320–322	CaCO$_3$	8750 ± 130
CAMS 2117		320–322	TOC	9050 ± 80
CAMS 2118		320–322	TOC	9000 ± 70
Beta 29806	M86-22P	76–98	TOC	1950 ± 200
Beta 29807		799–812	TOC	6320 ± 240
Beta 30637		874–920	CaCO$_3$	7600 ± 140
GU 2628	M86-24P	170–190	TOC	5160 ± 300
NCSU		260–261	CaCO$_3$	7043 ± 70
NCSU		380–381	CaCO$_3$	8562 ± 74

Beta = Beta Analytical Inc.; CAMS = Center for Accelerator Mass Spectrometry, Lawrence Livermore National Lab; GU = Scottish Universities Research and Reactor Centre; NCSU = North Carolina State University Stable Isotope Lab.

RESULTS

Interpolated sediment ages within the carbonate intervals were assigned using a least squares linear fit between ^{14}C dates determined from calcium carbonate and organic matter (Table 2; Finney and Johnson, 1991). This gave sedimentation rates within the carbonate intervals of 1.2 mm/year for core 12P, 1.5 mm/year for core 22P, and 0.6 mm/year for core 24P (Fig. 2). There is a hiatus or dramatic decrease in sedimentation rate near the top of core 12P. It will be shown that δ^{18}O peaks of core 12P correlate chronostratigraphically with δ^{18}O peaks in core 24P, substantiating this conclusion.

The weight percent carbonate content in the calcite-rich intervals of cores 12P, 22P, and 24P range from 0% to approximately 17% CaCO$_3$ (Fig. 3). The missing values in core 22P (from 874 to 920 cm depth) are the result of the core having been sampled earlier for a conventional ^{14}C date.

Cores 12P and 24P have δ^{18}O values ranging from –0.2 to –1.3‰, whereas in core 22P δ^{18}O ranges from 2 to –1‰ (Fig. 4). The oldest section of cores 12P and 24P are characterized by multiple positive peaks from about 7.5 to 9.0 kyr B.P. There is a broad single positive peak from 6.2 to 7.0 kyr B.P., and towards the top of the carbonate interval δ^{18}O values become more positive.

The range of δ^{13}C values increases from core 24P to 12P to 22P (Fig. 5). These yδ^{13}C values vary between approximately 1.2 and 13‰. Core 22P has no δ^{13}C trends which are paralleled in the δ^{13}C curves of the other two cores. Both cores 12P and 24P exhibit strongly fluctuating δ^{13}C values in the older sections and less variability in the younger sections.

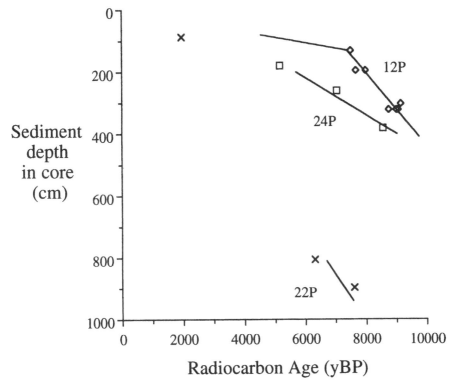

Figure 2. [14]C dates for cores 12P (diamonds), 22P (x), and 24P (squares) and averaged sedimentation rates between ^{14}C ages. [14]C dates are given in Table 2. The lines are the least-squares fits to the ^{14}C dates and were used to assign ages to the sediment.

DISCUSSION

The carbonate horizons in the cores have been interpreted as correlating with a lowstand of the lake 100 to 150 meters below the present level (Johnson and Ng'ang'a, 1990; Finney and Johnson, 1991). Authigenic calcite makes up about 0.5% of the sediment currently settling through the water column of Lake Malawi (Pilskaln, 1989). This small amount of calcite is not detected in core-top sediment. The water chemistry of the lake (reported by Gonfiantini et al., 1979; Table 3) was used to determine the degree of saturation with respect to calcite for the water column. This was carried out using the PCWATEQ program provided by Shadoware Inc. (Redmond VA), where the degree of calcite saturation is expressed as the ratio of ion activity product (IAP) of calcium and carbonate ions, divided by the saturation coefficient of calcite (K). Only the top few meters of the modern water column are supersaturated with respect to calcite (Fig. 6). The anoxic region below the chemocline has higher levels of CO_2 than the surface waters, making the deeper waters more corrosive to calcite (Finney and Johnson, 1991) resulting in calcite dissolution and preventing calcite accumulation in the modern lake sediments. A 125 meter drop in lake level would reduce the volume of the lake by 41%. Simply assuming that a drop in water volume of 41% results in a 170% increase in salinity, the top 300 meters of the water column would become oversaturated with respect to calcite. Under these simplified circumstances calcite could be

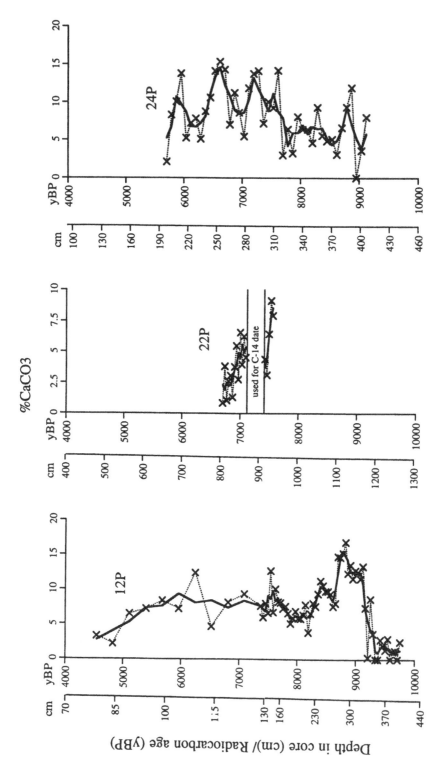

Figure 3. Weight percent calcite in the sediment from cores 12P,22P, and 24P. The solid line is a 3-point running mean of the measured data points. This span of the cores is interpreted to represent lake levels 100 to 150 meters below the present level.

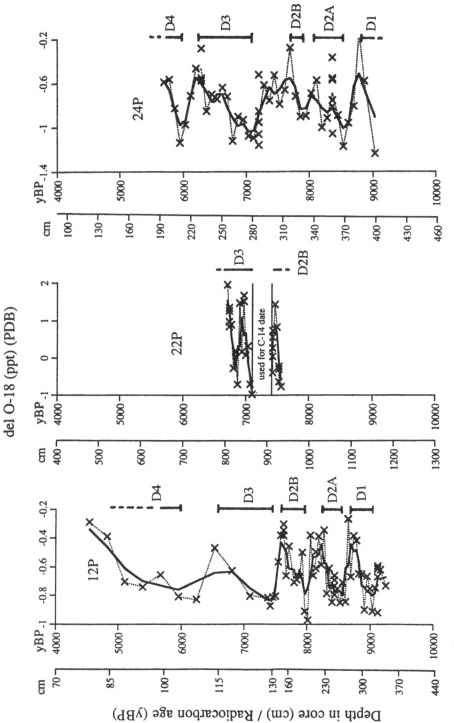

Figure 4. δ^{18}O values from cores 12P, 22P, and 24P. The solid line is a 3-point running mean of the measured data points. Levels marked D1, D2A, D2B, D3, and D4 are interpreted as representing "drying events" (see text).

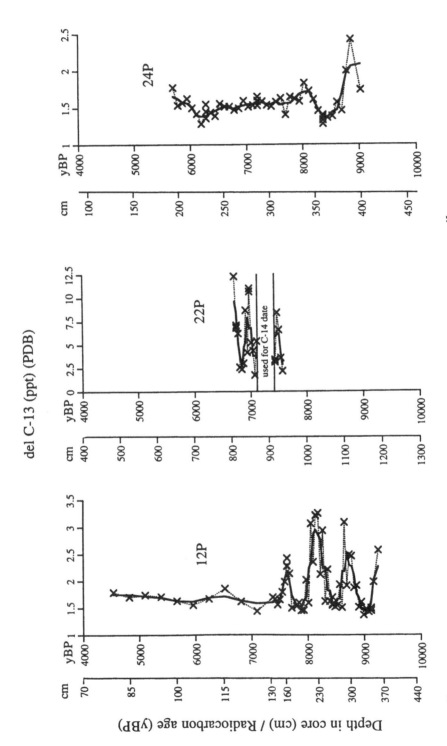

Figure 5. $\delta^{13}C$ values from cores 12P, 22P, and 24P. The solid line is a 3-point running mean of the measured data points. The $\delta^{13}C$ values in core 22P are interpreted as reflecting a mixture of authigenic carbonate and diagenetic carbonate formed in a methanogenic environment. Fluctuations of $\delta^{13}C$ values in cores 12P and 24P are interpreted as shifts in lacustrine primary productivity in the epilimnion.

Table 3. Water Chemistry Data for Lake Malawi (Gonfiantini et al., 1979)

Depth (m)	Temp. (°C)	pH	O_2 (mg/l)	Ca^{++} (meq/l)	Mg^{++} (meq/l)	Na^+ (meq/l)	K^+ (meq/l)	HCO_3^- (meq/l)	Cl^- (meq/l)	SO_4^- (meq/l)
1	25.62	8.0	7.90	0.90	0.61	0.88	0.16	2.36	0.14	0.13
100	23.38	7.75	4.72	0.94	0.63	0.90	0.16	2.40		0.09
200	22.87	7.7	0.36	0.98	0.63	0.90	0.17	2.41		
300	22.70	7.5	0.00	1.01	0.63	0.89	0.17	2.44	0.21	0.13
400	22.72	7.35	0.00	0.99	0.63	0.90	0.17	2.51		0.12
500	22.76	7.3	0.00	0.99	0.63	0.89	0.17	2.51		
600	22.75	7.3	0.00	1.00	0.63	0.90	0.17	2.49	0.28	0.13
640	22.70	7.3	0.00	0.99	0.63	0.90	0.17	2.50	0.28	0.14

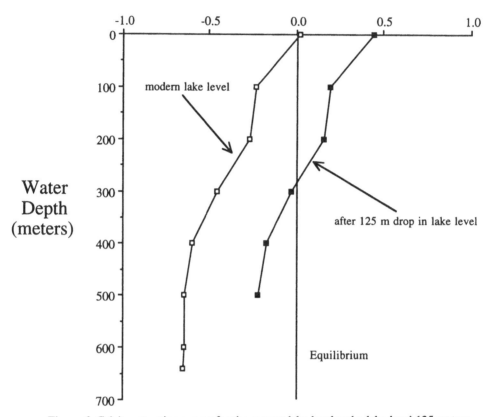

Figure 6. Calcite saturation curves for the current lake level and a lake level 125 meters below the current level. The saturation curve for a –125 meter lake level shows the probable lake water chemistry conditions for the carbonate analyzed in this study. The modern lake is mostly undersaturated with respect to calcite while the –125 meter lake is oversaturated with respect to calcite in the top several hundred meters of the water column.

deposited at the paleo-depths of the three cores (Fig. 6). These calculations support the contention that the carbonate horizons present in the cores represent low lake level conditions.

X-ray diffractometry analyses indicates that the magnesium content (Goldsmith and Graf, 1958) of the calcite in cores 12P and 24P varies down core between 2 and 5 mole percent. The carbonate in core 22P is magnesium calcite (4.5% Mg) and possibly another unidentified carbonate mineral. As mentioned previously, the sample pretreatment for isotopic analysis of the carbonates involves heating to 325°C. This step also effectively removes the unidentified carbonate mineral and displaces the calcite peak towards lower Mg values for samples from core 22P. These effects are not observed in the other cores. The presence of very positive $\delta^{13}C$ values in core 22P strongly suggests that at least a portion of those carbonates formed during burial diagenesis in an environment of methanogenesis.

Modern lake surface water has a $\delta^{18}O$ value of about 0.69‰ (SMOW) (T. Cerling, pers. comm.). Therefore calcite precipitated in isotopic equilibrium with the modern water should have a $\delta^{18}O$ value of −1.74‰ (PDB) at 27°C (Friedman and O'Neil, 1977). Most $\delta^{18}O$ values found in cores 12P and 24P are within +0.57 and +1.47‰ of this value, indicating that the calcite could have precipitated from water which was cooler or isotopically heavier.

If the calcite deposited during the early Holocene lowstand precipitated in the water column, variations in the $\delta^{18}O$ values could reflect past changes in lake water temperature or oxygen isotopic composition, as expressed by the equation (Craig, 1965):

$$t\,[°C] = 16.9 - 4.2(\delta c - \delta w) + 0.13(\delta c - \delta w)^2,$$

where t is water temperature and δc and δw are the $\delta^{18}O$ values of the calcite and water, respectively. Assuming the paleo-$\delta^{18}O$ value of the surface water was the same as today ($\delta^{18}O$ = 0.69‰; T. Cerling, pers. comm.) yields a temperature range of 21–24.2°C for the calcite $\delta^{18}O$ values found in core 12P and 20.5–25.5°C for core 24P. While these temperature ranges are approximately equal to the seasonal surface water temperature changes of 4°C found today (Eccles, 1974), one cannot rule out the possibility of a change in the isotopic composition of the water being responsible for producing the observed variability in the isotopic composition of the calcite, especially since one would not expect the paleo-$\delta^{18}O$ value of the surface water to remain constant after a 100 to 150 meter drop in lake level.

The theoretical response of $\delta^{18}O$ values of Lake Malawi water to two possible scenarios for fluctuations in lake level is rather disconcerting (Fig. 7). Following the hydrological modeling of Owen et al. (1990), three different possible climate regimes are used to vary lake level for the $\delta^{18}O$ theoretical calculations. The first climate regime, "dry," is characterized by a 50% decrease in rainfall compared to modern conditions, a 90% decrease in river inflow, and a 7% increase in evaporation flux. This regime leads to a drop in lake level of 1.2 meters per year until lake level has dropped 200 meters. The second climate regime, "steady-state," is characterized by a 15% decrease in rainfall compared to modern conditions, a 29% decrease in river inflow, and a 2% increase in evaporation. Using this regime, inputs and evaporation are balanced, therefore lake level remains constant. The third regime, "modern," is characterized by present day conditions. Modern conditions lead to an increase in lake level until there is outflow down the Shire River at about the current lake level. These three conditions are used in different sequences to cause two distinct histories of lake level fluctuation (Fig. 7). The top graph starts with dry conditions which drop lake level 75 meters. Then a combination of dry, steady-state, and modern conditions cause lake level to vary between 75 meters and 125 meters below current lake level. The bottom graph starts with dry conditions to drop lake level 125 meters. Steady-state and modern conditions then bring the lake level back up to the current level, with plateaus at 100 and 75 meters below current lake level.

The theoretical $\delta^{18}O$ values of lake water resulting from these conditions were calculated using the following equation from Phillips et al.(1992):

$$(\partial \delta_L)/(\partial t) = [B - \delta_L(\partial V/\partial t)]/V,$$

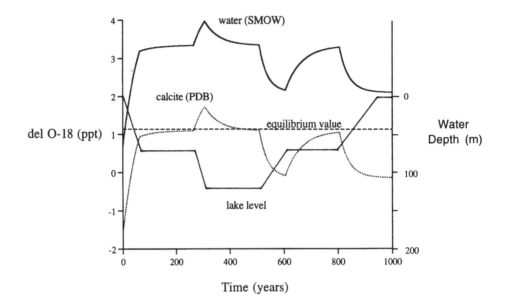

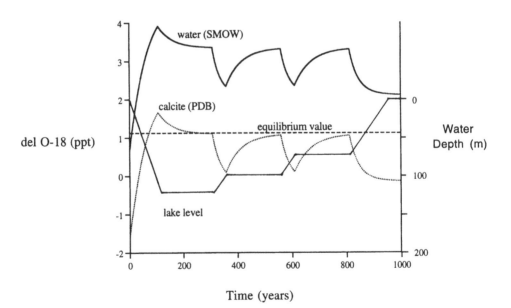

Figure 7. Theoretical $\delta^{18}O$ values for lake water and calcite using present day conditions as a starting point. The top graph illustrates the effects of a fall in lake level to 75 meters below the present level and then a further drop to 125 meters. The bottom graph illustrates the effects of a fall in lake level to 125 meters below the present level and then a stair-step rise to present conditions. The corresponding $\delta^{18}O$ curves are similar for these dissimilar lake level fluctuations.

in which V is volume of the lake, t is time, δ_L is the isotopic composition of the lake waters, and B is the isotopic budget for the lake:

$$B = \delta_I Q_I - \delta_E Q_E + \delta_C Q_C - \delta_O Q_O,$$

in which I represents input from rainfall and rivers into the lake, E represents gross evaporation, C represents condensation, O represents outflow, Q is the flux of the corresponding variable, and δ is the $\delta^{18}O$ value of the variable. Fluxes are determined using data from Owen et al. (1990) for the appropriate climate regime ("dry," "steady-state," or "modern" conditions) (Table 4) and the surface area of the lake at different volumes. The surface area of the lake at specific volumes are determined from a hypsographic curve for Lake Malawi. A value of $-7\%o$ is chosen for Lake Malawi precipitation because it is similar to the weighted mean values of rainfall measured at Ndola and Harare, which are to the west and south of lake Malawi (Rozanski et al., this volume). The $\delta^{18}O$ value for river input is set at $-5\%o$ to reflect the isotopic evolution of the river water as it flows to the lake.

δE is calculated using the following equation (Phillips et al., 1992):

$$\delta_E = \left(\frac{((1 + 10^{-3}\delta_L)(1 - k))}{((1 + 10^{-3}\varepsilon^*)(1 - kh))} - 1 \right) 10^3,$$

in which h is humidity and is 70% for modern conditions, and set at 60% for steady-state conditions and 50% for dry conditions, and k is a parameter dependent on wind speed which describes the transport of the isotopic species away from the water surface (for $H_2{}^{18}O$, $k = 0.002$ when wind speeds are between 6 and 7 m/s (Merlivat and Jouzel, 1979), a reasonable approximation for Lake Malawi). ε^* is the equilibrium enrichment factor and is equal to (Bottinga and Craig, 1969):

$$10^3 \ln(1 + \varepsilon^*) = 1.534(10^6 \, T^{-2}) - 3.206(10^3 \, T^{-1}) + 2.644,$$

where T is temperature in degrees Kelvin; therefore, at $27°C$, $\varepsilon^* = 9.0\%o$.

The isotopic composition of the condensation is given by (Phillips et al., 1992):

$$\delta_C = \varepsilon^*[1 + (\delta_A/10^3)] + \delta_A,$$

in which δA is the isotopic composition of the atmospheric moisture and is determined assuming that it was in isotopic equilibrium with the rainfall in the area ($\delta_A = -16\%o$ at an air temperature of $27°C$).

The flux of condensation to the lake's surface is:

$$Q_C = (hQ_E)/x,$$

in which x is the chemical activity of water, which is equal to 1 for the dilute lake water.

Evaporation fluxes reported in Owen et al. (1990) are net evaporation fluxes. Therefore the gross evaporative flux must be calculated from:

$$Q_E = Q_C + Q_{E\,net}.$$

δ_O is set equal to lake isotopic composition. The initial isotopic composition of the lake water is taken to be the current surface water composition ($0.69\%o$; T. Cerling, pers. comm.).

Table 4. Water Fluxes for the Different Climate Regimes Used
in the Theoretical Calculations of $\delta^{18}O$

	Modern	Steady-state	Dry
Net evaporation (mm/year)	1872	1909	2003
River inflow (mm/year)	1000	710	100
Rainfall on lake (mm/year)	1414	1202	707

The model does not account for the effect that a large lake such as Lake Malawi has on the isotopic composition of condensing water vapor forming over the lake. Even with this shortcoming the model does serve as a first approximation for the fluctuations in isotopic composition of lake water that a large lake would undergo.

The two climatic scenarios yield subtly different isotopic responses in the lake. Isotopic profiles do not track lake level. There is a shift in isotopic composition towards heavier values as lake level falls (and vice versa), but once the lake has achieved a steady level, the isotopic composition of the water shifts exponentially back towards an equilibrium value that depends on the isotopic budget of the lake rather than on the lake level (Fig. 7). The time required to achieve the equilibrium value is comparable to the residence time for water in the lake, or in the case of Lake Malawi, about 100 years. Thus an isotopic record for a lake such as Malawi that has undergone significant shifts in lake level in closed-basin status will be quite peaked in appearance, with the height of the peaks impacted by both the magnitude of the shift in lake level, and the rate of lake level change. A very gradual shift in lake level, taking place over the course of several water residence times, would be recorded as a low-amplitude isotopic shift because the isotopic equilibrium value for closed-basin status could be nearly maintained. On the other hand, a very rapid change in lake level will yield a shift in isotopic composition with an amplitude corresponding to the amplitude of the change in lake level. The model results all assume that the isotopic composition of the rainfall and other sources of fresh water are invariant. If there is a significant shift in atmospheric circulation such as one might anticipate over east Africa from glacial to interglacial times (Demenocal and Rind, 1993), this will also have a large impact on the isotopic composition of the water and the calcite.

The patterns of isotopic profiles for cores 12P and 24P show some similarities to those in the model results presented in Fig. 7. The profiles are quite peaked in appearance, each spans the order of a century, and the amplitude of the shifts is variable and up to 1‰. One problem with interpreting the profiles quantitatively is that an "equilibrium value" is not apparent from the shapes of the profiles, so although discriminating between positive and negative excursions, representing falling and rising lake levels, respectively, is possible, quantifying their amplitudes is not.

Given the limited precision of the radiocarbon dates, it is best to focus on the smoothed isotope curves in Fig. 4 to delineate a paleoclimatic history for the period between 10 and 5 kyr B.P. There appear to have been on the order of four periods when oxygen isotopic values increased. We refer to these simply as "drying events." Other factors such as cooling of the lake or isotopically heavier rainfall could also account for the shifts towards heavier isotopic values, but these two factors tend to work against one another. Assuming that our present age assignments are correct, there were significant drying events between 9 and 8.5 kyr B.P. (D1), between 8.5 and 7.5 kyr B.P. (D2), and between 7.2 and 6.5 kyr B.P. (D3). We suspect that the lake did not achieve open-basin status during intervening wet periods since carbonate continued to accumulate and there were no significant excursions of the isotopic profiles to lighter $\delta^{18}O$ values. After 6.5 kyr B.P. the record is too poorly constrained by radiocarbon dates to assign ages to the isotopic events. There was at least another lake level drop yielding the relatively heavy isotopic values at the tops of the profiles in cores 12P and 24P (D4), then presumably a very rapid rise in lake level to open-basin conditions at around 5–5.5 kyr B.P., at which time carbonate accumulation ceased.

The $\delta^{13}C$ values for the three cores show less distinct trends than those of the $\delta^{18}O$ values. Peaks common to both cores 12P and 24P are seen in the older sections of the carbonate intervals, and occur at approximately 8.0 kyr B.P. and 8.8 kyr B.P. In cores 12P and 24P the carbonate sections younger than 7.5 kyr B.P. both have relatively invariant $\delta^{13}C$ values. An increase in primary productivity can increase $\delta^{13}C$ values of calcite precipitated from lake waters by depleting the near-surface water in ^{12}C, which is then removed to deeper waters as the organic matter sinks (McKenzie, 1985). Therefore the 8.0 to 9.0 kyr B.P. period may have been a time of widely fluctuating productivity levels, with peaks at approximately 8.0 and 8.8 kyr B.P. These peaks correlate well with periods of higher $\delta^{18}O$ values, indicating that salinity and productivity increased as the lake was drying. From 5.0 to 7.5 kyr B.P. there may have been a time of relatively constant productivity levels. Core 12P has generally higher $\delta^{13}C$ values than those in core 24P, possibly indicating that the western side of the basin was more productive than the eastern side during the 5 to 10 kyr B.P. lowstand. In modern times the interaction of wind patterns over the lake can make the western shore of the lake more productive than the eastern shore (Eccles, 1974). This interpretation is supported by the average $\delta^{18}O$ values in the two cores: those in core 12P are about 0.3‰ heavier than those in core 24P, suggesting cooler waters that would be associated with upwelling.

The $\delta^{13}C$ values in 22P are generally much more positive than those for the other two cores, and probably indicate diagenetic formation of calcite coincident with methane production in the pore waters. Even with decreased lake level the core site would be close to or below the chemocline (Finney and Johnson, 1991). Organic matter in Lake Malawi sediment has $\delta^{13}C$ values of –20 to –26‰ (Johnson and Ng'ang'a, 1990) which is similar to the values found for other aquatic organic matter (–20‰; Deines, 1980). Organic matter which undergoes acetate fermentation methanogenesis would form isotopically light methane (approximately –59‰; Woltemate et al., 1984; Whiticar et al., 1986) and isotopically heavy CO_2 (approximately –2.5‰ with $\alpha = 1.06$; Woltemate et al., 1984; Whiticar et al., 1986). Since the fractionation factor between calcium carbonate and CO_2 is 1.010 (at 25°C; Bottinga, 1968), the carbonate formed from this CO_2 would form calcite with a $\delta^{13}C$ value of approximately 7.4‰. The average value of $\delta^{13}C$ from core 22P is 5.3‰, with some values above 10‰. $\delta^{13}C$ values in the carbonates from cores 12P and 24P are rarely above 3‰. The $\delta^{13}C$ values of carbonates from core 22P may be a result of a mixture of calcite from the water column and calcite formed from CO_2 left after methanogenesis. Methanogenesis would not be an explanation for the relatively heavy $\delta^{18}O$ values from core 22P since fractionation caused by methanogenic bacteria would not significantly affect oxygen iso-

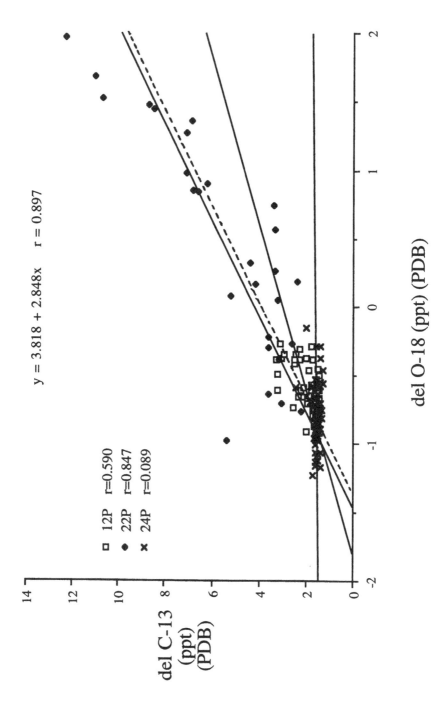

Figure 8. Plot of $\delta^{18}O$ vs. $\delta^{13}C$ for the three cores. Correlation is high ($r = 0.897$; dotted line) when all cores are considered together. Cores 12P and 24P, which are unaffected by diagenesis, have r values which are less than the value hypothesized by Talbot (1990) for closed basins ($r > 0.7$).

topic values (Talbot and Kelts, 1990) but the heavy values do reflect formation in colder bottom waters instead of surface waters.

Plots of $\delta^{18}O$ vs. $\delta^{13}C$ values for carbonates in cores 12P and 24P do not exhibit the strong linear trend Talbot (1990) found for a variety of closed basins (Fig. 8). Neverless, when all the data are considered, the resultant r value of 0.90 is above the criteria for a closed basin lake ($r > 0.7$). The influence of core 22P on this result, which may have been affected by methanogenesis, raises the question of what causes the linear relationship between the two isotopes in lacustrine carbonates, and to what extent it is truly indicative of a closed basin.

CONCLUSIONS

Analyses of $\delta^{18}O$ and $\delta^{13}C$ values for carbonates deposited during an early Holocene lowstand in Lake Malawi gives some indication of climate and productivity fluctuations. $\delta^{18}O$ values and sedimentological data from three cores are consistent with the hypothesis that there was a drop in lake level, which lasted from 10 to 5 kyr B.P. A closer analysis of the data revealed that the $\delta^{18}O$ fluctuations in authigenic calcite cannot clearly define the level of the lowstand because of their strong dependence on water temperature, the isotopic composition of the source waters, and the inability to detect an equilibrium value of $\delta^{18}O$ around which the profiles fluctuate. Neverless the data do indicate the timing of a few significant climatic events during the 4000 year period of carbonate accumulation. In particular, there were probably four relatively dry periods causing drops in lake level. The $\delta^{13}C$ values in the authigenic calcite in two of the cores suggest that primary productivity may have fluctuated between 8.0 to 9.0 kyr B.P. and then stabilized after 7.5 kyr B.P. The $\delta^{13}C$ values for carbonates in a third core indicate that diagenetic processes such as methanogenesis altered the carbonate after deposition. Even with the difficulties in interpretation, the isotopic measurements from the three cores illustrate the possible uses of ^{18}O and ^{13}C from authigenic carbonates in paleoclimate reconstructions.

ACKNOWLEDGMENTS

We thank the governments of Malawi and Tanzania for logistical assistance and for research permits issued to Project PROBE of Duke University. Tom Davis and John Graves supervised collection of most of the cores used in this study. We thank Peter Howd, Paul Baker, and Mary Mungai for helpful discussion and Keith Sturgeon for patience and assistance in the laboratory. We also thank Paul Baker and Guy Lister for their helpful reviews. Funding for this research was provided by the Bradley Foundation and U.S. AID grant number HRN-5600-g-00-2056-00 and NSF grant ATM88-16615.

REFERENCES

Beadle, L.C., 1981, The inland waters of tropical Africa, 2nd ed.: New York, Longman, 475 pp.

Bottinga, Y., 1968, Calculation of fractionation factors for carbon and oxygen isotopic exchange in the system calcite-carbon dioxide-water: Journal of Physical Chemistry, v. 72, pp. 800–808.

Bottinga, Y., and Craig, H., 1969, Oxygen isotope fractionation between CO_2 and water, and the isotopic composition of marine atmospheric CO_2: Earth and Planetary Science Letters, v. 5, pp. 285–295.

Cerling, T.E., Bowman, J.R., and O'Neil, J.R., 1988, An isotopic study of a fluvial-lacustrine sequence: the Plio–Pleistocene Koobi-Fora Sequence, east Africa: Palaeogeography, Palaeoclimatology, Palaeoecology, v. 63, pp. 335–356.

Chao, G.Y., 1969, 2-theta (Cu) table for common minerals: Geological Paper 69-2, Carleton University Department of Geology, 42 pp.

Craig, H., 1957, Isotopic standards for carbon and oxygen and correction factors for mass spectrometric analysis of carbon dioxide: Geochimica Cosmochimica Acta, v. 12, pp. 133–148.

Craig, H., 1965, The measurement of oxygen isotope paleotemperatures, *in* Stable Isotopes in Oceanographic Studies and Paleotemperatures: Spoleto, July 26–27, 1965. Consiglio Nazionale delle Richerche, Laboratorio di Geologia Nucleare, Pisa, pp. 1–24.

Deines, P., 1980, The isotopic composition of reduced organic carbon, *in* Fritz, P., and Fontes, P.C., eds., Handbook of Environmental Isotope Geochemistry I. The Terrestrial Environment: Amsterdam, Elsevier, pp. 329–406.

DeMenocal, P.B., and Rind, D., 1993, Sensitivity of Asian and African climate to variations in seasonal insolation, Glacial ice cover, sea surface temperature, and Asian orography: Journal of Geophysical Research, vol. 98, no. D4, pp. 7265–7287.

Eccles, D.H., 1974, An outline of the physical limnology of Lake Malawi (Lake Nyasa): Limnology and Oceanography, v. 19, pp. 730–742.

Finney, B.P., and Johnson, T.C., 1991, Sedimentation in Lake Malawi (East Africa) during the past 10,000 years: a continuous paleoclimatic record from the southern tropics: Palaeogeography, Palaeoclimatology, Palaeoecology, v. 85, pp. 351–366.

Friedman, I., and O'Neil, J.R., 1977, Compilation of stable isotope fractionation factors of geochemical interest, *in* Fleisher, M., ed., Data of Geochemistry, chapter KK U.S. Geol. Surv. Prof. Paper, 440-KK, 12 pp.

Goldsmith, J.R., and Graf, J.L., 1958, Relations between lattice constants and composition of the Ca–Mg carbonates: American Mineralogist v. 43, pp. 84–101.

Gonfiantini, R., Zupi, G.M., Eccles, D.H., and Ferro, W., 1979, Isotope investigation of Lake Malawi, in Isotopes in Lake Studies: Internal Atomic Energy Agency Panel Proceedings Series STI/PUB/5111, pp. 195–207.

Halfman, J.D., 1996, CTD-transmissometer profiles from Lakes Malawi and Turkana, *in* Johnson, T.C., and Odada, E.O., eds., The Limnology, Climatology and Paleoclimatology of the East African Lakes: Toronto, Gordon and Breach, pp. 169–182.

Johnson, T.C., and Ng'ang'a, P., 1990, Reflections on a rift lake: AAPG Memoir 50, pp. 113–135.

Jones, G.A., and Kaiteris, P., 1983, A vacuum-gasometric technique for rapid and precise analysis of calcium carbonate in sediments and soils: Journal of Sedimentary Petrology, v. 53, pp. 655–660.

Kelts, K., and Talbot, M.R., 1989, Lacustrine carbonates as geochemical archives of environmental change and biotic–abiotic interactions, *in* Tilzer, M.M., and Seruya, C., eds., Ecological Structure and Function in Large Lakes: Madison, Wisconsin, Science and Technology Publishers, pp. 290–317.

Lister, G.S., 1988, A 15,000-year isotopic record from Lake Zurich of deglaciation and climatic change in Switzerland: Quaternary Research, 29, pp. 129–141.

Lister, G.S., 1989, Reconstruction of Palaeo air temperature changes from oxygen isotopic records in Lake Zurich: the significance of seasonality: Eclogae Geol. Helv., 82/1, pp. 219–234.

Lister, G.S., Kelts, K., Zao, C.K, Yu, J., and Niessen, F., 1991, Lake Qinghai, China: closed-basin lake levels and the oxygen isotope record for ostracoda since the latest Pleistocene: Palaeogeography, Palaeoclimatology, Palaeoecology, v. 84, pp. 141–162.

McKenzie, J.A., 1982, Carbon-13 cycle in Lake Greifen: a model for restricted ocean basins, *in* Schlanger, S.O., and Cita, M.B., eds., Nature and Origin of Cretaceous Carbon-Rich Facies: New York, Academic Press, pp. 197–207.

McKenzie, J.A., 1985, Carbon isotopes and productivity in the lacustrine and marine environment, *in* Stumm, W., ed., Chemical Processes in Lakes: New York, Wiley, pp. 99–118.

McKenzie, J.A., and Eberli, G.P., 1987, Indications for abrupt Holocene climatic change: late Holocene oxygen isotope stratigraphy of the Great Salt Lake, Utah, *in* Berger, W.H., and Labeyrie,

L.D., eds., Abrupt Climatic Change: Evidence and Implications: Dordrecht, Holland, D. Reidel Publishing Co., pp. 127–136.

Merlivat, L., and Jouzel, J., 1979, Global climatic interpretation of the deuterium–oxygen 18 relationship for precipitation: Journal of Geophysical Research, v. 84, pp. 5029–5033.

Owen, R.B., Crossley, R., Johnson, T.C., Tweddle, D., Kornfield, I., Davison, S., Eccles, D.H., and Engstrom, D.E., 1990, Major low levels of Lake Malawi and implications for speciation rates in cichlid fishes: Proceedings of the Royal Society of London, B 240, pp. 519–553.

Phillips, F.M., and Campbell, A.R., Kruger, C., Johnson, P., Roberts, R., and Keyes, E., 1992, A reconstruction of the response of the water balance in western United States lake basins to climatic change: volume 1: Technical Completion Report, Project Numbers 1345662 and 1423687, New Mexico Water Resources Research Institute, 167 pp.

Pilskaln, C.H., 1989, Seasonal particulate flux and sedimentation in Lake Malawi, East Africa: EOS, v. 70, p. 1130.

Rozanski, K., Araguás-Araguás, L., and Gonfiantini, R., 1996, Isotope patterns of precipitation in the East African region: in Johnson, T.C., and Odada, E.O., eds., The Limnology, Climatology and Paleoclimatology of the East African Lakes: Toronto, Gordon and Breach, pp. 79–93.

Stuiver, M., 1970, Oxygen and carbon isotope ratios of fresh-water carbonates as climatic indicators: Journal of Geophysical Research, v. 75, no. 27, pp. 5247–5257.

Stumm, W., and Morgan, J.J., 1981, Aquatic Chemistry: An Introduction Emphasizing Chemical Equilibria in Natural Waters, 2nd Ed: New York, John Wiley and Sons, 780 pp.

Talbot, M.R., 1990, A review of the palaeohydrological interpretation of carbon and oxygen isotopic ratios in primary lacustrine carbonates: Chemical Geology., v. 80, pp. 261–279.

Talbot, M.R., and Kelts, K., 1990, Paleolimnological signatures from carbon and oxygen isotopic ratios in carbonates from organic carbon-rich lacustrine sediments: AAPG Memoir 50, pp. 99–112.

Whiticar, M.J., Faber, E., and Schoell, M., 1986, Biogenic methane formation in marine and freshwater environments: CO_2 reduction vs. acetate fermentation-isotope evidence: Geochimica et Cosmochimica Acta, v. 50, pp. 693–709.

Woltemate, I., Whiticar, M.J., and Schoell, M., 1984, Carbon and hydrogen isotopic composition of bacterial methane in a shallow freshwater lake: Limnology and Oceanography, v. 29, no. 5, pp. 985–992.

Wüest, A., Piepke, G., and Halfman, J.D., 1996, Combined effects of dissolved solids and temperature on the density stratification of Lake Malawi, in Johnson, T.C., and Odada, E.O., eds., The Limnology, Climatology and Paleoclimatology of the East African Lakes: Toronto, Gordon and Breach, pp. 183–202.

Late Quaternary Lake-Level Changes of Lake Malawi

B.P. FINNEY   *Institute of Marine Science, University of Alaska, Fairbanks, United States*

C.A. SCHOLZ   *Rosenstiel School of Marine and Atmospheric Science, University of Miami, Miami, Florida, United States*

T.C. JOHNSON   *Large Lakes Observatory, University of Minnesota, Duluth, United States*

S. TRUMBORE   *Center for Accelerator Mass Spectrometry, Lawrence Livermore National Laboratory, Livermore, California, United States*

Abstract — Reconstructions of lake-level variations in Lake Malawi, East Africa, indicate the lake was about 200 to 300 m below present during most of the period from ~40,000 to 28,000 years before present (BP), and 100 to 150 m below present from 10,000 to 6000 years BP. These fluctuations are generally out of phase with most African lake-level records north of Malawi. General Circulation Models (GCMs) simulate such an opposite response due to the effect of insolation variations on the intensity of the northern and southern hemisphere monsoon. These results suggest that climatic gradients in this region of tropical east Africa were much larger than today during significant periods of the past.

INTRODUCTION

The levels of many lakes fluctuate as climate changes. Maps of lake-level data for times in the past reveal changing climatic patterns and can infer shifts in atmospheric circulation (Street-Perrott and Harrison, 1985). While the coverage of lake-level fluctuations in Africa during the late Quaternary is better than other continents, very little paleoclimatic information exists for the region between about 9 to 20°S (Street-Perrott et al., 1989). Lake Malawi (9°30'–14°30'S, 34°30'E; also known as Lake Nyasa) is the fifth largest lake in the world (by volume) (Herendorf, 1982) and the largest lake south of 9°S (Fig. 1). It is the southernmost lake in the east African rift valley and at least 5 million years old (Johnson and Ng'ang'a, 1990). The lake has a volume of ~8000 km^3, a maximum depth of over 700 m, and is anoxic below 250 m (Eccles, 1974; Gonfiantini et al., 1979). Direct rainfall (~1400 mm/yr) over the lake surface mainly during the summer monsoon season (December–April) accounts for about 62% of the water input, while evaporation (~1870 mm/yr) controls about 82% of the water loss (Owen et al., 1990). Outflow from the Shire River regulates the present lake level.

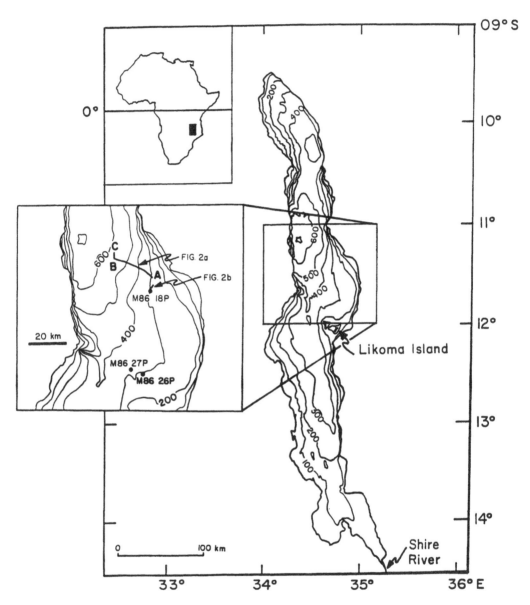

Figure 1. Bathymetry (m) and location of Lake Malawi in Africa. The large inset shows piston core locations and track lines of seismic profiles north of Likoma Island.

GEOLOGIC SETTING

Studies of high-resolution and multi-channel seismic reflection data have revealed a lake-wide depositional sequence up to 70 m thick which overlies an extensive erosional surface in water depths less than 450 m (Scholz and Rosendahl, 1988). This sequence, called the Songwe Sequence, is the shallowest part of a sedimentary package as much as 5 km thick (Flannery, 1988). Most of the piston cores (5–10 m long) recovered from the Songwe

Sequence have sedimentation rates of 0.5 to 1.5 mm/yr, and thus span the past few thousand to 20,000 years (Johnson and Ng'ang'a, 1990; Finney and Johnson, 1991). The Songwe Sequence is much thinner (<20 m) on the embayment north of Likoma Island in central Lake Malawi (Fig. 2a) probably due to limited terrigenous input from the small adjacent drainage area. Seismic reflection data reveal a well-developed erosional surface and truncated reflections beneath the Songwe Sequence in the shallower parts of this embayment (Fig. 2a).

Cores M86 18P, 26P and 27P were recovered from this region and can be correlated among one another as well as with the seismic data (Figs. 2b and 3). The 20 to 35% reduction in porosity observed near the base of each core corresponds to the basal Songwe unconformity. Another low porosity zone is observed at 0.5 to 3 m depth and corresponds to a horizon of Fe-rich sediment. This zone also correlates with a high-amplitude reflection in the high-resolution seismic data (Fig. 2b), and strengthens stratigraphic correlations among cores separated by as much as 40 km. The highly continuous nature of reflections within the Songwe Sequence in this region indicate a relatively uncomplicated sedimentary environment, undisturbed by turbidites which are common in the deepest basins of the lake, and without apparent influence of contour or seiche-driven currents as seen in southern Lake Malawi (Johnson and Ng'ang'a, 1990; Johnson and Davis, 1989; Scott, 1988) or southern Lake Tanganyika (Tiercelin et al., 1988). Thus the sediments from this region are ideal for pre-Holocene paleoenvironmental analysis.

CHRONOLOGY

Stratigraphically consistent correlations among cores, based on sedimentology, carbonate content, diatom assemblage, biogenic silica content and bulk chemistry, have been used to subdivide the cores into 6 units (Fig. 3). Radiocarbon dating of both authigenic calcite and organic matter has been used to establish a chronologic framework for the cores. These results (Table 1) indicate that the lower sections of the cores are near the limit of the accelerator mass spectrometry technique (AMS) and therefore significant errors can result from minor contamination by modern carbon. Age estimates between radiocarbon dated intervals have been interpolated assuming constant sedimentation rates between the dated intervals. For core M86 27P, mean ages for the 170 cm (33,260 yr BP) and 362 cm (36,000 yr BP) depths were used. In addition, a core top age of 10,000 yr BP is assumed as the core was lacking Unit 6, the well characterized Holocene section (Finney and Johnson, 1991).

In core M86 18P the carbonate dates (22,050 and 28,110 yr BP) in the upper calcite-rich horizon are somewhat younger than the organic carbon date from the same horizon (33,250 yr BP) and are not stratigraphically consistent with each other. The 33,250 yr BP date is nearly identical with the mean radiocarbon age of the correlated interval in M86 27P (170 cm; 33,260 yr BP) which shows good agreement between three dates from carbonate and organic samples (Table 1). Contamination by only 1.5 to 5% modern carbon would alter the radiocarbon age of a 33,000 year old sample to 28,000 to 22,000 yr BP. However, contamination by 46 to 74% radiocarbon-dead carbon would be required if the sample dated at 33,250 yr BP had a true age between 22,000–28,000 yr BP. Therefore, we use the older date because it seems more likely that some post-recovery recrystallization of calcite contaminated the carbonate dates. This implies only minor calcite neoformation; if 5% of the calcite in sediments containing 5% calcite is modern, this would amount to 0.25% of the bulk sediment. There is excellent agreement (within 1,000 yr) in age estimates for unit boundaries based on each core's independent timescale.

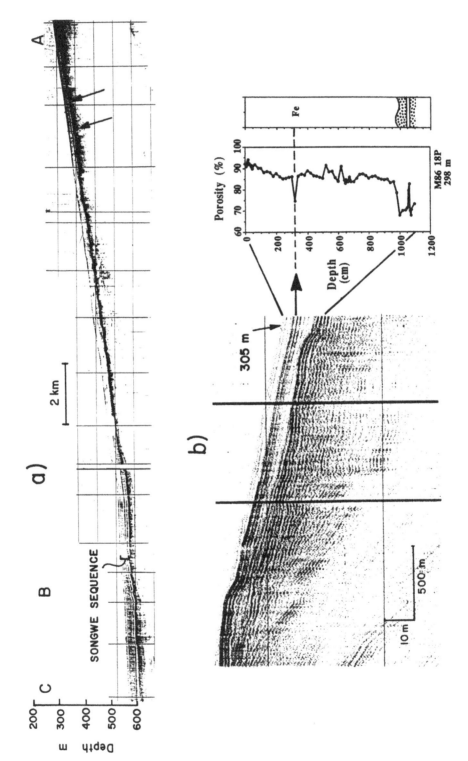

Figure 2. A) 1 kHz high-resolution seismic profile along transect A–B–C. Note condensed sedimentary section near A. The arrows point to truncated reflectors under the basal Songwe unconformity; B) detail of high-resolution seismic profile transect at core site M86 18P and correlation of high amplitude reflections to porosity and lithology.

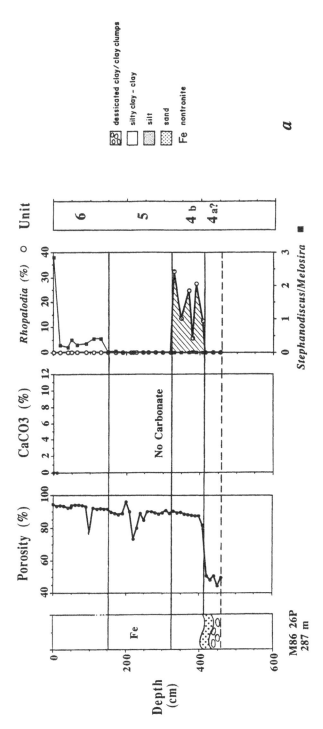

Figure 3. Downcore changes in lithology, porosity, calcium carbonate and diatom assemblage in cores a, M86 26P; b, M86 18P; and c, M86 27P. Two changes in diatom assemblage are shown. Enhanced abundance of periphytic diatom *Rhopalodia* (hatchured area) indicates lower lake-levels. The ratio of planktonic diatoms *Stephanodiscus* to *Melosira* is useful for lake-wide stratigraphic correlations (Finney and Johnson, 1991); relatively higher ratios are observed in Units 2 and 6. Details on radiocarbon dating are listed in Table 1.

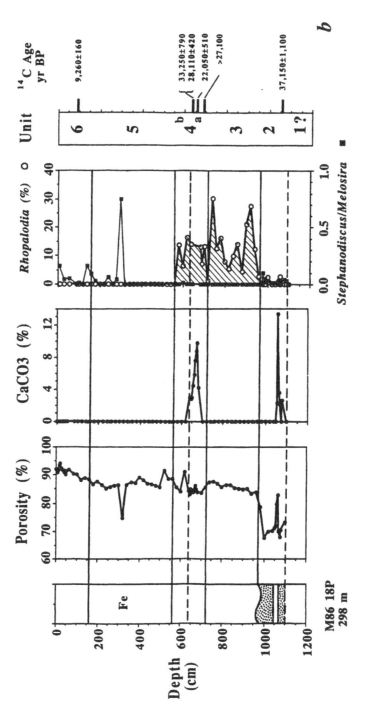

Figure 3, cont'd

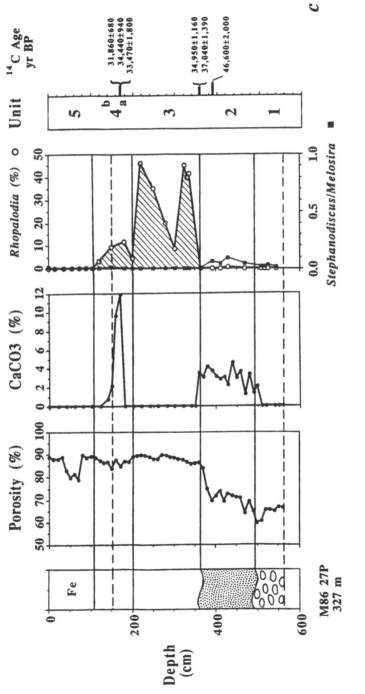

Figure 3, cont'd

Table 1. AMS Radiocarbon Ages of Lake Malawi Piston Core Samples

Core	Depth (cm)	Material[a]	Age	Lab. no.[b]
M86 18P	85–105	Organic carbon[a]	9,260 ± 160	GU2626
	637–638	Organic carbon	33,250 ± 790	LL
	637–638	Carbonate	28,110 ± 420	LL
	640–680	Carbonate[a]	22,050 ± 510	B33164
	690–700	Organic carbon[a]	>27,100	GU2627
	1060–1061	Carbonate	37,150 ± 1,100	AA5589
M86 27P	168–170	Carbonate	34,440 ± 940	LL1756
	168–170	Organic carbon	31,860 ± 680	LL1978
	170–171	Carbonate	33,470 ± 1,800	LL
	361–363	Carbonate	34,950 ± 1,160	LL1757
	361–363	Organic carbon	37,040 ± 1,390	LL1977
	390–391	Carbonate	46,600 ± 2,000	LL

[a]Conventional radiocarbon date. [b]GU = Scottish Universities Radiocarbon Dating Laboratory, LL = Lawrence Livermore National Laboratory, B = Beta Analytic Inc., AA = University of Arizona NSF Accelerator Facility.

PALEOENVIRONMENTAL RECONSTRUCTION

The oldest sediments recovered (Unit 1) consist of silty muds lacking carbonate, with a diatom assemblage dominated by *Melosira* (Fig. 3). This genus is an indicator of water with low salinity and alkalinity, as in the present-day lake (Owen et al., 1990; Haberyan, 1988; Hecky and Kling, 1987; Gasse, 1896; Gasse et al., 1983). The characteristics of this unit, undated but older than 47,000 yr BP, suggest a moderate to deep lake. The dried-out desiccated nature of this unit suggest it was subaerially exposed during a subsequent regression, the maximum extent of which may be represented by the erosional surface at the base of Unit 2 in core M86 27P.

Units 2 through 4 range from more than 40,000 to about 28,000 yr BP and show strong evidence for lake-level much lower than present. Unit 2 (>36,000 yr BP), a silt to clayey silt, contains micritic calcite authigenically precipitated from surface waters (Finney and Johnson, 1991). This indicates higher salinity than present, closed-basin conditions, and hence relatively lower lake-level (Finney and Johnson, 1991; Kelts and Hsu, 1978). The dominance of silts suggests a nearshore environment; there is no evidence from either core or seismic data for turbidites. The diatom assemblage is dominated by *Melosira* with trace to common benthic diatoms and planktonic *Stephanodiscus*. These characteristics are not unlike those found today just offshore of shallow, coarse grained facies (Haberyan, 1988). The presence of calcite and the erosional surface at the top of Unit 2 suggest falling lake levels during the later stages of this unit.

Unit 3 (ca. 36,000 to 34,000 yr BP) is an organic-rich, silty clay. The abundance of *Rhopalodia* range from less than 5% to more than 40%, with relatively lower values observed in the middle part of this unit. The intervals with high abundance of this periphytic diatom, many of which are intact, indicate deposition within the euphotic zone (Gasse, 1986; Haberyan, 1988; Haberyan and Hecky, 1987; Gasse et al., 1989), and hence water-levels as much as 250 m below present. Low *Rhopalodia* abundance in the middle portion of Unit 3 suggest relatively higher lake-levels, centered at about 35,000 yr BP. While the low lake-levels indicated by abundant *Rhopalodia* imply closed-basin conditions, carbonate was only detected in all one sample from this interval. Although carbonate precipitation would be expected during such conditions (Finney and Johnson, 1991), its absence in sediments from a small region may simply reflect the complexities of carbonate preservation in a large, deep lake. For example, the lack of carbonates may indicate post-depositional dissolution within the organic-rich substrate (Emerson and Bender, 1981), or a shift in the foci of carbonate precipitation to river mouths as in present day Lake Kivu (Haberyan and Hecky, 1987; Hecky, 1978).

An abrupt decrease in *Rhopalodia* at the transition between Units 3 and 4 coincides with a distinct drop in diatom abundance and preservation. Unit 4 (ca. 34,000 to 28,000 yr BP) is characterized by silty clays with few diatoms, many of which are *Rhopalodia*, indicating the core sites were within the euphotic zone. A spike of micritic calcite in Unit 4a is consistent with closed-basin sedimentation, and probably a period of negative water balance. An ^{18}O value of +2.7 on calcite from this unit in core M86 18P is further evidence for much higher salinities than present (Stuiver, 1970; Talbot, 1990), as carbonates precipitating in equilibrium from the modern lake should have ^{18}O values of about −0.5 (Ricketts and Johnson, this volume). Nearshore sands in the shallowest core M86 26P, which overly muds that appear to have been desiccated after deposition, may correlate with Unit 4a. The difference in water depth between this core and the deepest core is about 40 m, roughly the depth of the euphotic zone. Thus if benthic diatoms lived in relatively high abundance at the deeper site, the shallower core site would have been a very shallow water or subaerial environment during and prior to the time of Unit 4a. This is consistent with the stratigraphy of core M86 26P (nearshore sands overlying desiccated lacustrine mud). This interpretation indicates a lake-level about 290 m below present during Unit 4a. Unit 4b, a *Rhopalodia*-rich, calcite-poor silty clay, represents the shallow water phase of the subsequent transgression. At this time, water depths were sufficient for sedimentation to resume at core site M86 26P.

Unit 5 (ca. 28,000 to 10,000 yr BP) is a carbonate-free clay dominated by *Melosira*. Such sediments, similar to those currently deposited at these sites, indicate relatively deep water and suggest that open-basin conditions were established quickly. Unit 5 can be correlated with several cores in other regions of the lake. The Holocene section (Unit 6) consists of silty clays that are *Melosira*-rich and have variable *Stephanodiscus* abundance. Carbonate is not detected in the cores discussed here, but was preserved in many shallower cores during the early Holocene (Finney and Johnson, 1991). Evidence for lake-level variations during this period are more apparent in cores with faster sedimentation rates from shallower water. Our previous work (Finney and Johnson, 1991) suggests lake-levels 100 to 150 m below present during the early Holocene (6000–10,000 yr BP), followed by generally open-basin conditions. Lake level dropped briefly to about 100 m below present within the past five centuries, and again between 1150 and 1250 A.D. (Owen et al., 1990).

The pronounced unconformity seen in these and other seismic data (Scholz and Rosendahl, 1988; Johnson and Davis, 1989; Scholz and Finney, 1994) (Fig. 2a) is correlated with the porosity drop near the base of the cores (Figs. 2b and 3). The age of the upper surface of the unconformity is probably time transgressive over the basin; it correlates with the boundary between Units 2 and 3 in cores M86 18P and 27P and possibly to the Unit 4a–4b boundary in shallower core 26P. Thus it developed prior to ~30,000 yr BP in this region. The older age limit for this low lake-level stage is uncertain, but probably greater than 78,000 yr BP (Scholz and Finney, 1994). The angular discordance at this boundary suggests a significant amount of missing section in some localities.

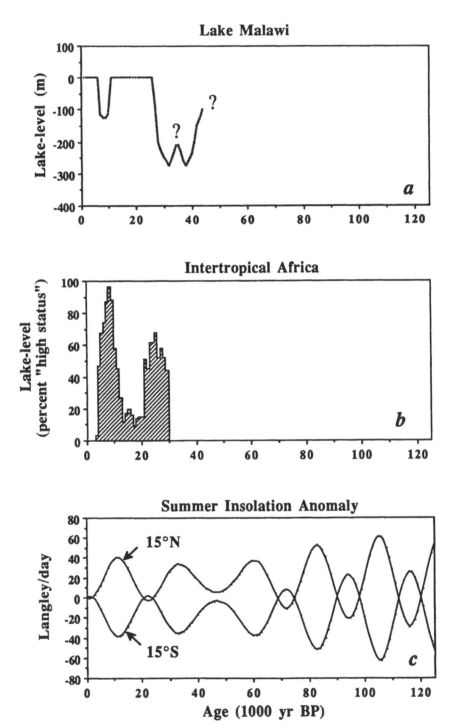

Figure 4. A) Estimated lake-level variations (m below present) for Lake Malawi from this work and Finney and Johnson (1991); B) percentage of basins in intertropical Africa with high lake-level status (modified from Street and Grove (1979); data primarily from northern hemisphere basins); and C) southern hemisphere (15°S) and northern hemisphere (15°N) summer insolation anomalies (Berger, 1978).

PALEOCLIMATIC IMPLICATIONS

These severe and relatively rapid lake-level variations are most probably controlled by changes in the precipitation/evaporation balance. A hydrologic model (Owen et al., 1990) demonstrates the lake's sensitivity to precipitation changes; a decrease in precipitation by 30% relative to present results in a 300 m lake-level drop within about 400 years. Similarly, a return to present day precipitation refills the basin in 300 years. Such changes in precipitation are reasonable in light of what is known from paleohydrologic estimates (Street-Perrott and Harrison, 1985; Haberyan and Hecky, 1987; Hastenrath and Kutzbach, 1983; Nicholson, 1989; Street, 1979) and GCM simulations (Kutzbach and Street-Perrott, 1985; COHMAP, 1988). The timing of the major lowstands in Lake Malawi is distinctly different from what has been reported for African lakes north of Malawi (Street-Perrott et al., 1989; Street-Perrott and Harrison, 1984; Butzer et al., 1972; Owen et al., 1982; Talbot et al., 1984; Gasse, 1977) (Fig. 4). In general, the northern lakes record high levels from ca. 40,000 until 20,000 yr BP, followed by an arid low-lake phase which lasted until about 12,000 yr BP. Higher levels were then reestablished until about 5000 yr BP followed by generally lower levels. Some paleoclimatic records from southern Africa (Deacon and Lancaster, 1988; Shaw and Cooke, 1986; Lancaster, 1989; Scott, 1989; Vogel, 1989) and Madagascar (Burney, 1987) indicate changes in water balance more similar to those reconstructed for Lake Malawi.

The Malawi lake-level variations are generally compatible with GCM simulations (COHMAP, 1988, Street-Perrott et al., 1990). Opposite responses of southern versus northern hemisphere precipitation are simulated by GCMs as a result of the effect of orbitally induced insolation variations on monsoonal circulation. The southern hemisphere response appears to be displaced significantly south (~9°) of the equator because the levels of Lake Tanganyika (Haberyan and Hecky, 1987; Gasse et al., 1989) and other lakes just to the north of Malawi (Street-Perrott et al., 1989) have varied similarly to African lakes north of the equator. The timing of the lake-level fluctuations in Lake Malawi appears to be similar to insolation variations, but the extent of lake-level change is not simply related to the intensity of the insolation anomaly (Clemens et al., 1991). If insolation variations are a major force driving changes in the water balance of Lake Malawi, then a period of severe fluctuations is suggested by insolation changes prior to about 70,000 yr BP (Fig. 4). Such fluctuations may have produced the erosional surfaces seen in the seismic records that predate the period recovered by the piston cores. Today, the ~23,000 year cycle in solar radiation is at a period of minimum contrast between northern and southern hemisphere summer insolation. Our results suggest that strong paleoclimatic gradients occurred in this region at times when the hemispheric summer insolation contrast was greater.

ACKNOWLEDGMENTS

We thank the governments of Malawi, Tanzania and Mozambique for logistical assistance and research permits. Thanks to Melissa Wenrich, Anne Hewes, Kate Green, Tom Davis, John Graves and Patrick Ng'ang'a for field and laboratory assistance. This work was supported by grants from NSF (ATM-9106558) and Texaco.

REFERENCES

Berger, A., 1978, Long-term variations of caloric insolation resulting from the Earth's orbital elements: Quaternary Research, v. 9, pp. 139–167.

Burney, D. A., 1987, Pre-settlement vegetation changes at Lake Tritrivakely, Madagascar: Paleoecology of Africa, v. 18, pp. 357–381.

Butzer, K. W., Isaac, G. L., Richardson, J. L., and Washbourn-Kamu, C., 1972, Radiocarbon dating of East African lake levels: Science, v. 175, pp. 1069–1076.

Clemens, S., Prell, W., Murray, D., Shimmield, G., and Weedon, G., 1991, Forcing mechanisms of the Indian Ocean monsoon: Nature, v. 353, pp. 720–725.

COHMAP members, 1988, Climatic changes of the last 18,000 years: observations and model simulations: Science, v. 241, pp. 1043–1052.

Deacon, J., and Lancaster, N., 1988, Late Quaternary Palaeoenvironments of Southern Africa: Oxford, England, Oxford University Press, 225 pp.

Eccles, D. H., 1974, An outline of the physical limnology of Lake Malawi (Lake Nyasa): Limnology and Oceanography, v.19, pp. 730–742.

Emerson, S., and Bender, M., 1981, Carbon fluxes at the sediment–water interface of the deep-sea: calcium carbonate preservation: Journal of Marine Research, v. 39, pp. 139–162.

Finney, B. P., and Johnson, T. C., 1991, Sedimentation in Lake Malawi (East Africa) during the past 10,000 years: a continuous paleoclimatic record from the southern tropics: Palaeogeography, Palaeoclimatology, Palaeoecology, v. 85, pp. 351–366.

Flannery, J. W., 1988, The acoustic stratigraphy of Lake Malawi, East Africa [M.S. thesis]: Durham, North Carolina, Duke University, 117 p..

Gasse, F., 1977, Evolution of Lake Abhé (Ethiopia and TFAI), from 70,000 b.p.: Nature, v. 265, pp. 42–45.

Gasse, F., 1986, East African diatoms: taxonomy, ecological distribution: Bibliotheca Diatomologica, v. 11, pp. 1–201.

Gasse, F., Talling, J. F., and Kilham, P., 1983, Diatom assemblages in East Africa: classification, distribution and ecology: Revue Hydrobiologie Tropical, v. 16, pp. 3–34.

Gasse, F., Lédéé, V., Massault, M., and Fontes, J. C., 1989, Water-level fluctuations of Lake Tanganyika in phase with oceanic changes during the last glaciation and deglaciation: Nature, v. 342, pp. 57–59.

Gonfiantini, R., Zuppi, G. M., Eccles, D. H., and Ferro, W., 1979, Isotope investigation of Lake Malawi, in Isotopes in Lake Studies: International Atomic Energy Agency, Vienna, pp. 195–207.

Haberyan, K. A., 1988, Phycology, sedimentology, and paleolimnology near Cape Maclear, Lake Malawi, Africa [Ph.D. thesis]: Durham, North Carolina, Duke University, 246 p..

Haberyan, K. A., and Hecky, R. E., 1987, The late Pleistocene and Holocene stratigraphy and paleolimnology of lakes Kivu and Tanganyika: Palaeogeography, Palaeoclimatology, Palaeoecology, v. 61, pp. 169–197.

Hastenrath, S., and Kutzbach, J. E., 1983, Paleoclimatic estimates from water and energy budgets of East African lakes: Quaternary Research, v. 19, pp. 141–153.

Hecky, R. E., 1978, The Kivu-Tanganyika basin: the last 14,000 years: Polskie Archiwum Hydrobiologii, v. 25, pp. 159–165.

Hecky, R. E., and Kling, H. J., 1987, Phytoplankton ecology of the great lakes in the rift valleys of central Africa: Ergebnisse Limnologie, v. 25, pp. 197–228.

Herdendorf, C. E., 1982, Large lakes of the world: Journal of Great Lakes Research, v. 8, pp. 379–412.

Johnson, T. C., and Davis, T. W., 1989, High resolution seismic profiles from Lake Malawi, Africa: Journal of African Earth Sciences, v. 8, pp. 383–392.

Johnson, T. C., and Ng'ang'a, P., 1990, Reflections on a rift lake, *in* Katz, B. J., ed., Lacustrine basin exploration — case studies and modern analogs: American Association of Petroleum Geologists Memoir, v. 50, pp. 113–135.

Kelts, K., and Hsu, K. J., 1978, Freshwater carbonate sedimentation, *in* Lerman, A., ed., Lakes: Geology, Chemistry, Physics: New York, Springer-Verlag, pp. 295–323.

Kutzbach, J. E., and Street-Perrott, F. A., 1985, Milankovitch forcing of fluctuations in the level of tropical lakes from 18 to 0 kyr B.P.: Nature, v. 317, pp. 130–134.

Lancaster, N., 1989, Late Quaternary paleoenvironments in the southwestern Kalahari: Palaeogeography, Palaeoclimatology, Palaeoecology, v. 70, pp. 367–376.

Nicholson, S. E., 1989, African drought: characteristics, causal theories and global teleconnections, *in* Berger, A., Dickinson, R. E., and Kidson, J. W., eds., Understanding Climate Change: American Geophysical Union Geophysical Monograph, v. 52, pp. 79–100.

Owen, R. B., Barthelme, J. W., Renaut, R. W., and Vincens, A., 1982, Palaeolimnology and archaeology of Holocene deposits northeast of Lake Turkana, Kenya: Nature, v. 298, pp. 523–528.

Owen, R. B., Crossley, R. Johnson, T. C., Tweddle, D. Kornfield, I., Davison, S., Eccles, D. H., and Engstrom, D. E., 1990, Major low levels of Lake Malawi and their implications for speciation rates in cichlid fishes: Proceedings of the Royal Society of London, v. B240, pp. 519–553.

Ricketts, R. D., and Johnson, T. C., 1996, Early Holocene changes in lake level and productivity in Lake Malawi as interpreted from oxygen and carbon isotopic measurements of authigenic carbonates, *in* Johnson, T. C., and Odada, E. O., eds., The Limnology, Climatology and Paleoclimatology of the East African Lakes: Toronto, Gordon and Breach, pp. 475–493.

Scholz, C. A., and Rosendahl, B. R., 1988, Low lake stands in Lakes Malawi and Tanganyika, East Africa, delineated with multifold seismic data: Science, v. 240, pp. 1645–1648.

Scholz, C. A., and Finney, B. P., 1994, Late Quaternary sequence stratigraphy of Lake Malawi (Nyasa), Africa: Sedimentology, v. 41, pp. 163–179.

Scott, D., 1988, Modern processes in a continental rift lake: an interpretation 28 khz seismic profiles from Lake Malawi, East Africa [M.S. thesis]: Durham, North Carolina, Duke University, 82 pp.

Scott, L., 1989, Climatic conditions in southern Africa since the last glacial maximum, inferred from pollen analysis: Palaeogeography, Palaeoclimatology, Palaeoecology, v. 70, pp. 345–353.

Shaw, P. A., and Cooke, H. J., 1986, Geomorphic evidence for the late Quaternary palaeoclimates of the middle Kalahari of northern Botswana: Catena, v. 13, pp. 349–359.

Street, F. A., 1979, Late Quaternary precipitation estimates for the Ziway-Shala Basin, southern Ethiopia: Palaeoecology of Africa, v. 11, pp. 135–143.

Street, F. A., and Grove, A. T., 1979, Global maps of lake-level fluctuations since 30,000 yr B.P.: Quaternary Research, v. 12, pp. 83–118.

Street-Perrott, F. A., and Harrison, S. P., 1984, Temporal variations in lake levels since 30,000 yr BP — an index of the global hydrologic cycle, *in* Hansen, J. E., and Takahashi, T., eds., Climate Processes and Climate Sensitivity: American Geophysical Union Geophysical Monograph 29, pp. 118–129.

Street-Perrott, F. A., and Harrison, S. P., 1985, Lake levels and climate reconstruction, *in* Hecht, A. D., ed., Paleoclimatic Analysis and Modeling: New York, John Wiley and Sons, pp. 291–340.

Street-Perrott, F. A., Marchand, D. S., Roberts, N., and Harrison, S. P., 1989, Global lake-level variations from 18,000 to 0 years age: United States Department of Energy, Technical Report TRO46, 213 pp.

Street-Perrott, F. A., Mitchell, J. F. B., Marchand, D. S., and Brunner, 1990, Milankovitch and albedo forcing of the tropical monsoons: a comparison of geological evidence and numerical simulations for 9,000 yr BP: Transactions of the Royal Society of Edinburgh, Earth Sciences, v. 81, pp. 407–427.

Stuiver, M. J., 1970, Oxygen and carbon isotopes of fresh-water carbonates as climatic indicators: Journal of Geophysical Research, v. 75, pp. 5247–5257.

Talbot, M. R., 1990, A review of the paleohydrological interpretation of carbon and oxygen isotopic ratios in primary lacustrine carbonates: Chemical Geology, v. 80, pp. 261–279.

Talbot, M. R., Livingstone, D. A., Palmer, P. G., Maley, J., Melack, J. M., Delebrias, G., and Gulliksen, S., 1984, Preliminary results from sediment cores from Lake Bosumtwi, Ghana: Palaeoecology of Africa, v. 16, pp. 173–192.

Tiercelin, J. J., Scholz, C. A., Mondeguer, A., Rosendahl, B. R., and Ravenne, C., 1989, Discontinuités sismiques et sédimentaires dans la série de remplissage du fossé du Tanganyika, rift Est-africain: Comptes Rendus de l'Académie des Sciences de Paris, v. 307, pp. 1599–1606.

Vogel, J. C., 1989, Evidence for past climatic change in the Namib desert: Palaeogeography, Palaeoclimatology, Palaeoecology, v. 70, pp. 355–366.

Structural Framework, Sequence Stratigraphy and Lake Level Variations in the Livingstone Basin (Northern Lake Malawi): First Results of a High-Resolution Reflection Seismic Study

M. DE BATIST and P. VAN RENSBERGEN   *Renard Centre of Marine Geology (RCMG), University of Ghent, Ghent, Belgium*

S. BACK   *Renard Centre of Marine Geology (RCMG), University of Ghent, Ghent, Belgium and Department of Geosciences, University of Potsdam, Potsdam, Germany*

J. KLERKX   *Royal Museum of Central Africa, Tervuren, Belgium*

Abstract — A first interpretation of a new grid of high-resolution reflection seismic profiles has revealed new insights in the sequence stratigraphic and structural framework of Livingstone Basin (northernmost Lake Malawi). Seven seismic sequences could be identified and were tentatively correlated with the Songwe and Mbamba Sequences of Pleistocene–Recent age. The acoustic substrate is interpreted to represent the top of the Nyasa/Baobab Sequence, correlative with the African Two Geomorphological Surface. The stacking pattern of the seven sequences represents a complete long-term lake level cycle, from a lake lowstand at appr. 320 m below the present level to the present-day lake highstand. Higher-frequency lake level oscillations controlled the succession of sedimentary facies within each of the sequences, which sometimes show classical passive-margin type sequence stratigraphic architectures. The observed structural features show evidence of a strongly variable tectonic behaviour. Some faults appear to be presently active and strongly influence the lake floor morphology, while the activity of others seems to have stopped a various stages throughout the Pleistocene. Large-scale tectonic movements, such as along the Livingstone Border Fault, have to a certain extent exerted an influence on the distribution of the sedimentary facies, as witnessed by the northward migration of the North-Kiwira delta lobes.

INTRODUCTION

The use of sequence stratigraphy concepts in the study of lacustrine deposits has aroused growing interest during the past years, mainly for its possible applications to paleoclimatological studies (e.g., Finney and Johnson, 1991) and to the development of exploration strategies for lacustrine reservoir rocks (e.g., Katz, 1990). Thanks to their long-term subsidence guaranteeing an expanded and complete stratigraphic record, large rift lakes are probably the best candidates to study the applicability of these concepts. Within the framework of the CASIMIR Project ("Comparative Analysis of Sedimentary Infill Mechanisms In Rifts") three large rift lakes have been selected for a comparative and interdisciplinary investigation: Lake Baikal in Siberia and Lake Tanganyika (Lezzar et al., in press) and Lake Malawi in the East African Rift System. This paper will present the results from a first field season on Lake Malawi.

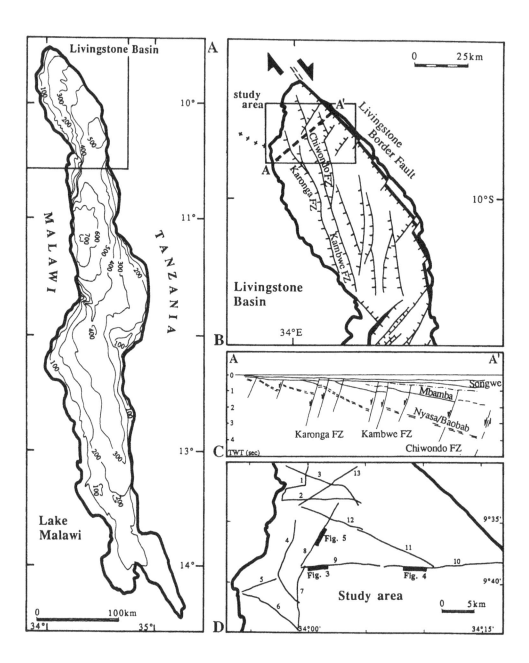

Figure 1. Location map. A) Bathymetry map of Lake Malawi. B) Structural map of Livingstone Basin (after Scholz, 1989) and indication of the study area. C) Interpreted line-drawing of PROBE multifold seismic profile (after Scholz, 1989) representing a typical cross-section through Livingstone Basin (for location see Fig. 1B). D) Localisation of the CASIMIR seismic profiles with indication of the position of Figs. 3, 4 and 5.

Lake Malawi (or Lake Nyasa) is more than 550 km long, has an average width of 50 km and a water depth of over 700 m (Fig. 1A). Many rivers, draining a total area of 65,000 km^2, are discharging into the lake, but only the Shire River accounts for the outflow. This shallow river is responsible for only 20% of the total water loss of the lake (an average level change of 0.45 m/yr); the rest (about 1.9 m/yr) is lost through evaporation (Pike, 1964). The lake is thus extremely sensitive to climate change and the lake level varies rapidly in response to only minor changes in rainfall, air temperature, etc., as documented by Johnson and Ng'ang'a (1990). The signal of these lake level fluctuations is thought to be recorded in the stratal pattern of the lacustrine sedimentary sequences, which can be determined in detail by high-resolution reflection seismic profiling.

Previous seismic investigations on Lake Malawi, such as those carried out in 1986–87 by Duke University within the PROBE Project (Scholz, 1989), already yielded a large amount of deep multi-channel seismic profiles and shallow 1 kHz Geopulse and 3.5 kHz subbottom profiles covering the whole of Lake Malawi. The multifold seismic data resulted in a large-scale tectonic model for the Malawi Rift (Ebinger et al., 1984; Rosendahl, 1987) and in a general stratigraphic description of its lacustrine infill (Scholz, 1989), and the high-resolution profiles added to the study of the recent sedimentation processes in the lake (Scott et al., 1991).

The CASIMIR high-resolution reflection seismic survey was organized on Lake Malawi in October 1992 and focused (Fig. 1D) on the delta area of the Songwe and North-Kiwira Rivers in the northern part of the Livingstone Basin (northern Lake Malawi). Thirteen seismic profiles (total length: 150 km) were acquired using a CENTIPEDE-sparker seismic source (frequency range: 150–1500 Hz, when operated at 300 J) and a single-channel streamer. Due to high-frequency noise the effective bandwidth had to be constrained to 200–800 Hz. The data were recorded on DAT-cassettes on board and subsequently digitized and processed at RCMG using the DELPH2 and PHOENIX VECTOR systems.

About 30% of the seismic grid bear little or no geological information due to the presence of gas in the surficial sediments preventing the penetration of the high-frequency acoustic waves (Fig. 6). Elsewhere, the seismic data yield a very good resolution (<1 m) and a penetration of generally more than 300 ms TWT. They allow testing of the applicability of sequence stratigraphic concepts, refining the stratigraphy of the upper layers (upper 200 m) of the sedimentary infill and adding information on the recent behaviour of the smaller tectonic features in the Livingstone Basin.

GEOLOGICAL SETTING

Lake Malawi extends over a large part of the Malawi Rift. This rift constitutes the southern segment of the Western Branch of the East African Rift System, which is connected to the northern segment through the Tanganyika–Rukwa–Malawi transcurrent fault zone (Tiercelin et al., 1988). It extends over 700 km from the Rungwe volcanic province in the North to the Urema Graben in the South and consists of a series of generally north–south trending half grabens of alternating polarity, which are typically 150 km long and 40 km wide (Ebinger et al., 1984; Specht and Rosendahl, 1989). Livingstone Basin is the northernmost and probably oldest of these half-graben basins presently covered by the lake (Fig. 1). Here, the rift shoulders rise more than 2000 m above the present lake level and the basin accommodates more than 4 km of sediment. Towards the South, the elevation of the rift shoulders decreases as does the depth of the sublacustrine basement. Sediment infill in the

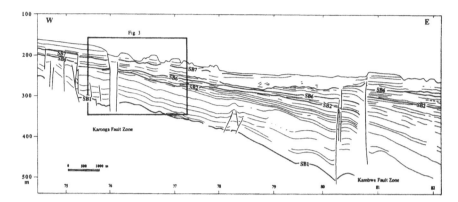

Figure 2. Interpreted line-drawing of profiles 9 and 10 representing a W–E cross-section through Livingstone Basin. Indication of the position of Figs. 3 and 4. Thick lines represent sequence boundaries, dotted lines represent transgressive surfaces, and dashed lines represent maximum flooding surfaces, as interpreted from the seismic sections.

southernmost basin — the Mwanjage-Mtakataka Basin — reaches a thickness of only about 800 m. It is generally believed that rifting was initiated at the northern end of the lake in Miocene times and that it migrated southwards through time.

Livingstone Basin is 50 km wide, 100 km long and has a maximum water depth of appr. 500 m (Fig. 1B). It is bordered by the steep, nearly rectilinear NW–SE Livingstone Border Fault (LBF) along the eastern bank of the lake and the gently sloping flexural margin at the western side. The LBF has a maximal vertical throw of more than 6 km and an important strike-slip component (Wheeler and Karson, 1989). The structural context of Livingstone Basin (Fig. 1B) was first established by Rosendahl (1987) and Scholz (1989) on basis of the PROBE multifold seismic data. They showed that the western flexural margin is marked by tilted blocks separated by N–S striking intra-basinal fault zones, which are synthetic to the main border fault. Transverse faults offset these intra-basinal faults and can reach far into the basin.

The stratigraphy of the lacustrine deposits in Livingstone Basin was also described by Scholz (1989), who applied the classical seismic stratigraphic methods to the PROBE data set. On basis of reflector termination and unconformity analysis, three regional depositional sequences were defined in Livingstone Basin: the Nyasa/Baobab Sequence, the Mbamba Sequence and the youngest Songwe Sequence (Fig. 1C). The Nyasa/Baobab Sequence overlies the bedrock and constitutes at least half of the sedimentary infill of the Livingstone Basin. The overlying Mbamba Sequence is separated from it by a high amplitude reflector, probably representing an abrupt lithological change or a major erosional surface, forming the top of uplifted blocks on the shoaling side of the basin. Both the Mbamba and Nyasa/Baobab Sequences were extensively eroded at the shoaling margins during a low-stand of the lake level about 78,000 yr. BP, when it fell appr. 350 m below its present level (Scholz and Finney, 1994). The Songwe Sequence, deposited during and after this major lowstand, is the youngest sequence observed in Lake Malawi. Off the eastern bank of the lake, the older Mbamba Sequence was completely eroded and the Songwe Sequence is directly overlying the Nyasa/Baobab strata at that location.

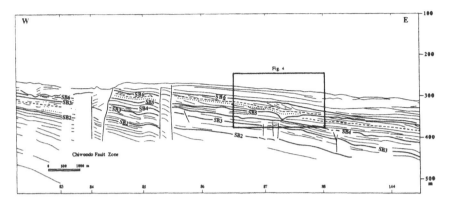

Figure 2, cont'd

DATA AND INTERPRETATION

Structure of Northern Livingstone Basin

The CASIMIR data in Livingstone Basin confirm its half-graben structure (Fig. 2). Tilted fault blocks are apparent on the shoaling flexural margin. Due to the large spacing of the seismic lines, the strike of different faults could not be mapped unambiguously, but the main faults seem to correspond with the Karonga, Kambwe and Chiwondo fault zones (Fig. 6), characterized by a N–S trend and by a typical 5–10 km spacing. The Karonga fault zone splits into two branches, which appear to continue onshore as the Songwe river and Mbaka river structural lineaments (Fig. 6).

In the vicinity of the Songwe delta, the flexural margin is characterized by an intensely faulted structural high, where the acoustic basement rises up to a depth of less than 100 ms TWT (appr. 80 m) beneath the lake floor (Fig. 2). Towards the East it rapidly dips below the penetration depth of the seismic signal (400 ms TWT or appr. 350 m). The top of the structural high is an erosion surface covered by appr. 40 m of relatively recent sediments. The feature probably corresponds to the crest of the tilted fault block between the Karonga and Kambwe fault zones.

Some of the observed faults, like those affecting the structural high off the Songwe River delta, seem to have been inactive for a relatively long period, while others show evidence of important recent displacements, which are even apparent at the lake floor. Many of the faults in the deeper parts of the basin coincide with areas of shallow gas accumulation (Fig. 3), which could indicate a hydrothermal origin. Similar observations have been reported from Lake Tanganyika (Tanganydro Group, 1992). The gas in the western coastal area is most probably mainly biogenic in origin as it is associated with the present-day deltas of the main rivers.

Subsidence along the LBF is responsible for reflector divergence to the East, which increases with depth but not in a constant way (Fig. 2). Abrupt changes in divergence between different reflector sets, which do not seem to punctuate or correspond to real unconformities, indicate the variable activity of the border fault through time.

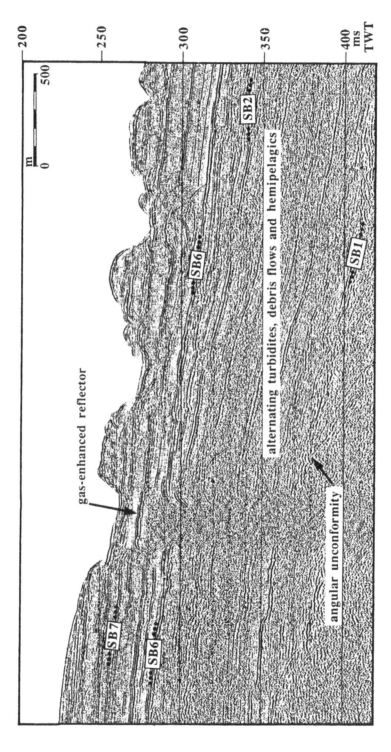

Figure 3. Part of seismic profile 10, showing aspects of sequences S4, S5 and S6. For location see Fig. 1D, for interpretation see Fig. 2.

Sequence Stratigraphy of Northern Livingstone Basin

On the CASIMIR high-resolution profiles, and especially on the west–east profile across Livingstone Basin (Fig. 2), 7 seismic sequences — sensu Van Wagoner et al. (1988) — were identified on basis of reflector termination patterns and were studied in detail. They were named S1 to S7 from old to young. Their thicknesses range from 0–10 m along the western margin to almost 100 m in the deeper part of the basin, adjacent to the LBF.

Due to the lack of boreholes or cores, there is no exact information on the age of the strata. The overall characteristics of the major sequence boundary SB6 (base of sequence S6), i.e., a distinct angular unconformity and an abrupt change in seismic facies, suggest that it might correspond to the base of the Songwe Sequence, as defined on deep PROBE data (Scholz, 1989). On basis of present-day sedimentation rates and thickness calculations, Scholz and Finney (1994) estimated the age of this boundary to be about 78,000 yr. BP. Through correlation with volcano-tectonic features onshore, Ebinger et al. (1993) propose a minimum age of 120,000 yr. BP for the base of the Songwe Sequence. Sequences 1 to 5 probably form the Mbamba Sequence and the acoustic substrate on the CASIMIR profiles (Fig. 2) may very well correspond to the top boundary of the Nyasa/Baobab Sequence. This boundary is probably correlative with the African Two Geomorphological Surface of King (1963), which was dated by Ebinger et al. (1993) to have an age of about 1.8 Ma.

Sequence S1 overlies the acoustic basement (Fig. 3). The reflector pattern is parallel at the western bank and becomes more divergent towards the East (Fig. 2). The base of the sequence is characterized by widely spaced parallel reflectors interbedded with reflection-free layers, interpreted as alternations of hemipelagic deposits, turbidites and debris flows (Fig. 3). Comparable facies have been described by Scholz and Rosendahl (1990). Towards the top of the sequence the facies changes into a pattern of closely spaced, high-amplitude reflections.

The major onlap surface SB2 and the distinct unconformity with the underlying strata mark the uplift of the structural high above lake level and a decrease of the subsidence along the LBF. The combined effect of these phenomena coincided with a major regressive phase with a strong basinward shift of the shallow-water depositional facies (sequences S2 and S3). Sequence S2 includes an incised valley fill on the upper slope of the flexural margin (appr. 250 m below present lake level) and a complete depositional sequence further downslope, which appears to be organized in backstepping transgressive deposits, overlain by downlapping progradational and aggradational highstand deposits (Fig. 2).

Sequence S4 witnesses the start of a major transgressive phase, related to a gradual rise of the lake level. In the deeper part of the basin (Fig. 4), it starts with a rather complexly organized lowstand unit, not unlike the subaqueous talus deposits described by Scholz and Rosendahl (1990). Towards the top it gradually develops into a typical prograding highstand deltaic complex (Fig. 4).

Sequence S5 displays the complete succession of prograding lowstand strata infilling incised depressions (appr. 320 m below present lake level), transgressive deposits and strongly prograding highstand deposits, separated by the transgressive and maximum flooding surfaces respectively (Fig. 4).

Sequence boundary SB6 separates shallow-water deposits of sequence S5 from deep-water sediments of sequence S6 (Fig. 4). The abrupt change in seismic facies reflects a sudden lake level rise, coinciding with but apparently not caused by a renewed subsidence along the LBF and reactivation of the Kwambe and Chiwondo faults, with displacements of over 30 m. As a result of this sudden transgression, fault-controlled channels at the

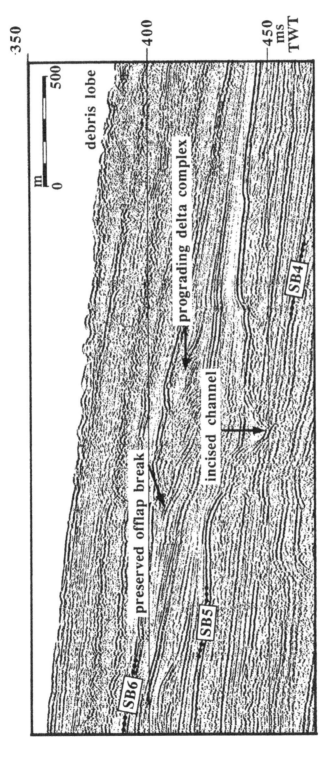

Figure 4. Part of seismic profile 9, showing aspects of sequences S1, S2, S6 and S7. For location see Fig. 1D, for interpretation see Fig. 2.

downthrown sides are filled in and the uplifted block on the western margin becomes flooded again at a present-day water depth of appr. 200 m (Figs. 2 and 3).

Two buried delta systems, the North-Kiwira and Songwe River delta systems, can be identified within sequence S6 along the western bank (Fig. 5). The latter was also studied by Scholz et al. (1993). The deltas are composed of a number of stacked lobes and characterized by complex packages of channel-fill deposits and of prograding and aggrading reflector sets, displaying the typical topset–foreset–bottomset facies (Fig. 5). Both systems are separated by the Karonga fault zone. The delta lobes are overlain by migrating channel-fill (delta plain?) deposits, which themselves are cut by a sharp erosive surface, probably representing a storm- or wave-base surface (Fig. 5). This surface corresponds to SB7 and defines the base of the present-day river deltas. The low amount of time-equivalent deposits in the deeper parts of the basin (Fig. 2) shows the backstepping effect associated with the rapid rise of the lake level throughout sequence S6 and into sequence S7.

DISCUSSION

The stratigraphic architecture of Livingstone Basin is strongly influenced both by tectonic activity and by variations in lake level. Subsidence in the half-graben system translates in reflector divergence towards the main border fault (Fig. 2). Changes in subsidence rate seem to punctuate discontinuities of only local significance. This observation leads to the conclusion that tectonics do not appear to be the major factor creating basin-wide unconformities — sensu Van Wagoner et al. (1988) — within the lacustrine strata. This was already suggested by Scholz and Rosendahl (1990). On the other hand, the CASIMIR seismic data have allowed the identification of a complex history of lake level fluctuations, with different orders of magnitude and frequency, through reflector pattern analysis.

In the stratigraphic section that was studied a complete long-term lake level cycle could be determined. In this cycle, sequences S2 and S3 would represent deposits formed during long periods of low lake levels and sequences S4 and S5 the early transgression. Sequence S6 — the Songwe Sequence of Scholz (1989) with an estimated age of about 78,000 yr. BP — would result from the rapid and massive rise of the lake level and sequence S7 would represent the present-day highstand situation. Within each of these sequences, higher-frequency oscillations controlled the distribution of lowstand, transgressive and highstand deposits. At present, quantification of the frequency of the different order cycles is not possible due to the lack of borehole information. A number of arguments are, however, at hand to quantify the amplitude of the lake level fluctuations.

Incised valleys in the stratigraphic section, such as the one associated with SB2 (Fig. 2), can be and have been used as quantitative indicators for past lake level lowstands (Gasse, 1989; Scholz and Rosendahl, 1988). Another and potentially more reliable lake level indicator can be derived — if seismic resolution allows to — from preserved toplap horizons and associated offlap breaks, as they represent fossil wave base levels. Sequence S5 (Fig. 4) offers a good example of a buried offlap break, indicating an original lake level that was at least 320 m below the present level. Comparable amplitudes have been reported by Scholz and Rosendahl (1988; 1990).

Also sequence S6 (Fig. 5) contains a number of fully preserved offlap breaks, within the two deltaic systems at the mouth of the North-Kiwira and Songwe Rivers. The observed shifting of these lobes (Fig. 6) illustrates a delicate interplay between gradual lake level rise, tectonic movement and sediment input. The southern Songwe delta system seems to have

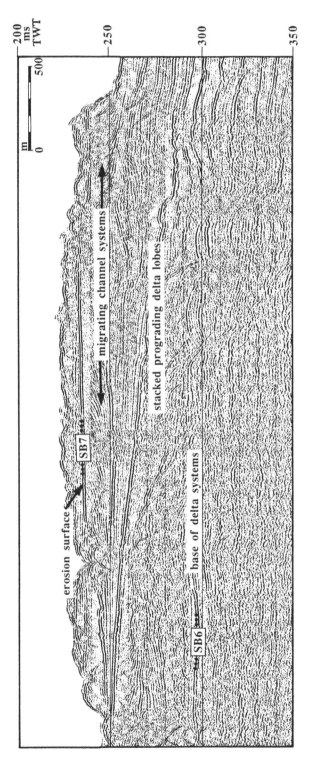

Figure 5. Part of seismic profile 8, showing aspects of sequences S6 and S7. For location see Fig. 1D.

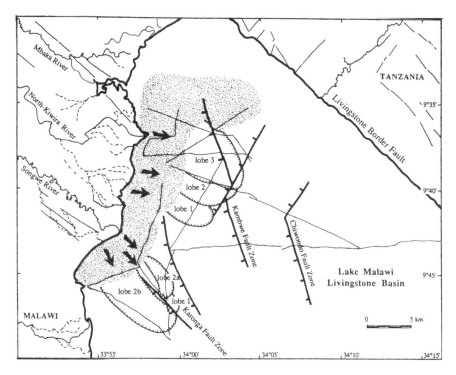

Figure 6. Geological interpretation map, showing the main faults of Scholz (1989) where they were identified on the CASIMIR profiles, the position of the North-Kiwira and Songwe delta lobes, and — onshore (after Delvaux and Hanon, 1993) — past and present river courses and structural lineaments interpreted from field and satellite data. The stippled pattern represents areas without seismic penetration due to gas blanking.

developed essentially in response to lake level rise and sediment input in a tectonically stable setting. The shifting of the lobes is therefore probably autocyclic. The North-Kiwira delta system on the other hand, gradually shifted towards the Northeast through time (Fig. 6). This observation seems to correspond to the observed northward shift of the river courses on the northwestern bank of the lake, as interpreted from satellite images and from land observations (Delvaux and Hanon, 1993), and is related to the overall subsidence of the basin along the LBF. The switching of the delta lobes is in this case probably largely allocyclic or tectonic in origin.

CONCLUSION

The good quality and high resolution of the CASIMIR seismic data reveal a wealth of stratigraphic information, which should — when placed in an interdisciplinary context and backed-up with the necessary borehole information — contribute to a better knowledge of the processes, and their forcing factors and time-scales, controlling the sedimentation in rift lakes.

ACKNOWLEDGMENTS

The CASIMIR Project is funded by the Belgian Science Policy Office and benefits from a grant of the Belgian Ministry of Environment. The development of RCMG's high-resolution seismic system was supported by the Flemish Ministry of Education (Concerted Research Action "Marine Geology"). We thank the Tanzanian Government (UTAFITI) for permission to conduct the CASIMIR survey and M. Johansen and the crew of the Nyanja for their logistic support. We also acknowledge E. Van Heuverswyn, W. Versteeg and M. Fernandez-Alonso for the acquisition of the data and W. Versteeg and P. Vanhauwaert for the seismic processing. Comments and suggestions by D. Rea and C. Scholz improved the quality of the paper. MDB is senior research assistant of the National Fund for Scientific Research (NFWO) and PVR is funded by the Flemish Government through an IWONL-IWT grant.

REFERENCES

Delvaux, D., and Hanon, M., 1993, Neotectonics of the Mbyea area, SW Tanzania. Musée royal de l'Afrique centrale, Tervuren (Belg.), Département Géologie et Minéralogie, Rapport annuel 1991–1992. 87–97.

Ebinger, C., Crow, M.J., Rosendahl, B.R., Livingstone, D., and LeFournier, J., 1984, Structural evolution of Lake Malawi. Nature, 308, 627–629.

Ebinger, C.J., Deino, A.L., Tesha, A.L., Becker, T., and Ring, U., 1993, Tectonic controls on rift basin morphology: evolution of the northern Malawi (Nyasa) Rift. Journal of Geophysical Research, 98, 17821–17836.

Finney, B.P., and Johnson, T.C., 1991, Sedimentation in Lake Malawi (East Africa) during the past 10,000 years: a continuous paleoclimatic record from the southern tropics. Paleogeography, Paleoclimatology, Paleoecology, 85, 351–366.

Gasse, F., Lédée, V., Massault, M., and Fontes, J.-C., 1989, Water-level fluctuations of Lake Tanganyika in phase with oceanic changes during the last glaciation and deglaciation. Nature, 342, 57–59.

Johnson, T.C., and Ng'ang'a, P., 1990, Reflections on a Rift Lake. In: Katz, B.J. (Ed.). Lacustrine basin exploration. American Association of Petroleum Geologists Memoir, 50, 113–135.

Katz, B.J., 1990 (Ed.), Lacustrine basin exploration. American Association of Petroleum Geologists Memoir, 50, 340 pp.

King, L.C., 1963, South African Scenery. Oliver and Boyd, Edinburgh. 699 pp.

Lezzar, K.E., Tiercelin, J.J., De Batist, M., Cohen, A.S., Bandora, T., Van Rensbergen, P., Mifundu, W., and Klerkx, J., in press, New seismic stratigraphy and Late Tertiary history of the North Tanganyika Basin, East African Rift system, deduced from multifold and high-resolution seismic data and piston core evidence. Basin Research.

Pike, J.G., 1964, The hydrology of Lake Nyasa. Journal of the Institute of Water Engineering, 18, 542–564.

Rosendahl, B.R., 1987, Architecture of continental rifts with special reference to East Africa. Annual Review of Earth and Planetary Sciences, 15, 445–503.

Scholz, C., 1989, Seismic Atlas of Lake Malawi (Nyasa), East Africa. Project PROBE, Geophysical Atlas Series, Vol 2, 116 pp.

Scholz, C., Johnson, T.C., and McGill, J.W., 1993, Deltaic sedimentation in a rift valley lake: New seismic reflection data from Lake Malawi (Nyasa), East Africa. Geology, 21, 395–398.

Scholz, C.A., and Finney, B., 1994, Late Quaternary sequence stratigraphy of Lake Malawi (Nyasa), Africa. Sedimentology, 41, 163–179.

Scholz, C.A., and Rosendahl, B.R., 1988, Low lake stands in Lakes Malawi and Tanganyika, East Africa, delineated with multifold seismic data. Science, 240, 1645–1648.

Scholz, C.A., and Rosendahl, B.R., 1990, Coarse-Clastic Facies and Stratigraphic Sequence Models from Lakes Malawi and Tanganyika, East Africa. In: Katz, B.J. (Ed.). Lacustrine basin exploration. American Association of Petroleum Geologists Memoir, 50, 151–168.

Scott, D.L., Ng'ang'a, P., Johnson, T.C., and Rosendahl, B.R., 1991, High-resolution character of Lake Malawi (Nyasa), East Africa and its relationship to sedimentary processes. International Association of Sedimentologists Special Publication, 13, 129–145.

Specht, T.D., and Rosendahl, B.R., 1989, Architecture of the Lake Malawi Rift, East Africa. Journal of African Earth Sciences, 8, 355–382.

Tiercelin, J.J., Chorowicz, J., Bellon, H., Richert, J.P., Mwanbene, J.T., and Walgenitz, F., 1988, East African rift system — Offset, age, and tectonic significance of the Tanganyika–Rukwa–Malawi intracontinental transcurrent fault zone. Tectonophysics, 148, 241–252.

Tanganydro Group, 1992, Sublacustrine hydrothermal seeps in northern Lake Tanganyika, East African Rift: 1991 Tanganydro Expedition. Bull. Centres Rech. Explor.-Prod. Elf-Aquitaine, 16(1), 55–81.

Van Wagoner, J.C., Posamentier, H.W., Mitchum, R.M., Vail, P.R., Sarg, J.F., Loutit, T.S., and Hardenbol, J., 1988, An overview of fundamentals of sequence stratigraphy and key-definitions. In: Wilgus, C.K., Hastings, B.S., Kendal, C.G.St.C., Posamentier, H.W., Ross, C.A., and Van Wagoner, J.C. (Eds.) Sea-level changes — An integrated approach. Society of Economic Paleontology and Mineralogy, Special Publication, 42, 39–45.

Wheeler, W.H., and Karson, J.A., 1989, Structure and kinematics of the Livingstone Mountains border fault zone, Nyasa (Malawi) Rift, southwestern Tanzania. Journal of African Earth Sciences, 8, 393–413.

Recent Ecosystem Changes in Lake Victoria Reflected in Sedimentary Natural and Anthropogenic Organic Compounds

E. LIPIATOU   *Gray Freshwater Biological Institute, University of Minnesota, Navarre, Minnesota, United States*

R.E. HECKY   *Freshwater Institute, Department of Fisheries and Oceans, Winnipeg, Manitoba, Canada*

S.J. EISENREICH   *Gray Freshwater Biological Institute*

L. LOCKHART, D. MUIR and P. WILKINSON   *Freshwater Institute, Department of Fisheries and Oceans*

Abstract — Aliphatic and aromatic hydrocarbons and organochlorines in a recent sediment core (Site 103) of Lake Victoria (East Africa) support the hypothesis that the lake has increased productivity since 1900, as reflected in the cyanobacteria population increase. This is evident by the increase in the accumulation rate of algal n-alkanes and organic carbon. n-alkanes indicative of terrestrial higher plants show an increased or stable accumulation rate during the same period. Good preservation of organic matter and organic indicators in the sediment suggest anoxic conditions have existed for the last 200 years. Increased human population of the basin and subsequent deforestation is indicated by the increase of the PAH derived from low temperature combustion processes. The hydrocarbon composition suggests an absence of petroleum inputs in the lake. The accumulation rates of PAH and organochlorines, although three times higher than 1900, remain low and are typical of remote environments.

INTRODUCTION

Organic compounds preserved in sediments provide information about the biological and anthropogenic inputs to sediments and their depositional environment over long periods of depositional history (Reed, 1977; Wakeham, 1993; Meyers and Ishiwatari, 1993). Local environments determine the relative importance of different sources and the degree of preservation of organic matter. Anoxic conditions and high sedimentation rates decrease alteration of organic matter in sediments resulting in better preservation (Didyk et al., 1978; Talbot and Livingstone, 1989; Wakeham, 1990).

Lake Victoria's ecosystem, one of the African Great Lakes, has undergone dramatic changes during the last three decades, including rapid eutrophication, increased anoxia, and alteration of the overall ecology of the lake (Talling, 1965; Ogutu-Ohwayo and Hecky, 1991; Hecky, 1993).

In order to delineate the causes and evolution of these limnological events with time, we examined the sediment record of a ^{210}Pb-dated core from Lake Victoria using organic molecules as indicators of ecosystem change. Examples of organic compounds analyzed are aliphatic and aromatic hydrocarbons of natural and anthropogenic origin. The core was also analyzed for nutrients (P, N, C, Si) and phytoplankton species (diatoms, chlorophytes, cyanobacteria) (Hecky, 1993). Our objectives were to infer biogenic sources to the sedimen-

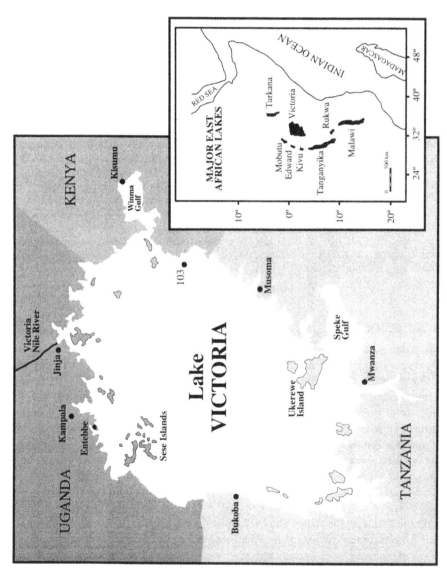

Figure 1. Map of Lake Victoria and of the major lakes of East Africa. Site 103 is the coring station sampled in 1990.

tary organic matter by using organic chemical fossils, and to determine the accumulation of anthropogenic chemicals in the lake. Organic compounds such as the aliphatic hydrocarbons n-C15 and n-C17 are derived from phytoplankton whereas n-C27, n-C29 and n-C31 are from terrestrial higher plants (Eglington and Hamilton, 1967; Han et al., 1968; Gelpi et al., 1970). Pesticides (DDT), polychlorinated biphenyls (PCBs), and polycyclic aromatic hydrocarbons (PAHs) are derived primarily from anthropogenic activities (use of fertilizers, combustion processes) (Laflamme and Hites, 1978; Hites and Eisenreich, 1987). Analysis of this core for the above compounds provides insight as to the preservation of organic matter in the sediments, comparison of autochthonous vs. allochthonous input, and information on the contaminant accumulation in the lake. Although the information obtained is limited to one core collected in the main depositional basin of the lake, these are the first data on organic chemicals in Lake Victoria and are some of the first organic geochemical data in the East African Lakes.

ANALYTICAL METHODS

Site Description and Sampling

Lake Victoria is located between Kenya, Uganda and Tanzania (~32°E, 0–5°S) at an altitude of 1137 m. Lake Victoria has a surface area of 68469 km^2 a mean depth of 40 m, and a maximum depth of 92 m (IDEAL, 1990). The sediments of the lake are rich in organic matter with an organic carbon content between 10 and 20% in the depositional basins and embayments but much less in the sandy sediments of the non-depositional areas (Mothersill, 1976; Kendall, 1969). A sediment core was collected in November 1990 in 55 m of water off western Kenya at site 103 (Ochumba and Kibara, 1989; Fig. 1).

Analysis of ^{210}Pb and ^{137}Cs

The core was analyzed for ^{210}Pb and ^{137}Cs activity to determine sediment chronology by a modification of the method reported in Eakins and Morrison (1978). The sediment was spiked with a known activity of internal standard (^{209}Po) and digested in concentrated HCl, and the Po isotopes plated onto silver planchettes. The daughter ^{210}Po activity was measured using conventional alpha spectroscopy to determine the total ^{210}Pb content. The ^{137}Cs activity was determined by gamma spectroscopy using Ge(Li) detection.

Extraction and Analysis

The sediment was stored frozen until lyophilized. The dry sediment (~15 g) was spiked with deuterated internal standards consisting of naphthalene, acenaphthalene, phenanthrene, chrysene, perylene. Lipid extraction was performed with dichloromethane (DCM) in a Soxhlet apparatus for approximately 16 hours. The extract was split in two equal parts. One part was used for analysis of non-aromatic hydrocarbons (NAHs) and PAHs, and the other for organochlorines pesticides and PCBs. The extract for hydrocarbon analysis was cleaned up by gel permeation chromatography and liquid–solid chromatography with 3% water-deactivated silica. Analyses were performed by high resolution glass capillary gas chromatography on a DB-5 column with electron capture detection. The hydrocarbon fraction was reduced to 2 ml in a rotary evaporator and 2 ml of hexane were added. The DCM/hexane was percolated through activated copper to remove sulfur. The extract was reduced to 1 ml, added to the top of a chromatographic column packed with BioBeads and eluted with 250 ml hexane:DCM (1:1). The eluate was evaporated to ~0.5 ml and added to the head of a

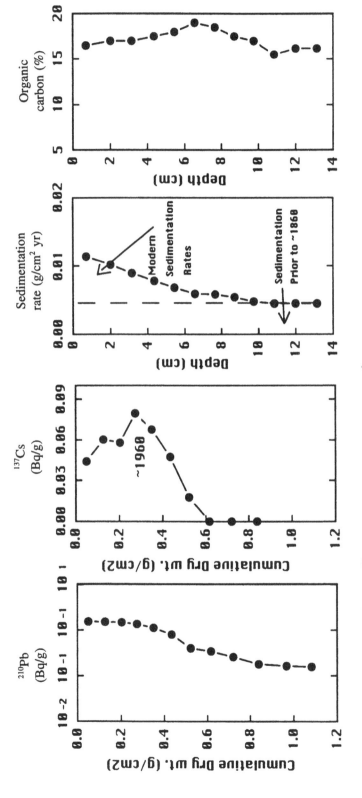

Figure 2. Profiles of ^{210}Pb and ^{137}Cs in Bq/g, sedimentation rate in g/cm^2 yr and organic carbon content (% of dry weight sediment) of Lake Victoria sediments in the site 103.

chromatographic column (1 cm × 15 cm) for the hydrocarbon fractionation. The column was packed with 11 g of silica (bottom) and 1 g of 5% deactivated alumina (top) and 1 cm anhydrous sodium sulfate on the top. The NAH fraction was eluted first with 23 ml of hexane and the PAH fraction second with 25 ml hexane:DCM (1:1). The NAH fraction was evaporated to 3 ml. The PAH fraction was evaporated to and exchanged with 1 ml of toluene. Both fractions were analyzed by gas chromatography–mass spectrometry (GC–MS). NAHs were analyzed in the total scan mode and the PAHs were analyzed in the selective ion monitoring mode. Quantification was performed by the isotope dilution method for PAHs or by external standards for NAHs (deuterated n-tetracosane).

The organochlorines analysis is described in detail elsewhere (Muir et al., 1993). In general, the extract was taken up in hexane and chromatographed on a Florisil column (1.2% deactivated with water). PCBs (hexane eluate) were separated from p,p'-DDT and other organochlorine pesticides (DCM:hexane, 15:85). Sulfur was removed using activated copper. PCBs and organochlorine pesticides were quantified using external standards (Muir et al., 1993). PCBs were confirmed by gas chromatography–mass spectrometry using an HP 5971MSD operated in the selected ion mode for P and P+2 masses of di- to deca-chloro-biphenyls.

RESULTS AND DISCUSSION

The sediment core was analyzed for ^{210}Pb and ^{137}Cs activity to determine sediment chronology for organic carbon and specific organic chemicals in order to quantitatively determine the recent environmental history of the lake.

Figure 2a,b shows the ^{210}Pb and ^{137}Cs profiles of core 103, the sediment accumulation rate and the organic carbon content. The sedimentation rate determined using ^{210}Pb and the CRS (Constant Rate of Supply; Oldfield et al., 1984) model (0.005 to 0.011 g/cm^2 yr) agreed well with the ^{137}Cs profile in that the ^{137}Cs peak occurred at ~1960 as determined by ^{210}Pb chronology. An apparent ^{210}Pb zone affected by mixing extended to ~5 cm, but chemical and radionuclide stratigraphy showed resolution in this depth range (shown later). This phenomenon may be attributed to focusing of fine sediment containing ^{210}Pb from non-depositional areas, and mixing processes operative on decadal time scales, but incomplete in shorter periods (Edgington and Robbins, 1990; Eisenreich et al., 1989). The sediment accumulation rate increased by a factor of two between 1830 (0.005 g/cm^2 yr) and recent years (0.011 g/cm^2 yr) (Fig. 2c). This increase, which indicates important changes in the lake's ecosystem probably results from increases in aquatic productivity of the lake in recent years and increased erosion of the watershed. The organic carbon content is high and stable throughout the core (16% to 19%) and reflects high productivity and good preservation of the natural organic matter (OM) reaching the bottom sediment (Fig. 2d), and a lack of erodable inorganic sediment. The sediment interface of core 103 is anoxic most of the year thereby inhibiting most of the microbial activity in sediments and resulting in better preservation of organic matter than in oxic environments (Didyk et al., 1978; Talbot and Livingstone, 1989). Since anoxia now affects up to 50% of the lake's bottom area for prolonged periods of time (Hecky, 1993), our results may be representative of these areas.

Aliphatic Hydrocarbons as Source Indicators of Natural OM

To understand the increase in the sedimentation rate as an ecosystem change in the highly organic sediments of Lake Victoria we examined the sources (autochthonous and allochthonous) of natural organic matter and their changes with time.

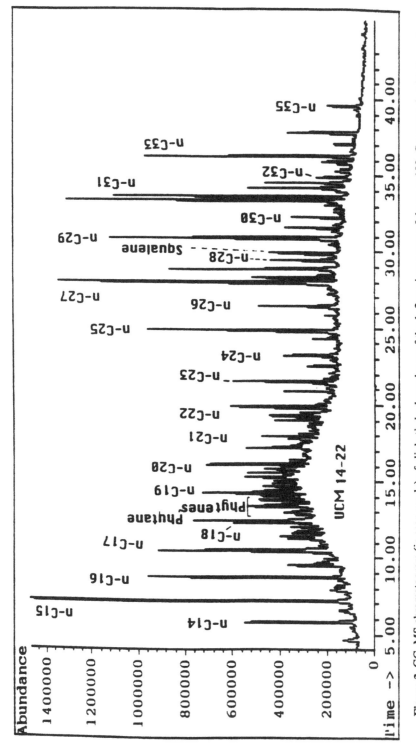

Figure 3. GC–MS chromatogram (in scan mode) of aliphatic hydrocarbons of the 1–2 cm increment of the core 103. Compounds with the code "n-" are straight chain alkanes. UCM = unresolved complex mixture.

Figure 3 shows a representative GC–MS chromatogram of non-aromatic hydrocarbons obtained from the sediments of the core 103 (1–2 cm depth). The distribution is dominated by the n-alkanes with carbon chain length of 14 to 35 (n-C14 to n-C35). A bimodal distribution of n-alkanes is seen with one mode dominated by n-C14 to n-C22 compounds maximizing at n-C15 and n-C17, and the second mode from n-C23 to n-C35 hydrocarbons (Fig. 3, Table 1). The chromatogram profile shows an Unresolved Complex Mixture (UCM) in the range from n-C14 to n-C23 hydrocarbons, which is generally attributed to microbial alteration of algal material (Dastillung, 1976). The absence of UCM in the range n-C23 to n-C35 hydrocarbons indicates negligible petroleum inputs to the lake (Mazurek and Simoneit, 1984).

The n-C14 to n-C18 components are indicators of autochthonous or in-lake inputs of organic matter (Leenheer and Meyers, 1983; Meyers and Ishiwatari, 1993). Significant amounts of n-C15 and particularly n-C17 are usually interpreted as indicators of a direct algal detritus input (Han et al., 1968; Gelpi et al., 1970). In fact the n-C17 has been found as a major component of cyanobacteria (Winters et al., 1969). Hecky (1993) reports that cyanobacteria are a dominant algal group in Lake Victoria that has undergone recent increases in abundance.

Isoprenoid-saturated hydrocarbons are considered as diagenetic derivatives of the phytol sidechain of the chlorophyll a (Tornabene et al., 1979). The isoprenoid hydrocarbon, phytane (2,6,10,14-tetramethyl-hexadecane), was identified in important concentrations in all sediment layers whereas pristane (2,6,10,14-tetramethyl-pentadecane), a related isoprenoid, is present in traces. Pristane is a byproduct of zooplanktonic digestive processing of chlorophyll a (Blumer et al., 1971). Its absence in the layers studied reflects the absence of copepods in the food chain of Lake Victoria (Meyers and Ishiwatari, 1993). Unsaturated isoprenoids such as phytenes and squalene ($C_{30}H_{50}$) were also identified in the core (Fig. 3). The preservation of such polyunsaturated compounds indicates the low degree of diagenesis of organic matter. In fact phytenes have been tentatively identified in methanogenic bacteria (Langworthy et al., 1982).

The straight chain alkanes from n-C23 to n-C35 show a pronounced odd-to-even predominance (Carbon Preference Index range: 3.69–5.12) illustrative of allochthonous inputs from terrestrial higher plants (Giger and Schaffner, 1977).

Similar types of chromatograms with a bimodal distribution and important concentrations in the low molecular weight compounds have been published for other lacustrine environments. Giger et al. (1980) published a similar chromatogram for Greifensee, a highly eutrophic Swiss lake with bottom anoxia and good organic matter preservation. Albaiges et al. (1984) show similar distributions from a coastal lagoon (Ebro Delta, Spain), a eutrophic environment with anoxic conditions. However, a unimodal NAH pattern dominated by the high molecular weight hydrocarbons, the lower molecular weight compounds being virtually absent, was observed in Lake Superior (US), an oligotrophic lake with oxic sediments (Lipiatou et al., in prep.). Both higher aquatic productivity in eutrophic environments and better preservation of OM in anoxic sedimentary conditions (Didyk et al., 1978; Talbot and Livingstone, 1989) exert a marked influence not only in the quantity of organic matter preserved but also in its composition at the molecular level evident in the non-aromatic hydrocarbon distribution.

Table 1. Concentrations of Individual n-Alkanes and Phytane (in ng/g) Measured in Sediments of Core 103

Compound	0–1 cm	1–2 cm	2–3 cm	3–4 cm	4–5 cm	5–6 cm	6–7 cm	7–8 cm	8–9 cm	9–10 cm	10–11 cm	11–12 cm
n-C14	44.8	534.0	1177.0	1225.0	1114.0	991.0	838.0	861.0	1269.0	725.0	540.0	538.0
n-C15	171.0	1398.0	1833.0	1830.0	1774.0	1686.0	459.0	1468.0	1838.0	1298.0	1023.0	1010.0
n-C16	107.0	586.0	1084.0	1160.0	1161.0	1094.0	1228.0	680.0	1012.0	542.0	373.0	390.0
n-C17	92.6	876.0	644.0	816.0	800.0	910.0	640.0	627.0	297.0	362.0	328.0	384.0
n-C18	44.6	226.0	263.0	341.0	280.0	375.0	349.0	271.0	259.0	201.0	213.0	228.0
Phytane	62.4	317.0	229.0	271.0	215.0	249.0	363.0	0.0	136.0	0.0	0.0	0.0
n-C19	46.8	300.0	379.0	526.0	508.0	660.0	327.0	284.0	250.0	160.0	184.0	245.0
n-C20	45.5	260.0	288.0	375.0	382.0	405.0	347.0	336.0	286.0	219.0	191.0	234.0
n-C21	31.3	133.0	205.0	386.0	364.0	472.0	373.0	468.0	331.0	316.0	247.0	239.0
n-C22	39.1	192.0	209.0	314.0	278.0	348.0	318.0	351.0	251.0	238.0	210.0	244.0
n-C23	55.1	263.0	399.0	613.0	566.0	730.0	698.0	915.0	676.0	701.0	569.0	608.0
n-C24	30.4	129.0	195.0	287.0	262.0	358.0	348.0	534.0	327.0	374.0	348.0	335.0
n-C25	121.0	570.0	790.0	1064.0	996.0	1181.0	1185.0	1436.0	1201.0	1276.0	957.0	1156.0
n-C26	45.5	189.0	266.0	389.0	376.0	509.0	499.0	634.0	502.0	530.0	455.0	469.0
n-C27	204.0	999.0	1343.0	2321.0	2298.0	2812.0	2685.0	3183.0	2419.0	2551.0	2798.0	3273.0
n-C28	42.3	147.0	211.0	299.0	307.0	547.0	369.0	478.0	367.0	394.0	363.0	383.0
n-C29	253.0	994.0	1268.0	1805.0	1737.0	1964.0	1911.0	2321.0	1650.0	1921.0	1596.0	2025.0
n-C30	56.0	176.0	220.0	315.0	313.0	438.0	431.0	509.0	378.0	415.0	452.0	521.0
n-C31	242.0	794.0	970.0	1235.0	1073.0	1368.0	1362.0	1587.0	1315.0	1350.0	291.0	393.0
n-C32	38.3	130.0	170.0	206.0	146.0	189.0	203.0	202.0	164.0	208.0	248.0	426.0
n-C33	261.0	765.0	83.4	1002.0	1067.0	1377.0	972.0	1165.0	967.0	967.0	1038.0	1249.0
n-C34	0.0	0.0	0.0	0.0	0.0	0.0	0.0	0.0	0.0	0.0	0.0	0.0
n-C35	70.8	138.0	168.0	198.0	194.0	291.0	262.0	344.0	264.0	270.0	393.0	472.0
Sum n-alkanes	2041.0	9798.0	12164.0	16708.0	15995.0	18704.0	15802.0	18654.0	16021.0	15015.0	12818.0	14823.0

Accumulation of Autochthonous and Allochthonous OM in the Sediments

The accumulation rate of autochthonous inputs (n-C15 and n-C17) in the Lake Victoria core has increased since 1900 by a factor of 1.8 and achieved maximum rates in 1960–70 (16 and 9 ng/cm^2 yr) (Fig. 4a,b). This behavior indicates an increase in the aquatic organic inputs to the lake sediments partly reflected in the increased dominance of cyanobacteria. These results are consistent with the increase in nutrient (phosphorous and nitrogen) accumulation during the same period (Hecky, 1993). Figure 4 shows that allochthonous organic matter, as reflected in the n-C25 and n-C29 hydrocarbon (Fig. 4c) accumulation profiles (~7 and 12 ng/cm^2 yr, respectively), show small changes with time and decrease since 1960. The n-C17/n-C29 ratios, indicative of in-lake versus terrestrial hydrocarbon inputs, show the increasing importance since 1900 of autochthonous organic matter (in-lake production) in the sedimentary organic carbon (Fig. 4d).

Polycyclic Aromatic Hydrocarbons: Combustion Processes

Polycyclic aromatic hydrocarbons (PAHs) are ubiquitous compounds in the environment produced during the pyrolysis of fossil fuels or organic-rich materials. Pyrolytic processes, can be either natural, such as forest fires (Wakeham et al., 1980; Ramdhal, 1983) or anthropogenic such as high temperature combustion of fossil fuels (coal, oil) (Laflamme and Hites, 1978; Larsen et al., 1986). PAHs are also introduced into the environment by release of crude oil or refined oil products (Jones et al., 1986). Finally some PAHs are produced from natural organic matter during in situ aromatization (Tan and Heit, 1981; Venkatesan, 1988). PAH abundance and distribution patterns provide insight as to their origin and mode of transport.

Table 2 provides concentrations of individual PAHs in the sediment increments, and Fig. 5 depicts a representative distribution of PAHs in the 1–2 cm increment. The distribution pattern is dominated by non-alkylated PAHs mainly of pyrolytic origin (Hites et al., 1980). The dominant compounds are phenanthrene (approx. 50% of the total PAH), a relatively volatile three ring PAH, produced in low temperature combustion processes, to a lesser extent naphthalene, a volatile two ring PAH and perylene, a naturally produced five ring PAH (Venkatesan, 1988). PAH distributions with dominance of lower molecular weight (MW) compounds (<4 condensed rings) were observed from biologic material collected in sediment traps (Prahl and Carpenter, 1979; Baker et al., 1991; Lipiatou et al., 1993) and atmospheric gas and water dissolved phase PAHs (McVeety and Hites, 1988; Baker and Eisenreich, 1990).

The PAH accumulation rate in core 103 shows an increase since 1860 (Fig. 6a) from 2.9 ng/cm^2/yr (background) to 10 ng/cm^2/yr in 1990 indicating an increase of 3.3 times over background levels. However these values are low and representative of remote environments influenced only from atmospheric deposition, such as Lake Superior (Gschwend and Hites, 1981; Lipiatou et al., in prep.), Isle Royale (McVeety and Hites, 1988), and the open Mediterranean Sea (Lipiatou and Saliot, 1991; 1992). The increase in the PAH accumulation rate is mainly due to the increase of phenanthrene whereas the higher molecular weight PAHs show small changes or remain constant with depth and time (Fig. 6b,c).

In Fig. 6d, retene (methyl-1 isopropyl-7 phenanthrene) shows an increase in accumulation rate in the bottom sediment since 1960. Retene is formed by diagenesis of the diterpenoid abietic acid, essentially found in resins of coniferous trees (Laflamme and Hites, 1978). This compound is commonly found in soils and the increase in the accumulation rate in this core

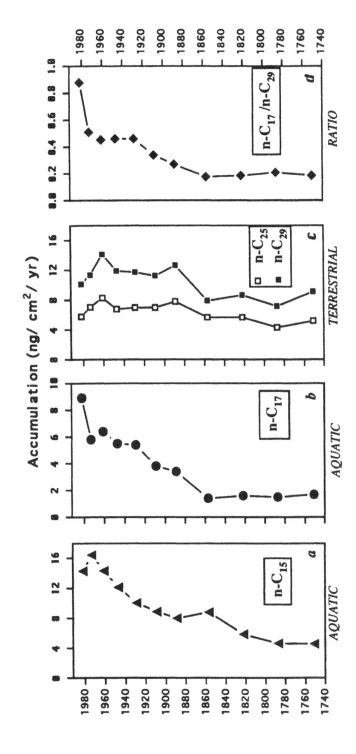

Figure 4. Sediment accumulation profiles in ng/cm² yr of a) C15 n-alkane, b) C17 n-alkane, c) C25 and C29 n-alkanes, and d) the ratio C17/C29, in the core 103.

Table 2. Concentrations of PAHs (in ng/g) Measured in Sediments of Core 103

Compound						Increment						
	0–1 cm	1–2 cm	2–3 cm	3–4 cm	4–5 cm	5–6 cm	6–7 cm	7–8 cm	8–9 cm	9–10 cm	10–11 cm	11–12 cm
Naphthalene	90.7	118.0	124.0	127.0	124.0	152.0	197.0	159.0	147.0	120.0	141.0	131.0
Acenaphtylene	4.8	13.7	4.4	11.6	10.9	14.3	18.1	15.7	13.5	10.7	18.9	6.0
Acenaphthene	15.6	20.3	17.0	18.7	16.0	18.7	31.4	20.8	19.8	13.1	14.7	15.3
Fluorene	51.2	78.2	60.7	67.1	62.9	67.8	117.0	73.5	67.1	44.6	48.4	52.2
Phenanthrene	411.0	475.0	257.0	294.0	330.0	240.0	567.0	256.0	249.0	125.0	129.0	159.0
Anthracene	50.7	55.7	31.7	36.3	39.7	31.7	54.7	31.5	34.0	19.3	22.7	26.5
Fluoranthene	84.9	78.1	66.7	73.5	66.1	86.7	119.0	95.0	91.2	81.3	86.2	82.9
Pyrene	54.3	54.8	38.8	39.5	48.6	40.9	61.7	38.6	38.6	33.1	32.4	30.4
Retene	32.6	40.1	31.8	24.0	51.9	32.6	34.6	10.2	18.1	10.1	6.7	9.2
Benzo(a)anthracene	4.9	5.1	5.5	5.8	6.5	7.1	6.7	6.5	6.2	6.1	5.3	6.7
Chrysene+triphenylene	8.9	10.0	10.0	10.0	11.3	14.0	13.8	13.8	13.8	12.8	12.5	13.9
Benzo(b)fluoranthene	18.4	20.2	63.8	30.0	36.6	40.4	41.8	45.8	95.6	50.9	43.1	51.0
Benzo(k)fluoranthene	26.9	27.3	3.1	37.0	36.3	38.9	40.0	41.8	4.4	43.3	40.0	47.4
Benzo(a)pyrene	7.0	6.1	7.3	6.6	6.0	6.3	6.2	6.2	5.7	6.1	6.1	6.2
Perylene	117.0	131.0	183.0	210.0	228.0	258.0	239.0	288.0	361.0	447.0	384.0	434.0
Indeno(1,2,3-cd)pyrene	27.1	36.4	41.7	45.0	49.9	54.8	51.0	54.6	55.9	63.2	45.4	58.1
Dibenzo(a,h)anthracene	0.0	0.0	3.0	3.2	3.4	3.6	0.0	0.0	0.0	0.0	0.0	0.0
Benzo(g,h,i)perylene	21.5	25.7	29.8	31.0	28.8	32.9	28.2	33.8	31.4	33.2	30.4	36.4
Coronene	45.3	47.4	66.9	67.0	72.8	79.3	81.7	87.7	100.0	94.2	83.1	98.4
Sum PAH	878.0	1027.0	764.0	836.0	897.0	850.0	1355.0	892.0	673.0	663.0	677.0	723.0

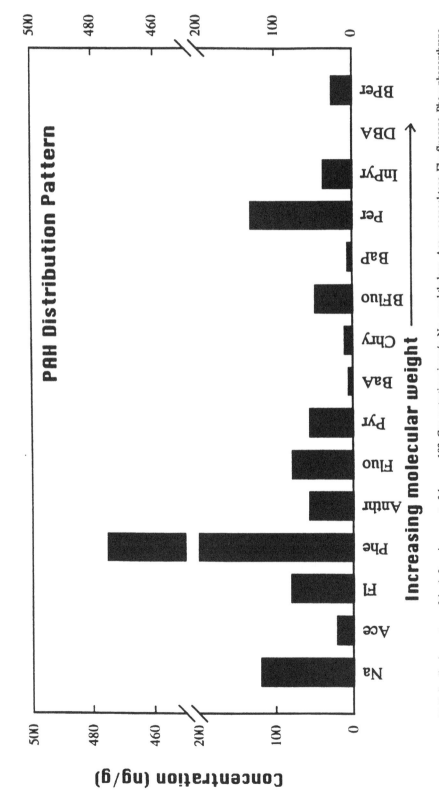

Figure 5. PAH distribution pattern of the 1–2 cm increment of the core 103. Concentrations in ng/g. Na = naphthalene, Ace = acenaphtene, Fl = fluorene, Phe = phenanthrene, Anthr = anthracene, Fluo = fluoranthene, Pyr = pyrene, BaA = benzo(a)anthracene, Chry+Tri = chrysene+triphenylene, BFluo = benzo(b)fluoranthene+benzo(k)fluoranthene, BaP = benzo(a)pyrene, Per = perylene, InPyr = indeno(1,2,3-cd)pyrene, DBA = dibenzo(a,h)anthracene, BPer = benzo(g,h,i)perylene.

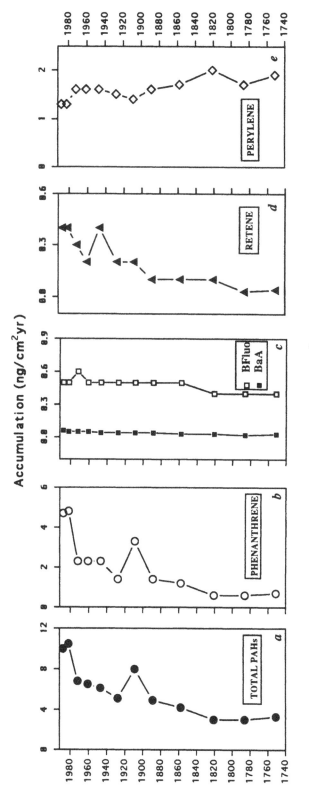

Figure 6. Sediment accumulation profiles for individual and total PAHs in ng/cm² yr, of the core 103. Total PAHs is the sum of compounds listed in the Fig. 5 except retene and perylene. Abbreviations as in Fig. 5.

may record increases in soil erosion as settlement and farming increased in the watershed. However, as retene could also be formed by the thermal degradation of abietic acid, wood combustion (Ramdhal, 1983) can be the cause of the increase in low molecular weight PAHs in the core.

What is the origin of phenanthrene and how is it transported? In this case the atmosphere is the mode of phenanthrene transport to the lake. Phenanthrene and the lower molecular weight PAHs exist in the atmosphere mainly in the gas phase (>90%; Bidleman, 1988); as the partitioning between gas and particle phases is influenced by temperature, higher temperatures such as those encountered in the equatorial region increase the gas phase concentrations (Baker and Eisenreich, 1990; Foreman and Bidleman, 1990). Phenanthrene and other low MW PAHs likely enter the lake via air–water exchange with subsequent incorporation into phytoplankton and the vertical flux ladder (Baker and Eisenreich, 1990). As the degree of diagenesis is low (see above), PAHs reaching the sediments are well preserved.

Perylene accumulation rate, opposite to the other PAHs, decreases in accumulation towards recent times (Fig. 6e). A diagenetic origin of perylene from terrestrial (Prahl and Carpenter, 1979) or marine precursors such as diatoms (Venkatesan, 1988) has often been noted in the literature. Therefore, a natural origin seems likely in the sediments of Lake Victoria.

PCBs and Organochlorine Pesticides

Sediment core 103 was analyzed for a series of organochlorine compounds such as poly-chlorinated biphenyls (PCBs) and pesticides such as toxaphene, DDT and metabolites, and chlordane and hexachlorocyclohexane (HCHs) species (Table 3). PCBs and organochlorine pesticides are ubiquitous in the global atmosphere and in aquatic and terrestrial ecosystems. PCBs, a mixture of ~100 chlorinated biphenyl compounds distributed in the environment, have been produced since 1929, used in hundreds of commercial and industrial applications and have now been banned in North America and elsewhere. Although released from urban and industrial centers, they are distributed worldwide through atmospheric transport and deposition. DDT was produced first in 1939 as an agent against many insects, achieved widespread use and peaked in usage worldwide in the 1960s. Environmental burdens are now generally decreasing (although data from developing countries are lacking). Toxaphene, a mixture of polychlorinated camphenes, has been used as an insecticide on cotton and soybean pests and is used to kill rough fish in developed countries. Toxaphene production was banned in late 1982 in the US (existing stockpiles could be used) and its use worldwide has dropped dramatically. The HCHs (active agent is g-HCH, Lindane) are still used as an effective pesticide in many countries. These compounds are semi-volatile in character, enhancing their atmospheric mode of transport, and are hydrophobic, meaning they partition readily into lipids of natural organic matter and living organisms (Schwarzen-bach et al., 1993). The net effect of these characteristics is that HCHs are transported globally by winds, deposited in aquatic ecosystems, become associated with organic rich particles, and accumulate in bottom sediments.

Table 3 shows the concentrations of PCBs and organochlorine pesticides found in Lake Victoria sediment core 103. PCBs exhibit concentrations much higher than other remote uncontaminated large lakes at values of 100–240 ng/g dry weight in the upper 5 cm (Eisenreich, 1987) and are dominated by the low MW PCB congeners. The PCB homologue distribution is not similar to known commercial mixtures in international use. This behavior

Table 3. Concentrations of Organochlorines (in ng/g)
Measured in Sediments of Core 103

Compound	Increment (cm)				
	0–1	1–2	2–3	4–5	11–12
Sum chlorobenzene	8.4	10.8	5.2	7.2	3.3
Sum hexachlorohexane	1.6	1.0	1.4	0.0	0.8
Sum chlordane	2.6	1.3	1.3	0.03	0.17
Sum DDT	5.1	4.9	4.3	2.9	1.0
Sum toxaphene	0.4	1.9	1.4	0.0	0.0
Sum mono-dichlorobiphenyl	136.0	81.0	73.2	45.5	5.7
Sum trichlorobiphenyl	48.4	52.0	25.5	27.9	10.6
Sum tetrachlorobiphenyl	27.8	35.3	14.4	18.4	5.3
Sum pentachlorobiphenyl	17.2	22.4	7.9	13.0	1.8
Sum hexachlorobiphenyl	5.6	7.3	4.0	8.0	1.7
Sum heptachlorobiphenyl	0.8	1.0	1.0	4.6	0.6
Sum octachlorobiphenyl	0.1	0.2	0.1	0.4	0.7
Sum polychlorobiphenyl	236.0	199.0	126.0	118.0	26.4

suggests that PCBs are weathered during transport to Lake Victoria, in processing by wet deposition during the rainy season, and continuously by air–water exchange. There is no direct evidence of local sources based on the PCB homologue distribution. Lake Superior, an atmospherically driven oligotrophic system in the center of the North American continent, has sediment concentrations of 10 to 20 ng/g dry wt. However, when PCB concentrations are normalized to the organic carbon (OC) content of each site (LS = ~2%; LV = ~15–20%), concentrations are nearly identical implicating natural organic matter (and its preservation) as a dominating factor in chlorinated hydrocarbon accumulation. That is, PCBs are likely incorporated into settling organic matter and reach the sediments where they undergo processing by chemical, physical and biological vectors. If bottom anoxia is a characteristic of the system, then organic matter is preserved along with much of its chlorinated hydrocarbon burden.

The organochlorine pesticides exhibit low sediment concentrations. Values are reported for the upper part (0–5 cm) and deep in the core corresponding to periods prior to production and use. DDT values of 3 to 6 ng/g dry wt or 15 to 30 ng/g OC, Toxaphene values of 0.4 to 2 ng/g or 4 to 10 ng/g OC, and HCH values of 1 to 2 ng/g or 5 to 10 ng/g OC are comparable to those of other uncontaminated environments (Eisenreich, 1987) and reflect regional usage patterns.

Accumulation rates of the organochlorines in the surface sediments of Core 103, calculated by subtracting the deepest sediment increment value (blank) from the surface sediment concentration and multiplying by the mass accumulation rate of 0.011 g/cm^2/yr are as follows:

Compound	Concentration (ng/g)	Accumulation rate (ng/cm^2/yr)
Sum chlorobenzenes	8.4	0.056
Sum HCHs	1.6	0.009
Sum chlordane	2.6	0.026
Sum DDT	5.1	0.045
Sum toxaphene	0.4	0.0044
Sum PCBs	236	2.3

The presence of chlorinated hydrocarbons in the deep layers of the cores indicate contributions to the chemical blank values from sampling and storage activities and analytical processing in the laboratory. Accumulation rates of the organochlorine compounds do not indicate a significant reduction of inputs to Lake Victoria in recent years but more sediment cores are needed to verify this observation.

CONCLUSIONS

The organic chemical data from core 103 in Lake Victoria support the hypothesis that the lake has experienced increased productivity since 1900 as evidenced by an increase in the accumulation rates for n-alkanes typical of in-lake production and an increase in the mass and organic carbon accumulation rate. Meanwhile, accumulation rates of n-alkanes typical of terrestrial higher plants have decreased or remained constant over the same period. Good preservation of natural organic matter, especially that typical of in-lake biologic production, suggests anoxia has existed at the sediment–water interface at this site for the last 200 years. Anoxic sedimentary conditions influence not only the quantity of organic matter preserved in the sediments but also its composition. Increased human population of the basin and subsequent deforestation is supported by the large increase in phenanthrene accumulation rate derived from combustion of fossil fuels at low temperature and the increase in retene accumulation attributed either to wood combustion or soil erosion. The aliphatic and aromatic hydrocarbons typical of oil pollution are absent from core 103 suggesting the absence of recent or chronic petroleum pollution. Organochlorines such as PCBs and selected pesticides occur at low concentrations typical of atmospherically driven inputs. Accumulation rates of organochlorines and combustion-derived PAHs are low and typical of remote environments of the earth. These organic markers are effective in delineating the changes in the Lake Victoria ecosystem over the last 200 years, and are accompanied by changes in nutrient and biologic species changes over the same period. To be definitive, more sediment cores from all of the depositional basins need to be analyzed for these compounds and others (e.g., fatty acids) to document the status and historical trends of the lake's ecosystem.

ACKNOWLEDGMENTS

We are grateful to Brian Billeck for assistance in the analytical procedure. We would also like to thank Shawn Schottler for helpful discussions, Joe Hallgren for assistance in figure preparation, Bert Grift for analytical work on the organochlorines and P. Meyers and L. Brazzi for very valuable comments on the manuscript. E.L. was supported by a grant from the Large Lakes Observatory (University of Minnesota) and Legislative Commission on Minnesota Resources.

REFERENCES

Albaiges, J., Algaba, J., and Grimalt, J., 1984, Extractable and bound neutral lipids in some lacustrine sediments. Org. Geochem., 6: 223–236.

Baker, J.E., and Eisenreich, S.J., 1990, Concentrations and fluxes of polycyclic aromatic hydrocarbons and polychlorinated biphenyls across the air–water interface of Lake Superior. Environ. Sci. Technol., 24: 342–352.

Baker, J.E., Eisenreich, S.J., and Eadie, B.J., 1991, Sediment trap fluxes and benthic recycling of organic carbon, polycyclic aromatic hydrocarbons, and polychlorobiphenyls congeners in Lake Superior. Environ. Sci. Technol., 25: 500–509.

Bidleman, T.F., 1988, Atmospheric processes controlling semi-volatile organic compounds. Environ. Sci. Tech.,

Blumer, M., Guillard, R.R.L., and Chase, T., 1971, Hydrocarbons of marine plankton. Mar. Biol., 8: 183–190.

Dastillung, M., 1976, Lipids des sediments recents. Thèse de Doctorat d'État, Université Louis Pasteur de Strasbourg, Strasbourg, 147 pp.

Didyk, B.M., Simoneit, B.R.T., Brassell, S.C., and Eglington, G., 1978, Organic geochemical indicators of paleoenvironmental conditions of sedimentation. Nature, 272: 216–222.

Eakins, J.D., and Morrison, R.T., 1978, A new procedure for the determination of lead-210 in lake and marine sediments. International Journal of Applied Radiation and Isotopes, 29: 531–536.

Edgington, D.N., and Robbins, J.A., 1990, Time scales of sediment focusing in large lakes as revealed by measurement of fallout Cs-137. Ecological Structure and Function, MM Tilzer (Serruya ed.), 1990, Springer-Verlag.

Eisenreich, S.J., 1987, Chemical limnology of non polar organic contaminants in large lakes: PCBs in Lake Superior. *In* Sources and Fates of Aquatic Pollutants. Hites, R.A., and Eisenreich, S.J., eds., Advances in Chemistry Series, #216, American Chemical Society: Washington, DC, 394–469.

Eisenreich, S.J., Capel, P.D., Robbins, J.A., and Bourbonniere, R.A., 1989, Accumulation and diagenesis of chlorinated hydrocarbons in lake sediments. Environ. Sci. Technol., 23: 1116–1126.

Eglington, G., and Hamilton, R., 1967, Leaf epicuticular waxes. Science, 156: 1322–1335.

Foreman, W. T., and Bidleman, T. F., 1990, Semivolatile organic compounds in the ambient air of Denver, Colorado. Atmos. Environ., 24A: 2405–2416.

Gelpi, E., Schneider, H., Mann, J., and Oro, J., 1970, Hydrocarbons of geochemical significance in microscopic algae. Phytochem., 9: 603–612.

Giger, W., and Schaffner, C., 1977, Aliphatic, olefinic and aromatic hydrocarbons in recent sediments of a highly eutrophic lake. *In* Advances in Organic Geochemistry 1975, Campos, R., and Goni, J., eds., pp. 375–390, Enadimsa, Madrid.

Giger, W., Schaffner, C., and Waheham, S.G., 1980, Aliphatic and olefinic hydrocarbons in recent sediments of Greifensee, Switzerland. Geochim. Cosmochim. Acta, 44: 119–129.

Gschwend, P.M., and Hites, R.A., 1981, Fluxes of polycyclic aromatic hydrocarbons to marine and lacustrine sediments in the northeastern United States. Geochim. Cosmochim. Acta, 45: 2329–2367.

Han, J., McCarthy, E. D., Van Hoeven, W., Calvin, M., and Bradley, W.H., 1968, Organic geochemical studies II. A preliminary report on the distribution of aliphatic hydrocarbons in algae, bacteria and a recent lake sediment. Proc. Nat. Acad. Sci. U.S.A., 59: 29–33.

Hecky, R.E., 1993, The eutrophication of Lake Victoria. Verh. Int. Verein. Limnol., 25 (in press).

Hites, R.A., and Eisenreich, S.J., eds., 1987, Sources and fates of aquatic pollutants. Adv. Chem. Series #216. American Chemical Society, Washington, DC.

Hites, R.A., Laflamme, R.E., and Windsor, J.G. Jr., 1980, Polycyclic aromatic hydrocarbons in marine/aquatic sediments: their ubiquity. Adv. Chem. Ser., 185: 289–311.

IDEAL, 1990, An International Decade for the East African Lakes. T.C. Johnson et al. Workshop Report 1 on the Paleoclimatology of African Rift Lakes, Bern, Switzerland, 39 pp.

Jones, D.M., Rowland, S.J., Douglas, A.G., and Howells, S., 1986, An examination of the fate of Nigerian crude oil in surface sediments of the Humber Estuary by gas chromatography and gas chromatography–mass spectrometry. Int. J. Environ. Anal. Chem., 24: 227–247.

Kendall, R.L., 1969, An ecological history of the Lake Victoria Basin. Ecolog. Monographs, 39: 121–176.

Langworthy, T.A., Tornabene, T.G., and Holzer, J., 1982, Lipids of archaebacteria. Zbl. Bakt. Hyg. I. Abt. Orig., C3: 228–244.

Laflamme, R.E., and Hites, R.A., 1978, The global distribution of polycyclic aromatic hydrocarbons in recent sediments. Geochim. Cosmochim. Acta, 42: 289–303.

Larsen, P.F., Gadbois, D.F., and Johnson, A.C., 1986, Polycyclic aromatic hydrocarbons in Gulf of Maine sediments: distribution and mode of transport. Mar. Environ. Res., 18: 231–244.

Leenheer, M.J., and Meyers, P.A., 1983, Comparison of lipid composition in marine and lacustrine sediments. Advances in Organic Geochemistry, 309–316.

Lipiatou, E., and Saliot, A., 1991, Fluxes and transport of anthropogenic and natural polycyclic aromatic hydrocarbons in the western Mediterranean Sea. Mar. Chem., 32: 51–71.

Lipiatou, E., and Saliot, A., 1992, Biogenic aromatic hydrocarbon geochemistry in the Rhone River Delta and in surface sediments from the open Northwestern Mediterranean Sea. Estuar. Coast. Shelf Sci., 34: 515–531.

Lipiatou, E., Marty, J.C., and Saliot, A., 1993, Sediment trap fluxes of polycyclic aromatic hydrocarbons in the Mediterranean Sea. Mar. Chem., 44: 43–54.

Lipiatou, E., Simcik, M., and Eisenreich, S. J. Accumulation and sources of organic chemicals in sediments of Lake Superior (USA) (in preparation).

Mazurek, M.A., and Simoneit, B.R.T., 1984, Characterization of biogenic and petroleum-derived organic matter in aerosols over remote, rural and urban areas. In Identification and analysis of organic pollutants in air, Keith, L.H., ed., Ann Arbor Science/Butterworth Publishers, Boston, 353–370.

McVeety, B.D., and Hites, R.A., 1988, Atmospheric deposition of polycyclic aromatic hydrocarbons to water surfaces: a mass balance approach. Atmos. Environ., 22: 511–536.

Meyers. P.A., and Ishiwatari, R., 1993, Lacustrine organic geochemistry — an overview of indicators of organic matter sources and diagenesis in lake sediments. Org. Geochem., 20: 867–900.

Mothersill, J.S., 1976, The mineralogy and geochemistry of sediments of northwestern Lake Victoria. Sedimentology, 23: 553–565.

Muir, D.C.G., M.D. Segstro, P.M. Welbourn, D. Toom, S.J. Eisenreich, C.R. Macdonald and D.M. Whelpdale, 1993, Patterns of accumulation of airborne organochlorine contaminants in lichens from the upper great lakes region of Ontario. Environ. Sci. Technol., 27: 1202–1210.

Ochumba, P.B.O., and Kibara, D.I., 1989, Observations on blue-green algal blooms in the open waters of Lake Victoria, Kenya. Afr. J. Ecol., 27: 23–34.

Ogutu-Ohwayo, R., and Hecky, R.E., 1991, Fish introductions in Africa and some of their implications. Can. J. Fish. Aquat. Sci., 48: 8–12.

Oldfield, F., and Appleby, P.G., 1984, Empirical testing of ^{210}Pb-dating models for lake sediments. In Lake sediments and environmental history, Hawarth, E.V., and Lund, J.W.G., eds., Univ. of Minnesota Press, Minneapolis, USA.

Prahl, F.G., and Carpenter, R., 1979, The role of zooplankton fecal pellets in the sedimentation of polycyclic aromatic hydrocarbons in Dabob Bay, Washington. Geochim. Cosmochim. Acta, 43: 1959–1972.

Ramdahl, T., 1983, Retene — a molecular marker of wood combustion in ambient air. Nature, 306: 580–582.

Reed, W.E., 1977, Biochemistry of Mono Lake, California. Geochim. Cosmochim. Acta, 41: 1231–1245.

Schwarzenbach, R.P., Gschwend, P.M., and Imboden, D.M., 1993, Environmental Organic Chemistry. Wiley-Interscience. John Wiley & Sons, Inc., New York, 681 pp.

Talbot, M.R., and Livingstone, D. A., 1989, Hydrogen index and carbon isotopes of lacustrine organic matter as lake level indicators. Paleogeography, Paleoclimatology, Paleoecology, 70: 121–137.

Talling, J.F., 1965, The photosynthetic activity of phytoplankton in East African lakes. Int. Revue ges. Hydrobiol. 50: 1–32.

Tan, Y.L., and Heit, M., 1981, Biogenic and abiogenic polynuclear aromatic hydrocarbons in sediments from two remote Adirondack lakes. Geochim. Cosmochim. Acta, 45: 2267–2279.

Tornabene, T.G., Lavgworthy, J., Holzer, G., and Oro, J., 1979, Squalenes, phytanes and other isoprenoids as major neutral lipids of methanogenic and thermoacidophilic "Archaebacteria." J. Mol. Evol., 13: 73–83.

Venkatesan, M.I., 1988, Occurrence and possible sources of perylene in marine sediments; A review. Mar. Chem., 25: 1–27.

Wakeham, S.G., 1990, Algal and bacterial hydrocarbons in particulate matter and interfacial sediment of the Cariaco Trench. Geochem. Cosmochem. Acta, 54: 1325–1336.

Wakeham, S.G., 1993, Reconstructing past oceanic temperatures. Environ. Sci. Technol., 27: 29–33.

Wakeham, S.G., Schaffner, C., and Giger, W., 1980, Polycyclic aromatic hydrocarbons in Recent lake sediments, II. Compounds derived from biogenic precursors during early diagenesis. Geochem. Cosmochem. Acta, 44: 415–429.

Winters, K., Parker, P.L., and Van Baalen, C., 1969, Hydrocarbons of blue-green algae: Geochemical significance. Science, 163: 467–468.

The Formation of Vivianite-Rich Nodules in Lake Victoria

J.S. MOTHERSILL *Royal Roads University, Victoria, British Columbia, Canada*

Abstract — During a survey of northwestern Lake Victoria in the 1970s vivianite-rich nodules were encountered in the lake-bottom sediments of Kome Channel. The deposit is located at the lakeward end of the channel at water depths of 45–50 m. The nodules, which are spherical to discoidal in shape and range from 0.3 to 1.5 mm in diameter, form less than 5 percent of the silty clay sediments of the lake-bottom. The nodules were identified as vivianite ($Fe_3P_2O_8 \cdot 8H_2O$) by X-ray diffractometry and their chemical composition was checked by atomic absorption and X-ray fluorescence methods. Thin section photomicrographs show the nodules to have the typical divergent fibrous structure of vivianite. The vivianite nodules appear to have been formed as a result of solubilization/redeposition in an ionic diffusion gradient process within the sediments.

INTRODUCTION

Lake Victoria, which is a monomictic lake, occupies an area of 67,800 km^2 in a drainage basin of approximately 263,200 km^2, is very shallow with a maximum depth of only 79 m (Temple, 1966). The lake overlies a pre-Middle Pleistocene drainage system that flowed westward. Uplift of the area to the west of Lake Victoria, associated with the development of the western rift system, resulted in reversal of drainage during the Middle Pleistocene (Bishop and Posnonsky, 1960). The lake basin, ponded between the upward on the west and the headwaters of the westward flowing drainage system on the east, was originally suggested by Wayland (1929).

The source material for the sedimentary fill of Lake Victoria consists predominantly of Precambrian granites, granite gneisses and argillites. The nearshore sediments consist of pebbly sand comprised predominantly of quartz with minor K-feldspar, and the basinal sediments which cover in excess of 95 percent of the lake bottom consist of dark yellowish brown silty clay comprised predominantly of quartz with subordinate K-feldspar, plagioclase, kaolinite and illite and minor vermiculite (Mothersill, 1976). Studies of the benthic macroinvertebrates of northwestern Lake Victoria were carried out by Mothersill et al. (1980).

The vivianite-rich nodule deposit was encountered during a survey of a 2600 km^2 area to the north of the Sese Islands utilizing a Ugandan Ministry of Works launch (Fig. 1). The occurrence of vivianite in lake bottom sediments has been reported by a number of researchers including Swain and Prokopovich (1957) and Dell(1973) in Lake Superior; Manning et al. (1991) in Narrow Lake, Alberta; Kjensmo (1968) in Lake Svinsjoen; Rosenquist (1970) in Lake Asrum; and Müller and Förstner (1973) and Kalindekafe (1993) in Lake Malawi; Renaut et al. (1986) in Lake Bogoria. The geochemistry of the formation of vivianite has been studied by Palache et al. (1955), Rosenquist (1970), Moore (1971) and Nriagu (1972) among others.

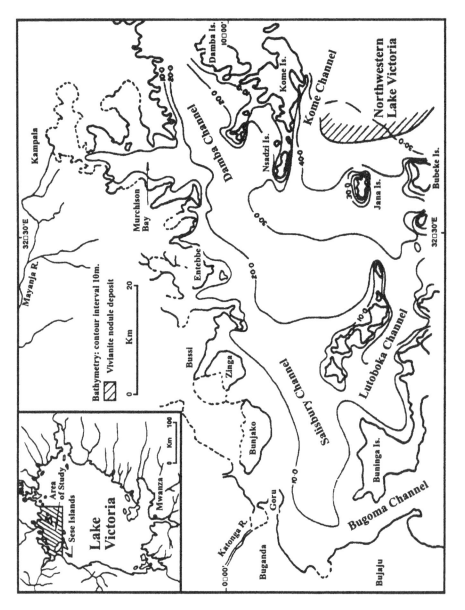

Figure 1. Location and vivianite deposit map.

PROCEDURE

Cores and grab samples were taken on an 8 km grid utilizing a Phleger corer and Shipek grab sampler. An echo sounding survey, utilizing a Kelvin Hughes MS 39/A echo sounder, was carried out between sampling stations. The pH and Eh of the bottom interface sediments were measured in a cut of the grab sampler in situ by a Corning Model 12 pH meter as soon as the sample was brought aboard.

The mineralogy of the lake sediments and the nodules was determined by X-ray diffractometry methods using an iron tube and focusing monochromator and based on charts from Berry (1972) and Smith (1967) for identification. Qualitative determinations of the elements present in the nodules were carried out by X-ray fluorescence techniques. Representative 1.0 g sample cuts of the nodules were then prepared for quantitative determinations by atomic absorption methods. These samples were taken twice to fumes of perchloric in a 10 ml solution of concentrated HF, HNO_3 and $HClO_4$ in the ratio of 15:5:2, and then dissolved in 10 ml of 30% HCl and made up to 100 ml with 3% HCl.

RESULTS

The vivianite-rich nodules were found at three sampling stations at the lakeward end of Kome Channel, which marks a change from channel to open-lake conditions (Fig. 1). Although the water depth of the stations where the nodules were found ranged from 45 to 50 m, the deposit was encountered at the eastern end of the sedimentological survey and could therefore be more extensive lakeward.

Eh measurements of cores and grab samples taken in the area of the nodule deposit showed a typical syndiagenetic sequence with an upper oxidized zone (initial phase) (+5 to +50 mv) forming a thin veneer (<5 mm), and underlain by a lower reduced zone (early burial phase) (−10 to −300 mv). The pH of the oxidized zone ranged from 6.95 to 7.30 and that of the reduced zone ranged from 7.10 to 5.50.

Vertical cuts made into the grab samples show that the nodules are concentrated just below the initial phase/early burial phase boundary at a depth of 0.5–1.0 cm below the water/sediment interface. The nodules do not form a segregated layer but form a unit from 1–3 cm thick mixed with the silty clay of the depositional sequence.

The nodules which are single spherulites originating from a single nucleus are spherical to discoidal in shape and range from 0.3 to 1.5 mm in diameter. The nodules were white to colourless when first examined in the grab sampler but within several hours changed to an intense blue colour. After several months the outer rim of the nodules became buff coloured. X-ray diffractometry analyses of powdered samples indicate that the nodules are formed of vivianite ($Fe_3P_2O_8 \cdot 8H_2O$). All the d-spacings of vivianite were recorded although the intensities, relative to the major peak at 6.80Å, tend to be lower than those noted by Berry (1972). The chemical composition of the nodules, checked semi-quantitatively by X-ray fluorescence methods and determined accurately by atomic absorption methods, confirmed that the mineral was vivianite. Thin section photomicrographs of the nodules showed the typical divergent fibrous structure of vivianite noted by Berry and Mason (1959). Atomic absorption analyses of the outer layer of the nodules which was stripped away by a 10% HCl solution showed that most of the Mn, Ca, Zn, Sr and Cr and all of the Cu and Ni occurred in the outer layer.

The silty clay sediments consisted of major amounts of quartz and subordinate amounts of K-feldspar, plagioclase, kaolinite and illite and minor vermiculite. The concentrations of

Table 1. Concentration of Elements in the Vivianite Nodules,
Sediments Within the Area of the Deposit, and the Average
for Northwestern Lake Victoria (ppm)

Element	Vivianite nodules (5 samples)	Sediments in deposit area (5 samples)	Average for northwestern Lake Victoria (30 samples)
Fe	335×10^3	89×10^3	56×10^3
P	124×10^3	259	669
Mn	358×10^2	52×10^2	443
Ca	14×10^2	413	35×10^2
Sr	23	14	–
Cu	4	24	43
Ni	20	34	47
Cr	28	56	67
Zn	110	89	91

selected elements in the sediments of the study area and for northwestern Lake Victoria were determined by atomic absorption methods (Table 1). The sediments in the area of the vivianite deposit contain considerably higher concentrations of Fe ($\times 1.5$) and Mn ($\times 12$), and lower concentrations of Ca ($\times 0.12$) and P ($\times 0.39$), than the average concentrations for the sediments of northwestern Lake Victoria.

DISCUSSION OF RESULTS

Vivianite has been reported in lake sediments in other East African lakes. Müller and Förstner (1978) reported vivianite crusts and hemispheres partially covering the surface of pebbles at the water/sediment interface under reducing and slightly alkaline conditions and Kalindekafe (1993) described vivianite-rich nodules that were found at the boundary between sands and muds at the sediment/water interface in Lake Malawi. Renaut et al. (1986) reported the occurrence of authigenic vivianite in the sediments of Lake Bogoria, Kenya.

   Although the vivianite-rich nodules described by Kalindekafe (1993) appear to be similar to those found in Lake Victoria they are located in a slightly different sedimentary setting associated with sands at the water/sediment interface, whereas the Lake Victoria nodules occur just below the water/sediment interface within a silty clay sequence. Both Müller and Förstner (1993) and Kalindekafe (1993) propose that the vivianite-rich nodules originate from the dissolution of C-phosphate from fish debris within the sediment and its redeposition in the uppermost sediment layers under reducing and slightly alkaline conditions. It is suggested that a similar origin could account for the vivianite-rich nodules of Lake Victoria with the Fe^{2+}, HPO_4^{2-}, Mn^{2+} and Ca^{2+} originating from the sediments. They are solubilized under the reducing and acid conditions of the early burial phase sediments and migrate upward through an ionic diffusion gradient. They are redeposited under the slightly reducing and alkaline conditions of the uppermost part of the early burial stage. The Mn^{2+} and Ca^{2+} found in the nodules probably occur as a substitution for Fe^{2+} in vivianite as noted by Berry and Mason (1959). The Sr^{2+} found in the nodules was probably a result of isomorphism of calcium and strontium. The other trace elements which have a higher abundance in the outer rim of the nodules probably are a result of surface absorption.

There would not appear to be a clear-cut conclusion as to whether the nodules were formed in situ or transported into the site. The lower values of HPO_4^{2-} in the associated silty clay of the deposit area relative to outside the deposit area could suggest an in situ formation.

Under reducing conditions the colour of the vivianite is white to colourless as were the nodules when first examined in the grab sampler. However after a few hours exposure to air the nodules changed to an intense blue colour, which is the partially oxidized form of vivianite called kertschenite by Palache et al. (1955). The buff coloured outer rim which formed on the nodules after several months was both opaque and X-ray amorphous and appears to be an intermediate stage in the alteration of vivianite to strengite as described by Nriagu (1972).

CONCLUSIONS

Vivianite-rich nodules found in the uppermost part of the early burial stage of the sedimentary sequence in the mouth of Kome Channel probably resulted from solubilization/redeposition in an ionic diffusion gradient process established in the sediments. It is not certain whether the nodules were formed in situ or transported to their present site.

REFERENCES

Berry, L.J., 1972, ed., Selected powder diffraction data for minerals, Joint Com. Powder Diffraction Standards, Swarthmore, 1081 pp.

Berry, L.J., and Mason, B., 1959, Mineralogy. W.H. Freeman and Co. San Francisco, 630 pp.

Bishop, W.W., and Posnonsky, M., 1960, Pleistocene environments and early man in Uganda. Uganda J., v. 24, pp. 44–61.

Dell, C.I., 1973, Vivianite: an authigenic phosphate mineral in Great Lakes sediments. Proc. 16th Conf. Great Lakes Res., pp. 1027–1028. Internat. Assoc. Great Lakes Res.

Kalindekafe, L.S., 1993, The mineralogy of Lake Malawi ferromanganese nodules. J. African Earth Sci., v. 17, pp. 183–192.

Kjensmo, J., 1968, Late and post-glacial sediments in the small meromictic Lake Svinsjoen. Arch. Hydrobiol., v. 65, pp. 125–141.

Manning, P.G., Murphy, T.P., and Prepas, E.E., 1991, Intensive formation of vivianite in the bottom sediments of mesotrophic Narrow Lake, Alberta. Can. Mineral., v. 29, pp. 77–85.

Moore, P.B., 1971, The $Fe_3^{2+}(H_2O)n(PO_4)_2$ homologous series: crystal–chemical relationships and oxidized equivalents. Am. Mineral., v. 56, pp. 1–17.

Mothersill, J.S., 1976, The mineralogy and geochemistry of the sediments of northwestern Lake Victoria. Sedimentology, v. 23, pp. 553–565.

Mothersill, J.S., Freitag, R., and Barnes, B., 1980, Benthic macroinvertebrates of northwestern Lake Victoria, East Africa: abundance, distribution, intra-phyletic relationships and relationships between taxa and selected element concentrations in the lake-bottom sediments. Hydrobiologia, v. 74, pp. 215–224.

Müller, G., and Förstner, V., 1973, Recent iron ore formation in Lake Malawi, Africa. Mineral. Deposita, v. 3, pp. 278–290.

Nriagu, J.O., 1972, Stability of vivianite and ion-pair formation in the system $Fe_3(PO_4)_2$–H_3PO_4–H_2O. Geochim. et Cosmochim. Acta, v. 36, pp. 459–470.

Palache, C., Berman, H., and Frondel, C., 1955, Dana's system of mineralogy, 2, John Wiley, 1124 pp.

Renaut, R.W., Tiercelin, J.J., and Owen, R.B., 1986, Mineral precipitation and diagenesis in the sediments of the Lake Bogoria basin, Kenya rift valley. *In* Sedimentation in the African rifts, Frosti, L.E., ed., p. 159–175. Blackwell Scientific.

Rosenquist, I.Th., 1970, Formation of vivianite in Holocene clay sediments. Lithos, v. 3, pp. 327–334.

Smith, J.V., 1967, ed., X-ray powder data file sets 1–5 (revised). Am. Soc. for Testing and Materials, Philadelphia, 635 pp.

Swain, F.M., and Prokopovich, N., 1957, Stratigraphy of the upper part of sediments of Silver Bay area, Lake Superior. Geol. Soc. Am. Bull., v. 68, pp. 527–542.

Temple, P.H., 1966, Evidence of changes in the level of Lake Victoria and their significance. Thesis, University of London.

Wayland, E.J., 1929, Rift valleys and Lake Victoria. Compte Rendu, XVth Session, Int. Geol. Congr. (Pretoria), v. 11, sect. 6, pp. 323–353.

Late Pleistocene Lake-Level Fluctuations in the Naivasha Basin, Kenya

M.H. TRAUTH   *Geologisch-Paläontologisches Institut, University of Kiel, Kiel, Germany*

M.R. STRECKER   *Department of Geophysics, Stanford University, Stanford, California, United States*

Abstract — A late Pleistocene lacustrine succession exposed in the Ol Njorowa Gorge, Naivasha Basin, Central Kenya Rift, contains the record of five lake-level highstands of Lake Naivasha between about 400 and 20 kyr. Near the southern rim of the Naivasha Basin unaltered tuff layers and diatomite beds up to 3.3 m thick characterize freshwater depositional phases, while severely altered silicic volcanic glass, authigenic silicate mineral phases (chabazite, clinoptilolite, analcime), and sedimentary fluorite signal lake-level lowstands associated with higher alkalinity. In addition to a well-documented lake-level highstand at 8–9 kyr, five other late Pleistocene lake-level highstands are interpreted as a response to enhanced African-Asian monsoons probably due to low-latitude climate forcing.

INTRODUCTION

In unglaciated tropical environments climatic changes are recorded in lacustrine sediments reflecting changes in water level and hydrochemistry. While low water levels coupled with increased alkalinity and salinity indicate an arid environment, high levels associated with freshwater conditions suggest a humid climate (Roberts, 1990). Long-term variations in tropical lake levels are thought to result from cyclic changes in orbital insolation parameters (Kutzbach and Street Perrott, 1985). Various studies show that tropical African lake levels respond primarily to precessional low-latitude monsoon forcing with periods of 19–23 kyr (e.g., Rossignol-Strick, 1983). In addition, North Atlantic sea-surface temperature anomalies may also influence rainfall variability and lake levels in tropical East Africa (e.g., Street-Perrott and Perrott, 1990).

Most studies concerning the climatic history as recorded in tropical African lakes have concentrated on the last 30 kyr (Gasse and Street-Perrott, 1978; Littmann, 1989), but little is known about earlier lake-level fluctuations. In the Kenya Rift diatomaceous sediments and stromatolites preserved high above present lake levels attest to multiple changes in lake levels during the last 1.8 Ma. Although radiometric age determinations are available for some of these highstand episodes (e.g., Sturchio et al., 1993, and references therein), detailed analyses of sediment variability associated with long-term fluctuations and inter-basinal correlations are still scarce. Such detailed comparisons would be particularly important in tectonically active rifts because both alkaline dominated and freshwater dominated basins often occur in close vicinity to each other. As the basins are bounded by faults or constructional volcanic relief their sedimentary deposits are commonly transient or incomplete. Therefore, in the light of tectonism and associated erosion, alternating freshwater and alkaline conditions could possibly reflect changes of local hydrologic conditions rather than regional climatic events.

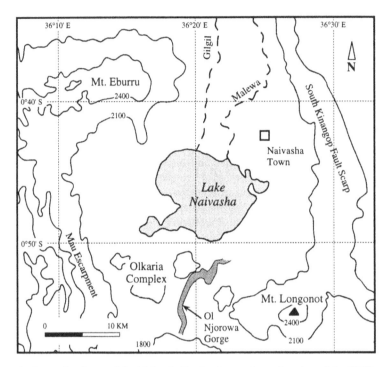

Figure 1. Generalized map of the Naivasha Basin. Arrow shows location of the Ol Njorowa Gorge.

This paper presents new paleohydrological data derived from exceptionally well pre-served lake deposits in the Naivasha Basin exposed in the Ol Njorowa Gorge (Fig. 1), south of Lake Naivasha in the Kenya Rift (0°55′S, 36°19′E). Apart from a well studied Holocene highstand at 8–9 ^{14}C kyr B.P. (e.g., Washbourn-Kamau, 1977) the succession documents five additional Pleistocene freshwater lake-level highstands of Lake Naivasha that may corre-spond to highstands documented in other basins in East Africa.

PHYSIOGRAPHY AND GEOLOGICAL SETTING

The Lake Naivasha Basin is one of the sedimentary sub-basins along the tectonically active axis of the Central Kenya Rift (Fig. 1) and was created by normal faulting between 900–700 kyr (Strecker et al., 1990). Situated at an elevation of 1890 m, Lake Naivasha is the highest freshwater body of the rift. To the west the Naivasha Basin is bounded by the Mau Escarpment and the volcano Mt. Eburru; to the east the basin is defined by faults limiting the intrarift Kinangop Plateau. To the north the basin is limited by fault scarps, whereas the southern boundary is composed of the trachytic volcano Mt. Longonot and the Olkaria Volcanic Complex (Fig. 1). The region receives 750 mm/yr rainfall, potential evapotranspi-ration is about 1600 mm/yr (Clarke et al., 1990). The lake is fed by the Malewa and Gilgil streams draining the Kinangop Plateau and the 4000 m-high Aberdare Range to the east (Fig. 1), where rainfall exceeds 1200 mm/yr (Clarke et al., 1990).

   The comenditic Olkaria Volcanic Complex consists of numerous lava flows and individ-ual eruptive centers younger than the foundation of the 400 kyr-old Mt. Longonot. The majority of the deposits at Olkaria are younger than 20 kyr (Clarke et al., 1990) and cover the predominantly lacustrine section exposed within the Ol Njorowa Gorge. The gorge is the site of an outlet of the high Holocene lake and was cut by headward erosion. The lacustrine succession is 60 m thick and unconformably overlies the oldest silicic lava flows and pyroclastics of the Olkaria Volcanic Complex (Trauth, 1992). The base of the profile is at 1840 m, i.e., 50 m below the present lake level. Similar lake sediments in the center of the basin are only exposed in subsurface channels. The deposits in the gorge are mainly yellow to buff colored waterlaid tuffs, characterized by altered pumice lapilli. Diatomite and laminated siltstone are also common, as are intercalations of coarse clastic fluvial sediments and pyroclastic deposits indicating intermittent subaerial conditions.

METHODS

An assessment of lake-level fluctuations in the Ol Njorowa Gorge, based on the analysis of stable isotopes, is hampered by the absence of suitable carbonate minerals in the entire profile. For this reason the late Quaternary paleoecology of the basin was reconstructed using diatom assemblages, sedimentary structures, and characteristic authigenic silicates. The presence of silicic pyroclastic deposits throughout the profile allows an evaluation of lake-level lowstands by means of the lateral zonation of authigenic silicates, which should reflect alkalinity and salinity of interstitial waters (e.g., Utada, 1966; Sheppard and Gude, 1973). Vertical variations of authigenic silicates in the Ol Njorowa Gorge profile thus could reflect lake-level lowstands through time. The mineral assemblages throughout the profile are presented in Fig. 2. Freshwater phases are characterized by diatomite, unaltered volcanic tuffs, and the absence of authigenic silicates. In contrast, silicic glass with perlitic cracks, glass shards with rims of montmorillonite, opal CT, occasional chabazite (Fig. 3a) and phillipsite represent the transition to alkaline conditions with a pH > 9 (e.g., Fisher and Schmincke, 1984). Increasing alkalinity causes the formation of clinoptilolite (Fig. 3b) associated with chabazite, and highly alkaline pore waters lead to the precipitation of analcime (Fig. 3c). Higher alkalinity is further indicated by fluorite concretions, often up to 5 cm in diameter. The concretions are interpreted to result from the dissolution of Ca^{++}-rich volcanic glass in the presence of fluorine released from hot springs into the lake (e.g., Sheppard and Mumpton, 1984). Sporadic halite occurs as a cementing agent for diatomite.

Bulk mineralogy was obtained from X-ray diffraction using standard techniques. Selected samples were also examined by scanning electron microscopy and X-ray spectroscopy to support the identification of diagenetic mineral phases. Diatom assemblages were studied and interpreted according to principles outlined in Gasse (1986).

RESULTS

The oldest deposits in the Ol Njorowa Gorge are partly waterlaid airfall tuffs, pyroclastic flow deposits (0–5.3 m), and fluvial sediments (5.3–6.8 m) that onlap the silicic lava flow in the southern gorge. The fluvial sediments are comprised of reworked pyroclastic deposits that become diatomaceous toward the top and grade into a diatomite bed, 3.3 m thick (6.8–10.1 m), which corresponds to the oldest lake-level highstand (V) in this <400 kyr-old succession. The diatom assemblages are dominated by *Epithemia zebra, E. sorex,* and *Fragilaria construens* (*Staurosira construens*) and indicate a shallow freshwater lake with a medium pH (e.g., Gasse, 1986). In addition, deeper water indicators such as *Nitzschia ancettula* and *Aulacosira granulata* occur as minor constituents. The diatom flora and facies

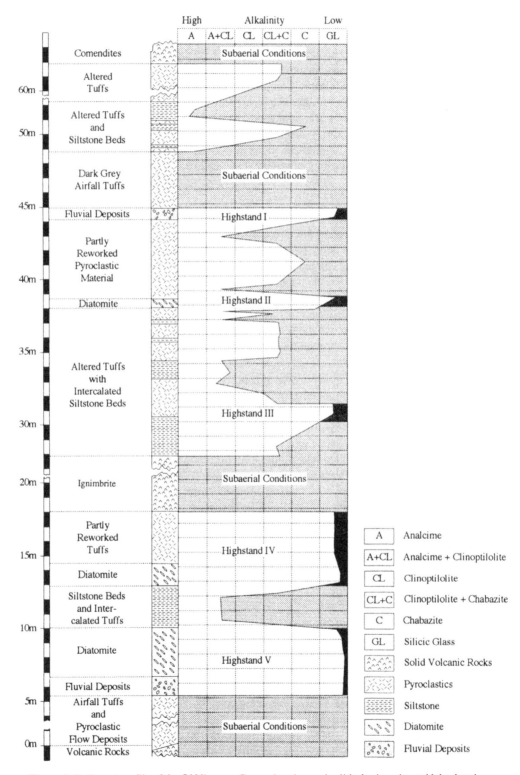

Figure 2. Sediment profile of the Ol Njorowa Gorge showing major lithologic units and lake-level fluctuations based on diagenetic mineral phases and diatomites.

Figure 3. a) Scanning electron micrograph of chabazite "cubes" or "rhombs" (CH). The crystals form polycrystalline crusts or strings of crystals. Chabazite is surrounded by bean-shaped aggregates of erionite (E). Scale bar = 5 μm. b) Scanning electron micrograph of clinoptilolite. Scale bar = 2.5 μm. c) Scanning electron micrograph of analcime. Scale bar = 5 μm.

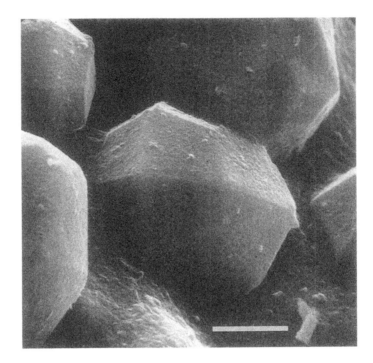

Figure 3, cont'd

patterns are interpreted to indicate a marginal environment of a large lake that deepened northwards.

The diatomite is overlain by a 2.8 m-thick succession of thinly bedded and laminated siltstone with layers of altered and partly reworked tuffs (10.1–12.9 m). The transition between the diatomite and this unit is characterized by abundant broken diatom frustules, sponge spicules and phytoliths (*Cyperaceae*) indicating a shallow sandy shore environment during regression. The interpretation of a major drop in lake level is further substantiated by the presence of authigenic clinoptilolite and analcime. The disappearance of authigenic analcime and the appearance of chabazite toward the top of the unit, however, suggests a trend toward less alkaline conditions in a very shallow water body at the end of this lowstand.

The tuffaceous unit is overlain by diatomite, 1.6 m thick, with tuff layers containing fresh glass (12.9–14.5 m). This unit signifies a second lake-level highstand (IV). Diatom assemblages, dominated by *Gomphonema angustatum* and *G. gracile*, indicate a large shallow freshwater lake with a pH < 8 (e.g., Gasse, 1986; Gasse et al., 1989). The diatomite bed is overlain by several beds of partly reworked pyroclastic material without any evidence of alteration (14.5–18.1 m), reflecting persisting freshwater conditions during deposition. However, in the upper part of this unit perlitic cracks in glass shards, abundant phytoliths and erosional unconformities herald a major regression. Subaerial conditions appear to have been attained at least along the periphery of the basin before an ignimbrite, 9.8 m thick was deposited (18.1–27.9 m).

The ignimbrite is covered by a 10.2 m-thick sequence that comprises tuffs and lacustrine siltstones representing fluctuating hydrologic conditions. Laminated siltstone at the base of the unit contains chabazite and clinoptilolite (27.9–30.6 m). These horizons are overlain by partly reworked pyroclastic deposits with unaltered volcanic glass (30.6–31.6 m). In comparison with severely altered glass in other layers, the pristine condition of the glass indicates deposition in a freshwater lake, equated with lake-level highstand (III). The absence of diatomite beds during this highstand may be a function of pronounced volcanic eruptions and high sedimentation rates.

Deposits associated with a superseding lake-level lowstand contain chabazite, clinoptilolite, and analcime (31.6–38.1 m). Finely laminated siltstones and siltstones with mud cracks and impact marks of airfall pumice lapilli suggest alternating subaerial and shallow water conditions.

Highstand (II) is manifested by three diatomite beds, up to 16 cm thick (38.1–38.7 m). The diatom assemblage is dominated by *Fragilaria construens* (*Staurosira construens*), *F. brevistriata* (*Pseudostaurosira brevistriata*), and *F. pinnata* (*Staurosirella pinnata*), which reflect shallow freshwater conditions along the southern basin margin. The presence of *Aulacosira granulata* and related species hints at a connection with a deeper lake farther north. However, in contrast to the older diatomites the occurrence of *Thalassiosira faurii*, *Mastogloia smithi*, *M. elliptica,* and *Anomoeoneis sphaerophora* suggest slightly alkaline conditions with a pH just above 8 during this highstand.

The deposition of diatomite was terminated by increasing volcanic activity represented by several tuff layers, up to 5.5 m thick (38.7–44.2 m). Clinoptilolite and analcime in the lower part of this pyroclastic section document rapidly increasing alkalinity in the course of lake regression. Numerous fluorite concretions attest to an increasing influence of hot-spring activity along the lake shores. Intermittent, less alkaline phases are represented by the substitution of analcime by chabazite within partly reworked tuff layers. However, severe alteration and formation of analcime above these layers shows a return to strongly alkaline conditions and a pronounced lake-level lowstand.

The upper part of the 5.5 m-thick pyroclastic sequence and the reworked tuffs above (44.2–45.0 m) contain unaltered glass shards, representing a renewed freshening of the lake (pH < 9), which is interpreted to correspond to lake-level highstand (I).

Above this unit follow subaerially deposited grey airfall tuffs, 3.8 m thick (45.0–48.8 m), with an increasingly yellow coloration in the highest horizons. These horizons contain analcime, similar to an overlying waterlaid tuff. The following pyroclastic and siltstone beds (48.8–52.2 m) have variable amounts of different authigenic silicates. Yellow tuffs in excess of 10 m thickness in the upper part of this sequence (>52 m) attest to a pronounced period of high alkalinity during a major drop in lake-level.

By about 20 kyr the final Pleistocene regression of the lake shore was complete (Clarke et al., 1990), and extreme volcanic activity created the comenditic flows and eruptive centers of the Olkaria Volcanic Complex, which were partly inundated by the lake-level highstand at 8–9 kyr, leaving behind well-preserved shoreline features and sediments (Washbourn-Kamau, 1970, 1977).

DISCUSSION

The sedimentary sequence in the Ol Njorowa Gorge represents five alternating phases of lake-level highstands and lowstands, with the record of highstands being less pronounced in the younger deposits. The described profile is laterally continuous throughout the 4 km-long gorge, and altered horizons containing authigenic silicates are also laterally persistent, which excludes alteration due to localized hot-spring activity. The depositional history of the lake basin began after the formation of the volcanic barrier in the southern part, which isolated the Naivasha Basin from other sub-basins of the Central Kenya Rift. The sediments contain modern, i.e., younger than middle Pleistocene diatom taxa (e.g., Gasse and Street-Perrott, 1978; Fourtanier and Gasse, 1988); the volcano-tectonic and erosional history of the basin and the surrounding drainage area indicates that this region has acted as a closed sedimentary basin for the last 400 kyr (Strecker, 1991). All important present tributaries were adjusted to Lake Naivasha during its existence. The fluvial history of the drainage area excludes major stream captures and resulting important freshwater input causing apparent high lake levels. The sedimentary succession, therefore, seems to be the long-term expression of major climate-controlled hydrologic changes affecting the basin and its surrounding drainage areas.

Biologic and lithologic evidence from various African tropical lacustrine successions (e.g., Gasse and Street-Perrott, 1978) show that the humid period and resulting high lake levels between 8–9 ^{14}C kyr B.P. (9–10 sidereal kyr B.P.; Radiocarbon v. 35, 1993) were in phase with maximum summer insolation at 10.4 kyr B.P. and thus were related to low-latitude monsoon forcing (Kutzbach and Street-Perrott, 1985; Clemens et al., 1992; Rossignol-Strick, 1983). The widely documented Holocene highstand in Kenya (e.g., Washbourn-Kamau, 1970, 1977) is in line with these observations. Because of the cyclic character of the Pleistocene lake-level fluctuations recorded in the Ol Njorowa Gorge we hypothesize that these hydrologic events may also be linked to Milankovitch mechanisms. With precision radiometric age determinations (Strecker, Trauth, and Deino, in prep.) it will be possible to interpret this continuous lacustrine succession in comparison with other East African sedimentary records, and thus provide a basis for testing the lake-level fluctuations in the light of enhanced African-Asian monsoons due to low-latitude climate forcing.

CONCLUSIONS

Lacustrine sediments in the Ol Njorowa Gorge in the southern Naivasha Basin provide new paleohydrological data for late Pleistocene time. Diagenetic mineral phases and the presence of intercalated diatomites allow an assessment of the paleohydrology and indicate pronounced variations in Pleistocene lake levels. Apart from a Holocene high lake level, there are five lake-level highstands in the Naivasha Basin between about 400 and 20 kyr.

The alternating phases in lake sedimentation reflect climatic changes and do not seem to be related to intrinsic basin dynamics. Although no precise age information exists for the lake sediments as yet, we expect that the lake-level fluctuations were in phase with variations in other East African lake basins and may correlate with enhanced African-Asian monsons due to low-latitude climate forcing.

ACKNOWLEDGMENTS

The authors thank the Deutsche Forschungsgemeinschaft (D.F.G.) for support. We thank the government of Kenya and the Kenya Wildlife Service for research permits. We are grateful to B. Owen and A. Grove for their reviews. We thank G. Eisbacher, F. Gasse, G. Haug, G. Muchemi, W. Smykatz-Kloss, E. Sittig, R. Tiedemann, and K. Winn for discussions. We also thank A. Brannath, A. Kamilli, E. Karotke, and B. Steiner for help with X-ray diffraction and spectroscopy measurements. Sonderforschungsbereich Karlsruhe (SFB 108) Contribution 409.

REFERENCES

Clarke, M.C.G., Woodhall, D.G., Allen, D., and Darling, G., 1990, Geological, volcanological and hydrogeological controls on the occurrence of geothermal activity in the area surrounding Lake Naivasha, Kenya: Ministry of Energy, Nairobi, Kenya, 138 pp.

Clemens, S., Prell, W.P., Murray, D., Shimmield, G., and Weedon, G., 1991, Forcing mechanisms of the Indian Ocean monsoon: Nature, v. 353, pp. 720–725.

Fisher, R.V., and Schmincke, H.U., 1984, Pyroclastic rocks: Springer, Berlin, 472 pp.

Fourtanier, E., and Gasse, F., 1988, Premier jalons d'une biostratigraphie et évolution des diatomées lacustres d'Afrique depuis 11 Ma: C. R. Acad. Sci. Paris, v. 306, pp. 1401–1408.

Gasse, F., 1986, East African diatoms. Taxonomy, ecological distribution: Berlin, J. Cramer, 201 pp.

Gasse, F., Lédée, V., Massault, M., and Fontes, J.C., 1989, Water-level fluctuations of Lake Tanganyika in phase with oceanic changes during the last glaciation and deglaciation: Nature, v. 342, pp. 57–59.

Gasse, F., and Street-Perrott, F.A., 1978, Late Quaternary lake-level fluctuations and environments of the northern rift valley and Afar region (Ethiopia and Djibouti): Palaeogeogr. Palaeoclimatol. Palaeoecol., v. 24, pp. 279–325.

Kutzbach, J.E., and Street-Perrott, F.A., 1985, Milankovitch forcing of fluctuations in the level of tropical lakes from 18 to 0 kyr. B.P.: Nature, v. 317, pp. 130–134.

Littmann, T., 1989, Spatial patterns and frequency distribution of Late Quaternary water budget tendencies in Africa: Catena, v. 16, pp. 163–188.

Roberts, N., 1990, Ups and downs of African lakes: Nature, v. 346, p. 107.

Rossignol-Strick, M., 1983, African monsoons, an immediate climate response to orbital insolation: Nature, v. 304, pp. 46–49.

Sheppard, R.A., and Gude, A.J., 1973, Zeolites and associated authigenic silicate minerals in tuffaceous rocks of the Big Sandy Formation, Mohave County, Arizona: U.S.G.S. Prof. Pap. 830, 36 pp.

Sheppard, R.A., and Mumpton, F.A., 1984, Sedimentary fluorite in a lacustrine zeolitic tuff of the Gila Conglomerate near Buckhorn, Grant County, New Mexico: J. Sed. Petrol., v. 54, pp. 853–860.

Strecker, M.R., 1991, Das zentrale und südliche Kenia-Rift unter besonderer Berücksichtigung der neotektonischen Entwicklung (habilitation thesis): Karlsruhe, Universität Karlsruhe, 182 pp.

Strecker, M.R., Blisniuk, P.M., and Eisbacher, G.H., 1990, Rotation of extension direction in the central Kenya Rift: Geology, v. 18, pp. 299–302.

Street-Perrott, F.A., and Perrott, R.A., 1990, Abrupt climate fluctuations in the tropics: the influence of Atlantic Ocean circulation: Nature, v. 343, pp. 607–612.

Sturchio, N.C., Dunkley, P.N., and Smith, M., 1993, Climate-driven variations in geothermal activity in the northern Kenya rift valley: Nature, v. 362, pp. 233–234.

Trauth, M.H., 1992, Tektonik und Sedimentation im Olkaria-Gebiet, südlich Lake Naivasha, Kenia (diploma thesis): Karlsruhe, Universität Karlsruhe, 119 pp.

Utada, M., 1966, Zeolites in sedimentary rocks, with reference to the depositional environments and zonal distribution: Sedimentology, v. 7, pp. 237–257.

Washbourn-Kamau, C.K., 1970, Late Quaternary chronology of the Nakuru-Elmenteita Basin, Kenya: Nature, v. 226, pp. 253–254.

Washbourn-Kamau, C.K., 1977, The Ol Njorowa Gorge, Lake Naivasha Basin, Kenya, in Greer, D.C., ed., Desertic Terminal Lakes: Utah Water Research Laboratory, pp. 297–307.

Comparative Paleolimnology in a System of Four Shallow Tropical Lake Basins

D. VERSCHUREN *Limnological Research Center, University of Minnesota, Minneapolis, Minnesota, United States; Laboratory of Animal Ecology, University of Ghent, Ghent, Belgium*

Abstract — Late-Quaternary sediment records from closed-basin lakes across tropical Africa are testimony to the complexity of climate history at time scales ranging from millennia to years. High-resolution reconstruction of this history has been complicated by the individualistic response of lakes to short-term climatic change, by basin-specificity of mechanisms governing the incorporation of climatic signals into the sediment record, and by deficiencies in traditional methods of sediment-core collection and dating. Future progress is expected to come from improved field methods, from consistent use of AMS radiocarbon dating, and from an integrated approach to climatic interpretation of multi-proxy sediment records. The study presented here attempts to improve high-resolution interpretation of complex stratigraphies by comparing sedimentary proxy records of lake response to recent climatic change in the four sedimentologically diverse but hydrologically interconnected lake basins of the Naivasha–Sonachi system in central Kenya.

INTRODUCTION

Late-Quaternary hydroclimatic oscillations had so strong an impact on closed-basin lakes in tropical Africa that most of them experienced a succession of distinct lake phases, each characterized by particular limnological and sedimentological behavior. In the past 30,000 years, many of the now shallow lakes and wetlands in the Eastern Rift have ranged from deep meromictic lakes or polymictic freshwater lakes to swamps, ephemeral ponds, or saline playas. Efforts to reconstruct Africa's climatic history from lake sediments have traditionally focused on recognition of the most obvious transitions between these lake phases, which in the sediment record are often clearly delineated as major lithostratigraphic units. At the millennium-scale resolution of these reconstructions, the late-Quaternary history of African climate is explained by orbital forcing through insolation-driven variations in paleo-monsoon activity (Kutzbach and Street-Perrott, 1985). The well-documented period of generally high lake levels during the latest Pleistocene and early Holocene is punctuated at ca. 11,000–10,000 yr BP and 8,000–7,000 yr BP by two arid intervals, which have been linked to the Younger Dryas event and the deglaciation of the Arctic through a mechanism that involves partial shutdown of the thermohaline circulation transporting heat from the Southern Hemisphere oceans to the North Atlantic (Street-Perrott and Perrott, 1990). Paleoceanographic reconstruction of surface-ocean salinity variations (Duplessy et al., 1992) has provided direct evidence for this partial suppression of NADW formation, during the two early-Holocene arid episodes as well as during a third at ca. 4,500–3,000 yr BP. The latter interval also coincides with widespread aridity on the African continent (e.g., Butzer et al., 1972).

In order to allow testing of the various mechanisms that may have forced African climate at time intervals shorter than the Milankovitch cycles (primarily those involving Atlantic Ocean circulation, El Niño–Southern Oscillation, and solar irradiance variations, but also volcanic aerosols and the anthropogenic impact on climate), sediment records of lake response to climatic change now need to be analysed and interpreted at higher resolution. However, correct high-resolution interpretation of the typically complex sediment strati-graphies demands special attention to the mechanisms governing the formation and preser-vation of sedimentary climate proxies. Apparent differences between climatic histories reconstructed from the sediment records of two lakes can be due to 1) lake-specific lag times and magnitudes of their responses to climatic change as determined by basin hydrology and local lake–groundwater interactions, 2) basin-specific preservation and resolution of sedi-mentary proxy signals as determined by mixing regime, taphonomy, and diagenesis, 3) sedimentation-rate changes and unrecognized disconformities, which may create spurious "events" and result in divergent interpolation between available radiocarbon dates, or 4) genuine regional differences in paleoclimate. Conceptual understanding of a particular lake's functioning sufficient to recognize how these various complicating factors may affect the climatic interpretation of proxy data from its sediment record can be achieved through the paleolimnological method, by validation and calibration of the sedimentary proxies in an investigation of lake response to documented recent climatic variability (Smol et al., 1991). Fine-interval study of dated core samples of recent sedimentation yields clues to exactly how local conditions of sedimentation and taphonomy affect the incorporation of a short-term climatic signal in the sediment record, and invites appreciation of the associated limits in temporal resolution. In a broader context, studying the recent history of closed-ba-sin lakes in tropical Africa may also promote understanding of their functional limnology and dynamic behavior as continually fluctuating wetland ecosystems. Lakes Naivasha and Sonachi in the Eastern Rift, which together form a complex of four ecologically and sedimentologically distinct but hydrologically interconnected lake basins, are a natural laboratory uniquely suited for this type of investigation. In the following account, I review relevant aspects of the limnology and recent history of the Naivasha–Sonachi system, argue for its suitability as object of study, and present results of preliminary analyses on a set of undisturbed short cores.

LIMNOLOGY AND RECENT HISTORY OF LAKES NAIVASHA AND SONACHI

Lakes Naivasha and Sonachi are located just south of the Equator, at ca. 1885 m a.s.l. in the central valley of the Eastern (Gregory) Rift in Kenya. The drainage basin of Lake Naivasha encompasses 2,378 km^2 of the Rift Valley and adjacent flanks of the Kinangop Plateau and Nyandarua Range on the east and the Mau Escarpment and Eburru Range on the west. Lake Sonachi occupies the floor of a 0.84 km^2 crater 3 km west of Lake Naivasha, well within the larger Naivasha catchment. Much of the monsoonal rainfall destined for the central Rift Valley is intercepted by the surrounding highlands, resulting in a strongly negative hydro-logical balance near the lakes. The main basin of Lake Naivasha (Fig. 1) and its satellite basins Lake Oloidien, Crescent Island Crater, and Lake Sonachi are maintained by river input primarily from the Malewa River, which drains the Kinangop Plateau and wet highlands in the Nyandarua Range.

Lake Naivasha is a fairly large (ca. 115 km^2 open-water surface area), shallow (Z_m averaged 7 m in the period 1963–93), and wind-exposed freshwater lake. Its drainage basin is topographically closed but the lake itself is hydrologically open. Water lost by groundwa-

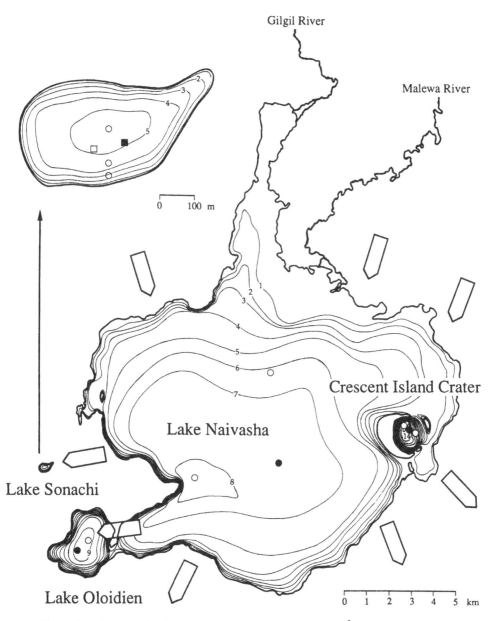

Figure 1. Bathymetry of Lake Naivasha (in 1983, at 1885.8 m a.s.l.; Åse et al., 1986) and Lake Sonachi (in 1990, at ca. 1884.2 m; Damnati et al., 1991), with location of core sites. Dots: short piston cores; squares: freeze cores. Closed symbols represent core sites of the profiles in Figure 3. Bold arrows show direction of groundwater flow (Gaudet and Melack, 1981).

ter seepage is estimated to average 12% of total water loss, the major output route being lake-surface evaporation with 80% (Gaudet and Melack, 1981; Darling et al., 1990). In the past 110 years lake level has fluctuated between 1881.5 m and 1896.5 m a.s.l. (Fig. 2, lower panel), causing relatively large changes in water depth (Z_m ranged between 3 and 18 m) and

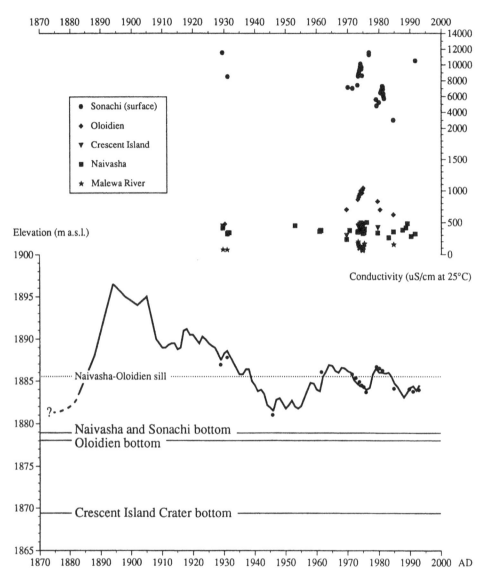

Figure 2. Lower panel: Lake Naivasha lake-level record 1883–1993 (heavy line; modified after Åse et al. (1986), with recent data from Åse (1987), Harper et al. (1993), and staff-gauge readings at Sulmac Cooperative, Naivasha) and depth soundings in Lake Sonachi (dots; data from Beadle (1932), Jenkin (1936), Melack (1976), MacIntyre and Melack (1982), Njuguna (1988), Clark et al. (1989), Damnati et al. (1991), and new fieldwork), in relation to lake-bottom elevation at the August 1991 core sites. For simplicity, the gradual increase of bottom elevation due to basin infilling is assumed negligible. Upper panel: Conductivity data for Lake Naivasha, Lake Oloidien, Crescent Island Crater, Lake Sonachi, and Malewa River (see legend; data from Beadle (1932), Jenkin (1936), Talling and Talling (1965), Kilham (1971), Melack (1976), Milbrink (1977), Gaudet and Melack (1981), MacIntyre and Melack (1982), Brierley et al. (1987), Barnard and Biggs (1988), Burgis and Mavuti (1988), Njuguna (1988), Clark et al. (1989), Harper et al. (1993), and new fieldwork). Alkalinity measurements (Beadle, 1932; Jenkin, 1936) were converted to conductivity via linear regression of all available pairwise measurements. All data are standardized to specific conductance at 25°C using a temperature coëfficient of 2.3% °C^{-1} (Marlier, 1951; Talling and Talling, 1965).

open-water surface area (ca. 100 to >200 km^2) but only minor changes in lakewater salinity (Fig. 2, upper panel). Conductivity measurements for the period 1929–91 range between 233 and 499 μS cm^{-1}, with a single measurement of 448 μS cm^{-1} from during the historical lowstand of the 1940s–1950s.

In contrast with the main basin, Lake Oloidien (5.5 km^2) is hydrologically closed. Water-budget calculations indicate that groundwater output is negligible, and water losses are due entirely to evaporation (Gaudet and Melack, 1981). The strongly negative local water balance is compensated for by seepage from the main basin into Lake Oloidien through the sill which separates the two basins (Fig. 1). Aerial photographs from 1946 and 1948 show that this flow is sufficient to keep both basins at similar level when disconnected, at least on interannual to decadal time scales. When lake level rises above ca. 1885.5 m a.s.l., Lake Oloidien becomes confluent with the main basin and input into "Oloidien Bay" is then direct. Fluctuations in lake level cause relatively large changes in water depth (Z_m range 4–19 m since 1883 AD) and open-water surface area (4.0 to 7.5 km^2), and also affect lakewater salinity (see Fig. 2, upper panel). Conductivity measurements for the period 1929–91 vary between 472 and 1040 μS cm^{-1}. No direct measurements are available from the mid-century lowstand, but the composition of fossil invertebrate assemblages deposited at that time suggest a maximum of 2,000–4,000 μS cm^{-1} (Verschuren, 1994).

Crescent Island Crater is a small (2.1 km^2) and deep (average Z_m: 17 m) inundated crater basin along the eastern shore of Lake Naivasha (Fig. 1). The crater is hydrologically open due to groundwater throughflow. Compared to the shallow main basin and Lake Oloidien, the 15 meter total range of historical lake-level fluctuations caused a lesser relative change in the water depth (Z_m range 12–27 m since AD 1883) and surface area (1.7 to 2.2 km^2) of Crescent Island Crater. Groundwater seepage and direct exchange of lakewater with the main basin keep the crater basin fresh, with conductivity measurements ranging between 291 and 491 μS cm^{-1} (Fig. 2).

Lake Naivasha's main basin, Lake Oloidien, and Crescent Island Crater can be classified according to Lewis (1983) as continuously warm polymictic, characterized by a daily stratification cycle in which a thermal gradient develops during calm morning hours but is later destroyed by strong afternoon winds and nocturnal cooling (Beadle, 1932). At current low to intermediate lake level, the main basin and Lake Oloidien probably mix completely almost every night. The usually more pronounced daytime surface-water heating in Lake Oloidien, together with the occasional occurrence of near-bottom oxygen depletion (Melack, 1979), indicates that the Oloidien basin is less prone to wind stress than the main basin. In Crescent Island Crater, the deepest and most sheltered basin of the system, nightly convective mixing extends down to at least 10–12 m depth and reaches the bottom frequently enough to keep it oxygenated. However, the occurrence during 1983 of a strongly developed oxygen stratification with complete anoxia below the depth of night-time mixing (Brierley et al., 1987) shows the potential of Crescent Island Crater to remain stratified for several weeks when Z_m is 17 m or more. It follows that during the high lake levels earlier in the 20th century, Crescent Island Crater may have been discontinuously polymictic with frequent near-bottom anoxia and its attendant consequences for the preservation of sedimentary proxies.

The fourth basin in the Naivasha–Sonachi system is Lake Sonachi (Fig. 1). It is sheltered from wind by its crater rim and is chemically stratified, with the chemocline typically at 4–5 m depth. The close match of Lake Sonachi depth soundings with the lake-level record of Lake Naivasha (Fig. 2, lower panel), supplemented by direct observation of synchronous lake-level changes (MacIntyre and Melack, 1982), support the notion of a groundwater connection and suggest that, at least on interannual to decadal time scales, the lake-level record of Lake Naivasha can be extrapolated to Lake Sonachi. Density stratification at the chemocline resists convective mixing and results in permanent near-bottom anoxia. Although Lake Sonachi is classified as meromictic (MacIntyre and Melack, 1982; Lewis, 1983), permanent stratification must have been disrupted during the lowstand of the 1940s–1950s when Z_m repeatedly fell to less than 3 m (Fig. 2). Low lake levels prevailing since 1984 until today (1993) probably also prevent chemical stratification from persisting year-round, with complete mixing likely at the seasonally lowest lake levels in February–March. Epilimnetic conductivity is variable, with high values during lake-level decline and lower values during lake-level rise (Fig. 2).

COMPARATIVE PALEOLIMNOLOGY OF THE NAIVASHA–SONACHI SYSTEM

As described above, all four basins of the Naivasha–Sonachi system are hydrologically connected to each other by surface water or via groundwater seepage so that lake level in the three satellite basins goes up and down in synchrony with the main basin. The lake-level record of Lake Naivasha itself (Fig. 2) has been linked to short-term trends in both the local Rift Valley climate and rainfall in the mountainous headwater regions of the Malewa River, and these in turn reflect interannual variation in the strength of the easterly monsoons (from the Indian Ocean) and the penetration of equatorial westerlies (from the Atlantic Ocean) onto the East African plateau (Vincent et al., 1979; Åse et al., 1986). When the Naivasha–Sonachi system is subjected to climate-driven fluctuations in lake level, its four basins are differently affected by 1) change in open-water surface area, because of differences in basin morphometry, 2) salinity change, because of differences in basin hydrology, and 3) disturbance of sedimentation by resuspension and bioturbation, because of differences in water depth and mixing regime. It can further be inferred from the lake-level record and bottom elevation in each basin (Fig. 2) that the 15 meter range of historical fluctuations must have caused mixing regimes and sedimentation patterns in each of the basins to switch between distinctly different states in just the past 110 years of documented history (Table 1). Thus, the Naivasha–Sonachi system is uniquely suited for a study in comparative paleolimnology of the mechanisms by which climate proxies are incorporated and preserved in the sediment record of lake basins with different sedimentary environments but subjected to the same recent events of short-term climatic change. The hydrological connectedness is an added advantage, because more independent lake–groundwater interaction would result in different lag times of lake response to climatic change and thus to different lake-level records and a priori different sediment records.

Fieldwork was conducted in August of 1991 and 1993, when (on both occasions) Lake Naivasha stood at ca. 1884.5 m a.s.l. Nine short cores with undisturbed sediment–water interface were collected with a rod-operated, single-drive piston corer (Wright, 1980) from central and transect locations in each of the four basins (Fig. 1). The cores were extruded upright in the field and sectioned in increments 1 cm thick with lightweight equipment designed specifically for on-site processing of highly porous recent sediments (Verschuren, 1993). Two freeze-cores (Renberg, 1981) were collected in Lake Sonachi and brought back to the U.S. intact for fine-interval stratigraphic analysis. Selection of multiple core sites (Fig. 1) to investigate within-basin variability of the sediment record is a compromise between maximization of horizontal distance, and avoidance of marginal areas where sedimentation

Table 1. Historical (1883–1993) Extremes in Water Depth, Mixing Regime,
Open-Water Surface Area, and Conductivity Measured or Estimated for
the Four Basins of the Naivasha–Sonachi System

Basin	Z_m (m)	Mixing regime	Area (km^2)	Conductivity (μS cm^{-1} at 25°C)
Naivasha	3 to 18	Wind-stressed to polymictic	ca. 100 to >200	ca. 250 (233) to 500 (499)
Oloidien	4 to 19	Wind-stressed to polymictic	ca. 4.0 to 7.5	ca. 250 (472) to 2–4000 (1040)
Crescent I	12 to 27	Continuously polymictic to discontinuously polymictic	ca. 1.7 to 2.2	ca. 250 (291) to 500 (491)
Sonachi	3 to 18	(Dis-)continuously poly-mictic to stably meromictic	ca. 0.07 to (?)0.20	(epilimnetic range 3000–11551)

Conductivity values between parentheses are the minimum and maximum recorded values. At the lake-level maximum of 1896.5 m a.s.l. (in 1894), exchange of water between the then broadly confluent Naivasha, Oloidien, and Crescent Island Crater basins is assumed to have resulted in a uniform water chemistry.

may have been discontinuous with episodes of non-deposition or erosion at low lake level. Loss-on-ignition profiles for the full set of eleven cores indicate that within-basin variability is small, with the exception of cores from Naivasha's main basin (Verschuren, 1996). Profiles of representative cores are used here to illustrate between-basin variability. All analyses were performed at contiguous 1-cm intervals throughout the profiles. Freeze-core NS93.2-F (Fig. 3) was analysed at contiguous intervals of variable thickness guided by fine structure of the stratigraphy. For uniformity, the data from this core were here transformed as if sampled at 1-cm intervals. Sediment chronologies are based on lead-210 dating, with sediment age calculated according to the constant-rate-of-supply (c.r.s.) model (Appleby and Oldfield, 1978). Laboratory methods are to be described in detail elsewhere.

PRELIMINARY RESULTS AND DISCUSSION

Sedimentary profiles in Figure 3 represent the last ca. 120 years of sedimentation in each basin of the Naivasha–Sonachi system. The profiles start in a dry period immediately before 1883 AD, when lake level stood lower than at any time in the 20th century (Fig. 2; Åse et al., 1986).

Volumetric water content (= porosity) of the recent sediments generally exceeds 95% in all four basins and is as high as 99% in surface sediments from Lake Oloidien, Crescent Island Crater, and Lake Sonachi. In the two hydrologically closed basins Oloidien and Sonachi, water content decreases to 90% in low-organic clays deposited during low lake level phase in the 1870s. The profiles of loss-on-ignition (Fig. 3) display signs of recent change in the depositional environment at the core sites that correlate with the major climate-driven events in the basins' recent history. Organic-matter content is typically low to intermediate (12–20%) in sediments deposited during the 1870s lowstand, at the bottom of the profiles. In all four basins this first unit is followed by a thick second unit of very dark brown high-organic muds (30–50%), deposited during the extended period of high lake level between about 1890 and 1930. Organic-matter content decreases during the progressive lake-level decline of the 1920s and 1930s to reach a new level of intermediate concentration (20–30%) by about 1940. In the main basin this third unit extends to the sediment surface,

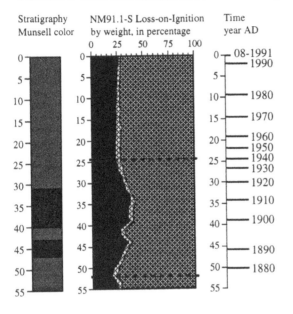

NM91.1-S Lake Naivasha, main basin   5.60 m depth

Stratigraphy   NM91.1-S Loss-on-Ignition   Time
Munsell color   by weight, in percentage   year AD

Figure 3. Visual stratigraphy, loss-on-ignition, and ^{210}Pb-derived sediment age of representative short-core profiles from central locations in each basin of the Naivasha–Sonachi system. Downcore depth in cm. Sediment chronologies are preliminary and subject to minor revision.

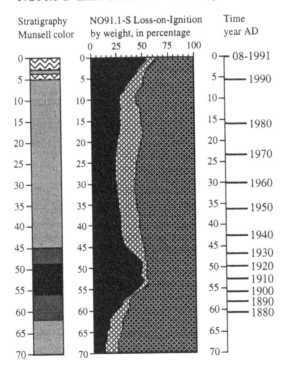

NO91.1-S Lake Oloidien   6.15 m depth

Stratigraphy   NO91.1-S Loss-on-Ignition   Time
Munsell color   by weight, in percentage   year AD

flocculent clayey mud
olive brown 2.5Y 4-5/4

clayey mud
dark brown 10YR 3/3

organic clayey mud
very dark brown 10YR 2-3/2

high-organic mud
black 10YR 2/1

laminated sediments

% Organic

% Inorganic

% CaCO$_3$

Figure 3, cont'd

NC91.1-S Crescent Island Crater   15.00 m depth

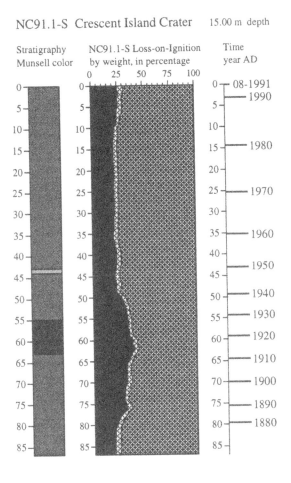

NS93.2-F Lake Sonachi   4.25 m depth

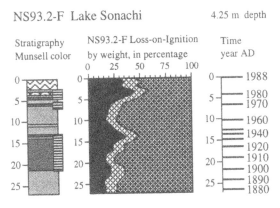

Island Crater) or strong (Lake Oloidien, Lake Sonachi) increases in organic matter near the top.

Based on comparison with core NM93.1-S from the deepest part of the main basin (Fig. 1), a disconformity must be inferred at 24 cm depth in core NM91.1-S (upper dashed line in the loss-on-ignition profile, Fig. 3). Constancy of organic-matter content in sediment deposited above this disconformity indicates that the local depositional environment has remained virtually unchanged for the past 40 years, notwithstanding lake-level fluctuations causing water depth at the core site to vary between 4 and 8 m (Fig. 2). Within this range of water depths the main basin is probably always continuously polymictic, with adequate oxygenation of the water column and a fully developed bottom fauna. Bioturbation, possibly augmented by wind-induced physical resuspension of surface sediments at the lowest lake levels, evidently formed a zone of mixing at the sediment–water interface in which potential proxy signals of lake-level fluctuation were smoothed. Even at water depths of 8 to (maximum) 18 m such as were common during the early 20th-century highstand, the main basin probably remained continuously polymictic with bottom oxygenation at least as adequate as in Crescent Island Crater today. Thus it is unlikely that the high organic-matter content of sediments deposited during that time reflect lower rates of organic decomposition in poorly oxygenated bottom waters. Similar dry-sediment accumulation rates during periods of high or low lake level further indicate that the changes in organic matter cannot be explained solely by concentration or dilution due to a variable influx of allogenic mineral components. Consequently, the higher organic-matter content of sediments deposited during high lake level most likely reflects increased organic deposition, for instance of plant material derived from desintegrating mats of swamp vegetation.

In core NO91.1-S from Lake Oloidien (Fig. 3), the early 20th-century highstand stands out clearly both in visual stratigraphy and organic-matter content. In Oloidien, deposition of high-organic sediments during the period 1880–1930 coincides with low rates of dry-sediment accumulation. Similarly, the slight increase in organic matter at about 20 cm depth coincides with a decrease in sediment accumulation rate during the 1970s and early 1980s (Verschuren, 1994). This inverse relationship between the sediment's organic-matter content and accumulation rate suggests that the organic-matter profile may be explained at least in part by dilution of organic matter with mineral sediment components during periods of low lake level. Throughout the early-20th century period of high lake level, Lake Oloidien was confluent with the main basin of Lake Naivasha and fresh (cf. overlapping conductivity values in 1929–30, Fig. 2). Reduced sedimentation rates in Lake Oloidien during this freshwater period can be attributed to the probable development of a papyrus-swamp fringe similar to that in the main basin of Lake Naivasha today. Emergent littoral vegetation limited influx of erosional material from the surrounding landscape, thus reducing bulk sediment accumulation in the Oloidien basin and effectively concentrating its organic component. When the fringing swamp became permanently stranded during lake-level decline in the period 1920–45, the papyrus may have died off or was cleared for agriculture. The resulting larger influx of inorganic sediment components contributed to a drastic increase in accumulation rate and to the lower organic-matter content of sediments deposited during low lake level. The carbonate content of Lake Oloidien sediments ranges from 11–13% near the bottom half to 15–23% in the upper two thirds of the profile, with lower values at the sediment-water interface and in the subsurface organic-matter maximum. Sedimentary evidence of carbonate precipitation in Lake Oloidien is consistent with the supersaturation of Lake Oloidien lakewater with respect to calcite and its low Ca/Na ratio relative to water from the main basin (Gaudet and Melack, 1981; compare with the low carbonate content of main-basin sediments, core NM91.1-S in Fig. 3).

In Crescent Island Crater, sediment accumulates faster than anywhere else in the Naivasha–Sonachi system. The deep and steep-sided crater acts as a sediment trap for the northeastern corner of Lake Naivasha. Carbonate content of the recent sediment is low, similar to the main basin. In core NC91.1-S the three major sedimentary units are well-defined, with high-organic sediments deposited at high lake level and less organic sediments at low to intermediate lake levels before and after. A modest increase in organic-matter content at 10 cm depth may correspond to enhanced organic-matter deposition during the early 1980s, which has been inferred by Brierley et al. (1987) to be the cause of profundal anoxia at that time. During the early 20th-century highstand, Crescent Island Crater may have been discontinuously polymictic with more frequent occurrence of near-bottom anoxia. Fossil assemblages of the benthic fauna (Verschuren, in prep.) may help to decide between increased organic deposition or slower decomposition as the main reason for high percentages of organic matter in the corresponding sedimentary unit.

Undisturbed core samples of the highly porous (95–99%) surface sediments in Lake Sonachi show its recent sediment record to be partially laminated, similar in aspect to many late-Quaternary records from closed-basin lakes in arid regions worldwide. Three units of cyclically laminated sediment in freeze-core NS93.2-F (Fig. 3) correspond with periods when Z_m in Lake Sonachi was greater than 5 m, the average observed chemocline depth. A narrow interruption of varved (i.e., annual) lamination at 5.0–4.5 cm depth is particularly interesting, as it appears to coincide with a shortlived lake-level drop between 1975 and 1977 (Fig. 2). Following the very low lake levels of the 1940s and early 1950s, renewed preservation of fine cyclical lamination appears to have started only about 1970, more than a decade after water depth had increased above 5 m. Sediments deposited before that time are coarsely laminated, suggesting a holomictic regime of discontinuous polymixis with possible physical resuspension but in which bioturbation is limited due to prevalence of near-bottom anoxia. The loss-on-ignition profile is complex and difficult to interpret in terms of the alternating deep-water meromictic versus shallow-water holomictic lake phases. Still, similar to core profiles from the other three basins, the 1940s–1950s lowstand is characterized in Lake Sonachi by decreasing and low organic-matter content. After the re-establishment of meromixis sometime in the 1960s, organic-matter content increased again to reach maximum values in varved sediment of the early 1980s. The overall inverse correlation of organic-matter content with the rate of dry-sediment accumulation indicates that, similarly to the situation in Lake Oloidien, organic-matter content of Lake Sonachi sediments is governed in part by the rate of dilution of authigenic organic matter with allogenic mineral components.

CONCLUSION

In reconstructing paleoclimate from lake-sediment records, various alternative scenarios typically exist to explain patterns and changes in sedimentary proxies. The discussion above of organic-matter content exemplifies the manner of how comparative paleolimnology can help to identify different causes for similar patterns of change in a selected proxy. The challenge of comparative paleolimnology is to identify combinations of patterns in independent sediment proxies that yield singular paleoclimatic interpretations. Identification of these same pattern combinations in longer sediment records can then be better trusted to result in an internally consistent reconstruction of climatic history. The recent, documented history of the Naivasha–Sonachi system includes phases of lake behavior corresponding to windstressed freshwater swamps, polymictic lakes with and without near-bottom anoxia, unstratified brine pools, and chemically stratified meromictic lakes. Considerable between-basin variability of the sediment record points to the profound influence of physical limnology, hydrology, and taphonomy on the details of the lake-sediment archive that is to become the source of paleoclimate-related information. Comprehensive paleolimnological analysis of the full set of undisturbed short cores will yield further insight into basin-specificity of mechanisms of climate-signal incorporation into the sedimentary proxy record, and may so contribute to improved high-resolution interpretation of late-Quaternary lake-sediment archives for the reconstruction of continental paleoclimate and environmental change in tropical Africa.

ACKNOWLEDGMENTS

This study is funded by NSF-RTG 90-14277-01 and the Quaternary Paleoecology Program at the University of Minnesota, and by OOA 120-50-790 to Dr. H. J. Dumont (University of Ghent). The fieldwork was conducted with research permission from the Office of the President of the Republic of Kenya to Dr. K. M. Mavuti (University of Nairobi). I wish to thank Dr. Mavuti, Dr. D. M. Harper (University of Leicester), and officials of the Olkaria Geothermal Co. for logistic support, the Lake Naivasha Riparian Owners Association for access to the lakes, Dan Engstrom for access to lead-210 dating facilities, Jo Verschuren, Nicola Pacini, and Fabienne Janssen for help in the field, and Herb Wright and Kerry Kelts for support. The author is research assistant with the National Fund for Scientific Research (Belgium), and acknowledges a Honorary Fellowship from the Belgian-American Educational Foundation. Contribution 469 of the Limnological Research Center, University of Minnesota.

REFERENCES

Appleby, P. G., and F. Oldfield, 1978. The calculation of lead-210 dates assuming a constant rate of supply of unsupported ^{210}Pb to the sediment. Catena 5: 1–8.

Åse, L.-E., K. Sernbo, and P. Syrén, 1986. Studies of Lake Naivasha, Kenya, and its drainage area. Forskningsrapport från Naturgeografiska Institutionen Stockholms Universitet 63: 75 pp.

Åse, L.-E., 1987. A note on the water budget of Lake Naivasha, Kenya. Geogr. Ann. 69A: 415–429.

Barnard, P. C., and J. Biggs, 1988. Macroinvertebrates in the catchment streams of lake Naivasha, Kenya. Rev. Hydrobiol. trop. 21: 127–134.

Beadle, L. C., 1932. Scientific results of the Cambridge expedition to the East African lakes, 1930–1931. 4. The waters of some East African lakes in relation to their fauna and flora. J. Linn. Soc. Zool. 38: 157–211.

Brierley, B., D. Harper, and R. Thomas, 1987. Water chemistry and phytoplankton studies at Lake Naivasha: short-term spatial and temporal variations. In D. Harper (ed.): Studies on the Lake Naivasha ecosystem 1982–1984. University of Leicester, Leicester.

Burgis, M. J., and K. M. Mavuti, 1988. The Gregory Rift. In M. J. Burgis and J. J. Symoens (eds.): African wetlands and shallow water bodies. Éditions ORSTOM: 331–340.

Clark, F., A. Beeby, and P. Kirby, 1989. A study of the macro-invertebrates of lakes Naivasha, Oloidien and Sonachi, Kenya. Rev. Hydrobiol. Trop. 22: 21–33.

Damnati, B., M. Taieb, M. Decobert, D. Arnaud, M. Icole, D. Williamson, and N. Roberts, 1991. Green Crater Lake (Kenya): chimisme des eaux et sédimentation actuelle. 3ième Conférence Internationale des Limnologues d'expression française (Hommage à F.A. Forel): 270–275.

Darling, W. G., D. J. Allen, and H. Armannsson, 1990. Indirect detection of subsurface outflow from a rift valley lake. J. Hydrol. 113: 297–305.

Duplessy, J.-C., L. Labeyrie, M. Arnold, M. Paterne, J. Duprat, and T. C. E. van Weering, 1992. Changes in surface salinity of the North Atlantic during the last deglaciation. Nature 358: 485–487.

Gaudet, J. J., and J. M. Melack, 1981. Major ion chemistry in a tropical African lake basin. Freshw. Biol. 11: 309–333.

Harper, D. M., K. M. Mavuti, and S. M. Muchiri, 1990. Ecology and management of Lake Naivasha, Kenya, in relation to climatic change, alien species' introductions, and agricultural development. Environ. Conserv. 17: 328–336.

Harper, D. M., G. Phillips, A. Chilvers, N. Kitaka, and K. Mavuti, 1993. Eutrophication prognosis for Lake Naivasha, Kenya. Verh. Internat. Verein. Limnol. 25: 861–865.

Jenkin, P. M., 1936. Reports of the Percy Sladen expedition to some Rift Valley lakes in Kenya in 1929. VII. Summary of the ecological results, with special reference to the alkaline lakes. Ann. Mag. Nat. Hist. Ser. 10, 18: 133–181.

Kelts, K., and K. Miller, in prep., Rapid change signatures in East African lakes. ?????

Kilham, P., 1971. Biogeochemistry of African lakes and rivers. Unpublished thesis, Duke University, Durham: 199 pp.

Kutzbach, J. E., and F. A. Street-Perrott, 1985. Milankovitch forcing of fluctuations in the level of tropical lakes from 18 to 0 kyr BP. Nature 317: 130–134.

Lewis, W. M., Jr., 1983. A revised classification of lakes based on mixing. Can. J. Fish. Aquat. Sci. 40: 1779–1787.

MacIntyre, S., and J. M. Melack, 1982. Meromixis in an equatorial African soda lake. Limnol. Oceanogr. 27: 595–609.

Marlier, G., 1951. Recherches hydrobiologiques dans les rivières du Congo oriental. Composition des eaux. La conductibilité électrique. Hydrobiologia 3: 217–227.

Melack, J. M., 1976. Limnology and dynamics of phytoplankton in equatorial African lakes. Unpublished thesis, Duke University, Durham: 453 pp.

Melack, J. M., 1979. Photosynthetic rates in four tropical African fresh waters. Freshw. Biol. 9: 555–571.

Milbrink, G., 1977. On the limnology of two alkaline lakes (Nakuru and Naivasha) in the East Rift Valley system in Kenya. Int. Revue ges. Hydrobiol. 62: 1–17.

Njuguna, S. G., 1988. Nutrient–phytoplankton relationships in a tropical meromictic soda lake. Hydrobiologia 158: 15–28.

Renberg, I., 1981. Improved methods for sampling, photographing and varve-counting of varved lake sediments. Boreas 10: 255–258.

Smol, J. P., I. R. Walker, and P. R. Leavitt, 1991. Paleolimnology and hindcasting climatic trends. Verh. Internat. Verein. Limnol. 24: 1240–1246.

Street-Perrott, F. A., and R. A. Perrott, 1991. Abrupt climate fluctuations in the tropics: the influence of Atlantic Ocean circulation. Nature 343: 607–612.

Talling, J. F., and I. B. Talling, 1965. The chemical composition of African lake waters. Int. Revue Ges. Hydrobiol. 50: 421–463.

Verschuren, D., 1993. A lightweight extruder for accurate sectioning of soft-bottom lake sediment cores in the field. Limnol. Oceanogr. 38: 1796–1802.

Verschuren, D., 1994. Sensitivity of tropical-African aquatic invertebrates to short-term trends in lake level and salinity: a paleolimnological test at Lake Oloidien, Kenya. J. Paleolim.

Vincent, C. E., T. D. Davies, and A. K. C. Beresford, 1979. Recent changes in the level of Lake Naivasha, Kenya, as an indicator of equatorial westerlies over East Africa. Climatic Change 2: 175–189.

Wright, H. E., Jr., 1980. Coring of soft lake sediments. Boreas 9: 107–114.

Impact of Man

Anthropogenic Threats, Impacts and Conservation Strategies in the African Great Lakes: A Review

A.S. COHEN *Department of Geosciences, University of Arizona, Tucson, United States*

L. KAUFMAN *Department of Biology, Boston University, Boston, Massachusetts, United States*

R. OGUTU-OHWAYO *Fisheries Research Institute, Jinja, Uganda*

Abstract — The African Great Lakes are unfolding human experiments on an enormous scale; even the most "pristine" of these lakes are undergoing rapid environmental change, the origins of which lie in regional human activities. Understanding changes in the lakes, and identifying possible remedial actions can only be done in the context of understanding changes in human demographics and land use in the region. Disturbance factors affecting the lakes are interactive, but can be categorized as follows: 1) Fishing activities, including overfishing and lack of effective management practices; 2) Discharge of pollutants; 3) Damage to watersheds, leading to cultural eutrophication and excess sedimentation problems; 4) Introduction of exotic species and translocations of species within lakes, leading to extinctions and drastically altered community structures, and; 5) Regional climate change.

There is a growing consensus that developing effective conservation strategies will require substantive changes in how the lakes are treated. These include legal reforms, renewal of monitoring and enforcement activities, development of incentive programs for improved land/lake use, management planning based on local ecological realities, more effective community participation in conservation planning and decision making and improvements in international cooperation. Development of protected reserves is a priority in all lakes both for economic reasons and for the maintenance of biodiversity. Long-term records of lake change are needed to assess the significance of our scanty historical data; this can be effected through greater use of paleolimnological proxy records. Measures are underway in all of these areas, but rapid progress is critical to prevent further environmental degradation of the lakes.

INTRODUCTION

Viewed from an airplane at 10 km altitude, the Great Lakes of Africa seem reservoirs of enormous ecological stability, far too large to succumb to the impacts of human activities around them. As for the African equatorial rainforest, the lakes' terrestrial counterpart, we now understand that this image of immutability was a chimera fostered by ignorance. The truth is that even the largest and most complex ecosystems, great lakes among them, have been severely altered by human interference. In this paper we review what is currently known about human impacts to the African Great Lakes. Other authors in this volume have addressed paleolimnologic and biotic histories. This story is different. It is changing. And the rate of change is faster than the scientific literature can track it, as lake ecosystems gyrate beneath the confluent impacts of anthropogenic and other forces.

Few things reveal the speed of these changes more clearly than the speed at which we have come to recognize them. In the 1970s there was doubt as to whether significant impacts even existed; by the 1980s the evidence was indisputable. The case was sealed by a stupendous rate of population growth within the region, which could only accelerate rates of change in the lakes. What remains uncertain today is the order of magnitude of these impacts, and consequently, the magnitude of risk that they pose to humanity and the rest of nature.

Small lakes have been frequently manipulated in experiments to test hypotheses of ecological dynamics (Schindler, 1990), laying the groundwork for theory on how human activities impact lake ecosystem structure and function. For example, small lakes have been perturbed to look at the effects of nutrient load enhancement, and the impact of removal or addition of species. Inadvertently, the large African lakes have been used as experimental systems, too. Unfortunately, the outcomes have often deviated from expectation, and scientific controls have been insufficient to figure out what had actually happened. Some lakes, like Victoria and Kyoga, have been heavily manipulated through species introductions and nutrient loading. Not surprisingly, their ecosystems are responding dramatically. The precise manner in which they are responding is indeed surprising ... frighteningly so. "Experiments" in such lakes as Tanganyika and Malawi/Nyassa are in much earlier stage, though even there signs of anthropogenic change are evident.

The African Great Lakes are frequently referred to as natural laboratories; this does not mean that human interference constitutes a laboratory experiment with a predictable outcome. The complexity of large lakes makes a multitude of ecological outcomes possible from a single set of starting conditions and input changes. This view is strongly supported both by empirical studies of multiple metastable community structures occurring under similar conditions, and by chaos theory. Thus, predictions about one lake based upon observations of the unfolding "experiments" in another, while worth indulging, must be taken with a grain of salt. In any event, apart from a discussion of possible impacts of changing climates on the lakes, we cannot prognosticate. Rather, our goal here is to synthesize what we know at present about the status of anthropogenically induced change in these lakes, and to review the various suggestions which have been made for regional conservation strategies.

A few generalizations are possible. One is that the African Great Lakes are acutely vulnerable to the impacts of pollution, much more so than one would guess on the basis of their size (Coulter and Jackson, 1981; Coulter, 1992; Hecky and Bugenyi, 1992; Bootsma and Hecky, 1993). The large volumes of these lakes, combined with their high evaporation and low outflow rates, make water (and pollutant) residence times exceedingly long, on the order of tens to hundreds of years. Additionally, their tropical setting fosters thermal stability, low oxygen saturation potential and high metabolic demand for oxygen, all of which make lakes extremely sensitive to the effects of eutrophying pollutants. More than twenty years have passed since Fryer (1972a,b) first called attention to this issue, and suggested that measures be considered to control such problems. Now this potential for pollution impacts is in fact being realized, particularly through cultural eutrophication. Fryer (1972a,b) also recognized the precarious position of the African fisheries and called for action to be taken to preserve the lake fish faunas both as economic resources and for their extraordinary biodiversity value. Subsequent research (e.g., Barel et al., 1985; Kaufman, 1992: Kaufman and Ochumba, 1993; Turner, 1993; Lowe-McConnell, 1993; Goldschmidt et al., 1993) has indicated that these warnings were well grounded; the African lake faunas

Table 1. Population Statistics for the African Great Lakes Basins

Lake basin	Basin population (mid-1990s)	1980s mean growth rate (%)	% basin population urban (1990)
Victoria	27.7	3.6	7
Tanganyika	6.2	3.4	12
Malawi/Nyassa	5.5	3.4	7
Kyoga	5	3.6	<2
Kivu	3	3.5	13
Edward	2?	3.5	<2
Turkana	2?	3.0?	<2
Albert	1?	3.4	<2

Source: World Bank and FAO statistics and projections (1990). Population statistics were calculated from regional population data within each country, using population density maps as additional information where necessary. Recent estimates for growth rates in the region have been revised downward sharply, as a result of the AIDS crisis. Estimates for Edward, Turkana and Albert Basins are very approximate.

are undergoing rapid change involving large-scale extinctions and alterations of ecosystem structure.

It is heartening that with time has come awareness: recognition of the challenges facing human communities around the African Great Lakes has greatly increased over the past twenty years. Unfortunately, time has not brought consensus on what to do about the problems, nor has it seen much energy directed toward solving them. There are ample reasons for the delay. Economic and social crises in many of the countries surrounding the lakes have impeded decision making and management for pollution control, reserve development and fisheries and wildlife conservation. Meanwhile, the rate of limnologic change has accelerated, making it ever more difficult to reverse degradation and conserve lake resources.

THE HUMAN FACTOR

Population and Land Use

Like any natural perturbation, human activity in the East African lake basins has a revealing historical context (Harris, 1993). The human population of the intralacustrine region of East and Central Africa (Table 1) is high, growing rapidly, and heavily concentrated near the lakes. Densities vary considerably among lake watersheds, from well over $100/km^2$ for the Victoria basin, to $<1/km^2$ for some regions around L. Turkana. Most of the people live in small villages and towns; urban areas represent only a small (albeit growing) fraction of the population. These dense rural human populations are often matched by equally high cattle populations, despite shifts away from a pastoral lifestyle (Bootsma and Hecky, 1993). Heaviest populations (both urban and rural) are concentrated within a short distance of the water, especially around Lakes Victoria, Kivu and the north end of Lake Tanganyika (see Bootsma and Hecky, 1993, Fig. 4). Thus, the cumulative impact of human activity is powerfully focussed upon these lakes, a fact with many ramifications for patterns of resource consumption (notably increased demands for potable water and fresh fish) and waste discharges, which are mostly into the lakes. Population density around the lakes is not merely high; it is growing at one of the highest rates in the world, averaging about 3.5% per year during the 1970s–80s (i.e., a doubling time of approximately 20 years). Unpublished data suggests that AIDS-related mortality has reduced the growth rate to 2% or less (Harris, pers. comm., 1993), though this should certainly not be regarded as relevant to long-term solution of population-related problems.

Rapid population growth has resulted in an equally rapid conversion of most of the watershed areas from forest and savannah woodland habitats (including areas of traditional slash and burn/shifting cultivation) to agricultural and range land. For example, lowland deforestation in Uganda and Burundi, both of which had extensive mesic forests or wood-land areas, is now complete but for small, isolated patches. Even these are subject to severe encroachment. The conversion has occurred most rapidly during the past 50 years. The history of this change provides an important context for understanding other problems in the lake basins.

Prior to the period of explosive population growth of the past few decades, most agriculturists throughout the intralacustrine watersheds practiced shifting cultivation, where small plots (usually communally held) were cleared of all but the largest trees for temporary cultivation (Hullsworth, 1987). This type of farming is still practiced in some low popula-tion-density portions of the lake watersheds (notably in eastern Zambia). Lal (1982) has shown that such practices, contrary to popular notion, are effective at controlling soil erosion, compared with intensively cropped areas with conventional soil conservation correctives. For example, in Nigeria he found that slash/burn plots which had been mulched with large trees left standing (traditional practice throughout most of the lake basins) lost on average 500kg of soil/ha/yr compared with ~30 tons/ha/yr on conventionally cleared plots which were protected by contour banks.

Traditional, intensive (i.e., nonshifting) agricultural practices in the Great Lakes region were largely restricted to groups who, for various reasons, found themselves limited to small plots, where shifting agricultural methods were not feasible. Examples include the Watengo people of Tanzania, who were restricted by warfare to farming a small mountainous area, the Wakara people of Ukara Island (Lake Victoria), and other groups in Ethiopia and Burundi (Hullsworth, 1987). In cases such as these, where land was limited and populations relatively stable, methods of soil conservation were derived (such as soil and grass terracing, fence gullying and embanking streams) which allowed for long-term utilization of steeply sloping farm lands. In most parts of the Great Lakes region however, soil terracing and other soil conservation techniques were not practiced, for the simple reason that low population pressures made them unnecessary (Anthony et al., 1979; Hullsworth, 1987). Thus, there are few East African counterparts to the soil conservation techniques developed over the long East Asian or Andean histories of high population pressure in mountainous areas.

As population densities in East and Central Africa have risen, rural land use practices have shifted towards more and smaller land holdings. This has forced a larger proportion of shifting-cultivation land to remain under "temporary" (intensive) cultivation. In a study in Western Kenya, Hullsworth (1987) found that farmers with "unrestricted" access to land retained 38% of their holdings under temporary crops; this figure rose to 60% when mean farm size dropped to under one hectare. A system of shifting agriculture becomes impractical as population densities rise above some threshold; this threshold has now been exceeded throughout most of the Great Lakes region except in Zambia and some parts of Zaire and Tanzania. The problem surfaced earlier and is most acute where soils are naturally poorer, such as the non-volcanic areas common in much of the western rift.

Unfortunately, the change from shifting to intensive agricultural methods has evolved so quickly in the Great Lakes basins that there has not been time to develop a tradition of palliatives to counteract the problems associated with intensive farming, soil erosion and loss of soil fertility. Terracing is practiced by only ~25% of farmers in western Kenya today (irrespective of farm size) and this is probably a maximum figure for the region. Ironically, the abandonment of traditional (and highly effective) methods for maintaining soil-fertility has proceeded most rapidly among younger, more literate farmers. These farmers are more likely to be swayed towards the use of fertilizers over traditional mulching, and also to abandon mixed cropping, in which the fields must be hoed by hand. Both of these trends have serious implications for the loss of nutrients in croplands through soil erosion and their subsequent discharge into the lakes (Hullsworth, 1987; Bizimana and Duchafour, 1991).

The use of fuel wood (and particularly charcoal) as a primary energy source among rural people in the Great Lakes region is a further contributory factor to both accelerating rates of deforestation and enhancing nutrient loading into the lakes (notably, by way of particulates carried by wind). At present 71% of total energy consumption in Subsaharan Africa is in the form of fuel wood, a figure which has remained remarkably constant in recent years (Davidson, 1992). In some countries (e.g., Tanzania, Malawi and Ethiopia) this figure is over 90%. Although afforestation and rural electrification programs exist in most of the intralacustrine countries, they have not reversed the growing gap between fuelwood consumption and production (Davidson, 1992).

Clearly, high population densities and growth rates are at the crux of the problems facing the African great lakes watersheds today. All potential solutions rest upon the ability of governments to halt these demographic trends and secondly to legislate rational land and water usage, and to gain the acceptance of such regulations by the people in the region. The problems of fuelwood shortage and high rural population density are especially acute and demand immediate attention if the long-term potential of these lakes is to be preserved.

MAJOR DISTURBANCE FACTORS AFFECTING THE LAKES

It is extremely difficult to ascribe any of the African Great Lakes' problems to a unique causal agent. Sometimes this is because of inadequate data (e.g., Kling, 1992; Kaufman, 1992); more often there are interactions among multiple causes of change. However, we can identify five primary categories of anthropogenic impacts on the lakes. The first two, fishing and species introductions, involve the artificial manipulation of species distribution and abundance patterns. The second two, deforestation and pollution, are related to lake inputs. Examples of these first four categories are well known. The fifth, anthropogenically caused climate change, is more controversial, though it is also likely to become more acknowledged in the future.

Fishing Activities

> Stocks can sustain considerably higher levels of exploitation indefinitely.
> Stoneman et al. (1973) after Bazigos (1972) in reference to Lake Malawi

> Over the past 30 years vital protein resources have been squandered by failure to manage (fishery) stocks properly.... It is a sad reflection on our ability to manage resources that the past decade has seen no amelioration of overfishing.
> Massinga (1990)

Fisheries represent one of the most important economic resources of the intralacustrine region. Fish is the least expensive form of animal protein available to the people of a region where protein deficiencies are a major health problem (Mwandu, 1992, cited in Hanek and Greboval, 1992). Fish comprises over 50% of the animal protein consumed throughout the region, and in some countries this figure rises to 70% (Hecky and Buugenyi, 1992; Ntakimazi, 1992; Ogutu-Ohwayo, 1992a; Tweddle, 1992). The fishes of the great lakes, particularly those of Lakes Malawi, Tanganyika and Victoria, also represent an enormous scientific and ecologic resource for Africa, given the importance that these endemic species flocks have played in understanding evolutionary processes and as hotspots of biodiversity for the planet. It is not surprising therefore that the quickening exploitation of fish resources in the lakes has been fraught with controversy, as short-term and long-term goals have collided.

Twenty years ago a sense of optimism was prevalent among many fisheries managers on the African Great Lakes. If all was not well with African lake fisheries, then at least all was on the right track, and the occasional critics of industrial fishery practices were castigated as doomsayers (cf. Fryer, 1972a; Stoneman et al., 1973; Jackson, 1973). This optimism was guided in large part by a rapidly expanding catch during the early phases of the industrial (or large-scale artisinal) fisheries on many of the Great Lakes. A useful example to consider comes from Lake Turkana, where rapid growth in catch rates cannot be attributed to species introductions (as in Lake Victoria). Impressive growth in the Lake Turkana gillnet fishery catches during the 1960s and early 70s led fishery advisers to predict maximum sustainable yields of between 30,000–225,000 tons per year for the lake (Wurtz and Simpson, 1964; Rhodes, 1966; Hopson, 1982; Kolding, 1992). Similarly impressive predictions, bolstered by more sophisticated data collection methods like acoustic surveys that estimate standing fish biomass, were produced for a number of the other Great Lakes (e.g., see review in Coulter, 1991, for Lake Tanganyika). Such impressive predictions naturally provided encouragement for both governments and the private sector to increase investment in fisheries, providing capital for both fishing fleets and fish processing facilities.

Subsequently, this early optimism has given way to a more sobering reality of currently declining fishery yields in most of the Great Lakes. Declining catch per unit effort data, even during the early phases of many of the Great Lake fisheries indicate that the initial predictions of fish production were unsustainable. Today the data call into question the entire concept of "maximum sustainable yield" as it has previously been applied by fisheries analysts, both in Africa, and elsewhere.

Declining catches per unit of fishing effort, observed in many lakes today (i.e., Roest, 1992, for Lake Tanganyika, Hanek and Greboval, 1992, for Lake Mweru) are, in part, indicative of the intensification of fishing activity around many of the lakes. In most African Great Lakes, the attraction of good fishing has brought about an uncontrolled growth in fishing effort, as people with few options for employment are drawn to a lucrative industry.

In both industrial and artisanal fisheries, this has resulted in growing numbers of people and fishing vessels pursuing fewer and fewer fish (e.g., for Lake Turkana, Kolding, 1992, Table 1; Lake Malombe, Malawi, D. Tweddle, pers. comm., 1993). In some areas, the large number of new fishing operations, particularly large-scale operations, has undermined traditional paths of authority which governed fishing rights (Yongo, 1991). Even where total yields have remained relatively stable over the past decade, as in Lake Mweru, the pressures from existing fishers to restrict fishing activities by newcomers are strong. In some cases, entire segments of the fishing industry have collapsed, as fishing has become no longer economically viable. This has been the experience of the industrial fishing fleet in northern Burundian waters over the past 5 years (C. Vrampas, pers. comm., 1992). It remains uncertain whether such collapses can be attributed strictly to overfishing or to more complex factors, such as shifts of production to more efficient or less costly artisanal methods (as has been proposed for Lake Tanganyika). Nevertheless, it is clear that in many of the African Great Lakes today, more fish are being removed than can be sustained through existing productivity; they are being overfished in the commonly understood sense of the word.

In addition to overfishing, other effects of fishing activities have included the selective damage to particular species or communities through certain types of fishing (for example, the damage caused by beach seining to shallow-water benthic fish communities or fishing with explosives), and the alteration of community structure caused by the size-selectivity of fishing nets (Kaningini, 1992; Lowe-McConnell, 1993).

COMPARATIVE FISHERIES IMPACTS ON AFRICAN GREAT LAKES ECOSYSTEMS

The following synthesis was developed from historical accounts of the fisheries on Lake Malawi/Nyassa (Turner, 1977; Magasa, 1988; Alimoso et al., 1990; Massinga, 1990; and Tweddle, 1992); Lake Tanganyika (Chapman, 1975; Roest, 1988, 1992; and Coulter 1992); Lake Turkana (Kolding, 1992); Victoria and Kyoga (Ogutu-Ohwayo, 1990; Lowe-McConnell, 1993); and Lakes Albert, Edward and George (Orach-Meza et al. 1990).

Fisheries in the African Great Lakes are, like the assemblages on which they are based, very diverse, representing a spectrum that ranges from cichlid-dominated resources like the Lake Malawi fishery, to those in which other families support the major commercial fisheries, as in Lake Albert and Lake Tanganyika. For convenience, the fisheries in each lake can be discussed in terms of three groups: ground, pelagic, and potadromous. Groundfish assemblages are in most lakes overwhelmingly dominated by cichlids, with centropomids, catfishes, lungfishes, mormyrids, and eels of varying significance. The pelagic assemblages consist of clupeids and/or cyprinids, planktivorous cichlids, and associated open water predators — cichlids, centropomids, and characoids. Potadromous fishes (the intralacustrine equivalent of anadromous species like salmon) are mostly cyprinids (Lowe-McConnell 1993).

Until the advent of large-scale mechanized fisheries, traditional fishing efforts were heavily concentrated within 1 or 2 km from the shorelines of all the African great lakes (Coenen, 1992; D. Tweddle, pers. com., 1993). Traps and beach seines were the major methods used to take groundfishes and potadromous species on their spawning runs. Pelagic fisheries were (and are) largely a canoe-based operation. The fishes are lured by fishing lamps and then taken in scoop nets, lift nets, or beach seines. Such methods still predominate in lakes Kyoga, Edward/George, Turkana, and Albert. However, on Lakes Tanganyika, Victoria, and Malawi, the traditional foundation of the African great lakes fisheries are being

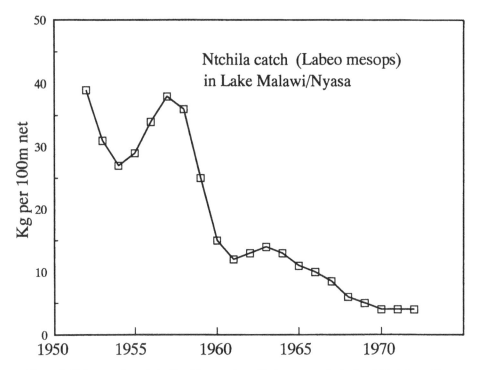

Figure 1. Fishery catch statistics for gillnet catches of *Labeo mesops* in the South East Arm of L. Malawi (From Alimoso et al., 1990).

obscured by a densely woven amalgam of 20th century changes: the introduction of modern fishing methods and improved "artisanal" methods, shifts in species assemblages, and changes in lacustrine environments.

In each of the lakes, a procession of modernization events was followed closely by serial collapse of the native fish stocks. Following their introduction in Lake Victoria, mass-produced flax gill nets led quickly to a precipitous decline in *Oreochromis esculentus* (at the time the most important commercial species in the lake); following this fishermen resorted to increasingly fine mesh sizes despite attempts at regulation (Ogutu-Ohwayo, 1990). Gill nets have also had a devastating impact around river mouths during spawning migrations of potadromous fishes (Soulsby, 1960; Ogutu-Ohwayo, 1990; Massinga, 1990; Tweddle, 1992; Fig. 1). Ringnets were brought to Lake Malawi/Nyassa to target *Oreochromis*, later to be used to catch the pelagic *Engraulicypris sardella* ("usipa") in 1974, and finally upon small haplochromine cichlids ("utaka") in 1975. The effects of such serial targeting are unknown. Until the mid-20th century, the open waters of the African great lakes were unexploited. This phase ended with the advent of a purse-seine fishery in Lake Tanganyika in the mid-1950s. A modernized lift-net fishery for the pelagic *Rastrineobola argentea* has recently developed in the Mwanza Gulf region of Lake Victoria, and there is strong interest in expanding it to other regions of this lake. The importance of pelagic cichlids as bycatch may pose a significant threat to these already endangered species. Uses of trawlers expanded greatly during the 1970s, first in Lake Malawi/Nyassa, then in Lake Victoria. In both lakes, the original targets were abundant benthic haplochromine cichlids (Kudhongania and Cordone, 1974; Lewis and Tweddle, 1990; Coenen, 1992; Tweddle, pers. com., 1993).

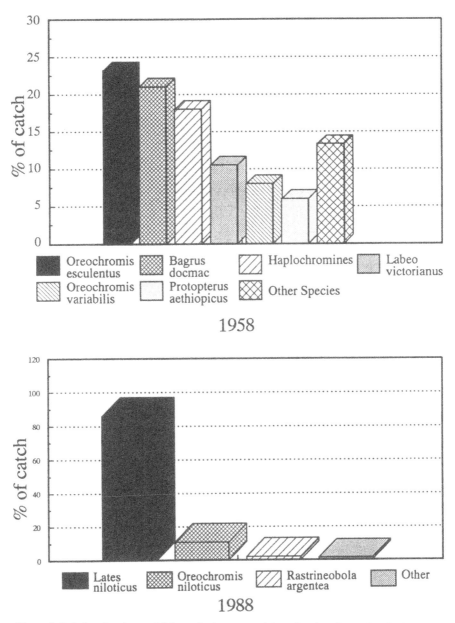

Figure 2. Relative abundance of fish species in commercial catches from Lake Victoria (Uganda waters) 1958 vs. 1988 (From Kudhongania, 1990). Notice the near complete disappearance of native species, and their replacement by Nile Perch (*Lates niloticus*) and Nile Tilapia (*Oreochromis niloticus*). By 1991, Rastrineobola catches had increased dramatically, with Nile Perch representing approximately 61% of the catch (Okaronon and Akumu, 1992).

With the collapse of the indigenous fauna in Lake Victoria (Barel et al., 1991), a new open-lake trawl fishery developed for the introduced Nile perch with heavy foreign subsidy (Fig. 2).

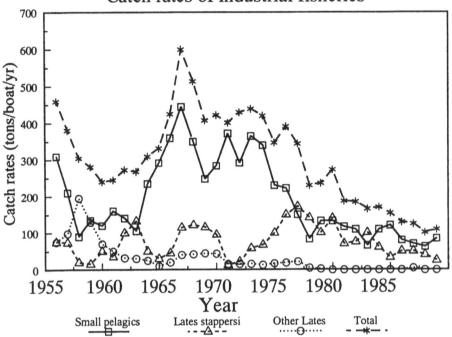

Figure 3. Catch rates of industrial fisheries in Burundi (From Roest, 1992).

Fishery impacts are usually thought of in terms of resource depletion. More far-reaching, however, are fisheries-driven ecosystem alterations (Coulter, 1991). For example, fisheries in Lake Tanganyika target six pelagic species: four centropomid predators (*Lates* spp.) and two small sardines. Large, predatory *Lates* constitute a large proportion of the biomass in unfished portions of the lake (Chapman, 1975; Roest, 1992). However, in heavily fished areas populations of the three largest *Lates* species have been reduced to very low levels, whereas populations of the smallest centropomid species (*L. stappersi*) initially increased, perhaps because of release from predation by the elimination of the largest species, *L. microlepis* (Fig. 3). Populations of the two clupeid species have been in oscillatory and apparently declining cycles during the same time period. These oscillations have been mirrored by out-of-phase cycles in *Lates stappersi*. Pearce (1988) suggested that marked oscillations in clupeid abundance resulted from predator–prey interactions between *Lates stappersi* and their clupeid prey.

That widespread declines in fisheries should be associated with a rapid increase in fishing pressure in extremely complex and volatile ecosystems is no surprise. However, the reasons for changes in the resource or the fishery are not straightforward. For example, the recent collapse of the industrial fishery in Lake Tanganyika may have more to do with advantages of small, easily landed craft than with any decline in catch–effort relationships (Ahayo,

1991; Elongo, 1991; Vrampas, pers. comm. 1992). Small boats can easily elude the array of regulations imposed on the industrial fleet.

The principal of Maximum Sustainable Yield, the foundation of modern fisheries management, has consistently failed in African Lakes for several reasons (Coulter, 1991; Kolding 1992). Although some of this failure may be attributable to poor statistics, better data are unlikely to solve the problem. It is important to note a widespread motivation to cleave to this explanation because it ensures continuation of programs and further research expenditures (see Twongo, 1990 for further discussion). Calculation of an MSY itself may be theoretically untenable, given the limnological and ecological changes imposed by the full panoply of anthropogenic impacts. Furthermore, numerous studies in both field and theoretical ecology have demonstrated that ecosystems may have several metastable states to which they can evolve following a perturbation. Thus it may not be possible to predict the course of effects following the introduction of a given intensity of fishing or of a given change in climate. This problem is compounded by the fact that, as ecosystems change state, inherent secondary productivity may go in or out of useable states for human consumption. A lake may retain a given level of inherent productivity, but for reasons of palatability or marketing, that productivity may not be useable. MSY estimates based on old or faulty data, irrelevant models, or which are applied in inherently unstable ecosystems, reinforce a complacency about the state of a key food resource for many African countries, a complacency that is surely not justified.

WATER POLLUTION AND SANITATION

Clean water from the African Great Lakes is a critical resource in regions like East and Central Africa, which are often water-poor (Hecky and Bugenyi, 1992). Maintaining high standards of water quality in the African Lakes is of considerable economic concern, both for human consumption (for domestic, industrial and agricultural uses) and for the role which water plays in public health. The health of the lake ecosystems themselves are also at risk from various forms of water pollution.

Tropical lakes are at particular risk from pollution hazards, owing to the lower oxygen saturation levels and higher oxygen consumption rates which occur at high temperatures. The large size and high evaporation rates of the African Great Lakes make pollutant retention time especially long, particularly in their hypolimnia. Bootsma and Hecky (1993) have modelled the retention time phenomenon for several large lakes. They estimated pollutant concentrations at various times both during the phase of pollutant input and following its cessation. Their results for Lakes Victoria and Malawi are shown in Fig. 4. Lake Victoria is relatively shallow and better mixed than Lake Malawi/Nyassa. Its inflow is approximately equal to outflow (both 20 km^3/yr), approximately 0.7% of total lake volume (2760 km^3). In this lake, the pollutant would tend to accumulate rapidly, reaching over 20% of the inflow concentration over the 40 year input period. Subsequently, following cessation of input, the concentration of pollutant would decrease relatively rapidly, but asymptotically, such that 100 years after cessation of pollution, the concentration in the lake would still be about 15% (0.7 µg/l) of the original input.

Lake Malawi/Nyassa is considerably deeper than Lake Victoria (690 m vs. 79 m max.), with a correspondingly larger volume (8,400 km^3). Outflow (11 km^3/yr) is considerably smaller than inflow (29 km^3/yr) and both make up a much smaller proportion of lake volume than in L. Victoria. Lake Malawi/Nyassa is also well stratified, which slows the mixing rates

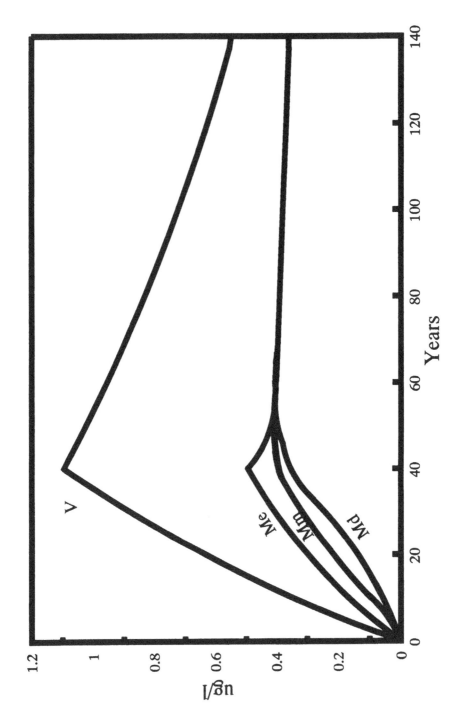

Figure 4. Pollution scenarios for Lakes Victoria and Malawi. Based on input of 5 μg/l of a pollutant introduced continuously for 40 yrs. from all influents, followed by complete cessation of pollutant input. V = Lake Victoria, Me = Lake Malawi epilimnion, Mm = Lake Malawi metalimnion, Md = Lake Malawi monimolimnion (from Bootsma and Hecky, 1993).

between epilimnetic and monimolimnetic waters considerably. Thus, in Lake Malawi/ Nyassa the rate of increase in lake water concentration of the pollutant is considerably slower than for Victoria. However, once a particular concentration of pollutant is achieved in the lake, its subsequent elimination from the lake via outflow is considerably retarded. Epilimnetic waters show a small immediate decline after pollution cessation, but this is offset by continued recharging from the monimolimnion. Concentrations in monimolimnetic waters actually continue to rise for several years after pollution ceases, as a result of slow mixing with the more concentrated surface waters. One hundred years after cessation of pollution, pollutant concentrations would have dropped only a small amount in the lake waters.

Production and use of organic and metallic pollutants is much lower in the intralacustrine nations of Africa than in the industrialized countries, so chemical pollutant levels are lower in the African lakes in comparison with European or North American large lakes (Table 2). In some lakes the impact of some pollutants may be mitigated by the reduction in chemical activity (through chelation) for toxic heavy metals in waters with high ionic strength (Bugenyi, 1990). However, pollution problems are compounded for the African Great Lakes by pollutant disposal. At present, most urban areas around the Great Lakes have either very limited or no sewage treatment capacity; raw sewage is discharged directly into the lakes (Nriagu, 1992). The lakes with the greatest chemical pollution problems have major urban centers located directly on their shorelines (notably Lakes Victoria and Tanganyika), although increasingly, transport-related pollution (harbor and bilge discharges) are likely to become more important on the lakes as North–South trade increases in the region (Cohen et al., 1992). Where agroindustries are being developed, or where spraying is used to control disease-carrying insects, threats from biocide pollutants is significant (Dejoux, 1988). This highlights a further problem; many compounds, especially organometallic compounds, which are banned in many parts of the world because of their known toxicity are routinely applied around the African Lakes (Dejoux, 1988; Nriagu, 1992). Finally, mine discharges and petrochemical spills are likely to play an increasing role as lake pollutants in Africa.

There is a dearth of information concerning current water quality on the African Great Lakes, as is evident from the compilation of Table 2. Unlike fisheries statistics, water quality data has been collected on an almost entirely ad hoc basis on the African lakes. Except for water quality monitoring stations located at municipal water intakes, there is little routine monitoring of water quality on the lakes. Thus the medium- and long-term evolution of water quality in various lakes has had to be deduced from patchy data, which may or may not be representative. In some areas there are abundant major element water-chemistry data, collected over long time periods. However, these data have been collected by so many investigations, each using slightly different methodologies, that their integration and interpretation in time series is nearly impossible (P. Coveliers, pers. comm., 1992). Because of their costs, analyses of trace metals and organic compounds in African lakes have rarely been performed outside of the context of short term studies.

An alternative method of monitoring aquatic pollution which holds great promise for Africa is the use of bioindicators (Dejoux and Deelstra, 1981; Dejoux, 1988; Mubamba, 1991; Caljon, 1992). Sinderman (1988) discussed the various approaches to using bioindicators, some of which could be utilized in African lakes. Where routine limnologic analyses are not feasible there is a need for establishing bioindicator monitoring programs as indices of incipient pollution problems.

Table 2. Table of Existing Water Pollution Statistics for African Lakes

	Lake Victoria	Lake Tanganyika	Lake Rutanzige/Edward	Lake George	Lake Nakuru	Background values for rivers and unpolluted sediment (10–12)	Proposed standards (1)
Inorganic parameters							
Arsenic (water)	0.002–0.008 (7)				0.006 (4)	0.0017	0.05
Arsenic (fish)	0.04–0.12 (13)	0.04–0.06 (±0.03) (6)			1.80 (4)		
Arsenic (sediment)	0.55–1.02 (9)				35 (4)		
Cadmium (water)	0.007–0.094 (7)		0.011–0.009 (±0.004) (3)	0.005–0.006 (±0.002) (3)	0.021 (4)	0.00002	0.005
Cadmium (organisms)		0.03–0.27 ± 0.1 (6)	0.614 (±0.318) (2,3)	0.355 (±0.299) (2,3)	0.26 (4)		0.2
Cadmium (sediment)			2.7 (±0.4) (3)	3.8 (±0.5) (3)	0.27 (4)	0.11	
Lead (water)			0.0011 (8)	0.006 (8)	0.005 (4)	0.003	0.05
Lead (fish)	0.4–1.1 (13)	0.01–0.04 (±0.02) (6)			0.84 (4)		
Lead (sediment)	6.02–77 (9)				34 (4)	0.019	1.5
Mercury (water)					0.001 (4)		0.001
Mercury (fish)		<0.05 (6)			0.22 (4)		0.31
Mercury (sediment)					0.05 (4)	0.05–0.3	
Copper (water)	0.005–0.0576 (7)		0.015–0.130 (±0.006) (3)	0.09–0.11 (±0.02) (3)	0.002 (4)	0.007	1.0
Copper (organisms)	0.15–0.53	1.7–3.2 (±0.4) (6)	0.039 (±0.029) (2,3)	0.0532 (±0.0326) (2,3)	10 (4)		10
Copper (sediment)	0.96–78.6 (9)		37 (±8) (3)	102 (±10) (3)	6.2 (4)	13–33 (15)	
Zinc (water)	0.025–0.125 (7)				0.049 (4)	0.02	5.0
Zinc (fish)	2.21–7.02 (13)	21–134 (±18) (6)			110 (4)		150
Zinc(sediment)	2.54–265 (9)				140 (4)	95	
Iron (water)			0.089 (8)	4.83 (8)		0.04	0.3
Iron (fish)	0.53–4.65 (13)	35–200 (±52) (6)					
Iron (sediment)	1180–52900 (9)		5500 (8)	69000 (8)		41000	
Mangance (water)	0.05–3.28 (7)				0.024 (4)	0.007	0.1
Manganese (fish)	0.22–0.74 (13)	5–17 (±12) (6)			19 (4)		
Manganese (sediment)	53.1–616 (9)				550 (4)	770	

Organic parameters

Parameter				
Dieldrin (water)	0.014–0.068 (16)			<0.0001 (4)
Dieldrin (fish)				0.02 (4)
Dieldrin (sediment)				<0.001 (4)
DDT (water)		0.02–0.13 (14)		<0.0001 (4)
DDT (fish)				<0.01 (4)
DDT (sediment)			0.004	<0.001 (4)
Lindane (water)				
Lindane (fish)		<0.01–0.04 (14)		<0.0001 (4)
DDE (water)	0.010–0.025 (16)			0.074 (±0.051) (5)
DDE (fish)		0.04–0.14 (14)		5.75 ± 7.57 (5)
DDE (birds)				<0.001 (4)
DDE (sediment)				<0.0001 (4)
DDD (water)				0.01 (4)
DDD (fish)		<0.005–0.010 (14)		<0.001 (4)
DDD (sediment)				<0.001 (4)

Sources are as follows: (1) WHO (1984); (2) Bugenyi (1981); (3) Bugenyi (1982); (4) Greichus et al. (1978); (5) Lincer et al. (1981); (6) Sindayigaya et al. (in press); (7) Ochieng (1987); (8) Bugenyi (1982); (9) Onyari and Wandiga (1989); (10) Burton (1976); (11) GESAMP (1982); (12) Salomons and Forstner (1984); (13) Wandiga and Onyari (1987); (14) Sindayigaya et al. (1990); (15) Bugenyi (1979); (16) Koeman and Penning (1970).

COMPARATIVE POLLUTANT IMPACTS TO AFRICAN GREAT LAKES ECOSYSTEMS

Pollution problems in Lakes Rutanzige and George exist as a result of uncontrolled contamination from copper mines at Kilembe (Bugenyi, 1979, 1981, 1982, 1990). Surface water concentrations of cadmium (1–9 µg/l) and copper (70–130 µg/l) are both high in this region, well above the relevant US EPA guidelines (though compliant with those of the World Health Organization (1984)). Both copper and cadmium decline with distance from the mines (i.e., they are higher in Lake George: Bugenyi, 1979; Nriagu, 1992). Although the effects of the high surface water copper concentrations on fish mortality may be somewhat mitigated by chelation in the alkaline waters of these lakes (Bugenyi, 1990), the levels are still very high (Nriagu, 1992).

Lake Victoria is subject to industrial and urban pollution around its main port cities and affluent rivers. Generally, waters analyzed from Lake Victoria fall within the WHO (1984) drinking water standards for metals, although locally Cd, Pb and Mn are too high (Okello, 1992). River delta and other coastal areas in heavily populated and/or industrialized regions are often much more polluted. These include the Rivers Nyando, Nyamasaria, Yala and Nzoia, Kasat, and all of Winam Gulf (Wandiga and Onyari, 1987; Onyari and Wandiga, 1989; CIFA, 1991; Okello, 1992). Onyari and Wandiga (1989) measured relatively high levels of copper and lead from sediments, even at stations which were distant from the shore (37 and 77 µg/g respectively). Much of this offshore contamination by heavy metals is probably derived from boat discharges.

Mining activities within the Lake Victoria Basin may pose a serious threat to the lake in future years. The Musongati Nickel Deposit, located on the Lake Victoria/Lake Tanganyika watershed divide in Burundi, is likely to be developed in the near future (Lavreau and Nkanira, 1990; D. Didas, pers. comm., 1992). Where discharges have been uncontrolled, nickel mining has been a notorious and extremely dangerous source of pollution for lakes in northern Europe (O. Lindqvist, pers. comm., 1992). The fact that Burundi and Rwanda do not occupy any of the shoreline of Lake Victoria illustrates the importance of involving all countries within a watershed in lake management compacts (rather than just those lying on the lake's shoreline).

Elsewhere, moderately high heavy metal concentrations have been documented, notably Cd and Hg from sediments, fish and birds of Lake Nakuru (Greichus et al., 1978). Presently, concentrations of heavy metals in urbanized portions of Lake Tanganyika are low (Deelstra, 1977, 1985; Meybeck, 1985; Ndayizeye, 1985; Ndabigengesere, 1986; Gasana, 1988; Niyabona, 1988; Sahiri, 1991; Sindayigaya, 1991; Sindayigaya et al., 1990, in press; Caljon, 1992). However, increasing quantities of urban waste from numerous industries, including textile, beer, battery and plastic fabrication factories as well as household waste, are currently discharged directly into the lake, with no pretreatment.

The central Kenyan rift lakes lie in an area of relatively intensive agricultural development. Their comparatively small sizes, coupled with the intensive land usage around them (including the town of Nakuru) make them quite vulnerable to pollution. This concern is magnified by the critical role both of these lakes play as migratory bird refuges along the East African flyway. Industrial, agricultural and urban discharges from Nakuru town currently flow untreated into Lake Nakuru, despite the latter being a National Reserve. Organic pesticide residues in fish tissues increased 50–100% during the 1970s (Koeman et al., 1972; Greichus et al., 1978; Dejoux, 1988).

Chlorinated pesticide residues have declined in northern Lake Tanganyika since the 1970s (Deelstra et al., 1976; Sindayigaya et al., 1990 and submitted), apparently as a result of a recent switch by cotton growers along the lakeshore to organophosphate and pyrethrenoid pesticides. However, serious accidental spills of DDT in the Kigoma harbor (Alabaster, 1981), and fuel oil leakages in the Mpulungu harbor, highlight the general absence of environmental regulations governing the handling and transport of toxic substances in and around the lake. Although its use in Burundi ended in the mid 1980s, DDT is still routinely applied near the shoreline in the Zambian portion of the lake (Sindayigaya et al., 1990; M. Pearce, pers. comm., 1992).

Pesticide residues have been recorded in both water birds and fish from Lake Victoria, although the magnitude of this problem is unclear (Koeman and Penning, 1970). Aketch et al. (1992) report that most of the pesticides in current use in the Victoria Basin do not create toxicity problems (although they note that both aldrin and altracine occur in analyzed water samples). Aldrin and dieldrin are both considered to be high-risk, persistent toxic compounds in lake environments (Bicknell, 1992). In Lake Malawi/Nyassa, very heavy application of copper fungicides is reported near Nbozi, in the Tanzanian portion of the watershed (Alabaster, 1981). An old proposal to build a paper pulp mill at Lake Malawi/ Nyassa has been resurrected. Such a plant could lead to disastrous consequences, as has occurred in Lake Baikal (Stewart, 1990; IUCN, 1991).

At Lake Tanganyika, there is a serious potential risk posed by petroleum exploration, both in the lake and within the immediate vicinity. Reflection seismic surveys were conducted by Project PROBE during the early 1980s, financed by a consortium of oil companies (Rosendahl, 1987, 1988). This work, as well as the recognition that natural oil seeps exist in the lake, prompted several oil companies to obtain exploration leases in the region. In 1986, AMOCO drilled 2 wells on the Ruzizi Plain, north of the lake, both of which proved to be dry holes (i.e., no commercially exploitable quantities of oil were found). Subsequent exploration has concentrated on areas west of the lake, as current oil prices are too low to make drilling in the lake itself profitable, although this situation could easily change. If oil were discovered and produced in the region, risks of oil spills would arise from well accidents, cross-lake transport and harbor spills. Baker (1992) has calculated the likely dispersion of a hypothetical mid-lake spill, based on a discharge of 30,000 tons. Figure 5 illustrates one scenario, based on a spill occurring between June and August, when winds are dominantly from the south. Assuming average wind and current speeds, an oil slick derived from the middle portion of the lake, can be expected to reach the north end (accumulating in the most densely populated portion of the lake basin) within 20 days. No rapid action plan currently exists to mitigate petrochemical or other hazardous waste spills on Lake Tanganyika, nor on any other African lake.

ENHANCEMENT OF NUTRIENT AND SEDIMENT LOADS

Dramatic changes in the trophic status of Lake Victoria have coincided with the rapid deterioration of its watershed, demonstrating the vulnerability of even the largest tropical lake to eutrophication. Watershed deforestation has also resulted in rapid increases in sediment yield to the lakes, the consequences of which are only now beginning to be understood.

The vulnerability of the large lakes to eutrophication and excess sedimentation is, in large part, controlled by how their watersheds are exploited. Cooking and land clearing fires can greatly enhance the levels of nutrients transported through the atmosphere and then depos-

After
1 day   10 days   20 days

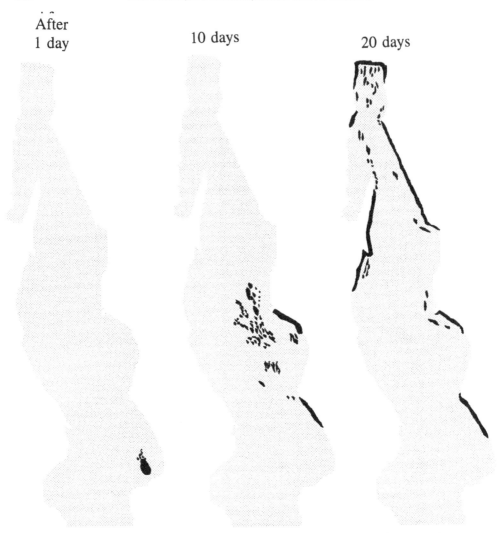

Figure 5. Speculative scenarios for oil spill dispersal on Lake Tanganyika, assuming a 30,000 ton spill in a mid lake area during June–August period of south winds (From Baker, 1992).

ited in the lakes (Hecky and Bugenyi, 1992; Bootsma and Hecky, 1993). Erosion increases the deposition of soil-bound nutrients in the lakes. Basin topography is important in determining a lake's susceptibility to nutrient or sediment enhancement impacts (Bootsma and Hecky, 1993). Low relief catchments, like those of Lake Victoria, are prone to rapid population growth, although their erosion rates may be low. The most densely populated lake margins are also the areas where sediment input and eutrophication problems are most serious.

Understanding eutrophication in the African Great Lakes has been hampered by a general uncertainty as to how, and which, nutrients limit productivity in the tropics (Talling and Talling, 1965; Melack et al., 1982; Hecky and Bugenyi, 1992).

Unlike temperate lakes, where P is most commonly the limiting nutrient, nutrient limitations are more complex (and poorly understood) in tropical lakes. For many lakes it is likely that limiting nutrients either alternate or work in concert in controlling primary productivity (Evans, 1961; Healey and Hendzel, 1980; Hecky, 1991; Bootsma and Hecky, 1993). All of the important types and sources of nutrients can be influenced, either directly or indirectly, by human activities. For example, enhanced loads of P and Si from soil erosion are small on an annual basis, but could affect the budgets of these nutrients in a lake over the decadal time scale (Bootsma and Hecky, 1993). Biological nitrogen fixation can be influenced indirectly by creating conditions favorable to the growth of N-fixing cyanobacteria. The impact of increased productivity on a lake will depend on the form it takes (Bootsma and Hecky, 1993). If the increase is not readily useable by zooplankton and fish, a high proportion of primary production will undergo bacterial decomposition, resulting in eutrophication. However, if enhanced productivity can be easily used by zooplankton and fish and indirectly by humans, then symptoms of eutrophication may not occur.

COMPARATIVE EUTROPHICATION IMPACTS ON AFRICAN GREAT LAKES ECOSYSTEMS

Because of its basin's high population density and intensive land-use, Lake Victoria has undergone a rapid increase in nutrient loading during the past 30 years, from agricultural and urban waste (Aketch et al., 1992; Bootsma and Hecky, 1993). Based on both studies within the basin and analogous areas in Africa, the increases in phosphorus loads in Lake Victoria basin waters probably results from livestock, burning and soil erosion (Ashton, 1981). Bootsma and Hecky (1993) have found that rainfall concentrations of phosphorus along the north shore of the lake and P deposition rates in the lake have increased dramatically in the past three decades, whereas nitrogen and sulfur levels in rain have remained static over the same period.

These enhanced nutrient loads have transformed Lake Victoria's water chemistry and ecosystem at least since the 1950s (Hecky and Bugenyi, 1992). A major decline in Si concentrations in the inshore waters of the lake has been accompanied by a major increase in chlorophyll concentrations (Talling, 1965; Hecky and Bugenyi, 1992). Mavuti and Litterick (1991) and Bootsma and Hecky (1993) have found that eutrophication in Lake Victoria is most severe in the Winam Gulf area, the region of highest human and cattle population density. Lake Victoria has also undergone progressive deoxygenation. Before the 1960s, Lake Victoria appears to have been at least seasonally oxygenated to its greatest depth (84 m). By 1992, oxygen concentrations of <1 $\mu g/l$ occur over 50% of the lake bottom during the stratified season; the lake is even more deoxygenated during seiche events (Hecky, 1992; Gophen et al., 1992). The oxycline now averages between 20–30 m (Ochumba et al., 1992). Eutrophication and shoaling of the surface mixed layer has increased the lake's bacterial biological oxygen demand (Bootsma and Hecky, 1993).

Numerous biotic impacts have been associated with the changes in Lake Victoria's water quality (Ochumba and Kibaraa, 1989; Hecky, 1992). Phytoplankton productivity has increased at least two times and phytoplankton biomass four to five times during the past 30 years (Hecky, 1992). We don't know at present whether these changes are strictly the result of cultural eutrophication, or represent natural variation in the lake's ecosystem (Hecky, pers. comm, 1992). However, between 1960 and 1992, Mugidde (1992) recorded a doubling in daily phytoplankton productivity for both nearshore and offshore areas. Some of this change may result from the competitive advantage of more productive N-fixing cyanobacteria, which have come to dominate the algal community. From dated cores, Hecky (1992) found that by the 1920s, the algal community began to transform itself (simultaneous with increases in C and N concentrations in the cores); a *Melosira*/chlorophyte community was replaced by filamentous, N-fixing cyanobacteria and thin-shelled *Nitzchia* diatoms. Increased algal productivity in the lake has greatly reduced transparency. Secchi disk depths are now 25% of their values in the 1950s, and the area of the lake floor within the photic zone has been halved (Hecky, 1992). Eutrophication has increased the frequency of algal blooms and fish kills after mixing events (Ochumba, 1990; Bugenyi and Magumba, 1992). At times the oxycline has risen to 5 m, resulting in extensive fish and snail mortality on the lake floor (Ochumba et al., 1992).

Major changes in Lake Victoria's zooplankton and secondary consumer community structure include a major trophic simplification compared with the pre-perch era (Mwebaza-Ndawula, 1992). Today new algal production is consumed by Nile Tilapia or zooplankton. Zooplankton are eaten by juvenile tilapia, *Chaoborus,* and *Rastrineobola.* Excess production generates benthic detritus which forms the basis for a second pathway, in which benthic bacteria decompose the accumulated organic matter. Detritus and bacteria are consumed by *Caridina* shrimp and other invertebrates. These in turn are consumed by juvenile Nile perch, which themselves are eaten by adult perch. *Caridina* can forage below the oxycline and form dense schools in deep waters where O_2 levels are low, and *Rastrineobola* can not reach. It is unclear to what extent benthic haplochromine cichlids (now eliminated by Nile Perch predation) might have taken advantage of this detritus pathway, since the oxygen requirements of these fish were not well studied prior to their disappearance.

Bootsma and Hecky (1993) noted that there has been a faster increase in Lake Victoria fish catches than in photosynthetic rates. This may imply that the new ecological system developing in Lake Victoria has become more energetically efficient, or it may simply reflect greater efficiency in fishers, as opposed to fisheries. A limited historical record and the rapid rate of change in this system, make all predictions for Lake Victoria's future evolution dubious (Harris, 1992).

Cultural eutrophication in the remaining African Great Lakes is much more localized than in Lake Victoria. In both Lakes Malawi/Nyassa and Tanganyika, in situ N fixation accounts for the bulk of total N requirements, whereas phosphorus and silica are mostly supplied by vertical mixing (Bootsma and Hecky, 1993). Over short time intervals it may be more difficult to change the nutrient dynamics of these voluminous lakes than Lake Victoria. However, on a decadal scale additions of large amounts of Si and P could have a eutrophying effect. There is some indication of eutrophication in the Bujumbura Bay region of Lake Tanganyika, where high population densities of people and cattle have existed for many years (Ndayizeye, 1985; Gasana, 1988; Niyabona, 1988; Caljon, 1992).

INCREASED RATES OF SEDIMENTATION

The role of eroded sediment in degrading water quality in the African Great Lakes is only now beginning to be understood (Dejoux, 1988). The practice of clear cutting vast quantities of land without erosion control measures, now common throughout the Great Lakes region, has greatly increased discharge rates of sediments. Rapid headward erosion, and stream incision are common by-products of this erosion, which, once started, often continues until all soil cover is denuded. The problem is particularly acute within the steeply sloping rift basins.

Erosion rates are known from only a few areas around the Great Lakes and even less is known about discharge rates into the lakes. On slopes of 28%, Bizimana and Duchafour (1991) recorded erosion rates of up to 27 tons/ha/yr for cropland in Burundi. These alarming rates are a consequence of monocultural farming without erosion control and are probably typical of the intensively farmed areas around the north end of Lake Tanganyika. Where slopes are higher (49%), soil loss rates rise to as high as 100 tons/ha/yr! Serious erosion problems have also been observed around Lake Malawi/Nyassa (Tweddle, 1992). Steep watersheds often given way to steep, rocky lake bottoms at the shoreline. This generates very high sedimentation rates in precisely those areas which when undisturbed, are typically sediment-free.

Many biological impacts can results from excess sedimentation. These include reductions in light penetration, affecting photosynthetic rates (Grobbelaar, 1985), reductions in the nutritional value of detritus (Graham, 1990), physical damage or abrasion to organisms (Bruton, 1985), interference with suspension feeding and respiration (Cairns, 1968), changes (reductions or increases) in nutrient loading, through release of nutrients from soil particles, or adsorption of nutrients from the water column (Golterman 1977, 1991), and loss of spatial heterogeneity required by habitat specialists (Cohen et al., 1993a). For most of the African lakes, only anecdotal information exists as to the magnitude of these problems.

COMPARATIVE SEDIMENTATION IMPACTS ON AFRICAN GREAT LAKES ECOSYSTEMS

Sediment pollution is probably the most serious environmental problem facing Lake Tanganyika at the present time (Bizimana and Duchafour, 1991; Nsabimana, 1991; Cohen et al., 1993a,b). Erosion is enhanced in the watershed by high population densities on marginal lands, the abandonment of terracing, and poor road and trail-building practices. Deforestation is essentially complete within the Burundi and northern Zaire portions of the watershed (Cohen et al., 1993b), and land clearing using uncontrolled large fires is proceeding at an alarming rate further south. Very large amounts of sediment eroded from deforested drainages are now being shed into the lake. In high sedimentation areas, anoxic, laminated muds are accumulating at depths where the water mass itself is well-oxygenated, suggesting that sedimentation is proceeding at a rate which is too high to allow organic matter to be consumed or oxidized (Cohen et al., 1993a). Under similar dissolved oxygen conditions in low sedimentation zones of the lake, sediments are oxidized and bioturbated.

Excess sedimentation in Lake Tanganyika has resulted in local reductions in species diversity of up to 60% for ostracode crustaceans (Fig. 6; Cohen et al., 1993b). Fish are also showing large magnitude reductions in diversity (Fig. 7; Cohen et al., in prep.). Specialized, benthic-algal browsers are disproportionately reduced compared to other trophic groups, suggesting that reductions in light intensity and habitat complexity are having a direct impact on the benthic fauna.

In the Lake Malawi/Nyassa watershed rapid land degradation has occurred in recent years. Major declines in large potadromous cyprinids and catfish have been recorded (Tweddle, 1992). These fish require relatively clear water streams to spawn (Jackson and Coetzee, 1982). Although once common (and important commercially), they have now been largely eliminated from Malawian waters through a combination of siltation and fishing pressure. They are still common in lightly populated areas of the Mozambiquan coast, but even there spawning grounds are beginning to show signs of serious deterioration (Massinga, 1990).

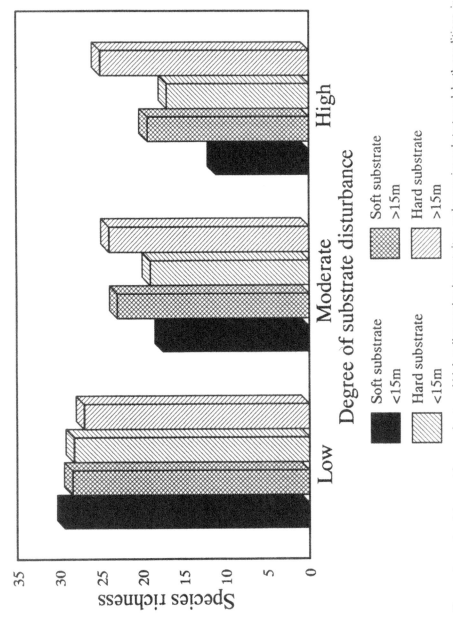

Figure 6. Ostracode species richness at low, moderate and high sedimentation impact sites, under varying substrate and depth conditions, in Lake Tanganyika (From Cohen et al., 1993b).

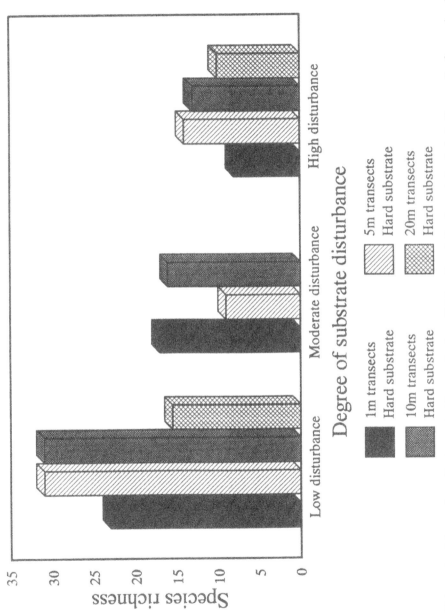

Figure 7. Fish species richness along dive transects at low, moderate and high sedimentation impact sites in Lake Tanganyika for varying depths (from Cohen et al., in prep.).

DAMS AND WATER DIVERSION SCHEMES

The precarious hydrologic balance of many African lakes make them vulnerable to disturbances caused by water diversion or impoundment projects. Elimination of swamps through water diversion or damming is likely to cause declines in diversity of indigenous fish through habitat loss, destruction of refugia and faunal mixing (Chapman et al., 1992). Impoundments have often lead to short-term eutrophication problems in African reservoirs, although these typically dissipate over time (Gaudet et al., 1981). In the Lake Malawi/Nyassa watershed the Dwanga River is completely closed off during the dry season and diverted to sugar cane fields. This river was originally an important locality for potadromous fish runs; such spawning runs have now ceased completely with the annual diversion (Tweddle, 1992).

The ecological impacts of the damming of Ripon Falls at the Nile outlet from Lake Victoria are unknown. Plans to dam Kabelega Falls (the barrier between Lakes Albert and Victoria) have been discussed in the past, but the falls are currently protected in Kabelega National Park (R. Lowe-McConnell, pers. comm. 1993).

HUMAN HEALTH PROBLEMS

Poor sanitation in the heavily populated areas around several of the lakes has resulted in serious disease transmission problems, particularly via lake fish. At Lake Tanganyika, cholera epidemics have been spread at many locations along the lake shore as a result of the practice of drying the small clupeid sardines (*Ndagala/Kapenta*) directly on beaches which are also used for defecation (Cohen et al., 1992). Major cholera outbreaks in the Zambian copperbelt (500 km from Lake Tanganyika) can be tied directly to the transport of infected kapenta from the lake.

INTRODUCED SPECIES

The introduction of alien fish species is the most controversial of all environmental issues facing the Great Lakes of Africa. Whether by accident or design, most of the large African lakes host alien fish species (Fig. 8) (Moreau, 1979; Moreau, 1988); all told, over 200 separate intentional introductions, representing more than 50 species, have occurred in African inland waters. The majority of introductions have been to streams or stock ponds for game and aquaculture. Only about 20 fish species have been introduced into the Great Lakes themselves, but the very term "watershed" connotes the threat represented by all introductions that lie within a lake's drainage basin. Including such accidental introductions from escape and downstream intrusion, only Lakes Turkana and Malawi/Nyasa have thus far escaped the introduction of extralacustrine fish species.

Exotic fishes have been introduced intentionally for angling, insect and snail pest control, and most importantly, as food (often in intensive pisciculture programs) (Moreau, 1979). Introductions for angling largely ended with the colonial era, whereas the repeated introductions of the mosquito fish (*Gambusia affinis*) to Africa have been largely unsuccessful (though the guppy, *Poecilia reticulata*, is widespread). Food fish introductions, especially for aquaculture, continue on a wide scale. Without denying the genuine need for fish protein, it must be recognized that the spread of aquaculture species has been spurred by the existence of "packaged" technologies for specific introductions, in which all of the details (manuals, food, etc.) have been worked out for the pond owner or fisheries agent. The appropriateness of a particular species for a given locality is rarely addressed adequately until after the fact of its introduction.

In addition to introduced fishes, exotic water plants and invertebrates also pose a risk to the lakes. Of particular importance are the water hyacinth (*Eichhornia crassipes*, Twongo, 1992) and the North American crayfish (*Procambrus clarkii*, Welcomme, 1988).

Species introductions can be disastrous, as illustrated unambiguously by the introduced zebra mussel and sea lamprey in the American Great Lakes. In the African Great Lakes, however, the balance sheet has been a subject of intense debate, particularly for the well publicized introductions of Nile Perch (and to a lesser extent various tilapia) into Lakes Victoria and Kyoga. In these, and similar cases, the rationality and moral validity of the introductions has depended on one's a priori views of the relative importance of foreign exchange weighed against local food availability and species and ecosystem preservation (e.g., Orach-Meza et al., 1990; Ogutu-Ohwayo, 1992a; Welcomme, 1992). Regardless, the evidence is overwhelming that tilapiine and *Lates* introductions have led to trophic simplification and species extinctions in several large African lakes (Barel et al., 1985; Goldschmidt et al., 1993; Lowe-McConnell, 1993). Some fishes are fecund and eurytopic, traits that predispose them to successful introduction, threatening native species that are often more stenotopic and less fecund with extinction (Witte et al., 1992; Lowe-McConnell, 1993; A. Ribbink, pers. comm., 1993). In all of the lakes where introductions have been both economically lucrative and ecologically damaging (Victoria, Kyoga and Nabugabo), there are rapid changes occurring, both limnologically and in apparently declining fisheries production. How these introductions will be viewed by fisheries management officials and socioeconomists, even 10 years from now, is not at all clear. The history of fish introductions in East Africa reads like good fiction, and says much about the limits of either our ability or our willingness to anticipate consequences (Welcomme, 1966; Bruton, 1990; Ogutu-Oh-wayo, 1992b,c, 1993; Kaufman 1992; Lowe-McConnell, 1993). Two examples are given here: the sardine, and the "Savior." In 1959, an effort was made to introduce *Stolothrissa tanganyicae*, one of Lake Tanganyika's two endemic sardines, to Lake Kivu, which lacked this type of fish and the popular fishery that it might support (Mahy, 1979; Reusens, 1988; Hanek et al., 1990). The effort failed; the other of the two sardine species, isambaza (*Limnothrissa miodon*), became established instead (Bruton, 1990). This species did well, with official landings of 200 to 400 tons/yr and actual landings perhaps ten times higher (Hanek et al., 1990). The only problem was, the fish from Lake Kivu tasted like sulfur while those from Lake Tanganyika did not. The latter, even dried, commands a higher price in local markets than fresh fish of the same species from Lake Kivu. Meanwhile, the introduction has had its impact on Lake Kivu, with significant changes in the zooplankton community, including the extinction of some local species (Bruton, 1990; J. Lehman, pers. comm. 1993). *L. miodon* has also been introduced into several artificial reservoirs. It was established in Lake Kariba by 1968, from which it escaped downstream to Lake Cabora Bassa; it has also been introduced to Lake Itezhi-Tezhi (Marshall, 1985, 1988; Mubamba, 1992). *L. miodon* currently contributes approximately 80% of the total fish catch (~30,000 t/a) in Lake Kariba (Chifamba, 1992; Coulter, pers. comm., 1994).

The introductions of *L. miodon* into these reservoirs has elicited little controversy. Ogutu-Ohwayo (1992a) notes that intensification of fishing effort for *L. miodon* in L. Kariba coincided with a decline in *Hydrocynus* populations, although the exploitation of gravid female *Hydrocynus* during their breeding migrations was probably a contributing factor (Marshall, 1985). The most serious consequence of these successful introductions has been

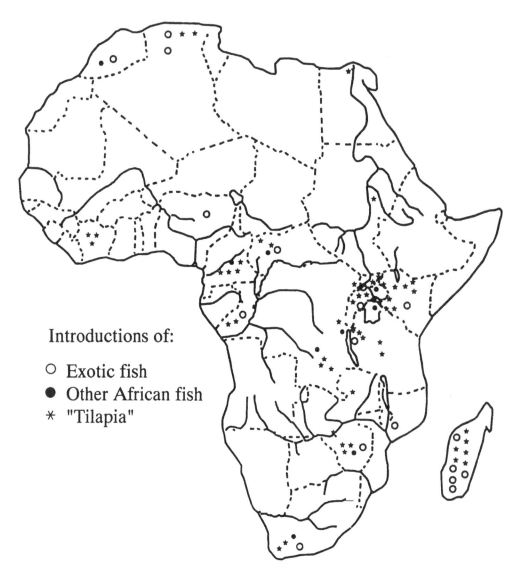

Figure 8. Species introductions in continental Africa: a) introductions between 1950 and 1960; b) introductions after 1960 (from Moreau, 1988).

the perseverance of efforts to gain support for introduction of pelagic clupeids into other African Great Lakes with more complex community structures.

Right after the more or less successful introductions of *L. miodon* into Lakes Kivu and Kariba, Turner (1982) proposed the same for Lake Malawi/Nyassa. Turner observed that fish production per unit area in Lake Malawi/Nyassa was much lower than in Lake Tanganyika, and argued that this was so because cladocerans and larvae of the midge *Chaoborus* were underexploited in the absence of a clupeid predator, these organisms being scarce in Lake Tanganyika where two such predators abound. An introduction of clupeids into Lake

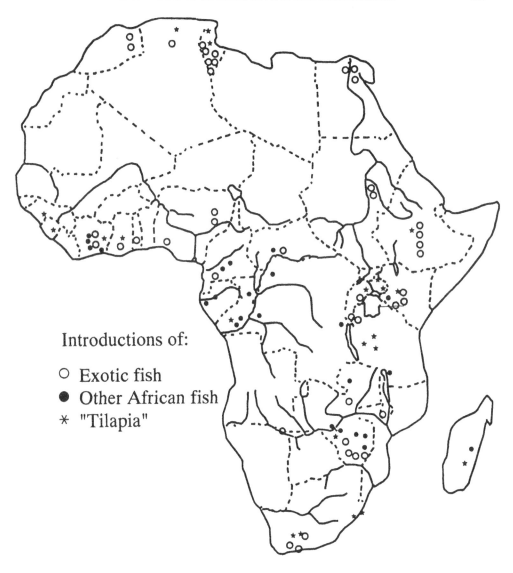

Introductions of:

○ Exotic fish
● Other African fish
* "Tilapia"

Figure 8, cont'd

Malawi/Nyassa, might therefore increase the proportionate transfer of secondary productivity into harvestable fish by supplying a more efficient trophic link than that apparently provided by the indigenous zooplanktivorous fishes in Malawi/Nyassa. Turner's suggestion did not go unchallenged. Eccles (1985) noted that other African lakes, notably Lake Turkana, which lack *Chaoborus* also lack clupeids, calling into question the validity of Turner's original thesis. Furthermore, Lake Malawi/Nyassa, unlike Lake Kivu, already has a complex pelagic ecosystem, including an extraordinary array of zooplanktivorous fishes. The effects of introducing an exotic clupeid into this assemblage would be impossible to predict, and could well lead to the mass extinction of species without succeeding in replacing it with something more valuable, thus threatening an existing fishery, and raising moral issues as well (Eccles, 1985; McKaye et al., 1985).

Nongwa (1986) reviewed Malawian government policy on clupeid introductions into the lake: it had no plans to do so and would not even consider the proposition unless the outcome could be reliably predicted. Even then, Malawi would convene an international conference of the three riparian countries before undertaking an introduction. Recent investment in the development of fisheries based on indigenous pelagic fishes in the lake shows that this policy remains in effect as of 1993. No official policy statements have been issued by the governments of the other countries surrounding the lake concerning introductions.

Lake Victoria is at the nexus of both research and controversy concerning impacts of exotic species introductions on African lakes. The mass extinction of hundreds of indigenous species following the population irruption of Nile Perch is the sine qua non example of the hazards of species introductions into previously species-rich ecosystems. It has occasioned resolutions at several international ichthyological and conservation conferences calling for a halt to further introductions into the African Great Lakes (Balon and Bruton, 1986; Cohen, 1991; Kaufman et al., 1993; SIAL, 1993).

In the 1950s exotic tilapiine cichlids led the vanguard of exotic introductions to Lake Victoria (*Oreochromis niloticus*, *O. leucosticus*, and *Tilapia zilli*). Next in line was the Nile Perch (*Lates niloticus*) (Welcomme, 1964, 1966; Ogutu-Ohwayo, 1992a). Some introduced species were not actually intended for Lake Victoria. North American black bass (*Micropterus salmoides*) has spread down the Sondu-Miriu River, and may be in the main lake by now (Moreau et al., 1988; Ochumba et al., 1992), and water hyacinth (*Eichornia crassipes*) have recently invaded Lake Victoria from ornamental populations in Rwanda. Although it has been established in this lake (as well as Lake Kyoga) for less than 15 years, it has proliferated around almost the entire lake shoreline (Twongo, 1992). Water hyacinth are well known agents of local deoxygenation and habitat alteration.

The dynamics of introduction in Lake Victoria owe much to its being part of a larger Lake Basin System (LBS). The system includes minute lagoons, backponded river mouths (Katonga, Yala, Mara), and satellite lakes (Nabugabo, Kanyaboli, Kyoga) that rim the main lake on all sides, and communicate periodically with it through river flow and flooding (Kaufman and Ochumba 1993). Many of these waterways are separated from each other by the barest rise of land or stretch of swamp. The swamps may act as valves, alternately providing avenues of dispersal at high water, and serving as hypoxic barriers to the movements of some fishes, such as Nile perch, when water levels subside. Behind such barriers are refugia for indigenous species now rare or absent in the main lake (Kaufman and Ochumba 1993). For example, two tilapiines were endemic to the LBS: the mbiru (*Oreochromis variabilis*), and the ngege (*O. esculentus*), once the most important food fish in East Africa. The mbiru is now rare, perhaps due to competition on nursery grounds from *Tilapia zillii*, and hybridization with other introduced species (Welcomme, 1964, 1966). The ngege actually appears to be extinct in Lake Victoria; relict populations elsewhere in the LBS are threatened by hybridization with *O. niloticus* (Kanyaboli) and are in other cases severely stunted (Ogutu-Ohwayo 1993; Kaufman and Ochumba, 1993). The last known healthy wild population of ngege in the world inhabits Lake Kayugi, a tiny satellite pond in Uganda (Ogutu-Ohwayo, 1993). Protection from upstream migration of Nile perch is not sufficient to ensure a future for these fishes in the wild, as illustrated by our discovery that *Tilapia zilli* from upstream fish farms wash into Lake Kayugi in big rainstorms. Fortunately, this species is unlikely to hybridize with the ngege, but it may be only a matter of time until *O. niloticus* finds its way in.

While there seemed to have been little objection to the tilapiine introductions, Fryer (1960) challenged the first planned introductions of Nile perch on the grounds that the notion that it would bring about increased yields of food fish was unjustified (Fryer, 1960). The opposition lost, and by the mid-1980s Nile perch was undergoing a population explosion in the lake (Figs. 9, 10 and 11 from Ogutu-Ohwayo, 1992a, Witte et al. 1992 and Witte et al., 1992, respectively). This rapid increase was mirrored by an equally precipitous decline in native haplochromine cichlids. Current estimates suggest that as many as 65% of the endemic cichlid species in Lake Victoria were eliminated as a result of the introduction of the Nile Perch (Goldschmidt et al., 1993; Kaufman and Ochumba, 1993). The patterns in Lake Kyoga and Nabugabo were similar, Kyoga a bit ahead, and Nabugabo a bit behind (Ogutu-Ohwayo, 1993; Namulemo, 1992; Olowo, 1993).

The evidence is strong that species introductions had major impact on Lake Victoria's ecosystem, but it was as one part in a fugue that included fishing pressure, deforestation, siltation, eutrophication, changes in the lake's physical dynamics, oxygen depletion of the lake's deeper waters, and even climate change (Barel et al., 1985; Acere, 1986; Orach-Meza et al., 1990; Ogutu-Ohwayo, 1992a; Witte et al., 1992; Kaufman, 1992; Kaufman and Ochumba, 1993; Lowe-McConnell, 1993). Whether the outcome of these introductions was "good" or "bad" remains a point of contention (Orach-Meza, 1990; Welcomme, 1992). Simultaneous with the dramatic extinctions of haplochromines came major increases in fish production among three species — the Nile perch, the Nile tilapia and a native pelagic cyprinid *Rastrineobola argentea*. The result was a vast increase in the yield from the fishery as compared to the pre-Nile Perch era (Ogutu-Ohwayo, 1992a). Where Nile perch had been formerly disdained by many, it was being referred to by the late 1980s as *mkombozi* (savior) in Tanzania (Ligtvoet, 1988). Supporters of the Nile perch introduction emphasized the importance of this increased catch to the local economy and diet (Reynolds and Greboval, 1988; Orach-Meza, 1990). It will remain a question for speculation whether native tilapiines (notably the highly prized *O. esculentus*) might have undergone similar vast increases in population sizes had they not been previously eliminated by a combination of forces. Supporters of the Nile perch introduction may yet be forced to revise their opinions. Yields of Nile Perch began to decline during the late 1980s–early 1990s in the Nyanza Gulf, Kenya (Rabuor and Polvina, 1992) and in Uganda waters (Okaronon and Akumu, 1992). There is considerable doubt as to the long-term sustainability of the Nile Perch "boom" at Lake Victoria, and *Rastrineobola* catches have rapidly increased as a proportion of total catch in the past 5 years (cf. Ogutu-Ohwayo, 1992b, Harris, 1992). Furthermore, much of the Nile Perch production from the lake is exported out of the region and local profits from the catch are limited. In Uganda, marketing decisions for perch are made at a high management level, through a very rigid hierarchy; people at the grassroots level, especially local fishermen and consumers are not involved. This raises the question of how "valuable" Nile Perch actually is as a resource for local people (Naikoba et al., 1992). Whether the changes in catches represent extreme fishing pressure on the Nile Perch, or are a manifestation of continued changes in the trophic status of Lake Victoria, is unknown at present. If the Nile perch boom proves to be as short-lived as many predict, only those few who directly benefited from the introduction could possibly find it justified in the face of the massive and permanent scars that it left on the ecosystem (Chitamwebwa, 1993).

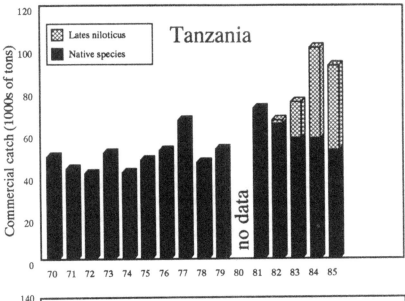

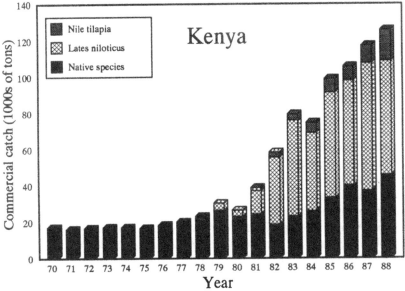

Figure 9. Lake Victoria fisheries catches for the three surrounding countries, plus total catches. Totals are only shown through 1985 because of incomplete data in Tanzania and Kenya. (From Ogutu-Ohwayo, 1992.)

HUMAN-INDUCED CLIMATE CHANGE

Perhaps the greatest unresolved question concerning human impacts on the African lakes is the effect of climatic change. Although the impact of global climate change on lakes has been considered in a temperate context (Carpenter et al., 1992), little consideration has been given as to how such changes might affect the large lakes of Africa. The sensitivity of

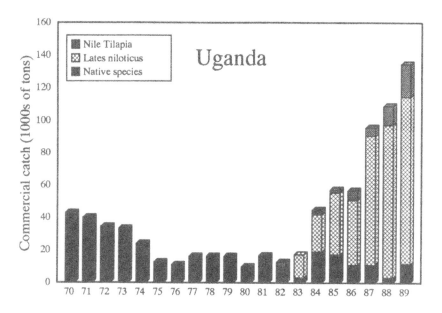

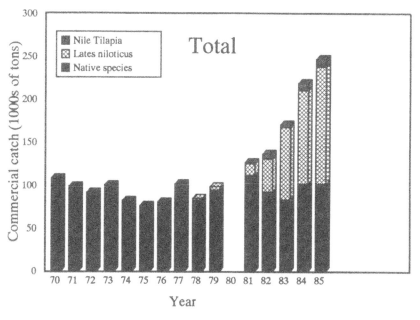

Figure 9, cont'd

African lakes to climate change lies mostly in their delicate balance of precipitation and evaporation. Several of the lakes (e.g., Tanganyika and Malawi/Nyassa) remain open basins only by very slight excesses of precipitation in their drainage basins. Slight drops in precipitation or increases in either temperature or average wind speeds could convert these lakes to closed basins (Owen et al., 1990). Climate change could cause significant changes

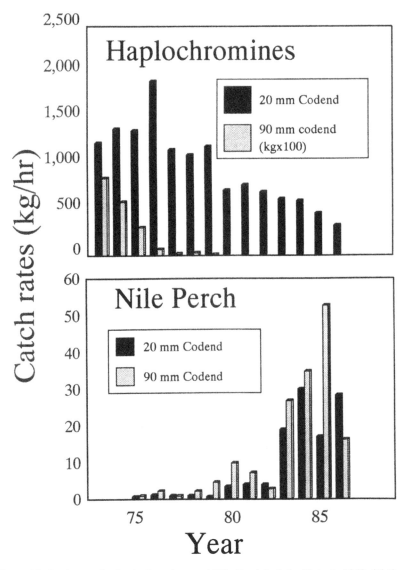

Figure 10. Catch rates for haplochromines and Nile Perch in Lake Victoria 1973–86 (from Witte et al., 1992).

in the thermal stability of a lake's water mass and its mixing dynamics (Hecky and Bugenyi, 1992). Even minor warming events would increase the stability of the water column and decrease vertical mixing in the large lakes which are currently stratified, and could push other marginally stratified lakes (Victoria, Edward, Albert) into states of permanent stratification.

Changes in precipitation and evaporation may also cause lake water chemistry to vary, particularly when a lake changes between open and closed basin conditions. Geological evidence for such changes in Africa during the Quaternary is abundant, particularly for those lakes which are, today, in relatively arid regions (Eugster, 1980; Cohen and Nielsen, 1986;

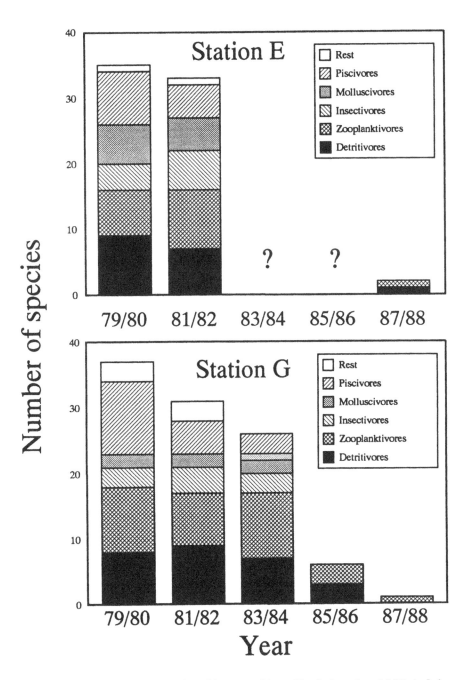

Figure 11. Species diversity and trophic composition of haplochromine cichlids in Lake Victoria, 1979–88 (From Witte et al., 1992).

Table 3. Global Circulation Model (GCM) Results for Various East African Lakes

| Lake | Temperature results (mean annual change in temperature in °C) Model | | | | Precipitation results (mean annual change in precipitation in mm) Model | | |
	CCC (highest spatial resolution)	UKHI	GFDL		CCC	UKMO gradual CO_2 increase model	GFHI
Malawi	+2 to +4	+2 to +4	+2.5		−200 to −300	−200 to −300	+50 to +100
Tanganyika/Kivu	+2 to +4	+2 to +4	+2.5		−50 to −100	no change	no change
Victoria	+2 to +4	+2 to +4	+2		−150 to −200	no change	no change
Turkana	+2 to +4	+2 to +4	+2.5		−300 to −400	+50 to +100	no change

Values for both precipitation and temperature change are based on 1× CO_2 model to 2× CO_2 model comparisons, calculated from maps in Gates et al. (1992). Results from three high-resolution GCMs are presented because the interpretations for equatorial Africa in some cases conflict.

Cassanova and Hillaire-Marcel, ms.). Many aquatic organisms are sensitive to major ion chemistry and concentration changes; it can be expected that such changes would lead to dramatic extinctions during a period of increasing alkalinity (Cohen et al., 1983). An increase in ionic strength would also affect the utility of lake waters for human consumption and agriculture; in some lakes water is currently potable but solute concentrations are too high for irrigation. Some of the potential effects of climate change over short time intervals are illustrated by the historic lake level changes observed in several African lakes. Lake Turkana underwent a major decline in fisheries productivity following a decline in lake level of several meters during the late 1970s and 1980s (Fig. 3) (Lindqvist and Beveridge, 1987; Kolding, 1992). The fisheries boom and bust events observed in Lake Turkana are also typical of African reservoirs, which show short term increases in production early after their formation, as nutrient-rich soils are inundated, followed by rapid declines (Gaudet et al., 1981; Kolding, 1992).

The most recent Ocean-Global Circulation Model simulations (CCC, UKHI and GFDL) for the entire African Great Lakes region during the next 70 years all suggest increases of between 2–4°C over simulated modern temperatures, based on a doubling of atmospheric CO_2 over a 60 year period (2050 A.D. $2 \times CO_2 = 1990 \ 1 \times CO_2$, Table 3) (Gates et al., 1992). To date these results have been robust to changing assumptions about rates of CO_2 increase, atmosphere–ocean coupling, incorporation of trace gas components and spatial resolution refinements in the models. Changes in precipitation predicted by model results are more mixed. For Lake Turkana, model results range from major decreases to moderate increases in precipitation. Lakes Tanganyika, Kivu and Victoria all showed moderate to major decreases in the CCC model (which had the highest spatial resolution of any of the models considered in the study), but no change for the other models. Finally, for Lake Malawi/ Nyassa the CCC and UKMO models suggested major decreases in precipitation (approximately 30% of present rates), whereas the GFHI model predicted a modest increase.

A 30% decrease in precipitation at Lake Malawi/Nyassa (even assuming no increase in temperature) would produce a lake level decline of approximately 50 meters, according to the hydrologic model of Owen et al. (1990) (Fig. 12), which would drop Lake Malawi well below its outlet level, and result in increased salinity. With a +2 to +4°C temperature rise, the actual lake level decline might be considerably greater, accompanied by more stable lake stratification. Such a lake level fall would have serious consequences for a region which, over the same 60 yr time period, may undergo a significant population increase.

CONSERVATION STRATEGIES AND MEASURES IN PROGRESS IN THE AFRICAN GREAT LAKES

Coulter (1989) and Ntakimazi (1992) have reviewed the numerous values attached to conserving the African Great Lakes. Their value as economic (water, fish, tourism, transport, energy), biodiversity, scientific, cultural and aesthetic resources is abundantly evident. A consensus has been growing in recent years among African scientists that in order address all of these values, conservation strategies for the lakes should concentrate on five key areas: 1) organizational structure and the law, 2) international cooperation, 3) developing local constituencies for conservation efforts, 4) establishing protected areas, and 5) obtaining long-term records of ecologic change.

Several authors have noted the weakness of existing legal protection for the lakes and the need for clearer chains of authority in monitoring and policing lake-basin environmental regulations (e.g., Lake Victoria, Rukuba-Ngaiza, 1992; Lake Tanganyika, Mutahinga, 1991,

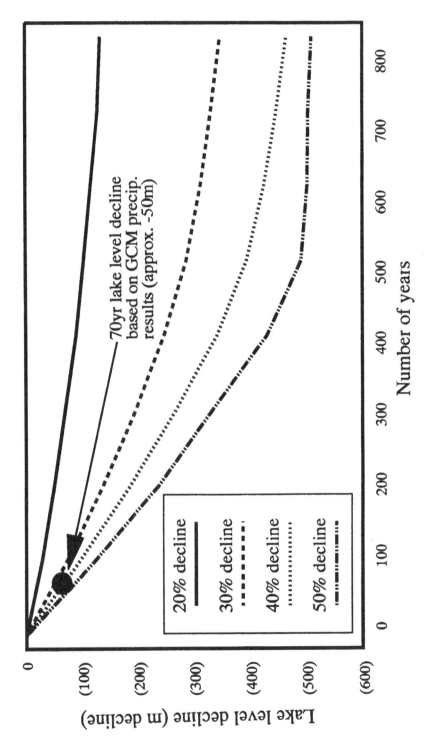

Figure 12. Model lake-level responses for decreased rainfall at Lake Malawi/Nyassa. A 30%[+] decline is predicted by two of the three GCMs considered in this paper. Hydrologic model modified from Owen et al. (1990)

Pearce, 1991, Mwasaga, 1991). Mwasaga (1991) has argued that incentive systems for protecting the lakes and their watersheds, coupled to fair and effectively enforced penalties for non-compliance, are essential. Initial targets for legal reforms might concentrate on easily monitored problems; these might include discharge standards for industrial and large-scale agricultural wastes. Getabu (1992) has suggested that concerted effort needs to be made to reestablish mesh size regulations on lakes where these are either nonexistent, or have been allowed to lapse. Programs developed in continuous collaboration with local communities to develop creative incentives for compliance with environmental standards (with less emphasis on penalties) have proven effective in terrestrial settings in East Africa and could serve as models for the lakes and their watersheds (MTNRE, 1991; WWF, 1991a,b). For example, Kabazinya (1991) has proposed that programs for intensive cultivation, agroforestry using native species and the use of traditional forest products should replace monocultural reforestation efforts. Global Environmental Facility programs now in the final stages of development for Lakes Tanganyika, Victoria and Malawi/Nyassa will provide major funding to strengthen the institutional infrastructure of communication and environmental monitoring in the countries around these lakes.

Work by Coulter (1991) and Kolding (1992) suggest that there is an urgent need to move beyond inapplicable Maximum Sustainable Yield models and to develop a new theoretical framework for fisheries management in the African Great Lakes, one based on principles of tropical hydrobiology, ecosystem disequilibria and multispecific fisheries. Coulter and Mubamba (1993) have suggested that fishing methods and regulations need to be much more closely tailored to the specifics of the ecology of commercially important species in each lake. Such regulations should be developed in consultation with the local fishers who will have to abide by them and efforts to develop self-policing and incentive schemes should be intensified, as they have proven effective in wildlife management (e.g., Mwenya et al., 1990; TEEP, 1991; United Republic Of Tanzania, 1991). Elongo (1991b) has demonstrated how community-based Nongovernmental Organizations (NGOs) dealing with fishing or environmental issues may be effective agents for developing such dialogues between government officials and fishers. Based on work at Lake Kivu, Kaningini (1992) has shown that well-organized public education campaigns organized by local NGOs can be highly effective in deterring fishers from using popular, but highly destructive fishing methods.

Recommendation reports of many recent conferences and African governmental working groups concerned with the African Great Lakes have called for significant increases in the level of international cooperation between riparian countries concerning watershed/lake issues over that which now exists (FAO, 1989; IAC, 1990; Cohen, 1991; Kaufman et al., 1993; PTA, 1993). Programs of this sort are beginning to take shape for fisheries development (for example, the Lake Tanganyika Research Program and a planned Lake Victoria Fisheries Service Commission), but have yet to be developed in other aspects of lake management and conservation. However the planned Global Environmental Facility Projects for Lakes Victoria, Tanganyika and Malawi/Nyassa all provide mechanisms for increasing international cooperation between riparian countries on lake-environmental concerns.

Conservation efforts on the African Great Lakes have been hampered by an absence of well-defined local constituencies which can exert pressure to ensure that environmental regulations are emplaced and enforced (Kaningini, 1991; United Republic Of Tanzania, 1991). The highest priority for developing such a constituency is a system of environmental education, particularly in the primary and secondary schools (TEEP, 1991). Schools could become involved in cooperative education programs to teach children in families involved in fishing and agriculture about the needs for creating community buffer zones to control erosion, or the use of more selective fishing gear (Coulter et al., 1986). Governmental and National N.G.O. policy statements from several of the riparian countries have argued that in order to be successful, these programs must be tailored to accommodate local community concerns, and establish better lines of communication with regional authorities (Anon., 1991; United Republic of Tanzania, 1991). A good example of successful community involvement in the establishment of an underwater reserve comes from the development of the Mafia Island National Park, in coastal (Indian Ocean), Tanzania (WWF, 1991b). Similarly, the Village Afforestation Program of Tanzania is showing promising signs of improving land use practices in its demonstration programs.

Fishing pressure and watershed impacts make both the establishment of restricted fishing zones and underwater reserves critical components of any lake environmental protection program (Anon, 1991; Cohen, 1992; Coulter and Mubamba, 1993; Ogutu-Ohwayo, 1993). The positive impact of the Hol Chan National Underwater Park in Belize on local fishing yields in the surrounding area (Carter et al. in press) shows how such reserves can have strong economic and biologic justification. Pearce (1985a,b) has suggested that Nsumbu Park waters in Lake Tanganyika may be serving a similar, though unintended, purpose as a nursery areas for commercial fish species, thus explaining higher fisheries yields in areas adjacent to park.

Reserves need to incorporate watershed as well as lake areas, and require easily boundaries which can be easily identified by local fishers without navigation equipment or maps (Coulter and Mubamba, 1993). Reinthal (1993) has argued that in lakes such as Malawi/ Nyassa, with large geographic variations in fauna, multiple reserves should be identified, to preserve significant, representative elements of the lake's biodiversity. Mechanisms should be identified, in concert with the local community, to create substitute sources of income for people displaced by the creation of the reserve; this might include small-scale tourism or creation of sport-fishing concessions controlled by local villages (Cohen, 1991, Conference Recommendations).

Several authors have demonstrated the need to protect refugia for populations of lake endemics in areas where these species are endangered (Ogutu-Ohwayo 1992a, Namulemo, 1992; Olowo, 1992). Generally, habitat protection is greatly preferable to captive breeding as a strategy for species protection, for reasons of cost and potential sustainability, although captive breeding can be useful for scientific and educational purposes (SIAL, in prep., Conf. Recommendations, but see Ribbink, 1987 for arguments favoring captive breeding).

There have been a number of recent expansions and creations of underwater reserves and protected areas in the Great Lakes. These include the Lake Malawi National Park (Bootsma, 1992); the Nkhotakota Game Reserve, Lake Malawi, for the protection of spawning grounds of Mpasa (*Opsaridium microlepis*) in the Bua River (Tweddle, 1992); the restricted fishing zones of Lake Victoria, including Ndere Island, Mfangano Island, and the Nyando River Delta; and the Nsumbu National Park (Zambia), Gombe Stream National Park (Tanzania), Mahali Mountains National Park (Tanzania), and Ruzizi Delta National Park (Burundi), at Lake Tanganyika (Cohen, 1991).

The Lake Victoria Research and Conservation Program has been established in Lake Victoria under the joint auspices of the IUCN Captive Breeding Specialists Group, the American Association of Zoological Parks and Aquariums and the European Union of

Aquariums, with 30 participating institutions and hatcheries in Kenya and Uganda (Kaufman 1989; Andrews and Kaufman, in press). This program is intended to reintroduce native food fish (*Labeo victorianus* and *Oreochrmis esculentus*) into the Lake and promote worldwide public awareness through major aquariums. Haplochromine species in the breeding program are providing the basis for laboratory study of the recent paleoecology of Lake Victoria's indigenous fishes. Under a proposed Global Environmental Facility program for Lake Malawi/Nyasa a hatchery would be built to regenerate the populations of *Ospharidium*, which have been seriously reduced as a result of damage to spawning grounds (D. Tweddle, pers. comm., 1993).

Many recent conferences and government reports have recognized the critical need to establish environmental monitoring programs at fixed representative stations on each of the Great Lakes for the regular collection of routine water chemistry and physical limnological data and for the collection of bioindicators (e.g., Anon, 1991; Kaufman et al., 1993). Such stations need to be established with care; to assure their sustainability, the most critical measurements must be identified and provided higher funding priority. Where choices of methods are possible, an emphasis should be placed on using equipment and supplies which are easily obtained and inexpensive; there is an unfortunate history of long-term monitoring programs on African lakes being abandoned for lack of funds to purchase materials. Bioindicators should be sought out for key problems in each lake because of their potentially low costs. All of the GEF projects planned for the Great Lakes contain provisions for long-term monitoring programs.

The absence of long-term historical records of limnologic change on the African lakes has seriously hampered the ability of investigators to assess the meaning of short-term changes, for example those observed in the last 30 years in many lakes. Data is needed to identify the range of natural variability in limnologic parameters and to provide baselines of rates of change in undisturbed systems. The most promising approach to this problem is through paleolimnology. Although numerous paleolimnological studies have been undertaken in African lakes since the 1960s, their purposes have been largely to understand climatic change during the Quaternary; hence, the temporal resolution and time intervals studied have mostly been inappropriate for resolving baseline data for modern anthropogenic impacts. The potential for such investigations is enormous. For example, the Paleoecological Investigation of Recent Lake Acidification (PIRLA) project of northeastern North America has played a crucial role in understanding the timing of onset of acid precipitation and its biological impacts on U.S. and Canadian lakes (Whitehead et al., 1990). In Lake Victoria, diatom paleoecology and sedimentation records have already been used to identify a baseline of change in trophic systems prior to the last 30 years of historical records (Hecky, 1992). A similar approach is now being taken using ostracode, pollen and sedimentation records to understand the history of sedimentation impacts on the biodiversity of Lake Tanganyika (Cohen et al., in prep.). Creative uses of this approach (e.g., use of fossil fish scales and otoliths where possible, to identify changes in community structure among pelagic fish) should be encouraged, as the costs of collecting and analyzing such samples need not be high.

CONCLUSIONS

The African Great Lakes today encompass a wide range of human-impact intensities. Some, like Lake Victoria, have been seriously disturbed by human activities, whereas others, like Turkana, have experienced much less pressure. None have been immune to anthropogenic disturbance, and with anthropogenic climate changes looming in the near future (if not already here), all will be exposed to major limnologic perturbations. The directions that these African lake "experiments" will take in future years is far from clear. Even with a total cessation of pollution inputs (a highly unlikely prospect in any event), it is unclear whether lakes such as Victoria will return to their pre-impact state, or whether quasi-permanent structural changes will have been set in place, moving such lakes into new ecosystem states (cf. Schindler, 1990).

One thing is clear; the signals of change are not encouraging. Declining water quality and fishing yields, and species extinctions, in a region which has achieved great benefits from its water, fisheries and biodiversity, all signal a need to move the Great Lake experiments in a new direction. There is an urgent need today to reform the fisheries, water and watershed-use practices throughout the intralacustrine region. There is an urgent need for lake-basin management plans, which are both well founded scientifically **and** enforced. Without such reforms, these Great Lakes, upon which so much of the regions culture, economy and aesthetic values are based, face a grim future.

ACKNOWLEDGMENTS

We thank Pere Anadon, George Coulter, Debra Gevirtzman, Rosemary Lowe-McConnell, Ellinor Michel, and Tony Ribbink for their many suggestions on this manuscript. We also thank Fred Bugenyi, Gashagaza Masta Mukwaya, Steve Njuguna, Gaspard Ntakimazi, and Dennis Tweddle for sharing ideas on conservation and fisheries of African Lakes. Funding for this project was provided by the National Underseas Research Center (USA)(A.C.), the Institut de Ciencies de la Terra "Jaume Almera" (CSIC-Spain)(A.C.) and the Pew Scholars Program for Conservation and the Environment (L.K.). Research in Africa by A.C. was authorized by the Ministière de l'Enseignmement Supérieur et de la Recherche Scientifique of the Govt. of Burundi and the Ministère de la Recherche Scientifique et Technologie of the Govt. of Zaire.

REFERENCES

Ahayo, K.N., 1991, La diminution de la production dans le nord-ouest du Lac Tanganyika. *In* Cohen, A. (ed.), Report on 1st Intl. Conf. on Conservation and Biodiversity of Lake Tanganyika. Biodiversity Support Prog. Washington, DC, pp. 38–40.

Acere, T., 1986, Nile Perch, *Lates niloticus* (Linne); the scapegoat for the decline/disappearance of the indigenous fish species of Lake Victoria. UFFRO seminar on Current State and Planned Development Strategies of the Fisheries of Lakes Victoria and Kyoga, Jinja, 1986, 21 pp.

Aketch, M., Calamari, D., and Ochumba, P., 1992, A rapid survey of the watershed impacts to the Kenyan portion of Lake Victoria. in Biodiversity, Fisheries, and the Future of Lake Victoria, Conference Abstracts, pp. 2–3.

Alabaster, J., 1981, Review of the state of aquatic pollution of East African inland waters. FAO/CIFA Occas. Paper 9, 36 pp.

Alimoso, S., Magasa, J and van Zalinge, N., 1990, Exploitation and management of fish resources in Lake Malawi. *In* Fisheries of the African Great Lakes (ed. anon.), Intl. Agricultural Center, Wageningen, Netherlands, Occas. Paper 3:83–95.

Andrews, C., and Kaufman, L. (in press), The role of captive breeding in the conservation of endangered freshwater fish species. 6th World Congress on Breeding Endangered Species In Captivity. ms.

Anonymous, 1991, Actes de la Consultation Nationale des O.N.G. du Zaire. Bukavu, Zaire 16–20 Dec., 1991, Preparatory Document for U.N. Conference On The Environment and Development, Rio de Janeiro, 1992.

Anthony, K., 1979, Agricultural Change in Tropical Africa. Cornell Univ. Press., 326 pp.

Ashton, P., 1981, Nitrogen fixation and the nitrogen budget of a eutrophic impoundment. Water Research 15:823–833.

Baker, J., 1992, Oil and African Great Lakes. Mitt. Internat. Verein. Limnol. 23:71–77.

Balon, E., and Bruton, M., 1986, Introduction of alien species or why scientific advice is not heeded. Envir. Biol. of Fishes 16:225–230.

Barel, C., Dorit, R., Greenwood, P., Fryer, G., Hughes,N., Jackson, P.B.N., Kawanabe, H., Lowe-McConnell, R., Nagoshi, M., Ribbink, A., Trewavas, E., Witte, F., and Yamaoka, K., 1985, Destruction of fisheries in Africa's lakes. Nature 315:19–20.

Barel, C., Ligtvoet, W., Goldschmidt, T., Witte, F., and Goudswaard, P., 1991, The haplochromine cichlids in Lake Victoria: an assessment of biological and fisheries interests. Pp. 258–279 *in* Keeleyside, M. (ed.), Cichlid Fishes, Behaviour, Ecology and Evolution, Chapman and Hall, London.

Bazigos, G., 1972, Report to the Promotion of Integrated Fishing Development Project FAO/MLW 16 Malawi Fisheries Dept., Zomba, Malawi, 23 pp.

Bicknell, D., 1992, Ranking Great Lakes persistent toxics. Water Envir. and Technol. 4:50–55.

Bizimana, M., and Duchafour, H., 1991, A drainage basin management study: The case of the Ntahangwa River Basin. *In* Cohen, A. (ed.), Report on 1st Intl. Conf. on Conservation and Biodiversity of L. Tanganyika. Biodiv. Support Prog., Washington, DC, pp. 43–45.

Bootsma, H., 1992, Lake Malawi National Park: An overview. Mitt. Inter. Verein. Limnol. 23:125–128.

Bootsma, H., and Hecky, R.E., 1993, Conservation of the African Great Lakes: A limnological perspective. Conserv. Biol. 7:644–656.

Bruton, M., 1985, The effects of suspensoids on fish. Hydrobiologia 125:221–241.

Bruton, M., 1990, The conservation of the fishes of Lake Victoria, Africa: an ecological perspective. Envir. Biol. of Fishes 27:161–175.

Bugenyi, F., 1979, Copper ion distribution in the surface waters of Lakes George and Idi Amin. Hydrobiologia 64:9–15.

Bugenyi, F., 1981, The aquatic environmental pollution in Uganda. Current status and conservation measures. Jinja mtg. 1–13.

Bugenyi, F., 1982, Copper pollution studies in Lakes George and Edward, Uganda: The distribution of Cu, Cd and Fe in the water and sediments. Envir. Pollution (Ser. B.) 3:129–138.

Bugenyi, F., 1990, Likely effects of salinity on acute copper toxicity to the fisheries of the Lake George-Edward Basin. Hydrobiologia 208:39–44.

Bugenyi, F., and Magumba, K., 1992, The annual cycle of stratification in Lake Victoria revisited. *In* Lakes Victoria, Kyoga and Nabugabo. Biodiversity, Fisheries and the Future of L. Victoria, Jinja, abstr., pp. 3–4.

Burton, J.D., 1976, Basic properties and processes in estuarine chemistry. *In* Burton, J., and Liss, P. (eds.), Estuarine Chemistry, Acad. Press, pp. 1–36.

Cairns, J., 1968, Suspended solids standards for the protection of aquatic organisms. Purdue Univ. Eng. Bull. Part 1. 129:16–27.

Caljon, A., 1992, Water quality in the Bay of Bujumbura (Lake Tanganyika) and its influence on phytoplankton communities. Mitt. Internat. Verein. Limnol. 23:55–65.

Carpenter, S., Fisher, S., Grimm, N., and Kitchell, J., 1992, Global change and freshwater ecosystems. Ann. Rev. Ecol. Syst. 23:119–139.

Carter, J., Gibson, J., Carr, A., and Azueta, J., in press, Creation of the Hol Chan Marine Reserve in Belize: A grass roots approach to barrier reef conservation. The Environmental Professional.

Casanova, J., and Hillaire-Marcel, C., 1993, Carbon and oxygen isotopes in African lacustrine stromatolites: Palaeohydrological interpretation. In Swart, P. (ed.), Climate Changes in Continental Isotopic Records. AGU Monograph 78:123–133.

Chapman, D., 1975, Fishery development in Lake Tanganyika. UN FAO Report FIA:DP/URT/71/012/21:1–8.

Chapman, L., Chapman, C., Kaufman, L., and Liem, K., 1992, The role of papyrus swamps as barriers to fish dispersal: A case study and implications for fish diversity in the Lake Victoria basin. in Biodiversity, Fisheries, and the Future of Lake Victoria, Conference Abstracts, p. 6.

Chifamba, P., 1992, The life history of Limnothrissa miodon in Lake Kariba. In Coenen, E. (ed.), Report on the symposium on biology, stock assessment and exploitation of small pelagic fish species in the African Great Lakes. UN/FAO/Finnida Report GCP/RAF/271/FIN-TD/06, p. 10.

Chitamwebwa, D., 1993, Socio-economic gains versus biodiversity losses following Nile Perch explosion in Lake Victoria. In J.-F. Guégan, Teugels, G., and Albaret, J.-J. (eds.), Symp. International sur la Diversité Biologique des Poissons d'Eaux Douces et Saumâtres de l'Afrique. Dakkar, 1993, pp. 11–12.

CIFA, 1991, Report of the third session of the working party on pollution and fisheries, Accra, Ghana, Nov. 1991, FAO Fisheries Report 471, 43 pp.

Coenen, E., 1992, Report on the symposium on biology, stock assessment and exploitation of small pelagic fish species in the African Great Lakes region. FINNIDA/FAO Report GCP/RAF/271/FIN-TD/06, 29 pp.

Cohen, A. (ed.), 1991, Report on 1st Intl. Conf. on Conservation and Biodiversity of Lake Tanganyika. Biodiversity Support Prog. Washington DC, 128 pp.

Cohen, A., 1992, Criteria for developing viable underwater natural reserves in Lake Tanganyika. Mitt. Int. Verein. Limnol 23:109–116.

Cohen, A., Dussinger, R., and Richardson, J., 1983, Lacustrine paleochemical interpretations based on East and South African ostracodes: Paleogeog., Paleoclimat., Paleoecol. 43:129–151.

Cohen, A., and Nielsen, C., 1986, Ostracodes as indicators of paleohydrochemistry in lakes: A Late Quaternary Example from Lake Elmenteita, Kenya. Palaios 1:601–609.

Cohen, A, Coulter, G., and Cardy, F., 1992, Global Environmental Facility Planning Mission Report. Pollution Control and other measures to protect biodiversity in Lake Tanganyika., United Nation。 Development Programme, GEF Program for Africa, 87 pp.

Cohen, A., Bills, R., Gashagaza, M., Michel, E., Tiercelin, J., Martens, K., Soreghan, M., Coveliers, P., West, K., Ntakimazi, G., Mboko, S., and Kimbadi, S., 1993a, Preliminary observations of sedimentation impacts on benthic communities and biodiversity using an ROV submersible in Lake Tanganyika. Symp. Intl. Diversité Biologie des Poissons d'Eaux Douce et Saumatres d'Afrique (PARADI), Dakar, Senegal, abstr. w/prog. pp. 12–13.

Cohen, A.S., Bills, R., Cocquyt, C., and Caljon, A., 1993b, The impact of sedimentpollution on biodiversity in Lake Tanganyika. Conservation Biology. 7:667–677.

Cohen, A., Bills, R., Gashagaza, M., Michel, E., Tiercelin, J., Martens, K., Soreghan, M., Coveliers, P., West, K., Ntakimazi, G., Mboko, S., and Kimbadi, S., in prep., First ROV observations in Lake Tanganyika, Africa indicate sedimentation impacts on benthic communities and biodiversity.

Coulter, G., 1989, A matter of value. Soc. Int. Limnol. Symp. on Resource Utilization and Conservation of the African Great Lakes, Bujumbura, Burundi, 7 pp.

Coulter, G. (ed.), 1991, Lake Tanganyika and its Life. Oxford Univ. Press, Oxford, 354 pp.

Coulter, G.W., 1992, Vulnerability of Lake Tanganyika to pollution, with comments on social aspects. Mitt. Int. Verein. Limnol 23:67–70.

Coulter, G.W., Allanson, B., Bruton, M., Greenwood, P., Hart, R., Jackson, P., and Ribbink, A., 1986, Unique qualities and special problems of the African Great Lakes. Envir. Biol. Fish. 17:161–183.

Coulter, G.W., and Jackson, P.N.B, 1981 Deep Lakes. ch. 9 *in* Symoens, J, Burgis, M., and Gaudet, J. (eds.), The Ecology and Utilization of African Inland Waters. UNEP Reports and Proceedings Series 1, pp. 114–124.

Coulter, G., and Mubamba, R., 1993, Proposal for designing reserve management plans for Sumbu National Park, Lake Tanganyika. Conservation Biology 7:678–685.

Davidson, O., 1992, Energy issues in Sub-Saharan Africa: Future directions. Ann. Rev. Energy and the Envir. 17:359–404.

Deelstra, H., 1977, Danger de pollution dans le Lac Tanganyika. Bull. Soc. Belge Etude. Géograph. 46:23–35.

Deelstra, H., 1985, Millieuproblemen in ontwikkelingslanden. Chemische pollutie in Afrika. Meded. Zitt. K. Acad. overzeese Wet. 29:179–194.

Deelstra, H., Power, J., and Kenner, C., 1976, Chlorinated hydrocarbon residues in the fish of Lake Tanganyika. Bull. Envir. Contamination and Toxicology, 15:689–698.

Dejoux, C., and Deelstra, H., 1981, Pollution. *In* Symoens, J. Burgis, M., and Gaudet, J. (eds.), The Ecology and Utilization of African Inland Waters. UNEP Reports and Proceedings Series 1:149–161.

Dejoux, C., 1988, La Pollution Des Eaux Continentales Africaines. ORSTOM Trav. et Doc. 213, 513 pp.

Eccles, D., 1985, Lake flies and sardines — A cautionary note. Biol. Conservation 33:309–333.

Elongo, S., 1991a, Situation des prises dans la partie Zairois du Lac Tanganyika. *In* Cohen, A. (ed.), Report on 1st Intl. Conf. on Conservation and Biodiversity of Lake Tanganyika. Biodiversity Support Prog. Washington, DC, pp. 56–59.

Elongo, S., 1991b, CADIC Environnement-Developpement, Rapport Annuel, 1991, B.P. 386, Bujumbura, Burundi, 13 pp.

Eugster, H., 1980, Lake Magadi and its precursor. *In* Nissenbaum, A. (ed.), Hypersaline brines and evaporitic environments. Devel. in Sedimentology v. 28, Elsevier, Amsterdam, pp. 195–230.

Evans, J., 1961, Growth of Lake Victoria phytoplankton in enriched cultures. Nature 189:417.

FAO (Committee For Inland Fisheries Of Africa), 1989, Report of the Second Session of the Working Party On Pollution and Fisheries, Nairobi, Kenya, Oct., 1989. FAO Fisheries Rept. 437, 24 pp.

Fryer, G., 1960, Concerning the proposed introduction of Nile perch into Lake Victoria. E. Afr. Agr. Jour. 25:267–270.

Fryer, G., 1972a, Conservation of the Great Lakes of East Africa: A lesson and a warning. Biol. Conserv. 4:256–262.

Fryer, G., 1972b, Some hazards facing African lakes. Biol. Conserv. 4:301–302.

Gasana, J.C., 1988, Etude sur la qualité des eaux du lac Tanganyika dans la baie de Bujumbura. Univ. du Burundi, Bujumbura, 140 pp.

Gates, W., Mitchell, J., Boer, G., Cubasch, U., and Meleshko, V., 1992, Climate modelling, climate prediction and model variation. *In* Thoughton, T., Callander, B., and Varney, S. (eds.), Climate Change 1992, Cambridge Univ. Press, London, pp. 101–134.

Gaudet, J., Mitchell, D., and Denny, P., 1981, Macrophytes (Aquatic Vegetation). *In* Symoens, J. Burgis, M., and Gaudet, J. (eds.), The Ecology and Utilization of African Inland Waters. UNEP Reports and Proceedings Series 1:27–36.

GESAMP, 1982 (IMO/FAO/UNESCO/WMO/IAEA/UN/UNEP) Joint group of experts on the scientific aspects of marine pollution. The health of the oceans Rep. Study GESAMP 15:108 pp, and UNEP Reg. Seas Rep. Stud. 16:108 pp.

Getabu, A, 1992, Growth parameters and total mortality in *Oreochromis niloticus* (Linneaus) from Nyanza Gulf, Lake Victoria. Hydrobiologia 232:91–97.

Goldschmidt, T., Witte, F., and Wanink, J., 1993, Cascading effects of the introduced Nile Perch on the detritivorous/phytoplanktivorous species in the sub-littoral areas of Lake Victoria. Conservation Biology, 7:686–700.

Golterman, H., 1977, *in* Viner, A., and Lee, G. (eds.), Sediment/Freshwater Interactions. Devel. in Hydrobiol. 9, Junk, The Hague, Netherlands

Golterman, H., 1991, Influence of sediments on lake water quality. Int. Conf. on Land–Water Interactions, New Delhi, India, Dec. 1991, abstr. p. 4.

Gophen, M., Ollinger, U., Ochumba, P., 1992, Limnological studies in Lake Victoria. in Biodiversity, Fisheries, and the Future of Lake Victoria, Conference Abstracts, pp. 5–6.

Graham, A., 1990, Siltation of stone-surface periphyton in rivers by clay-sized particles from low concentrations in suspension. Hydrobiologia 199:107–115.

Greichus, Y., Amman, B., and Hopcraft, J., 1978, Insecticides, polychlorinated biphenyls and metals in African lake ecosystems. III Lake Nakuru. Bull. Envir. Contamination and Toxicology 19:454–461.

Grobbelaar, J., 1985, Phytoplankton productivity in turbid waters. Jour. Plankton Res. 7:653–663.

Hanek, G., Baziramwabo, T., and Lamboeuf, M., 1990, La peche de l'Isambaza (*Limnothrissa miodon*) au Lac Kivu. *In* Fisheries of the African Great Lakes (ed. anon.), Intl. Agricultural Center, Wageningen, Netherlands, Occas. Paper 3:33–43.

Hanek, G., and Greboval, D., 1992, Report on the meeting of Project Managers for the coordination of stock assessment work on East African Lakes. FINNIDA/FAO Rept. GCP/RAF/271/FIN-TD/07, 51 pp.

Harris, C., 1992, Future prospects for the fisheries of Lake Victoria: Three scenarios. in Biodiversity, Fisheries, and the Future of Lake Victoria, Conference Abstracts, pp. 27–28.

Harris, C., 1993, Anthropogenic causes of changes in water quality of Lake Victoria. Intl. Decade of East Afr. Lakes Symp., Jinja, Uganda, March, 1993, abstr. w/prog. pp. 13–14.

Healey, F., and Hendzel, L., 1980, Physiological indicators of nutrient deficiency in lake phytoplankton. Can. Jour. Fish. Aquatic Sci. 37:442–453.

Hecky, R., 1991, The pelagic ecosystem. ch. 5 *in* Coulter, G. (ed.), Lake Tanganyika and Its Life. Oxford, pp. 90–110.

Hecky, R., 1992, Historical evidence of eutrophication in Lake Victoria. in Biodiversity, Fisheries, and the Future of Lake Victoria, Conference Abstracts, pp. 1–2.

Hecky, R., and Bugenyi, W., 1992, Hydrology and chemistry of the African Great Lakes and water quality issues: Problems and solutions. Mitt. Int. Verein. Limnol. 23:45–54.

Hopson, A. (ed.), 1982, Lake Turkana. A report on the findings of the Lake Turkana Project 1972–1975. v. 1–6, Overseas Devel. Admin, London, 1614 pp.

Hullsworth, E.G., 1987, Anatomy, Physiology And Psychology Of Erosion. Wiley, N.Y. 176 pp.

IAC (International Agricultural Centre), 1990, Report on the International Symposium on Resource Use and Conservation of the African Great Lakes. Bujumbura, Burundi, Jan. 1990, Int. Agr. Centre Occas. Paper 2, 50 pp.

IUCN, 1991, Lake Baikal on the Brink. IUCN East European Programme, Environmental Research Series 2, 36 pp.

Jackson, P.B.N., 1973, The African Great Lakes: Food source and world treasure. Biol. Conserv. 5:302–304.

Jackson, P.B.N., and Coetzee, P., 1982. Spawning behaviour of *Labeo umbratus* (Smith) (Pisces: Cyprinidae) S. Afr. Jour. Sci. 78:293–295.

Kabazinya, K., 1991, La foret d'Itombwe a l'ouest du Lac Tanganyika. *In* Cohen, A. (ed.), Rept. on 1st Intl. Conf. on the Conservation and Biodiversity of L. Tanganyika. Biodiv. Support Program, Washington, DC, pp. 61–65.

Kaningini, M., 1992, Problématique de la protection des ressources halietiques dans les lacs du Kivu et Tanganyika. Actes de la Consultation Nationale des O.N.G. du Zaire, Bukavu, Zaire 16–20 Dec., 1991, Preparatory Document for U.N. Conference On The Environment and Development, Rio de Janeiro, 1992, pp. 72–74.

Kaufman, L., 1989, Challenges to fish faunal conservation programs as illustrated by the captive biology of Lake Victoria cichlids. *In* Dresser, B., Reece, R., and Marushka, E. (eds.), 5th World Congress on Breeding Endangered Species in Captivity, pp. 105–120.

Kaufman, L., 1992, Catastrophic change in a species rich freshwater ecosystem. Bioscience 42:846–858.

Kaufman, L., and Ochumba, P., 1993, Evolutionary and conservation biology of cichlid fishes as revealed by faunal remnants in northern Lake Victoria. Conservation Biology 7:719–730.

Kaufman, L., Ochumba, P., and Ogutu-Ohwayo, R. (eds.), 1993, People, Fisheries, Biodiversity and the Future of Lake Victoria. Edgerton Research Lab., New England Aquarium, Boston, MA, Report 93–3, 68 pp.

Kling, G., 1992, Limnological change in Lake Victoria; contrasts in structure and function between temperate and tropical great lake ecosystems. Biodiversity, Fisheries, and the Future of Lake Victoria, Conference Abstracts, pp. 7–8.

Koeman, J., and Penning, J., 1970, An orientational survey of the side effects and environmental distribution of insecticides used in tse-tse control in Africa. Bull. Envir. Contam. Toxicol. 5:164–170.

Koeman, J., Penning, J., De Goeij, J., Tijoe, P., Olindo, P., and Hopcraft, J., 1972, A preliminary survey of the possible contamination of Lake Nakuru in Kenya, with some metals and chlorinated hydrocarbon pesticides. Jour. Appl. Ecol. 9:411–416.

Kolding, J., 1992, A summary of Lake Turkana: An ever-changing mixed environment. Mitt. Internat. Verein. Limnol. 23:25–35.

Kudhongania, A., 1990, The future prospects for the Lake Victoria fishery. in Roest, F. (ed). Fisheries of the African Great Lakes. S.I.L. Conf. Internat. Agr. Center, Occas. Paper 3:75–82.

Kudhongania, A., and Cordonne, A., 1974, Bathyo-spatial distribution patterns and biomass estimates of the major demersal fishes in Lake Victoria. Afr. Jour. Trop. Hydrobiol. Fish. 3:15–31.

Lal, R., 1982, Annual Rept. Inst. of Tropical Agriculture, Ibadan, Nigeria.

Lavreau, J., and Nkanira, T., 1990, Gitologie et spectroradiométrie du gisement nickelifere de Musongati (Burundi). *In* Lavreau, J. (ed.), H. Spectrometrie a Haute Resolution *In Situ* et Analyse d'Image LANDSAT et SPOT au Burundi. Musée Roy. de l'Afrique Centrale, Tervuren, Belgium, Ann. Ser. 8, Sci. Geol. 98:95–108.

Lewis, D., and Tweddle, D., 1990, The yield of Usipa (*Engraulicypris sardella*) from the Nankumba Peninsula, Lake Malawi (1985–86). Collected Reports of Fisheries Research in Malawi, Occas. Papers, 1:57–66.

Lincer, J., Zalkind, D., Brown, L., and Hopcraft, J., 1981, Organochlorine residues in Kenya's Rift Valley lakes. Jour. Appl. Ecol. 18:157–171

Lindqvist, O., and Beverridge, M., 1987, Project review of experimental tilapia culture at Lake Turkana-KEN 043, Lake Turkana Fisheries Management Project. Report to NORAD, Feb., 1987.

Ligtvoet, W., 1988, The Nile Perch in Lake Victoria: a blessing or a disaster, HEST report 42.

Lowe-McConnell, R., 1993, Fish faunas of the African Great Lakes: Origins, diversity and vulnerability. Conservation Biology, 7:634–643.

Magasa, J., 1988, A brief review of the fish stocks and dependent fisheries of Lake Malawi. *In* Lewis, D. (ed.), Predator–Prey relationships, population dynamics and fisheries productivities of large African lakes. FAO Comm. for Inland Fisheries of Africa (CIFA), FAO-Rome, Occas. Paper 15:45–52.

Mahy, G., 1979, Biologie et écologie du Lac Kivu. Etudes Rwandaises, 12.

Marshall, B., 1985, Changes in abundance and mortality rate of tigerfish in the Eastern Basin of L. Kariba. Jour. Limnol. Soc. South Africa 11:46–50.

Marshall, B., 1988, Seasonal and annual variations in the abundance of pelagic sardines in Lake Kariba, with special reference to the effect of drought. Arch. Hydrobiol. 112:399–409.

Massinga, A., 1990, River impact on the cyprinid fishery of the Mozambican side of Lake Nyassa/ Malawi, with particular reference to the genus *Osparidium* and the importance of long-term data collection. *In* Fisheries of the African Great Lakes (ed. anon.), Intl. Agricultural Center, Wageningen, Netherlands, Occas. Paper 3:96–108.

Mavuti, K., and Litterick, M., 1991, Composition, distribution and ecological role of the zooplankton community in Lake Victoria, Kenya waters. Int. Assoc. Theor. Appl. Limnol. Proc. 24:1117–1122.

McKaye, K., Makwinja, R., Menyani, W., and Mhone, O., 1985, On the possible introduction of non-indigenous zooplankton-feeding fishes into L. Malawi. Biol. Conservation 33:289–307.

Melack, J., Kilham, P., and Fisher, T., 1982, Responses of phytoplankton to experimental fertilization with ammonium and phosphate in an African soda lake. Oecologia 52:321–326.

Meybeck, M., 1985, Evaluation préliminaire de la pollution du Lac Tanganyika. UNEP, Nairobi, 46 pp.

Moreau, J., 1979, Introductions d'Especes Etrangeres dans les Eaux Continentales Africaines. S.I.L. Conf. Document, Nairobi, Ecole Nat. Super. Agron., Tolouse, 22 pp.

Moreau, J., 1988, Introductions d'especes etrangeres dans les eaux continentales africaines. Interets et limites. *In* Leveque, C., Bruton, M., Ssentongo, G. (eds.), Biologie et ecologie des poissons d'eau douce africains. ORSTOM Travaux et Documents 216:395–425.

MTNRE (Ministry of Tourism, Natural Resources and Environment, United Republic Of Tanzania), 1991, The Proposed Mafia Island Marine Reserve — Proposals from the Steering Committee. Dar es Salaam, 14 pp.

Mubamba, R., 1991, A survey of littoral invertebrates in the Zambian waters of Lake Tanganyika. *In* Cohen, A. (ed.), Rept. on 1st Intl. Conf. on the Conservation and Biodiversity of L. Tanganyika. Biodiv. Support Program, Washington, DC, p. 80.

Mubamba, R., 1992, Introduction of Lake Tanganyika sardines into Itezhi-Tezhi Reservoir, Zambia. in Coenen, E., Report on the symp. on biology, stock assessment and exploitation of small pelagic fish species in the African Great Lakes region. FINNIDA/FAO Document GCP/RAF/271/FIN-TD/06, p. 20.

Mugidde, R., 1992, The increase in phytoplankton photosynthesis and biomass in Lake Victoria. in Lakes Victoria, Kyoga and Nabugabo. Biodiversity, Fisheries and the Future of L. Victoria, Jinja, abstr. p. 13.

Mutahinga, M., 1991, Quelques aspects de la legislation de la peche au Zaire et son application. *In* Cohen, A. (ed.), Rept. on 1st Intl. Conf. on the Conservation and Biodiversity of L. Tanganyika. Biodiv. Support Program, Washington, DC, pp. 81–84.

Mwandu, D., 1992, End of Assignment Report, Vitamin A Deficiency Project for Luapula Valley, Govt. of Zambia.

Mwasaga, B., 1991, The status of laws or regulations pertaining to conservation and national parks: A Tanzanian case study. *In* Cohen, A. (ed.), Rept. on 1st Intl. Conf. on the Conservation and Biodiversity of L. Tanganyika. Biodiv. Support Program, Washington, DC, pp. 85–86.

Mwebaza-Ndawula, L., 1992, Distribution, abundance and changes in relative abundance of zooplankton in northern Lake Victoria. Biodiversity, Fisheries and the Future of L. Victoria, Jinja, abstr. pp. 14–15.

Mwenya, A., Lewis, D., and Kaweche, G., 1990, ADMADE: Policy, background and future. World Wildlife Fund, 17 pp.

Naikoba, R., Odongkara, O., and Harris, C., 1992, Structure and process in the Uganda Lake Victoria fishery. Biodiversity, Fisheries and the Future of L. Victoria, Jinja, abstr. pp. 26–27.

Namulemo, G., 1992, The state of the native cichlids of Nabugabo Lake. Biodiversity, Fisheries and the Future of L. Victoria, Jinja, abstr. p. 20.

Ndabigengerese, A., 1986, La charge pollutante du lac Tanganyika par les arrivées dans la baie de Bujumbura. Univ. du Burundi, Bujumbura, 202 pp.

Ndayizeye, P., 1985, Influence des déchets industriels sur la qualité de l'eau du Lac Tanganyika. Université du Burundi, Bujumbura, 159 pp.

Niyabona, J.C., 1988, Quelques aspects de la pollution dans le Lac Tanganyika. Univ. du Burundi, Bujumbura, 108 pp.

Nongwa, G., 1986, African Lakes. Nature 322:679.

Nriagu, J., 1992, Toxic metal pollution in Africa. The Science of the Total Environment 121:1–37.

Nsabimana, S., 1991, Soil erosion and pollution of Lake Tanganyika in Burundi. In Cohen, A. (ed.), Rept. on 1st Intl. Conf. on the Conservation and Biodiversity of L. Tanganyika. Biodiversity Support Prog., Washington, DC, pp. 91–94.

Ntakimazi, G., 1992. Conservation of the resources of the African Great Lakes: Why? An overview. Mitt. Internat. Verein. Limnol. 23:5–10.

Ochieng, E., 1987, Limnological aspects and trace element analysis of Kenyan natural inland waters. M. Sc. thesis, Univ. Nairobi, 294 pp.

Ochumba, P., 1990, Massive fish kills within the Nyanza Gulf of Lake Victoria, Kenya. Hydrobiologia 208:93–99.

Ochumba, P., and Kibaraa, D., 1989, Observations on blue-green algal blooms in the open waters of Lake Victoria, Kenya. Afr. Jour. Ecol. 27:23–34.

Ochumba, P., Gophen, M., and Kaufman, L., 1992, Changes in oxygen availability in the Kenyan portion of Lake Victoria: Effects on fisheries and biodiversity. in Lakes Victoria, Kyoga and Nabugabo. Biodiversity, Fisheries and the Future of Lake Victoria, Jinja, abstr. pp. 4–5.

Ogutu-Ohwayo, R., 1990, The decline of the native fishes of Lakes Victoria and Kyoga (East Africa) and the impact of introduced species, especially the Nile Perch, Lates niloticus, and the Nile Tilapia, Oreochromis niloticus. Envir. Biol. Fishes 27:81–96.

Ogutu-Ohwayo, R., 1992a, The purpose, costs and benefits of fish introductions: With specific reference to the Great Lakes of Africa. Mitt. Int. Verein. Limnol. 23:37–44.

Ogutu-Ohwayo, R., 1992b, Stability in fish stocks and in life history parameters of the Nile Perch Lates niloticus L. in Lakes Victoria, Kyoga and Nabugabo. Biodiversity, Fisheries and the Future of L. Victoria, Jinja, abstr. pp. 23–24.

Ogutu-Ohwayo, R., 1992c, The impact of predation by Nile Perch, Lates niloticus L. on the fishes of Lake Nabugabo: with suggestions to conserve native cichlids. Biodiversity, Fisheries and the Future of L. Victoria, Jinja, 1992 abst., p. 19.

Ogutu-Ohwayo, R., 1993, The effects of predation by Nile Perch, L. niloticus on the fish of L. Nabugabo, with suggestions for conservation of endangered cichlids. Conservation Biology 7:701–711.

Okaronon, J., and Akumu, J., 1992, The increased fishing pressure and the future of the fish stocks of Lake Victoria. In Kaufman, L. et al. (eds.), Biodiversity, Fisheries and the Future of Lake Victoria, Jinja, 1992 abst., p. 26.

Okello, E., 1992, Heavy metal concentrations in the great rivers of the Lake Victoria Basin. In Kaufman, L. et al. (eds.), Biodiversity, Fisheries and the Future of Lake Victoria, Jinja, 1992 abst., p. 32.

Olowo, J., 1992, Some observations on surviving native species of Lakes Victoria, Kyoga and Nabugabo. Biodiversity, Fisheries and the Future of L. Victoria, Jinja, 1992 abst., p. 20.

Onyari, J., and Wandiga, S., 1989, Distribution of Cr, Pb, Cd, Zn Fe and Mn in Lake Victoria sediments. East Africa. Bull. Envir. Contam. Toxicol. 42:807–813.

Orach-Meza, F., Coenen, E., and Reynolds, J., 1990, Past and recent trends in the exploitation of the Great Lakes fisheries of Uganda. *In* Roest, F. (ed.), Fisheries of the African Great Lakes. Intl. Agricultural Center, Wageningen, Netherlands, Occas. Paper 3:109–121.

Owen, R., Crossley, R., Johnson, T., Tweddle, D., Kornfield, I., Davison, S., Eccles, D., and Engstrom, D., 1990, Major low levels of Lake Malawi and their implications for speciation rates in cichlid fishes. Proc. Roy. Soc. London B, 240:519–553.

Pearce, M., 1985a, The deepwater demersal fish in the south of Lake Tanganyika. Report of Fisheries Dept., Govt. of Zambia, Lusaka, 163 pp.

Pearce, M., 1985b, Some effects of *Lates* species on pelagic and demersal fish in Zambian waters of Lake Tanganyika. UNDP/FAO, CIFA Document SAWG/85/WP2, FAO Rome, 18 pp.

Pearce, M., 1988, Some effects of *Lates* spp. on pelagic and demersal fish in Zambian waters of Lake Tanganyika. *In* Lewis, D. (ed.), Predator–Prey Relationships, Population Dynamics and Fisheries Productivities Of Large African Lakes. UN FAO Committee for Inland Fisheries of Africa (CIFA) Occas. Paper 15:69–87.

Pearce, M., 1991, Utilization and legislation pertaining to the Zambian section of the Lake Tanganyika Basin. *In* Cohen, A. (ed.), Rept. on 1st Intl. Conf. on the Conservation and Biodiversity of L. Tanganyika. Biodiversity Support Prog., Washington, DC, pp. 105–107.

PTA (Preferential Trade Area For Eastern And Southern Africa), 1993, Proposal For The Establishment Of A Regional Centre For Fisheries Research And Management (Eastern/Central/Southern Africa Region). PTA, Lusaka, 31 pp.

Rabuor, C., and Polovina, J., 1992, Multispecies and multigear analyses of fisheries data from the Kenya waters of Lake Victoria. Biodiversity, Fisheries and the Future of Lake Victoria, Jinja, 1992. abstr. pp. 24–25.

Reinthal, P., 1993, Evaluating biodiversity and conserving Lake Malawi's cichlid fish fauna. Conservation Biology 7:712–718.

Reusens, M., 1988, Debut d'etude d'evaution du stock de *Limnothrissa miodon* au Lac Kivu. *In* Lewis, D. (ed.), Predator–Prey Relationships, Population Dynamics and Fisheries Productivities Of Large African Lakes. UN FAO Committee for Inland Fisheries of Africa (CIFA) Occas. Paper 15:88–103.

Reynolds, J., and Greboval, D., 1988, Socio-economic effects of the evolution of Nile Perch fisheries in Lake Victoria: A review. FAO, CIFA Tech, Paper 17.

Ribbink, A.J., 1987, African lakes and their fishes: conservation scenarios and suggestions. Environmental Biol. of Fishes. 19:3–26.

Roest, F., 1988, Predator–prey relations in northern Lake Tanganyika and fluctuations in the pelagic fish stocks. *In* Lewis, D. (ed.), Predator–Prey Relationships, Population Dynamics and Fisheries Productivities Of Large African Lakes. UN FAO Committee for Inland Fisheries of Africa (CIFA) Occas. Paper 15:104–129.

Roest, F., 1992, The pelagic fisheries resources of Lake Tanganyika. Mitt. Internat. Verein. Limnol. 23:11–15.

Rhodes, D., 1966, Fisheries development possibilities, Report to the government of Kenya. UNDP/FAO Rep. No. TA 2144, 77 pp.

Rosendahl, B., 1987, Architecture of continental rifts with special reference to East Africa. Ann. Rev. Earth and Planetary Sci. 15:445–503.

Rosendahl, B. (ed.), 1988, Seismic atlas of Lake Tanganyika, East Africa. Duke Univ. Project PROBE Geophysical Atlas Series 1, 82 pp.

Rukuba-Ngaiza, N., 1992. A review of the laws and institutions on fish conservation in Uganda. in Biodiversity, Fisheries and the Future of L. Victoria. Conference Abstracts, Jinja, 1992, p. 30.

Sahiri, C., 1991, La pollution chimique du Lac Tanganyika dans les eaux voisines de la ville de Bujumbura. *In* Cohen, A. (ed.), Rept. on 1st Intl. Conf. on the Conservation and Biodiversity of L. Tanganyika. Biodiversity Support Prog., Washington, DC, pp. 108–109.

Salomons, W., and Forstner, U., 1984, Metals in the hydrocycle. Springer-Verlag, Berlin, 349 pp.

Schelske, C., 1988, Historic trends in Lake Michigan silica concentrations. Int. Rev. ges. Hydrobiol. 73:559–591.

Schindler, D., 1990, Experimental perturbations of whole lakes as tests of hypotheses concerning ecosystem structure and function. Oikos 57:25–41.

SIAL (Speciation In Ancient Lakes), in prep., Conference Proceedings from Symposium at Mont-Rigi, Belgium, March, 1993.

Sindayigaya, E., Deelstra, H., and Dejonckheere, W., 1990, Evolution de la contamination des poissons du Lac Tanganyika par les residus de pesticides organochlores. Med. Fac. Landbouww. Rijksuniv. Gent. 55:1361–1368.

Sindayigaya, E., 1991, Quelques données concernant la surveillance de la contamination des poisons de Lac Tanganyika. *In* Cohen, A. (ed.), Rept. on 1st Intl. Conf. on the Conservation and Biodiversity of L. Tanganyika. Biodiversity Support Prog., Washington, DC, p. 110.

Sindayigaya, E., Van Cauwenbergh, R., Robberecht, H., and Deelstra, H. (in press) Copper, zinc, manganese, iron, lead, cadmium, mercury and arsenic in fish from Lake Tanganyika, Burundi, Africa. The Science Of The Total Environment, 20 pp., ms.

Sinderman, C., 1988, Biological indicators and biological effects of estuarine/coastal pollution. Water Res. Bull. 24:931–939.

Soulsby, J., 1960, Some Congo basin fishes of Northern Rhodesia. Part II. North. Rhodesia Jour. 4:319–334.

Stewart, J. 1990, The great lake is in peril. New Scientist, 30 June, 1990, pp. 58–62.

Stoneman, J., Mecham, K., and Mathotho, A., 1973, African Great Lakes and their fisheries potential. Biological Conservation 5:299–302.

Talling, J., 1965. The photosynthetic activity of phytoplankton in East African Lakes. Int. Revue ges. Hydrobiol. 50:1–32.

Talling, J., and Talling, I., 1965, The chemical composition of African lake waters. Int. Revue ges. Hydrobiol. 50:421–463.

TEEP (Tanzania Environmental Education Programme), 1991, A Mission Report. WWF, Dar es Salaam, 134 pp.

Turner, G., 1993, Species composition changes as the result of fishing in Lake Malawi. Speciation In Ancient Lakes Symp., abstr. w/ prog. p. 43, Mont-Rigi, Belgium, March, 1993.

Turner, J., 1977, Some effects of demersal trawling in Lake Malawi (Lake Nyassa) from 1968 to 1974. Jour. Fish. Biol. 10:261–271.

Turner, J., 1982, Lake flies, water fleas and sardines. in Biological studies on the pelagic ecosystem of Lake Malawi FAO Technical Report FI:DP/MLW/75/019, 1:165–173.

Tweddle, D., 1992. Conservation and threats to the resources of Lake Malawi. Mitt. Internat. Verein. Limnol. 23:17–24.

Twongo, T., 1990, On dangers to rational exploitation and development of the resources of the Great Lakes of Africa. *In* Roest, F. (ed.), Fisheries of the African Great Lakes. Intl. Agricultural Center, Wageningen, Netherlands, Occas. Paper 3:122–130.

Twongo, T., 1992, The spread of water hyacinth on Lakes Victoria and Kyoga and some implications for aquatic biodiversity and fisheries. Biodiversity, Fisheries and the Future of Lake Victoria Conference Abstracts, Jinja, 1992, p. 16.

United Republic Of Tanzania, 1991, National Report For The 1992 United Nations Conference On Environment And Development (UNCED), Rio de Janeiro, July, 1991, Executive Summary, 17 pp.

Wandiga, S., and Onyari, J., 1987, The concentration of heavy metals: Mn, Fe, Cu, Zn, Cd, and Pb in sediments and fish from the Winam Gulf of Lake Victoria and fish bought in Mombasa town markets. Kenya Jour. Sci. 8:5–18.

Welcomme, R., 1964, Notes on the present distribution and habits of the non-endemic species of *Tilapia*, which have been introduced into Lake Victoria. Rep. E. AFr. Freshwater Fish. Res. Org. 1962/63:36–39.

Welcomme, R., 1966, Recent changes in the stocks of *Tilapia* in Lake Victoria. Nature 212:52–54.

Welcomme, R., 1988, International introductions of inland aquatic species. FAO Fish. Tech. Paper 294, 318 pp.

Welcomme, R., 1992, The moral dilemma of development — The case of Lake Victoria. Biodiversity, Fisheries and the Future of Lake Victoria Conference Abstracts, Jinja, 1992, pp. 22–23.

Whitehead, D., Charles, D., and Goldstein, R. 1990, The PIRLA project (Paleoecological Investigation of Recent Lake Acidification): An introduction to the synthesis of the project. Jour. Paleolimnol. 3:187–194.

WHO (World Health Organization) Anonymous, 1984, Guidelines for drinking water quality. Recommendations. 130 pp.

Witte, F, Goldschmidt, T., Wanink, J., van Oijen, M., Goudswaard, K., Witte-Maas, E., and Bouton, N., 1992, The destruction of an endemic species flock: quantitative data on the decline of the haplochromine cichlids of Lake Victoria. Envir. Biol. of Fishes 34:1–28.

Wurtz, A., and Simpson, H., 1964. Limnological survey of Lake Rudolf (British East Africa) Verh. Internat. Verein. Limnol. 15:149.

WWF, 1991a, ADMADE program document, Lusaka, Zambia.

WWF, 1991b, Mafia Island Reserve Planning Document, Dar-es-Salaam, Tanzania.

Yongo, E. 1991, Socioeconomic aspects of fish utilization and marketing. Kisumu Kenya Marine Fish. Res. Inst. document.

Anthropogenic Perturbations on the Lake Victoria Ecosystem

A.W. KUDHONGANIA, D.L. OCENODONGO and J.O. OKARONON
Fisheries Research Institute, Jinja, Uganda

Abstract — Anthropogenic impact on the environment of Lake Victoria has caused a dramatic decline in fish species diversity accompanied by an intriguing phenomenon of higher annual fish yields. Future prospects for fish production, biodiversity and environmental sustainability are currently unpredictable and of concern. Prompt intervention through appropriate research and management protocols are required to stimulate and sustain the socioeconomic viability of the ecosystem.

INTRODUCTION

Lake Victoria is a large (69,000 km^2), relatively shallow basin having mean and maximum depths of 40 and 80 m, respectively. The lake lies at an elevation of 1135 m in the central part of the great plateau stretching between the western and eastern arms of the East African Rift Valley. The lake may be as old as Late Pliocene or Early Pleistocene (Wayland, 1931). Another lake is believed to have existed much earlier during the Miocene (25 million years BP) in the region now occupied by Lake Victoria, although it later completely dried out (Greenwood, 1966).

The present lake basin is believed to have originated from Pliocene uplift of the western arm of the Rift Valley and ponding of westward flowing rivers (Wayland, 1931; Worthington, 1954; Kendall, 1969). Recent hydrological data of Lake Victoria suggest that between 13,000 and 15,000 years BP, the lake was no more than 26 m deep (Stager et al., 1986) and perhaps had dried up completely. Its present coastline of 3440 km is very irregular, especially in the north and south ends. Water income is mostly (80–90%) from direct rainfall (1450 mm per year) and is nearly balanced by water loss by evaporation. The rest of the water budget (10–20%) is due to the inflow from the streams dominated by the Kagera River and by nearly equal outflow through the Victoria Nile (Talling, 1965). The three major bottom types of the lake are *hard* (sand, gravel and rock), *soft* (silt, mud, humus or clay) and *mixed*. On the basis of other ecological parameters, however, Lake Victoria has been described as many lakes within one lake (Beauchamp, 1958).

With regard to the fish species, *Lates* and *Polypterus* are believed to have existed in the Miocene lake basin (Greenwood, 1951, 1966) but disappeared with the original lake. The genus *Tilapia* first appeared in the Pleistocene fossil records (Trewavas, 1937). The endemic ichthyofauna of the modern fish genera in Lake Victoria were derived from the fishes of swamps and rivers that existed in the basin at the time of tectonic ponding (Corbet, 1960). This accounts for the many fish species that could not quickly and fully adapt to the lacustrine regime and retained various degrees of anadromesis (Whitehead, 1959; Corbet, 1961).

According to Greenwood (1966), the indigenous fishes of Lake Victoria fell into 12 families derived from 6 cichlid and 14 non-cichlid genera. There were 50 non-cichlid fish species (Lowe-McConnel, 1975). The cichlids consisted of two tilapiine species (Graham, 1929) and, through endemic explosive speciation (Greenwood, 1965), more than 300 species of haplochromines (van Oijen et al., 1981). It has been suggested that the radiation of the haplochromine species flock was achieved in the geologically short time span of about 750,000 years (Greenwood, 1984). However, since there are indications of the lake having dried substantially more recently (Stager et al., 1986) the time span for species radiation may have been even shorter. It is also possible that the evolution of the cichlid fish species was closely related to the evolutionary changes of the lake environment. Comprehensive paleo-limnological data will be needed to assess this possibility.

A generalized lake-wide survey carried out between 1969 and 1971 revealed the depth distribution and abundance of the major demersal fish species present (Kudhongania and Cordone, 1974). The high species diversity declined with increasing depth and at least 80% of the trawlable demersal ichthyomass, estimated at 700,000 tons, was composed of haplochromines.

The endemic ichthyofauna of Lake Victoria (originally more than 350 species) demonstrated the species diversity which some tropical ecosystems could support. Since the fauna is no longer as diverse, Lake Victoria appears to have succumbed to the impact of human activities on its limnology, productivity and biodiversity. This paper summarizes the anthropogenic perturbations on the Lake Victoria ecosystem.

ANTHROPOGENIC EFFECTS ON THE RESOURCES

Lake Victoria has a densely peopled watershed and there is strong evidence to indicate that its ecosystem has been impacted by man's activities, including fishery exploitation, exotic fish introductions, eutrophication and introduction of ornamental macrophytes. The lake has undergone significant change during the past 30 years (Hecky and Bugenyi, 1992) and is now in a transient state that causes considerable concern.

Fisheries Exploitation

One of the human activities known to have had significant impact on the Lake Victoria fauna is the exploitation of a multispecies fishery. The traditional fishery was artisanal, lucrative and exploited by simple fishing gear, including basket traps, hooks and spine nets of papyrus. But fishing intensified with urban development along the lake shore, the introduction of flax gill-nets in 1905, the arrival of the railway at the Nyanza Gulf in 1908, the introduction of the nonselective beach seines early in the 1920s (Graham, 1929), use of the more efficient synthetic fibre gill-nets since 1952 and the arrival of the outboard engine in 1953 (Mann, 1969). Industrial trawl fishing was introduced in Mwanza Gulf in 1973 (Witte and Goudswaard, 1985) while trawling started in the northern part of Lake Victoria in 1987 (Kudhongania and Colenen, 1991). In addition, industrial exploitation of the *Rastrinebola argentea* fishery, using seine nets, has been developing during the last decade (Wandera, 1990). Other types of fishing gear such as cast nets were also added to the fishery.

In the absence of any effective controls for the gill-net mesh sizes, entry of fishermen, canoes or fishing gear per canoe, fishing pressure eventually became detrimental to the fish stocks. Localized over-fishing was first observed in 1927 (Graham, 1929). Beverton (1959) predicted that the continued use of smaller mesh gill-nets would decrease both the immediate and long-term catches of several fish species. This was confirmed by Garrod (1961).

Intensive gill-netting of gravid *Labeo victorianus* during breeding migrations wiped out the *Labeo* fishery (Garrod, 1961; Cadwalladr, 1965). Over-fishing for *Labeo* affected 13 other anadromous or anadromous-like fish species (Whitehead, 1959). A lake-wide bottom trawl exploratory survey conducted between 1969 and 1971 indicated that most of the table fish species exploited by the traditional gear were declining to alarmingly low levels (Kudhongania and Cordone, 1974). Where trawling occurred on a commercial scale (for example, Mwanza Gulf), fishing had more damaging effects on the haplochromine stocks than the alleged predation by Nile perch (Kudhongania et al., 1992).

There is no doubt that the prolonged lack of control on the use of mesh-size gill-nets, beach seines, cast nets, lift nets, and trawl nets, as well as unlimited fishing pressure, had detrimental effects on the fish stocks (e.g., Mainga, 1985; Wanjala and Marten, 1974; Ogari and Asila, 1990; Wandera, 1990; Okaronon et al., 1985), including the capture of immature fish and interference with their breeding and nursery strategies (Welcomme, 1964). Although the use of different types of fishing gear was meant to exploit the multispecies fishery more effectively, the capture of non-target components of the stocks was inevitable and significant, leading to significant decline in fish population densities and species diversity.

Introduction of Exotic Fish Species

When the two indigenous tilapia species declined from the commercial catches, four exotic tilapiines were introduced into Lake Victoria in the 1950s (EAFFRO, 1964) in order to stimulate the commercial fishery. The Nile perch (*Lates niloticus* L.) was also introduced into the lake in 1954 (Amara, 1986) in order to convert the abundant haplochromine biomass into table fish. These exotic fish introductions have had significant impact on the ecology and biodiversity of Lake Victoria.

The introduction of four exotic tilapiine species into the lake suddenly increased inter-specific competition with the two indigenous species for food, spawning sites and nursery grounds (Fryer, 1961; Welcomme, 1966 and 1967). Competition and possible hybridization appear to have been instrumental in accelerating the decline in the tilapiine stocks in favour of only one exotic species (*Oreochromis niloticus* L.) among a total of six tilapia species. *O. niloticus* is is the only tilapiine species which has managed to coexist with the Nile perch in Lakes Victoria and Kyoga (Balirwa, 1992).

Lates became established in the Uganda, Kenya and Tanzania sectors of Lake Victoria in 1975, 1977 and 1978, respectively (Ssentongo and Welcomme, 1985). The establishment of the Nile perch fishery coincided with the decline of haplochromine stocks. There has been ample evidence to indicate that predation by *Lates* was largely responsible for the dramatic collapse of the haplochromine diversity (Okedi, 1970; Ogari, 1984; Ogutu-Ohwayo, 1985; Okaronon et al., 1985; Witte and Goundswaard, 1985; Moreau, 1982; etc.). At least two thirds of the original species may be extinct (Witte et al., 1992). The decline in haplochromine stocks was intriguingly accompanied by tremendous increases in fish landings constituted mainly by Nile perch and Nile tilapia (Ogutu-Ohwayo, 1985; Okaronon et al., 1985; Ssentongo and Welcomme, 1985). From the average annual catch of about 100,000 tons during the 1970s, the annual harvest rose to more than 400,000 tons in the 1980s. Although fish yield from Lake Victoria has increased several fold, the decline in haplochromine species has been the most rapid loss of vertebrate species diversity in recent years in the East African region. Haplochromine cichlids had dominated the fish fauna of Lake Victoria in terms of species numbers, biomass and ecological diversity (Witte et al., 1992).

The severe reduction in fish species diversity has led to modifications in the trophic patterns of the Lake Victoria ecosystem, including alterations in faunal and floral composition, and reduced grazing pressure on phytoplankton. It has been suggested that reduced grazing pressure may be partly responsible for the increased phytobiomass and widespread anoxia in the lake (Mugidde, 1992; Hecky and Bugenyi, 1992).

Cultural Eutrophication

Agricultural developments around Lake Victoria have involved vegetation clearing, deforestation, bush fires and draining of swamps for growing various crops, all leading to accelerated soil erosion (Bugenyi, 1987). In addition, pesticides, fertilizers and other agrochemicals are increasingly applied and some eventually find their way into the lake system.

Urban and industrial expansions along the lake shores have also increased the volumes of sewage, garbage and other effluents which end up in the lake (Bugenyi, 1984 and 1987). These add to the nutrient loading and are a source of water pollution.

Increased nutrient inputs from rain (Bugenyi, 1987), industrial and domestic sewage, and agricultural runoff have doubled the biological productivity of Lake Victoria during the last 30 years (Mugidde, 1992; Hecky and Bugenyi, 1992). Higher biological productivity is largely responsible for the increased algal biomass leading to deoxygenation of the deep waters. Lake Victoria is especially sensitive to oxygen depletion because tropical lakes hold less oxygen than those in colder climates. Given the extensive catchment area of 184,000 km^2 and flushing time of 140 years (Hecky and Bugenyi, 1992), the retention capacity of the nutrients is so high that eutrophication and anoxia may become long-term phenomena since an advanced stage of eutrophication under these conditions is virtually irreversible (Jorgensen and Volenweider, 1988). Coupled with prolonged thermal stratification which may develop due to the global temperature changes by the greenhouse effect (Hecky and Bugenyi, 1992), the productivity and biotic diversity of benthic organisms and demersal fish species could be greatly impaired.

Introduction of Ornamental Macrophytes

Wetland ecotones are known to have the highest biological activity and productivity in lake ecosystems since they comprise both aquatic and terrestrial components of production (Jorgensen and Loffler, 1990). They are essential nurseries for many organisms.

In the Lake Victoria basin, with a coastline of 3440 km, the wetland zone is extensive enough to adequately stimulate production and biodiversity of the lake. Unfortunately the zone was invaded by the water hyacinth (*Eichhornia crassipes* Maritus) in 1990. While the introduction of the weed into the lake may have been accidental, the water hyacinth was originally transported from its native home in South America to Africa to be used as an ornamental plant.

E. crassipes has spread to many parts of the lake shores and proliferated in inshore areas with high nutrient content (Twongo et al., 1992). These are the same areas where many fish species breed, nurse, shelter and feed. Proliferation of the water hyacinth leads to reduced oxygen due to its respiration and shading of solar radiation, thereby reducing photosynthesis. It also results in reduced pH due to increased carbon dioxide, increased water clarity due to reduced phytoplankton, reduced floral and faunal diversity due to low oxygen levels, and increased loss of water from the lake by evapotranspiration.

The weed is distributed by currents, winds, boats, and rafts. Its prolific vegetative reproduction by stolons and its long-lived seeds add further to its rapid expansion in the lake (Harley, 1990; Twongo et al.. 1992). The potential for further proliferation and spread of the weed in Lake Victoria is very high. It is no wonder that *E. crassipes* has been described as the world's worst aquatic weed (Harley, 1990). The weed imposes serious uncertainties for future fish production and biodiversity in Lake Victoria.

SUMMARY

Lake Victoria, more than any other large lake in East Africa, provides a striking example of the negative consequences of human activities on ecosystem dynamics and aquatic biodiversity. Out of the original 350 or more fish species, probably less than one third still survive, while only three species occur in commercially viable quantities. Other components of the biota and the water environment have also been changing over the past 30 years. These ecological changes are the result of a complex web of human interventions:

1. Over-exploitation of fish populations due to the uncontrolled types of fishing gear and fishing pressure. Over-fishing affected mainly the traditional commercial fish species populations by interference with their recruitment processes and breeding and nursery strategies.

2. Introduction of exotic fish species. The presence of four exotic tilapiines introduced interspecific competition for food, breeding and nursery grounds so that out of the total six tilapiines only one (*O. niloticus*) has remained commercially viable. Predation by the Nile perch, on the other hand, was largely responsible for reducing the haplochromine diversity so that less than one third of the original number of species (Witte et al., 1992) still survive in sparse densities. Although the reduction in fish species diversity has been followed by a dramatic increase in the annual fish yield, the main worry is whether the higher catches can be sustained under the existing bio-ecological and exploitation regimes. The reduction in fish species diversity, at the same time, has contributed to the increase in phytobiomass due to the reduced herbivory.

3. Cultural eutrophication is mainly due to urban, industrial and agricultural activities, and have increased nutrient inputs into Lake Victoria. This has resulted in increased productivity and phytobiomass, alterations in primary and secondary community structures, and widespread anoxia in the deep waters of the lake. The mode and tempo of eutrophication will determine the viability of the water environment for future productivity and biodiversity.

4. The presence of *Eichhornia crassipes* (water hyacinth) in Lake Victoria has introduced more ecological problems for the lake. The weed proliferates in the inshore zone where productivity has been high and where many desirable organisms would normally breed and feed. The future of Lake Victoria is of great concern because extensive anoxia is limiting benthic organisms and demersal fish in the deep waters of the lake, while the biota in the pelagic zone are restrained during mixing, and *E. crassipes* is ravaging the inshore area. Appropriate research and management interventions are promptly needed to stimulate the viability of the ecosystem.

ACKNOWLEDGMENTS

I am grateful to the International Development Research Centre (Ottawa) for financing the on-going research studies which have spearheaded the general understanding of the Lake Victoria ecosystem. I also thank Mr. J.S. Balirwa (EAFFRO) for his very useful comments on the original draft, and Mrs. F. Balirwa and R. Byekwaso for their secretarial services.

REFERENCES

Amara, J. Ofulla, 1986, The truth about the Nile perch: The Standard (Kenya), Tues., July 22, 1986.

Balirwa, J.S., 1992, The evolution of the fishery of *Oreochromis niloticus* (Pisces: Cichlidae) in Lake Victoria: Hydrobioligia, v. 232, pp. 85–89.

Beauchamp, R.S.A., 1958, Utilizing the natural resources of Lake Victoria for the benefit of fisheries and agriculture: Nature, v. 18, pp. 1634–1689.

Beverton, R.J.H., 1959, Report on the state of the Lake Victoria fisheries: Unpublished Report, Fisheries Laboratory, Lowestoft.

Bugenyi, F.W., 1984, Potential aquatic environmental hazards of agro-based chemicals and practices in a developing country, Uganda: Proct. International Conference on the Environmental Hazards of Agrochemicals, v. 1, pp. 461–473.

Bugenyi, F.W., 1987, Human activities and the detrimental aspects inflicted on the Nile basin and its resources in Uganda, Unpublished Report, Uganda Freshwater Fiseheries Research Organization.

Cadwalladr, D.A, 1965, The decline in the *Labeo victorianus* Boulenger (Pisces: Cyprinidae) fishery of Lake Victoria and an associated deterioration in some indigenous fishing methods in the Nzoia river, Kenya: East African Agriculture and Forestry Journal, v. 30, pp. 249–256.

Corbet, P.S., 1960, Breeding sites of non-cichlid fishes in Lake Victoria: Nature, v. 187, pp. 616–617.

Corbet, P.S., 1961, The food of non-cichlid fishes in Lake Victoria basin with remarks on their evolution and adaptation to lacustrine conditions: Proceedings of the Zoological Society of London, v. 136, pp. 1–101.

EAFFRO, 1964, East African Freshwater Fisheries Research Organization: Annual Report, 1964.

Fryer, G.E., 1961, Observations on the biology of the cichlid fish, *Tilapia variabilis* Boulenger in the northern waters of Lake Victoria (East Africa): Revue Zoologie Botanie Africa, v. 64, pp. 11–33.

Garrod, D.J., 1961, The rational exploitation of the *Tilapia esculenta* stocks of the Northern Buvuma Island area of Lake Victoria: East African Agriculture and Forestry Journal, v. 27, pp. 69–76.

Graham, M., 1929, The Victoria Nyanza and Its Fisheries: A Report on the Fishery Survey of Lake Victoria, 1927–1928: Crown Agents, London.

Greenwood, P.H., 1951, Fish remains from Miocene deposits of Rusinga Island and Kavirondo Province, Kenya: Annual Magazine of Natural History, v. 2, pp. 1192–1201.

Greenwood, P.H., 1965, Explosive speciation in African lakes: Proceedings of the Royal Institute of Zoology, v. 184, pp. 256–269.

Greenwood, P.H., 1966, The Fishes of Uganda: The Uganda Society, Kampala, 127 pp.

Greenwood, P.H., 1984, African cichlids and evolutionary theories, *in* Echelle, A.A., and Kornfield, I., eds., Evolution of Fish Species Flocks: University of Maine at Orono Press, Orono, pp. 141–154.

Harley, K.L.S., 1990, The role of biological control in the management of water hyacinth, *Eichhornia crassipes*: Biocontrol News and Information, v. 11, pp. 11–22.

Hecky, R.E., and Bugenyi, F.W., 1992, Hydrology and chemistry of the African Great Lakes and water quality issues: Problems and solutions: Internationale Vereinigung Limnologie, v. 23, pp. 45–54.

Jorgensen, S.E., and Vollenweider, R.A., eds., 1988, Guidelines of Lake Management. Vol. 1. Principles of Lake Management, ILEC and UNEP publication.

Kendall, R.L., 1969, An ecological history of the Lake Victoria basin: Ecological Monographs, v. 39, pp. 121–176.

Kudhongania, A.W., and Cordone, A.J., 1974, Batho-spatial distribution patterns and biomass estimates of the major demersal fishes in Lake Victoria: African Journal of Tropical Hydrobiological Fisheries, v. 3, no. 1, pp. 15–31.

Kudhongania, A.W., and Colenen, E.J., 1991, Trends in fisheries development, prospects and limitations for Lake Victoria (Uganda): Seminar on the management of the fisheries of Lake Victoria, Aug. 1991, Jinja, Uganda.

Kudhongania, A.W., Twongo, T., and Ogutu-Owayo, R., 1992, Impact of the Nile perch on the fisheries of Lake Victoria and Kyoga: Hydrobiologia, v. 232, pp. 1–10.

Lowe-McConnell, R.H., 1975, Fish Communities in Tropical Freshwaters: Their Distribution, Ecology and Evolution: Longmans, London and N.Y.

Mainga, O.M., 1985, Preliminary results of an evaluation of fishing trends in the Kenya waters of Lake Victoria: FAO Fishery Report, v. 335, pp. 110–116.

Mann, M. J., 1969, A resume of the evolution of the Tilapia fisheries of Lake Victoria up to the year 1960: EAFFRO Annual Report, pp. 21–27.

Moreau, J., 1982, Expose synoptique des donnees biologiques sur la perche du Nile, *Lates niloticus* L.: FAO Symposium Peches, v. 132, 44 pp.

Mugidde, R., 1992, Changes in phytoplankton primary productivity and biomass in Lake Victoria (Uganda): Master of Science Thesis, University of Manitoba.

Ogari, J., 1984, Distribution, food and feeding habits of *Lates niloticus* in Nyanza Gulf of Lake Victoria (Kenya): FAO Fishery Report, v. 335, pp. 68–80.

Ogari, J., and Asila, A., 1990, The state of the Lake Victoria fisheries, Kenya waters: FAO Fishery Report, v. 430, pp. 18–23.

Ogutu-Ohwayo, R., 1985, The effects of predation by Nile perch introduced into Lake Kyoga (Uganda) in relation to the fisheries of Lakes Kyoga and Victoria: FAO Fishery Report, v. 335, pp. 18–39.

Okaronon, J., Acere, T., and Ocenodongo, D., 1985, The current state of the fisheries in the northern portion of Lake Victoria: FAO Fishery Report, v. 335, pp. 89–98.

Okedi, J.O., 1970, Further observations on the ecology of Nile perch (*Lates niloticus* Linne) in Lakes Victoria and Kyoga: EAFFRO Annual Report, pp. 42–55.

Ssentongo, G.W., and Welcomme, R.L., 1985, Past history and current trends in the fisheries of Lake Victoria: FAO Fishery Report, v. 335, p. 123–135.

Stager, J.C., Reinthal, P.N., and Livingstone, D.A., 1986, A 25,000-year history of Lake Victoria, East Africa, and some comments on its significance for the evolution of cichlid fishes: Freshwater Biology, v. 16, pp. 15–19.

Talling, J. F., 1965, The photosynthetic activity of phytoplankton in East African lakes: International Revue Ges. Hydrobiology, v. 50, no. 1, pp. 1–32.

Trewavas, E., 1937, Fossil cichlid fishes of Dr. L.S.B. Leaky's expedition to Kenya in 1934–1935: Annual Ag. Natural History, v. 19, pp. 381–386.

Twongo, T., Bugenyi, F.W., and Wanda, F., 1992, The potential for further proliferation of water hyacinth in Lakes Victoria and Kyoga and some urgent aspects for research: Proceedings of the CIFA Sub-Committee for Lake Victoria Meeting, Feb. 1992, Jinja, Uganda. van Oijen, M.P.J., Witte, F., and Witte-Maas, E.L.M., 1981, An introduction to ecological and taxonomic investigations on the haplochromine cichlids from the Mwanza Gulf of Lake Victoria: Netherlands Journal of Zoology, v. 31, pp. 149–174.

Wandera, S.B., 1990, The Dagaa fishery of the Uganda northern portion of Lake Victoria: Proceedings of the Mwanza Workshop, 8–10 March 1990.

Wanjala, B., and Marten, G., 1974, Survey of the Lake Victoria fishery in Kenya: EAFFRO Annual Report.

Wayland, E.J., 1931, Summary of progress of the Geological Survey of Uganda for the Years 1919 to 1929: Entebbe, Government Printer.

Welcomme, R.L., 1964, The habitat and habitat preference of the young of the Lake Victoria Tilapia (Pisces: Cichlidae): Revue Zoology and Botany of Africa, v. 70, pp. 1–28.

Welcomme, R.L., 1966, Recent changes in the stocks of Tilapia in Lake Victoria: Nature, v. 212, p. 52–54.

Welcomme, R.L., 1967, Observations on the biology of the introduced species of Tilapia in Lake Victoria: Revue Zoology and Botany of Africa, v. 76, pp. 249–279.

Whitehead, P.J.P., 1959, The anadromous fishes of Lake Victoria: Revue Zoology and Botany of Africa, v. 59, pp. 329–363.

Witte, F., and Goudswaard, P., 1985, Aspects of the haplochromine fishery in Southern Lake Victoria: FAO Fishery Report, v. 335, pp. 81–88.

Witte, F., Goldschmidt, T., Wanink, J., van Oijen, M., Goudswaard, K., Witte-Maas, E., and N. Bouton, 1992, The destruction of an endemic species flock: quantitative data on the decline of the haplochromine cichlids of Lake Victoria: Environmental Biology of Fishes, v. 34, pp. 1–28.

Worthington, E.B., 1954, Freshwater organisms, in A discussion on the problems of distribution of animals and plants in Africa: Proceedings of the Linnaeus Society of London, part 1, pp. 67–74.

Growing Impact of Water Hyacinth on Nearshore Environments on Lakes Victoria and Kyoga (East Africa)

T. TWONGO *Fisheries Research Institute, Jinja, Uganda*

Abstract — Water hyacinth, *Eichhornia crassipes*, has become an imposing and dominant component of the marginal ecotones of sheltered shores of Lakes Kyoga and Victoria as well as along the River Nile, within less than three years after the weed was first reported in 1988 and 1990 on the shores of the respective lakes. The negative impacts of the weed on nearshore environments and on the attendant human interests have just began to show and have yet to be studied and quantified. In this contribution, the rapid proliferation of water hyacinth on Lake Victoria and associated Lake Kyoga is illustrated. It confirms earlier predictions of a looming impact of infestation by the weed on the water environment, aquatic biodiversity and socioeconomic activities, if control is not effected.

WATER HYACINTH INTRODUCTIONS

The natural lake systems of East Africa (Kenya, Tanzania, and Uganda) have been free of the scourge of water hyacinth, *Eichhornia crassipes*, until the mid 1980s when the weed appears to have invaded Lakes Victoria and Kyoga. Water hyacinth is, however, said to have been kept for ornamental purposes in Nairobi and Mombasa, Kenya as early as 1957. The first serious infestation with the weed recorded in the region was on River Sigi near Tanga, Tanzania in 1955. It subsequently appeared in River Pangani near Kologwe in 1959 (Harley 1991). The water hyacinth may have been introduced in neighboring Zaire (River Congo) as early as 1952 (Gopal 1987) and in Sudan (River Nile) in the 1950s. The earliest African host countries for water hyacinth were probably Egypt where the weed appears to have been introduced between 1879 and 1892 (Gopal and Sharma 1981) and the Republic of South Africa where invasions of about 1910 have been reported (Gopol 1987).

Water hyacinth was first recorded on lakes in East Africa at about the same time: Lake Kyoga, Uganda in May 1988 (Twongo 1991a) (Fig. 1); Lake Naivasha, Kenya in 1988 (Harley 1991) and Lake Victoria, where infestation was already widespread along the north western shoreline in June 1990 (Freilink, 1991) (Fig. 3). The weed entered Lake Albert via the River Nile in 1990. While the origin of the infestation on lake Kyoga is still unknown, Lake Victoria is believed to have received the weed from River Kagera which originates in Burundi, flows through Rwanda and Tanzania, and enters the lake in Uganda. The original source of water hyacinth infestation for Lake Naivasha and possibly for Lake Victoria was ornamental plant hobbyists in Kenya and Rwanda, respectively.

SPREAD ON EAST AFRICAN LAKES

Proliferation of water hyacinth on the infested lakes in East Africa has been extremely rapid. The weed spread throughout the littoral areas of Lake Naivasha within three years after it was first recorded (Harley 1991). In this lake, water hyacinth competes with *Salvinia*

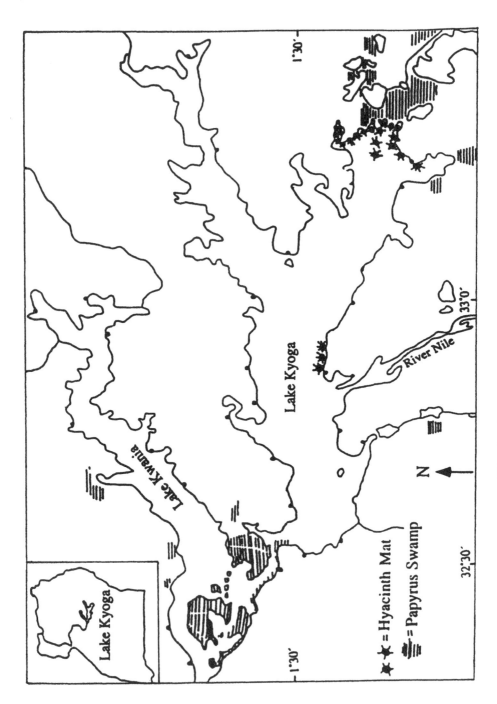

Figure 1. Water hyacinth on Lake Kyoga, May 1988 (after Twongo, 1991).

molesta. The weed was firmly established in almost all suitable areas around Lake Kyoga and the southern portion of adjoining Lake Kwanin by November 1991 (Twongo 1991b) (Fig. 2). In addition to the efficient flotation mechanism, rapid spread of water hyacinth on Lake Kyoga was enhanced by the general east to west water flow through the lake. The weed subsequently descended River Nile which flows through about half the lake, and by the end of 1990 was present in the quiet zones of most of the Kyoga Nile and Albert Nile, as well as the sheltered areas of Lake Albert.

The rate of spread of water hyacinth around the shores of Lake Victoria has been extremely rapid in view of the vast size of the lake (68460 km^2) and the convoluted formation of its shores, particularly on its northern, western and southern margins. In Uganda, the first survey for water hyacinth in June 1990 showed the weed largely confined to a portion of the northwestern shore west of Entebbe (Fig. 3).

However, by January 1992 the weed was firmly established in most suitable littoral environments along the entire shoreline of Uganda (Twongo et al., in press). Water hyacinth had already been reported in Tanzania (AAPS, 1990) as well as in the sheltered bays in Kenya (Ochumba, personal communication). Distribution across this immense lake was greatly aided by strong winds which swept large masses of weed from the bays along the northwestern shores across the lake, to the northern and possibly northeastern shores. It is most probable that water hyacinth was transported to the southern shores in Tanzania in a similar manner. Large islands of the weed were dispersed by strong waves into small units which then floated to various destinations. Wave action caused the tangled stolons of the weed to break up into numerous pieces which floated about the lake only to sprout into daughter plants in suitable environments. During the first ground survey for the weed on the Ugandan side of Lake Victoria, many new colonies of the weed were observed to start from short stolons, many of which had only one internode (Twongo et al., in press).

PROLIFERATION POTENTIAL

Observations during the ground surveys on water hyacinth infestation around Lake Kyoga (Twongo 1991 a and b) and Lake Victoria (Twongo et al., in press) indicated that the weed thrives in bays and inlets which are sheltered from strong offshore and along-shore wind action; have flat or gently sloping, relatively shallow shores (rarely deeper than six meters); and have a muddy bottom rich in organic matter. It was further observed on Lake Victoria that shorelines with the above characteristics supported an emergent macrophyte flora of papyrus (*Cyperus papyrus*) with patches of *Vossia*. Stands of *Typha* were often scattered among the papyrus or formed a background to it. The floating and submerged macrophytes found in such bays and inlets were mostly Nile lettuce, *Pistia stratiotes*, *Potamogeton* sp., *Ceratophyllum demersum*, *Mylliophylum* sp., and *Nymphaea* sp. These macrophytes were therefore used as indicators of environments with a potential for water hyacinth infestation and indeed, this prediction has been upheld by later infestations by the weed in Lake Victoria.

The emergent swamp vegetation, dominated by papyrus, which borders the shallow bays and inlets are well marked on Admiralty charts for Lake Victoria, 1956 edition. These charts were used to identify and map environments suitable for hyacinth infestation around the entire lake (Fig. 4). It is interesting to note that the potential distribution patterns for water hyacinth shown in Fig. 4 approximate closely what is now present in Tanzania and Kenya (Bwathondi and Ochumba, personal communication). The infestation potential indicates

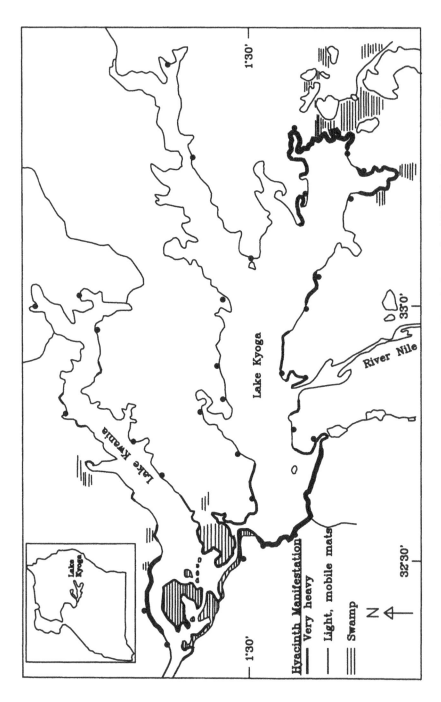

Figure 2. Water hyacinth infestation on Lakes Kyoga and Kwania, November 1991 (after Twongo, 1991).

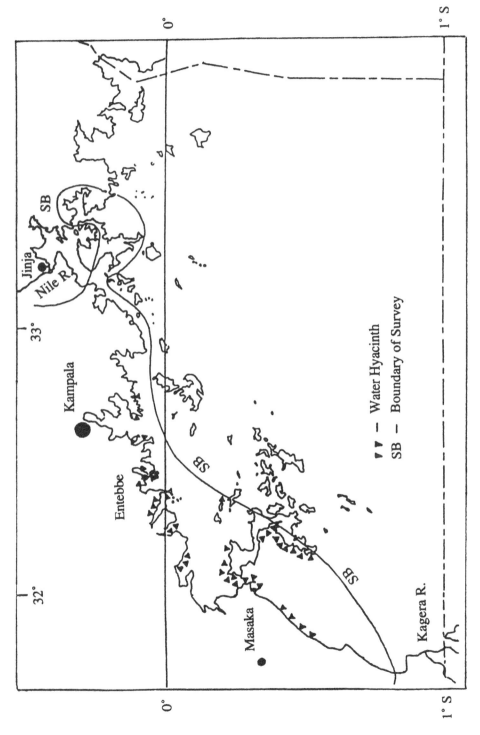

Figure 3. Water hyacinth on Lake Victoria, June 1990 (after Frielink, 1990).

also that the nearshore environments of Uganda are most susceptible to hyacinth proliferation among the riparian states of Lake Victoria. Later visits to some of these shorelines in Uganda by UFFRO scientists have indicated extremely rapid spread of the weed to entire bays, e.g., Namirembe Bay at the mouth of River Katonga.

IMPACT OF WATER HYACINTH PROLIFERATION

Assessing the impact of water hyacinth proliferation in aquatic environments such as the lakes in East Africa require documentation of baseline data on water quality, biodiversity and socioeconomic activities. This data is currently lacking. It is regrettable, for instance, that so little was known about the ecology and dynamics of the immense and diverse stocks of haplochromines in Lake Victoria when they disappeared over the past 20 years, largely under the influence of introduced Nile perch. Secondly, detailed assessment of the impact of water hyacinth infestation on a given ecosystem and its resources might justify the huge economic tag that may be required to control the vast expanses of the weed around the shores of Lake Victoria. For instance, in Uganda today the outcry about water hyacinth infestation is getting louder because the costly impact of weed infestation on various interests of society, notably fisheries, is becoming evident.

On the other hand, it is often assumed that many studies have already been made on water hyacinth in various countries, particularly in the United States and in Asia, and that there is a lot of information on the weed. It would appear, however, that little detailed quantitative information actually exists in the literature on specific changes to natural aquatic environments and their biodiversity due to infestation by the water weed. Besides, the climatic and environmental differences, as well as the diversity of target benefits to be considered, reduce the comparative value of the socioeconomic impact of water hyacinth infestation in the Tennessee Valley, USA, or in the irrigation channels of Indonesia, with that on the fishing communities of Lakes Victoria and Kyoga.

A major aspect of the impact of hyacinth infestation in aquatic environments is its potential to disrupt fishing activities, transportation and the functioning of various installations, such as domestic water purification and hydroelectricity generation plants. This aspect of impact, which is of direct socioeconomic nature, is rapidly gaining momentum on Lakes Victoria and Kyoga, as well as on the River Nile. Fish landings at infested shorelines now constantly remove masses of water hyacinth from their sites, and those with narrow approaches, like many along the south shore of Lake Kyoga, are periodically sealed off from the lake by thick plugs of the water weed. A quick survey of the major fish landing sites on the Ugandan portion of the Lake Victoria mainland in January 1992 indicated that most of the landings (at least 70% of those surveyed) (Fig. 4) were located in areas already infested with the water weed. However, their location may not be changed because of their shelter from rough waters and proximity to productive fishing grounds. Major economic installations in Uganda are also threatened by water hyacinth. Mats of the weed now arrive at the Owen Falls hydroelectric generation facility at the source of River Nile, with increasing frequency. Authorities at the plant have reported increased fouling of the screens to the turbines by organic debris. The National Water and Sewage Corporation is greatly concerned about the rapid expansion of mats of water hyacinth close to the water intake point for the City of Kampala in Murchison Bay. Increased organic matter in the water, partly attributed to rotting water hyacinth, has led to higher water treatment bills (Arebahona, personal communication). Encroachment of water hyacinth on the "quiet water" areas of the River

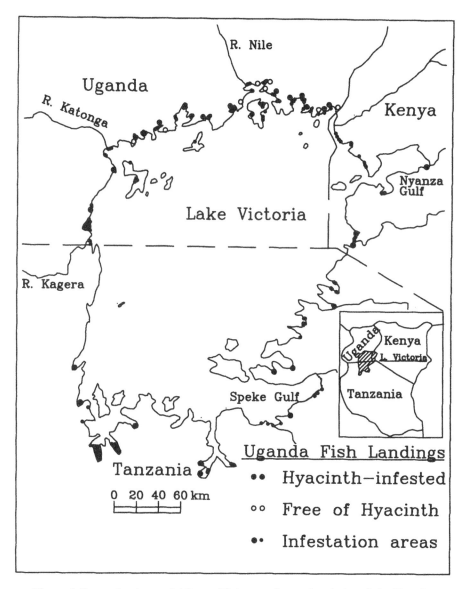

Figure 4. Extrapolated potential for establishment of water hyacinth on Lake Victoria.

Nile in the Murchison Falls National Park is displacing hippos and crocodiles from their traditional sites (Game Warden, Masindi District, personal communication). This displacement is unwelcome news to the tourist industry which offered these sites for game viewing.

The spread of water hyacinth in Uganda has fueled speculation about a likely resurgence of diseases from schistosomiasis whose vectors may flourish in shallow water environments similar to those infested by the water weed. While studies have yet to be made in Uganda to investigate the association of water hyacinth with various vector borne diseases, large numbers of snails are almost always found among water hyacinth in Lakes Victoria and Kyoga. The taxonomy and diversity of this mollusc population has not been investigated.

However, Dazo et al. (1966) reported a positive association between the bilharzia vectors *Bulinus trancatus* and *Biomphalaria alexandrina*, and various aquatic macrophytes, including water hyacinth in Egypt. It is probable, therefore, that a resurgence of schistosomiasis may occur in hyacinth-infected areas of Uganda, particularly in parts of northwestern Uganda along River Nile, where incidences of bilharzia have been common (Ongom and Bradley, 1972). Clearly there is a need to monitor the impact of water hyacinth infestation on incidence of water-associated, vector-borne diseases.

Twongo et al. (in press) give preliminary observations made under small hyacinth mats in sheltered environments of Lake Victoria to illustrate depressed dissolved oxygen concentrations and lowered pH. The thick, extensive cover developed by the weed over a water surface has the effects of light-shading, restriction of gaseous exchange, and hinderance to water mixing. The high rate of respiration and decomposition of dead organic matter would add to the depletion of oxygen. Uptake of certain nutrients from the environment would also be accelerated in view of the rapid growth rate of the water weed. In this connection, it is of interest to note that the inlets of most affluent rivers into the Uganda sectors of Lake Victoria are ideal for infestation by water hyacinth. In the event that catchment nutrient sources are important in the nutrient dynamics of Lake Victoria, a massive build up of the water weed could form a significant sink to influence localized nutrient levels, particularly in sheltered bays. Detailed studies on physicochemical changes and how they affect biodiversity under water hyacinth mats on these natural aquatic systems are, therefore, urgently required.

Information on biodiversity in the submerged portion of the riparian strip of Lakes Victoria and Kyoga where the water hyacinth thrives is scanty. Hence, there is almost complete ignorance about the likely impact of water hyacinth on aquatic biodiversity of these inshore environments. The preliminary observations and inference made during the surveillance field trips cited above are outlined below, if only to emphasize the urgency of baseline and impact assessment studies on aquatic biodiversity before water hyacinth totally engulfs the sheltered inshore environments.

As indicated earlier, environments with a potential for the establishment of water hyacinth are usually associated with floating and submerged macrophytes rich in planktonic and encrusted algae. However, it is unlikely that much of this flora would survive under large, established hyacinth mats. The higher Secchi disc readings recorded under water hyacinth as compared to those just outside the weed mat during preliminary surveys (Twongo et al., in press) indicate lower phytoplankton biomass under the mat. This would be expected due to light-shading and competition for nutrients which, together with superior physical crowding by the weed, would also overpower other aquatic macrophytes such as *Cerato-phyllum*, *Myriophyllum*, *Potamogeton*, *Nymphaea*, and *Pistia*. Loss of original submerged and floating macrophytic flora at the expense of water hyacinth has been observed under thick mats of the weed on Lakes Victoria and Kyoga by this author.

Insufficient knowledge about the invertebrate fauna, including the benthic component, associated with the riparian strip susceptible to water hyacinth infestation, limits discussion on the impact of the water weed on invertebrate faunal diversity. However, it is to be expected that reduced algal growth will lead to a reduction in algal grazers, consequently disrupting populations of dependent fauna in the food chain. Secondly, anoxia under the mat excludes fauna with high oxygen demand including fish like Nile perch and Nile tilapia.

In Lakes Victoria and Kyoga, large populations of a limited number of invertebrate species, notably leeches, dragonfly nymphs, and several mollusc species, have been found associated with hyacinth mats. Among the vertebrate fauna, the lungfish *Protopterus aethiopicus* and some snakes appear to thrive in the mats. The environmental conditions associated with water hyacinth appear to favor some animals particularly, those which tolerate low oxygen environments. Such organisms tend to develop large populations. The factors and mechanisms favoring the above and other fauna, while apparently excluding most indigenous biodiversity, provides interesting contrast to an apparent resurgence of several haplochromine species in Lakes Victoria and Kyoga at the edge of hyacinth mats (Witte; Wandera, personal communication).

The impact of water hyacinth on the commercial fish stocks of Lakes Victoria and Kyoga and River Nile is likely to be most severe on the Nile tilapia, currently the most important commercial fish species in Lake Kyoga. This water fish thrives best in less than 10 meters of water (Kudhongania and Cordone, 1974) and frequents environments which are susceptible to hyacinth infestation for its food and shelter, particularly at the fry and juvenile stages. Although Nile tilapia is said to breed on relatively shallow sandy or gravely shores which are not suitable for establishment of hyacinth, observations by this author in Lake Kyoga indicate that the species breeds well on muddy bottoms that are ideal for the weed. Welcomme (1967) suspected *O. niloticus* to breed and nursery in similar environments in Lake Victoria. However, Nile tilapia has been reported to avoid very low oxygen environments (Welcomme 1967) and the species has not been able to penetrate the anoxic conditions of papyrus swamps. Nile tilapia is thus unlikely to thrive under water hyacinth mats given the low oxygen regime and poor supply of food. Furthermore, the mats form an effective barrier to the food in the emergent vegetation along the shore where Nile tilapia is known to feed, especially during the rainy season.

The impact of hyacinth infestation on Nile perch will be felt by juveniles which frequent indigenous submerged and floating macrophytes in search of food and shelter. While juvenile perch prey heavily an nymphs of *Odonata* which are plentiful among water hyacinth mats, it is unlikely that the fish will exploit this prey under anoxic conditions. Young Nile perch will, however, feed on the nymphs at the edge of the weed mats. In the final analysis, the effect of water hyacinth infestation on Nile perch will depend on the ability of Nile perch juveniles to find alternative prey in the open water and on shores which are not infested with hyacinth. The fish appears to be well adapted to use a variety of prey species (Ogutu-Ohwayo 1985).

RECOMMENDATIONS

In view of the mounting impact of water hyacinth on Lake Victoria and Kyoga and the River Nile, and the great potential the weed has to spread to other aquatic environments in the region, it is recommended that:

1. Control of water hyacinth in the Upper Nile catchment be given priority consideration.

2. A regional strategy to control water hyacinth on Lake Victoria be agreed upon without delay.

3. Riparian countries around Lake Victoria take a common stand on biological control, possibly the only long-term measure available for the control of water hyacinth in this complex environment.

4. Baseline data on water quality, biodiversity and production dynamics in shoreline aquatic systems of Lake Victoria and Kyoga should be assembled for impact assessment purposes.

5. Impact of water hyacinth proliferation on nearshore environments and socioeconomic interests be systematically evaluated to facilitate impact assessment.

ACKNOWLEDGMENTS

I wish to thank Mr. Sowobi (UFFRO) who drew all the figures in this paper and Florance Bazanya for typing the manuscript.

REFERENCES

All Africa Press Service, 1990, Harmful hyacinth resists elimination: Science and Technology Feature of the All Africa Press Service, January 1990, Nairobi, Kenya.

Dazo, B.C., N.G. Hairstone, and I.K. Dadwood, 1966, The ecology of *Bu. trancatus* and *Bi. alexandrina* and its implications for the control of bilharziasis in the Egypt 49 Project Area: Bulletin of the World Health Organization, v. 35, pp. 339–359.

Freilink, A.B., 1991, Water hyacinth in Lake Victoria: a report on an aerial survey on 12–13 June 1990, in Thompson, K., ed, The Water Hyacinth in Uganda. Ecology Distribution Problems and Strategy for Control. Workshop: 22–23 October, 1991. FAO TCP/UGA/9153/A.

Gopal, B., 1987, Water Hyacinth. Elsvier, Amsterdam.

Gopal, B., and K.P. Sharma, 1981, Water Hyacinth (*Eichhornia crassipes*): The Most Troublesome Weed of the World, Hindasia, Delhi.

Harley, K.L.S., 1991, Survey project on exotic floating African water weeds: Commonwealth Science Council Survey Report.

Kudhongania, A.W., and A.J. Cordone, 1974, Batho-spatial distribution patterns and biomass estimate of the major demersal fishes in Lake Victoria. African Journal of Tropical Hydrology and Fisheries, v. 3, pp. 15–31.

Ogutu-Ohwayo, R., 1985, The effects of predation by Nile perch *Lates niloticus* (Linne), introduced into Lake Kyoga (Uganda) in relation to the fisheries of Lake Kyoga and Lake Victoria. Food and Agricultural Organization Fisheries Report, v. 335, pp. 18–41.

Ongom, V.L., and D.J. Bradley, 1972, The epidemiology and consequences of *S. mansoni* infestations in West Nile, Uganda, Part I. Field studies of a community at Panyigoro: Transactions of the Royal Society of Tropical Medicine, v. 66, pp. 835–851.

Twongo, T., 1991a, The water hyacinth on Lake Kyoga: Special report no. 2, October 1989, *in* Thompson, K., ed., The Water Hyacinth in Uganda. Ecology Distribution, Problems and Strategies for Control. Workshop: 22–23 October, 1991. FAO TCP/UGA/9153/A.

Twongo, T., 1991b, Water hyacinth on Lakes Kyoga and Kwania. Special report no. 3, November 1991, in Thompson, K., The Water Hyacinth in Uganda. Ecology Distribution, Problems and Strategies for Control. Workshop: 22–23 October, 1991. FAO TCP/UGA/9153/A.

Twongo, T., F.W.B. Bugenyi, and F. Wanda, in press, The potential for further proliferation of water hyacinth in Lakes Victoria and Kyoga and some urgent aspects for research. At CIFA Sub-Committee for Management and Development of the Fisheries of Lake Victoria, Sixth session, 10–13 Feb., 1992. Jinja, Uganda, in press.

Welcomme, R.L., 1967, Observation on the biology of the introduced species of *Tilapia* in Lake Victoria: Review of the Zoology and Botany of Africa, v. 76, pp. 249–279.

Preliminary Studies on the Effects of Water Hyacinth on the Diversity, Abundance and Ecology of the Littoral Fish Fauna in Lake Victoria, Uganda

N.G. WILLOUGHBY and I.G. WATSON   *Fisheries and Aquatic Resources Group, Natural Resources Institute, Chatham Maritime, Kent, United Kingdom*

T. TWONGO   *Uganda Freshwater Fisheries Research Organisation, Jinja, Uganda*

Abstract — Electrofishing was used to take samples of fish from 30 sites representing 6 habitats along the northern and western shorelines of Lake Victoria (Uganda). The habitats sampled were open shore, rocky shore, *Vossia*, *Typha*, papyrus and *Eichhornia* (water hyacinth). 2,860 specimens were caught, weighing 36.1 kg, and representing 57 species. Haplochromines contributed the bulk of the species (45 spp = 79%) most individuals (2,514 = 88%) and most biomass (26.4 kg = 73%). The reductions in the numbers of specimens and in the biomass around *Eichhornia* habitats compared with the other habitats were significant at the 1 and 5% levels of probability respectively. There was also a statistically significant reduction (at the 10% level) in species diversity around *Eichhornia* compared with all but one of the other habitats. Data on sex ratios, states of maturity and food and feeding habits in relation to the *Eichhornia* and other habitats are also given. It is concluded that the presence of water hyacinth in Lake Victoria is probably a threat to its biodiversity and shoreline fisheries, but that further sampling is needed to confirm this.

INTRODUCTION

Water hyacinth (*Eichhornia crassipes* (Mart.) Solms) is a native of South America, but it has been transferred by man to other freshwater habitats around the world, usually as an ornamental plant. It has passed into local waterways, and although it is seldom a problem in South America, it has proliferated to such an extent in the waters of Africa, Asia, Australia and North America that is now probably the world's major aquatic plant pest. Its status as a pest arises from the fact that it can form dense floating masses of weed, sometimes hundreds of metres in extent, which can block waterways or blanket lake shores, causing problems in transport, hydropower generation, health and fisheries.

Literature is divided over whether *Eichhornia* has an adverse or beneficial effect on fisheries. Possible benefits include the findings that it forms a substrate for invertebrate food organisms (Green et al., 1976), and that its roots provide shelter for juvenile fish (Hickley and Bailey, 1986). Data from Malawi indicate that despite difficulties in setting fishing gear during the gradual encroachment of the weed through areas of swamp, total catches are reported to have increased (Terry, 1991). Blocking access to fishing grounds is a major problem (Singh and Nasiruddin, 1980). The oxygen depletion around and under large *Eichhornia* mats would be expected to deter fish (McVea and Boyd, 1975), though its introduction in small quantities is sometimes used as a fish aggregating device in Asian swamps (P. Edwards, pers. comm.).

Water hyacinth has been spreading steadily across Africa's river and lake systems for decades. It was first reported in Uganda (Lake Kyoga) in 1988 (Twongo, 1991), and was first seen in the Ugandan sector of Lake Victoria in 1990 (Freilink 1991), though Harley (1991) suggests that low levels of infestation may have been there for ten years prior to that. The source of the weed in Lake Victoria was probably the Kagera River, on the western side of the lake (Harley, 1991). Fringing mats of the weed as large as 50–60 m across may now be found in many places along the shoreline of Ugandan Lake Victoria, and one mat of 200 m in length was seen floating in the western part of the lake.

The difficulties of fishing in *Eichhornia* have resulted in very sparse quantitative data on catch rates or species diversity in and around the weed. The possibilities for comparing data from *Eichhornia* with those from other habitats are therefore very limited indeed. This work reports on an attempt to compare the fish populations of *Eichhornia* beds with those of the natural shoreline habitats of Lake Victoria.

MATERIALS AND METHODS

Equipment

The fishing method used was pulsed direct current electricity, at 200 V, 100 Hz, and fishing at 2–5 A. The cathode was a piece of expanded wire mesh of approximately 1.5 m². This needed to be larger than that which would usually be required, but was necessary to increase current flow in the low conductivity water (100 υS) of Lake Victoria. Two circular copper anodes were used, these having 50 cm diameter heads mounted on 2.0 m fibreglass poles. They were fitted with mesh across the face so that they doubled as catch nets.

This system had the capacity to draw fish from dense weed cover into open water before they were stunned and netted. The limitations of the electro-fishing method will be discussed later, but no other method would have had the capacity to draw fish from the inaccessible parts of their habitats in this way, and it provided comparable samples quickly and effectively under most shallow water conditions.

Habitat Types

Sites belonging to six habitat types were fished. These were: open beaches (4 sites), rocky shores (4), stands of papyrus (*Cyperus papyrus*, 4) stands of hippo grass (*Vossia*, 4) stands of bullrush (*Typha*, 4) and areas affected by *Eichhornia* mats (10). The latter were divided into "shallow" sites (5) where the weed had been blown against sandy or muddy beaches, but would probably be moved away at a later time by changes in wind direction; and "deep" sites (5) usually associated with papyrus swamp, where the *Eichhornia* mats were trapped within bays, and were probably more permanent. Pure stands of each of the vegetation types were chosen, and all sampling sites were adjacent to the shore. (A habitat would be considered pure if more than 90% of its length was of the type required. This posed no problem for open or rocky shores or for *Typha* beds, but some micro-habitats consisting of other vegetation were present in some of the *Eichhornia*, *Vossia* and papyrus.) Where water or mud were too deep for wading, electrofishing was conducted from a boat. Sampling was carried out at sites between Lambu, near Masaka in the southwestern part of the lake and MacDonald Bay in the eastern part (see Fig. 1).

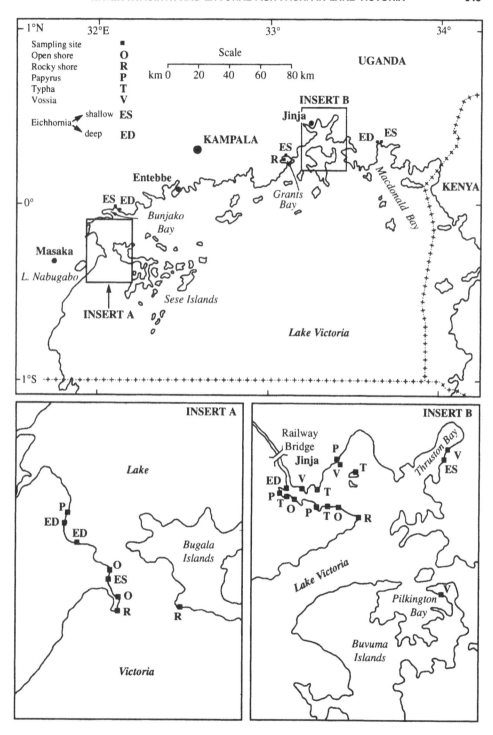

Figure 1. Sampling sites for fish in Lake Victoria (Uganda).

METHODOLOGY

Each sampling site was first surrounded by a stop-net 50 m long and 5 m deep with 16 mm (stretched) mesh to prevent fish escaping. This was set parallel to the shore and within 2 m of the weed fringe for most of its length. For rocky shores or open beaches the outer edge of the net was approximately 5 m from the shore. Electrofishing was then carried out within the stop net for three 15 minute sessions, and the fish from each session held separately to assess depletion of the populations as a result of the sampling.

The fishing was carried out by the same personnel (NGW and IGW) on all occasions to reduce bias. Each held an anode, and the system worked only when both anodes were in the water and switched on. The operators either waded along the fringes of the weed, or were paddled along it in the boat. Fishing was also carried out as far into the weed beds as far as possible. Each 15 minute session usually provided time for 3 traverses along the length of the 50 m net. Fish were attracted to the anodes from a distance of approximately 1 m, becoming stunned as they approached closely. They were scooped into 2 plastic bins during fishing, the contents of which were merged at the end of each 15 minute session, thus producing 3 samples from each sites

The sampling was carried out over a 6 week period during February and March 1993, with fishing taking place between 0800 and 1800 hours. The weather varied from full sunshine to overcast, with no rain falling during the sampling period.

Provisional identification of the fish was made in the Fisheries Research Organisation laboratories in Jinja, or at the field base near Masaka, after which the fish were measured and weighed, and their sexes and diets checked. Definitive identification of the haplo-chromines is still being carried out in UK.

RESULTS AND DISCUSSION

Assessment of Capture Efficiency

In the depletion method, the catches from a few of the second and third periods of fishing were greater than those from the first, inferring infinite populations. However, when the results from 28 of the 30 sampling sites were combined (2 merged in error), they provided numbers of fish from samples 1–3 in the proportions 44:32:24. Graphical analysis of these figures suggests that approximately 50% of those fish available to electro-fishing were caught during the triplicate sampling.

Species Abundance

57 species of fish, 45 of which were haplochromines, were found during the survey. Owing to the complexities of haplochromine taxonomy, the identification of some species must await further analysis, so some aspects of the results presented here are tentative.

The species abundance by numbers and weights are summarised in Table 1. The non-haplochromines contributed 12% by number and 27% by weight of the catches, while the haplochromines contributed 88% by weight and 73% by number. The shallow water habitats of the lake are therefore still populated predominantly by haplochromines, both in terms of number and weight, despite the introduction of the cichlid *Oreochromis niloticus* and the predatory Nile perch, *Lates niloticus*.

Table 1. Species Abundance by Number and Weight (all sites combined)

	Number caught	% total	Weight caught (g)	% total
A. Non-Haplochromines				
Protopterus annectens	3	–	265	1
Rastrineobola argenteus	46	2	72	–
Aplocheilichthys pumilus	12	–	5	–
A. eduardensis	5	–	1	–
Clarias liocephalus	8	–	89	–
Oreochromis niloticus	82	3	4,217	12
O. esculentus	10	–	622	2
O. variabilis	4	–	800	2
O. leucostictus	17	1	1,663	5
Tilapia zillii	18	1	917	2
Mastacembalus frenatus	24	1	191	1
Lates niloticus	34	1	852	2
Tilapia/Oreochromis fry	83	3	54	–
Total Non-Haplochromines	346	12	9,748	27
B. Haplochromines				
Astatoreochromis allaudi	24	1	565	2
Astatotilapia barberi	15	1	235	1
Astatotilapia nubila	43	2	843	2
A. 'robust nubila'	208	7	4,694	13
Enterochromis (3 spp)	67	3	374	–
Gaurochromis sp.	21	1	619	2
'Haplochromis' (24 spp)	296	10	5455	15
Lipochromis obesus	14	–	582	2
Lipochromis sp. 2	1	–	27	–
Prognathochromis (3 spp)	58	2	1020	3
Psammochromis acidens	293	10	1,128	3
Psamm. riponianus	22	1	731	2
Psamm. saxicola	242	8	6,244	17
Psammochromis (5 spp)	28	–	211	–
Haplochromine fry, unident.	1,202	42	3,653	10
Total Haplochromines	2,514	88	26,381	73
Grand Total, all species	2,860	–	36,129	–

Only 3 species, all haplochromines, contributed more than 5% to the total numbers, and only 3 species, including *Oreochromis niloticus*, contributed more than 5% to the total weight. *Lates niloticus*, did not contribute significantly in terms of either number or weight.

Table 2. Fish Diversity and Biomass in Shallow Water Habitats of
Lake Victoria (Uganda)

Habitat	No. of samples	No. of fish per sample		Biomass per sample (g)		No. of species per sample	
		Mean	Range	Mean	Range	Mean	Range
Eichhornia (shallow)	5	28	3–53	375	26–1274	5	2–7
Eichhornia (deep)	5	38	14–95	313	55–748	4	1–8
Open shore	4	23	1–52	432	2–1007	4	1–8
Rocky shore	4	332	61–716	1800	1031–3631	7	4–9
Papyrus	4	100	18–190	1391	135–2,720	6	2–10
Typha	4	103	23–199	3112	846–6283	9	7–10
Vossia	4	75	25–120	1437	73–2936	5	4–7

Species Diversity and Distribution

The catch data in terms of numbers, biomass and species diversity for each habitat type are summarised in Table 2.

The numbers of fish/sample for the shallow and deep *Eichhornia* sites were compared with those from the combined other sites. An analysis of variance test using ln-transformed data is given in Table 3. The difference between the two groups (*Eichhornia* vs. the rest) is statistically significant at the 1% level.

The differences between the levels of biomass of fish/sample in the two groups of habitats were also tested by analysis of variance using square root transformed data (Table 4). The difference between *Eichhornia* and the rest of the habitats is statistically significant at the 5% level.

Detailed analysis of fish species diversity in different habitats showed that the *Eichhornia* sites have lower diversity than the others, but that the variability of the results among habitats was high. However, when the open water results are excluded from this analysis (for which there are good reasons which will be presented later) the difference between *Eichhornia* and the remainder becomes statistically significant at the 10% level.

Sex Ratios and Breeding Condition

The sexes and states of maturity of 518 specimens belonging to 25 key species were determined, the results being presented in Table 5. A significant feature was the total absence of any male specimens from 6 species of haplochromine, despite the examination of a total of 61 specimens. It is not known whether this is a result of a sampling anomaly, sex-specific habitat selection, or misidentification of the males of these species.

Most of the female haplochromines (17 out of 20 species) were in or were coming into breeding condition when caught, suggesting that the inshore areas are their preferred breeding habitats. While the same was true of several of the tilapias (and also of *Clarias* and *Mastacembalus*) no breeding specimens of the species which is now of primary commercial

Table 3. Analysis of Variance on ln-Transformed Total Catch Data (mean numbers per sample) Comparing *Eichhornia* Sites with All Other Habitats

	df	SS	Mean square	F
Water hyacinth vs. others sites	1	6.269	6.269	4.59*
Other site comparisons	5	19.302	3.860	
Site effect	6	25.571	4.262	
Error	23	31.386	1.365	
Total	29	56.957		

*$p = 0.5$.

Means	*Eichhornia*	Other	SE of Diff
Ln scale	3.096	4.073	0.452
Retransformed	22.1	58.7	

The difference between the two groups (*Eichhornia* vs. the rest) is statistically significant at the 1% level.

Table 4. Analysis of Variance on Square Root-Transformed Total Catch Weights at Each Site Comparing *Eichhornia* with All Other Habitats

	df	SS	Mean square	F
Water hyacinth vs. others sites	1	2398.814	2398.814	9.63**
Other site comparisons	5	2626.163	525.233	
Site effect	6	5024.977	837.496	
Error	23	5731.703	249.204	
Total	29	10756.680		

**$p = 0.1$.

Means	*Eichhornia*	The rest	SE of Diff
Square root scale	16.19	35.45	6.113
Retransformed	262.1	1256.7	

The difference between the two groups (*Eichhornia* vs. the rest) is statistically significant at the 5% level.

importance in the lake, *L. niloticus*, were found in the sampling areas. Amongst the tilapias it is interesting to note the preponderence of male *O. niloticus* and *Tilapia zillii* and of female *O. leucostictus* in the inshore areas. Only juvenile *Lates* were caught in the inshore habitats.

Table 5. Sex Ratios of Key Species

Species	No. of specimens	No. of adult specimens	Sex ratio male:female maturing (yes/no)
Non-Haplochromines			
Oreochromis niloticus	82	21	N 20 : 1 N
O. esculentus	10	8	Y 6 : 2 N
O. leucostictus	17	16	N 1 : 15 Y
Tilapia zillii	24	13	N 12 : 1 Y
Lates niloticus	34	–	–
Haplochromines			
Astatorechromis allaudi	24	17	Y 8 : 9 Y
Astatotilapia barberi	15	11	N 1 : 10 Y
A. nubila	43	43	Y 22 : 21 Y
A. 'robust nubila'	208	77	Y 44 : 33 Y
Enterochromis sp 1	12	7	– 0 : 7 N*
Enterochromis sp 3	49	5	– 0 : 5 Y*
Gaurochromis sp	21	20	– 0 : 20 Y*
Haplo. 'few inner'	15	11	– 0 : 11 Y*
Haplo. 'green–gold'	30	30	Y 17 : 13 Y
Haplo. 'many inner'	20	11	– 0 : 11 Y*
Haplo. 'Masaka oranjefin'	26	17	Y 8 : 9 Y
Haplo. 'Masaka redfin'	60	30	Y 14 : 16 Y
Haplo. 'olive'	11	7	– 0 : 7 Y*
Haplo. 'olive–yellow'	11	11	N 4 : 7 N
Lipochromis obesus	14	12	N 4 : 8 Y
Prognath. prognathus	44	43	N 10 : 33 Y
Psammochromis acidens	293	17	Y 7 : 10 Y
Psamm. 'redflank'	11	11	Y 8 : 3 N
Psamm. riponianus	22	19	Y 10 : 9 Y
Psamm. saxicola	242	61	Y 42 : 19 Y

* = species for which no males were found; Y = yes, maturing; N = no, not maturing.

Food and Feeding Habits

The results of the stomach contents analysis are shown in Table 6. Only 230 (45%) of the 518 stomachs examined, contained identifiable contents, indicating that the majority of fish had not been feeding for several hours prior to capture. The tilapias relied more on algaceous mud and vegetable matter (51 out of the 61 with stomach contents = 84%) as a food source than the haplochromines (49 out of 139 = 29%), while the latter consumed many more chironomid and insect larvae (87 out of 139 = 62%) than the tilapias (10 out of 61 = 16%). The insect nymphs taken were predominantly damselfly and dragonfly nymphs. The appar-

Table 6. Diets of Selected Species as Indicated by Number of Stomachs Containing a Dominant Item

Species	Number examined	Empty	Mud	Detritus	Macrophyte plant	Chironomid larvae	Other ins larv	Prawns	Fish
Non-Haplochromines									
Clarias liocephalus	9	2	1	–	–	2	4	–	–
Oreochromis niloticus	38	6	26	2	–	3	1	–	–
Oreochromis variabilis	4	–	3	1	–	–	–	–	–
Oreochromis lecosticta	15	5	9	1	–	–	–	–	–
Tilapia zillii	16	1	5	3	1	6	–	–	–
Mastacembalus frenatus	15	10	–	–	–	1	3	1	–
Lates niloticus	28	10	–	3	–	–	2	10	3
Subtotals	**125**	**34**	**44**	**10**	**1**	**12**	**10**	**11**	**3**
Haplochromines									
Astatoreochromis allaudi	20	12	1	2	–	4	1	–	–
Astatotilapia barberi	10	7	–	–	–	1	2	–	–
Astatotilapia nubilis	12	8	–	–	–	4	–	–	–
A. 'robust nubilis'	91	58	4	6	–	22	–	1	–
Enterochromis spp. (2)	7	4	–	2	–	–	1	–	–
Gaurochromis sp.	19	14	–	–	–	2	3	–	–
'Haplochromis' spp. (6)	93	58	3	5	–	15	10	2	–
Prognathochromis spp. (2)	33	22	2	4	–	1	4	–	–
Psammochromis acidens	16	11	–	–	–	5	–	–	–
Psammochromis riponianus	17	12	–	2	–	–	3	1	–
Psammochromis saxicola	58	36	1	17	–	2	2	–	–
Psammochromis spp. (2)	17	12	–	–	–	1	4	–	–
Subtotals	**393**	**254**	**11**	**38**	**0**	**57**	**30**	**3**	**0**
Totals	**518**	**288**	**55**	**48**	**1**	**69**	**40**	**14**	**3**

ent absence of lakefly larvae in the stomach contents suggests that the species were feeding in the littoral area of the lake, rather than moving into the open water to seek prey.

The only piscivore noted during this work was *Lates*, though 3 other species also consumed the caridean prawns found around the lakeshore. (Several of the fish were found to have haplochromine eggs or fry in their stomachs. These were almost certainly consumed while the fish were being held in the bins after capture — alive but under stress — and was not taken to indicate a piscivorous diet.)

DISCUSSION

Capture Efficiency

Bohlin et al. (1991) using electrofishing for stock assessment, primarily in temperate rivers and streams, asserted that fish population sizes are always underestimated and that the accuracy varied from poor to fair. The present study was not intended to measure the population sizes or the fishing efficiency of the equipment, but to take samples from a variety of habitats in the most standardised fashion available. The critical question relates to whether the samples obtained from the different habitats are comparable.

Zalewski and Cowx (1991) list 23 factors which can affect electrofishing catches. These are grouped into environmental features (abiotic, habitat and seasonality); biological features (community and population structure) and technical features (personnel, equipment and organisation).

Of the environmental features, water conductivity, quality and clarity can be taken as the same for all sites, as can water velocity (nil), temperature and weather. From a technical viewpoint, the personnel, equipment and standardisation of effort were also the same for all sites. This leaves the biological features (community and population structure) and some abiotic features (habitat dimensions and substrate) as factors which might cause apparent differences in the efficiency of sampling between sites.

The biological biases inherent in electrofishing are worthy of some consideration. Several genera were not represented or were only poorly represented in these catches, though there were reasons for expecting their presence. Species which generate their own electical fields, such as the mormyrids, are very difficult to attract using electrofishing, and were not caught at all, though they were taken from these habitats by other methods (J. Balirwa, pers. comm.). Species which are dark all over such as *Protopterus* and to some extent *Clarias*, were difficult to see coming to the anode and so would be underrepresented, while silvery species, such as most haplochromines would be readily caught. The charcin genus *Alestes* was probably present, but these species have a flight reaction which would take them away from the area and to safety while the stop net was being set, rather than into the weeds and to capture. However, all these caveats apply to the entire sampling programme, and while some of the above groups would have been expected to be more abundant in some of the habitats sampled than others, no sampling technique will provide perfect representation of the populations it is addressing.

Another area of particular interest is the question of whether the fishing efficiency around the *Eichhornia* differed significantly from other sites. Analysis of variance of the population estimates using Zippin's method (Zippin, 1956) indicated that the mean proportion of the population caught from the *Eichhornia* sites was slightly greater than those from the other sampling sites, thus tending to minimise any differences in the results between these sites and the others.

Further questions relates to the open shore sampling, its low biomass and species diversity levels. Although it might be expected that this habitat would be the easiest to sample, it has been noted elsewhere (Zalewski and Cowx, 1991) that shelter tends to have a concentrating effect on the distribution of fish and therefore enhances the likelihood of their capture. Furthermore, the behaviour patterns of some species living along open shores mean that they are significantly more adept at avoiding capture by electrofishing (eg *Alestes*) than those which seek shelter, thus reducing the apparent efficiency of the gear more in "open water" habitats than in others (Bohlin et al., 1991). For these reasons the open shore results have not been included in the comparison between *Eichhornia* habitats and the other sheltered habitats when calculating the species diversity obtained around the lakeshore.

Species Abundance and Distribution in Inshore Habitats

The haplochromines are still the dominant group in inshore habitats despite the population expansions of the introduced species, *Oreochromis niloticus* and *Lates niloticus*. The haplochromines were also breeding at or around the time of sampling these inshore areas, unlike the introduced species.

The results suggest that of the habitats sampled, only the open shoreline was as poor in abundance and species diversity as the *Eichhornia* fringes, and that all 4 of the other habitats (rocky shore, *Vossia*, *Typha* and papyrus) were richer in all aspects of their fish faunas. The open shores where electro-fishing was carried out were sites where beach seining by local fishermen could also be conducted and were therefore disturbed habitats. This disturbance and the paucity of vegetative or rocky cover are probably sufficient to explain the low yields from the open shores.

Most (35 out of 45) of the haplochromine species were only taken from a single habitat, though this will be due, in part, to low population densities of the species concerned. Of the apparently habitat-specific haplochromines, 8 were taken only around rocks, 7 around each of *Vossia*, *Typha* and *Eichhornia*, and 3 around each of papyrus and open water. Thus 7 "uncommon" species were taken during sampling at 10 *Eichhornia* sites, while 28 such species were taken from 20 other sites, suggesting a higher preponderance of uncommon species in the pre-*Eichhornia* habitats. Only one of the 12 non-haplochromine species, *Aplocheilichthys pumilus*, was apparently restricted to a single habitat, this species being found only around *Eichhornia*. Further sampling would clarify whether these 36 species were truly associated with the habitat with which they have currently been identified, or were occupying a micro-habitat within an impure stand of vegetation, or were merely unfortunate in passing through the area during the sampling period.

The deoxygenation which results from the dense *Eichhornia* cover, and the associated lowering of the pH, become discernable even in measurements made within a metre or so of the fringing edges of the weed mat (Willoughby et al, 1993). The only fish species which would be able to inhabit areas under the mat and more than a few metres from its edge would be those having accessory breathing organs such as *Protopterus* or *Clarias*, which are pre-adapted to life in deoxygenated waters such as those of papyrus swamps. In practice, the only fish taken within the *Eichhornia* mats were either such species, or were found in small clear areas of water within the mat.

The argument that the fringes of the weed provide unusually beneficial concentrations of food items is not supported by this work, as the proportion of fish from *Eichhornia* habitats with food items in their stomachs was not significantly different from that for the overall population (49 vs. 44%).

Another case which has been put forward is that the fringes of the weed provide protection for juvenile or small fish. In the case of the haplochromines, this does not seem to be valid, as only 153 of the haplochromine fry (13% of the total of 1202) came from *Eichhornia* sites, which made up 33% of the sampled habitats. Interestingly, the weed may be a more protective habitat for tilapia fry, as 60 of these (72% of the total of 83) were taken around *Eichhornia*.

Habitat Change and Biodiversity

It is not possible at present to assess the proportion of Uganda's Lake Victoria shoreline which is affected by *Eichhornia*, but it might be as high as 15%, especially in the western parts. There are already areas, such as around Port Bell at the northern end of Murchison Bay, which are severely affected in terms of access for commercial shipping. The length of shoreline blanketed by the weed will almost certainly increase, making movement on the lake and fishing along the shoreline progressively more difficult and less profitable. Vegetated habitat types which have evolved with the lake such as *Typha*, papyrus and *Vossia*, though they are unlikely to vanish as a result of the presence of *Eichhornia*, are likely to become deoxygenated as a result of the hyacinth development between them and open water, making them unsuitable habitats for the majority of fish species. This should be expected to reduce the numbers and biomass of inshore species, and, as their preferred habitats become less available, the biodiversity of the inshore environments.

The figures presented here show the adverse effects of *Eichhornia* mats on both the biomass and species diversity of fish associated with inshore habitats of the lake. The results suggest strongly that water hyacinth is likely to bring very little benefit to the fisheries of Lake Victoria, as species number, biomass and diversity are reduced, the former two very significantly, in its vicinity. On the basis of the data presented here, *Eichhornia* can only be considered to be a major long term threat to the catch levels and the diversity of the shoreline fisheries of Lake Victoria, though a longer period of sampling should be undertaken to confirm this.

ACKNOWLEDGMENTS

We wish to express our thanks to Mr. W. Kudhongania of the Ugandan Fisheries Research Organisation in Jinja for his agreement to allow this study to take place, his permission to use FRO facilities and his active personal interest in the work. Many FRO staff assisted in the work, particularly Dr. F. Bugenyi, Mr. G. Mbahinzireki, Mr. F. Wanda, Mr. M. Magumba and Mr. I. Musana. Thanks are also due to the British High Commission in Kampala for their assistance in clearing equipment and in the provision of a vehicle. Ms. S. Green and Mr. C. Gay of NRI helped with statistical analyses. The project was funded by the Overseas Development Adminsitration, UK, through the Natural Resources and Environment Department budget it provides to NRI.

REFERENCES

Bohlin, T., T.G. Heggberget, and C. Strange, 1991, Electric fishing for sampling and stock assessment. *In* Fishing with electricity; applications in freshwater fisheries management. Cowx, I.G., and Lamarque, P., eds., Fishing News Books, Oxford, pp. 112–139.

Freilink, A.B., 1991, Water hyacinth in Lake Victoria. A report on an aerial survey on 12–13 June 1990. *In* Thompson, K., ed., Workshop on water hyacinth in Uganda: ecology, distribution, problems and strategies for control. 22–23 October 1991. Kampala, Uganda.

Green J., S.A. Corbet, E. Watts, and O.B. Lan, 1976, Ecological studies on Indonesian lakes. Overturn and restratification of Ranu Lamongan. J. Zool. Lond., 180: 315–354.

Harley, K., 1991, Survey project on exotic floating African water weeds. Unpub CSIRO rept., 30 pp.

Hickley, P., and R.G. Bailey, 1986, Fish communities in the perennial wetlands of the Sudd, southern Sudan. Freshwater Biology 16: 695–709.

McVea, C.M., and C.E. Boyd, 1975, Effects of water hyacinth on water chemistry, phytoplankton and fish in ponds. J. Env. Qual. 4.3: 375–378.

Singh, S.R., and Nasiruddin, 1980, Limnlogical investigations on Dah Lake (Ballia). I. The fishery. Ind. J. of Zootomy, XXI: 59–66.

Terry, P.J., 1991, Water hyacinth in the Lower Shire, Malawi, and recommendations for its control. Unpub report to ODA, London. Long Ashton Research Station, Univ of Bristol, 64 pp.

Twongo, T., F.W.B. Bugenyi, and F. Wanda, 1992, The potential for further proliferation of water hyacinth in Lakes Victoria and Kyoga, and some urgent aspects for research. CIFA sub-committee for management and development of Lake Victoria fisheries, Feb. 1992, Jinja, Uganda.

Willoughby, N.G., I.G. Watson, S. Lauer, and I.F. Grant, 1993, The effects of water hyacinth on the biodiversity and abundance of fish and invertebrates in Lake Victoria, Uganda. Final technical report to ODA/NRI, 48 pp.

Zalewski, M., and I.G. Cowx, 1991, Factors affecting the efficiency of electric fishing. In Fishing with electricity; applications in freshwater fisheries management. Cowx, I.G., and Lamarque, P., eds., Fishing News Books, Oxford, pp. 89–111.

Zippin, C., 1956, An evaluation of the removal method of estimating animal populations. Biometrics 12: 163–189.

Historical Note

Early Research on East African Lakes: An Historical Sketch

E.B. WORTHINGTON *C.B.E.*

Abstract — This is concerned with fishery surveys and limnology of Lake Victoria (1927), Lakes Albert and Kioga (1928), and lakes studied by the Cambridge expedition to African Lakes, 1930–31, on Lakes Naivasha, Baringo, Turkana (Rudolf) in Kenya; and in Uganda — Lakes Edward and George, Bunyoni and Nabugabo. Other lakes farther south are mentioned briefly, namely Tanganyika, Rukwa, Bangweulu, and Malawi (Nyasa).

INTRODUCTION

I was asked to write of early experiences and the surprise of discovery on early research studies in East Africa, so I start from Mombasa in July 1927. The three East African Governors of Kenya, Uganda and Tanganyika Territory had asked for a fishery survey of Lake Victoria from which supplies of "ngege" were decreasing following the introduction of gill-nets in 1910. Michael Graham of the British Government's fisheries laboratory at Lowestoft had been commissioned to do the job and I joined as his assistant. We had reached Mombasa by sea with a ton of fishing and biological equipment and proceeded up country by rail. That first introduction to Africa is unforgettable: in the evening the tropical smells, fire flies and the music of cicadas. At dawn, sitting above the cow-catcher in front of the locomotive, the first sights of Africa's fantastic richness of wild life at that time. After a day or two in Nairobi for briefing we rejoined the railway to Kisumu which was the rail head.

LAKE VICTORIA, 1927

For six months the SS Kavirondo, a converted tug, was our headquarters, in company with Lt. Stevenson as skipper and Dick Dent, Kenya's fish warden, part time. In the SS Kavirondo we cruised all round the lake with frequent stations using varied fishing gear, taking samples of fauna and flora, observing water temperature and quality at different depths, visiting many islands and crossing the lake on several transects. We threw out many labeled bottles to obtain an idea of currents. At one station in heavy weather the flat-bottomed tug nearly rolled herself over. There were also land excursions to talk with local fishermen, and to augment our food supply with buck, a few guinea fowl or duck. We measured innumerable fish, examined stomach contents and gonads, and did post-mortems on fish-eating birds and crocodiles.

At that time, little was known of Lake Victoria's biology except from a few collections which had reached the Natural History Museum in London. The most important was W.A. Cunnington's expedition at the beginning of the century. He had been on Lake Tanganyika to check the hypothesis of Moore that it was the relict of a Jurassic sea, and collected from Lake Victoria on the way home. In addition to the fish themselves I became deeply interested in the indigenous native fishing methods and was suprised at their variety, all made from local materials such as papyrus for rope, reeds for baskets, adapted to what was a clear understanding of the fish themselves. The Luo fishermen we employed had a better eye for a species than we had and pointed out that the "ngege," as served for breakfast in Nairobi, was in fact new to science and was named by Graham, *Tilapia esculenta*. Up to that time, it had been confused with *T. variabilis*. One of my own concerns was with zooplankton. I wanted to know whether daily migrations between deep and surface water, such as I had previously studied on Lake Lucerne in Switzerland, took place also in the tropics. Although the species were different, their behaviour was remarkably similar, as revealed by closing plankton nets fished in vertical series over a 48 hour period at a single station.

Although a great deal of research has been carried out on Lake Victoria since 1927, some of our more casual observations may still have significance. For example, among memories which may relate to present conditions, I recall one day in the northern area well away from land, when we steamed for miles through a calm sea carpeted with blue-green algae; and another occasion in the western area, some thousands of dead fish, one floating in every 2–3 square metres, nearly all *Bagrus*. Around most shores were many crocodiles of all sizes to about 4 metres long. They were rarely hunted by tribesmen until their skins became valuable when an American leather company discovered how to deal with the "buttons." In spite of their abundance, swimming was excellent when away from land with a water temperature nearly constant at 24°.

At the end of the Lake Victoria survey came recommendations of which three are worth recalling: one was intended to avoid over-fishing for *Tilapia*, by allowing growing fish at least one chance to breed — no gill-nets of mesh less than 5 inches should be legal. This was accepted and applied for some years, but it collapsed as the catches with 5 inch mesh nets diminished. Another recommendation was about the suggested introductlon of *Lates*, which was being advocated by keen anglers and as a means of utilizing the abundant small sized species: we recommended that it should not be attempted until and unless the results could be predicted by further research. A third recommendation was to establish a permanent research centre on Lake Victoria which should assist in the development of freshwater fisheries in East Africa, not only those of Lake Victoria. This was reinforced in reports on Lakes Albert and Kioga and by the Cambridge expedition later. It was ultimately acted on after a quarter century, when funds became available from the Colonial Development & Welfare Act enacted during World War II. The plan of the original laboratory was copied from the Rockefeller Yellow Fever Laboratory at Entebbe, and R.S.A. (Bobby) Beauchamp was appointed director in 1946.

Before leaving Lake Victoria, I should mention also the cooperation between the three countries which we arranged from1949 through the Lake Victoria Fisheries Board. It did a good job for a number of years until broken up shortly before Independence. It is good news that a new Lake Victoria committee or board is to be set up to coordinate research and advise on fisheries management for the lake as a whole. Today it would be advisory to the three governments, not executive as was its predecessor.

LAKE ALBERT, 1928

I was asked by Uganda's Governor to study Lakes Albert and Kioga after Graham returned to England. This presented a different problem, not only in its general ecology and fauna, but also in its facllities for research. Stevenson was allocated to assist me, but Butiaba, the only port, sported only one powered boat other than the mail steamer, namely an old steam launch that was available for only ten days. I needed more than ten days to get all round and across the lake, up the Victoria Nile to Murchison Falls, and down the Albert Nile. So we acquired the largest row-boat there, 16 feet long, and made of metal. Outboard motors had not yet reached East Africa, so Steve took one look at her and said "We'll fit her with sail." He did this with all the khaki drill in the one local duka. With fishing and limnological equipment, tent, a cook, two fishermen, Steve and myself, there was not much freeboard. We had some exciting times in her, but Steve was a master sailor and saw us through. On one occasion near Murchison Falls, a hippo, probably defending its calf, rushed at us, lifted the boat, smashed the rudder and nearly turned us over. One night under sail when crossing the lake, we were overtaken by a sudden storm. Water poured over the gunwale, we baled like crazy and had to dump a lot of equipment, although I saved my microscope. Just as we were giving up, a bright light suddenly shone through deluging rain at the spar's tip. "St. Elmo's fire," called Steve, "we'll come through." And we did.

At that time no one had recorded any soundings on Lake Albert except near ship landings, and like other lakes it was said to be bottomless. However, the deepest sounding recorded on our bathymetric map was 45 metres. The possibility of fishing development was, of course, our main objective. Apart from *Lates*, a target of anglers who occasionally visited the lake, its fish were quite unknown. I was surprised therefore to find hardly any species other than airbreathers such as *Protopterus* and *Clarias* which lived both above and below Murchison Falls. There seemed to be no record of this in the accounts of previous travellers, and it was a fascinating exercise to fit our data from stomach contents, these truly "Nilotic" fishes, into a food web with *Lates*, crocodiles and mankind at its summit. *Lates*, which we took readily all round the lake on long-lines and large-mesh gill-nets, appeared to offer the best opportunity for economic fishing. However the only fishery of significance at that time was for *Citherinus*, gill-netted by Belgians near the Semliki River mouth, to help feed workers at the Kilomoto gold mines.

LAKE KIOGA, 1928

Lake Kioga was easier to work on than Albert, with a motor boat available to live on. Soon after boarding her at Masindi Port, I succumbed to an acute bout of malaria, confirmed later from a blood smear on my whisky flask (the only glass readily available). Quinine was then the only anti-malaria drug so I swallowed a near-lethal dose thinking of Alfred Russell Wallace somewhere in the tropical jungle emerging from a particularly high bout of fever with his thoughts suddenly condensed into "survival of the fittest." My head was buzzing with the reasons why the fauna above Murchison Falls were so different from those below. I hoped for a similar revelation when regaining *compos mentis*; but none came.

We collected what I think were the first lot of fishes from the swamps and water-lilies of Lake Kioga, to compare with those of Lake Victoria and thereby assess the degree of isolation caused by Ripon Falls. We also admired the ingenuity and fishing methods of the Bakenzi tribesmen who lived on floating islands of papyrus and reeds.

CAMBRIDGE EXPEDITION OF 1930–31

Back in Cambridge after working on Lakes Albert and Kioga, I prepared for another expedition to Africa. I recruited Leonard Beadle who later was Professor of Biology at Makerere Collgege in Kampala for many years. I also enlisted Vivian (Bunny) Fuchs who was just finishing a degree in Geology and was later to become Director of the British Antarctic Survey. We were joined by Stella, my wife, who was studying geography at Newnham College, and in Kenya, by Dick Dent.

Lake Naivasha

We made base at Lake Naivasha where Dick Dent lent us his house on the shore. The first limnologist to work on tropical African lakes, Penelope Jenkin, had been here for a year or two earlier as a member of Louis Leakey's archaeological expedition, so we had a previous study to build on. We started by dropping our outboard motor, one of the first to reach East Africa, into 15 feet of weed-clad water. It took two days to recover.

As part of the great prehistoric lake which had once inundated Lakes Elmenteita and Nakuru as well as Naivasha, the latter had but a single indigenous fish. However, a *Tilapia* had already been introduced from the Tana River and Dick Dent was negotiating about introducing the Black Bass from America. We studied plankton and fish populations before other introductions took place, notably Coypu which decimated the magnificent water lilies and seemingly reduced to about a half the enormous number of bird species identified at Naivasha by John Williams. Still later the American prawn, when introduced, multiplied prodigiously for some years.

Lake Baringo

The expedition's main objective in Kenya was Lake Rudolf (Turkana), but we had two short spells also on Lake Baringo which was almost equally unknown in 1930. There was no road or even track there from Nakuru, but we made it in two Ford box-body cars. We had many tire punctures by thorns along the way. One car arrived with two tires stuffed with grass, and a broken spring bound up with a block of wood and rawhide.

There was no boat or canoe on Baringo, so we crossed to the islands with their volcanic fumaroles in the unique boat-shaped rafts made from bundles of Ambatch wood bound together by the Njemps fishermen. The lake seemed to be over-populuted with *Tilapia nilotica* of small size. The Njemps caught them by an ingenious form of fly fishing — tying dragonflies onto barbless hooks dangled over the fishes' nests. The Njemps disregarded crocodiles, which were small but numerous in the shallow waters. One boy was brought to the camp, however, nearly dead with ghastly wounds. We took him to the hospital at Nakuru and he rejoined us on a later visit. Bunny Fuchs spent several days walking round the lake studying geology, but unfortunately caught relapsing (tick) fever and was out of action in hospital for our main period on Lake Rudolf.

Lake Turkana (Rudolf)

The first obstacle to research on Turkana was to get a serviceable safe boat to the lake where there was not even a canoe. We were having a flat-bottomed boat built at Nakuru for use in Uganda, but that would be quite unsuitable on Lake Rudolf where high winds and heavy seas were expected. The expedition was being run on a shoe string financially, but several departments of government had offered to help. At a dinner party arranged by Archie Ritchie, the Chief Game Warden, my wife Stella sat between the Director-General of the Kenya and Uganda Railway and the Colonel of the Kings African Rifles, with instructions to vamp them into lending a boat and transporting it to Lake Rudolf. She succeeded, and in due course a lifeboat of the SS Clement Hill at Kisumu was carried by military lorry and a squad of troops to Ferguson Gulf on Lake Turkana. We built a temporary house/laboratory on the shore of the Gulf. Our outboard engine, resurrected from the weedy depths of Lake Naivasha, was fitted to the lifeboat which we dubbed "The Only Hope." We crossed the lake several times, recorded its first soundings (maximum depth 73 metres) and went through the limnological routines and fishing trials. Stella surveyed by plane table the first map of Central Island and parts of the Lake Turkana shore. The whole area was "closed" owing to raids by "shifters" from Ethiopia, but the Turkana were very helpful. Stella, as the first white woman they had ever seen, caused quite a stir.

Our study of the fishes showed close similarity to the fauna of Lake Albert, but taxonomy showed several divergencies from the Nilotic fauna, supposedly the result of Lake Turkana's isolation since the last Pluvial Period when it drained to the Nile via the Sobat River. We were, of course, much concerned with the changes in level of Lake Turkana since its outlet ceased to exit, and the quality of its waters was of particular interest to Beadle. To the practical person, the concentration of sodium bicarbonate made it almost undrinkable until citric acid was added, when it frothed up like Eno's Fruit Salts, with similar results. Laundry was not needed, because luxuriating in the lake fully clothed for a few minutes resulted in a clean suit.

What surprised us most about Lake Turkana was the size of the fish. We caught a *Lates* of 214 lb. on a long line with unbated hooks. The *Tilapia nilotica* were often 4 or 5 times the weight of those from Lake Albert. In a report to the Kenya government I recommended initiating a commercial fishery, provided transport could be established. In due course my head Luo fisherman, Pangrassio by name, was sent up to start one. Now, 60 years later, the restaurants in Nairobi serve "Nile Perch from Lake Turkana," and discerning sportsmen and tourists regard the lake as a special place to visit.

Lake Edward and Lake George

The Cambridge Expedition moved to Uganda early in 1931. Scientifically, the main object was to study Lakes Edward and George, connected to Lake Albert by the River Semliki. We were to assess how their fauna and its evolution fitted into the lake complex, and to assess possible fishery development. On Lake Edward there was already a small gill-net fishery for *Tilapia nilotica*, connected to the salt industry at the extinct crater lake of Katwe. At the opposite southwest end was another small fishery started by the Belgians at Kamande. It was already known that the fish fauna was somewhat limited, but similar in some respects to that of Lake Victoria, and wholly without crocodiles and major predator fish such as *Lates* and *Hydrocyon*.

On reaching the Kazinga Channel by road at Namasagali between the two lakes, we launched our boat. We attached a large dug-out canoe in tow, and set off under the power of our outboard engine to the astonishment and cheers of local fishermen. On reaching Lake Edward, we observed elephants, hippos and various other wild animals on both sides of the channel. We made a base on the Mweya peninsula, on the site which was later developed as the main center for the Queen Elizabeth National Park. From there we had views far across Lake Edward, to Katwe and the Kazinga channel nearer by. We proceeded to our routine of geographic, biological and fishery research.

When making the first bathometric map of Lake Edward, it was rather surprising to find a maximum depth of 117 metres, greater than either Lakes Albert or Turkana. We also found a pronounced chemocline near what is now the Zaire shore, with a substantial volume of anoxic water below. Stella and I recorded the Semliki River's flow near our camp by floating bananas for a measured distance down a reach where we had measured the cross section. (This was the first discharge record for the Semliki in H.E. Hurst's great work of seven volumes on the Nile.) Meanwhile, Beadle and Fuchs on several days foot safari had identified the Semliki Falls and Rapids. Small though they are, they had prevented mixing between the Nilotic fish fauna of Lake Albert and the Victorian fauna of Lake Edward, and had prevented the invasion of Lakes Edward and George by crocodiles.

However the period of isolation of the Victorian from Nilotic fauna was not very long, because Fuchs found fossil vertebrae of *Lates* and also of *Crocodilus* in the ancient lake beds along the eroded sides of the Kazinga channel. Years later fossils of *Lates* and other Nilotic fishes were identified in lake beds near the shores of Lake Victoria. Such discoveries impacted subsequent hypotheses of African aquatic evolution, which on occasion have broken into arguments about sympatric versus allopatric speciation.

It was clear that the gill-net fisheries on Lake Edward at Katwe and Kamande were capable of considerable expansion, and that Lake George offered even better opportunity for a new Tilapia fishery. However, regulation would be needed if the experience of reduced production as experienced in the Kavirondo Gulf was to be avoided.

The abundance of wild life around these lakes was at that time intense, partly because the whole African population, other than in a few settlements, had been removed to reduce sleeping sickness epidemics. Jumbos were all around. We fed the camp on buffalo; only tongue and tail were edible in my experience. The biggest python I've ever seen slithered into the lake from our camp site on Lake George.

Lake Bunyoni

Stella and I had a fortnight's relaxation at the end of the expedition on beautiful Lake Bunyoni. It is dammed by an ancient lava flow. We found it to have a unique simple food chain in the open water of invertebrates → frogs (*Xenopus lacois*) → otters (*Lutra maculicolis*). This is now obliterated, I believe by various introductions of fish and the slaughter of otters. The *Xenopus* were very small and had large eyes. Dr. Parker at the Natural History Museum was about to describe them as a new species when he found that their eye sockets were packed with metacircaria of a fluke of which the adult stage was believed to reside in the otters.

CONCLUSION

When homeward bound we flew from Port Bell to Cairo with one of the early flights of Imperial Airways. The trip included two days in a flying boat down the Nile to Wadi Halfa, and two more days on to Cairo in a Handley Page biplane. We and our luggage were weighed in prior to departure. Among the dozen passengers, one was so large he had to pay overweight on himself before weighing his bags. The flying boat's greatest challenge was to avoid landing on a hippo.

It was a great privilege to have been, like the Ancient Mariner, "the first that ever burst upon that silent sea," but it is difficult, after half-a-century, to keep pace with the variety and quantity of research projects which are now proceeding on these lakes.

Printed and bound by CPI Group (UK) Ltd, Croydon, CR0 4YY

24/10/2024

01778291-0012